Mechatronics for Complex Products and Systems

Mechatronics for Complex Products and Systems

Project-Based Design Approaches for Robots, Cyber-Physical Systems, Digital Twins, and Other Emerging Technologies

Zhuming Bi
Civil and Mechanical Engineering
Purdue University Fort Wayne
Fort Wayne, IN, USA

WILEY

Published by John Wiley & Sons, Inc., Hoboken, New Jersey.
Published simultaneously in Canada.

Library of Congress Cataloging-in-Publication Data Applied for:

Hardback ISBN: 9781394209590

Cover Design: Wiley
Cover Image: © IM Imagery/Shutterstock

Set in 9.5/12.5pt STIXTwoText by Straive, Pondicherry, India

Contents

Preface

In this book, *Project-Based Design Approach* is a methodological framework to integrate a number of multidisciplinary design methods, including *Model-Based System Engineering* (MSBE), to support needs-driven designs of *Robots*, *Cyber-Physical Systems* (CPSs), *Internet of Things* (IoT), *Human-Cyber Physical Systems* (HCPS), and beyond. This book aims to provide an alternative textbook on Mechatronics with the improvement on (1) the theoretical foundation of mechatronics, (2) a methodological framework to enable needs-driven designs of *multi-*, *inter-*, *intra-*, or *cross*-disciplinary products or systems, and (3) the evolution of mechatronics in the digital era.

The author has worked and taught in the field of Mechatronics for over three decades. His professional development has been accompanied and greatly influenced by the continuous evolution of Mechatronics in three decades. As an emerging discipline, Mechatronics has been broadened rapidly due to its applicability and compatibility to many new technologies including *Digital Twins* (DT), CPS, *Human-Robot-Interaction* (HRI), IoT, *Artificial Intelligence* (AI), *Blockchain Technologies* (BCT), and *Metaverse*. In contrast, the evolution of *Engineering Education* has lagged significantly to catch up with the educational needs, especially in the field of Mechatronics. This book-writing proposal is highly motivated by the author's desire to expand the coverage of existing mechatronic textbooks beyond conventional mechanical designs, electrical and electronic designs, control designs, and robotics. Existing textbooks used to be excellent sources for students in their respective engineering disciplines to explore multidisciplinary solutions to real-world design problems from scratches. However, these textbooks are gradually outdated and become not suitable for next-generation students. With the rapidly developed technologies, the innovations on modern products depend mostly on learning and understanding customers' needs, identifying and selecting off-the-shelf *Mechatronic Functional Modules* (MFMs), assembling selected MEMs as an integrated solution, and verifying and validating design solutions thoroughly before a finalized design solution is turned into its physical model to meet customers' needs. Available textbooks cover mostly on disciplinary knowledge for designs of integrated mechatronic components; they lack a theoretical foundation to explain what innovations of mechatronics are and why the thoughts on mechatronics are so critical to design successful products. In addition, the contents are outdated and yet cover numerous newly developed technologies such as DT, CPS, and HRI. Therefore, this author believes that instructors in Mechatronics can have some better options in guiding next-generation engineers to seek their own answers to some fundamental questions such as, *what is the philosophy of mechatronics?*, *how to make a mechatronic product or system innovative?*, *what is the theoretical fundamental to make mechatronics unique?*, *how to integrate AI and multidisciplinary design methods in mechatronic design?*, *what is the role of mechatronics in the digital era?* and *how to use a mechatronic design approach to develop ever-complicating products and systems?*

The book is organized comprehensively in five innovative ways: (1) discuss main-stream multi-disciplinary design methods including MBSE, *Concurrent Design Optimization* (CDO), *Axiomatic Design Theory* (ADT), and *Virtual Verification and Validation* (VVV) with the emphases on their roles in creating the innovations of mechatronic systems. (2) Generalize the concept of a mechatronic system as an integrated solution by assembling and interacting a set of *static* and *dynamic*, *sole-disciplinary* and *multi-disciplinary*, *virtual* or *physical* MFMs that add values to system in their own ways. Students will be educated mainly on how to interpret customers' needs into system-level *Functional Requirements* (FRs), decompose high-level FRs into low-level sub-FRs, and finally to select right MFMs to meet sub-FRs, respectively. Knowledge, experience, and skills to design and implement functional modules are secondary. (3) Emphasize the innovations of a mechatronic product on selection, integration, coordination, collaboration, and cooperation of MFMs to achieve system-level objectives rather than innovations on individual MEMs. (4). Develop a methodological framework for students to utilize appropriate multidisciplinary methods at different development phases of mechatronic products. (5) Broaden mechatronics to cover most relevant and influential technologies including DT-I, *Digital Triad* (DT-II) CPSs, HRI, HCPSs, IoT, Cloud Computing (CC), BCT, and Metaverses.

The book is structured as a textbook for (1) a core undergraduate course on Mechatronics and Robotics, (2) a graduate-level course in combined mechanical, electrical, and computer engineering program, and (3) a graduate-level elective course in Manufacturing Engineering (ME), System Engineering (SE), and Industry Engineering (IE). It also serves as a comprehensive reference and a design manual for engineers in developing sustainable mechatronic products and systems in the applications of automotive, aerospace, defense, agriculture, services, and healthcare industries.

About the Companion Website

This book is accompanied by a companion website:

www.wiley.com/go/bi/mechatronics

For the purpose of being used as a text book for Robotics, Mechatronics, Cyber-Physical Systems, or Digital Manufacturing, this product is accompanied by PPT presentations for all 15 chapters.

1

Introduction

1.1 Introduction

Mechatronics is an engineering discipline that brings multiple conventional disciplines including *mechanical, electrical, electronic*, and *information engineering* together to optimize the solutions to various engineering problems. Originally, the concept of Mechatronics was coined by a Japanese Engineer Tetsuro Mori from the words of **mecha**nical engineering and elec**tronics** in 1969. Nowadays, the coverage of modern mechatronics has gone far beyond from the integration of conventional engineering disciplines to the extension to many new disciplines such as *artificial intelligence* (AI), *telecommunications*, and *cybersecurity* as long as emerging disciplines can be integrated to enhance the capabilities of mechatronic systems.

In comparison with conventional systems, a mechatronic system consists of a set of multiple mechatronic components that exhibit multidisciplinary behaviors. Therefore, the design of a mechatronic system must be performed concurrently, so that design constraints in multiple disciplines can be modeled, analyzed, and satisfied simultaneously. From this perspective, mechatronics is also viewed as *a philosophy* where a system design associated with multiple disciplines is performed concurrently to seek integrated solutions to complex engineering problems. *Mechatronics* becomes a growing discipline that has been, and it will be evolve continuously with emerging technological advancements in *Materials, Science, Processes, Engineering, Integration technologies*, and *Information Technologies* (ITs). Most of the classic books on mechatronics have lagged to reflect recent advancements, especially in ITs. In the following sections, the trends of the developments in engineering designs, integration technologies, and ITs are discussed with a focus on their impact on Mechatronics.

1.2 Growing Complexity of Engineering Designs

Engineering design is to formulate *customer's requirements* (CRs) into a design problem with specified constraints and objectives and develop a *design solution* (DS) that can satisfy CRs optimally. A complete engineering design usually includes *design for manufacturing* (DfM) and *design for assembly* (DfA) where the constraints of manufacturing or assembling processes are taken into considerations at the phases of manufacturing and assembling, respectively. By DfM and DfA, a virtual model can be used to analyze system behaviors, predict system outcomes, and verify if all design

Mechatronics for Complex Products and Systems: Project-Based Design Approaches for Robots, Cyber-Physical Systems, Digital Twins, and Other Emerging Technologies, First Edition. Zhuming Bi.

Companion website: www.wiley.com/go/bi/mechatronics

constraints can be satisfied. These reduce the needs of iterations when design defects are identified and fixed at late phases of system development.

Since the information and knowledge about a product or system are accumulated gradually when its design process proceeds, the constraints involved in later design stages cannot be verified until relevant information becomes available. This becomes an obvious reason why an engineering design process is naturally iterative. In other words, the constraints that are ignored in early design phases must be verified later. In such a way, a design space with tentative solutions should be continuously refined to satisfy more and more constraints until all of them are fully satisfied.

It is desirable that less number of iterations is needed to transform a virtual model into a physical model. This implies that design iterations all occur in the virtual world with no additional cost on physical prototyping. The finalized virtual model is converted into its physical model correctly at the first time. It is referred as *First-Time Right* (FTR) practice (Bi and Wang 2020). The methodologies for engineering designs are being advanced continuously to cope with the growing complexity of products or systems in their lifecycles from design to manufacturing, assembly, application, and to disposal. Figure 1.1 shows the trend of increasing complexity of engineering designs from the perspective of manufacturing (Alkan et al. 2018). The growth of the complexity of a manufacturing system can be observed in the aspects of *products*, *enabling technologies*, and *business environments*.

Figure 1.2 shows the dimensions of complexity in engineering designs that are dependent on those of products, technologies, and degrees of dynamics and uncertainties. The complexity of each aspect could be further decomposed when the solutions to corresponding functional

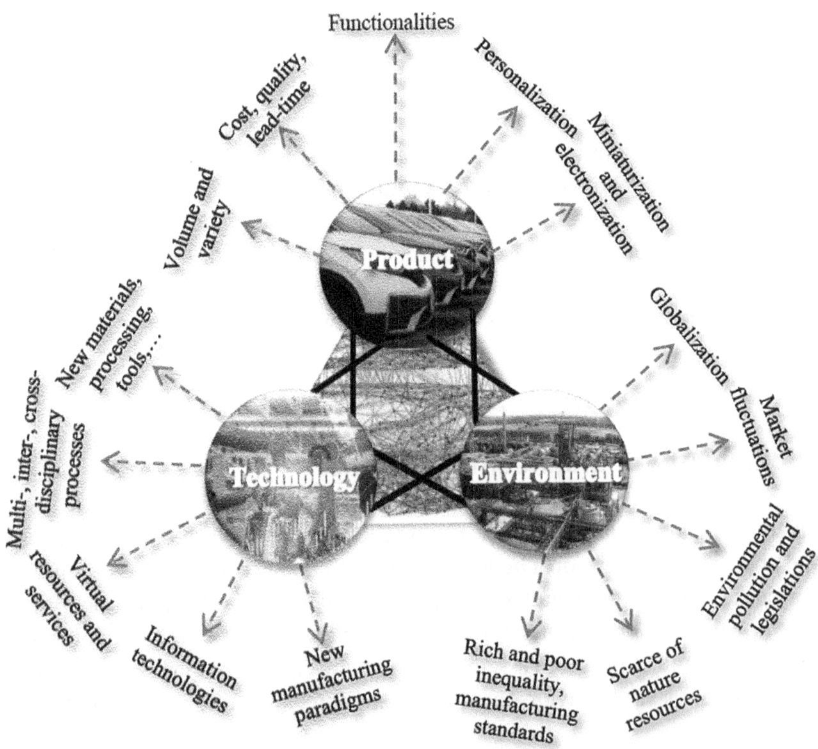

Figure 1.1 Dimensions of growing complexity of engineering designs

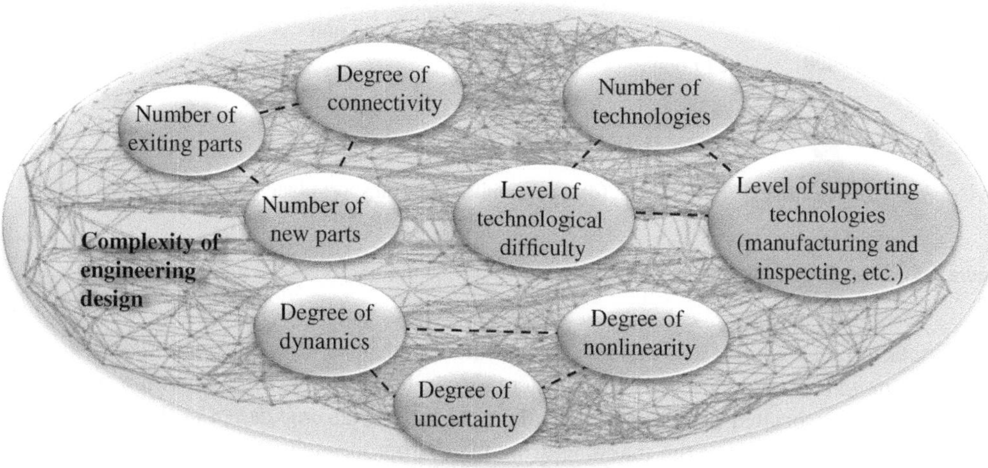

Figure 1.2 The dimensions of complexity in engineering design: products, technologies, and degrees of dynamics and uncertainties

requirements are not available. Accordingly, the complexity of products depends on many factors including the number of parts and assemblies, the degrees of connectivity, nonlinearity, and dynamics, the number of accessible technologies, and the levels of technological difficulties and assistive technologies.

1.2.1 Products

The complexity of a product has been measured by numerous factors such as *types* and *numbers* of constitutive parts and components, *types* and *numbers* of the processes to manufacture parts and assemble parts into components, *variants* and *volumes* of products, and system performance criteria such as *quality*, *lead-time*, *cost*, *lifespan*, and *after-sales services* of products (Orfi et al. 2011). Researchers agree that the scale and the complexity of modern products have been increasing greatly. Figure 1.3 shows some examples of main variables that affect the complexity of products (i.e., lawn mowers, grand pianos, cars, and airplanes); some factors such as *numbers*, *types*, and complexity of constitutive components contribute to the complexity of products directly. It seems clear that a product with a high-level complexity involves a high number and types of simple or complex parts and components.

The growing complexity of modern products can be evidenced by the evolution of various product families. Adamsson (2005, 2007) used the examples of wiring harnesses to show an increase in the complexity of automotive products. An automobile in 1949 had ~60 contact points with ~40 wires. An automobile in 1990 had around 3800 contact points and used approximately 1900 wires with a total length of ~3 km. An automobile in 1999 used 110 electric motors and 60 electronic control units (ECUs). Three data bus systems were used to support information integration and data exchanges. BeyondPLM (2018) discussed the trend of the ever-increasing complexity of modern products; it was associated with the *number of configuration items* (NICs). A typical mechanical system, mechatronic system, and large-scale integrated system have typically less than 10^3, 10^3–10^5, and over 10^8 NICs, respectively.

Product complexity is related to numerous factors in manufacturing and production such as design, development, manufacturing, assembly, and supply chain management. As shown in

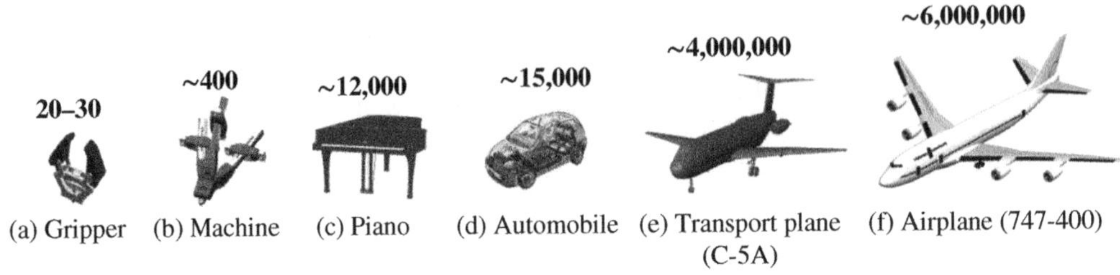

Figure 1.3 Examples of product complexity versus numbers of parts

Figure 1.4a, product complexity was modeled in the dimensions of *designs, manufacturing processes, functionalities*, and *varieties*. A manufacturer can be profitable only when the complexity of products and associated processes can be managed by *mechatronic design* at the design phase and by *mass customization* at the manufacturing phase appropriately. As shown in Figure 1.4b), the more products a company makes, the higher revenue the company can gain. On the other hand, making more product variants implies the increase in the complexity of the corresponding production system, thus affecting the productivity, lead time, and cost reduction. Production cost increases monolithically with the number of products and variants. Knowing customers' needs becomes a strategic resource to enterprises now. However, there is a limited business window for enterprises to make products to meet customers' needs in a profitable way. To expand a profitable business window, efforts can be made to increase the production revenue by making more products through mass customization and reduce the development cost by increasing system efficiency such as through mechatronic design.

With the need for more versatile and advanced products, the number and types of parts and the complexity levels of parts are expected to be increased continuously. The complexity of products due to other factors, such as volume and variety in enterprise and personalization, has been thoroughly discussed by other scholars (Bi et al. 2021a). The survey of over 246 engineers by Rowe (2019) concluded that the complexity of products was continuously increasing, and design

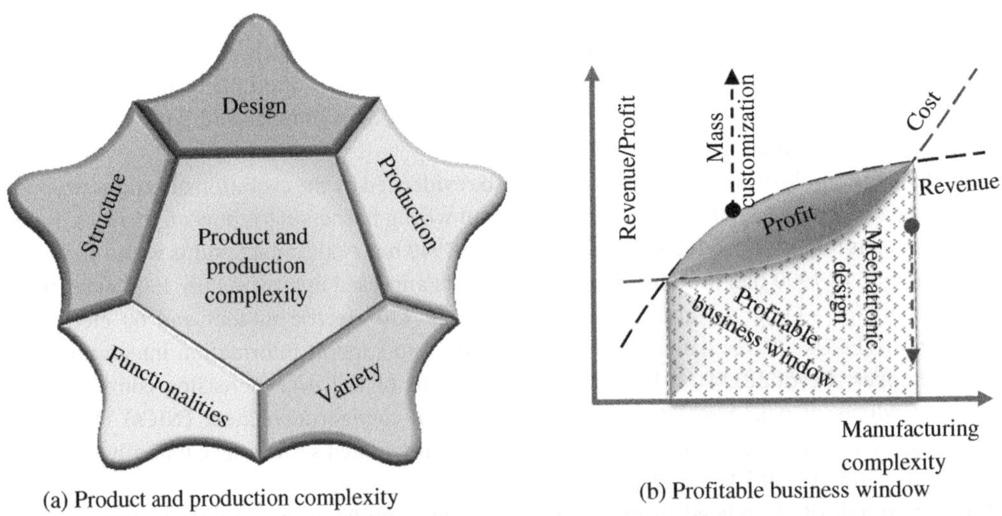

Figure 1.4 Main dimensions of product complexity and profitable business window

methodologies need to evolve to manage the complexity effectively. Ninety-two percent of the engineers reported that in the last five years, the products had increased the complexity in various aspects such as intricate mechanical designs, embedded electronics, and newly introduced materials and processes. It was found that the main causes of increased products' complexities were attributed by intricate mechanical design (57%), more electronics (47%), adoption of different materials (43%), reduced reductions (40%), system integration (30%), compacted sizes (30%), embedded software (26%), pressure for AI and smart things (17%), and more microprocessors (16%).

1.2.2 Manufacturing Technologies

Manufacturing technologies are the collections of methods, skills, techniques, and processes used to make goods or provide services to meet users' needs. Three primary factors of a manufacturing technology are *materials*, *manufacturing processes*, and *systems* (Groover 2020). More and more new alternatives for materials, processes, and systems become practically available, and this leads to high-level complexity in evaluating, selecting, and using advanced manufacturing technologies.

Advanced products or systems usually have high requirements of materials such as extremely high strengths, low weight, long durability, and better self-healing and resilience. Numerous research efforts have been made to explore high-performance materials such as new cements and composites, nanomaterials, optical materials, biomaterials, lightweight materials, and biodegradable materials (Hutagalung 2012; Grant 2013). Using advanced materials helps enhance the performance of products greatly. For example, an average percentage of composite materials in airplanes increased from 10% to 15% in 1980 to over 50% in 2015 (GeneralPlastics 2021). New methods have been developed to design, compose, and customize composite materials for given applications. Traditionally, materials' characterizations rely mainly on experiments, which are time consuming and expensive. The rapidly developed Computing Science offers the capabilities for virtual designs of new materials through *Machine Learning* (ML), *Big Data Analytics* (BDA), fuzzy logic, and Soft Computing (Babanli et al. 2019). Technologies for advanced materials are classified into basic sciences, generic technologies, and proprietary technologies; the innovations in advanced materials are critical since new materials (1) contribute to manufacturing applications greatly, (2) bring the opportunities to span new enabling technologies, (3) address key socio-economic challenges, (4) add high values to manufacturing businesses, and (5) be compatible to innovative infrastructure (Featherston and O'Sullivan 2014).

ITs play irreplaceable roles in (1) automating manufacturing processes and (2) supporting decision-making activities at all domains and levels of manufacturing businesses. With the advances in the computer hardware and software and information infrastructures, ITs have greatly proliferated from standalone mechanization to *computer-aided drawing*, *computer-aided design* (CAD), *computer-aided engineering* (CAE), *computer-aided manufacturing* (CAM), *supply chain management* (SCM), *enterprise resource planning* (ERP), computer-integrated manufacturing (CIM), argument reality (AR), digital twins (DT-Is), cloud computing (CC), *Big Data analytics* (BDA), and recently to *digital triad*-II (DT-II) and *Internet of Digital Triad Things* (IoDTT) (Bi et al. 2021a, 2021b).

The continuous evolution of automation and ITs has made it possible to advance manufacturing paradigms from *mass production* to *lean production*, *flexible manufacturing*, *CIM*, *mass customization*, and recently *sustainable manufacturing*. This provides the possibility for enterprises to select and implement certain manufacturing paradigms for their specific applications (Bi et al. 2014a, b). Moreover, *Internet of Things* (IoT) offers an effective mechanism for enterprises to utilize virtual manufacturing resources dynamically, and this adds a new dimension to the complexity of manufacturing systems.

1.2.3 Business Environments

Manufacturing environments used to be steady, in the sense that enterprises can define the boundaries of their manufacturing systems clearly, and any changes in operating their manufacturing systems are predictable. Manufacturing operations are planned, scheduled, and executed within manufacturing systems with few disturbances from environments.

However, today's manufacturing businesses become more and more closely related to manufacturing environments. Business environments have been gradually complicated and become dynamic greatly due to a number of factors: (1) global manufacturing capacities have exceeded the overall capacities required to meet users' needs of products; this has led to high diversifications and fluctuations of product demands and fierce competitions among manufacturers across nations and regions. (2) Environmental issues such as global warming, population, scarcity of natural resources, and deterioration of global living conditions become ever worsening. More and more regulations and laws have been developed for the industries to alleviate environmental challenges. Manufacturing enterprises are highly pressured to minimize the environmental impact by expanding manufacturing businesses over product lifecycles from raw materials, to design, manufacturing, assembly, and use of products, and finally to disposal and recycling (Bi 2011). (3) The development trends of human society have also increased the complexity of manufacturing businesses in the sense that full automation is impractical for most manufacturing enterprises. These trends include the aging of population, lacking of experienced workers, and growing inequality of the rich and poor.

1.2.4 Engineering Design

Mechatronic design is typically multidisciplinary and deals with multiple disciplines to meet economic, environmental, and social needs (Adamsson 2007; Strong 2021). Modern products and systems are mostly complex and sophisticated in terms of their intelligence levels and information-relevant contents. It becomes necessary to adopt some global and integrated approaches to deal with the design activities from a very early stage and throughout the whole design lifecycle (Defoort et al. 2012). Designs of a complex system are often open ended and ill defined, and the design content goes far beyond the scope of traditional individual disciplines. The trend of system complexity has evolved from *multidisciplinary* to *interdisciplinary* and to *transdisciplinary*. Engineers are required to work both within and outside the boundaries of their own disciplines. They should be trained to understand, synthesize, analyze, and apply knowledge and skills from disciplines other than their own (Beemt et al. 2020). Telenko et al. (2016) discussed the need to solve multidisciplinary problems throughout the curriculum, and they proposed the pedagogical underpinnings to integrate design experiences and multidisciplinary learning into engineering curricula. With the rapid development of ITs, mechatronic design turned out to be a vital approach to solve various multiple disciplinary engineering problems (Alciatore 2019).

As illustrated in Figure 1.5, the complexity of an engineering design depends on the *functional requirements* (FRs) associated with products, manufacturing processes, and business environments. The more complex the products, manufacturing processes, or business environments are, the more complex the corresponding engineering design is. Moreover, the integration of products, manufacturing processes, supply chains, and planning and controlling as a comprehensive manufacturing system brings new constraints for correspondences, compliances, and dependencies among system elements. The complexity of engineering design can be increased exponentially due to the comprehensive effects of the increasing complexities of products, manufacturing processes, and business

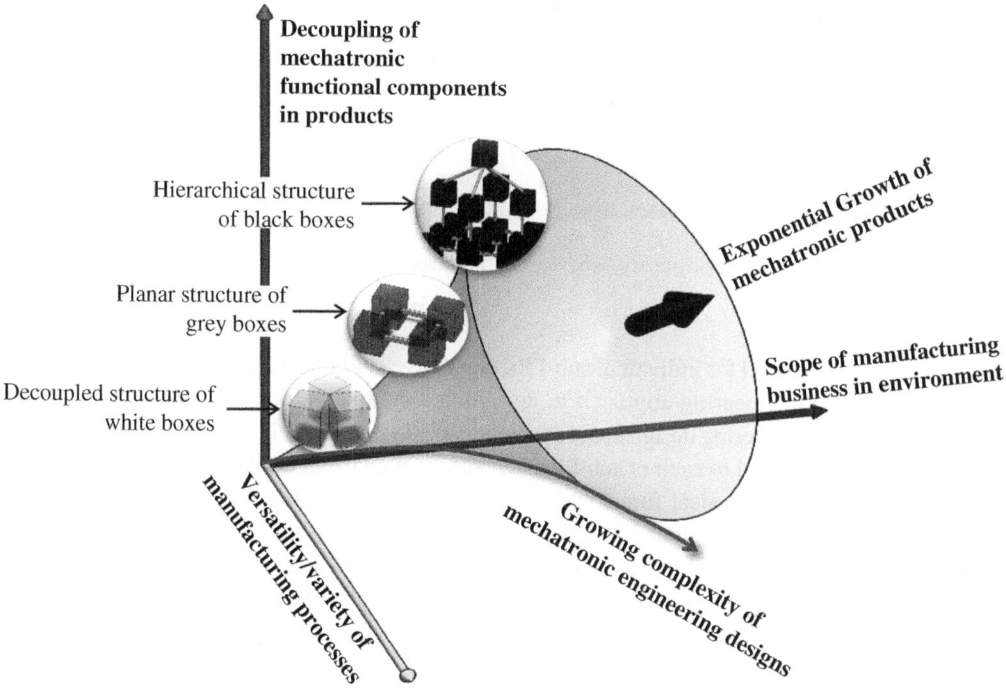

Figure 1.5 Growing complexity of mechatronic engineering designs

environments discussed in Sections 1.2.1, 1.2.2, and 1.2.3. Effective design methodologies are needed to deal with the ever-increasing complexity of modern engineering designs.

1.3 Integrated Engineering Design

In mechatronic design, the complexity of a mechatronic system can be measured by the numbers and types of FRs related to products, manufacturing processes, and business environments. Moreover, these multidisciplinary FRs are strongly coupled, and the DSs to these FRs have to be developed concurrently in such a way that (1) the design space for all potential solutions is not reduced due to a sequential design procedure and (2) all of the design constraints in these disciplines can be satisfied simultaneously (Bi 2002). As shown in Figure 1.6, raw materials are transformed into final products through a series of manufacturing processes; and the theories, design knowledge, and tools in different disciplines are utilized to implement manufacturing transformations. However, design constraints and outcomes in one discipline can strongly be coupled with those in other disciplines, since all of them are coupled to physical objects or processes in the material flow.

Three fundamental ways to deal with changes and complexity are (1) using an integral component with its flexibility of adapting certain changes, (2) using a modularized structure where different numbers and types of modules can be selected and assembled into various system configurations for different functions, and (3) using modularized structure with adjustable components as part of modules. Since a modularized structure has been proven in its great effective in using topological parameters to deal with changes, an integrated engineering design methodology should be capable of decomposing system-level FRs into loosely coupled sub-FRs, so that integrated

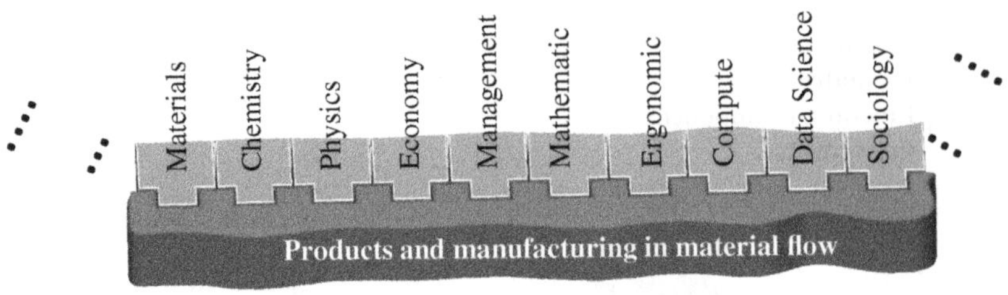

Figure 1.6 Couplings of multidisciplinary behaviors in products and manufacturing

solutions can be developed for individual sub-FRs, respectively. Figure 1.7 describes the scenario of using an integrated engineering approach to designing a complex product or system. Five main tasks of integrated engineering design are (1) decomposing system-level FRs into sub-FRs until corresponding DS is identified to each of sub-FRs, (2) performing virtual design to identify DSs for all sub-FRs, (3) developing physical functional modules for required virtual DSs, (4) selecting and assembling modules into a system configuration to meet the required FRs, and (5) developing the control system to integrate and coordinate the controls of decentralized modules in physical implementation. It is worth to note that conventional multiple-disciplinary designs emphasize on task 3 in Figure 1.7.

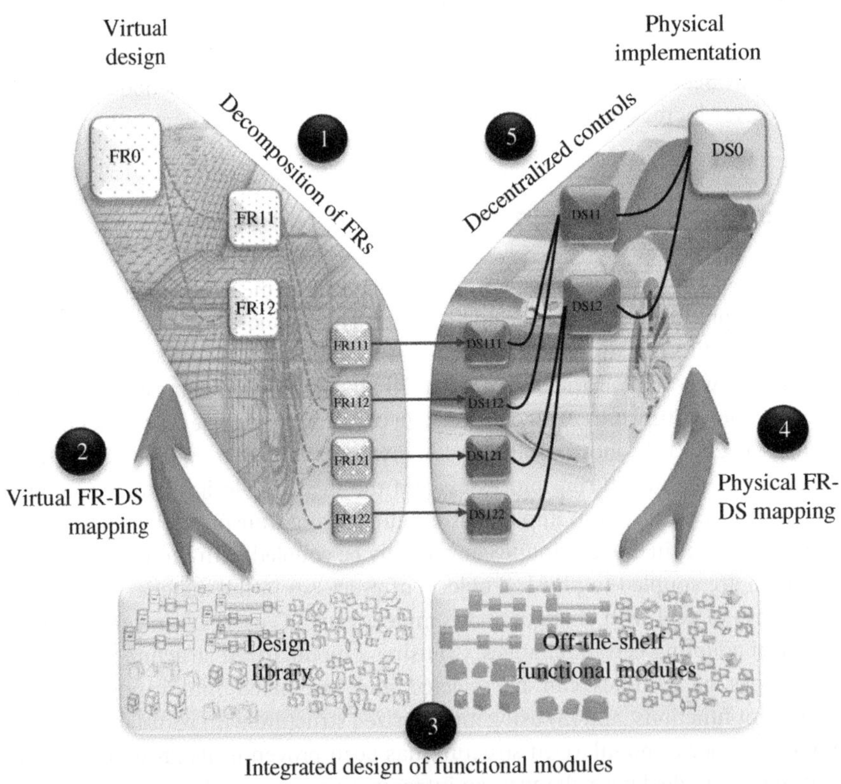

Figure 1.7 Scenario of using an integrated design method for complex product or system

1.4 Mechatronics for Multi- or Interdisciplinary Designs

With the saturation of globalized manufacturing capacities, companies are facing some fierce competition in advancing their products and systems. A mechatronic design helps reduce the complexity of objects significantly (AIMSek 2024). DE247 (2024) gave a few examples of using the mechatronic methodologies to deal with product complexity: (1) adopting the mechatronic design method reduced a half number of parts from Boeing 767 to Boeing 787 and the design was actually performed in a distributed and decentralized environment with 5400 suppliers in 38 states in the United States and 19 counties. (2) A conventional fuel nozzle with 18 parts was simplified and fabricated as a single component by additive manufacturing at the General Motors (GM). (3) An iPhone camera consisted of over 200 parts, and it was developed collectively by a team of over 800 members.

van der Roest (2017) discussed the evolution of enabling technologies in dealing with the increasing complexity of products. As shown in Figure 1.8, conventional mechatronic design is at one phase of technological development, where electric and electronic components are embedded in mechanical components as an integrated system. With the continuous increase of product complexity, the scope of mechatronic design has to be expanded greatly to integrate dynamic resources such as smart things, cyber-physical systems (CPSs), and dynamic services in distributed and decentralized environments.

Products were traditionally dominated by mechanical designs for motion and power transforms; however, more and more electronic and software components are integrated on mechanical structures as mechatronic products and systems. Taking the example of automobile industry, the number of electronic components has increased steadily. Over 80% of new functions were implemented by mechatronic components, and around one-third of the total cost of products was spent on

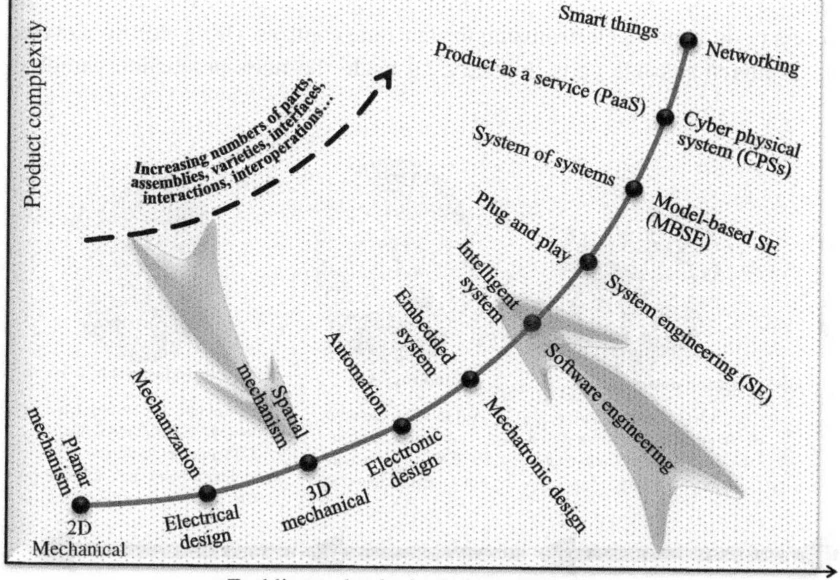

Figure 1.8 Relation of enabling technologies and product complexity

electronics. To make a product smart, electronic and software components play an increasingly important role in sensing internal and external changes, processing data, and controlling the products to respond to the changes adequately. In addition, more and more traditional mechanical components were gradually replaced by electronic or mechatronic systems. For example, a conventional gear transmission in an automobile has been replaced by electronic and software components. In the globalized business environment, enterprises are facing the challenges to increase product diversities with improved efficiency and productivity and the level of customers' satisfaction (Unido 2020).

The progress and the miniaturization we have seen in electronics during the last decades have allowed engineers to come up with new products and new engineering disciplines. Another factor that gives a booming to mechatronics applications is the continuously decreasing prices of electronic parts and the challenges to design very small systems. Taking an example, microprocessors with high performance are becoming very cheap, which encourages their use in computer-controlled systems (Boukas and Al-Sunni 2011). Figure 1.9 shows that enabling ITs has been evolved to support the design and control of complex manufacturing systems with the industry revolution from Industry 1.0 to 4.0.

1) Industry 1.0 was featured by the *mechanization* of manual operations. A machine was purposely constructed to perform tasks on objects. It began with the adoption of water-powered engines in the late 18th century. In over a century, Industry 1.0 had increased six times of average income per capita over the world. Two basic motions of machines were translational and rotational motions that were implemented by basic machine elements such as linkages, gears, cams and cranks, and pulleys and belts. Machine elements changed the speeds or types of motions, so that machinery was not man-powered; therefore, mechanization was synonymous with motorized machines (Wikipedia 2024).
2) Industry 2.0 was featured by *electrification* and *electronization* of machines. Electrification and electronization sparked the standardization and industrialization in the late 19th century. Electricity was developed as the new way of generating energy from petroleum; *eletronization* referred to a process or outcome of electronizing for machine controls. Using electric-powered machines to make interchangeable parts, Industry 2.0 made the mass production paradigm

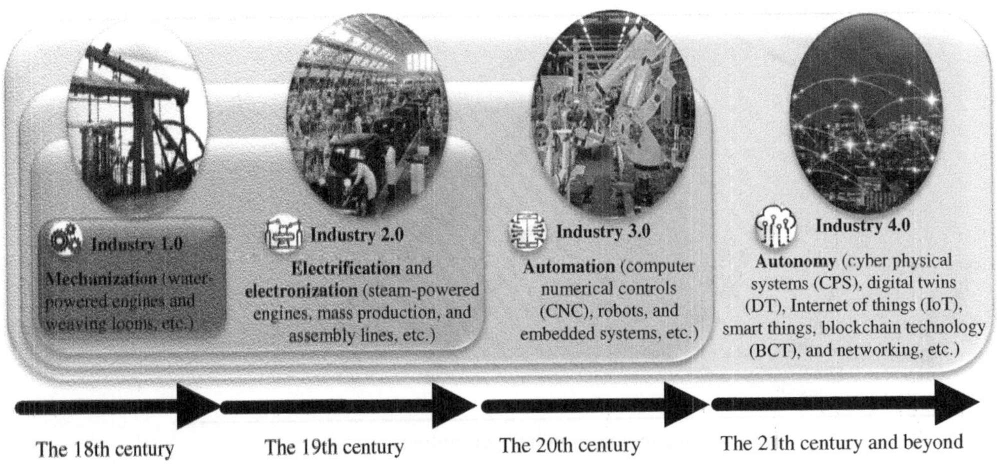

Figure 1.9 From mechanization to autonomy in Industry Revolution (IR)

possible, and this triggered the inventions of cars, airplanes, and telephones, which radically changed people's lives and working styles.

3) Industry 3.0 was featured by the *automation* of products or systems in the early 20th century. Products and systems were automated by the use of *electronics* and ITs. Embedded technologies, microprocessors, and computers were replacing humans' roles in controlling machines and systems; this increased productivity and lowered manufacturing costs. In Industry 3.0, logic processes and ITs were widely used to automate manufacturing processes with less and less human intervention (Venneslan 2009).
4) Industry 4.0 was featured by the *autonomy* of smart systems. Since 2018 with the incorporation of newly developed technologies such as CPS, DT-I, IoT, and *Blockchain Technologies* (BCTs), modern products are expected to become autonomous in the sense that they are capable of (1) sensing the changes and uncertainty in the environment, (2) processing and utilizing sensed data, and (3) reacting at the changes and uncertainties to optimize system objectives autonomously.

Industry 4.0 deals with the increasing levels of complexity, dynamics, uncertainty, distribution, and decentralization. Industry 4.0 is driven by four main factors: (1) the increases of data *Volume, Variety, and Velocity* (3 V); and the capabilities of computing power and networking; (2) the increasing demands of analysis, capabilities, and business intelligence; (3) the needs of the synergies of humans and machines; and (4) the emergence of innovative interactions of digital and physical worlds such as additive manufacturing, digital manufacturing, and robotics. Therefore, Industry 4.0 is characterized by many features linking digitization, optimization, customization, autonomy, adaptation, and human–machine interaction (Wiboqo et al. 2020). Bishop (2002) and Yan et al. (2020) gave a detailed discussion on the impact of emerging technologies on the development of mechatronic systems in Figure 1.10.

1.5 Mechatronic Design Examples

Modern products and systems are mostly mechatronic. Therefore, an engineering design in the real world mostly requires taking into consideration of mechanical, electrical, electronic, and control requirements simultaneously in developing a complete solution to meet the FRs of products. In this section, a few of capstone design projects by the author's students are introduced to show the necessity of using a mechatronic design approach in practice.

1.5.1 Development of Football Robot Team

The design aimed to build a football-playing robot team that would compete in the intercollegiate mechatronic game (Bi et al. 2017). As shown in Figure 1.11, different robots were designed to serve as the roles of a *quarterback*, *receiver*, *center*, or *runner*. These robots must satisfy geometric, weight, and power requirements specified in the regulations and rules of robotic football competitions. Some critical FRs of these robots included that (1) a center robot passed over a football to a quarterback in less than 20 seconds with a success rate higher than 75%; (2) a center, quarterback, or receiver robot was capable of accomplishing a complete pass with a success rate higher than 65%; (3) a receiver robot was capable of traveling 50 ft in 5 seconds, and a central robot was capable of traveling 50 ft in 8 seconds. Moreover, each robot should be able to identify teammates, acquire,

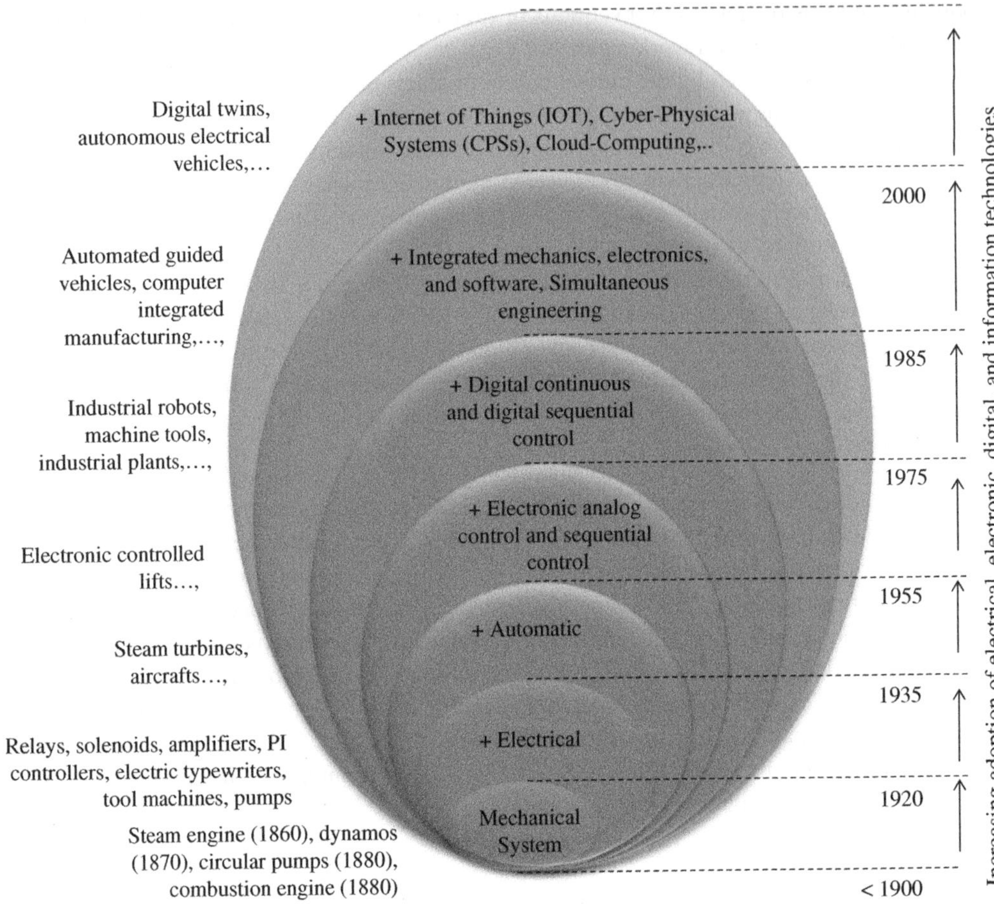

Figure 1.10 Historical development of mechatronic systems

share, and utilize sensed data in a real-time mode, and support its decision-making activities for optimized team performance.

As a system, all of the system elements must be integrated to achieve system-level functions with the desired performance. Figure 1.12 shows that a mechatronic design method is used to (1) decompose system-level FRs into module-level FRs, (2) develop physical solutions to each sub-FRs, and (3) develop the integration solution by which all functional modules are networked, interacted, coordinated, and collaborated to fulfill system-level FRs.

1.5.2 Reusing Robots to Unload Heat Sinks Automatically

The design aimed to adopt a lightweight robot to automate the unloading process of heat sinks at the exit of the heat treatment line (Bi et al. 2015). The expected system requirements include (1) unloading heat sinks without causing damages such as drops, scratches, or dents; (2) placing a part in a box at exits with the positional accuracy of 0.1 in.; (3) achieving a cycle time of operation less than 40 seconds to catch up with the moving speed of the conveyer up to 12.3 in./min; (4) showing the counts of heat sinks in box, alert a supervisor to replace a box, and allow human intervention in operation; and (5) meeting the safety requirements in running the physical system.

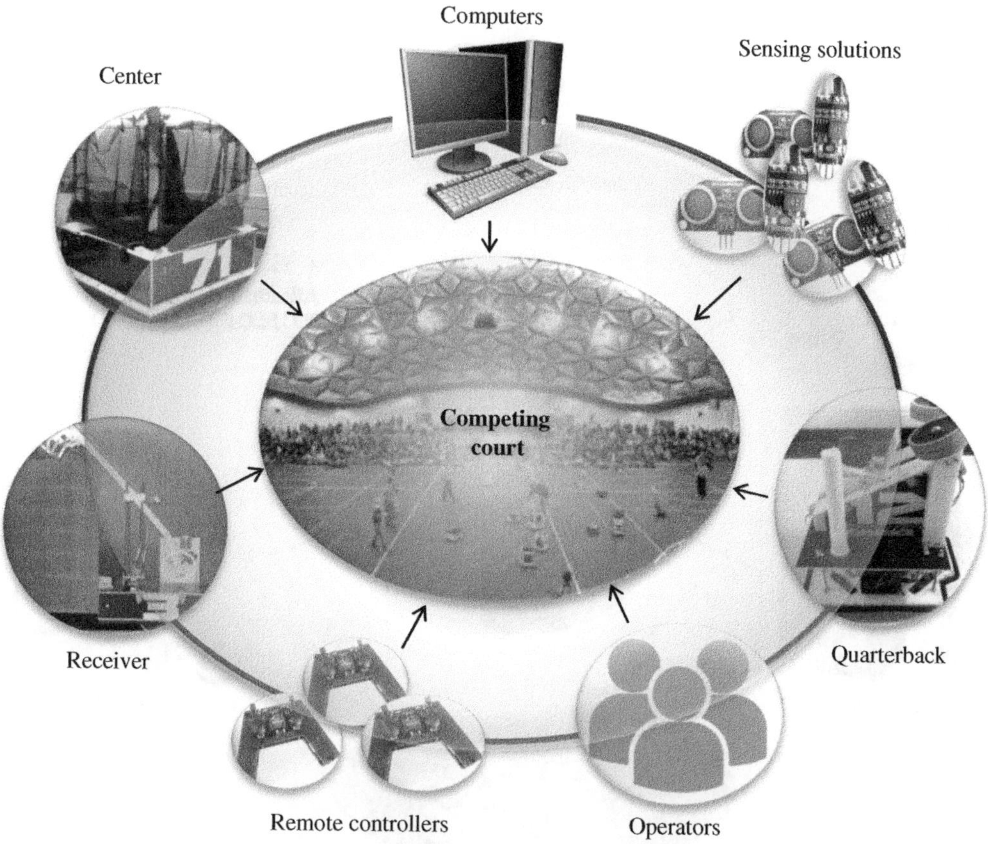

Figure 1.11 Development of robot team for football-playing competition (Bi et al. 2017)

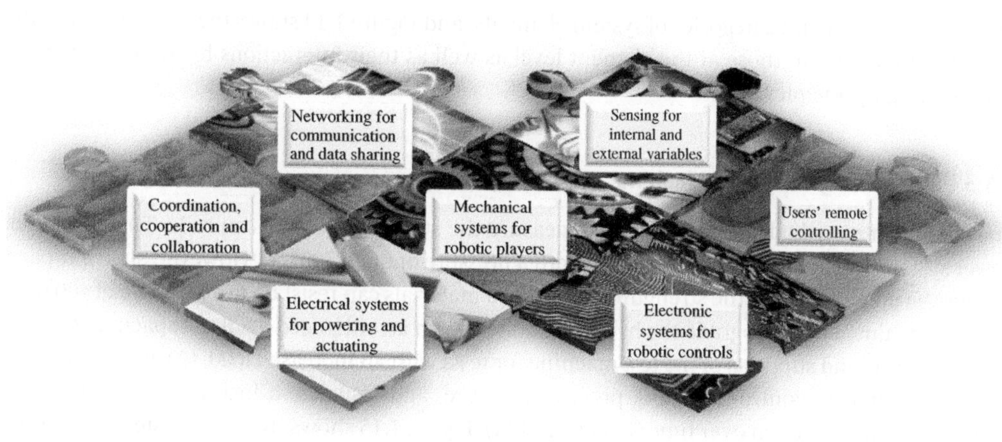

Figure 1.12 Main FRs of designing a football robot team

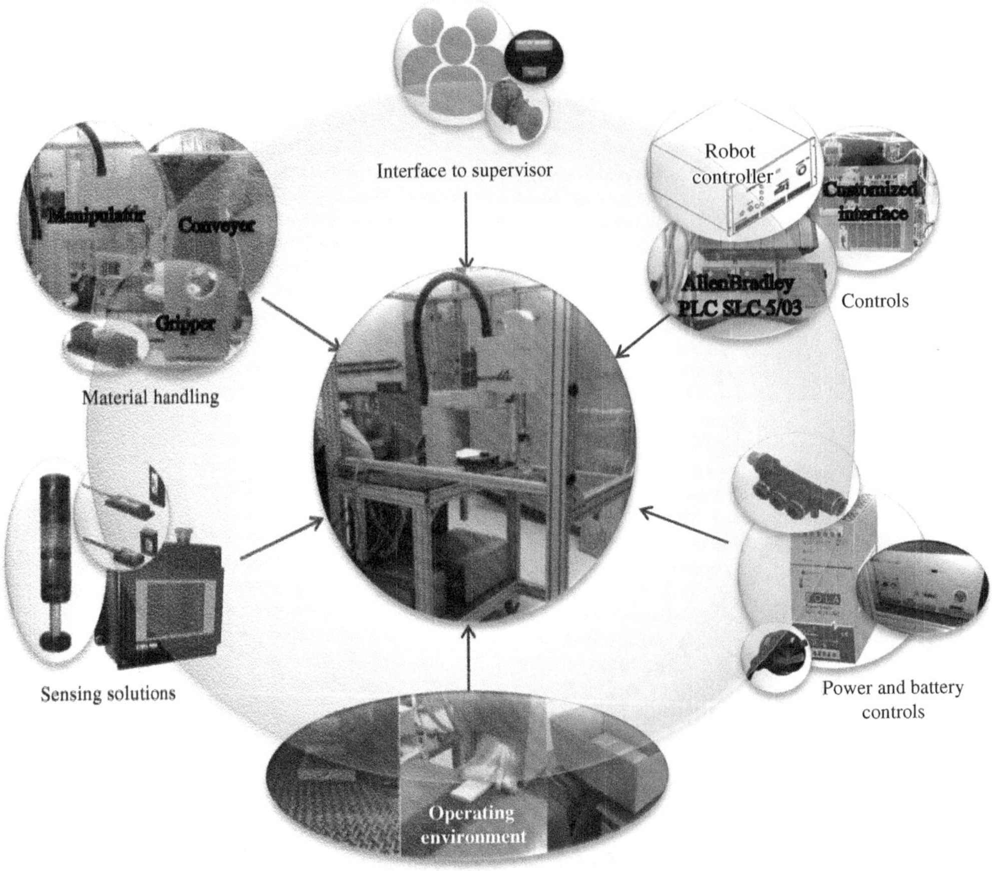

Figure 1.13 Automating heat sink unloading process by a used robot (Bi et al. 2015)

Figure 1.13 shows main categories of system elements, and Figure 1.14 shows that system-level FRs were decomposed into those at the element level as well as their interactions by following mechatronic design principles.

1.5.3 Rebuilding Rail Test Machine

The client company had a test machine to measure the strengths of welds on rails. The hydraulic power system was able to press a specimen with a pressure of up to 300 bar. However, such a capacity must be upgraded since the strengths of upgraded materials can resist up to a pressure of 400 bar. The project aimed to (1) upgrade the hydraulic system with an enhanced loading capacity; (2) reinforce the frame and supports of the test machine for tests with an increased load; (3) develop a user-friendly solution to control a testing process, measure, and visualize a load and corresponding deflection of the specimen over time (Bi et al. 2014a). Figure 1.15 shows five main system elements, that is, mechanical frame, energy absorption, hydraulic power supply, measurement and instrumentation, and testing controls and visualization. Figure 1.16 shows that the system-level FRs were decomposed into four sub-FRs including mechanical support and energy absorption, power supply, sensing system, and testing controlling.

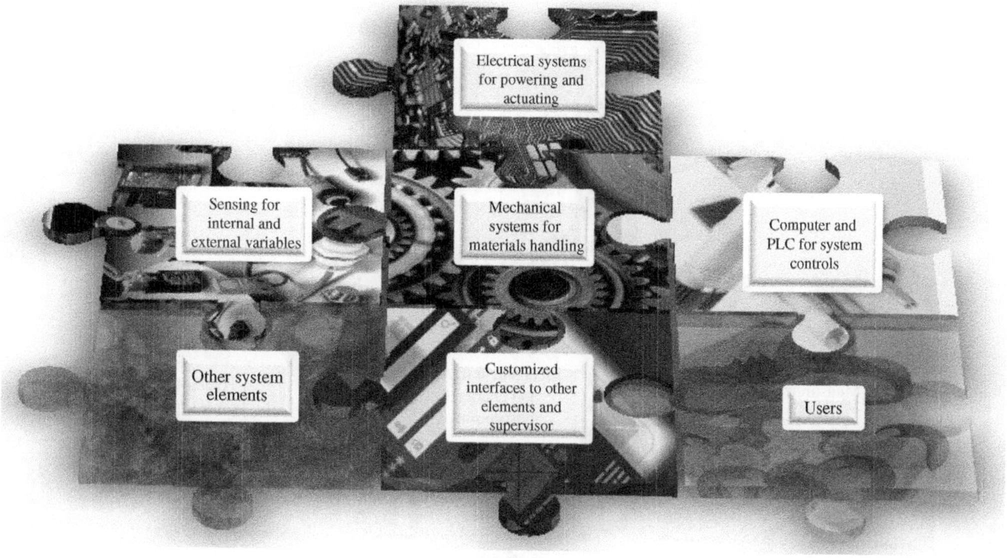

Figure 1.14 Main sub-FRs of automating a heat sink unloading process

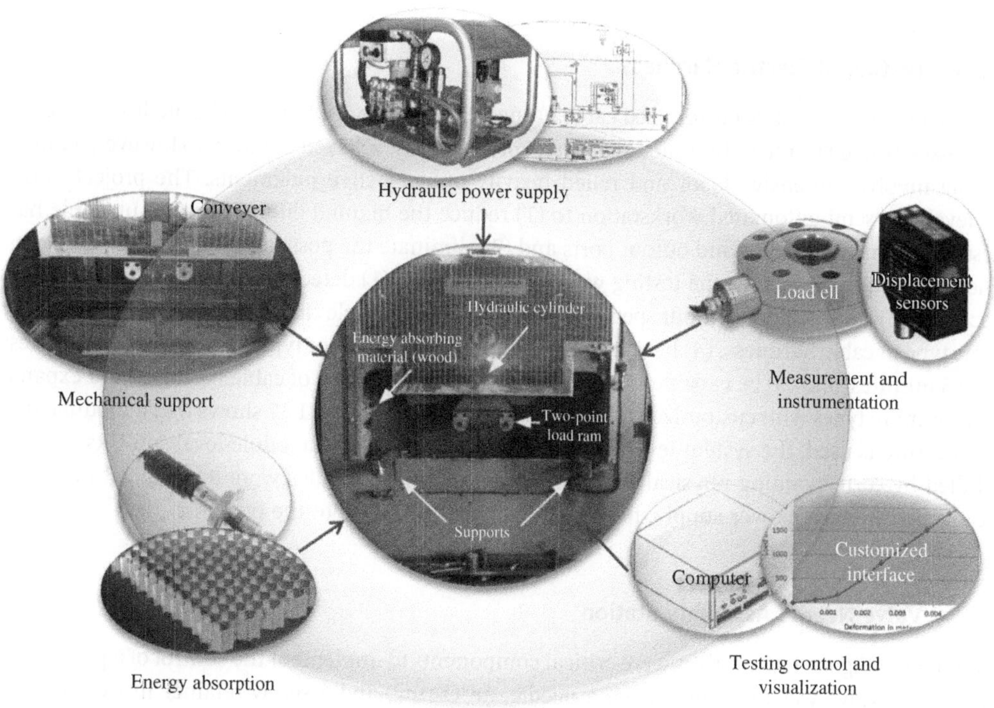

Figure 1.15 Rebuilding a test machine for rails (Bi et al. 2014a)

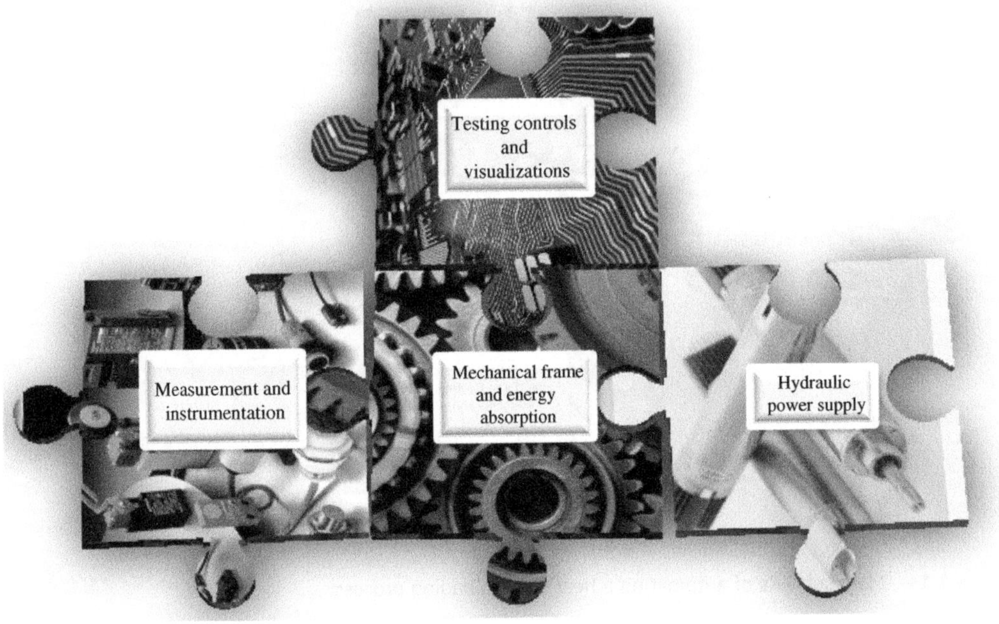

Figure 1.16 Main sub-FRs of rebuilt rail test machine

1.5.4 Testing of Electric Hardness

The client company made customized cable harnesses for home appliances. The quality of its cable harnesses was guaranteed by functional tests on randomly selected products. However, manual testing involved intensive labor and relied heavily on subjective judgments. The project aimed to develop a semi-automated workstation to (1) reduce the manual effort in setting up cable harnesses and mapping input and output ports and (2) eliminate the possibility of undetected defects. The main requirements for the testing workstation were to (1) detect both open and crossed connections, (2) assist testers without special training, (3) have a cycle time of less than 20 seconds for same type of cable hardness, (4) have a changeover time for different types of cable hardness of less than 5 minutes, and (5) be capable of testing over 20 different types of cable hardness and expandable for more types with customized adaptors (Bi et al. 2018). Figure 1.17 shows that a modularized architecture is used, the system-level FRs are decomposed into five module-level sub-FRs that are fulfilled by corresponding physical solutions. The five main modules were mechanical interface, electronic interface, power supply, testing control, and interface to the operator.

1.5.5 Valve Needle Assembly Station

As shown in Figure 1.18a, valves were critical components to implement the control of a pneumatic actuator. A valve was assembled from a needle, an O-ring, and a screw retainer in two steps as shown in Figure 1.18b, that is, (1) a pressing process to place the R-ring in the slot of the needle and (2) a threading process to thread the screw at the end of the needle. Both these processes were performed manually, and an operator easily felt vision or muscle fatigue due to the small sizes of parts at a scale of millimeters.

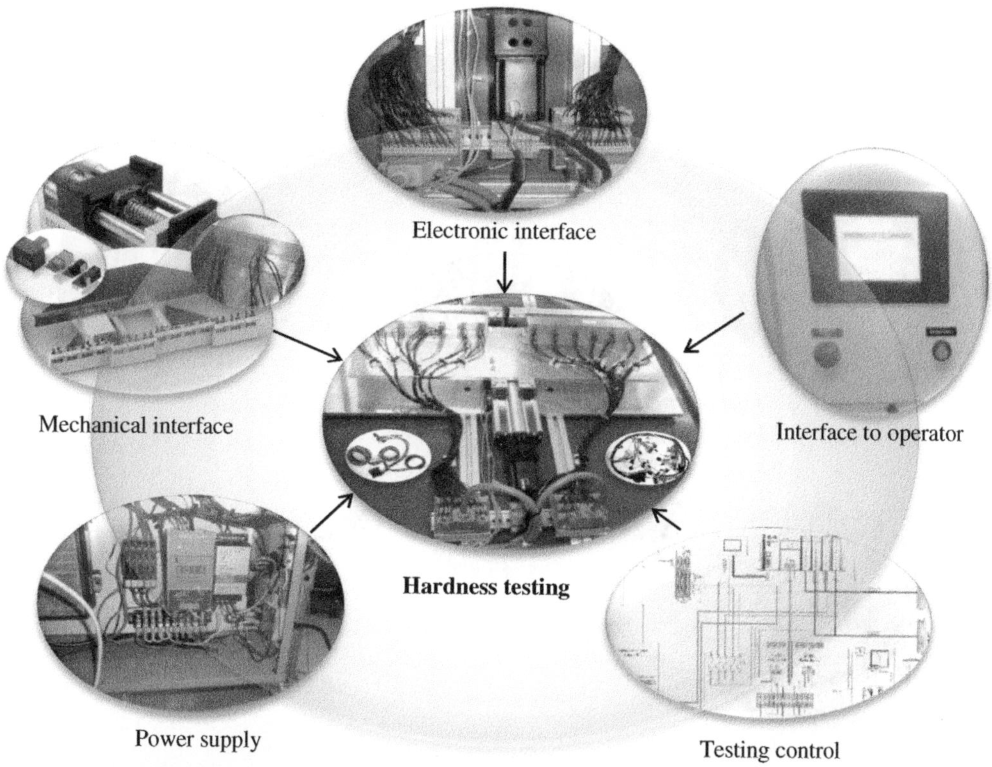

Figure 1.17 Main modules of a semi-automated cable hardness testing workstation (*Source:* Bi et al. (2018)/ MDPI/CC BY 4.0)

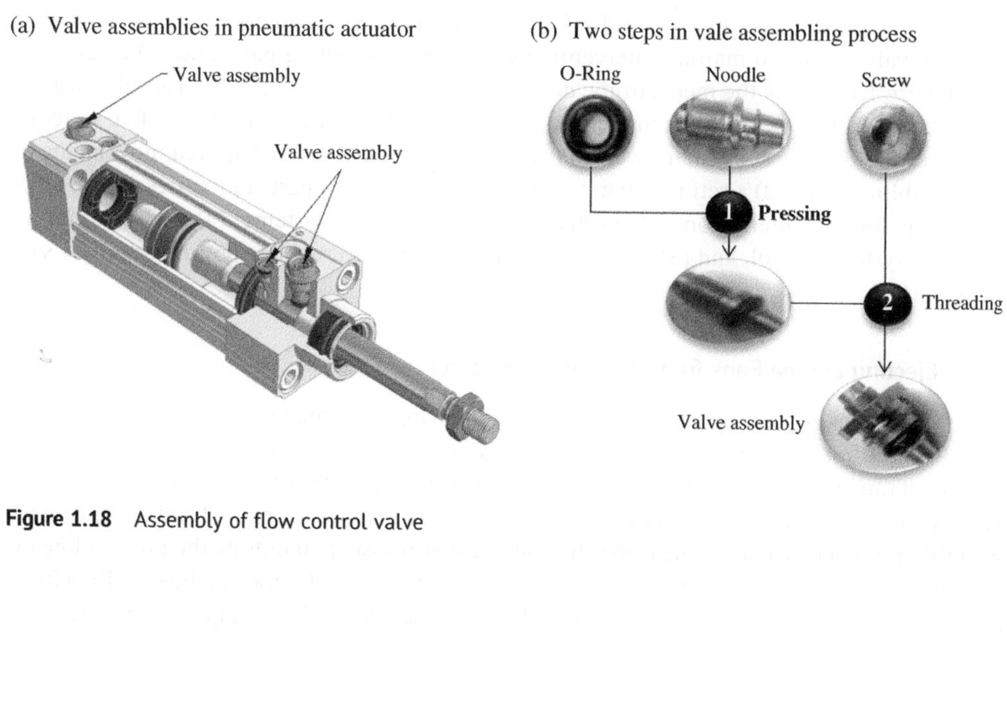

Figure 1.18 Assembly of flow control valve

Figure 1.19 Mechatronic design of semi-automated valve assembly workstation

The client requested to develop an assistive workstation where parts were placed manually at given locations and the pressing and threading processes were automated. The semi-automated system was required to produce one valve in less than five seconds, one setup for a batch of at least 1000 valves, and no manual intervention except for (a) feeding parts and (b) emergency stops. Figure 1.19 shows the mechatronic design of the semi-automated valve assembly works. The system-level FRs were decomposed into five sub-FRs, which were fulfilled by five systems, respectively. These five subsystems were (1) an actuating system for all required movements in two assembling tasks, (2) a setup system to position and orientate parts in fixtures, (3) an electric system for the interactions of mechanical and actuating systems, (4) a programming system for system control, and (5) a pneumatic power system to regulate power supply for system operation.

1.5.6 Ejecting Engine Fans from Performance Tester

The specifications of engine fans were measured by having performance tests for products randomly selected from a production line. The client company had a workstation to test engine fans. However, it ran into a number of technical issues that led to uncertain downtimes in a product line. In the existing system, a fan was lifted by a pair of parallel rails and was conveyed and ejected out on top of thin round belts to a mating conveyor, then further transported outside the product line by inertia force. Parts were easily damaged in the lifting process and the duration of the transition from one location to another, and fixtures were quickly worn or damaged due to improper changeovers.

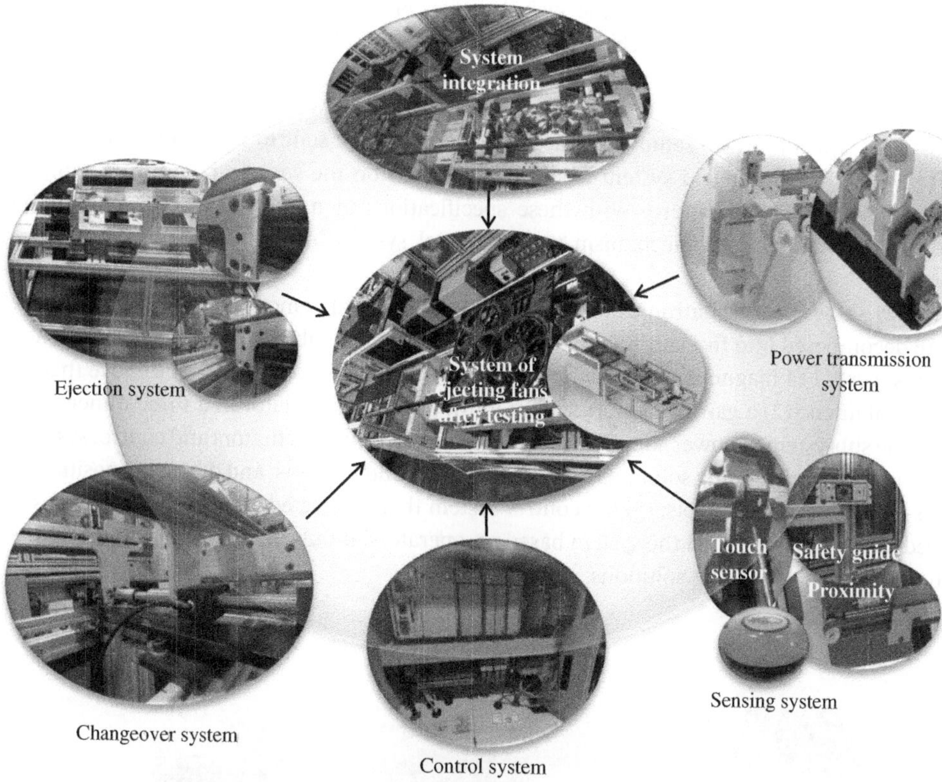

Figure 1.20 Six subsystems of the new engine fan ejection system

The project aimed to develop a new ejection system to meet the requirements of (1) automating fan-ejection process without any damage to fans, (2) being applicable to a variety of engine fans of up to 20 lb for both single- and dual-fan assemblies, (3) ejecting a fan in less than 3.5 cycle time, (4) being integrated with existing CompactLogix PLC, and (5) meeting system safety and reliability standards. Figure 1.20 shows the mechatronic design of a new fan ejection system. The system-level FRs were decomposed into six sub-FRs, which were fulfilled by six systems, respectively. These six subsystems were (1) a new ejection mechanism to lift and transfer a tested fan from the testing workstation to outside of product line in less than 3.5 seconds, (2) a changeover system, especially dealing with two identified technical issues, (3) a power transmission system to provide power for all moving components, (4) the sensing system that acquires the data of motion variables for closed-loop system controls, (5) a control system that uses sensed data to operate fan testing workstation, and (6) an integration system to connect all sub-systems seamlessly.

1.5.7 Demonstrator of Automated Spacer Removals in Truck Assembly Line

The client company produced 0.5–1 ton trucks at a rate of one truck per minute. The first step of a truck assembly line was to place the truck chassis at its entrance, and one essential task was to remove frame spacers from the chassis. Frame spacers were placed among the chassis so that individual chassis were protected and transported in stacked structures. When a chassis was placed on the assembly line, up to four frame spacers must be detected and removed from the chassis, and these operations were performed manually.

The company requested to explore the technical feasibility of automating the detection and removal of frame spacers from the chassis at this workstation. The automation was required to (1) locate spacers; (2) lift, remove, and transport spacers one by one to storage at the specified location; and (3) alert the operator for a manual operation if a spacer was sticking or at an abnormal position. If the system fails to remove a spacer, an alert must be sent to an operator for manual spacer removal. The automated system would be evaluated on the success rate of removing the spacers and its reaction to failure. With these specifications in mind, the mechatronic design focused on a spacer removal mechanism and a control system in order to achieve a working prototype.

Figure 1.21 shows the mechatronic design of the spacer removal demonstrator. The system-level FRs were decomposed into five sub-FRs that were fulfilled by five sub-systems, respectively. These subsystems were (1) a magnetic manipulator to pick spacers from truck frames and move them to specified containers; (2) a gantry support system to move the manipulator in a three-dimensional space; (3) the subsystem of power supplies for electric and pneumatic actuators and control systems; (4) the subsystem of all sensing solutions to identify spacers and humans, and detect the positions of actuators for closed loop controls; (5) the control system that connected all system elements as an integrated system and operated the system based on operators' inputs and real-time data acquired from the subsystem of sensing solutions.

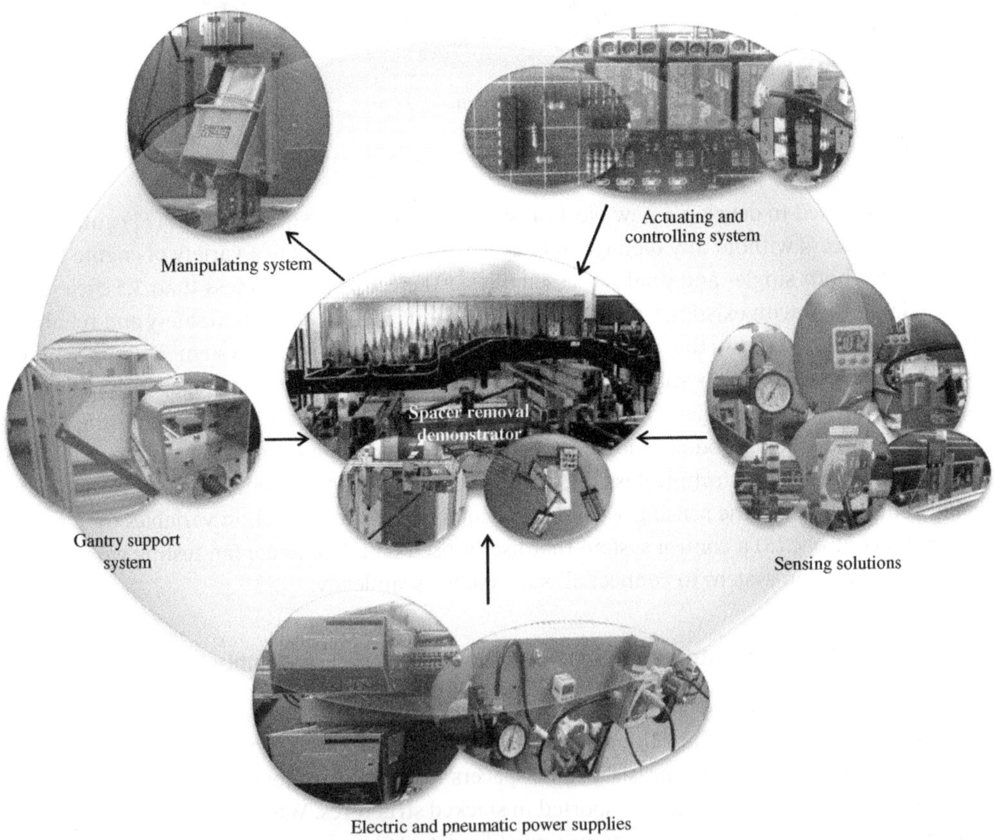

Figure 1.21 Subsystems of a spacer removal demonstrator

1.6 Group Technologies (GTs) for Mechatronic Designs

The complexity of a product or system can be measured by a *Shannon entropy* that was initially introduced to quantify the amount of energy in thermodynamics; it was then used to measure the randomness and disorders of the system as (Isik 2010; Vrabič and Butala 2012; Sönmez and Koç 2015; Fredendall and Gabriel 2019),

$$H(\chi) = \sum_{x \in \chi} p(x) \cdot \log_2 \left(\frac{1}{p(x)} \right) \tag{1.1}$$

where $H(\chi)$ is the system entropy for the amount of information relating to randomness and uncertainties of the interactions of system elements. χ is the set of possible events, that is, interactions of system elements. $p(x)$ is the probability of the occurring event $x \in \chi$. $\log_2$ is the function to normalize probability into "0" or "1."

Equation (1.1) shows that the entropy depends on two factors: (1) the number of possible events χ and (2) the probability $p(x)$ when event $x \in \chi$ occurs. Moreover, the entropy decreases when (1) the number of possible events decreases and (2) the probability $p(x)$ of an event $x \in \chi$ becomes more certain.

Since the number of possible events depends on the number of interactions among system elements, grouping system elements into components and sub-systems will reduce system complexity significantly (Bi et al. 2021b, 2021c). Taking an example of a system with a total number (n_S) of FRs is $(3 \times 4 \times 8) = 96$, if DSs would be defined by these FRs, respectively.

When one DS can satisfy one FR, that is, the number of FRs that are satisfied by one DS (n_G) is 1; then, the total number (n_I) of possible interactions among system elements becomes,

$$n_I = \left[\left(\frac{n_S}{n_G} \right)^2 - \left(\frac{n_S}{n_G} \right) \right] / 2 \tag{1.2}$$

where $\left(\frac{n_S}{n_G} \right)$ is the number of groups in the system.

As shown in Figure 1.22, when the number (n_G) of the satisfied FRs in each DS is set as 1, 3, 4, 6, 12, 24, 32, and 48, respectively, Eq. (1.2) determines the corresponding number of system interactions as 4560, 496, 276, 120, 66, 6, 3, and 1, respectively.

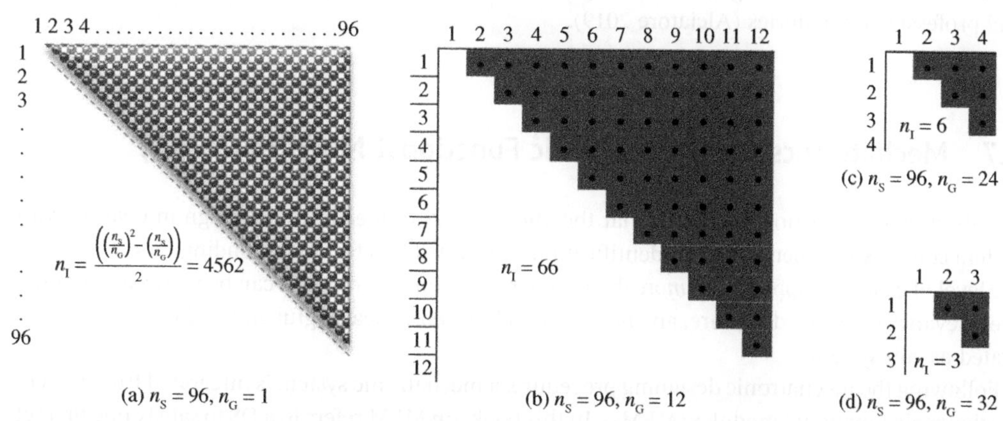

Figure 1.22 Decrease of system complexity (n_I) with an increase of group size (n_G)

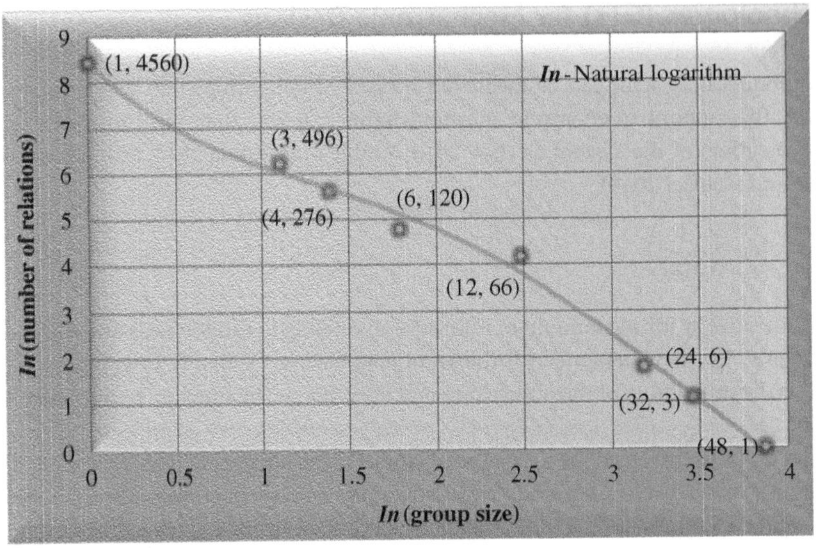

Figure 1.23 Example of the impact of group size on system complexity when the number of basic system elements is $n_s = (3 \times 4 \times 8) = 96$

Figure 1.23 shows that system complexity (n_I) is reduced monotonically with an increase of granularity (n_G) in a grouping.

Engineered products usually consist of components from different disciplines. Products can be developed in different ways that can be classified into three categories as shown in Figure 1.24. First, traditionally, *a sequential design process* is used to take into consideration of design constraints and optimization discipline by discipline. A high cost occurs due to the need of iterations for redesigning and retooling. Second, *disciplinary-specific concurrent designs*, such as DfA and DfM, are used to take into consideration of design constraints and optimizations in two or a few disciplines. Third, *mechatronic design* is used to take into consideration of all constraints and system-level objectives simultaneously. Traditional engineering disciplines, such as mechanical, electrical, chemical, and civil engineering, were developed in the 20th century; they have respective bodies of theories, knowledge, methodologies, and design tools in mutually exclusive intellectual and professional territories (Alciatore 2019).

1.7 Mechatronics and Mechatronic Functional Modules (MFMs)

The discussion in Section 1.6 shows that the effectiveness of mechatronic design in dealing with system complexity depends on the identification of grouped DSs for corresponding FRs. *The higher* the level of grouping applied, *the more* the number of system interactions can be reduced; *the closer* the relevance of grouped DSs are, and *the easier* and *the more* meaningful grouped DSs can be integrated as sub-systems.

Following the mechatronic designing procedure, a mechatronic system is integrated from a set of mechatronic functional modules (MFMs). In this book, *an* MFM refers to a DS to satisfy one or a set of system FRs. Depending on specified FRs, an MFM can be a collective and integrated solution of

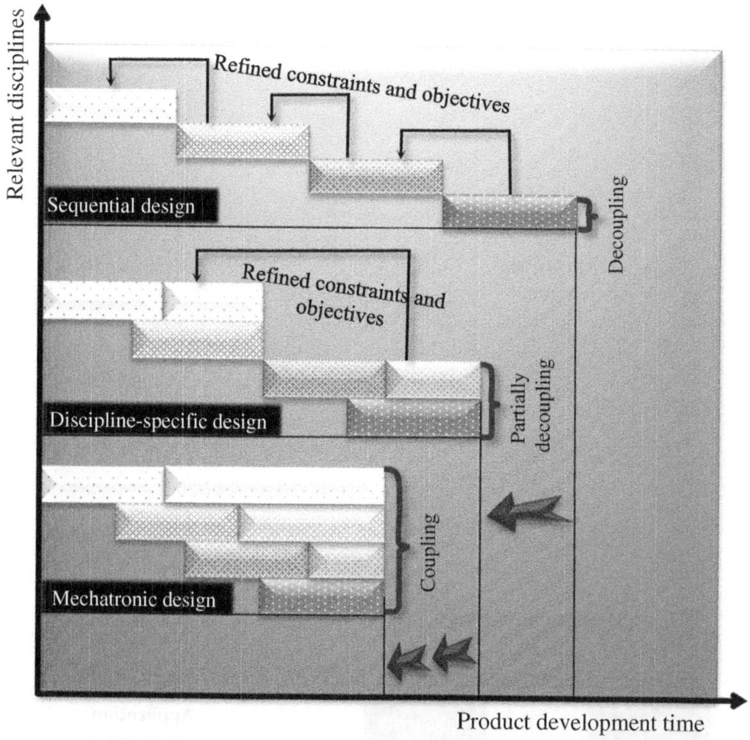

Figure 1.24 Differences among mechatronic, discipline-specific, and sequential designs

mechanical, electric, electronic, and control solutions. An MFM can also be an integration of a set of low-level MFMs.

Mechatronics, also called mechatronics engineering, is an interdisciplinary branch of engineering that focuses on the engineering of electronic, electrical, and mechanical engineering systems. In regard to relevant disciplines, mechatronics also closely relates to robotics, electronics, computers, telecommunications, systems, control, and product engineering. Note that mechatronics was originally intended to be nothing more than a combination of mechanics and electronics; hence, the name being a portmanteau of mechanics and electronics; however, as the complexity of technical systems continued to evolve, the definition had been broadened to include more technical areas (Wikipedia 2024) (Figure 1.25).

Figure 1.26 provides an external view of a product-oriented mechatronic system and its boundaries. Alongside the actual mechatronic system (the product), two additional important, interacting agents can be seen: the *user* and the *environment*. The rationale behind a mechatronic product was its *user-based purpose*. A mechatronic system never existed for its own sake; instead, it was designed to interact with users and environment through *man–machine interface*. In the simplest case, this consisted of a start button to enable a complex task that would be carried out by the mechatronic system automatically.

Figure 1.27 illustrates a mechatronic system from the structural perspective. A mechatronic system was used to replace or expand humans' abilities in (1) sensing environments; (2) performing tasks; (3) making decisions to respond to changes; and (4) performing comprehensive tasks with data collection, processes, and utilization for decision-making and execution.

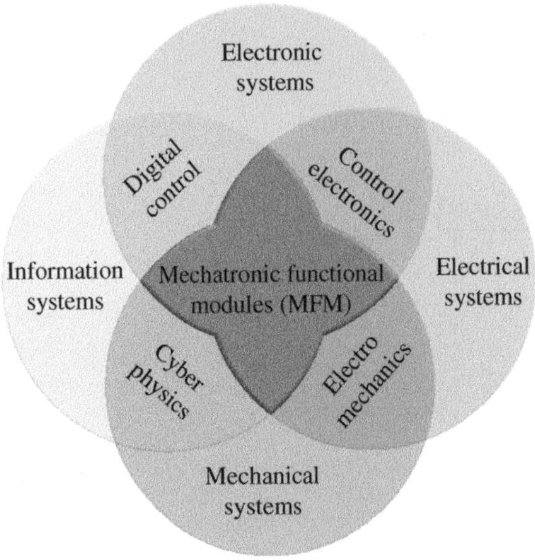

Figure 1.25 Integrated mechatronic system (*Source:* Craig and Stolfi (2002)/with permission of Elsevier)

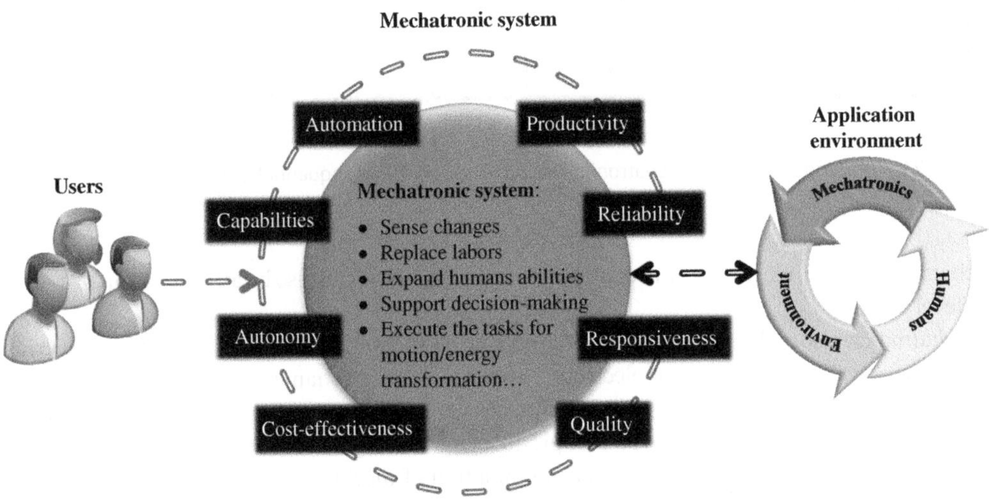

Figure 1.26 Mechatronic system from users' perspective

1.8 Mechatronic Design Methodologies

Mechatronic products and systems are not new. On the one hand, most of the engineered products are integrated mechanical, electrical, and information systems; on the other hand, these integrated products are not necessarily developed by engineers who have been trained for multidisciplinary designs. However, learning mechatronics helps engineers greatly to perform an effective and efficient process in defining, classifying, organizing, and integrating disciplinary solutions into a system-level solution. From this perspective, mechatronics should be taught more on *mechatronic design methodologies* rather than *mechatronics* itself; in addition, the scope of mechatronics has been broadened gradually over the years.

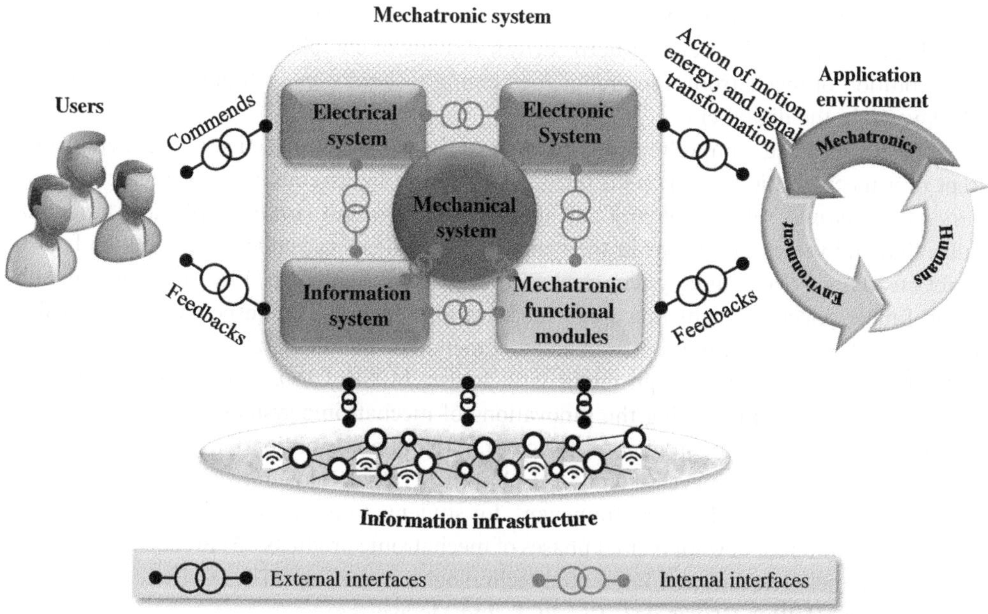

Figure 1.27 Mechatronic system from a structural perspective

Mechatronics is *a philosophy* that reflects the advance of human civilization in interdisciplinary science. Designers in any engineering discipline can benefit greatly by setting their minds on the synthesis of interdisciplinary knowledge to meet emerging engineering challenges. In fact, the concept of mechatronics itself came from the evolution of engineering design methodologies. These design methodologies have been integrated to advance mechatronic products by adopting more and more embedded components, computing resources, and ITs. Moreover, rapidly developed DT-I, CPSs, *bio-electro-mechanical systems*, *quantum computing*, *micro-electronic-mechanical systems* (MEMS), and *nanotechnologies* have brought new growth points of traditional mechatronics (Yan et al. 2020).

In the next chapter, mechatronic systems will be analyzed to identify more important characteristics such as modularity, integrability, coupling disciplines, concurrent deigns, decentralized controls, event-driven automations, adaptability, predictability, resilience, and continuous adaptation. Therefore, a mechatronic design process must be needs-driven, and different design methodologies such as *model-based system engineering* (MBSE), *axiomatic design theory* (ADT), *unified modeling language* (UML), *concurrent design optimization* (CDO), and *virtual verification and validation* (VVV) should be utilized to satisfy the FRs of mechatronic designs at a different phase. Finally, *the project-based mechatronic design* (PBMD) will be introduced as a platform to integrate all available computer-aided tools as long as they add values to the mechatronic system to be designed.

1.9 Organization of the Book

In this book, PBMD is introduced as a methodological framework to integrate a number of multidisciplinary design methods to support needs-driven designs of *Mechatronics*, CPSs, IoT, *Human Cyber Physical Systems* (HCPSs), and beyond. This book aims to provide an alternative book on mechatronics with the improvement on (1) the theoretical foundation of mechatronics; (2) the methodological framework to enable needs-driven designs of complex products or systems; and (3) the evolution of mechatronics to embrace emerging technologies including DT-I, CPSs, IoT, and Metaverse in the digital era.

The book is organized comprehensively to achieve the following objectives: (1) generalize the concept of a mechatronic system as an integrated solution by assembling and interacting a set of *static* and *dynamic*, *sole-disciplinary* and *multidisciplinary*, and *virtual* or *physical* functional modules that add values to systems in their own ways. Students are educated mainly on how to interpret customers' needs into system-level FRs, decompose high-level FRs into low-level sub-FRs, and finally select the right functional modules to meet sub-FRs, respectively. Knowledge, experience, and skills to design and implement functional modules are secondary. (2) Discuss mainstream multidisciplinary design methods including MBSE, CDO, ADT, and VVV with an emphasis on their roles in creating the innovations of mechatronic systems. (3) Emphasize the innovations of a mechatronic product on selection, integration, coordination, collaboration, and cooperation of MFMs to achieve system-level objectives rather than innovations on individual MFMs. (4) Develop a methodological framework for students to utilize appropriate multidisciplinary methods at different developmental phases of mechatronic products. (5) Broaden mechatronics to cover the most relevant and influential technologies including DT-I, DT-II, CPSs, human–robot interactions (HRIs), HCPSs, IoT, CC, BCT, and Metaverses.

Figure 1.28 shows the organization of the book. Chapter 1 gives an overview of the historical development of Mechatronics and discusses the need to teach mechatronic design methodologies rather than mechatronics itself. Chapter 2 introduces the common ways for innovations to characterize mechatronic systems from the design perspective, commonly used multidisciplinary design methodologies are discussed to explore their roles in creating innovations in mechatronic systems, and PBMD is presented as the methodological framework to integrate various multidisciplinary design methodologies for needs-driven designs. Chapters 3–8 discuss the designs of MFMs in power supplies, actuating, sensing, signal processing, cyber systems, and control systems. Chapters 9–15 introduce emerging technologies in the recent development of mechatronics including DT-Is, CPSs, IoT, Robotics, end-effector tools, Metaverse, and HRIs.

1.10 Summary

Since its birth, mechatronics has continuously evolved to deal with multidisciplinary designs with the ever-increasing complexity. Classic books focus on mechatronics itself and underestimate its importance as a design philosophy. There is a need to provide some alternatives in teaching mechatronic courses with two primary purposes to (1) understand mechatronic design methodologies and (2) incorporate newly developed technologies in mechatronics.

PBMD is proposed as a methodological framework to integrate various multidisciplinary methodologies to create innovations in mechatronic systems at different levels and aspects. The proposed book is expected to fill the gap of mechatronic educations with its features of (1) classifying innovations and converting these into FRs of mechatronic systems; (2) presenting PBMD as a methodological framework to integrate system modelling language (SysML), ADT, MBSE, CDO, and VVV in mechatronic design; (3) generalizing single or multidisciplinary system components as MFMs; (4) embracing emerging technologies DT-I, CPSs, Metaverse, HRIs, and HCPSs in mechatronics.

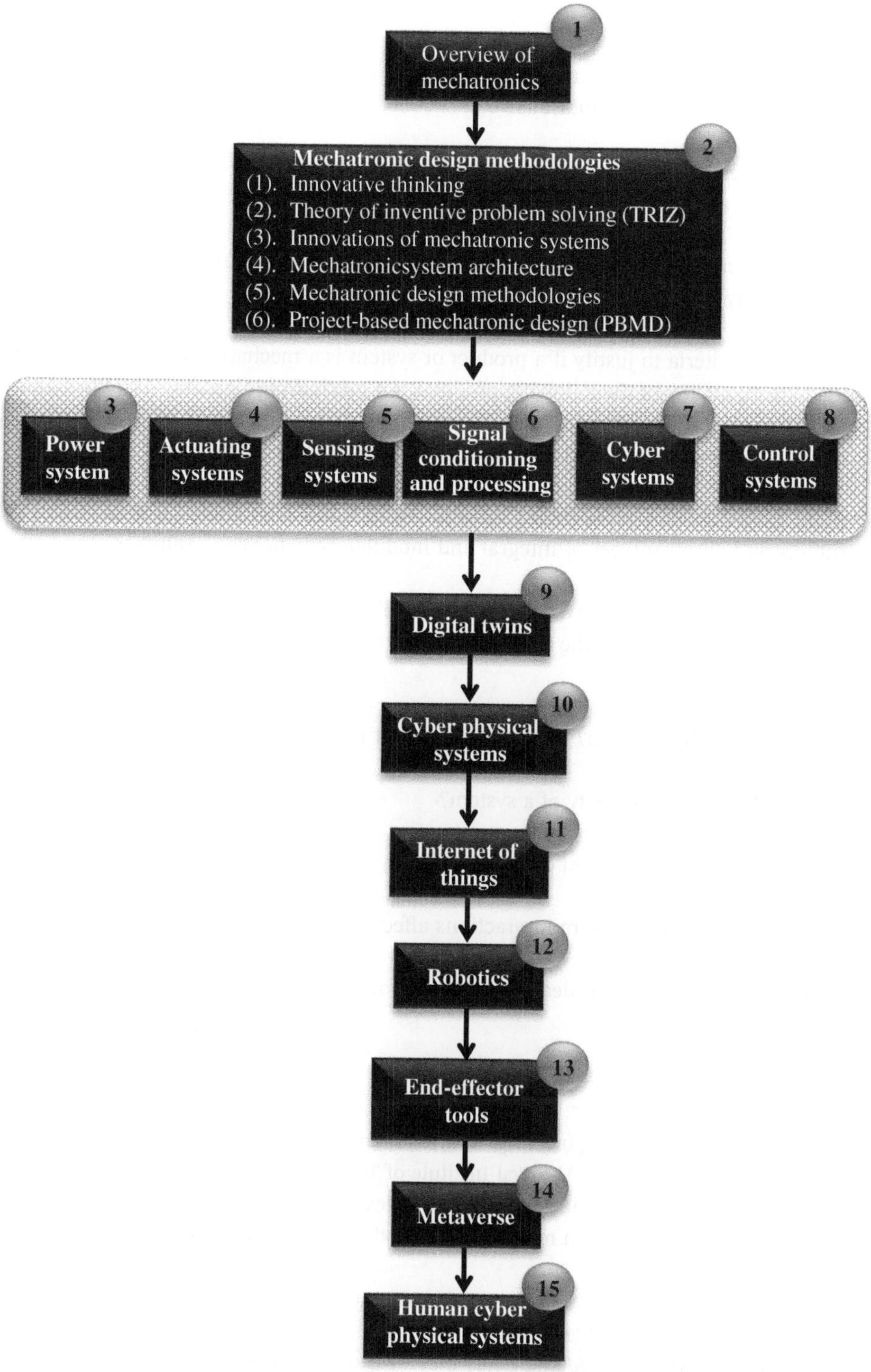

Figure 1.28 Organization of book

Problems

1.1 What are the basic functional requirements (FRs) of a mechatronic system? Give a few real-life examples to support your answers.

1.2 What are some essential components of a mechatronic system? Give a few real-life examples to support your answers.

1.3 What are your criteria to justify if a product or system is a mechatronic system? Use your criteria to justify if an industrial robot is a mechatronic system.

1.4 What are your criteria to justify if a product or system is a mechatronic system? Use your criteria to justify if any of the following products is a mechatronic system: (1) a mechanical watch, (2) a smartwatch, (3) a mercury thermometer, (4) an LED light, (5) a smartphone, (6) an app, (7) a microwave, and (8) a programmable logical controller.

1.5 What are the differences between an "integral" system and a "modular" system? Compare advantages and disadvantages of integral and modular systems with examples.

1.6 What are the differences between a "sequential" design and a "concurrent" design? Compare the advantages and the disadvantages of sequential and concurrent designs with examples.

1.7 What is a needs-driven design? Why is it important in mechatronic design?

1.8 How to measure the complexity of a system?

1.9 How does the number of system elements affect system complexity?

1.10 How does the number of system interactions affect system complexity?

1.11 How can mechatronic designs deal with system complexity effectively?

References

Adamsson, N. (2005). Mechatronics engineering: new requirements on cross-functional integration. Licentiate thesis. ISSN 1400-1179. KTH Royal Institute of Technology S-100 44 Stockholm, Sweden.

Adamsson, N. (2007). Interdisciplinary integration in complex product development – managerial implications of embedding software in manufactured goods. Doctoral thesis. Royal Institute of Technology, ISBN 978-91-7178-610-4.

AIMSek (2024). How does PLM help in managing next-generation complexity. https://www.aimdek.com/blogs/plm-help-managing-next-generation-product-complexity (accessed 12 September 2024).

Alciatore, D.G. (2019). *Introduction to Mechatronics and Measurement Systems*, 5e. McGraw-Hill Education.

Alkan, B., Vera, D.A., Ahmad, M. et al. (2018). Complexity in manufacturing systems and is measures: a literature review. *European Journal of Industrial Engineering* 12 (1): 116–150.

Babanli, M.V., Prima, F., Vermaut, P., et al. (2019). Review on the new materials design methods. https://link.springer.com/chapter/10.1007/978-3-030-04164-9_124 (accessed 12 September 2024).

den Beemt, A.V., MacLeod, M., der Veen, J.V. et al. (2020). Interdisciplinary engineering education: a review of vision, teaching, and support. *Journal of Engineering Education* 109: 508–555.

BeyondPLM (2018). Achilles heel of MBSE and PLM modeling. http://beyondplm.com/2018/08/14/achilles-heel-mbse-plm-modeling (accessed 12 September 2024).

Bi, Z.M. (2002). On adaptive robot systems for manufacturing applications. Doctoral thesis. University of Saskatchewan, Canada. https://harvest.usask.ca/handle/10388/etd-07302008-094853 (accessed 12 September 2024).

Bi, Z.M. (2011). Revisiting system paradigms from the viewpoint of manufacturing sustainability. *Journal of Sustainability* 3 (9): 1323–1340.

Bi, Z.M. and Wang, X.Q. (2020). *Computer Aided Design and Manufacturing*, 2020. Hoboken, NJ: Wiley-ASME Press Series.

Bi, Z.M., Pomalaza-Raez, C., Singh, Z. et al. (2014a). Reconfiguring machines to achieve system adaptability and sustainability: a practical case study. *The Proceedings of the Institution of Mechanical Engineers, Part B: Journal of Engineering Manufacture* 228: 1676–1688.

Bi, Z.M., Xu, L.D., and Wang, C. (2014b). Internet of things for enterprise systems of modern manufacturing. *IEEE Transactions on Industrial Informatics* 10 (2): 1537–1546.

Bi, Z.M., Yanfei, L., Baumgartner, B. et al. (2015). Reusing industrial robots to achieve sustainability in small and medium-sized enterprises (SMEs). *Industrial Robot* 42 (3): 264–273.

Bi, Z.M., Wang, G., Xu, L.D. et al. (2017). Witte C and Sidwell C (2017) IoT-based system for communication and coordination of football robot team. *Internet Research* 27 (3): 162–181.

Bi, Z.M., Pomalaza-Raez, C., Hershberger, D. et al. (2018). Automation of electrical cable harnesses testing. *Robotics* 7 (1): https://doi.org/10.3390/robotics7010001.

Bi, Z.M., Jin, Y., Maropoulos, P. et al. (2021a). Internet of things (IoT) and big data analytics (BDA) for digital manufacturing (DM). *International Journal of Production Research* https://doi.org/10.1080/00207543.2021.1953181.

Bi, Z.M., Zhang, W.J., Wu, C. et al. (2021b). Generic design methodology for smart manufacturing systems from a practical perspective, part I – digital triad concept and its application as a system reference model. *Machines* 9 (10): 207.

Bi, Z.M., Zhang, W.J., Wu, C. et al. (2021c). Generic design methodology for smart manufacturing systems from a practical perspective, part II – systematic designs of smart manufacturing systems. *Machines* 9 (10): 208.

Bishop, R.H. (2002). The mechatronics handbook, ISA-The Instrumentation Systems, https://www.cic.ipn.mx/~pescamilla/SensAct/Bishop2002.pdf (accessed 12 September 2024).

Boukas, E.K. and Al-Sunni, F.M. (2011). *Mechatronic Systems – Analysis, Design and Implementation*. Springer.

Craig, K. and Stolfi, F. (2002). Teaching control system through mechatronics: academic and industrial perspectives. *Mechatronics* 12 (2): 371–381.

DE247 (2024). By the number: product complexity. https://www.digitalengineering247.com/article/by-the-numbers-product-complexity (accessed 12 September 2024).

Defoort, S., Balesdent, M., Klotz, P. et al. (2012). Multidisciplinary aerospace system design: principles, issues, and onera experience. *Multidisciplinary Aerospace System Design* 2012 4: 1–15.

Featherston, C. and O'Sullivan, E. (2014). A review of international sector strategies and roadmaps: a case study in advanced materials. https://www.ifm.eng.cam.ac.uk/uploads/Resources/Featherston__OSullivan_2014_-_A_review_of_international_public_sector_roadmaps-_advanced_materials_full_report.pdf (accessed 12 September 2024).

Fredendall, L.D. and Gabriel, T.J. (2019). Manufacturing complexity: a quantitative measure, https://pdfs.semanticscholar.org/39c5/b53dd70414acae58532a0a8010193af221d3.pdf (accessed 12 September 2024).

GeneralPlastics (2021). Composite core: factors to consider when choosing core material for your application. https://www.generalplastics.com/technical-papers/composite-core-factors-to-consider-when-choosing-core-material-for-your-application (accessed 12 September 2024).

Grant, P. (2013). New and advanced materials. https://assets.publishing.service.gov.uk/government/uploads/system/uploads/attachment_data/file/283886/ep10-new-and-advanced-materials.pdf (accessed 12 September 2024).

Groover, M.P. (2020). *Fundamentals of Modern Manufacturing: Materials, Processes and Systems*, 7e. Wiley.

Hutagalung, S.D. (2012). *Materials Science and Technology*. Rijeka, Croatia: InTech.

Isik, F. (2010). An entropy-based approach for measuring complexity in supply chains. *International Journal of Production Research*, 2010 48 (12): 3681–3696.

Orfi, N., Termpenny, J., and Sahin-Sariisik, A. (2011). Harnessing product complexity: step 1 – establishing product complexity dimensions and indicators. *The Engineering Economist* 51 (1): 59–79.

van der Roest, D. (2017). Empowering quoting and program execution for automotive suppliers. https://www.vdr.com/the-practical-plm-blog/2017/fueling-new-business-effective-execution-for-automotive-suppliers-part-1-of-3 (accessed 12 September 2024).

Rowe, J. (2019). Managing product complexity, regulations and requirements. https://www.engineering.com/DesignSoftware/DesignSoftwareArticles/ArticleID/18471/Managing-Product-Complexity-Regulations-and-Requirements.aspx (accessed 12 September 2024).

Sönmez, O.E. and Koç, V.T. (2015). On quantifying manufacturing flexibility: an entropy based approach. *Proceedings of the World Congress on Engineering 2015 Vol II WCE 2015*, London, UK (1–3 July 2015).

Strong, D.S. (2021). Towards effective multidisciplinary engineering education: the multidisciplinary design stream at Queen's university. https://pdfs.semanticscholar.org/a6b2/e9c04f6d3e378391e24f5f46a91ea483d889.pdf (accessed 12 September 2024).

Telenko, C., Wood, K., Otto, K. et al. (2016). Designettes: an approach to multidisciplinary engineering design education. *ASME Journal of Mechanical Design* 138: 022001-1–022001-11.

Unido (2020). Industry 4.0 opportunities behind the challenges background paper. https://www.unido.org/sites/default/files/files/2020-06/UNIDO%20Background%20Paper%20on%20Industry%204.0_FINAL_TII.pdf (accessed 12 September 2024).

Venneslan, C. (2009). Electrification and industrialization: an assessment of the industrial breakthrough in Norway. *Scandinavian Economic History Review* 57 (2): 124–155.

Vrabič, R. and Butala, P. (2012). Assessing operational complexity of manufacturing systems based on statistical complexity. *International Journal of Production Research* 50 (14): 3673–3685.

Wiboqo, E.B., Legionosuko, T., Mahroza, J., and Jaya, Y.C. (2020). Industry 4.0 and opportunities in competency development for defense apparatus' human resources. *International Journal of Advanced Science and Technology* 29 (7): 45–60.

Wikipedia (2024). Mechatronics. https://en.wikipedia.org/wiki/Mechatronics (accessed 12 September 2024).

Yan, X.-T., Bradley, D., Russel, D., and Moore, P. (2020). *Reinventing Mechatronics – Developing Future Direction for Mechatronics*. Switzerland AG: Springer Nature.

2

Mechatronic Designs – Innovations, Theories, and Methods

2.1 Innovative Thinking

Science and technology is used to make products for the human society; artifacts are designed and made to meet humans' needs (Fiorineschi et al. 2018). Engineering design aims to conceive and develop an artifact as an engineering solution. Engineering design of an artifact usually consists of three basic tasks as shown in Figure 2.1: (1) recognizing and defining the needs in human society; (2) identifying science and technologies that are applicable to develop innovative solutions; and (3) designing and implementing the solution for identified humans' needs (Taura and Nagai 2017).

Since mechatronic design is a typical engineering design, and it is innovative in nature. The concept of mechatronic design was originally coined to describe an innovative process in dealing with the complexity of a multidisciplinary system. Innovations may be needed to accomplish the tasks at any stage of mechatronic design from conceptual design to the delivery of finalized artifacts. Taking an example of conceptual design, the demanded needs are analyzed and decomposed into manageable units in which specific solutions can be identified and developed; functional modeling is the key to the conceptual design of a complex design problem. *Function modeling* can provide an abstract functional representation of an artifact to describe the strategies for decomposing a problem, creating concepts, designing architecture and models, and organizing resources (Hirtz et al. 2001, 2002). Functional modeling leads to a functioning model with identified functional modules as well as their assemblies as the whole artifact.

It is critical to incorporate innovative thinking in mechatronic design. The innovation in mechatronic design is not just a collection of functions and modules; it is mainly the integration and synergizing of multidisciplinary modules to meet the needs of complex systems. Most modern products rely on close interactions of mechanics, electrics, electronics, controls, and software engineering. Mechatronics mainly refers to two trends, that is, spatial integration of multidisciplinary solutions, and integrated controls of multi-body systems (Gausemeier et al. 2011). In particular, mechatronic designs are playing more and more important roles in *computer*, *communication*, *control*, and *consumer* (4C) products. Hammadi et al. (2014) argued that mechatronic design should consider the interactions and integration of multidisciplines, multitasks, multiphysical simulations, multilevel abstractions, and multi-tool designs. For example, around 90% of all innovations relevant to automobiles were through multidisciplinary integration (Bradley and Russell 2010; Malmquist 2014). Figure 2.2 illustrates some representative

Mechatronics for Complex Products and Systems: Project-Based Design Approaches for Robots, Cyber-Physical Systems, Digital Twins, and Other Emerging Technologies, First Edition. Zhuming Bi.

Companion website: www.wiley.com/go/bi/mechatronics

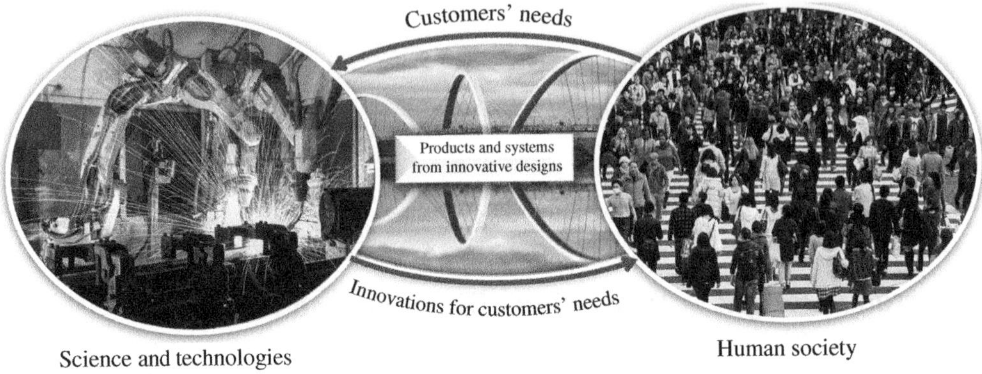

Figure 2.1 Artifacts to bridge science and technologies and humans' needs through innovations (*Sources:* (a) tum2282/Adobe Stock Photos; (b) MIDASIT Company Ltd. / https://www.midasbridge.com/en/blog/bridgeinsight/design-considerations-for-arch-bridges / last accessed on July 18, 2024; (c) Thomas La Mela / Shutterstock)

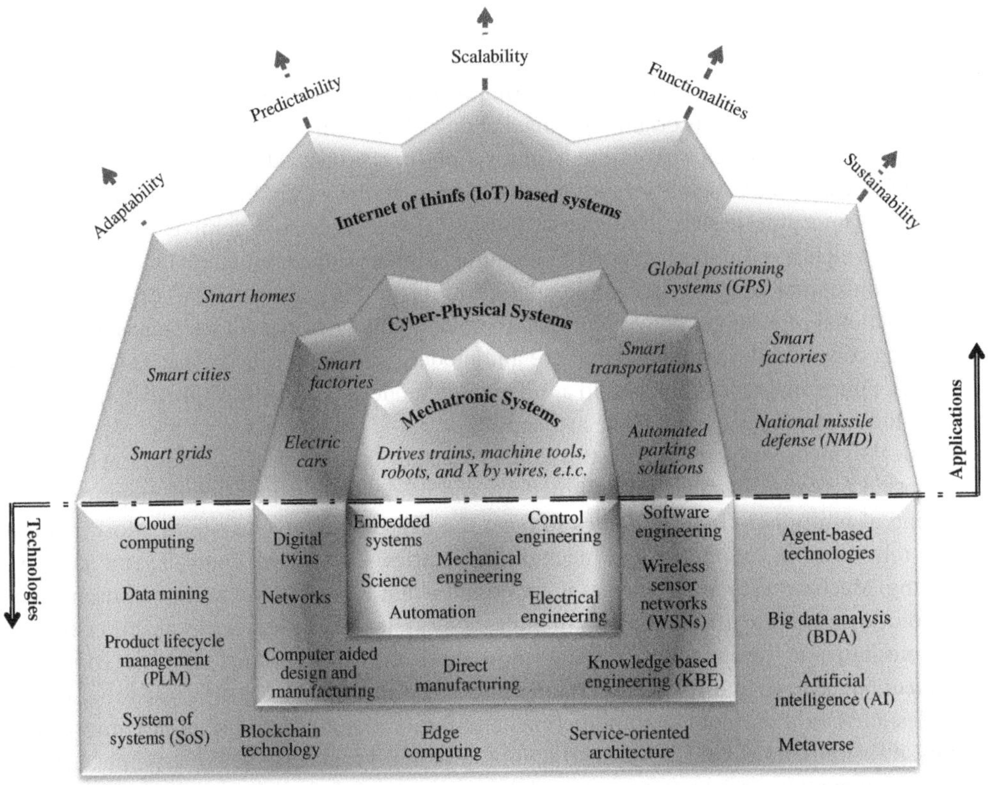

Figure 2.2 Transition from mechatronic systems to cyber-physical systems (CPSs) and to Internet of Things (IoT)

innovations and examples of when mechatronic design approaches have evolved from the traditional integration of mechanical and electrical systems, to *cyber-physical systems* (CPSs), and to *Internet of Things* (IoT). The interactions of system elements are playing more and more important roles in modern manufacturing such as *Industry 4.0, smart manufacturing*, and *cyber-physical production systems* (Hehenberger et al. 2016). Future system integration methods and models must support the integration of distributed, decentralized, and an increasing number of heterogeneous system elements in the network and over an extended time domain.

Although every innovation may be involved in a complex process with multiple iterations and feedback loops, common tasks are involved in an innovative process referred to as *an innovation spiral* (Nesta 2019). These common tasks are to (1) identify new opportunities; (2) define functional requirements (FRs); (3) define design space and constraints; (4) define design solutions (DSs); (5) integrate, verify, and validate a system as a whole; (6) scale up and upgrade a system when it is needed; and (7) reconfigure a system when it is needed (see Figure 2.3). In addition, an innovation can be made by *open-box thinking* or *in-box implementation*. In an in-box implementation, a solution is found by identifying, selecting, and integrating available resources; note that in-box implementation is used to develop the solutions to over 98% of the inventive problems. The in-box implementation contributes more and more along with the progress of development; in particular, mechatronic design principles are mostly applied to fulfill the tasks involved in stages 3–7 (Trabar 2014).

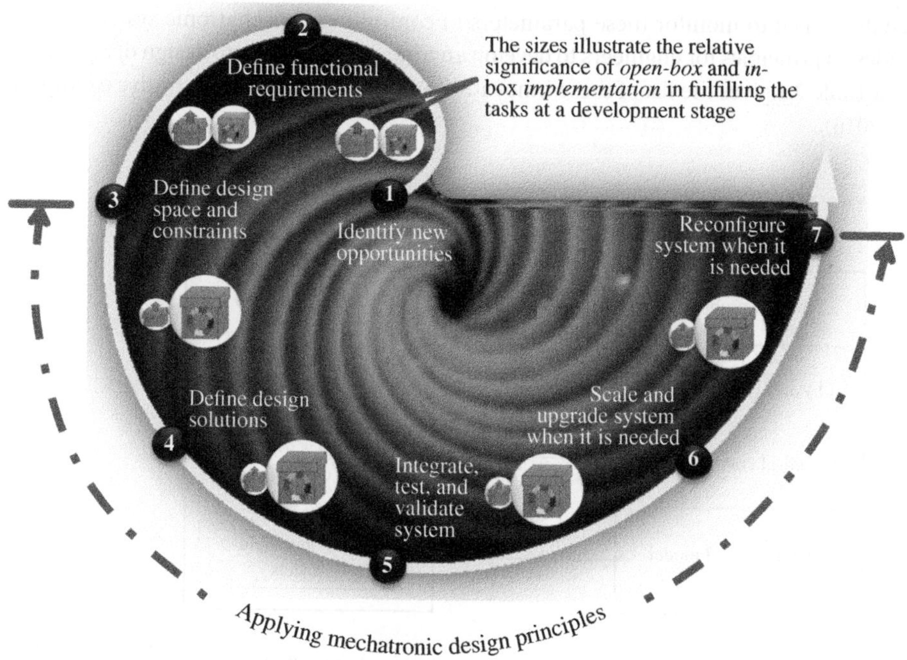

Figure 2.3 Role of mechatronic design in an innovation spiral

2.2 Theory of Inventive Problem-Solving (TRIZ) as Tactic Methodology

To convert an innovative idea into tangible mechatronic products, a systematic engineering methodology makes significant differences of efficiency and productivity in formulating a design problem, defining the scope of possible solutions, and optimizing the DS (Hollanders and van Cruysen 2009). The *theory of inventive problem solving* (TRIZ) is widely used to explore and implement the innovativeness of mechatronic products at a module level (Gadd 2011; Chang et al. 2016).

TRIZ was developed by G. Altshuller based on the findings from the analysis of existing patents. In applying TRIZ, (1) an innovative problem is formulated by contradictory statements on 39 parameters; (2) the space of DSs consists of 11 separation principles and 40 principles of innovations; and (3) a solution refers to the principles that resolve the contradictions adequately. The scope of a design problem by TRIZ is limited in terms of the number of contradictions. When an innovative problem becomes complex, separation principles are used to separate contradictions in a space domain and/or the time domain. Figure 2.4 shows the process of using TRIZ to solve an engineering problem.

Thirty-nine parameters are used to describe the contradictions of an innovative problem include *weight*, *length*, *area*, *shape*, *stability*, or *volume* of a stationary or moving object, *force*, *speed*, *energy*, and *power* of a moving object, a distribution of *stress*, *strength*, *intensity*, and *temperature* of an object, a duration of *action*, loss of *energy*, *substance*, *information*, and *time of action*, *quantity*, *accuracy*, and *precision of action*, *harmful factors*, *ease of manufacture*, *operation*, and *repair of an object*, *adaptability*, *versatility*, and *complexity* of object, *difficulty* of detecting and measuring, *automation*, and *productivity* of action (Gadd 2011). In sequential chapters, the measurement and instrumentation will be developed to monitor these parameters in controlling a mechatronic system.

TRIZ provides 40 principles for engineers to develop an innovative solution to design of a product or system, and Table 2.1 gives the list of these innovative principles as well as illustrative examples (Chang et al. 2016).

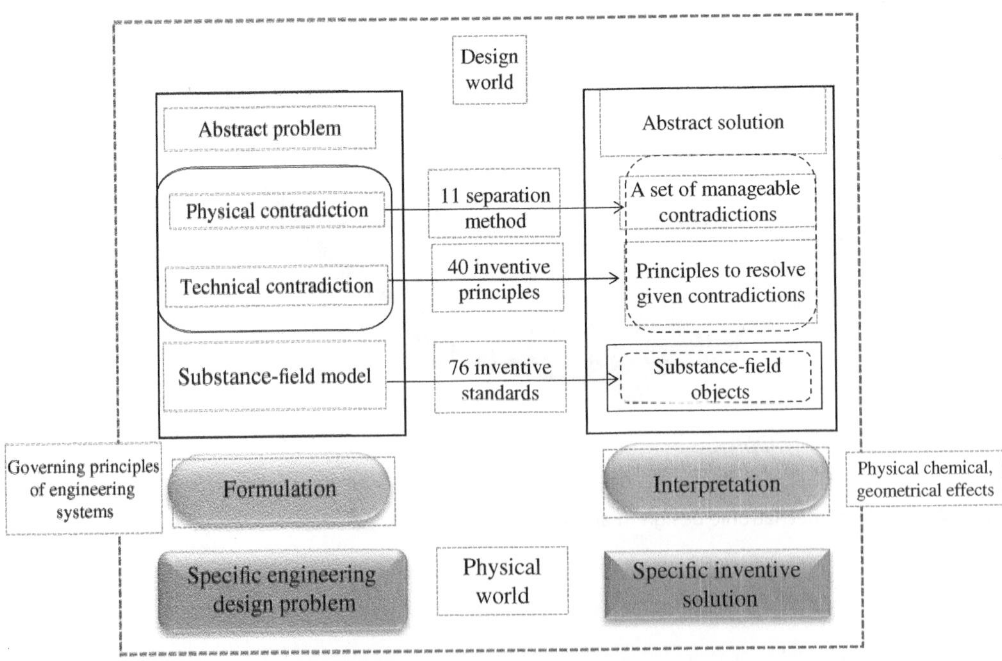

Figure 2.4 Problem-solving process by using TRIZ. (Rousselot and Renaud 2015; Russo and Duci 2015)

Table 2.1 Examples of using 40 principles for innovative mechatronic designs

	Principles	Examples	Illustrations
1	**Division**	Modularized architecture of a product or system	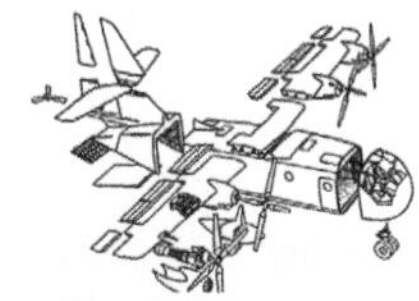
2	Taking out	Taking out of notoriously noisy components, such as power units, engines, and turbines, from systems	
3	Local quality	Luxury Edition (LE), Sport Edition (SE), Executive Luxury Edition (XLE), Limited, and Premium trims of a car brand	
4	Asymmetry	Assistive robots and personalized robots	
5	**Merging**	Reconfigurable systems, networking	
6	Universality	General-purpose drones, robots, and self-guided vehicles	
7	**Embedded structure**	Matryoshka doll, embedded circuits and controls, and drive-by-wire systems	
8	Counterweight	Anti-gravity counterbalance systems, fish bladders, and balloons	
9	Preliminary anti-action	Counting acting waves, anti-impact piezoelectric system in cutting, and anti-missile systems	
10	Prior action	Self-supervised deep reinforcement learning navigation, cleaning printing nozzles, resetting reading heads, and default settings.	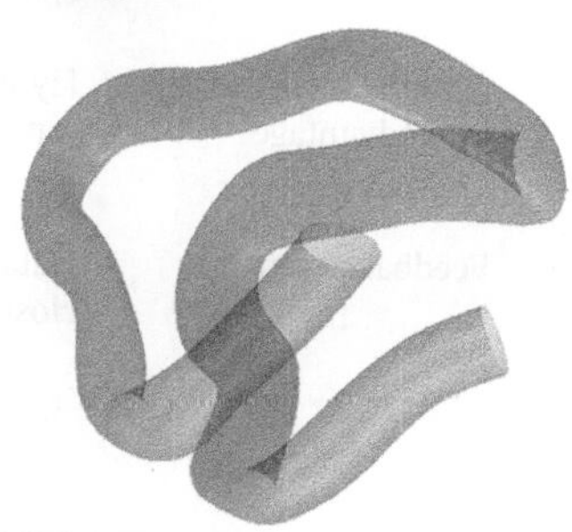

(*Continued*)

Table 2.1 (Continued)

	Principles	Examples	Illustrations
11	Beforehand cushioning	Airbag, masking before painting, and preventive actions	
12	Equipotentiality	Mechanism synthesis, dissolvable surgeon threads, and replaceable parts	
13	Inversion	Sensors and actuators, motors and generators	
14	Spheridality	Omni-vehicles, bearings, spherical robots, and rollocopters	
15	Dynamicity	Self-balancing robots, automatic air-locks, automatic fuel injections depending on driving conditions, automatic gears in cars, and autonomous driving	
16	Partial, overdone, or excessive actions	Dimensional margins of castings for machining, removing excessively sprayed paint, assembling for interference fits, trimming, and deburring processes	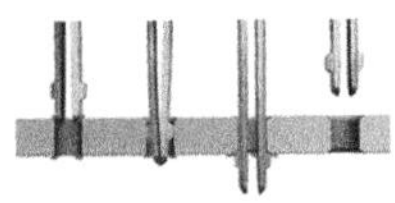
17	**Adding another dimension**	Augmented reality (AR) systems, metaverse	
18	Mechanical self-induced vibration (resonance)	Piezo- micro-electromechanical systems (MEMSs), vibration energy harvesting systems, mechanical oscillators, and concrete vibrators	
19	Periodic action, pulsed action	Smart lifts, hammer drills, pulsed amplifiers, and pulsed lasers	
20	Continuity of useful actions without an idle mode	Expert-guided machine learning, printing cartridges, stream turbines, and pump-storage power plants	
21	Skipping, quick mode, or pace of realization	Heat treatments, pico- or emio-second pulsed laser cutting, and rapid battery chargers	
22	Convert disadvantage into advantage	Hybrid cars, brakes, clutches, and renewable technologies	
23	Feedback principle	Data-driven controls, autopiloting, and closed-loop controls	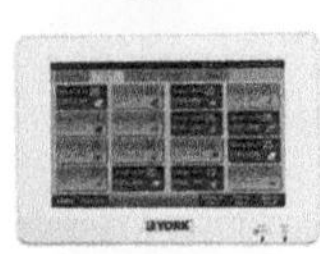

Table 2.1 (Continued)

	Principles	Examples	Illustrations
24	Intermediate means, mediating, and fitting principle	Agent-based controls, electric circuits for fittings of impendence, resistances, or power inputs, fittings of in-sense pressure, loading, transmitting, and reconfigurable systems	
25	Self-service principle	Self-healing materials and robots, self-repairing devices, and automated teller machines (ATMs)	
26	**Copying and imaging**	Ultrasounds, magnetic resonance mapping, mirrors, printers, and self-replicating living robots	
27	An inexpensive short-lived object and disposable solutions	Kitchen utensils, dishes, plastic bags, wrappers, disposable gloves, self-destructing origami robots	
28	**Substitutive mechanical systems**	Replacing gear sets with step motors, replacing fasteners with magnetic forces (magnetic couplings), and electrical discharge machining	
29	Use of a pneumatic or hydraulic construction	Pneumatic and hydraulic power generators, automobile brakes, and pneumatic or hydraulic controls	
30	Flexible films, foils, and membranes	Non-wettable films, inflatable balloons, domes, and barriers, wrapping packaging, air-pumped bubbles, anti-bullet vests, coatings, and paints.	
31	Use of porous materials	Self-lubricating bearings, aerated concrete, foams, and sponge materials	
32	Color change	Smart camouflage devices, lapping processes, and probing of phosphorescence distributions	
33	Homogeneity	Homogeneous robotic systems, products with homogenous materials, and behaviors, features, or other characteristics	

(Continued)

Table 2.1 (Continued)

	Principles	Examples	Illustrations
34	Discarding, recovering, rejecting, and regenerating parts	Dissolvable pills, rockets, cornstarch-based packages, resilient system architectures, and controls	
35	Changing states and properties of materials	High-temperature food processing, chemical products, heat treatments, vehicles, and microwaves.	
36	Phase transition	Vanadium dioxide light sensors, transition-edge sensors and phase-changing fixtures, refrigerators, and liquefied ammonia heat carriers	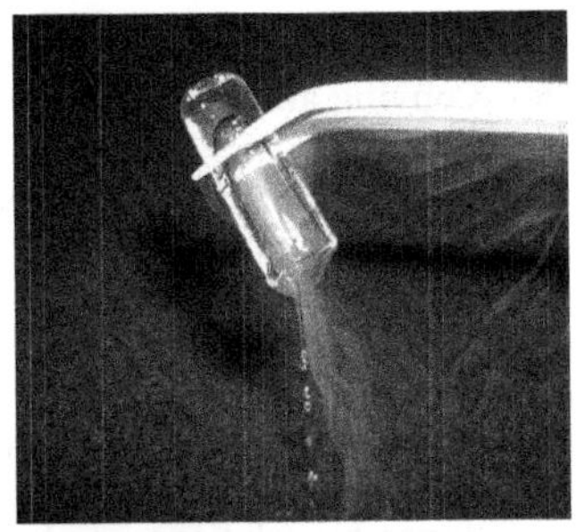
37	Thermal expansion	Relays, thermally activated MEMSs, thermometers, and thermal balancing	
38	Use of strong oxidizers	Oxygen (O_2), ozone (O_3), indirect vapor (H_2O), and high-velocity oxygen fuel coating	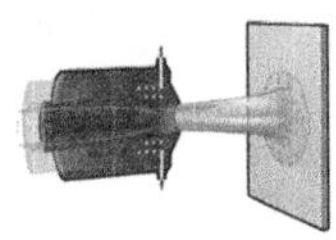
39	Neutral atmosphere, inert environments	CO_2 extinguishers, N_2 or He_2 protections, and shielding in arc welding.	
40	**Composite materials**	Titanium alloys, aluminum alloys, and composite materials on airplanes	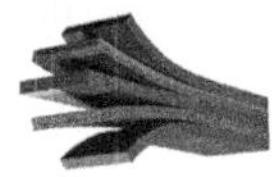

Bold terms are more closely relevant to Mechatronics.
Sources: Tesla / https://cdn.motor1.com/images/mgl/RvGyr/s1/tesla-model-3-y-prices-20210221.webp / Last accessed on July 19, 2024; Zeelool Inc / https://www.zeelool.com/goods-detail/FA0602-01 / Last accessed on July 19, 2024; Trumpet, LLC / https://blog.trumpetinc.com/dont-be-a-victim-to-all-in-one-ria-software / Last accessed on July 19, 2024; Elmo Motion Control Ltd / https://www.elmomc.com/product-tag/g-bee/ / Last accessed on July 19, 2024; DBolical Pty Ltd / https://www.moddb.com/mods/the-end-of-days/images/china-hq-61-anti-air-missile-launcher / Last accessed on July 19, 2024; Magnus Engineered Equipment / https://www.magnusequipment.com/industry/remanufacturing/ / Last accessed on July 19, 2024 AliExpress.com / https://www.aliexpress.com/item/1005004686271202.html / Last accessed on July 19, 2024; Autodesk Inc / https://www.autodesk.com/community/gallery/project/137361/spherical-robot / Last accessed on July 19, 2024; Innovation Origins / https://innovationorigins.com/en/the-brain-of-a-self-driving-car-learns-just-like-a-child/ / Last accessed on July 19, 2024; Gedema / https://www.gedema.com/en/products/deburring-tool-x-grat/deburring-tool-x-grat-oe-10-31-16-28-mm/20991/deburring-tool-x-grat-oe-13-87-14-30-mm-xg-13-87 / Last accessed on July 19, 2024; Maruti Sagar Engineerin / https://www.marutisagarengineering.co.in/petroleum-oil-testing-equipment.html / Last accessed on July 19, 2024; Bosch Limited /

https://www.rs-online.vn/p/impact-drill-gsb-19-2re-110v/6957101/ / Last accessed on July 19, 2024; Printer Test Page / https://printertest-page.com/how-to-test-printer-ink-and-toner-quality/ / Last accessed on July 19, 2024; Schumacher Electric Corporation / https://www.homedepot.com/p/Schumacher-Electric-15A-6-Volt-12-Volt-Fully-Automatic-Battery-Charger-and-Maintainer-with-Auto-Voltage-Detection-SC1323/315434899 / Last accessed on July 19, 2024; Laird Thermal Systems, Inc / https://www.telerex-europe.com/en-gb/thermoelectric-peltier-modules-ms-series-ms4--129--10--15--11--w8 / Last accessed on July 19, 2024; Johnson Controls / https://www.latam.johnsoncontrols.com/en/hvac-equipment/variable-refrigerant-flow-systems/controllers / Last accessed on July 19, 2024; Adapa A/S / https://www.metyx.com/wp-content/uploads/PDF_Files/METYX%202021%20Activities%20Report.pdf / Last accessed on July 19, 2024; Iconfinder ApS / https://www.iconfinder.com/icons/87409/atm_icon / Last accessed on July 19, 2024; BISINFOTECH / https://www.linkedin.com/posts/akshay-bellare-772a0b7_honeywell-starts-production-of-disposable-activity-6698593056184266753-xXTJ/ / Last accessed on July 19, 2024; Hyvair Hydraulic Pneumatic Fluid Power Components / https://hyvair.com/ / Last accessed on July 19, 2024; GlobalGilson.com / https://www.globalgilson.com/242in-latex-membranes-0012in-thick / Last accessed on July 19, 2024; Hebei Baorui Rubber Products Compony Ltd / https://www.alibaba.com/product-detail/0-48-8-inch-custom-soft_62287660853.html?spm=a2700.7724857.0.0.551183ddZQO1DS / Last accessed on July 19, 2024; SPC GERALD JAMES / Wikimedia Commons / Public Domain; ROBOTLAB Inc / https://www.robotlab.com/store/cubelets-boundless-builder-pack / Last accessed on July 19, 2024; Estes Industries LLC / https://estesrockets.com/products/destination-mars-mav / Last accessed on July 19, 2024; Deglr6328 / Wikimedia Commons / CC BY SA 3.0; Circuit Interruption Technology, Inc / https://www.citrelay.com/view_relay.php?series=A2H / Last accessed on July 19, 2024; R.C. Alderliesten (2005) / R.C. Alderliesten)

It is seen from Table 2.1 that a solution to an engineering problem is traditionally found in a dominant discipline such as mechanical or electrical engineering. With the ever-increasing complexity of modern products, the uses of TRIZ are more and more relied on the principles that are highlighted in the table such as *divisions*, *merging*, and *copying*. These principles focus on the innovations occurring in the integration and synergizing of multidisciplinary behaviors of products (Bonnema 2011; Mhenni et al. 2014).

2.3 Innovations of Mechatronic Systems

In comparison with traditional products in certain physical domains, a mechatronic product is characterized by the integration of multidisciplinary solutions. The innovations of mechatronic products can be achieved from many aspects such as *modularization*, *integrability, reconfigurability*, *decentralization*, and *distribution* for *versatility*, *adaptability*, *predictability*, and *resilience* of systems.

2.3.1 Modularization

Differing from an integral system, a mechatronic system is modularized in which system elements are developed and maintained as individual modules, respectively. Different types and the number of modules can be used to configure different systems for specific FRs in applications. Modularization allows engineers to solve a complex problem by developing sub-solutions to a set of simple problems. *The top-to-bottom approach* is used for modularization to deal with the complexity of mechatronic systems (Figure 2.5).

Modularity allows to maximization of the commonality in system configurations and provides more variants by selecting different numbers or types of modules, and assembling the selected modules in different ways (Bi and Zhang 2001a, 2001b; Bi 2002; and Holtta-Otto and de Weck 2007). A *design structure matrix* (DSM) represents mechatronic structure and modularization that are characterized by the dependence and encapsulation of modules. A singular value modularity index

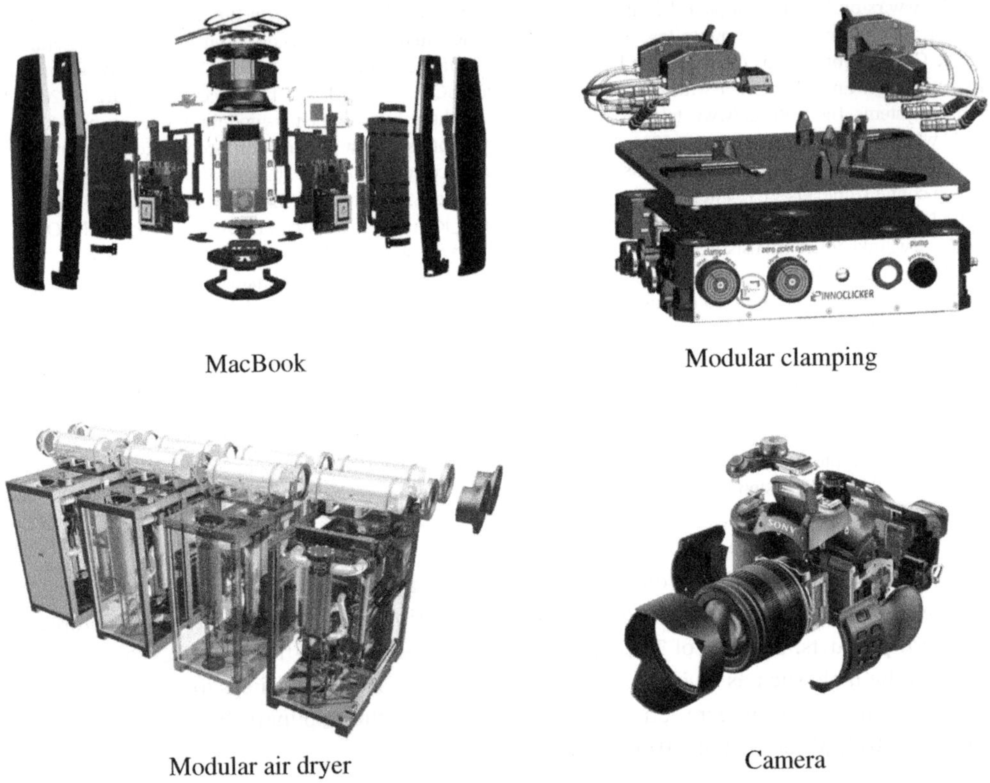

Figure 2.5 Examples of modularized mechatronic systems (*Sources:* (a) Apple Inc / https://www.parts-people.com/blog/2016/03/23/msi-takes-on-macbook-pros-cylindrical-design-with-new-vortex-gaming-pc / last accessed on July 04, 2024; (b) pneumatic Products / https://blog.mnteng.com/2018/03/modular-refrigerated-air-dryer-for.html / last accessed on July 04, 202; (c) Innoclamp GmbH / https://www.innoclamp.de/en-gb/innoclicker / last accessed on July 04, 2024; (d) Sony Electronics Inc)

(SVMI) was introduced to evaluate the degree of modularity of a system or product (Holtta-Otto and de Weck 2007). With a modularized architecture, a set of plug-and-play modules is selected and assembled into various system configurations to meet the needs of specific applications. Modularized architecture supports (1) *composability* where selected mechatronic functional modules (MFMs) can be connected physically and logically to form an MFM at an upper level; (2) *integrability* where multiple MFMs are integrated to fulfill expected tasks; (3) *interoperability* where power and data can be shared within a set of modules via appropriate connectors; (4) *granularity* by which MFMs at lower levels can be composed to scale up a system; (5) *openness* where MFMs are plug-and-play that allow them be connected to intended collaboration; and (6) *reusability* where specific MFMs can be assembled and disassembled in different system configurations (Bi 2002; Zou et al. 2022).

It was found that 80% of mechatronic systems adopted modularized architectures and the main components were reusable modules. This implied a new mechatronic system should begin with selecting available functional modules and then customizing their assemblies for specified requirements. Adopting standardized modules implied that a great variety of products could be made using different combinations of modules. Accordingly, these modular components could be made by mass production (Lettner et al. 2015). For example, a robot can be built by assembling a set of selected motors, linkages, joints, end-effectors, and controllers.

Modularization is a fundamental principle of mechatronic designs, and the benefits of system modularization are self-evidenced. *From designers' perspectives*, modularization allows to (1) assemble standardized modules into customized products for specific needs by users, (2) change a system configuration to accommodate changes, (3) simplify the modification and substitution of modules or assembling, (4) reduce the number of innovative designs, and (5) facilitate *concurrent engineering* (CE) for complex system designs. From the users' perspective, modularization allows to (1) upgrade modules and systems when needed, (2) improve reliability or resilience by upgrading or replacing malfunctioning modules readily, and (3) have cost-effective products since the modules with a large volume can be made by mass customization (Bi and Zhang 2001a, 2001b).

2.3.2 Integrability

Integrability refers to the capability of undergoing integration or of being integrated. Integrability is an ability of developing separate components, respectively, and constructing them together as a system appropriately (Kazman et al. 2020). A mechatronic system is usually versatile; the system-level functions are mostly fulfilled by integrating a number of functional modules. On the one hand, the performance of a mechatronic system can be measured by integrability at both the system level and module level. At the *system level*, the more the system configurations can be generated from a set of system modules, the higher the system integrability is. At the *module level*, the more the interfaces a functional module has to connect other modules, the higher the integrability of this module is. *The bottom-up approach* is used by integrability to deal with the complexity of a mechatronic system.

Figure 2.6 shows the types of interfaces to integrate elements into a system. Other than the types and the number of system elements, the interfaces of system elements are other main factors to determine the variants of an integrated system. The *heterogeneous architecture* in Figure 2.6a allows

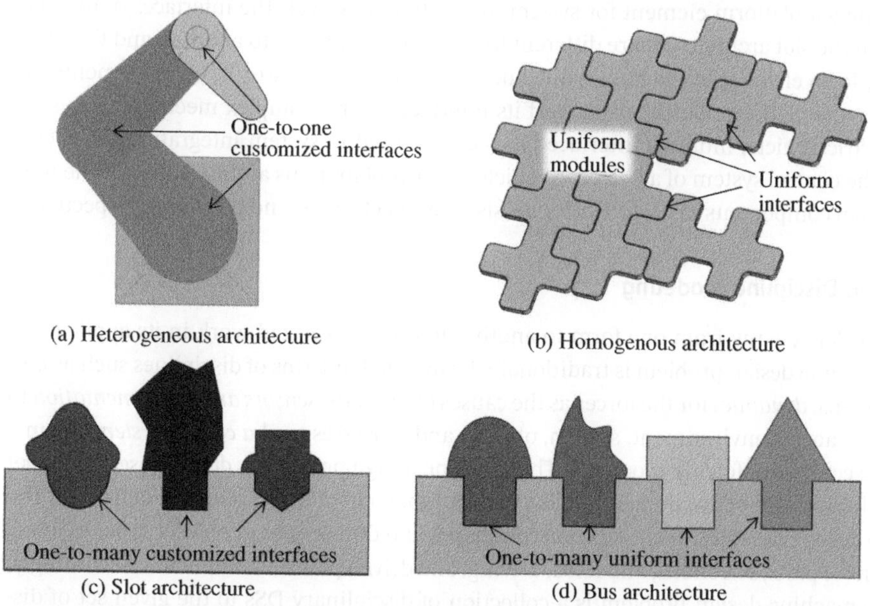

Figure 2.6 Interface types for system integration

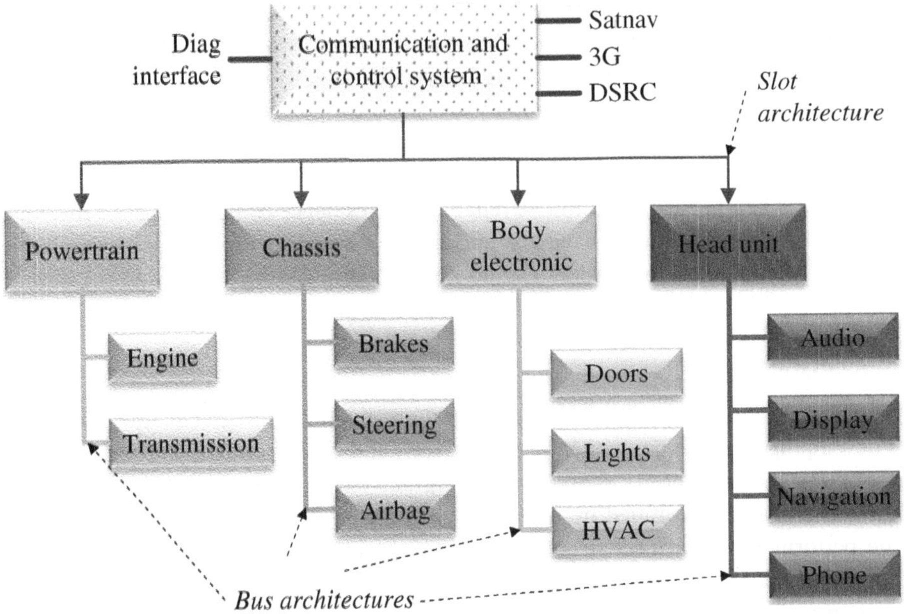

Figure 2.7 Examples of mixed interfaces in complex systems (*Source:* Adapted from Lokman et al. (2019))

one-to-one interaction of two elements, and the system configuration is specific to a given application. *Homogeneous architecture* in Figure 2.6b is applicable to a modular system with the same type of system elements; elements can be integrated with others in an arbitrary way to maximize the number of system configurations. Both *slot architecture* in Figure 2.6c and *bus architecture* in Figure 2.6d deploy a platform element for system integration; however, the interfaces of the platform element in the slot architecture are different from one element type to another, and the interfaces of the platform element in the bus architecture are same; in other words, other elements can be integrated to the platform element at any of its interfaces. For a complex mechatronic system such as an electric vehicle, multiple interface types can be used in system integration. As shown in Figure 2.7, the control system of an electric vehicle consists of four bus architectures for the integrations of system components in powertrain, chassis, body electronics, and head unit, respectively.

2.3.3 Coupled Discipline Modeling

A machine transfers energy from one form to another to do the expected work in its application. Therefore, a machine design problem is traditionally formulated in terms of disciplines such as *kinematics* for motions, *dynamics* for the forces as the causes of motions, *sensors and instrumentation* to understand the states of environment, system, objects, and processes, and *a control system* to run a machine and execute transferring processes. The machine design problem is decomposed into a set of the subproblems in respective domains; DSs to a subproblem in one discipline are defined by the corresponding design variables, such as geometric shapes and dimensions of objects, types of joints and connections of parts, selections, and fixed, driving, and driven parts in kinematics. Accordingly, a solution to a machine design problem is a collection of disciplinary DSs to the given set of disciplinary subproblems obtained in respective disciplines. As shown in the low portion of Figure 2.8, a machine design process is iterative to deal with the dependencies of multidisciplinary behaviors. Note that an iterative design process might lead to a convergence problem when the multidisciplinary behaviors become strongly coupled.

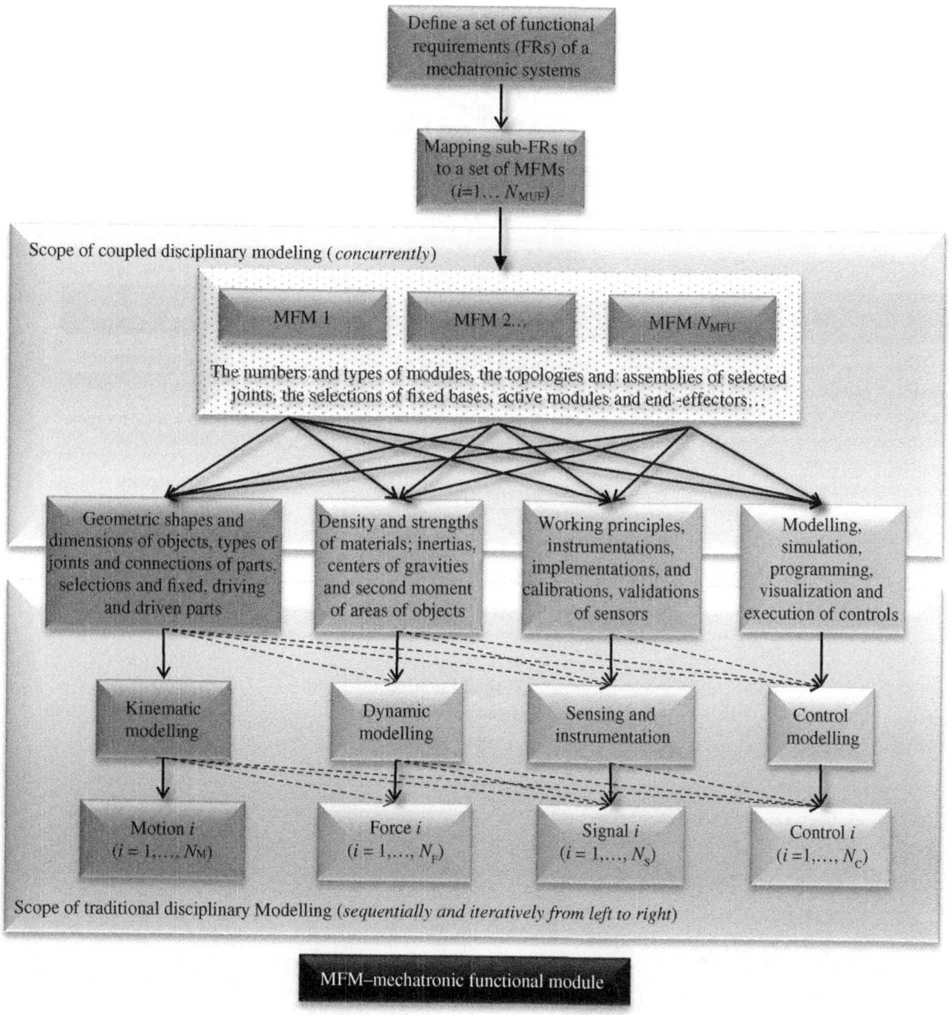

Figure 2.8 Coupled disciplinary modeling in a mechatronic system

A mechatronic system consists of a set of *mechatronic functional modules* (MFMs), and the design of a mechatronic system relates to the selections and assemblies of MFMs. Different from the design variables in an individual discipline of conventional machine design, the typical design variables in a mechatronic system are *the numbers and types of modules*, *the topologies and assemblies of* selected joints, *the selections of fixed bases*, and *active modules* and *end-effectors* as shown in Figure 2.8. Moreover, these design variables are strongly coupled in the sense that all of the design variables must be solved to meet the design constraints in all relevant domains simultaneously. Figure 2.9 shows the strong couplings of design variables of a modular robotic system with those of design variables in kinematics, kinetics, dynamics, and real-time controls.

2.3.4 Concurrent Design

A mechatronic system is involved with multiple disciplines, and each discipline brings a set of constraints to system design variables. Therefore, the design of a mechatronic system must take into consideration of the design constraints in all the disciplines simultaneously to fully explore a design

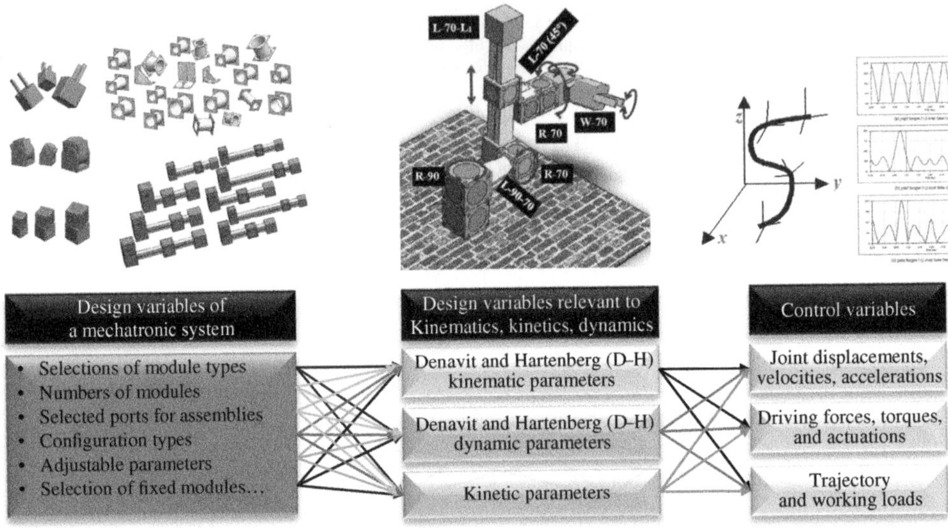

Figure 2.9 An example of strongly coupled mechatronic system

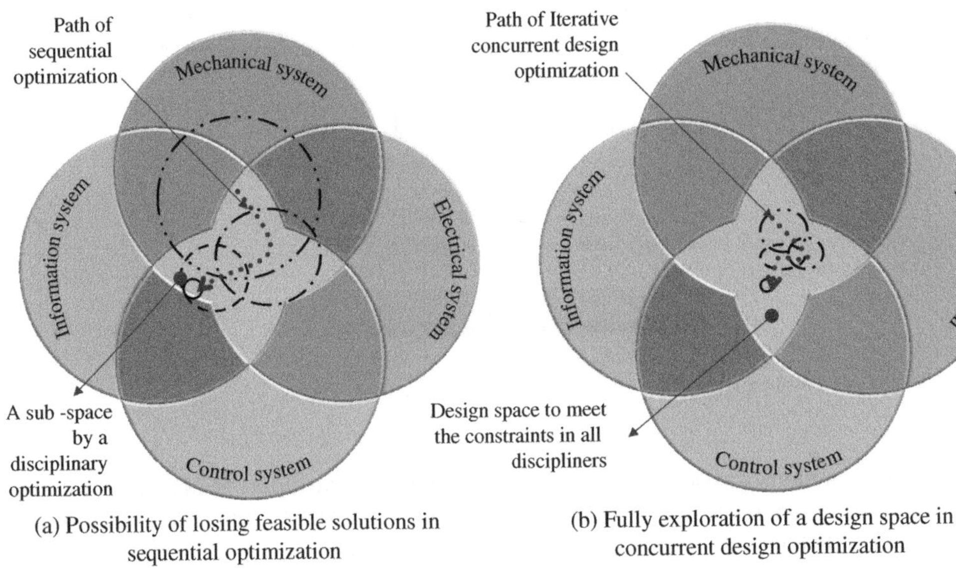

Figure 2.10 Concurrent design versus sequential design

space. As shown in Figure 2.10, the traditional sequential design brings the risk of missing feasible solutions due to a local optimization at each step without the consideration of the design constraints in sequential steps (Bi and Zhang 2001a, 2001b).

Concurrent engineering (CE) is a method where all of the FRs, design constraints, and optimization criteria in multiple disciplines are taken into consideration simultaneously to fully explore the design space and optimize a DS globally. Figure 2.10 shows the comparison of the path in seeking an optimal solution to a mechatronic system. *In a sequential design procedure*, the system design problem is divided into a sequence of subproblems in respective disciplines; the local optimizations in these disciplines bring the risks of losing feasible solutions in a global space; since the design constraints in other disciplines cannot be automatically satisfied when a sub-solution is sought in one discipline. *In a concurrent design procedure*, the design constraints and criteria in all disciplines are taken into consideration simultaneously, and this allows for exploring and optimizing a system solution in a full design space.

2.3.5 Decentralized Controls

Traditionally, a hieratical architecture is used to control a complex system. Figure 2.11a shows a layered control architecture in which the control units in the lower layer are coordinated by the system controller at the upper layer. However, different time frames are used to plan, schedule, and control activities in different layers. The inputs of a control unit at a specific layer are the outputs of these units at the upper layer, and the outputs of the control units at this layer are the inputs of the lower layer until these inputs become the execution commands of hardware resources at the physical layer. As shown in Figure 2.11a, an integrated control system consists of two data flows, that is, *top-down breakdown* for system operation and execution, and the *bottom-up feedback* for system updating. The rigidity of these two data flows limits the system's flexibility in dealing with the changes and uncertainties in the business environment; since real-time interactions occur only to the units within one layer.

The controllers of mechatronic functional units are embedded in their hardware systems, and the units can respond to environmental changes through predefined interfaces automatically. In other words, the system-level goals are achieved by coordinating the behaviors of available modules. Figure 2.11b shows the decentralized control architecture of a mechatronic system. The control system consists of two types of system elements, that is, the controllers at the module level and the coordinators at the system level. The system elements at both levels are closely networked, so that the behaviors of modules are coordinated effectively in a real-time mode to achieve system-level objectives efficiently. No time will be wasted for the mechatronic system to respond to environmental changes promptly.

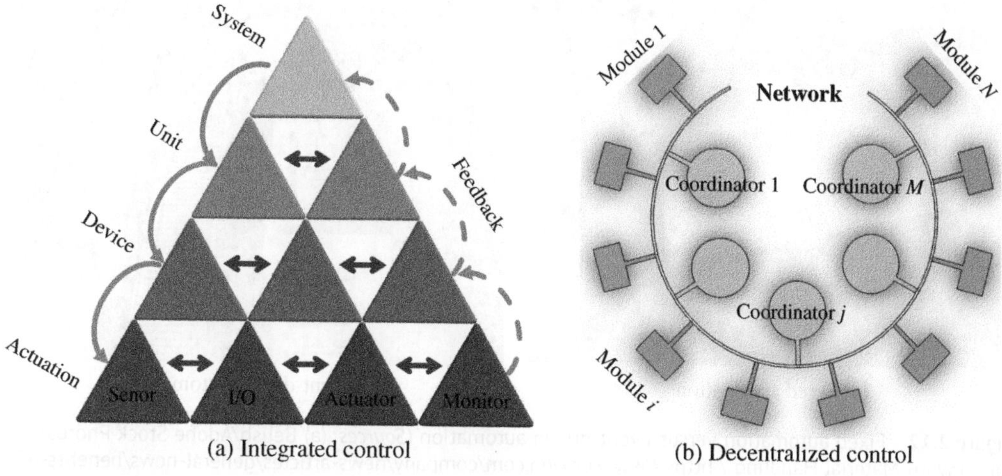

Figure 2.11 Integrated control versus decentralized control (*Source:* Adapted from Fan and Chang (2018))

2.3.6 Event-Driven Automation

A mechatronic system is reprogrammable in sense that the same hardware system can be used for different tasks based on how it is programmed. Depending on the automation level of programming or the capability of a certain program in dealing with changes and uncertainties in applications, automation in a mechatronic system can be either *fixed automation* or *event-driven automation* (Bi and Lang 2007). Figure 2.12 shows two examples of fixed and event-driven automation, respectively. The conveyer in Figure 2.12a is a type of fixed automation since the paths of the objects to be transferred are fixed that are determined when a conveyer is constructed. On the other hand, a robot in an automated warehouse in Figure 2.12b is a type of event-driven automation; since the motion path for a task depends on the detected origin and designation of the object to be transported. The application of a mechatronic system in an ill-structured and dynamic environment requires event-driven automation.

2.3.7 Adaptability and Re-configurability

Adaptability refers to a set of changes or task variants a mechatronic system can fulfill. On the one hand, system adaptability is evaluated at the interface of a mechatronic system with its environment based on the set of changes or task variants the system is able to cope with. On the other hand, the adaptability at the system interface is enabled by the re-configurability of the system itself. *Re-configurability* refers to the system's capability to satisfy new FRs or deal with distribution, uncertainties, or unanticipated changes. Adapting to any change in the environment requires re-configurability within the system, and the system re-configurability can be achieved by *software reconfiguration, hardware reconfiguration* (see Figure 2.13), or a combination of both (Bi 2002).

Adaptability and re-configurability relate to the changes of tasks in the environment or that of the configurations in the system; it will be desirable to identify and convert a set of changes into the set of FRs when a mechatronic system is designed. Unfortunately, task variants may or may not be

(a) Fixed automation (b) Event-driven automation

Figure 2.12 Fixed automation versus event-driven automation (*Sources:* (a) Belish/Adobe Stock Photos; (b) Gonvarri Material Handling / https://www.dexion.com/company/news-articles/general-news/benefits-of-investing-in-an-automated-warehouse/ last accessed on July 18, 2024)

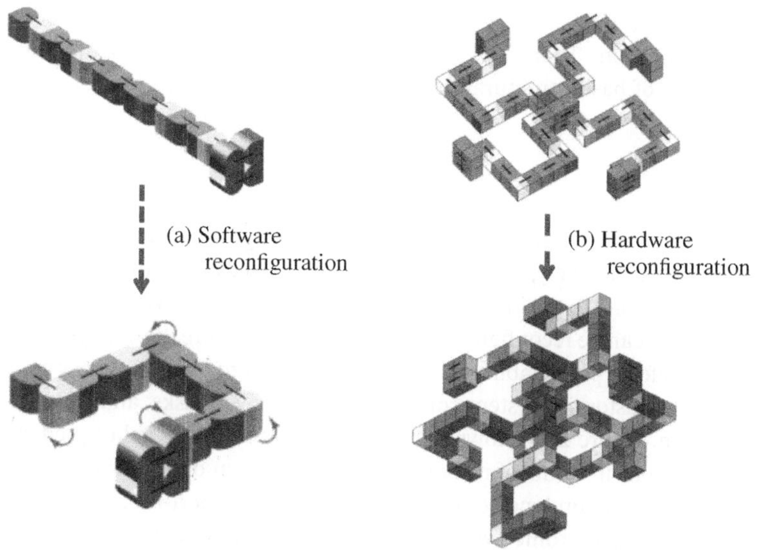

Figure 2.13 Example of software and hardware reconfiguration for system adaptability (*Source:* Aloupis et al. (2013)/with permission of Elsevier)

anticipated; this brings a great challenge in defining the FRs of a mechatronic system with less uncertainty and system re-configurability can be under- or over-defined.

To design a mechatronic system with its precise system re-configurability for given applications, it is ideal but impractical to determine its FRs from users' perspectives. Figure 2.14 shows two common ways to define the FRs in regards to system re-configurability. The first way in Figure 2.14a assumes that the hardware system is designed to be capable of fulfilling all task variants, and a change from one task to another is adopted by software reconfiguration, that is, programming. Therefore, the FRs of the system re-configurability are defined for the main characteristics of the hardware systems such as workspace, accuracies, and maximum loads, speeds, and accelerations. The second way in Figure 2.14b assumes that system structure or topology should be

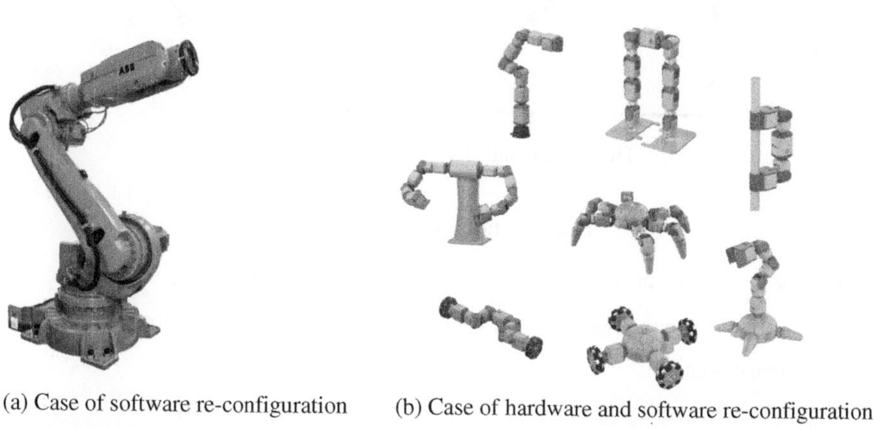

Figure 2.14 Two ad hoc ways of defining FRs of system re-configurability (*Sources:* (a) ABB (2024)/ABB; (b) Ozrobotics (2024)/Oz Robotics)

changeable to fulfill all task variants, and a change from one task to another is adopted by both hardware and software reconfiguration. Therefore, the FRs of the system's re-configurability are defined by specifying (1) a set of basic structural variants and (2) the specifications of main functional modules.

2.3.8 Predictability

A mechatronic system is mostly a closed system that can respond to changes adequately. When a change is detected and acknowledged passively, the mechatronic system needs some time to reconfigure its software or hardware to respond to the identified change correspondingly. When a change can be predicted actively, the system can be reconfigured and become ready before the change actually occurs and takes effect. Therefore, predictability is a critical measure to (1) improve system responsiveness in a short term, (2) customize system configurations for given tasks, and (3) optimize system design in a long term. Figure 2.15 shows the correspondence of the level of predictability and performance improvement in terms of responsiveness and system optimization.

System predictability depends on the abundance of the information about short-term and long-term changes and uncertainties of mechatronic systems and applications. System predictability cannot be fully utilized unless that a system possesses the capabilities to meet the identified changes

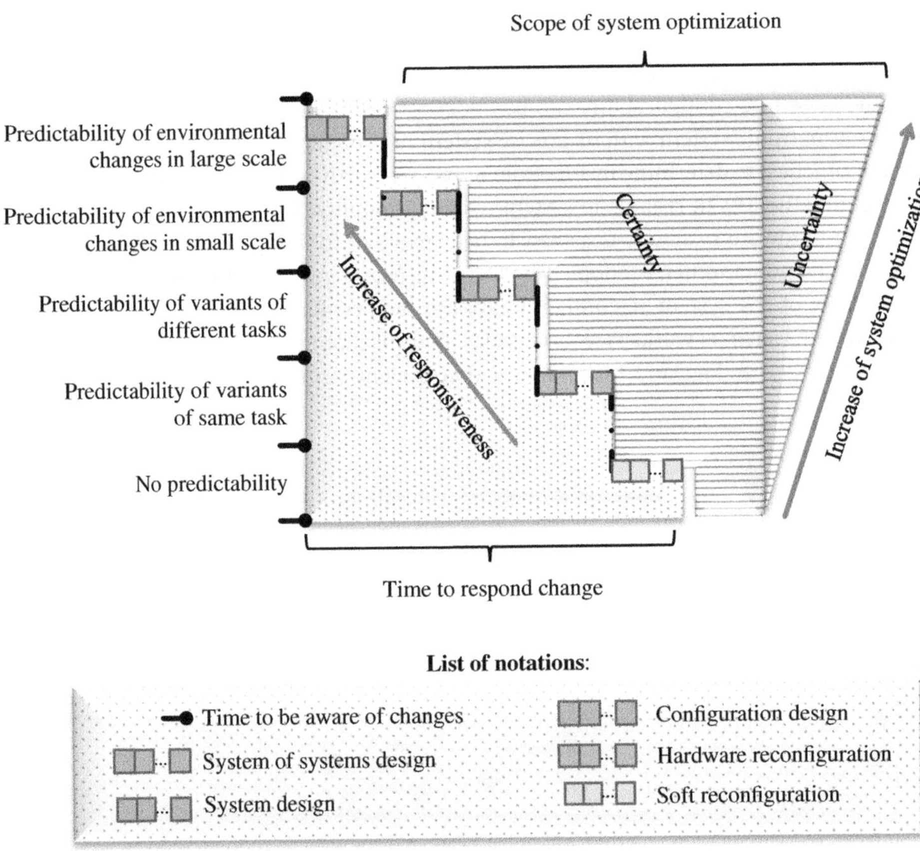

Figure 2.15 Level of predictability versus performance improvement (i.e., responsiveness and system optimization)

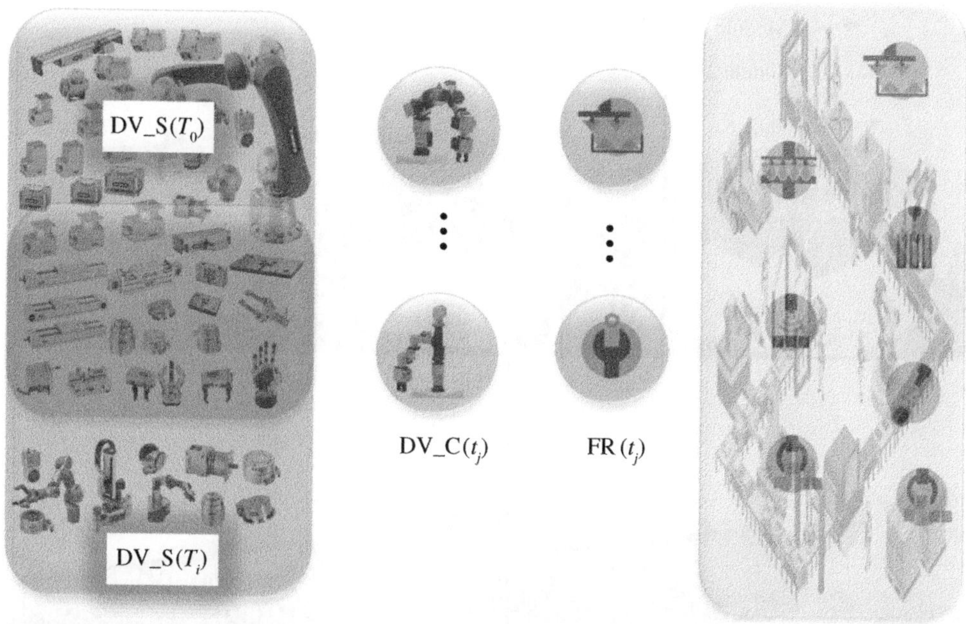

Figure 2.16 Modular systems as solutions to meet predicted changes

in a reasonable time frame. As shown in Figure 2.16, a mechatronic system usually adopts modular architecture. This implies that (1) numerous system configurations can be generated by selecting different modules and assembling them in different ways and (2) modules in a system can be added, modified, upgraded, or removed based on the needs of applications and technological readiness (Figure 2.17).

2.3.9 System Resilience

System resilience originally refers to the abilities of a system to continue to (1) operate under an adverse stress or a condition such as a debilitated or degraded state for basic system functionalities and (2) recover to an affective operational condition within a specific time frame that is consistent with mission needs (NIST 2024). The second ability relates to the self-healing capability of the system. It is the ability to predict potential problems and take correct actions to prevent system failures. System resilience was introduced into mechatronic systems by Zhang and van Luttervelt (2011), Zhang (2015), and Han et al. 2016); it was used to measure the capabilities of a mechatronic system to resume or recover the system operations when one or more mechatronic module(s) are malfunctioning. A resilient mechatronic system must be built on modular architecture and with functional redundancy (Zhang et al. 2017).

System resilience is achieved by reconfiguring a system to perform a certain task when a malfunction occurs in a system component. The main differences from the objectives of the system reconfigurability are that (1) the change that occurred to the system is internal and (2) no additional resources will be available when a malfunction occurs. Therefore, some modules in a mechatronic

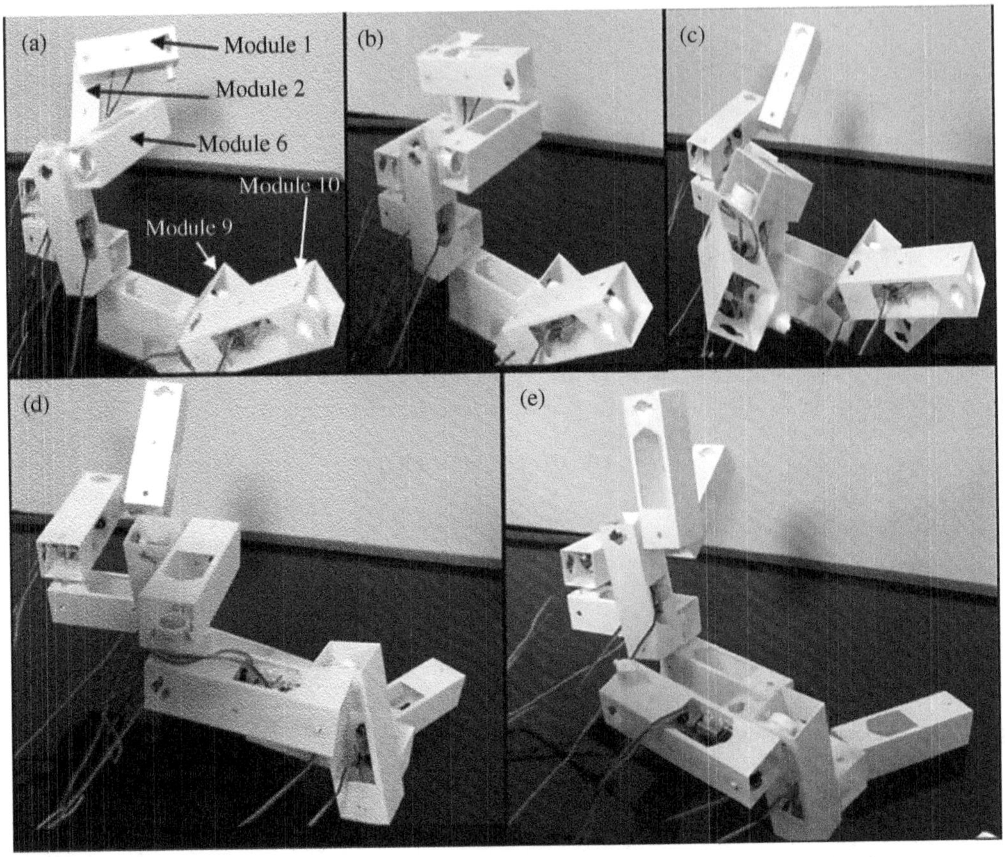

Figure 2.17 A self-reconfigurable robot for system resilience (*Source:* Zhang (2015)/Tan Zhang)

system with resilience must possess functional redundancy, so that basic functions can be still fulfilled when a part of them is malfunctioning. Zhang (2015) developed a self-reconfigurable robot that was able to reconfigure its modules to resume basic operations autonomously when one module was malfunctioning. The system showed two basic properties of a resilient system, that is, modular architecture and functional redundancy.

2.3.10 Continuous Adaptation (CA)

Due to the multidiscipline nature, concurrent design is essential for a mechatronic system for satisfying the design constraints in all disciplines simultaneously. However, one significant challenge in implementing a concurrent design methodology is that design constraints and objectives may be changed along with time; mechatronic solutions in applications should evolve with changes that occur.

A mechatronic system is required to be intelligent to deal with anticipated and unanticipated changes over time. In other words, the system should be flexible, reconfigurable, or adaptive to improve itself in fulfilling its missions in its applications. As shown in Figure 2.18, CA is the ongoing improvement of a mechatronic system through a series of system reconfigurations. CA seeks "incremental" improvement over time instead of a "breakthrough" improvement at once, as illustrated in Figure 2.19. From the perspective of the system life cycle, the dynamic data of the tasks for

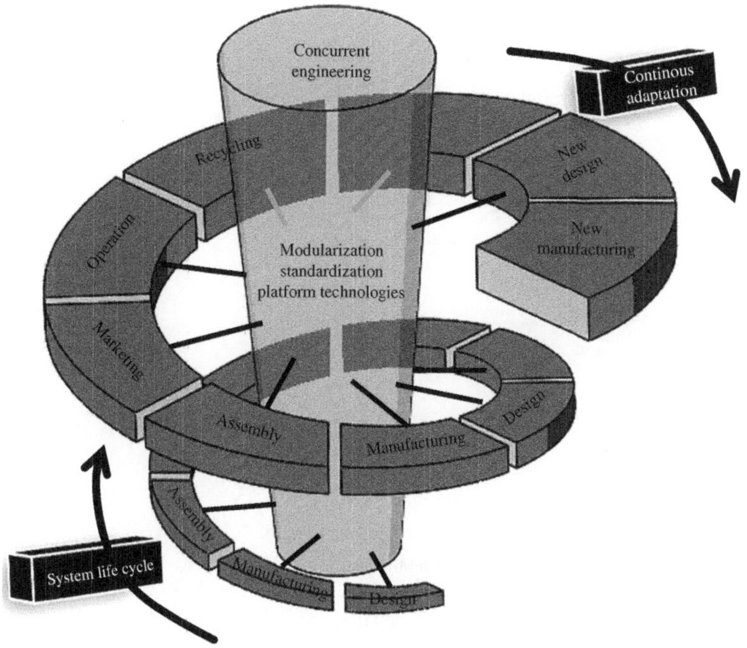

Figure 2.18 CE and *continuous adaptation* (CA) in a mechatronic manufacturing system

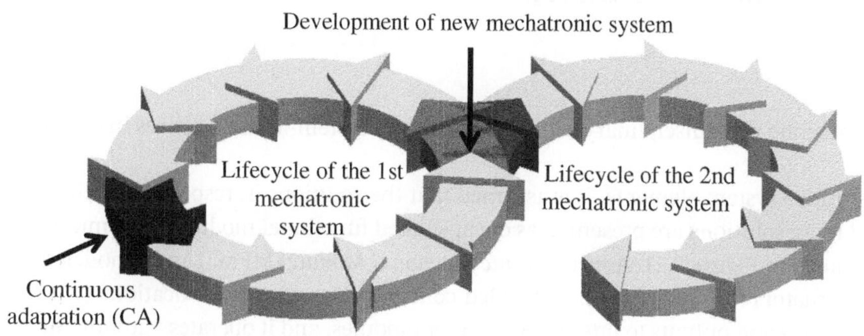

Figure 2.19 Evolution of a mechatronic system

a mechatronic system are collected until it is significant enough to trigger a system reconfiguration, and such a process is iterative as long as the mechatronic system can continuously fulfill the tasks satisfactorily.

2.4 Architecture of Mechatronic Systems

System architecture describes constitutive system elements, the relations of system elements, the changes of system elements and interactions, and the principles, rules, and methods to evolve a system over time. Figure 2.20 shows the architecture of a mechatronic system. Mechatronics aims

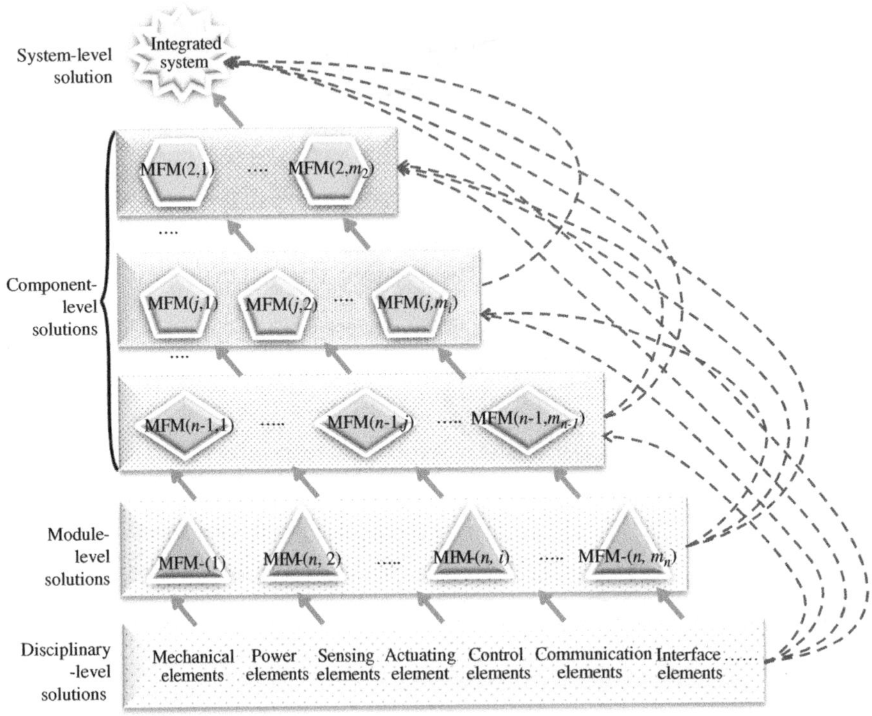

Figure 2.20 Generic architecture of a mechatronic system

to synergize and integrate multidisciplinary solutions to achieve system-level objectives in complex applications.

To model constitutive system elements, it is assumed that the solutions in respective disciplines are available, and these solutions are presented as encapsulated functional modules with multidisciplinary nature shown in Figure 2.21. A *mechatronic functional Module* (MFM) can be modeled as an integration of actuators (A), sensors (S), embedded controls (E), and communications (C). An MFM specifies its inputs and outputs to interact with other modules, and it operates autonomously in three types of resources, that is, material structure, information, and power. Figure 2.22 shows a generic model of MFM. It can be viewed as an encapsulated unit of mini-mechatronic subsystems. Accordingly, a comprehensive mechatronic system with a number of MFMs can be viewed as a system of mechatronic systems.

Similar to other engineering systems, an MFM is designed to implement one of four primary transformations. Therefore, an MFM can be abstracted as one of these primary transformations (Andreason 2022): (1) transforming materials in terms of geometries, dimensions, strengths, and quality); (2) transforming energy in terms of types, duration, forms, and quantities such as horsepower, amplitudes, and frequencies; (3) transforming information in terms of types, formats, amount, quality, and code; and (4) changing the states of biological objects such as people and animals. Figure 2.23 abstracts the functions of an MFM based on its transformations.

To model the relations of system elements, design of a mechatronic system is involved with (1) developing or selecting a number of MFMs that can fulfill respective FRs and (2) assembling

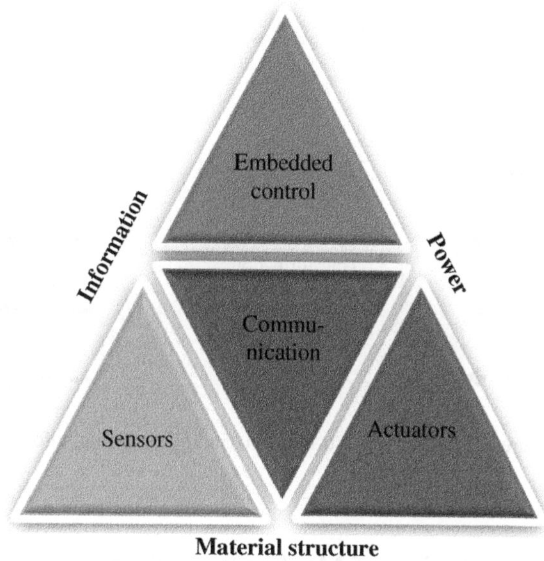

Figure 2.21 A typical encapsulated mechatronic unit (*Source:* Cheng et al. (2014)/ with permission of Elsevier)

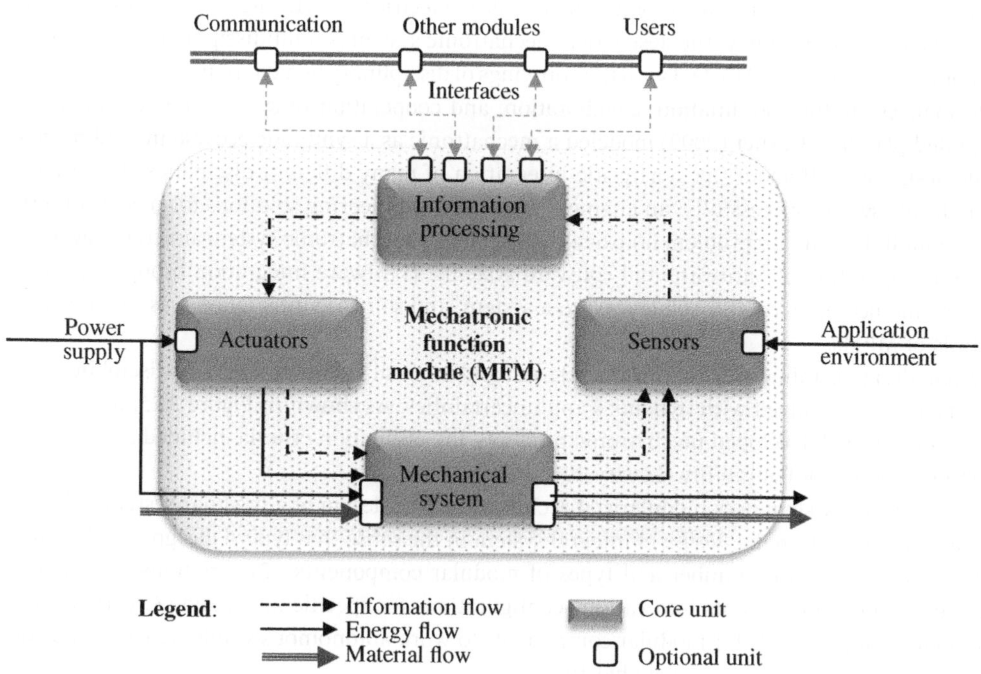

Figure 2.22 A generic model of mechatronic functional module (MFM)

selected MFMs as an integrated system that can fulfill given FRs at the system level. Qian et al. (2015) gave an example of a fish robot that was designed and built as a modular mechatronic system. Figure 2.24 shows that the fish robot is composed of three MFMs for motions, actuations, and controls to decouple the designs of respective MFMs to a great extent.

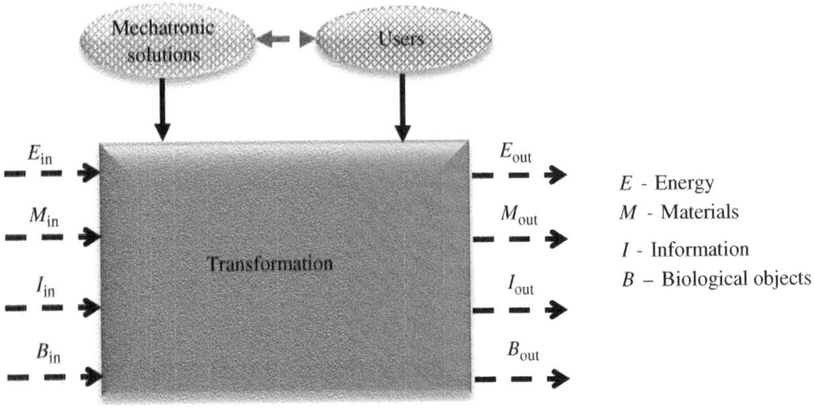

Figure 2.23 MFM as a transformation of materials, information, and biological objects

2.5 Design of Mechatronic Systems

A mechatronic system is an integration of mechanical, electrical, electronic, and information processing systems. Accordingly, the design of a mechatronic system is multidisciplinary, and systematic approaches are needed to deal with the couplings of disciplinary behaviors and optimize system performances via the coordination, collaboration, and cooperation of system components in an integrated platform. Garner (2009) modeled a mechatronic as a symbiotic ecosystem, and mechatronic design was all about the synergistic integration of many different disciplines. To design a mechatronic system successfully, an engineer developed engineering solutions to solve a comprehensive multidisciplinary problem holistically. Design of a mechatronic system aimed to develop a coherent integration of hardware and software systems. Hardware integration brought together sensors, microcomputers, and actuators into a system, and software integration was for advanced controls of MFMs.

System design is influenced by various external and internal factors in multiple disciplines, and these factors are involved with dynamics and uncertainties. As shown in Figure 2.25, to optimize the performance of developed mechatronic products, the evaluation criteria included adaptability, flexibility, and re-configurability (Levary 2006).

A mechatronic system uses modularized architecture. The design of a modular system is fundamentally different from the design of integral system in the sense that typical design variables in a design space are (1) the number and types of modular components, (2) variations to assemble selected modular components into various configurations for specific tasks, and (3) variations of coordination and controls for modular components to work autonomously, interactively, and collaboratively to achieve system-level objectives.

Figure 2.26 illustrates the differences between designs of mechatronic and integral systems. Mechatronic design requires a holistic approach to deal with coupled physical domains concurrently to reveal inexplicit synergies among these domains. Traditional ways to tackle with respective domains and then integrate them as a system-level solution usually need the iterations to find feasible solutions and expose the risk of not finding optimized solutions when the behaviors in these domains are closely coupled (Malmquist 2014). The procedure of designing a mechatronic system is featured by (1) exchanges of design data and models, (2) close interactions and

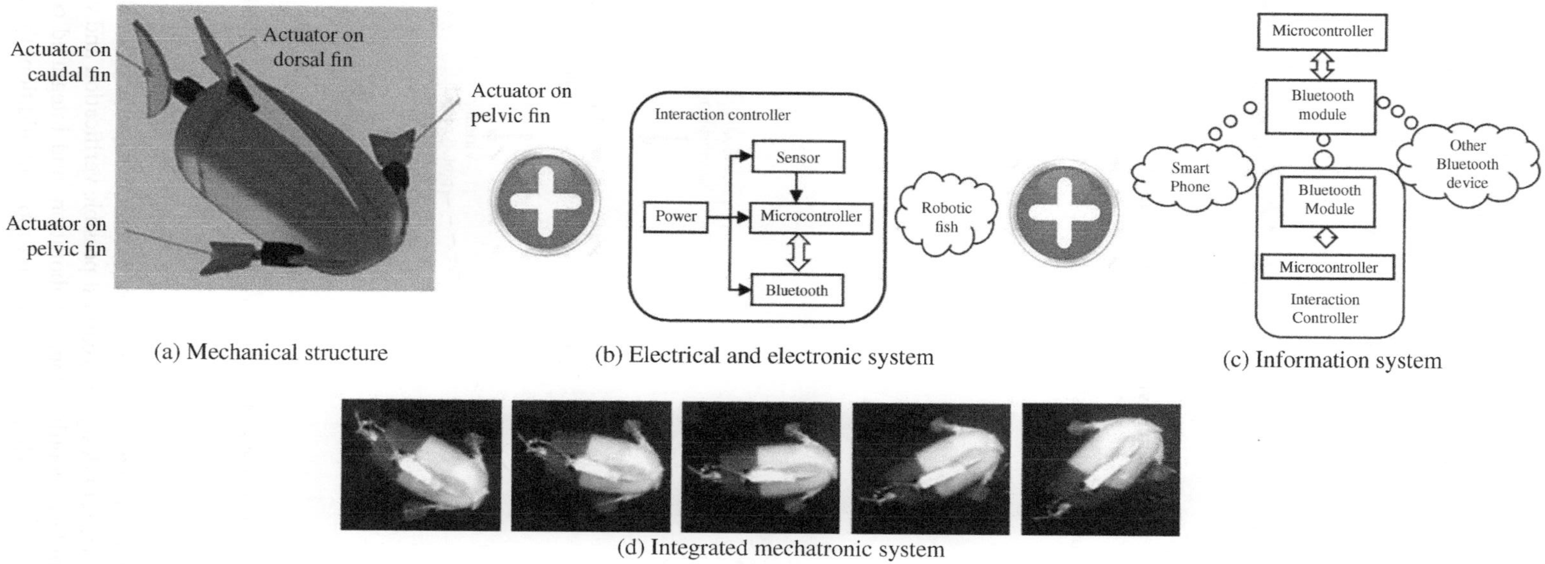

Figure 2.24 Example of combining a set of MFMs as an integrated solution (*Source:* Qian et al. (2015)/INASE)

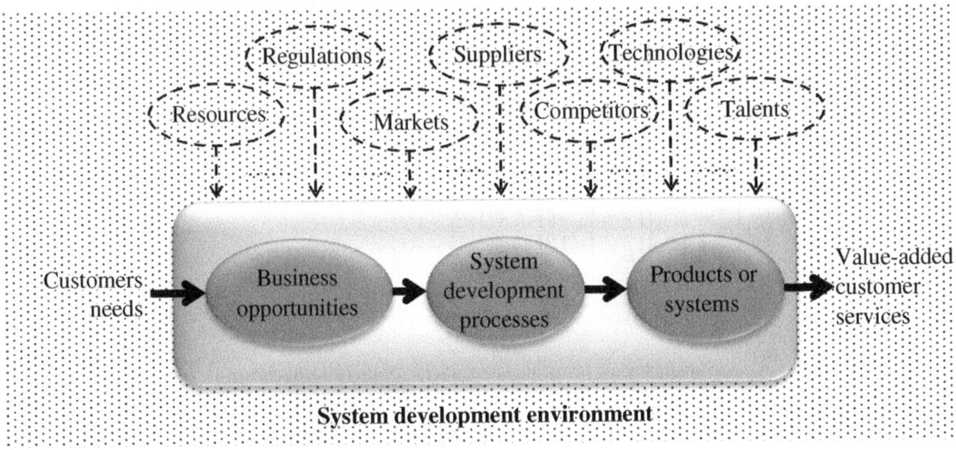

Figure 2.25 Mechatronic manufacturing systems for multidisciplinary objectives on adaptability

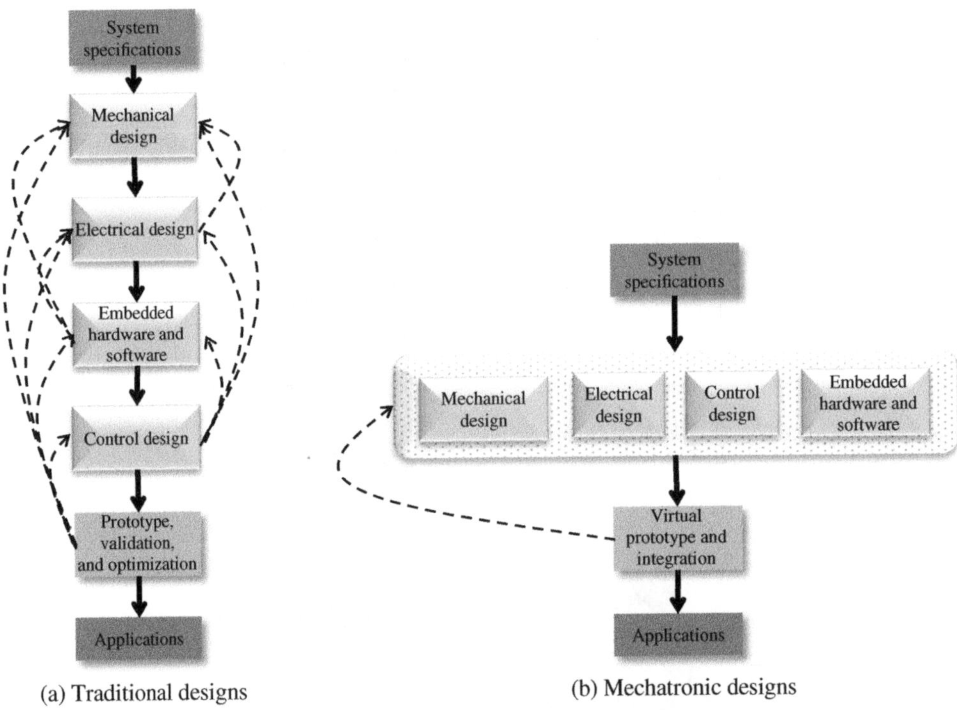

Figure 2.26 Comparison of traditional and mechatronic designs (Wits et al. 2012; Sell 2014; and Solidworks 2024)

collaborations of designers in respective disciplines, (3) multidisciplinary modeling and simulation, (4) multidisciplinary concurrent design, (5) earliest possible verification and validation, (6) methods and tools for system-level coordination, collaboration, and integrated controls, and (7) the needs of systematic design methodologies and control tools. In addition, the mechatronic design

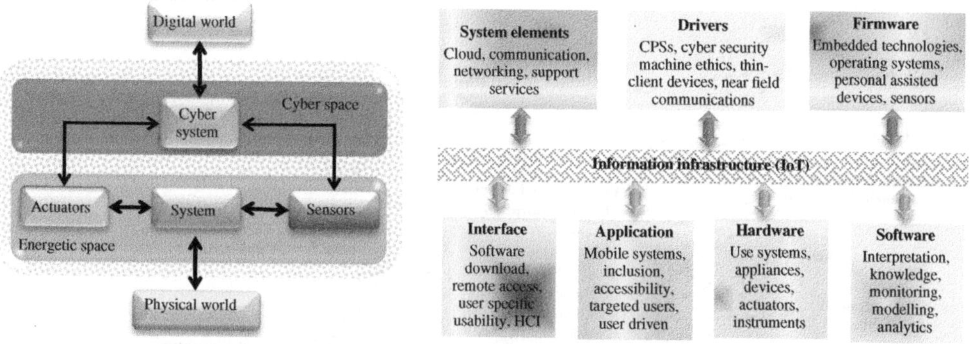

(a) Conventional mechatronics (b) Mechatronics and beyond

Figure 2.27 Multidisciplinary integrations for mechatronics and beyond

differs from conventional sequential multidisciplinary design optimization (MDO) in the sense that multiple iterations among disciplines could be eliminated by concurrent designs (Cheng et al. 2014).

Hehenberger et al. (2016) discussed the evolution of mechatronic systems; they argued that conventional integrated mechatronic systems had transited into *cyber-physical systems* (CPS) and cloud-based systems such as IoT. However, conventional mechatronic systems and elevated systems shared the commonalities in multidisciplinary developmental processes that were needed to deal with the interactions and integrations of disciplinary solutions (see Figure 2.27).

2.6 Mechatronic Design Methodologies

In Section 2.3, mechatronic systems have been characterized to identify the main features of systems. These features will be interpreted as measurable FRs in system development. For designs of mechatronic systems, numerous methodologies have been developed; while it can be beneficial to identify some effective methodologies that can tackle with main features of mechatronic systems and seek the right solutions to FRs systematically.

In this section, some popular methodologies that are used in different design phases of mechatronic systems are discussed. Figure 2.28 shows the correspondences of the discussed methodologies with the main features, or innovations, of mechatronic systems. These methodologies have been selected to persevere the innovations of mechatronic systems that have been elaborated in Section 2.3. More specifically, *System Modeling Language* (SysML) by the *National Institute of Standards and Technology* (NIST) is discussed to model mechatronic systems in Section 2.6.1. *Model-based System engineering* (MBSE) is introduced to deal with system design with a modular architecture in Section 2.6.2. *Axiomatic design theory* (ADT) is used to decompose the complexity of FRs at a system level into that at component level in Section 2.6.3. *Concurrent design optimization* (CDO) is used to tackle multidisciplinary couples of modular components at the assembly level in Section 2.6.4. *Virtual verification and validation* (VVV) will be used to minimize the iterations from digital models to physical systems in Section 2.6.5. Finally, *project-based design methodology (PBDM)* is introduced to integrate the aforementioned methodologies in customizing mechatronic systems for specific applications in Section 2.7.

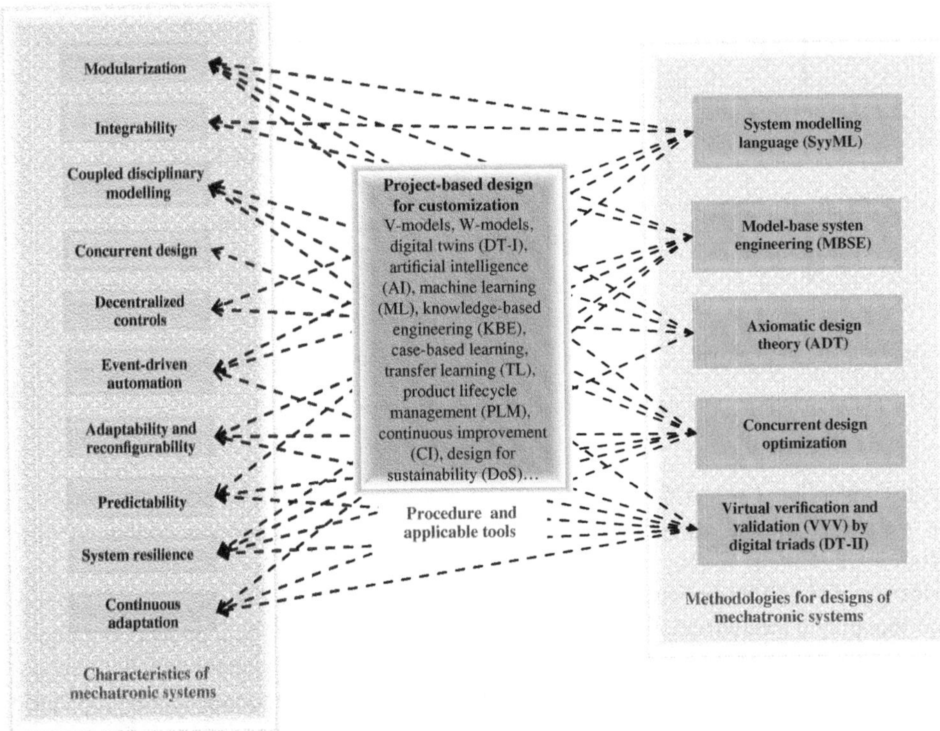

Figure 2.28 Methodologies for designs of mechatronic systems

2.6.1 System Modeling Language (SysML)

A mechatronic system is characterized by synergic interactions of multidisciplinary functional modules to perform advanced functionalities. The functionalities of a mechatronic system are not limited to the aggregation of the functionalities of individual modules (MHenni et al. 2014; Chami and Bruel 2015). The multidisciplinary nature of a mechatronic system provided opportunities for numerous functionalities; while this brought substantial challenges to the mechatronic design due to the aggregation of multiple engineering disciplines, business management fields, and the heterogeneity of modeling methods of mechatronic components in terms of modeling languages, the determinism of models, and representations of interactions. It is desirable that a mechatronic modeling method can support *multi-paradigm modeling* (Barisic et al. 2022).

System engineering (SE) methods should be used to describe a complex mechatronic system as a whole rather than individual system components. Many SE methods such as SysML by NIST have been developed to model complex mechatronic systems. SysML is used to establish a common system information basis on a comparatively abstract level for all participating stakeholders. SysML encompasses information such as requirements, structures, behaviors, and system parameters and variables (Nigischer et al. 2021). SysML is a general-purpose multi-view language for system modeling. SysML consists of two phases: (1) a black box analysis for a comprehensive set of FRs from users' perspective and (2) a while box analysis for the solutions of system elements, constitutions, and controls (MHenni et al. 2014). SysML was used to verify the executability of system behaviors (Barbau and Bock 2020).

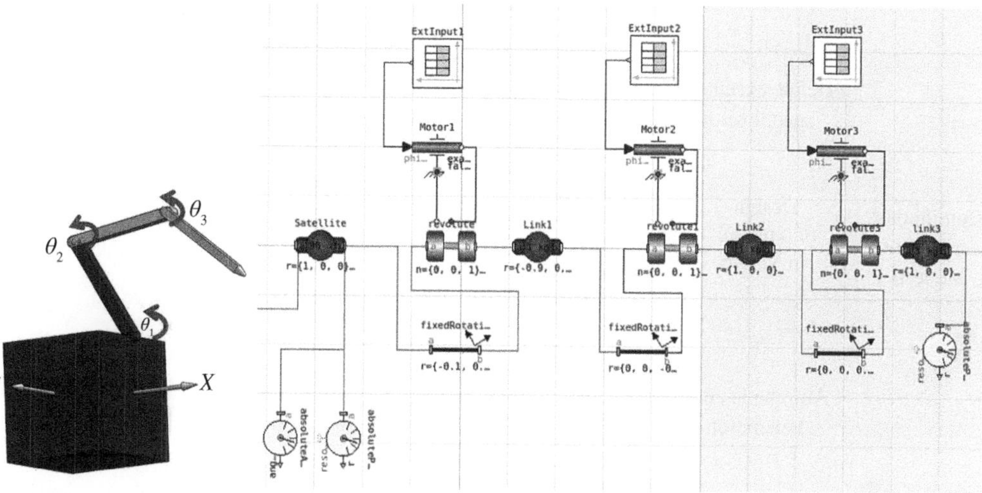

Figure 2.29 An example of using OpenModelica for the representation of a mechatronic system (*Source:* Santos et al. (2022)/with permission of Elsevier)

Flows and functions are defined as elements in a library for users to select the right ones to meet their requirements. Predefined flows in a Model Library include these for materials, energy, and signals. The attributes of these flows can be customized to specific models. Components in the library are defined for products and services, and they are represented as blocks in SysML. A block is augmented with ports that can be possibly connected with others. Components are identified by mapping incoming and outgoing flows to corresponding ports. Components in the Model Library are classified based on the flows of functions. When the flows are assigned to functions, the connections of the functions with others are established. A model can be used to represent *a virtual object* such as an algorithm, an app, or software or *a physical object* such as a robot or a machine tool. In defining a model for specific purposes, functional modeling is performed to analyze flows of materials, energy, and signals (Wolki 2013). Figure 2.29 shows an example where the open-source OpenModelica is used to represent a robot using SysML.

2.6.2 *Model-Based System Engineering (MBSE)*

A mechatronic system consists of modular components, each module represents a multidisciplinary solution to certain FR, and mechatronic design aims to select and assemble modular components into different configurations for specified tasks. Therefore, MBSE can simplify mechatronic designs in the sense that MBSE supports designs and communications across different engineering disciplines.

As shown in Figure 2.30, each MFM in a mechatronic system (see system architecture in Figure 2.20) is defined as a model. A model is an encapsulated system unit with specified inputs, outputs, and the functions to produce outputs based on given inputs. Typical tasks in designing, selecting, and operating a model are classified and include *defining executable specifications*, *modeling and simulations, automatic code generations*, and *continuous verifications and tests* (Lennon 2008; Barner et al. 2008). At the model level, each model is defined by an executable specification by uniquely describing its natural and controlled behaviors mathematically. At the system level,

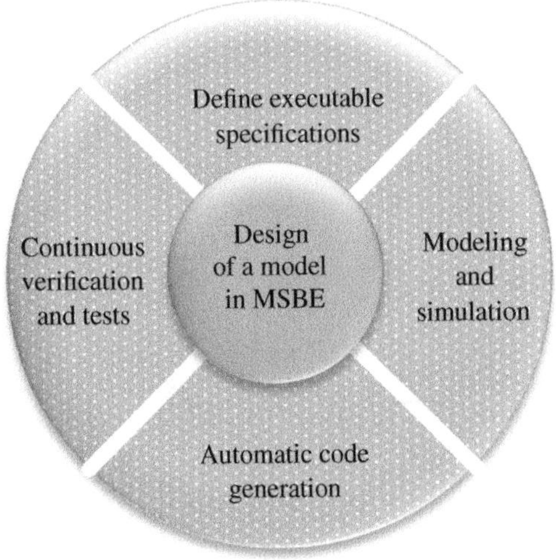

Figure 2.30 Tasks in developing a model in MBSE

developed models are simulated and executed to evaluate or control the system. In applications, a number of commercial and open-source software tools, such as Modelica, Matlab/Simulink, GNU/Octave, and Scilab/Xcos, are available to support MBSE (Lohofener and Merseburg 2013; Goseva-Popstojanova et al. 2016). For example, MBSE was integrated with SysML in representing the system architecture and implementing methods, tools, and processes (Wolki 2013). Qamar et al. (2011) modeled a mechatronic system as integration of a set of models at different phases in a system's life cycle. In Figure 2.31, the system-level mechatronic design is integration of models for

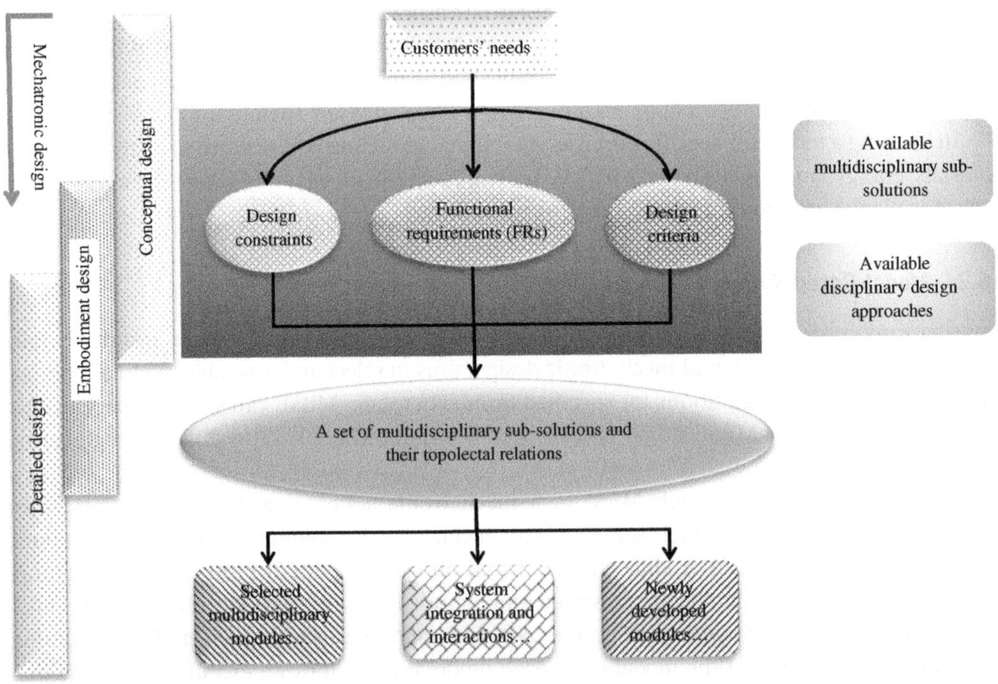

Figure 2.31 Integration of models in MBSE

conceptual design, embodiment designs, and detailed design. Each model is an encapsulated unit that can be designed, upgraded, or removed independently.

Since the interactions of hardware and software systems are most critical to mechatronic systems, MBSE tools are most effective to manage the ever-increasing complexity of mechatronic systems. Barner et al. (2008) developed a modular and expandable tool to support MBSE with an emphasis to maximize the reusability of modules in mechatronic development (Barner et al. 2008). In system integration, using MBSE tools can improve productivity and quality by shortening lead time and simplifying verification and validation (Goseva-Popstojanova et al. 2016). A mechatronic system is generally built from a set of functional modules, a function–behavior–stage model can be used to represent the relations of function modules. In determining modularized structure, a DSM can be used to identify the similarities of components and cluster them into modules (Van Beek et al. 2010). Akundi and Lopez (2021) discussed the industry-wide transition from document-centric systems engineering to MBSE and they found that MBSE emphasized more on modeling principles in defining the system's requirements, designing, analyzing, verifying, and validating the systems.

2.6.3 Axiomatic Design Theory (ADT)

A mechatronic system achieves the functions at the system level by coordinating, collaborating, and co-operating a set of constitutive modules that fulfill corresponding sub-functions at the module level. Accordingly, designing a mechatronic system requires to decompose the FRs at the system level into the sub-FRs at the module level. *Axiomatic design theory* (ADT) was introduced to manage the complexity of system decomposition effectively. ADT has been used as a standard guidance to understand *customers' requirements* (CRs), convert CRs into FRs, and seek and map *design solutions* (DSs) to FRs at solvable levels (Delaram and Valilai 2018).

ADT defines a DS as a set of constitutive modular components with associated interactions, and a mechatronic system solution should use as many off-the-shelf MFMs as possible; therefore, ADT promotes knowledge reuses in engineering design. A large amount of knowledge and information about these MFMs and their interactions can be fully utilized to reduce the cost and time of new system development. From this perspective, ADT has shown its effectiveness in mechatronic design since existing parts, components, products, or systems can be adopted, modified, and integrated into new mechatronic systems. ADT can also be readily integrated with other methods such as *case-based reasoning* (CBL) to reuse knowledge and span innovative designs using existing micro-electromechanical systems (MEMSs) (Bi 2002; Janthong et al. 2010).

A mechatronic system is modularized where ADT is applied to decompose a system into a set of modular components, and it integrates multiple disciplinary elements for successful coordination, collaboration, and cooperation in a system. ADT designs and analyzes a mechatronic system using two axioms, that is, (1) *independence Axiom* and (2) *information axiom*. The independence axiom aims to minimize the couplings of decomposed modules; a design is optimized when it is uncoupled and system modules can satisfy corresponding FRs independently without affecting other FRs or DSs. The information axiom aims to minimize the information content, and the information content is quantified as the sum of the logarithms ($\sum_{i=1}^{N} \log(p_i)$) of the probabilities of success for uncoupled functional modules to fulfill specified FRs as shown in Figure 2.32. Moreover, the actual and desired ranges of FR_i or DS_i should be close to each other as much as possible to avoid *under-* or *over-*engineering. The information content is related directly to system complexity, and it can be reduced by modularization and platforming technologies (Schuh et al. 2016; Bi and Wang 2020).

To use ADT in mechatronic design, FRs defined in respective disciplines such as mechanical engineering should be re-organized so that these disciplinary FRs can be mapped to corresponding

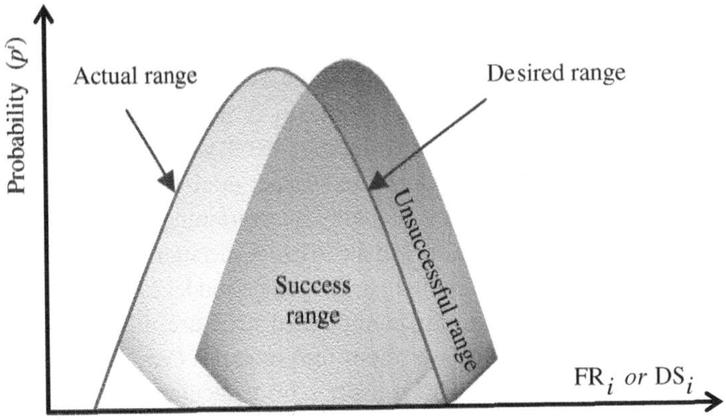

Figure 2.32 Definition of information content in ADT

MFMs. Schuh et al. (2016) introduced *generic disciplinary functions* (GDFs) to derive MFMs from FRs. If an *FR* can be modeled as a set of GDF_i ($i = 1,2, \ldots, n$) that have corresponding MFM_j as DSs, the design process can be represented by,

$$\text{FR}_i = \begin{Bmatrix} \text{GDF}_1 \\ \ldots \\ \text{GDF}_n \end{Bmatrix} \text{MFM}_j \tag{2.1}$$

Note that applying Eq. (2.1) to map FRs into MFMs is an interactive process. *Modular function deployment* (MFD), graphical models, matrix-based models, mathematical models, and modularity frameworks can be used to define MFMs based on the specified FRs in mechanical engineering, electrical and electronic engineering, and computer engineering (Askhoj et al. 2020). Schuh et al. (2016) suggested to select the number of GDFs reasonably. As shown in Figure 2.33, the decomposition of FRs should be refined iteratively until an MFM is identified to meet the group of eight GDFs satisfactorily.

It is common that not all FR_i or GDF_i can be satisfied by an existing MFM_i. In such cases, one or more *architecture elements* (AEs) can be introduced to fulfill the corresponding FR_i. The mappings of FR_i into AE_i can be coupled with each other; moreover, AE_i can be the solutions in respective disciplines other than in multiple disciplines as,

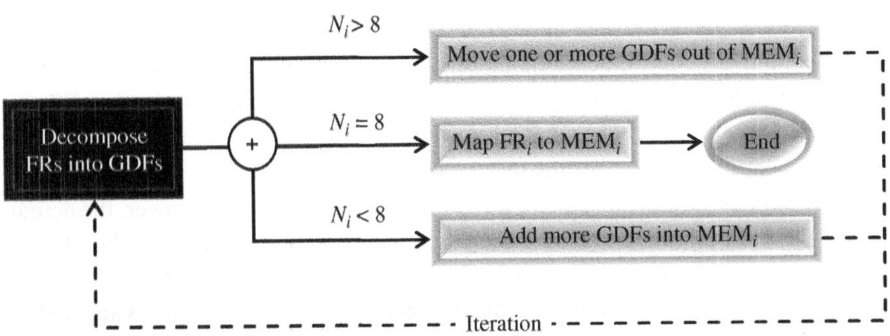

Figure 2.33 Mapping MFMs to FRs by defining a set of GDF_i

$$\begin{Bmatrix} FR_1 \\ FR_2 \\ FR_3 \\ FR_4 \\ FR_5 \end{Bmatrix} = \begin{bmatrix} \times & & & & \\ & \times & & & \\ & & \times & & \\ & & & \times & \times \\ & & & \times & \times \end{bmatrix} \begin{Bmatrix} MFM_1 \\ MFM_2 \\ MFM_3 \\ AE_4 \\ AE_5 \end{Bmatrix} \tag{2.2}$$

2.6.4 *Concurrent Design Optimization* (CDO)

As shown in Figure 2.34, a mechatronic system is modularized and consists of a number of MFMs. Mechatronic design is also referred to as configuration design that requires designers to develop a configuration by selecting and assembling from a set of available MFMs to meet the corresponding FRs of the system (Zheng et al. 2019; de Carvalho and da Hora 2021).

In an MFM, mechanical, electrical and electronic, and information system components are integrated. This leads to couplings of system behaviors in multiple physical domains. This brings a significant challenge to traditional sequential design methods to find a globally feasible solution by optimizing sub-solutions to corresponding FR_i from the upper to lower layers of system architecture. It is desirable to sustain a design space as large as possible by taking into consideration of all design variables and constraints concurrently. As shown in Figure 2.35, CDO was proposed (1) to

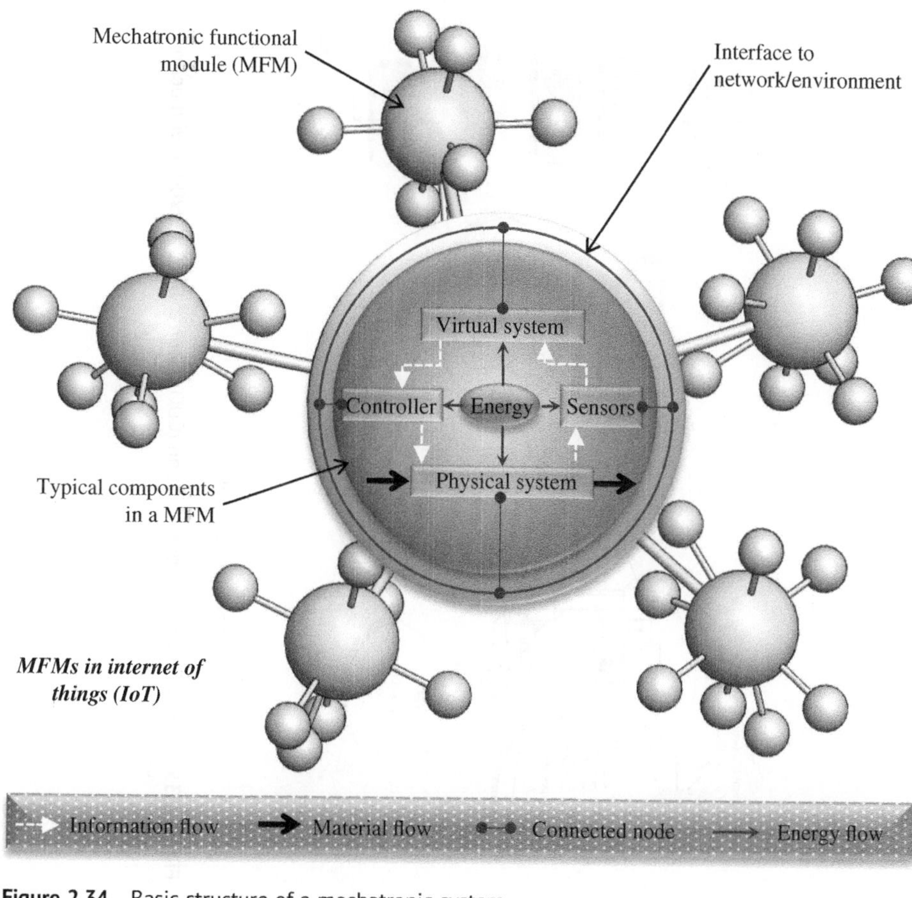

Figure 2.34 Basic structure of a mechatronic system

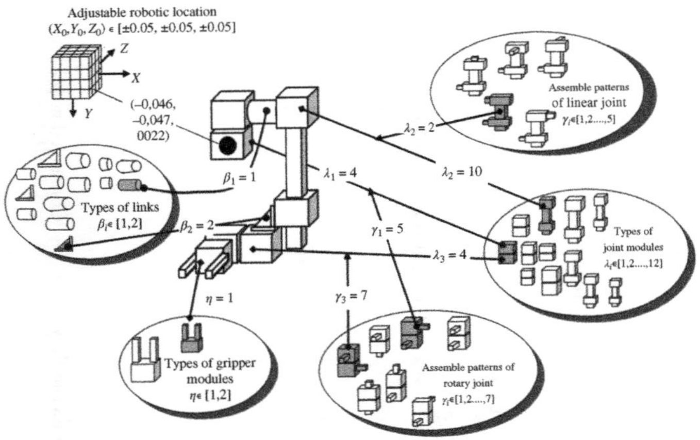

(a) Example configuration

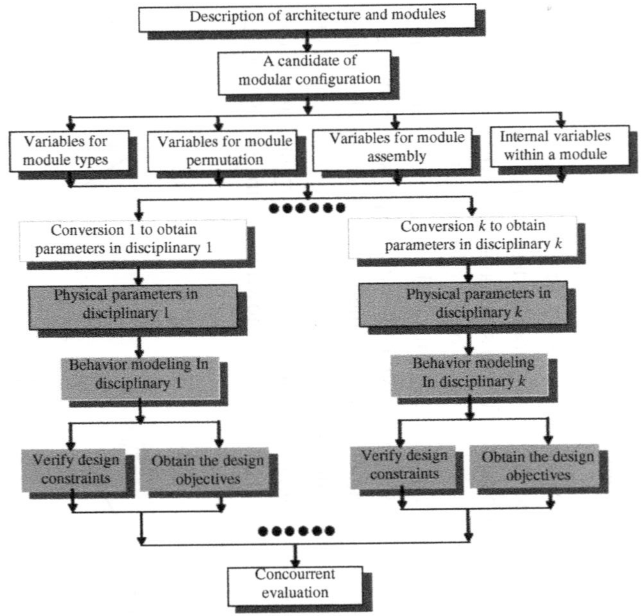

(b) Configuration design process

Figure 2.35 Concurrent design optimization (CDO) (*Source:* Bi et al. (2007)/with permission of Elsevier) formulate a mechatronic system design problem into an optimization problem where design variables include a number and types of modules, the topologies and assemblies of selected modules, and relative setups of the mechatronic system and the application environment; (2) analyze and compare feasible system configurations by taking the full consideration of design constraints in multidisciplines; and (3) search and improve a DS by applying heuristic algorithms.

In contrast to a conventional system in a specific domain such as a mechanical system or a software tool, a mechatronic system has extended its functionalities and capabilities by integrating multidisciplinary solutions from different domains as a holistic system solution. To manage system complexity, Balkhair (2019) proposed a multilayer *hierarchical design development* (DDP) process to define a *conceptual integrated model* (CIM) using standard components in model design libraries. Design libraries include DSs that can satisfy FRs such as functional blocks, simulation tools, and knowledge bases. Moreover, implementing DDP itself requires to integrate DT, V-model variations, MBSE, *function-behavior-state model* (FBSM), SE, *unified modeling language* (UML), and SysML.

2.6.5 Virtual Verification and Validation (VVV)

The design of a complex engineering system is usually iterative. On the one hand, the same set of FRs may be satisfied by numerous solutions; on the other hand, a solution must be rigorously verified and validated before it is transferred into a physical solution. The concept of *digital twins* (DT-I) was introduced to deal with the mappings of DSs from the virtual space to the physical space. However, it is our understanding that the importance of DT-I was undermined. DT-I can play a more important role in analyzing and comparing DSs in a vast design space and thoroughly verify and validate a finalized virtual model before it is converted into its physical model. This helps to explore more design alternatives, reduce design iterations, reduce prototyping costs, and shorten development times of new mechatronic systems. The deployment of VVV is also driven by the evolution of mechatronic systems (Hehenberger et al. 2016) so that more and more virtual and dynamic resources can be integrated into synthesized solutions of complex engineering problems as illustrated in Figure 2.36.

Moreover, VVV benefits mechatronic systems in other innovative aspects: (1) it becomes possible to verify if a DS meets various constraints in different disciplines at the earliest possible time; (2) it supports system reconfiguration by analyzing and comparing valid options when changes or events are detected so that a current system configuration can be transited to a future system configuration smoothly; (3) it improves system sustainability in the sense that the system can be upgraded at the earliest possible time before the changes become too much to be dealt with without a system revolution.

2.7 Project-Based Mechatronic Design (PBMD)

Innovations in mechatronic designs rely greatly on available enabling technologies, hardware and software resources, and MFMs, and design of a mechatronic system must be pulled by customers' needs rather than pushed by technologies (Bi and Wang 2020). Here, *project-based mechatronic design* (PBMD) is introduced as a methodological framework to integrate the methodologies in Section 2.6 for needs-driven mechatronic designs. For illustration, PBMD is used to design an evacuation device for a disabled person in a wheelchair to be evacuated over stairs from a public facility in an emergency situation (Bi 2024).

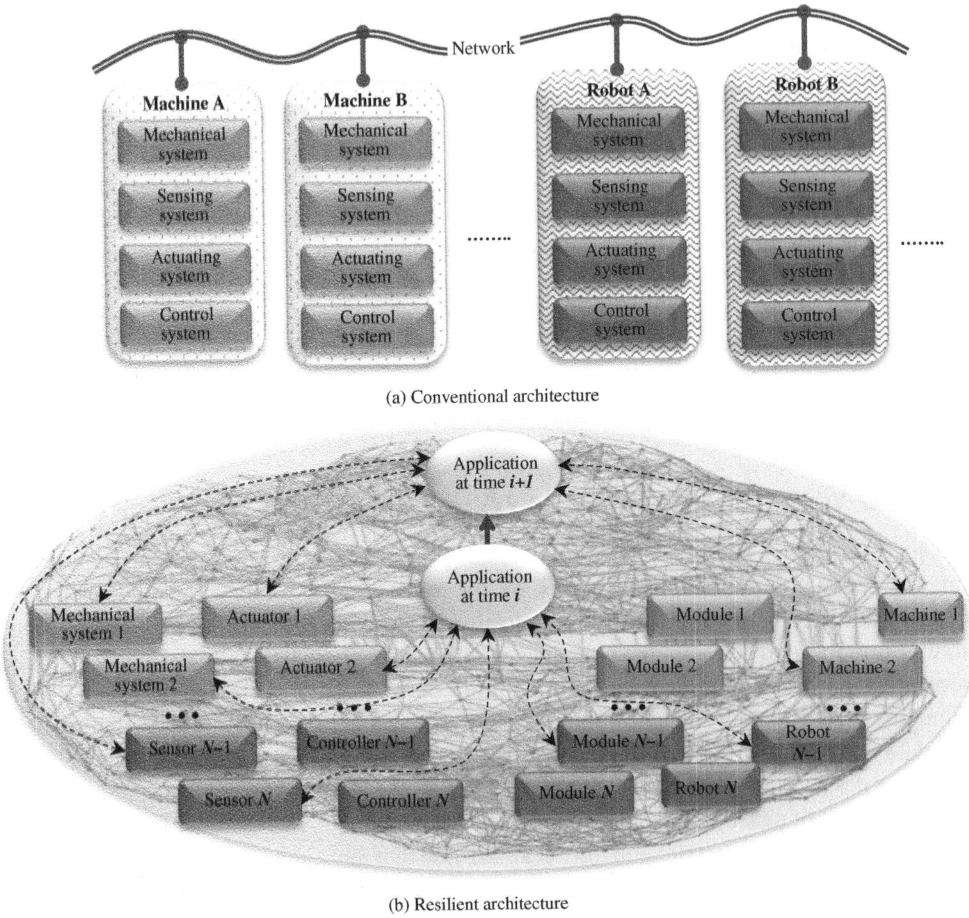

Figure 2.36 Virtual resources in integrated mechatronic solutions

2.7.1 Existing Assistive Evacuating Technologies

In the majority of not-for-profit organizations (NGOs), wheelchair users are manually evacuated by caregivers when an emergency situation occurs. As shown in Figure 2.37, depending on the weight of a loaded wheelchair, two, three, or even four caregivers are needed to navigate and evacuate a patient in a wheelchair from a multi-floor building to the ground safely. This may bring a number of issues in practice such as (1) the availability of caregivers, (2) the strengths of caregivers, (3) the reliability of human navigation, and (4) other safety concerns when caregivers collaborate to carry a heavy object on stairs.

Assistive technologies have been widely studied; a number of innovative products have been developed for the disabled persons to gain the mobility over stairs. As shown in Figure 2.38, existing assistive technologies for mobility over stairs are classified into five categories (Gelenbe and Wu 2012; Cooper 2013; Cornell University 2014; Bi and Gelenbe 2018; and Evacuation Chairs Australia 2024), that is, *intelligent wheelchairs*, *fixed stair-lifting devices*, *portable ramps*, *stair-climbing devices*, and *all-terrain robots*.

Figure 2.37 Manual evacuation of a disabled person over stairs (*Source:* Adam Scotti / https://www.huffpost.com/archive/ca/entry/justin-trudeau-carries-man-in-wheelchair-down-stairs-photo_n_5159721 / last accessed on July 18, 2024)

1) The products in the first two categories can be customized to impaired individuals for mobility or private access over stairs. A smart electric-powered wheelchair with the capability of stair climbing has attracted great attention in the past two decades. A few of the smart wheelchairs are commercially available; these products can assist patients with walking difficulties to move over stairs conveniently to participate in outdoor activities. On the other hand, fixed stair lifts are the most common ways to provide access across floors. However, lifts are generally expensive and occupy large space. While low-cost compact lifts are only applicable to small residences. For an emergency evacuation in a large public amenity, using smart wheelchairs or fixed stair lifts are simply cost-prohibitive.
2) An alternative solution for the first or the second category is to develop fixed ramps. On the other hand, building additional ramps for wheelchair evacuations in an existing infrastructure is involved with a high cost, and this affects space utilization and aesthetics greatly. Therefore, building additional ramps for wheelchair evacuation became infeasible for our client Turnstone.
3) The solutions for the third category are portable and applicable to a small number of stairs. However, these solutions are not adaptable to stair variations. Moreover, operating such devices takes a long setup time, especially for an evacuation on multiple floors.
4) The solutions for the fourth category are mostly at their prototype stages. A low market acceptance rate is attributed to a number of the limitations such as the complexity of system operations, the need for precautions to stabilize a wheelchair by caregivers, the limited adaptability to the variants of public amenities, and the limited intelligence in navigating the evacuation autonomously.

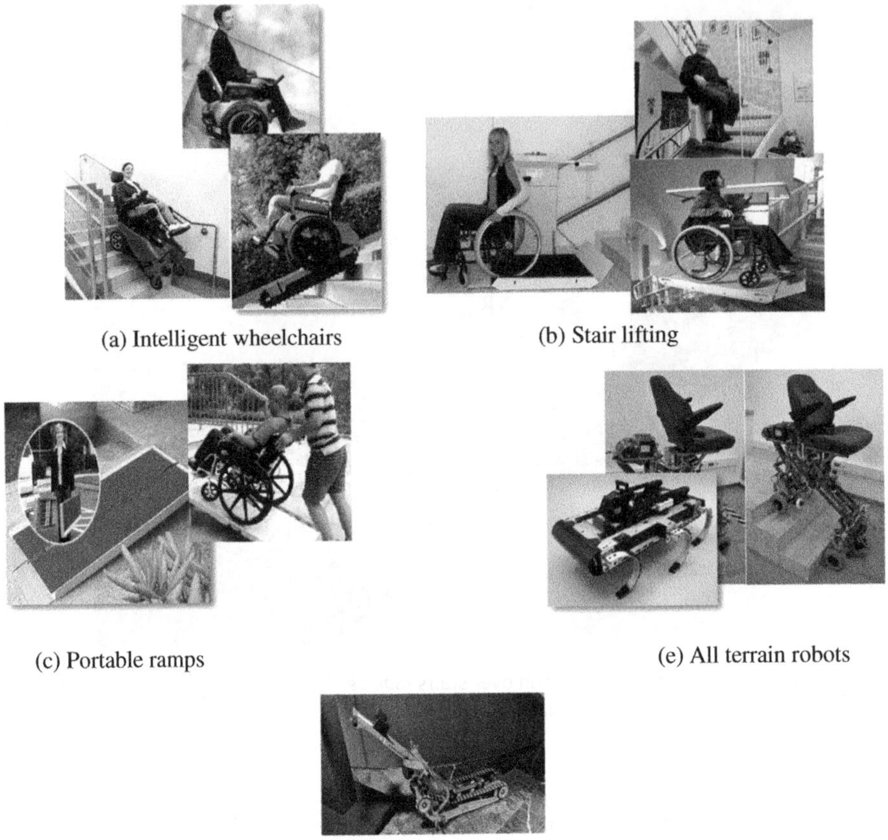

(a) Intelligent wheelchairs

(b) Stair lifting

(c) Portable ramps

(e) All terrain robots

(d) Portable stair climbing

Figure 2.38 Types of stair ascending/descending devices (*Source:* (a1) Image of a Scewo BRO wheelchair prototype climbing stairs. *Source:* © Scewo; (a2) Insider Inc / https://www.businessinsider.com/electric-stair-climbing-wheelchair-college-students-zurich-switzerland-scalevo-2015-10 / last accessed on July 18, 2024; (a3) Михаил Решетников/Adobe Stoc006B; (b1) Ingo Bartussek/Adobe Stock; (b2) Thirdleg Mobility Aids / https://www.indiamart.com/proddetail/wheelchair-stair-lift-4257807833.html / last accessed on July 18, 2024; (b3) Nationwide Lifts, Inc / https://www.home-elevator.net/apex-delta-ipl.php / last accessed on July 18, 2024; (c1) AliExpress.com / https://www.aliexpress.com/item/3256805254675101.html?_randl_currency=USD&_randl_shipto=US&src=google / last accessed on July 18, 2024; (d1) Mattes/Wikimedia Commons; (e1) Chocoteco et al. (2017) / MDPI / CC BY 4.0; (e3) Fastcompany.com / https://www.fastcompany.com/3020241/know-what-to-do-with-a-swarming-aerial-robot-tell-samsung-and-cash-in / last accessed on July 18, 2024.)

5) The solutions for the fifth category cover different mobile robots for transportation. Differing from conventional powered wheelchairs, a mobile robot integrates sensing, perceiving, decision-making, navigating, and controlling seamlessly to automate an evacuation process. The new functions of a mobile robot for emergency evacuation are (1) sensor-based controls of evacuations; (2) the adaptability to changes of moving speed, turning angle, and stairs; and (3) the capability of collision avoidance in an ill-structured environment with the presence of still or movable objects. By all means, robotic wheelchairs are very expensive in general, and the efforts

are limited to customize mobile robots as effective tools for emergency evacuations; this is less relevant to the development of a portable evacuation device (Rulik et al. 2022; Sivakanthan et al. 2022).

Stairs are widely used in multi-floor buildings due to their high spatial efficiency to connect areas of differing vertical elevations. However, stairs bring a big challenge to transport wheelchairs across floors. Especially, emergency evacuation of users with disabilities from a public amenity must follow specified guides (National Fire Protection Association 2022; NSF 2024). There is an emerging need to advance assistive technology to evacuate impaired users in wheelchair, especially in the presence of emergency circumference. It seems that a no vital solution is available to the emergency evacuation problems. For example, powered wheelchair users face challenges when going outdoors, including slopes, steps, uneven surfaces, and other environmental barriers. Very limited works have been reported on the prevention of wheelchair accidents due to unbalancing or instability. As shown by some examples of wheelchair accidents, there were over 100 000 emergency room visits resulting from wheelchair-related injuries, of which more than 65% were attributed to tips and falls (Sundaram et al. 2017).

2.7.2 Proposed Assistive Evacuation Device

We propose to adopt modularized architecture to decompose the evacuation system into a set of functional modules with manageable complexity, respectively. The implementation of each module is encapsulated for its specified sub-FR, and its interactions with other modules occur only by its interfaces.

Existing techniques for stair ascending or descending exhibit two main limitations. The first one is the effectiveness of stair walking for desired speeds and adaptability. The second one is the stability of mechanism while it moves on stairs. A normal person uses the legs as the stepping mechanism and the movement is elegantly controlled for a precise balance by the brain in conjunction with a number of sensory organs. The legs are equipped with high speed and high peak power output actuators in the form of muscles. The brain also acts on a combination of visual data to estimate stair location and height.

To overcome the identified limitations of existing stair descending products, we propose a comprehensive solution with an integration of recent research advancements in assistive technologies, mobile robots, human–machine interaction, sensors and actuations, and IoT-based navigation as shown in Figure 2.39. In addition, the system should be optimized based on the design criteria of *safety*, *cost*, and *operating time*.

As shown in Figure 2.40, the highest level of decomposition leads to three main components: (1) *mechanical components* including automatic brakes, steering systems, transmission mechanisms, and loading/unloading mechanisms; (2) *electric and electronic* components including power supplies, assistive controllers, stair detecting and Gyroscope sensors; (3) *navigation and control systems* including embedded micro-controllers, graphic user interfaces, and other smart things to enable IoT navigations.

2.7.3 Main Functional Requirements from Use Cases

Any mechatronic design can begin with the formulation of a design problem itself. The design of an evacuation device is viewed as a system design problem. It is formulated by defining system boundaries, converting users' needs into FRs at quantified measures, and identifying system parameters and constraints.

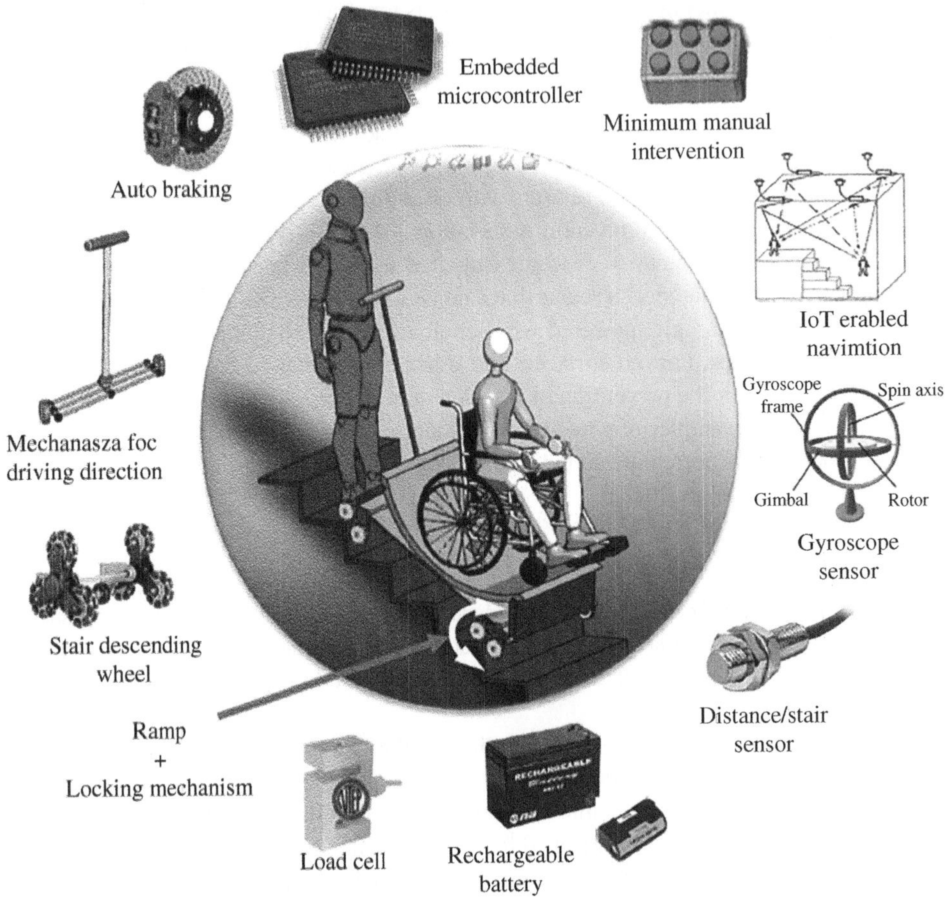

Figure 2.39 Conceptual design of an evacuation device (*Sources:* (a) Brembo / https://autotalent.com/products/brembo-gt-345x28mm-4-piston-rear-big-brake-kit-for-bmw-7-series-e65-e66-2002-2005 / last accessed on July 18, 2024; (b) Scanrail/Adobe Stock Photos; (c) Linen-Trolley / https://linen-trolley.com.au/rotatruck-stair-climber-conversion-kit-r2 / last accessed on July 18, 2024; (d) Coti Global Sensors Manufacturing / https://www.cotiglobal.com/products/load-cells/s-beam/ / last accessed on July 18, 2024; (f) AUTONICS / https://www.amazon.com/AUTONICS-PRL08-2DN2-Inductive-Sensing-Non-Shielded/dp/B00TXX6V1M / Last accessed on July 19, 2024)

Due to the complexity of the system, *axiomatic design theory* (ADT) is integrated with *use cases* in *model-based system engineering* (MBSE) to deal with the complexity and couplings of system components. System-level FRs are decomposed into a multilayer hierarchical structure until a number of physical solutions are defined to each sub-FR. This will decouple *design solutions* (DSs) for respective sub-FRs. In this way, the solution to individual functional modules can be implemented independently.

Figure 2.40 shows three main use cases that are defined to quantify FRs of mechanical components in an evacuation device; these use cases correspond to typical operations of mechatronic components in three predicted application scenarios. Figure 2.41 shows the specifications of stairs (Sumesh et al. 2021). Accordingly, the technical specifications of an evacuation device are specified as follows:

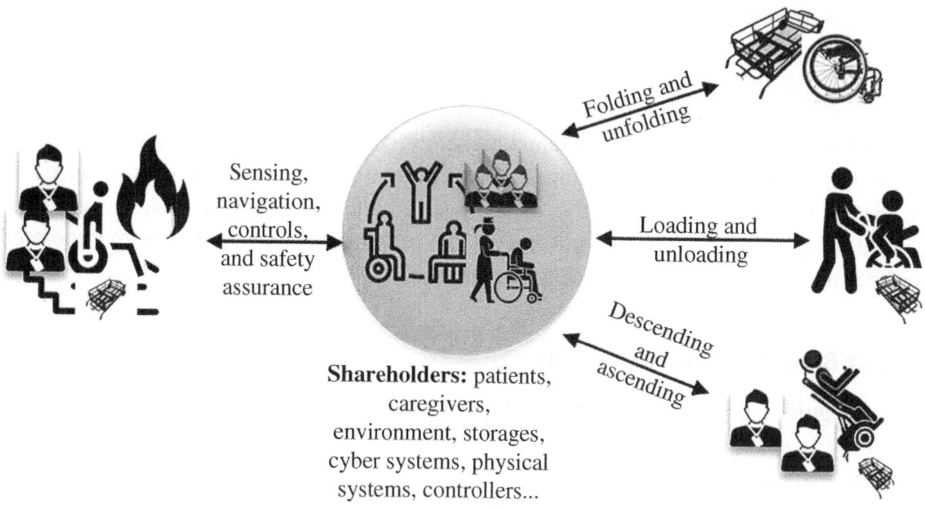

Figure 2.40 Quantifying FRs of mechanical systems based on operations in predicted scenarios of applications

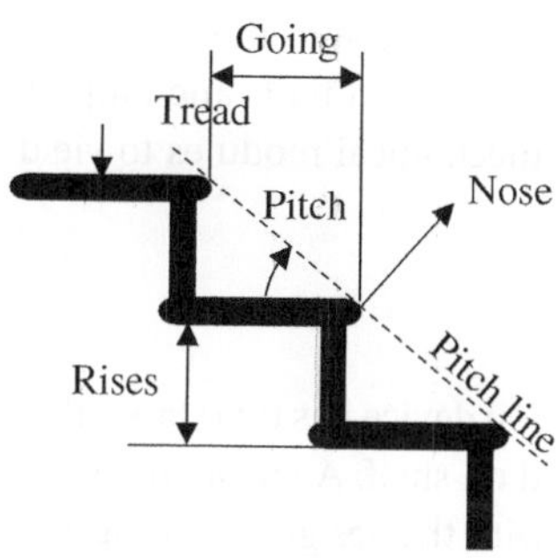

(a) Stair model

Standard	Rise (mm)	Going (mm)
CSN	150–180	210 +
DIN	140–200	230–270
BS	150–200	250–370

(b) Dimensions in different standards

Figure 2.41 Specifications of stairs

- setup with a wheelchair: <20 seconds,
- stair height and slope: <8 in. and 40°,
- travel mode: flat floor, down-stairs, and up-stairs,
- switch mode: <5 seconds,
- tolerance of measurement of stair height: <0.25 in.,
- travel on flat floor speed <30–60 steps/min,
- travel down-stairs speed <20–30 steps/min,
- load: > 500 lb,
- power battery life: 5 hours,
- responsive time of auto braking under an abnormal condition <1 μs,
- real-time self-detection and self-rebalancing >20 Hz,
- IoT of integrated navigation, and
- self-weight < 100 lb.

FRs of a subsystem should be decomposed further to develop the solutions to all expected tasks. Taking an example of a mechanical system, further decomposition of FRs includes the requirements to (1) adjust patient's posture, (2) travel over stairs, (3) autonomous brake for stability, (4) pack and fold, (5) assist descending, and (6) carry the required load with sufficient strengths. Moreover, the design criteria to be optimized include safety, compactness, weight, rigidity, and cost. Kinematic and dynamic models are developed to analyze stair-descending and braking mechanisms in a virtual environment. All of the mechanical components will be designed for safer human–machine interaction and faster movement. The braking mechanism will be incorporated to ensure the stability of the stair-descending mechanism.

2.7.4 Project-Based Mechatronic Designs

PBMD is used to guide a design process of a mechatronic product. PBMD allows one to use various multidisciplinary design methods at different developmental phases. In particular, MBSE is adopted here to manage the complexity of CPSs, map the FRs with DSs, and provide a complete view of system using the aids of formal semantics and rich visual modeling and tools. MBSE was the formalized application of modeling to design and analyze system requirements, verify and validate activities, and support all design activities in the whole life cycle from the conceptual design phase to deployment (Trehan 2018). MBSE helps to manage FRs, reduce ambiguity, and make it feasible to perform a formal verification and validation process in the system development (de Saqui-Sannes et al. 2022). MBSE makes a transition from document-centric to model-centric system engineering (Chen et al. 2022). Here, the idea of MBSE is used to define inputs and outputs for each mechanical module and develop the solution of respective mechanical modules to yield expected outputs optimally.

2.7.4.1 Folding and Unfolding Mechanism

The first use case in Figure 2.28 shows that in applications, an evacuation device has two states; that is, the folded state when the evacuation device is not in use and stored on shelf. A virtual model of the evacuation device must be created, and it is particularly involved with the design of folding and unfolding such as in Figure 2.42. The ranges of joints should be specified to determine the oversize of the evacuation device in unfolded and folded states. Figure 2.43 shows the overall dimensions of the evacuation device in the drawing.

2.7.4.2 Reaction Forces on Tracks for Structural Elements

The dimensions of structural elements in the evacuation device must be optimized to (1) minimize the overall weight and (2) assure the appropriate material strength for the required *factor of safety*.

Figure 2.44 shows an example of determining the dimensions of a structural member using Motion Simulation in SolidWorks. The structural member is used to transform the external load to the front track. The amount of force in this element varies when the front track is moved from flat ground to a stair. To calculate the force, a virtual motor is introduced in a Motion Study model, and the reaction force applied to the front track is determined based on the torque predicted in the Motion Study (see Figure 2.44b). Table 2.2 shows the obtained reaction forces, assuming that the length of the front track is 600 mm.

Moreover, computer-aided engineering can be integrated to take advantage of dynamic loads and evaluate the stress distribution of any structural member directly. Figure 2.45 shows the changes of stress distribution over the member of interest in Figure 2.44. It shows that the max stress exceeds

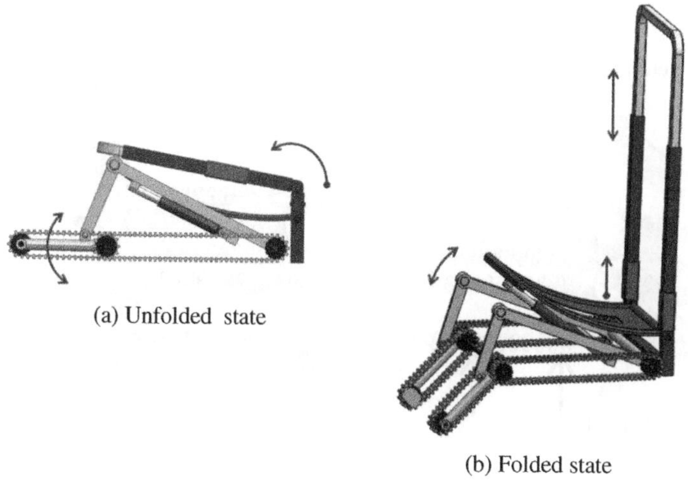

Figure 2.42 Evacuation device in an unfolded/folded state

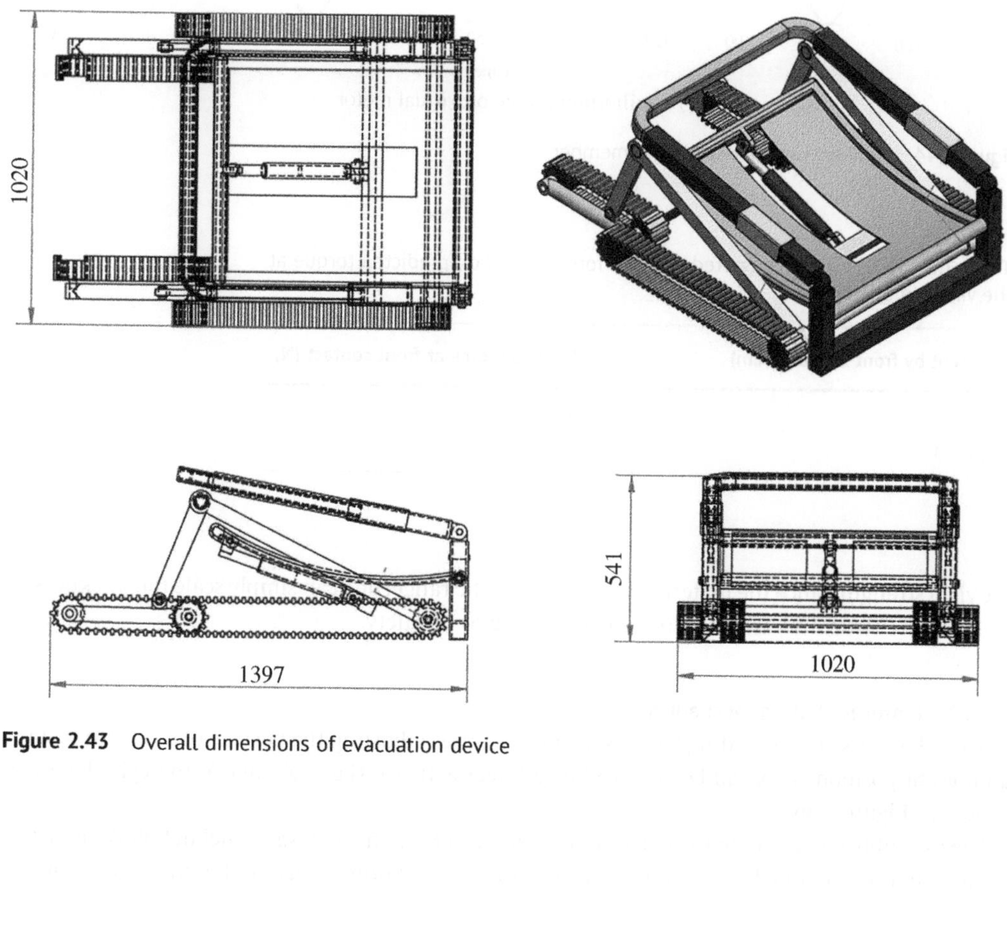

Figure 2.43 Overall dimensions of evacuation device

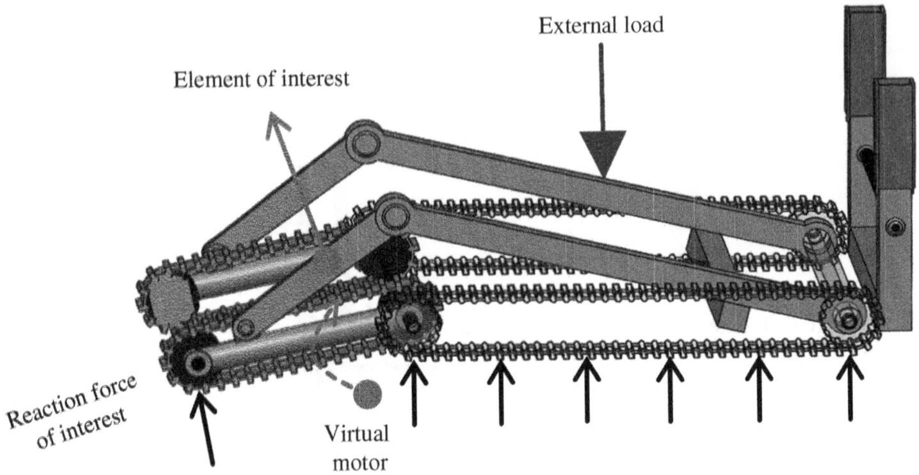

(a) Virtual motor for calculation of reaction force

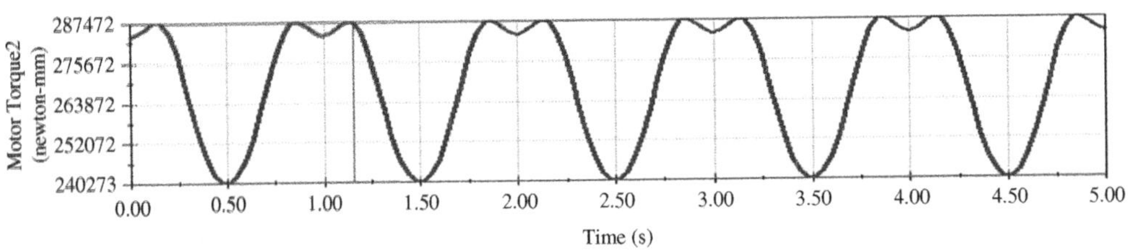

(b) Equilibrium torque on virtual motor

Figure 2.44 Optimization of a structural member

Table 2.2 Examples of converted reaction forces based on predicted torque at the virtual model

Moment by front support (N.m)	Reaction force at front contact (N)
287.472	479.12
240.273	400.46

the yield strength. Since the material works in its elastic range, one can simply scale the thickness or width of the structural member to increase the factor of safety.

2.7.4.3 Motor for Lifting Mechanism

The third use case for ascending and descending requires adjusting the posture of a rider to sustain an upright position. This can be achieved by a linear actuator (i.e., cylinder-A and cylinder-B) as shown in Figure 2.46.

The assembly model in Figure 2.46 is used to define a Motion Analysis model in SolidWorks, the linear actuator is defined as a motor, and the speed and working range of the linear actuator are

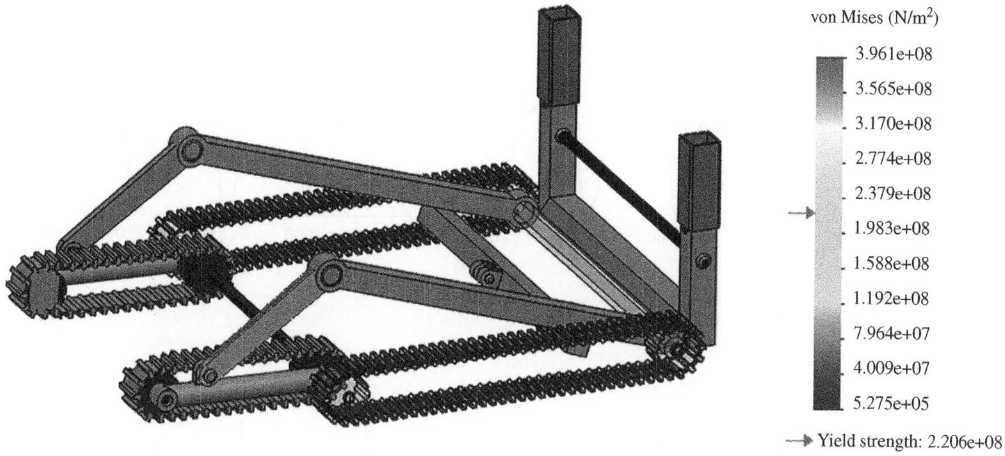

Figure 2.45 Dynamic stress analysis in motion analysis

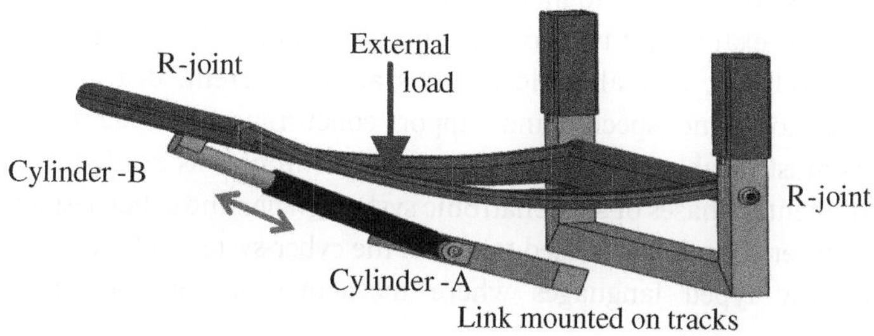

Figure 2.46 Cylinders to adjust gravity center during transportation

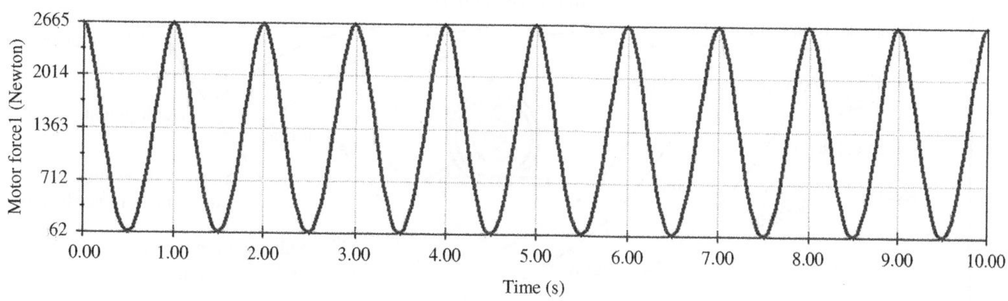

Figure 2.47 The required driving force over time

defined. The simulation on this model finds the required torque and power applied on the linear actually as shown in Figures 2.47 and 2.48, respectively. The ranges of the driving force and power are determined as (62–2665) N and (0–401) Watt, respectively. These can be used to select a linear actuator appropriately.

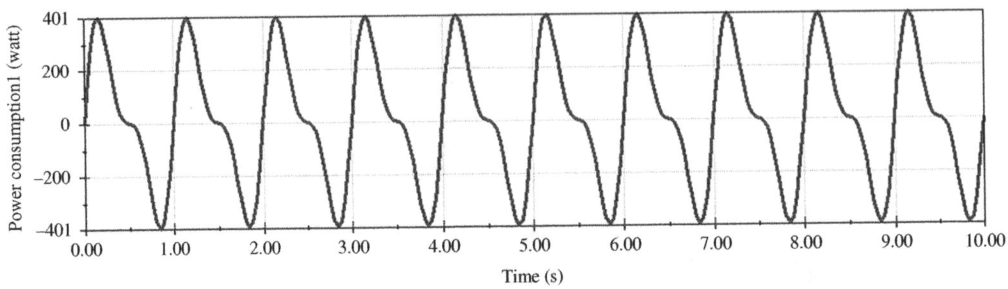

Figure 2.48 The required power over time

2.7.4.4 Control of Evacuation Device

MBSE can further be used to design and implement the cyber system of an evacuation system. To model a cyber-system, a hierarchical structure is adopted to model system elements, relations, and behaviors since a mechatronic system typically involves many entities including elements, their states, and events for states' transitions. MBSE helps to model system behaviors based on relations and interactions of system entities. For example, if two components are connected, the events at their interface are predictable. In addition, as real-world systems are concurrent, distributed, and decentralized, there is the need to define, specify, and support concurrency in designing a cyber-system. System components must be able to communicate and synchronize as needed.

Figure 2.49 shows all the developmental phases of a mechatronic system. When the cyber system is to be designed for a mechanical system, Modelica is used to model the cyber-system. Modelica is featured as (1) declarative statically typed languages where mathematical equations and

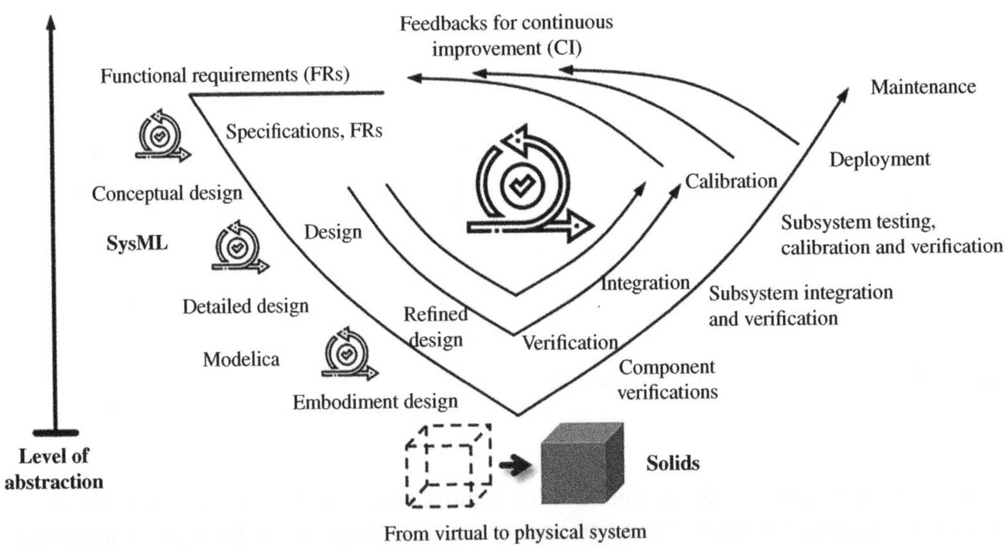

Figure 2.49 Developmental phases of a mechatronic system (*Source:* Marwedel (2021)/Springer Nature/CC BY 4.0)

functionals allow a causal modeling and provide the checks on high-level specification and static type; (2) supports modeling in multi-domains such as mechanical, electrical, hydraulic, thermodynamic, biological, and financial systems; (3) everything is treated as a class, it is an object-oriented language and uses the syntax similar to that of Java and Matlab; (4) it is visual component programming and possesses the capabilities to construct hierarchical structures; (5) it is non-proprietary and efficient that is comparable to C and supports advanced equation compilation for over 300,000 equations. It describes continuous-time behaviors by differential algebraic equations and discrete event dynamic system behaviors by event triggers.

2.7.4.5 PBMD in Mechatronic Design

PBMD has been introduced as a methodological framework to integrate various multidisciplinary methods in designing and implementing a mechatronic system. PBMD is used to support needs-driven mechatronic design. It begins with the definition of *customers' requirements* (CRs); CRs are then translated as a set of user cases to elaborate system-level FRs in all predicted use scenarios. PBMD is further used inclusively to seek a DS for each FR, respectively, until corresponding DSs are identified for all FRs satisfactorily. Since mechatronic systems are modularized, ADT can be integrated to decompose system-level FRs into sub-FRs at lower levels. SysML can be integrated to represent data flows and interactions among MFMs. MBSE can be integrated to analyze, compare, and select MFMs as the solutions to corresponding FRs at different levels and domains. CDO can be integrated to model and evaluate if one or an assembly of MFM(s) meet all constraints in multiple disciplines. Finally, VVV must be integrated to verify and validate DSs to reduce the iterations between the virtual world and the physical world.

2.8 Summary

Mechatronic systems are characterized by modularization, integrability, coupled-discipline modeling, concurrent design, decentralized controls, event-driven automation, adaptability, reconfigurability, predictability, system resilience, and CA. Innovative thinking helps improve mechatronic designs, and TRIZ can be utilized to develop innovative MFMs at different levels and domains of mechatronic systems.

The architecture of a mechatronic system is modularized; it consists of a set of standardized or customized MFMs. Design of a mechatronic system is involved with identifying and selecting MFMs, assembling selected MEMSs into integrated DSs, and fully verifying and validating DSs for expected FRs before a virtual DS is transferred into a physical solution in applications. Design of a mechatronic system begins with defining and translating CRs into system-level FRs, high-level FRs are decomposed hierarchically until DSs are developed to meet all sub-FRs satisfactorily. Mechatronic design is naturally multidisciplinary. Various multidisciplinary methods, such as SysML, ADT, MBSE, CDO, and VVV have been introduced to support the developments at different design phases. In particular, PBMD is proposed as a methodological framework to integrate these multidisciplinary methods for effective designs of mechatronic systems. In implementation, PBMD can be integrated with other design methods and tools such as *V-models*, *W-models*, *digital twins* (DT-I), *artificial intelligence* (AI), *machine learning* (ML), *knowledge-based engineering* (KBE), *case-based learning*, *transfer learning* (TL), *product life cycle management* (PLM), *continuous improvement* (CI), and *design for sustainability* (DoS).

Problems

2.1 Give two examples of mechatronic products and discuss their innovations.

2.2 Review the products in Figure 2.50 and prove that they are mechatronic products.

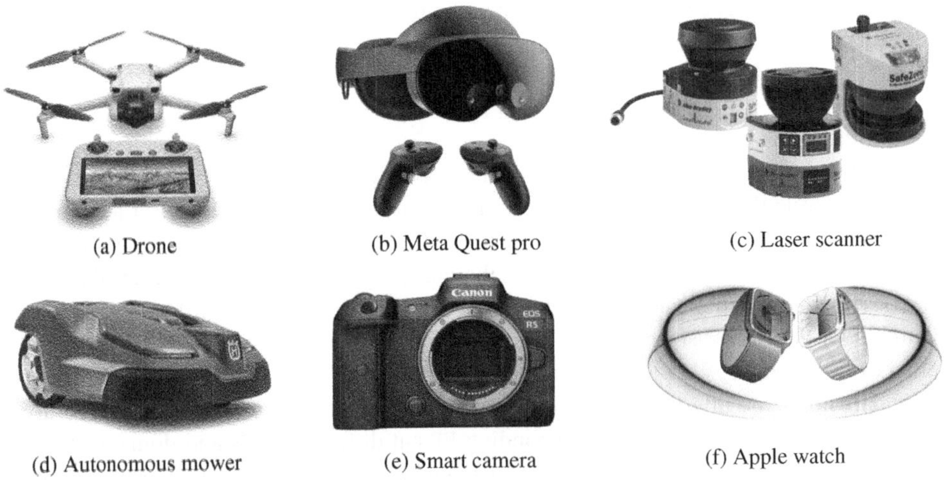

Figure 2.50 Examples of mechatronic products (*Source:* (a) Generic / https://www.amazon.in/Lightweight-Foldable-Tri-Directional-Obstacle-Photography/dp/B0CKPX2DTJ / Last accessed on July 19, 2024; (b) Future US, Inc / https://www.tomsguide.com/best-picks/best-vr-headsets / Last accessed on July 19, 2024; (c) Rockwell Automation / https://www.rockwellautomation.com/en-us/products/hardware/allen-bradley/safety-products/safety-presence-sensors.html / Last accessed on July 19, 2024; (d) Husqvarna AB / https://www.husqvarna.com/us/battery-lawn-mowers/ / Last accessed on July 19, 2024; (e) Canon / https://www.bhphotovideo.com/c/product/1547009-REG/canon_eos_r5_mirrorless_digital.html / Last accessed on July 19, 2024; (f) Apple Inc / https://www.theverge.com/2023/12/21/24010965/apple-watch-series-9-ultra-2-removed-from-online-sale-store / Last accessed on July 19, 2024)

2.3 What is innovative thinking? What innovative approaches are you aware of? What are the new innovative methods that you have learned from TRIZ?

2.4 What is an MFM? Debrief the functions of main components in an electric vehicle shown in Figure 2.51.

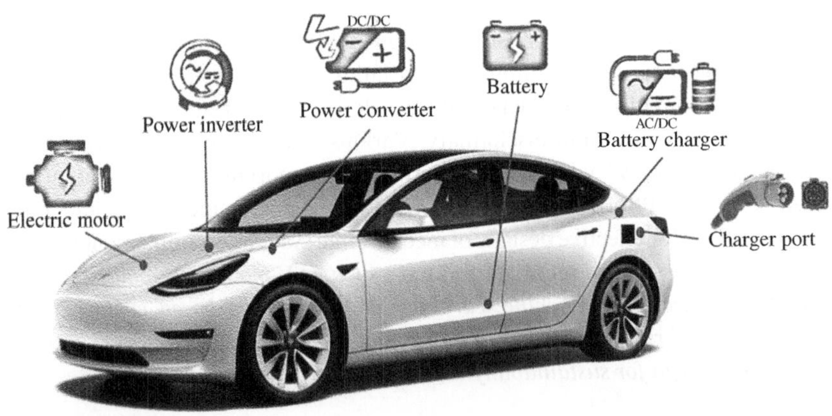

Figure 2.51 Main components of an electric vehicle (*Source:* Owl WiS / https://www.youtube.com/watch?v=JTOtU78c5So&ab_channel=OwlWiS / last accessed on July 04, 2024)

2.5 Discuss the importance of modular architecture for a mechatronic system.

2.6 Discuss the necessity of concurrent design optimization in mechatronic design using examples.

2.7 What is a project-based mechatronic design (PBMD)? What are the roles of use cases in applying PBMD for a mechatronic design? Use a mechatronic product (e.g., a foldable treadmill) as an example to show your understanding.

2.8 Design an all-terrain biomimetic robot (team project with 3–4 members) such as one of the exemplifying design concepts as shown in Figure 2.52; it should satisfy the following functional requirements:
- At least 3 Degrees of Freedom (DOF) locomotion
- All-terrain (height variant): $\pm$50 mm
- Max dimensions: L $\times$ W $\times$ H 500 $\times$ 300 $\times$ 100 mm
- Carrying load: >2 kg
- Self-weight: less than <10 kg
- Max speed: >1.0 m/s
- Power consumption: less than <100 W
- Battery life: >5 hour
- Budget cost: <$400

Figure 2.52 Examples of biomimetic robots (*Sources:* Virginia Tech / https://www.youtube.com/watch?v=SjURGAwz9Pw&lc=Ugxal9MzEzVGu1L77914AaABAg / Last accessed on July 19, 2024)

2.9 Design an assistive lift for emergency evacuation of disabled in wheelchair (team project with 3–4 members) such as an exemplifying design concept as shown in Figure 2.53; it should satisfy the following FRs:
- set up with a wheelchair: < 20 seconds,
- stairs height and slope: <8 in. and 40,
- travel mode: flat floor, down-stairs, and up-stairs,
- switch mode: < 5 seconds,
- tolerance of measurement of stair height: < 0.25 in.,
- travel on flat floor speed: <30–60 steps/min,
- travel down-stairs speed: <20–30 steps/min,
- load: > 300 lb,
- power battery life: 5 hours,
- responsive time of auto braking under an abnormal condition: <1 second,
- real-time self-detection and self-rebalancing: >20 Hz,
- IoT integrated navigation, and
- self-weight: < 100 lb.

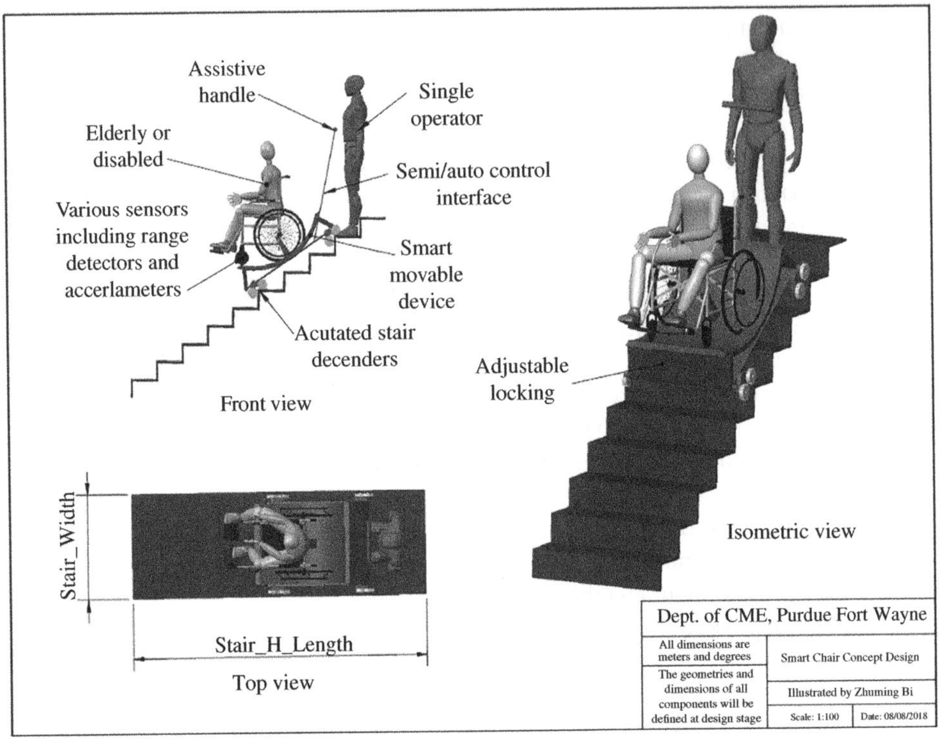

Figure 2.53 An example of the design concept of an evacuation device

2.10 Design a testing platform for electric riding lawn mowers. It aims to assess the vibration and fatigue life of the main driving shaft and the damage to electric contacts of battery and circuits, and it needs to acquire the data of driving torque and bending deflection of shaft over time (see Figure 2.54).

Figure 2.54 Instrumentation for testing of a shaft with a diameter of 2-in., a motor power of up to 20 horsepower (available length for instrumentation is less than 4-in.) (*Source:* Black & Decker / https://www.greenbuildingadvisor.com/article/lawn-mowing-season / Last accessed on July 19, 2024)

References

ABB (2024). Software products. https://new.abb.com/medium-voltage/digital-substations/softwareproducts. (accessed 12 September 2024).

Akundi, A. and Lopez, V. (2021). A review on application of model based systems engineering to manufacturing and production engineering systems. *Procedia Computer Science* 185 (2021): 101–108.

Aloupis, G., Benbermou, N., Damian, M. et al. (2013). Efficient reconfiguration of lattices-based modular robots. *Computational Geometry* 46 (8): 91–928.

Andreason, M.M. (2022). 41030 mechatronic engineering design, Technical University of Demark. https://fdocuments.in/document/mogens-myrup-andreasen-eafit-interactiva-application-of-mechanical-electronic.html (accessed 12 September 2024).

Askhoj, C., Lokkegaard, M., Bertram, C.A., and Mortensen, N.H. (2020). Identifying modularity practices across mechanics, electronics and software. *NorDesign 2020, Kgs.* Lyngby, Denmark (11–14 August 2020).

Balkhair, H. (2019). Development of a multi-layer V-model design process and computational tools for mechatronic conceptual design. Doctoral thesis. University of British Columbia, Vancouver, Canada.

Barbau, R. and Bock, C. (2020). Verifying executability of SysML behavior models using satisfiability modulo theory solvers. *NISTIR 8283*. https://doi.org/10.6028/NIST.IR.8283.

Barisic, A., Ruchkin, I., Savic, D. et al. (2022). Multi-paradigm modeling for cyber-physical systems: a systematic mapping review. *Journal of Systems and Software* 183 (2022): 111081.

Barner, S, Geisinger, M, Buckl, C, and Knoll, A (2008). Easylab: model-based development of software for mechatronic systems. *2008 IEEE/ASME International Conference on Mechatronic and Embedded Systems and Applications*, pp. 540–545. https://doi.org/10.1109/MESA.2008.4735652.

Bi, ZM (2002). On adaptive robot systems for manufacturing applications. Doctoral thesis. University of Saskatchewan Saskatoon, Canada.

Bi, ZM (2024). Model-based system engineering (MBSE) for design of mechatronic products. *The 11th IEEE International Conference on Cybernetics, Intelligent Systems, Robotics, Automation and Mechatronics (CIS-RAM)*, Hangzhou, China (8–11 August 2024).

Bi, H and Gelenbe, E. (2018). Emergency and normal navigation in confined spaces. https://arxiv.org/pdf/1310.2886.pdf (accessed 31 March 2018).

Bi, Z.M. and Lang, S.Y.T. (2007). A framework for CAD-and sensor-based robotic coating automation. *IEEE Transactions on Industrial Informatics* 3 (1): 84–91.

Bi, Z.M. and Wang, X.Q. (2020). *Computer Aided Design and Manufacturing*. Hoboken, NJ: Wiley-ASME Press Series.

Bi, Z.M. and Zhang, W.J. (2001a). Modularity technology in manufacturing: taxonomy and issues. *International Journal of Advanced Manufacturing Technology* 18 (5): 381–390.

Bi, Z.M. and Zhang, W.J. (2001b). Concurrent optimal design of modular robotic configuration. *Journal of Robotic Systems* 18 (2): 77–87.

Bi, Z.M., Zhang, W.J., Chen, I.M., and Lang, S.Y.T. (2007). Automated generation of the DH-parameters for configuration design of modular manipulators. *Robotics and Computer-Integrated Manufacturing* 23 (2007): 553–562.

Bonnema, G.M. (2011). The engineers' innovation toolkit. *Procedia Engineering* 9 (2011): 345–354.

Bradley, D. and Russell, D.W. (2010). *Mechatronics in Action – Applications and Education*. London Dordrecht Heidelberg New York: Springer.

de Carvalho, R.A. and da Hora, H. (2021). Fernandes R (2021) a process for designing innovative mechatronic products. *International Journal of Production Economics* 231: 107887.

Chami, M. and Bruel, J.M. (2015). Towards an integrated conceptual design evaluation of mechatronic systems: the SysDICE approach. *Procedia Computer Science* 51 (2015): 650–659.

Chang, Y.S., Chien, Y.H., Yu, K.C. et al. (2016). Effect of TRIZ on the creativity of engineering students. *Thinking Skills and Creativity* 19 (2016): 112–122.

Chen, J., Chen, Y., Hu, Z. et al. (2022). A semantic ontology based approach to support model-based systems engineering design for an aircraft prognostic health management system. *Frontiers in Manufacturing Technology* 2: 886518.

Cheng, C., Bricogne, M., Duigou, J.L., and Eynard (2014). Survey on mechatronic engineering: a focus on design methods and product models. *Advanced Engineering Informatics* 28 (3): 241–257.

Chocoteco, J.A., Morales, R., and Feliu-Batlle, V. (2017). Enhancing the trajectory generation of a stair-climbing mobility system. *Sensors* 17 (11): 2608. https://doi.org/10.3390/s17112608.

Cooper, R.A. (2013). Assistive technology: from research to practice. http://aaate.net/wp-content/uploads/sites/3/2013/12/AAATE2013.pdf (accessed 12 September 2024).

Cornell University (2014). Disability status report Indiana. http://www.disabilitystatistics.org/StatusReports/2014-PDF/2014-StatusReport_IN.pdf (accessed 12 September 2024).

Delaram, J. and Valilai, O.F. (2018). An architectural view to computer integrated manufacturing systems based on axiomatic design theory. *Computers in Industry* 100 (2018): 96–114.

Evacuation Chairs Australia (2024). Legal overview. www.evacuationchairs.com.au/legalities (accessed 12 September 2024).

Fan, Y.C. and Chang, J.Y.J. (2018). Equipment communication architecture for smart manufacturing. *Proceedings – 2018 IEEE International Conference on Smart Manufacturing, Industrial & Logistics Engineering SMILE 2018*, pp. 27–32, vol. 2018 (January 2018).

Fiorineschi, L., Frillici, F.S., and Rotini, F. (2018). Enhancing functional decomposition and morphology with TRIZ: literature review. *Computers in Industry* 94 (2018): 1–15.

Gadd, K. (2011). *TRIZ for Engineers: Enabling Inventive Problem Solving*, 1e. Wiley.

Garner, G.T. (2009). Mechatronics as a technological basis for an innovative learning environment in engineering. Doctoral thesis. Mechanical and Aerospace Engineering, University of Virginia.

Gausemeier, J., Dumitrescu, R., Kahl, S., and Nordsiek, D. (2011). Integrative development of product and production system for mechatronic products. *Robotics and Computer-Integrated Manufacturing* 27 (2011): 772–778.

Gelenbe, G. and Wu, F.J. (2012). Large-scale simulation for human evacuation and rescue. *Computers & Mathematics With Applications* 64: 3869–3880.

Goseva-Popstojanova, K., Kahsai, T., Knudson, M. et al. (2016). Survey on model-based software engineering and auto-generated code, NASA/TM-2016-219443.

Hammadi, M., Choley, J.Y., and Mhenni, F. (2014). A multi-agent methodology for multi-level modelling of mechatronic systems. *Advanced Engineering Informatics* 28 (3): 187–258.

Han, B., Liu, C.L., and Zhang, W.J. (2016). A method to measure the resilience of algorithm for operation management. *IFAC-PapersOnLine* 49 (12): 1442–1447.

Hehenberger, P., Vogel-Heuser, B., Bradley, B. et al. (2016). Design, modelling, simulation and integration of cyber physical systems: methods and applications. *Computers in Industry* 82 (2016): 273–289.

Hirtz, J.M., Stone, R.B., McAdams, D.A., et al. (2001). Evolving a functional basis for engineering design. *Proceedings of DETC'01 ASME 2001 Design Engineering Technical Conference and Computers and Information in Engineering Conference*, Pittsburgh, PA (9–12 September 2001), DETC2001/DTM-21688.

Hirtz, J.M., Stone, R.B., McAdams, D.A. et al. (2002). A functional basis for engineering design: reconciling and evolving previous efforts. NIST Technical Note 1447. https://www.nist.gov/publications/functional-basis-engineering-design-reconciling-and-evolving-previous-efforts (accessed 12 September 2024).

Hollanders, H. and van Cruysen, A. (2009). Design, creativity and innovation: a scoreboard approach. https://www.yumpu.com/en/document/view/10263247/design-creativity-and-innovation-a-scoreboard-approach/3 (accessed 12 September 2024).

Holtta-Otto, K. and de Weck, O. (2007). Degree of modularity in engineering systems and products with technical and business constraints. *Concurrent Engineering* 15 (2007): 113–126.

Janthong, N., Brissaud, D., and Butdee, S. (2010). Combining axiomatic design and case-based reasoning in an innovative design methodology of mechatronics products. *CIRP Journal of Manufacturing Science and Technology* 2 (2010): 226–239.

Kazman, R., Bianco, P., Ivers, J., and Klein, J. (2020). Integrability. Technical Report CMU.SEI-2020-TR-001, Carnegie Mellon University. https://resources.sei.cmu.edu/asset_files/TechnicalReport/2020_005_001_637385.pdf (accessed 12 September 2024).

Lennon, T. (2008). Model-based design for mechatronic systems. www.ee.co.za/wp-content/uploads/legacy/AutT_Mod.pdf (accessed 12 September 2024).

Lettner, D., Hehenberger, P., Nohrer, A. et al. (2015). Variability and consistency in mechatronic design. *Concurrent Engineering: Research and Applications* 23 (3): 213–225.

Levary, V. (2006). Model-based framework for the adaptive development of engineering systems. Doctoral thesis. https://www.semanticscholar.org/paper/Model-based-Framework-for-the-Adaptive-Development-L%C3%A9v%C3%A1rdy/ac896e777c407a8aa12f64c724bc444883d1915c (accessed 12 September 2024).

Lohofener, M. and Merseburg, H. (2013). Model-based design of mechatronic systems with open source software. *Proceedings to the 14th International Workshop on Research and Education in Mechatronics REM 2013*, Wien (5–6 June 2013). ISBN 978 390275 907-8.

Lokman, S.F., Othman, A.T., and Abu-Baker, M.H. (2019). Intrusion detection system for automotive controller area network (CAN) bus system: a review. *EURASIP Journal on Wireless Communications and Networking* 184: https://jwcn-eurasipjournals.springeropen.com/articles/10.1186/s13638-019-1484-3.

Malmquist, D. (2014). A tool for holistic optimization of mechatronic design concepts. Doctoral thesis. Department of Machine Design, KTH Royal Institute of Technology, SE -100 44 Stockholm, Sweden.

Marwedel, P. (2021). *Embedded System Design Embedded Systems Foundations of Cyber-Physical Systems, and the Internet of Things*, 4e. Spring International Publishing https://doi.org/10.1007/978-3-030-60910-8.

Mhenni, F., Choley, J.Y., Penas, O. et al. (2014). A SysML-based methodology for mechatronic system architecture design. *Advanced Engineering Informatics* 28 (3): 218–231.

National Fire Protection Association (2022). Emergency evacuation planning guide for people with disabilities. http://www.nfpa.org/disabilities (accessed 21 August 2022).

Nesta (2019). A compendium of innovation methods. www.nesta.org.uk/report/compendium-innovation-methods (accessed 12 September 2024).

Nigischer, C., Bougain, S., Riegler, R. et al. (2021). Multi-domain simulation utilizing SysML: state of the art and future perspectives. *Procedia CIRP* 100 (2021): 319–324.

NIST (2024). Information system resilience. https://csrc.nist.gov/glossary/term/information_system_resilience (accessed 12 September 2024).

NSF (2024). A roadmap for us robotics from internet to robotics. https://lulabi.live/download/4682620-a-roadmap-for-us-robotics-from-internet-to-robotics (accessed 12 September 2024).

Ozrobotics (2024). ModMi modular versatile robot for research, learning and experimentation. https://ozrobotics.com/shop/modmi-modular-versatile-robot-for-research-learning-and-experimentation (accessed 12 September 2024).

Qamar, A., Wikander, J., and During, C. (2011). Designing mechatronic systems: a model-integrated approach. *International Conference on Engineering Design (ICED11)* (15–18 August 2011), Technical University of Denmark.

Qian, Z., Liu, H., and Bi, Z.M. (2015). Design and experiment of miniaturized and low-cost robotic fish with customized electromagnetic actuation. *International Journal of Systems Applications, Engineering & Development* 9 (2015): 159–166.

Rousselot, F. and Renaud, J. (2015). On triz and case based reasoning synergies and oppositions. *Procedia Engineering* 131 (2015): 871–880.

Rulik, I. et al. (2022). Control of a wheelchair-mounted 6DOF assistive robot with chin and finger joysticks. *Frontiers in Robotics and AI* 22: https://doi.org/10.3389/frobt.2022.885610.

Russo, D. and Duci, S. (2015). From Altshuller's 76's standard solutions to a new set of 111 standards. *Procedia Engineering* 131 (2015): 747–756.

Santos, R.S., Rade, D.A., and da Fonseca, I.M. (2022). A machine learning strategy for optimal path planning of space robotic manipulator in on-orbit servicing. *Acta Astronautics* 191 (2022): 41–54.

de Saqui-Sannes, R., Vingerhoeds, R.A., Garion, C., and Thirioux, X. (2022). A taxonomy of MBSE approaches by languages, tools and methods. *IEEE Access* 10: 120936.

Schuh, G., Rudolf, S., and Breuning, S. (2016). Modular platform design for mechatronic systems using axiomatic design and mechatronic function modules. *Procedia CIRP* 50: 701–706.

Sell, R. (2014). *Model Based Mechatronic Systems Modelling Methodology in Conceptual Design Stage*. Lambert Academic Publishing.

Sivakanthan, S., Canliotti, J.L., Sundaram, S.A. et al. (2022). Mini-review: robotic wheelchair taxonomy and readiness. *Neuroscience Letters* 772 (16): 136482.

SolidWorks (2024). Incorporating mechatronic into your design process. White Paper. https://www.solidworks.com/sw/docs/Mechatronics_WP_ENG.pdf (accessed 12 September 2024).

Sumesh, J.V., Kulkarni, N., Anwaar, T. et al. (2021). Design of a powered multi-tracked stair climbing carrier with loading capability into automobiles. *9th RSI International Conference on Robotics and Mechatronics (ICRoM)*, pp. 479–484, Tehran, Iran, Islamic Republic of.

Sundaram, S.A., Wang, H., Ding, D., and Cooper, R.A. (2017). Step-climbing power wheelchairs: a literature review. *Topics in Spinal Cord Injury Rehabilitation* 23 (2): 98–109.

Taura, T. and Nagai, Y. (2017). Creativity in innovation design: the roles of intuition, synthesis, and hypothesis. *International Journal of Design Creativity and Innovation* 5 (3-4): 131–148.

Trabar, A. (2014). The use of CX-TRIZ cards for inventive problem solving. https://studylib.net/doc/14459512/cx-triz-cards--letter-x-designates-the-process (accessed 12 September 2024).

Trehan, V. (2018). A process model in platform independent and neutral formal representation for design engineering automation. Doctoral thesis. Birmingham City University. https://www.open-access.bcu.ac.uk/8873/7/PhD%20Thesis%20-%20Vibhor%20Trehan.pdf (accessed 12 September 2024).

Van Beek, T.J., Erden, M.S., and Tomiyama, T. (2010). Modular design of mechatronic systems with function modelling. *Mechatronics* 20 (2010): 850–863.

Wits, W.W., Bakker, H.M., and Chechurin, L.S. (2012). Towards multidisciplinary support tools for innovation tasks. *Procedia CIRP* 2 (2012): 16–21.

Wolki, S.J. (2013). Modular libraries for conceptual designs. Doctoral thesis. TH Zurich. https://ethz.ch/content/dam/ethz/special-interest/mavt/design-materials-fabrication/engineering-design-and-computing-lab/DoctoralTheses/PhD_Wo%CC%88lkl.pdf (accessed 12 September 2024).

Zhang, T. (2015). Towards a novel resilient robotic system. Doctoral thesis. University of Saskatchewan Saskatoon, Canada.

Zhang, W.J. and van Luttervelt, C.A. (2011). Toward a resilient manufacturing system. *CIRP Annals* 60 (1): 469–472.

Zhang, T., Zhang, W.J., and Gupta, M.M. (2017). Resilient robots: concept, review and future directions. *Robotics* 6 (4): 22.

Zheng, C., Qin, X., Benoit, E. et al. (2019). Interface model-based configuration design of mechanical systems for industrial manufacturing applications. *Robotics and Computer-Integrated Manufacturing* 59 (2019): 373–384.

Zou, Y., Kim, D., Norman, P. et al. (2022). Towards robot modularity – a review of international modularity standardization for service robots. *Robotics and Autonomous Systems* 148 (2022): 203943.

[illegible]

3

Power Generation, Storage, Supply and Transmission

3.1 Introduction

A mechanical system is a system of elements that transfers an input power in one form (i.e., motion and force) to the output power in another form to meet users' needs in applications. As shown in Figure 3.1, three basic element types of a mechanical system are (1) *driving elements* to provide input power, (2) *transmitting elements* to convert the input to the output power in terms of its motion and force, and (3) *driven elements* by which the output power is used to implement the functions of systems. For example, a furnace or steam boiler converts chemical energy into thermal energy, an automobile engine uses chemical energy to drive motion, an electrical generator produces electrical power using mechanical energy, and a gas turbine converts chemical energy into mechanical energy.

From the perspective of energy flow, a mechatronic system can be generally viewed as a power transmission system, and the system should be designed by (1) selecting the right energy source, (2) distributing and transmitting power through a set of system elements in right forms of energy, and (3) using the power by driven components to fulfill prescribed functions as expected. In this chapter, the forms of power are discussed to determine the *functional requirements* (FRs) of power transmission. The main types of mechanical solutions to power inputs, power transmission, and power uses in applications are introduced in respective sections.

3.2 Energy, Work, and Power

Any physical systems including mechatronic systems, are designed with certain functions. Fulfilling a function requires *work*; work is a measure of the transferred *energy* on an object, and *power* is the work done in a unit of time. For example, (1) when a force F is used to lift an object to a height of h in a time period of t, the amount of work ($W = F \times h$) is transferred to the object, and the corresponding power is determined as ($P = W/t$). (2) When a current (I) passes through a resistor (R), the amount of electric power ($P = I^2 \cdot R$) is transferred to dissipated heat energy.

Energy refers to the ability of doing work to change things. Energy exists in any matter, and it can be transferred in various forms such as *motions, phases, reactions, waves, sounds, heats*, and *lights*. However, an energy form can be of either a "potential" or a "kinetic" form. As shown in Figure 3.2 (EIA 2007), *potential energy* is stored in a matter in its state, and examples of potential energy are chemical energy, nuclear energy, gravity field, and mechanical energy. *Kinetic energy* exhibits in the

Mechatronics for Complex Products and Systems: Project-Based Design Approaches for Robots, Cyber-Physical Systems, Digital Twins, and Other Emerging Technologies, First Edition. Zhuming Bi.

Companion website: www.wiley.com/go/bi/mechatronics

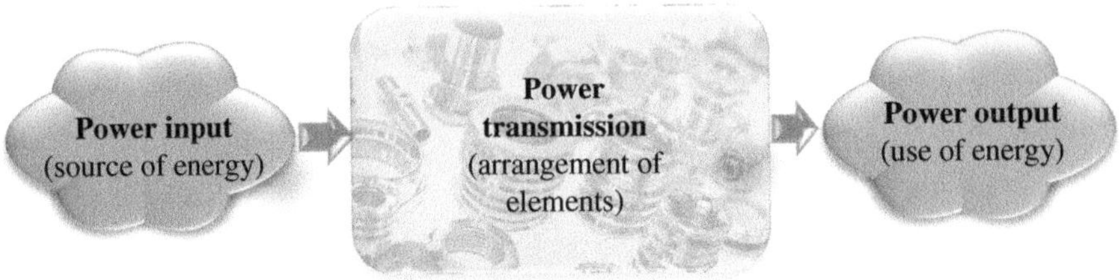

Figure 3.1 Power transmission in a mechanical system

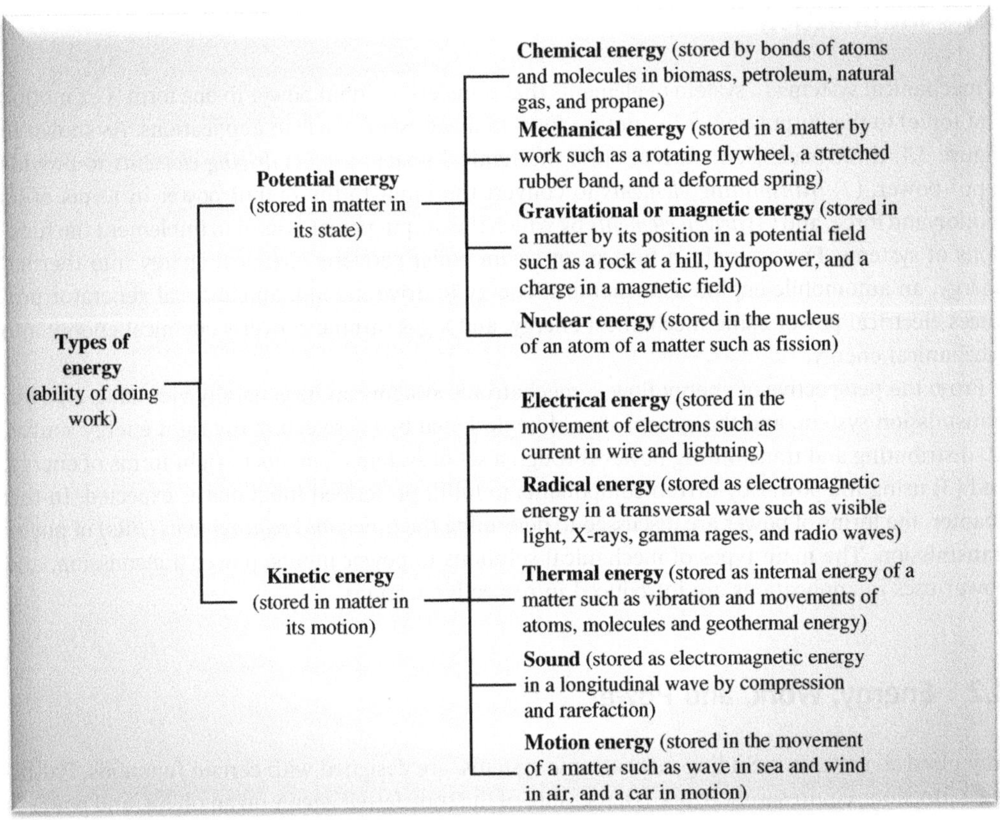

Figure 3.2 Types and examples of energy

form of motion of a matter such as electrons, atoms, molecules, gas, fluid, and solids; examples of kinematic energy are electric energy, radical energy, thermal energy, and motion of objects.

The advancement of mechatronic systems depends intrinsically on available energy. Energy sources should be safe, reliable, and affordable; although our human society is facing the growing challenge of accessing energy to meet people's high living standards, the global population is expected to grow from 7.0 billion today to 9.7 billion by 2050 (ExxonMobil 2024).

Example 3.1 An object with a mass of $m = 10$ kg is pulled up in a plane at an inclined angle of $\theta = 30°$ with the floor. The coefficient of friction (CoF) at a contact surface is $f = 0.1$. and the pulling force is $T = 100$ N as shown in Figure 3.3a. Assume that the initial speed $v_0 = 0$ (m/s) and it is moved by a distance of $L = 5$ m over the plane, calculate (1) the work done by the pulling force, the friction force, and gravity and (2) the speed of the object after it is moved for 5 m over the plane.

Solution

The work by the pulling force can be calculated as,

$$W_T = T{\cdot}L = (100)(5) = 500\ (\text{J}). \tag{3.1}$$

To calculate the consumed work by friction, one has to know the relation force (N) that can be calculated based on the force balance in the free-body diagram (FBD) of the object in Figure 3.3b as,

$$\sum F_{\text{Normal}} = 0 \quad \Rightarrow \quad N = mg\ \cos\theta = 84.91\ \text{N}. \tag{3.2}$$

The work by the friction force can be calculated as,

$$W_f = F_f{\cdot}L = (N{\cdot}f){\cdot}L = (84.91)(0.1)(5) = 42.46\ (\text{J}). \tag{3.3}$$

The work by the gravity force can be calculated as,

$$W_g = (mg){\cdot}L\ \sin\theta = (10)(9.81)(5)(\sin(30,°) = (84.91)(0.1){\cdot}(5) = 245.25\ (\text{J}). \tag{3.4}$$

To calculate the speed when the object is moved for 5 m, the energy conservation principle can be used, since the kinematic energy (U) of the mass also comes from the work by the pulling force as,

$$U = \frac{mv^2}{2} = W_T - W_f - W_g \Rightarrow v = \sqrt{\frac{2(W_T - W_f - W_g)}{m}} = \Rightarrow 6.516\frac{\text{m}}{\text{s}}. \tag{3.5}$$

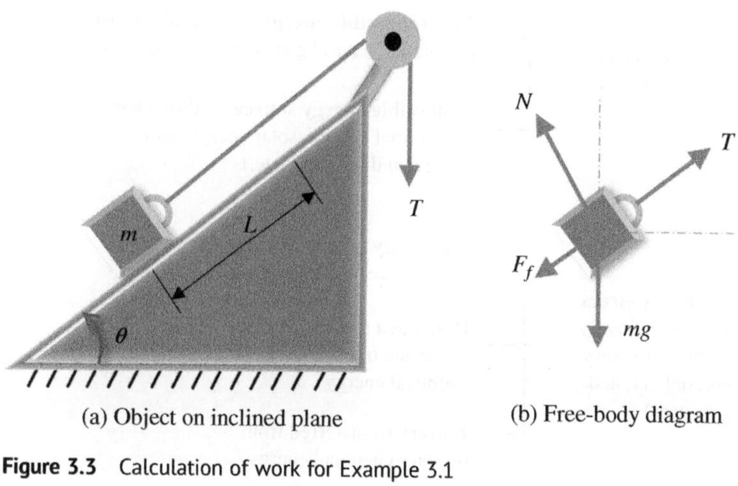

Figure 3.3 Calculation of work for Example 3.1

Example 3.2 An elevator lifts a mass of 1500 kg, and it is capable of accelerating the mass from rest (i.e., $v_0 = 0$ m/s) up to $a = 5.0$ m/s^2 in one second, what is the minimum power the elevator has for this capability?

Solution

According to the principle of energy conservation, the power spent in one second is converted into the potential energy of the mass as well as its kinematic energy as,

$$\text{Power} = \frac{mgh + \frac{m}{2}v^2}{t} = \frac{mg\left(\frac{at^2}{2}\right) + \frac{m}{2}(at)^2}{t} = \frac{(1500)(9.81)\left(\frac{(5)(1)^2}{2}\right) + \frac{(1500)}{2}((5)(1))^2}{1}$$
$$= 55{,}527.5.(\text{W}) \tag{3.6}$$

3.3 Energy Source

Energy is the driving power of human civilization, and energy exploitation and utilization are essential to develop and advance social and economic societies. With the over-exploitation of natural resources, how to utilize more renewable energy becomes a crucial concern in the research community and industries; a great deal of research effort is made to develop renewable, green, and environment-friendly energy from green sources such as waves, sunlight, wind, geothermal, water flows, nuclear, and biomass (Pandian 2010).

When energy is used in a mechatronic system, it is neither created nor destroyed according to the law of conservation of energy; instead, the energy is transferred from one form into another in fulfilling the functions of the system. Therefore, *a source of energy* must be determined in designing a mechatronic system. For example, gasoline is used as the source of energy by a car engine to convert chemical energy in gasoline into the kinematic energy required to move a car. The total amount of energy in our universe remains the same. Figure 3.4 shows a classification of energy sources by

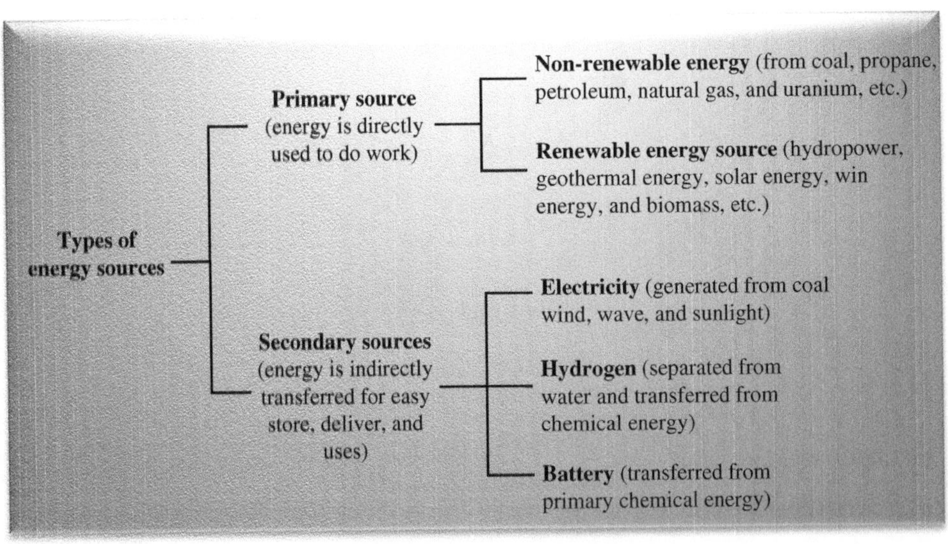

Figure 3.4 Classification of energy sources

which a primary or secondary energy source can be used by a driving component in a mechatronic system. The following two criteria can be used to select the right energy source.

One criterion in selecting an energy source is to see if the energy from the source is renewable or not. A *nonrenewable energy* source has limited supplies; for example, coal, natural gas, uranium, and petroleum are nonrenewable, and the excessive use of these nonrenewable energy sources has led to the scarcity of natural resources globally. A *renewable energy* source has sustainable supplies; for example, solar energy, hydropower, geothermal energy, biomass, and wind energy are renewable since they can be replenished in a short time. It should be noted that electricity and hydrogen are secondary energy sources that are prepared for easy storage, transmittance, delivery, and use of energy. When a renewable energy source is used, one has to know how much energy is needed to run the mechatronic system adequately. The evaluation of the energy harvesting capability of an energy source depends on the type of energy source.

The second criterion is the efficiency of energy transformation. When energy is transferred from one form to another, not all the amount of energy can be used to do work, and *energy efficiency* measures the amount of usable energy in contrast to the total amount of energy obtained from its source. Energy transformation is usually not very efficient; for example, the efficiency of a human body in transferring energy from food to mechanical energy in moving, thinking, and exercising is less than 5% in most of time (EIA 2007). Since any energy transformation involves a loss of usable energy, a primary energy source is advantageous in comparison with a secondary energy source since no energy loss occurs in an indirect energy transformation.

3.4 Driving Components – Functional Requirements (FRs)

Depending on the purpose and application of a certain mechatronic system, FRs for an energy source can be classified in terms of the form of power. Figure 3.5 shows a decomposition of FRs for an energy source. The overall functional requirement (FR0) is decomposed into a set of sub-FRs in the form of energy (FR11) (e.g., power by voltage (V) and current (I)), the rates and frequencies (FR12), the amplitudes (FR13), and the other requirements (FR1i) such as efficiency, cost, availability, and environmental factors. In addition, the form of power depends on its

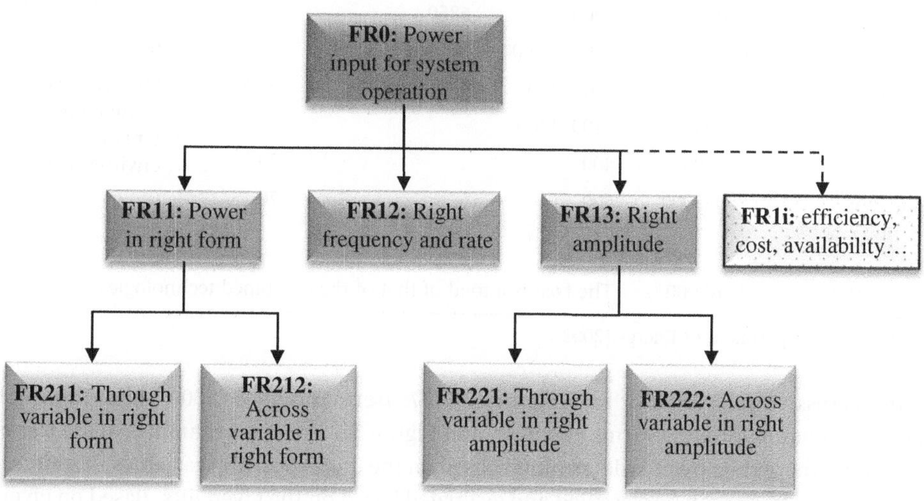

Figure 3.5 Functional requirements (FRs) of driving components

Table 3.1 Through and across variables in different forms of power

Form of power	Through variable	Across variable
Mechanical – translation	Force (F)	Linear velocity (V)
Mechanical – rotation	Torque (T)	Angular velocity (ω)
Electrical	Electric Current (I)	Electric voltage (V)
Hydraulic	Flow rate (Q)	Pressure (p)
Magnetic	Magnetic moment (m)	Magnetic field (B)
Thermodynamic	Entropy flow ($\dot{s}$)	Temperature (T)

Table 3.2 A comparison of distributed energy forms

Type	Power (kW)	Cost ($/kW)	Heat rate Btu/kWh)	Efficiency (%)	Environmental impact (lb/kWh)	
					NO_x	CO_2
Diesel engine	1–10,000	350–800	7800	45	0.017	1.7
Natural gas engine	1–5000	450–1100	9700	35	0.0059	0.97
Natural gas engine with combined heat and power (CHP)	1–5000	575–1225	9700	35	0.0059	0.97
Due-fuel engine	1–10000	625–1000	9200	37	0.01	1.2
Microturbine	15–60	950–1700	12,200	28	0.00049	1.19
Microturbine with combined heat and power (CHP)	15–60	1100–1850	12,200	28	0.00049	1.19
Combustion turbine	300–10,000	550–1700	11,000	31	0.0012	1.15
Combustion turbine with combined heat and power (CHP)	300–10,000	700–2100	11,000	31	0.0012	1.15
Fuel cell	100–250	5500	6850	50	0.000015	0.85
Photovoltaics	0.01–8	8000–13,000	Storage device virtually has no heat rate	N/A	0.0	0.0
Wind turbine	0.2–5000	1000–3000		N/A	Storage device virtually has no emission to environment	
Battery	1–1000	1100–1300		70		
Flywheel	2–1600	400		70		
Superconducting magnetic energy storage (SMES)	750–5000	600		70		
Hybrid energy systems	1–10,000	The cost is a total of that of the combined technologies.				

Source: Adapted from Department of Energy (2002).

through and across variables (Otter and Elmqvist 1997; Isermann 1999, 2000), and Table 3.1 shows these variables in different forms of power, and Figure 3.6 shows six examples of the forms of energy source where the power is determined based on the corresponding variables in sources.

Available energy sources can be classified and evaluated based on the given FRs. Based on given FRs, an energy source must be selected to have the same form of the required power in a

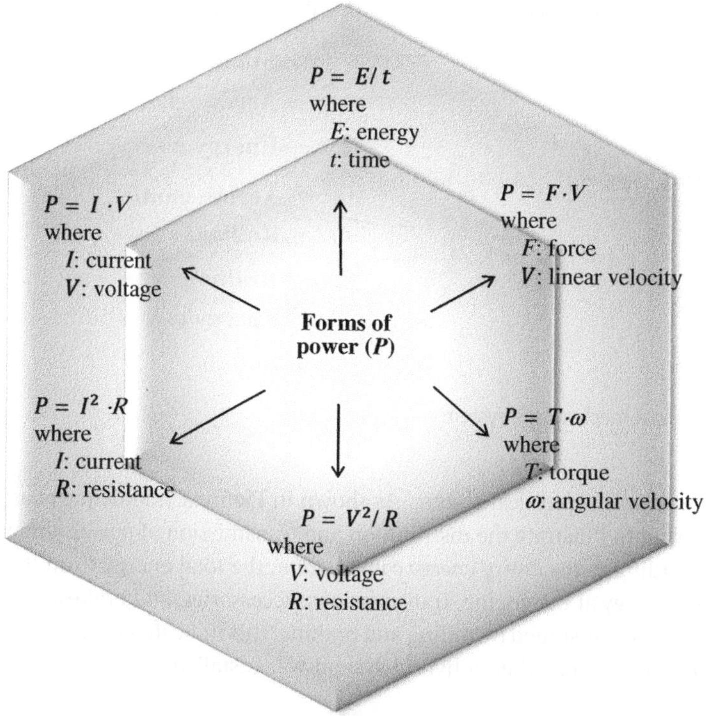

Figure 3.6 Forms of power (P)

mechatronic system. With the industrialization and infrastructure development, electrical power lines have become the most accessible secondary energy sources for machines; however, not all the mechanical systems can be connected to a power grid, and these systems are stand-alone, and they must be self-powered or equipped with *distributed energy sources* (DESs). A self-powered or distributed energy source is usually modular, small, and equipped with storage, and energy may come from sunlight, wind, fuel cells, gas, or battery. For example, Table 3.2 shows a comparison of available distributed energy sources with a capacity of power of up to 10 MW; the considered factors are capacity, cost, heat rate, efficiency, and environmental impact (Department of Energy 2002).

3.5 Power Transmission

A machine can be characterized by its energy flows. A mechatronic system is mostly to apply and control force and tools to fulfill given tasks; any energy sources such as electricity, hydropower, nuclear or thermal power, or wind power can be possibly used to drive a mechatronic system, and the mechatronic system functions to transform the power from the energy source to the right form to fulfill the tasks. However, mechatronic systems are highly diversified and complex in the sense that in a mechatronic system, the power from the energy source is distributed and flows through different system elements to driven elements where system functions are performed. The forms of power in machine elements and driven components are different in terms of through and across variables of power.

Otter and Elmqvist (1997) and Isermann (1999) argued that a mechatronic system could be modeled based on its energy flow, and the bond graph method and unified object-oriented language

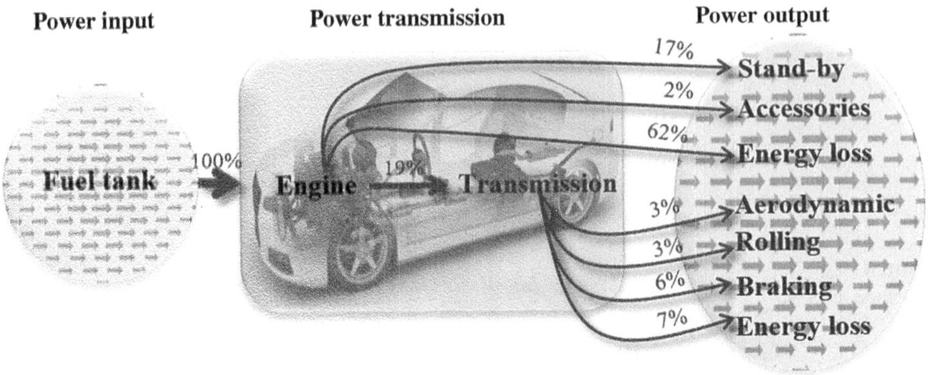

Figure 3.7 Example of energy flow in a mechatronic system

were used to represent an energy flow in a physical system. As shown in Figure 3.7, Maleque et al. (2012) modeled energy flows in a car to illustrate the distribution and transmission of power when the car was used in the city. According to *the law of energy conservation*, the total energy from the source was the sum of consumed energy at the engine, transmissions, accessories, and brakes, 75% of energy input from a wheel rotor was consumed by rolling and braking; this indicated the amount of consumed energy that could be harvested when a hybrid system was installed.

3.5.1 Functional Requirements (FRs)

A system element changes the form of power in a certain way. The FRs of a system element can be defined based on its role in power transmission accordingly. Some common FRs of power transmissions for system elements are exemplified in Figure 3.8. It shows that a system element may be used to *collect* power, *supply* power, *split* or *consume* power, *engage/disengage* a power flow, or *change* the form of power in terms of its type, direction, or amplitude of motion or force.

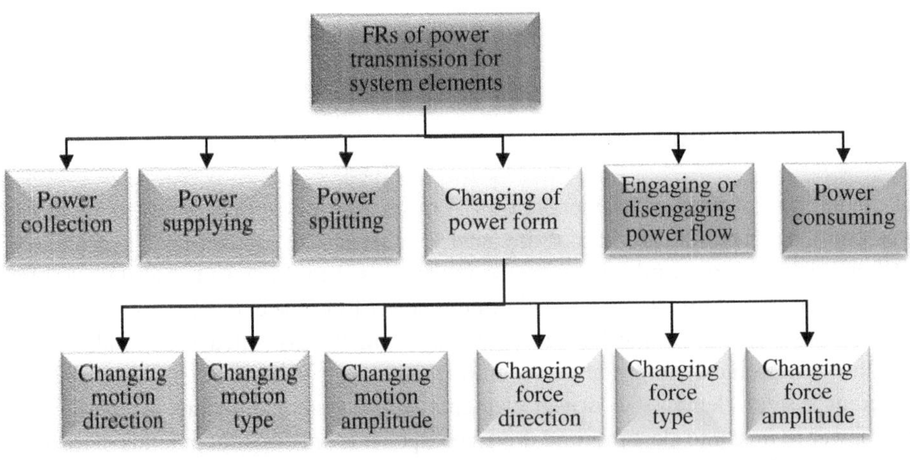

Figure 3.8 Exemplifying FRs of power transmission for system elements

3.5.2 Machine Elements for Power Transmission

The FRs of a system element can be defined according to its role of power transmission such as the collection, supply, and transmission as exemplified in Figure 3.9. Most of machine elements are used to transmit power by changing the form of power in terms of the type, direction, or amplitude of motion or force. As shown in Figure 3.9, the machine elements for power transmission are classified in terms of *used media* and degree of automation. On the one hand, examples of machine elements that transmit power by solid bodies are clutches, gearboxes, flywheels, belts, winches, pulleys, cams, connecting rods, couplers, spindles, cranks, chains, and brakes. Examples of machine elements that transmit power by fluid or gas fluid coupling are fluid converters, valves, gates, switches, reservoirs, solenoids, regulators, and pumps. On the other hand, the degree of automation of a machine element is critical since it determines the need of mechatronic design. A machine element for power transmission can be *manual*, fully *automated*, or *semi-automated* depending on the uses of electrical and electronic elements in its operation and controls.

3.5.3 Types of Machine Elements

In an engineering product such as automation drive and machinery, power transmission elements are used to transmit source power into the right form of power to fulfill the functions of products. Due to their broadened applications, machine elements are mostly standardized and made in mass production to reduce manufacturing costs. Common types of machine elements are introduced in Table 3.3.

3.5.4 Procedure in Designing or Selecting Machine Elements

The FRs of a machine element should be defined by a system in terms of external loads and *boundary constraints* (BCs), and the performance of a machine element should be evaluated based on system design criteria such as safety factor, cost, and lifespan. Note that the capability of a system,

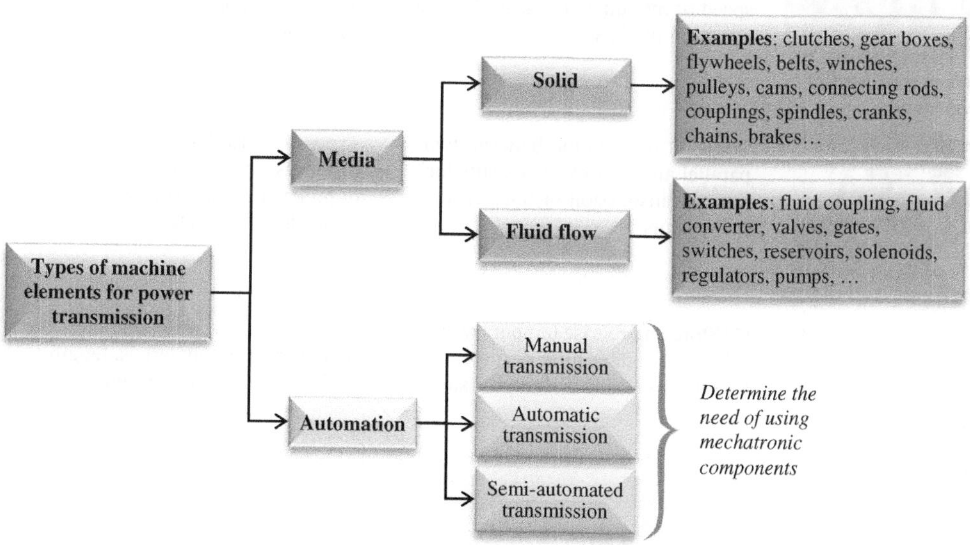

Figure 3.9 Types of mechanical power transmissions

Table 3.3 Common types of machine elements

Types	Functional requirements (FRs)	Design variables (DVs)
 ***Gear transmission***	(1) Change type, amplitude, direction, and speed of motion; (2) combine or divide power; (3) transmit power	Types, numbers of tenth, pitch diameters, materials, manufacturing grades, center distance, and pressure angles, spatial orientations, etc.
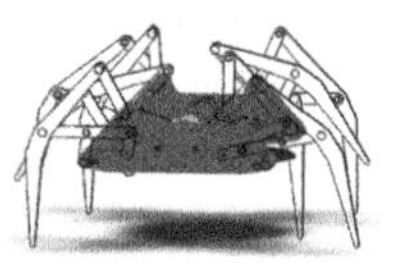 ***Linkage***	Transfer a motion/force in one form to another specified form for task execution	Numbers, geometric dimensions, assembly of rigid bodies and joints, selections of fixed bodies, active joints, etc.
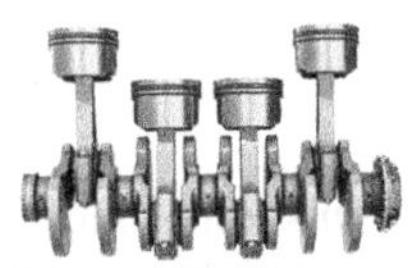 ***Cam mechanism***	Convert a rotation of a shaft into simple or complex reciprocating linear motion	Cam profile, geometric dimensions and supports of cam, rollers, shaft, follower, etc.
 ***Bearing***	(1) Reduce the friction at contacts; (2) provide supports to moving components; (3) prolong the lifespans of moving elements	Type, outside and bore diameters, widths, static and dynamic loading capacities, reliabilities, etc.
 ***Power screw***	(1) Gain mechanical advantage; (2) transfer a rotation to a translation; (3) perform precision positioning	Number of threads, pitch and pitch diameter, thread angles, rollers, collar, coefficients of friction, motion range, etc.
 ***Clutches and brakes mechanism***	(1) Reduce the motion of object; (2) engage two objects for the same speed of motion; (3) engage/disengage a motion of object	Configuration, geometric dimensions, braking patches, acting force or torques, supports of acting components, etc.
 ***Belts, ropes, and chains***	Transmit the rotation between two parallel and non-parallel shafts for power transmission or track relative movements	Types and profiles of belts, central distances, the geometric dimensions of pulley, number, configuration, preloads of belts, etc.
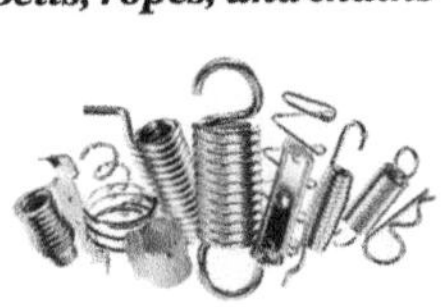 ***Springs***	(1) Store and release temporary energy; (2) measure deflection or force; (3) serve as elastic or damping supports	Number of coils, wire diameter, pitch diameter, pitch, materials, length, preload, end-conditions, etc.

Table 3.3 (Continued)

Types	Functional requirements (FRs)	Design variables (DVs)
 ***Fastenings, rivets, and welds***	(1) Join two or more objects temperately or permanently; (2) assemble components with different materials and manufacturing processes into structure; (3) support easy maintenance and upgradation	Crest and pitch diameters, pitch, threading direction, dimensionless stiffness rate, caps and washers, length, preload of bolts, etc.

Sources: alexlmx/Adobe Stock Photos; Gear Technology / https://www.geartechnology.com/articles/30409-abma-update / Last accessed on July 19, 2024; Firgelli Automations / https://www.firgelliauto.com/blogs/news/types-of-leadscrews-used-in-actuators-from-acme-to-ballscrews / Last accessed on July 19, 2024; ACTION CLUTCH / https://rcclin.cl/brand/action-clutch/ / Last accessed on July 19, 2024; Babcox Media Inc / https://www.tomorrowstechnician.com/curing-belt-slippage/ / Last accessed on July 19, 2024; CHINA SPRING INC. / https://www.chinaspringparts.com/about-us.html / Last accessed on July 19, 2024; Endeavor Business Media, LLC / https://www.machinedesign.com/mechanical-motion-systems/article/21835438/aviation-and-aerospace-fasteners-why-it-pays-to-buy-legit / Last accessed on July 19, 2024; AliExpress.com / https://www.aliexpress.com/item/1005002028197582.html / Last accessed on July 19, 2024

such as *production throughput, safety factor, loading capability,* or *reliability*, will be determined by that of the weakest machine element; therefore, a system is optimized when the capabilities of all system elements are equally utilized. As shown in Figure 3.10, a general procedure is followed to design or select a machine element. For example, a shaft design follows the following procedure:

(1) Draw a *FBD* by isolating the machine element from the system; any interface between the machine element and the system is replaced by either an external load or a displacement constraint on FBD. The common types of external loads are *axial loads* (*P*), *transverse loads* (*V*), torques (*T*), moments (*M*), or *body loads* (*B*). The common types of constraints are *fully* or *partially* fixed degrees of freedom (DoF) and *given displacement*.

(2) Establish a reference coordinate system for FBD, draw the bending moment diagrams in the x–y and x–z planes; determine some critical sections where the resultant internal moments are their peak values;

(3) Draw a torque diagram along the x-axis; note that torque in a power transmission element is fully balanced from its input to output, determine some critical sections where the torques are at their peak values.

(4) At all critical sections, calculate the normal stresses by moments and shear stresses by torques and evaluate the corresponding effective stresses such as *von Mises stresses* or *maximized shear stresses*, and determine the most critical positions by comparing the effective stresses at all critical sections.

(5) For ductile materials, use the *maximum-shear-stress theory* (MSST) or *the distortion-energy theory* (DET) to determine the design factor when a part dimension is given, or determine the part dimension when the design factor is given.

(6) For brittle materials, use *the maximum-normal-stress theory* (MNST), *Brittle Coulomb–Mohr Theory* (BCM), or *the modified Mohr theory* (MMT) to determine the design factor when a part dimension is given, or determine the part dimension when the design factor is given.

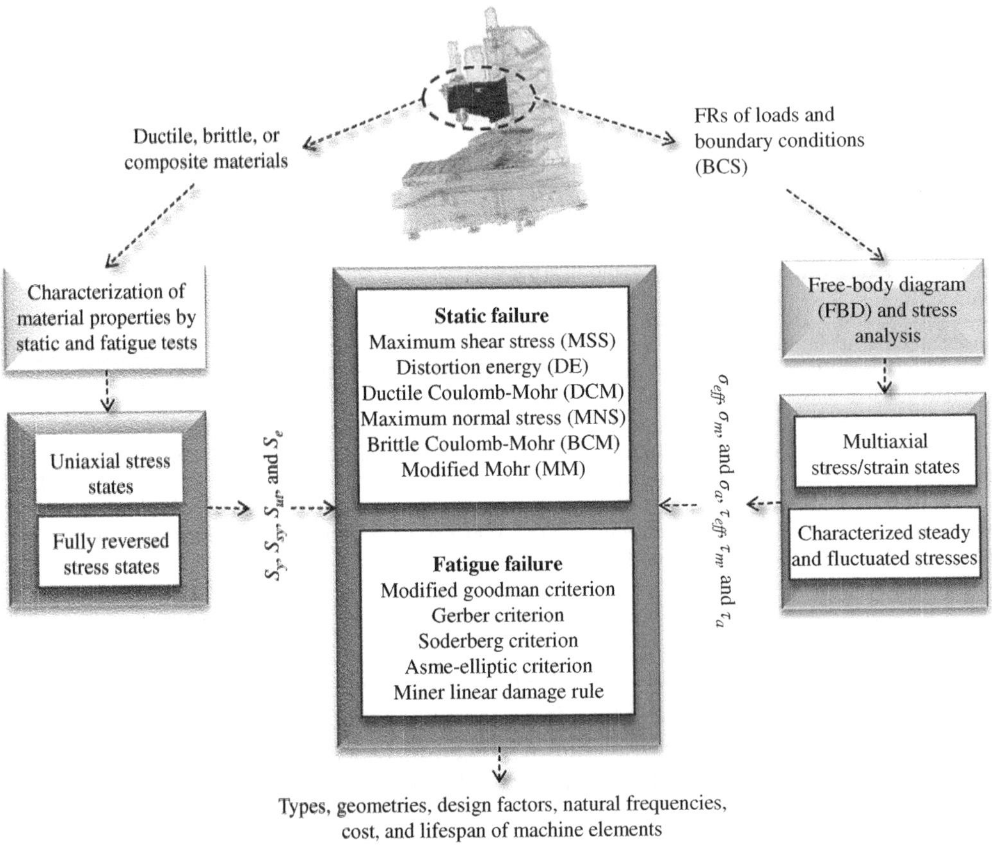

Figure 3.10 Procedure in designing or selecting a machine element

3.5.5 Machine Elements in Mechatronic Systems

Engineering design is open ended since there are numerous design solutions to meet the same set of FRs of a system. However, it is desirable to adopt as many standardized machine elements as possible to shorten the development cycle and reduce the cost of a mechatronic system. Under modularized architecture, the interfaces of a machine element with other system elements are most critical to an integrated mechatronic system. On the other hand, defining the inputs and outputs of a machine element based on standardized interfaces quarantines the flexibility and upgradability of individual machine elements. In Table 3.4, the main inputs/constraints, outputs/performance, and design variables of common machine elements are discussed for mechatronic engineers to define and select machine elements efficiently.

3.5.6 Mechanical Power Transmission Examples

Different types of machine elements can be selected and assembled to create different machines to transform power from one as shown in Table 3.5.

Table 3.4 Inputs, outputs, and design variables of common machine elements

Inputs and constraints	Machine element and main design variables	Outputs and performance
Locations, orientations, and supports of input and output shafts; central distance, input power, torque and speed, expected gear ratio, etc.	 ***Gear transmission***: gear transmission type, the number of gears, the number of teeth on gears, pitches, pressure angles, widths, materials, grades of gears, etc.	Output power, torque and speed, cost, weight, dimensions, design factors against fracture and fatigue failures, reliability, lifespan, etc.
Locations and orientations of input and output bars; input torque and motion profile, the function of output motion with respect to input motions, etc.	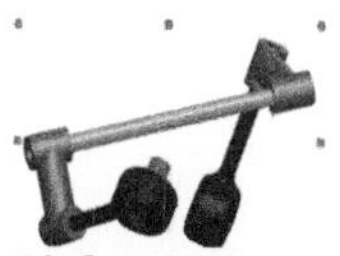 ***Linkage***: The numbers of links and joints, topological structures, the shapes and dimensions of links, selections of input and output links, motor, etc.	Output motion, power and torque, singularity positions, mechanical advantages, working ranges, efficiency, cost, weight, dimensions, cycle time, etc.
Locations, orientations, and supports of input, intermediate and output shafts; input power, torque and speed, and desirable motions and torques of outputs, etc.	 ***Belt system***: the numbers, types, materials, profiles, preloads, and lengths of belts and pulleys, the diameters and widths of pulleys, intermediate shafts, etc.	Output power, torque and speed, cost, weight, dimensions, mechanical advantages, efficiency, reliability, etc.
Locations, orientations, and supports of input, intermediate and output shafts; input power, torque and speed, and desirable locations of output motions and torques, etc.	 ***Chain system***: the numbers, types, materials, profiles, preloads, and lengths of chains and gears, the numbers of teeth on gears, designs of ideal gears, etc.	Output power, torque and speed, cost, weight, dimensions, mechanical advantages, efficiency, reliability, etc.
Locations, orientations, and supports of input and output shafts; input power, torque and speed, braking torque, etc.		Braking force, power, torque, braking or engaging cycle, time, cost, weight, size, lifespan, reliability, efficiency, etc.

(*Continued*)

Table 3.4 (Continued)

Inputs and constraints	Machine element and main design variables	Outputs and performance
	Clutch/brake: types, sizes, widths, and materials of disks, patch and strips, actuating mechanism, force, and supports, lubrication, etc.	
The number, the materials, the thicknesses of the members to be joined, the static and dynamic characteristics of external loads, accessible space, etc.	 ***Fastener***: type and configuration of fastening, the pitches, crest diameters, materials, grades, preloads of bolts, etc.	The cost, weight of fasteners or assembly, design factors against bending, shearing, tearing, and fatigue failures on bolts or members, etc.
Expected mechanical advantage, power, torque, and speed of power source, the location and geometry of lift object, load, speed, range of lifting, etc.	 ***Power screw***: type, pitch, crest diameter, number of threads of screw, the dimensions of collar and nut, the length of thread body, rolling elements, etc.	Output power, torque and speed, operating range, raising and lowering efficiencies, cost, weight, reliability, lubrication, wear, self-locking, etc.
Mean and alternative load, stroke of deflection, the location orientation and type of applied load, structural support, etc.	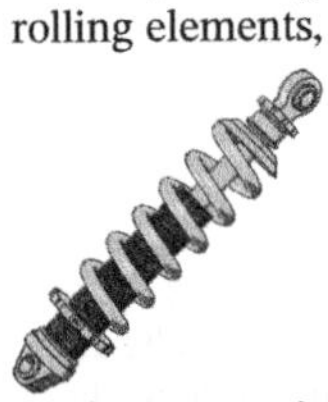 ***Spring***: type, pitch, crust diameter, wire diameter, number of coils, materials, end-conditioning, preload of spring, pin, etc.	Spring rate, output displacement and force, maximum load, cost, weight, stability, energy loss rate, lifespan, etc.

Source: AliExpress.com / https://www.aliexpress.com/item/1005002028197582.html / Last accessed on July 19, 2024.

3.6 Power Generation

A mechatronic system consumes energy to do work. Conventionally, a mechatronic system is powered by external sources such as batteries, electric power, gas, or fuel. However, it is a growing trend that more and more mechatronic systems, such as self-propelled drug deliveries (Zhang et al. 2021) and self-driven sensors (An et al. 2020; Pu et al. 2021a), become self-powered to achieve energy-independence and sustain long-term operations in extreme environments. Therefore, power generation and supply might be essential FRs of a mechatronic system. In this section, common types of energy sources as well as corresponding energy harvesting methods are discussed.

Table 3.5 Inputs, outputs, and design variables of mechanical power transmissions

Inputs	Machine	Outputs and performance
Velocity, acceleration, torque of rotation over time, etc.	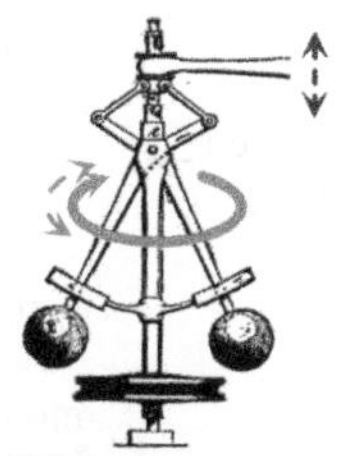 ***Water governor***: bearing, shelf, flywheels, lever, linkage, supports, etc.	Displacement, force, speed of output valve operation; cost, weight, dimensions, response time, reliability, and lifespan of machine
Volume rate and pressure of input flow, etc.	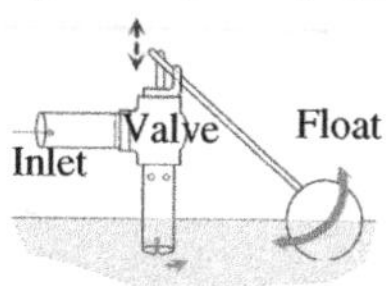 ***Float valve***: inlet and outlet, valve, float, pins, lever, container, etc.	Output pressure and rate of flow, displacement of valve operation; cost, weight, dimensions, response time, reliability, and lifespan of machine, etc.
Input current and atmosphere temperature over time, etc.	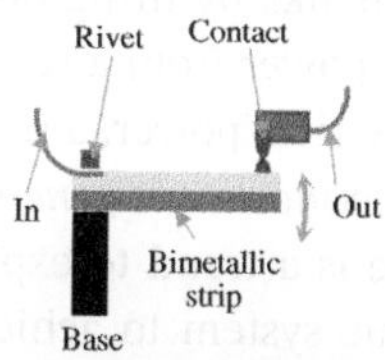 ***Thermal-induced switch***: the configuration of bimetallic strip, contact points, rivet, input and output wires, base support, etc.	Openness and closeness of circuit contact; cost, weight, dimensions, response time, reliability, and lifespan of machine
Volume rate and pressure of input flow, etc.	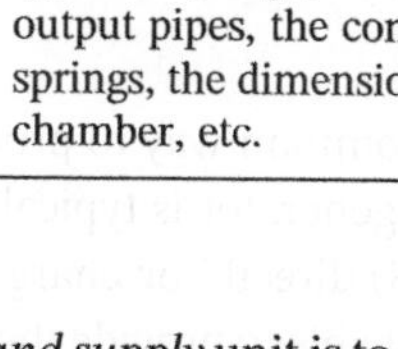 ***Spring energy storage***: number and types of springs, the sizes of input and output pipes, the configuration of springs, the dimensions of the pressured chamber, etc.	Output pressure and rate of flow, displacements of springs, stored energy; cost, weight, dimensions, response time, reliability, and lifespan of machine

A *power generation and supply* unit is to collect power from a source of energy and converts it into usable power to powered modules in a mechatronic system. A *source of energy* in the application environment can be (1) *mechanical* energy such as wind or wave, (2) *chemical* energy such as gastric acid, (3) *solar* energy radiated by a light source, and (4) *thermal* energy that transfers, converts, or radiates due to a temperature gradient. Figure 3.11 also shows that two basic functional modules are required to self-power a mechatronic system (Necula et al. 2014): *power generation* is to convert a source energy into a new form that can be usable by the mechatronic system by following certain physical principles of energy conversion. *Power storage* is to store the collected energy and power other functional modules for their appropriate uses.

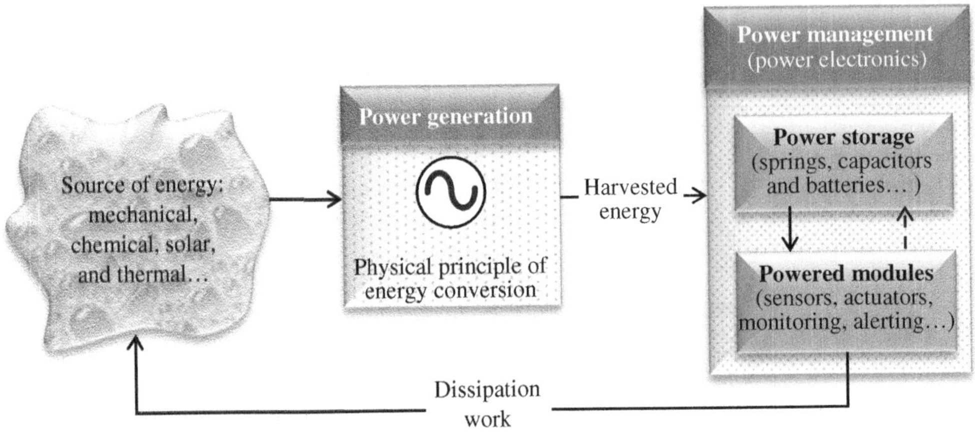

Figure 3.11 FRs of power generation and supply

The design of a *power generator* depends on the type of source energy. For example, an internal combustion (IC) generator is designed to charge battery banks by firing propane or natural gas. A number of power generators are developed to generate power from a renewable energy source such as *photovoltaics* (PVs) to convert solar power into electrical power, a *wind* or *hydro turbine* to convert wind or hydropower into electric power, and a *thermoelectric generator* (TEG) to convert geothermal energy into electrical power. Moreover, there is a trend to explore the possibility of directly integrating a power generator in the mechatronic system to achieve the independence of the system where direct power is inaccessible. Examples of these applications are (1) micro- and nano-systems used in in vivo environments and (2) ubiquitous sensing in highly unstructured environments.

Various energy sources are used to power machines, devices, or systems. As shown in Figure 3.12, other than electric power, coals, natural gases, propane, nuclear energy, and gasoline, renewable energy sources such as solar power, wind power, geothermal power, and hydroelectric power become more and more preferable to power sensors and mechatronic systems (Aurora 2015; Will et al. 2019).

3.6.1 Internal Combustion (IC) Generator

As shown in Figure 3.12a, *internal combustion* (IC) is a common way to produce power from propane, diesel, or natural gas. The rated power from an IC generator is typically 1–2.5 kW. It can be used to power a *road weather information system* (RWIS) directly or charge battery banks when other renewable power sources such as solar power are not able to provide the required power capability. In general, the capability of a power generator should be two to five times of the power required by the application.

Alternatively, fuel cells can be used as fossil fuel-based power sources without mechanical or combustive conversion. A fuel cell uses hydrogen as the fuel and generates electrical energy by combining hydrogen and oxygen in a ceramic cell membrane, the byproduct is water. Note that ceramic cell membranes are very fragile and they can be easily fractured in transportation; in addition, cell membranes can be corroded due to the impurity of the fuel; in addition, the fuel cell must remain operational in cold weather due to byproduct water. Two common sources of hydrogen for fuel cells are solid oxide and methanol.

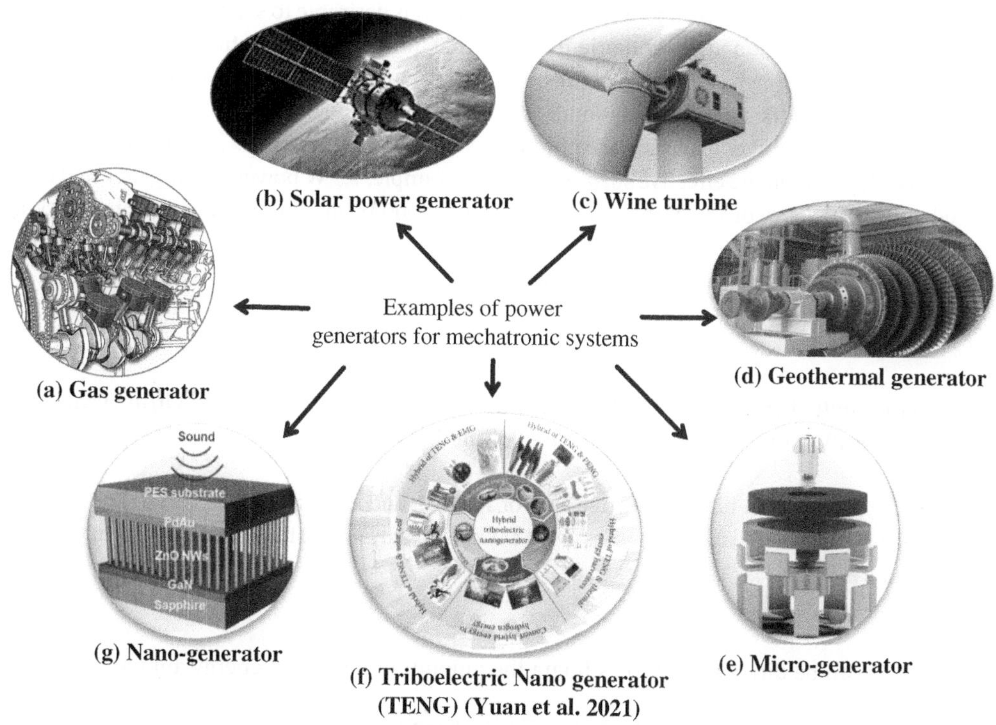

Figure 3.12 Examples of power generators for mechatronic systems (*Source:* (a) Swaroopvarma/Wikimedia Commons; (c) General Electric Company / https://www.windpowermonthly.com/article/1811518/ge-build-new-york-plants-haliade-x-offshore-wind-turbine-patent-work-around-nets-sufficient-orders / Last accessed on July 19, 2024; (d) MITSUBISHI HEAVY INDUSTRIES, LTD. / https://power.mhi.com/products/steamturbines/lineup/geothermal / Last accessed on July 19, 2024; (f) Yuan et al. (2021)/with permission of Elsevier)

3.6.2 Solar Power Generator

As shown in Figure 3.12b, a solar power generator harvests energy from sunlight. The power is stored in a battery bank. In the daytime under the sunlight, the battery bank is charged. Therefore, the capacity of a solar power generator depends on the size of solar panels, intensity of sunlight, and the efficiency of energy conversion. A solar panel has a typical efficiency of 11–15%, and the size of a solar panel should be customized based on the required power for an application. Taking the example of a solar power generator used in a remote metrological site, a typical solar PV module has a power of 0.2–0.9 kW, and 6 of these modules are able to output 12 or 24 V direct current (DC). Note that the actual power output or voltage depends on the sunlight on surfaces that relate to season, weather conditions, and hour of day. To optimize power use and storage, the charge is controlled to maintain the charging voltage above the nominal voltage when the output voltage of solar PV is lower than the nominal voltage.

Solar energy is presented as a type of electromagnetic radiation in an environment similar to radiofrequency (RF) radiation. The power harvested from electromagnetic radiation is widely used in many mechatronic systems. In particular, the technology of using PV to collect solar energy is very mature and widely used; the power from a PV system varies from milliwatt to megawatt for a broad application from a wristwatch to large-scale power grids.

PV systems are very attractive as renewable energy solutions to portable energy systems. Since solar energy comes from solar radiation that varies from one location to another due to the differences in longitude, latitude, and weather conditions, it is desirable to set up a PV system based on the optimized inclination angle and orientation to maximize its power output. The efficiency of power conversion is defined as the ratio of the output power and irradiance power over surface of a solar cell. Typically, indoor irradiance is small and around 3.5–20 W/m^2, it implies that the output power will be 0.35–2 W/m^2 considering the power conversion efficiency of 10%. The total power of a solar cell also depends on the effective surface area. For example, 1.5 W power can be generated by a solar cell with a efficiency of 15% if the solar irradiation is 1000 W/m^2 and the effective surface area is 0.01 m^2.

The solar power is solar irradiance calculated as,

$$E = H_y * r * A \tag{3.7}$$

where E is the emitted energy (kWh), H_y is the annual average solar radiation (kWh/m^2/year), r is the efficiency of PV panel (%), and A is the area of the PV panel (m^2).

Example 3.3 A PV panel has an effective area of 10.0 m^2, and an efficiency of 12.5%. The annual average solar radiation H_y is estimated as 1800 kWh/m^2/year, estimate the amount of harvested energy annually.

Solution

Equation (3.7) can be directly used to estimate the amount of harvested energy annually as,

$$E = H_y * r * A = (1800)(12.50\%)(10.0) = 2250 \text{ (kWh)}$$

When the power demand of an application is known, the size of a solar power system (m^2) can be determined coordinately as,

$$\left.\begin{aligned} E_d &= P * t \\ S &= E_d / \left(H_y/365 * r\right) \end{aligned}\right\} \tag{3.8}$$

where E_d is the daily consumed energy (kWh/day), P is the power of the mechatronic system (kW), t is the number of operating hours (hours), S is the size of the solar power system (m^2), H_d is the daily average solar radiation (kWh/m^2/year, r is the efficiency of the PV panel (%), and A is the area of the PV panel (m^2).

Example 3.4 A smart home is equipped with a refrigerator (0.5 kW) and a central air-conditioning system (2.0 kW) that are powered by a solar energy system. Assume that average daily solar radiation of 2000 kWh/m^2/year and a panel efficiency of 15%, determine the minimal size of the solar energy system to support 24-hour operations of these appliances.

Solution

Equation (3.8) can be used to estimate the size of the solar energy system as,

$$E_d = P * t = (0.5 + 2.0)(24) = 60 \text{ (kWh)}$$

$$S = \frac{E_d}{H_d/365 * r} = \frac{60 * 365}{2000 * 15\%} = 73.0 \text{ m}^2$$

Other electromagnetic energy can also be harvested to power a mechatronic system. For example, an ID card can be powered by a nearby source with high-power electromagnetic energy. An urban area has numerous RF sources such as radios, television channels, mobile telephony, and wireless networks; however, it is very challenging to collect from all these disparate sources and utilize energy in a useful way, since the working principle to collect electromagnetic energy is rectification by a rectifying antenna. The antenna consists of a Schottky diode at its dipoles, and it just produces too low energy to power an electronic device. Such types of generators will become more and more important when more and more future electronic devices with low-power consumption are developed, and advanced technologies are developed to recycle RF energy cost-effectively.

3.6.3 Wind Turbine Generator

A wind turbine generator transfers wind energy into electrical power. The wind energy is presented as an aerodynamic force over rotor blades. When the wind flows through a set of blades, the air pressure on one side of the blade is decreased, and it causes a pressure difference of the two sides of blade and creates both lift and drag. When the lift and drag forces are unbalanced, the rotor will spin. The generator is directly or indirectly connected to the rotor to create electricity under aerodynamic forces on blades. For the applications of remote weather information systems, a wind power generator can typically generate less than 100 W to charge batteries, the output is three-phase alternative current (AC), and it can be rectified as 12–48 DC. The voltage of the DC output is rectified as a constant so that it can be directly connected to battery charge controllers. In addition, it is not uncommon to integrate a wind power generator with solar PV; in such a case, the controller is advanced to manage the power from wind and solar PV optimally.

The amount of power harvested by a wind turbine is proportional to the effective area of the turbine and the cube of the wind speed. It implies that the harvested power will reduce greatly when the size is scaled down. Moreover, the wind speed is unstable due to its variation in altitude, time, gusts, air density, and direction. The harvested power can be estimated as (Sun 2017),

$$P_t = \frac{1}{2}\rho A v^3 c_p(\lambda, \beta) \tag{3.9}$$

where P_t is the harvested power (W), ρ is the air density, A is the effective area of the rotor, v is the wind speed, $c_p(\lambda, \beta)$ is the coefficient of power determined by the tip speed of the wind speed ratio (λ), and the blade pitch angle (β). $c_p(\lambda, \beta)$ is limited to its theoretical Betz limit as 0.59.

Example 3.5 Assume that the effective area of a wind turbine $A = 50.0\,\text{m}^2$, air density ρ is 1.225 kg/m, the average wind speed is 4.0 m/s, and the coefficient of power c_p is 0.3, determine the maximum output power of the wind turbine.

Solution

Equation (3.5) can be used to estimate the maximum output power of the wind turbine as,

$$P_t = \tfrac{1}{2}\rho A v^3 c_p(\lambda, \beta) = \tfrac{1}{2}(1.225)(50)(4.0)^3(0.3) = 0.588\ (\text{kW})$$

3.6.4 Geothermal Generator

A geothermal generator transfers geothermal energy into electrical energy. The amount of energy for a geothermal source depends on a number of factors such as the temperature, the flow rate of the resource, and the efficiency of the geothermal generator as,

$$P = m * h * \eta \tag{3.10}$$

where P is the power output in kilowatt (W), m is the mass flow rate of the steam or hot water (kg/s), h is the enthalpy of the steam or hot water (J/kg), and η is the efficiency of the geothermal generator.

Example 3.6 Assume that the mass flow rate (m) of steam in a geothermal generator is 2.0 (kg/s), the specific heat (c) of the cooling water is 4180 (W/kg·°C), the temperatures of the condenser cooling water at entry and exit are 15 °C and 30 °C, respectively, and the efficiency (η) of a geothermal generator is 20%, determine the output power of the geothermal generator.

Solution

Equation (3.6) can be used to estimate the output power of the geothermal generator. The enthalpy of the water flow (h) is calculated as,

$$h = c(T_{\text{in}} - T_{\text{out}}) = (4180)(30 - 15) = 62{,}700\ (\text{J/kg})$$

The output power can then be determined by,

$$P = m * h * \eta = (2)(62{,}700)(20\%) = 25.08\ (\text{kW}) \tag{3.11}$$

The raw materials can also be lava that is supplied by fluid cells, pump, or fluid distributor directly. A geothermal generator will typically charge the battery or output power at a rate of 20 energy units per tick (EU/t). To fill a multi-function electric storage unit (MFE) with a capacity of 2000 EU, one geothermal generator will take more than 50 minutes to consume 100 ticks if one bucket or cell of lava lasts for 30 seconds.

Thermal energy is usually converted into electricity before it can be used by a mechatronic system. A TEG is another type of generating electrical power from thermal energy; the raw materials are fuels such as propane and natural gas. A TEG is designed based on the Seebeck effect: one side of the thermoelectric material is heated by the burned fuel and the other side is exposed to ambient air to create a potential voltage difference due to the significant temperature differences of two sides of the thermoelectric material. Due to the need for fuel combustion, TEG is used in open or in a well-ventilated environment.

Since thermal energy is a form of energy commonly presented in the environment, TEG can also be used to harvest thermal energy from a machine, device, and a human or animal body. In such a case, the thermocouple in TEG produces a current that is proportional to the temperature difference at its two junctions. The current is generated based on the Seebeck effect; it is determined by the positive and negative Seebeck coefficients of p-type and n-type materials, respectively (Mateu and Moll 2005). A thermal energy flow can be formed in the thermocouple due to (1) the temperature gradient at the junction or (2) the absorption or dissipation of energy due to the Peltier effect. The upper limit of recovered thermal energy can be determined by the Carnot efficiency as,

$$\eta = \frac{Q_{\text{Hot}} - Q_{\text{Cold}}}{Q_{\text{Hot}}} \leq \left(1 - \frac{T_{\text{Cold}}}{T_{\text{Hot}}}\right) \times 100\% \tag{3.12}$$

where η is the efficiency of the thermocouple, Q_{Hot} and Q_{Cold} are the heat flows at the hot and cold junctions, and T_{Hot} and T_{Cold} are the temperatures at the hot and cold junctions, respectively.

3.6.5 Other Generators

Harvesting mechanical energy is the most practical way to self-power a mechatronic system. The harvesting principle is to use mechanical energy to produce certain movement or deformation of a smart structure, and the induced movement or deformation is then converted into electrical power.

Depending on the smart structure type, the power can be converted in a piezoelectric, electromagnetic, electrostatic, or magneto-strictive way. A power generator should be designed to maximize the efficiency of energy harvesting; note that the allowable size and weight of a power generator as well as available mechanical energy are usually specified as FRs. Among mechanical energy harvesters, piezoelectric components are widely used to transfer energy from mechanical vibration to electric voltage, it is then regulated and stored to power an electronic device such as a microsensor, tire pressure sensor, or medical electronic device. Mechanical energy exists in various forms such as fluctuating deformation or pressure, vibration, random movements, air or fluid flows, and ocean waves.

As shown in Figure 3.12e, a micro-generator produces power by converting kinetic energy from a magnetic rotor into an alternating magnetic field in a stator due to electromagnetic induction. The stator consists of multiple electrical windings to produce alternating voltage and current by the induction principle. As shown in Figure 3.12g, a nano-generator converts mechanical or thermal energy from a nano-scale physical change into electric power; a piezoelectric or triboelectric generator produces the power from mechanical motion, and a pyroelectric generator produces the power from thermal energy when the environment involves a fluctuation of temperature over time. A nano-generator can be possibly used as an independent or a secondary energy source to power a nano or micro device whose power consumption is low and kinetic energy is supplied continually in the application environment. For example, pH or UV nano-generators by Wang and Song (2001) and Wang et al. (2008) were able to generate 20–20 mV to power sensors. Most generators at nanoscale produce limited electric energy, and they are used as a secondary energy source to the battery. In this case, a nano-generator can be integrated with other energy harvesting devices such as biochemical energy and a solar cell; it has been used in developing smart dust (Niccolai et al. 2019). A good example of piezoelectric generator is piezoelectric fiber used in textiles for outfit; the kinetic energy by a human body's motion can be harvested by piezoelectric fibers to power portable electronic devices such as a system to monitor health conditions. An exo-shoe triboelectric nanogenerator (ES-Teng) by Yun et al. (2021) was able to generate 13 μW/g subjected to daily stepping motion of a person to charge wearable electronics such as a health condition monitoring system. Figure 3.12f shows some examples of triboelectric nanogenerators (TENGs) that harvest power by triboelectricity. Yuan et al. (2021) 3D-printed an acoustic TENG capable of harvesting 4.33 mW from an excitation of sound pressure with 100 dB; that power level was able to power a commercial calculator and up to light-emitting diodes (LEDs), and it was developed for edge-sensing for smart things in Internet of Things (IoT).

3.6.6 Selection of Power Source for Mechatronic System

Power is the source of energy to run a mechatronic system; it is critical to select the right power supply to optimize the mechatronic system. However, numerous solutions are available to transfer power from one form to another for a given application. With a growing concern for environmental sustainability and scarcity of nonrenewable energy, it becomes more attractive to utilize renewable energies such as solar energy, wave, wind, and geothermal energies; these types of energy resources are usually dispersed and suitable to power mechatronic systems in distributed environments (Arani et al. 2017). Different solutions to power generation and supplies must be compared based on the FRs of the application. For example, Aurora (2015) discussed alternative power sources for remote weather information systems and compared these power sources based on the criteria of capacity, cost, and fuel rate in Table 3.6.

Table 3.6 Example of comparison of alternative power sources for RWIS

Power source	Capacity	Cost ($)	Fuel rate
STO-135A solar PV module	135 W	105/module	—
WT10 wind turbine (24 V)	75 W at 12 V and 40 mph	4000	—
RP500 fuel cell	500 W at 12 V	30,000	0.12
EFOY pro 800 full cell	45 W at 12 V	21,000	0.24
Gentherm 5060 TEG	54 W at 12 V	6500	1.16

Source: Aurora (2015)/Iowa State University.

Necula et al. (2014) discussed various energy harvesting systems to acquire electric power for mechatronic systems from an ambient environment. For a portable mechatronic device, electrical supplies or batteries are still the most common choices. However, the need of extra power supplies may cause a number of issues in many applications where (1) the access to direct power source is prohibited or (2) the lifespan of device must be extended for a long-term use. Self-powering becomes the vital solution to these mechatronic systems; a self-powering device is capable of harvesting energy from the energy sources in the form of mechanical, thermal, electromagnetic, acoustic, wave, wind, and others. Self-powering mechatronic systems help to reduce human society's dependence on nonrenewable resources such as fossil fuels (Tiefenbeck et al. 2018).

A considerable number of mechatronic systems still use batteries as a direct power source, and batteries are characterized by volumetric and gravimetric energy density, number of cycle life, and the self-discharge rate per year. Mateu and Moll (2005) gave a comparison of four types of commonly used batteries in Table 3.7. Li-ion offers the best option of energy density and number of recharging cycles. Recently developed capacitor technology allows an ultra-capacitor to have a capacitance value in an order of kilo farad (kF), and such a large capacitor is able to present an energy density of 3 W-h/kg. Ultra-capacitors are widely used in the automotive industry due to a large peak power and number of cycles. It is worth to noting that the battery technologies have been developed more slowly than other electronic technologies. For example, the density of disk storage has been increased over 1200 times; while the density has been increased only about 3 times since 1990. The development of new battery technologies is mainly for low-capacity batteries for portable applications and smart devices in the Internet of Things (IoT). To increase battery capability, thin-film Li-ion and polymer cells can be applied, while a promising alternative is using fuel cells. A fuel cell uses chemical fuels such as methanol or hydrogen as a power source, and it can be

Table 3.7 Comparison of common battery types

Type	Volumetric energy density (Watt-hour/cm^3)	Gravimetric energy density (Watt-hour/kg)	Self-discharge rate per year (%)	Life recharging cycles (times)
Ni-Cd	100	30–35	15–20	300
Ni-MH	175	50	20	300
Li-ion	200	90	5–10	500
Alkaline	300	125	4	1

Source: Mateu and Moll (2005)/SPIE

expected to increase the gravimetric energy density by three to five times in comparison with a Li-ion cell. However, a fuel cell has to be refueled or carry a sufficient amount of fuel for the expected lifespan.

A battery system should be selected based on the storage capacity required by a mechatronic system as,

$$C = E_{\mathrm{d}} * N * V \tag{3.13}$$

where C is the battery capacity (Ah), E_{d} is the daily energy demand (kWh/day), N is the number of expected days in a recharging cycle (days), and V is the battery voltage (V).

Example 3.7 Assume that the daily energy demand of a mechatronic system is 0.025 (kWh), and it is expected to recharge its battery in 30 days, determine the minimal capacity of a 12 (V) battery.

Solution

Equation (3.13) can be used to estimate the minimal battery capacity as,

$$C = E_{\mathrm{d}} * N * V = (0.025) * (30) * (12) = 9\,(\mathrm{Ah})$$

Despite the great dependence on batteries and external power supplies, there is a trend to integrate power generation and supply in a mechatronic system for more and more mobile electronics. The external power supply is not always accessible, and conventional batteries are involved in some issues such as replacement, maintenance, and lifespan. A vital alternative to obtain the power needed to operate these mechatronic systems is to harvest energy from the ambient environment; self-powering by energy harvesting provides the unlimited solution to operate a mechatronic system in its full life cycle (Necula et al. 2014).

3.7 Requirements of Power Supplies and Storages

The generated power may not be directly used to power a mechatronic system. Take an example of car batteries used to power electronic systems: power comes from the engine, it recharges the batteries, and it is then consumed by electronic devices at the right voltage level. This implies that a power supply system is often needed to generate, store, and finally transfer the power into the right form for use. For most of portable mechatronic systems, batteries play an irreplaceable role in storing and supplying power; note that using batteries increases the size and weight of the system, and it also requires proper maintenance. This has become a main hurdle to advancing ubiquitous computing and IoT when thousands even millions of small systems require batteries to sustain their functions.

Since source power often requires to be stored and regulated for appropriate uses, traditional batteries can be advanced by integrating with a power generator and emphasizing their functions as energy reservoirs and regulators. Accordingly, power supply systems are classified by volumetric and gravimetric energy density and the output power is regulated or not as shown in Table 3.8.

3.7.1 Requirements of Power Supplies

In a mechatronic system, fulfilling any action consumes energy. Figure 3.13 shows that energy is supplied to both information subsystems for decision-making supports and physical subsystem for actions; traditionally, the powers in two subsystems are related to weak and strong electricity,

Table 3.8 Comparison of different power supplies

Types	Unregulated	Regulated		
		Linear	Switched	Battery
Advantages	• Simple circuitry • Durability • Low cost	*Constant voltage; high-quality power supplies with noise filtering, adjustable output, and precise tuning* • Safe and reliable • Small residual ripple • Less noise • Good line and load regulation • stable	• Small size and lightweight • Large voltage ranges • High efficiency • Less expensive than linear regulation	• Portable • No need of on-site power
Disadvantages	• Voltage changed with load current • Applied to fixed output	*Complex circuitry and high cost* • Poor efficiency • Large heat sinks • Large size and heavy	• Complex design • Pollution • Higher noise • Durability	• Fixed voltage • Short life • Output voltage drop when reserves are used.

Source: Adapted from Wavelength (2013).

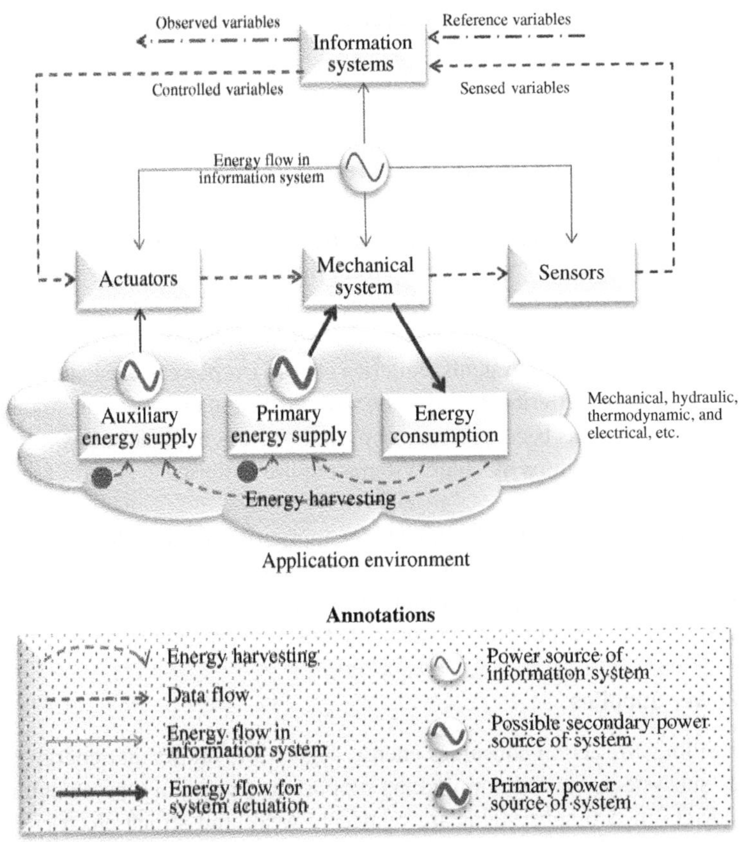

Figure 3.13 Energy flow in a mechatronic system

respectively. The power supply system in a mechatronic system can have one unified or multiple power sources based on accessibility, and the power can be directly supplied or harvested from the application environment. Since energy or power can be transformed from one form into another, the requirement of power supply can be generally characterized in its electrical form by voltage, current, and power and the power supply system is featured by its site, weight, power density, life-span, and wastes. A power supply system can be evaluated by its efficiency, safety class, output voltage, and no-load consumption (ITU and GeSi 2012).

The minimization affects the need of energy supplies of the information subsystem significantly. In a mechatronic system shown in Figure 3.13, the information about system statuses is acquired by various sensors such as tachometers, thermometers, cameras, range-finders, and acoustic sensors. The acquired data is processed and utilized by the information system to control and operate the physical subsystem in the application environment. From this perspective, a mechatronic system consists of a feedforward energy flow and a feedback information flow, and primary energy flows into the system, and it is either directly or indirectly consumed depending on if the form of input power should be transformed into another form in the applications (Isermann 2000).

The amount of power to operate a mechatronic system depends on the size and weight of matters or materials of its electronic system. Moore's law tells that the technology for integrated circuits (ICs) evolves with the size shrinkage of transistors, and the supply voltages to ICs are also reduced correspondingly due to reliability; therefore, a size reduction of parasitic components reduces the energy consumption of a mechatronic system (Ghazanfarian et al. 2021). For a reduced scale factor of α (i.e., $\alpha < 1$), the required energy to operate ICs with the same amount of computing is reduced by α^3. Note that when the miniaturization of a mechatronic system is accompanied by additional requirements such as processing an increased amount of data or increasing responsiveness, the required energy can be reduced by instead of α^3.

To reduce energy consumption, a mechatronic system usually has multiple power modes. In such cases, it is mandatory for its energy supply subsystem to manage the power modes effectively to (1) sustain an appropriate power level in a full operation, (2) refill energy reservoir when the system is in partial operation or inactive, and (3) save energy when the system is in an inactive status (Mateu and Moll 2005).

Two important developmental trends of mechatronic systems from their origins are (1) the increase of versatility and functionality and (2) the decrease of cost, size, and weight (Pandian 2010). For example, more and more portable products become available to assist people in sensing, interacting, supporting decision-making, and performing daily tasks; it becomes popular to further reduce the size and weight of a portable product as an embedded mechatronic subsystem in more complex mechatronic systems such as smart watches, glasses, and cloths. With the miniaturization of mechatronic systems, the energy required to operate a mechatronic system has been reduced drastically in past decades; especially, it has become feasible for some small mechatronic systems to harvest energy directly from the application environment to power themselves.

3.7.2 Classification of Energy Storage Systems

Energy storage serves two purposes: *the first one* is to store energy when it is generated or supplied excessively and release energy when the energy is consumed; the power supply and demand are mostly irregular depending on the change of external loads. Power storage is extremely critical to an integrated energy harvesting system since it is required to store energy for a continuous system operation duration, the time when no energy can be harvested. *The second one* is to deal with different levels of energy consumption when a mechatronic system has multiple optional modes. Take

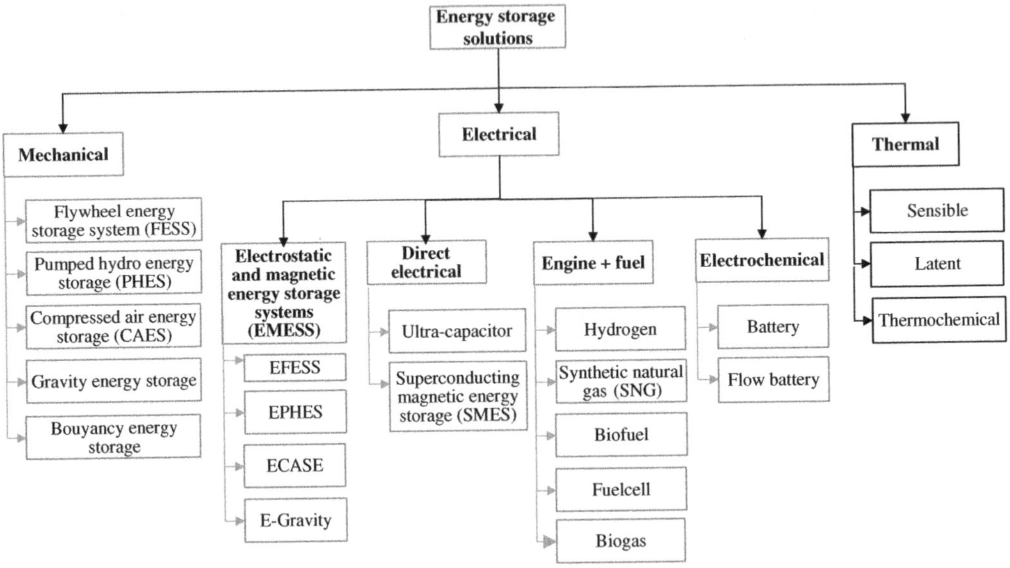

Figure 3.14 Types of energy storage solutions (Alami 2019; Olabi et al. 2021; Rosa et al. 2021; and Oreilly 2022)

an example of many wireless devices over the Internet, they usually have an *active* mode and an *inactive* mode, and high-level energy is consumed only when the data is sensed, processed, and transmitted. Both active and inactive modes provide the feedback to the power supply and storage system for an appropriate energy level for system operations.

Figure 3.14 classifies the types of energy storage systems into *mechanical*, *electrical*, and *thermal* systems (Olabi et al. 2021; Rosa et al. 2021). The common types of mechanical energy storage systems are *flywheel energy storage systems* (FESSs), *pumped heat electrical storages* (PHESs), *compressed air energy storage systems* (CAESs), and *gravity* energy storage systems. The common types of electrical energy storage (EES) systems are *electrostatic and magnetic*, *direct electrical*, *combined engine and fuel*, and *electrochemical* energy storage systems. The common types of thermal energy storage (TES) systems are *sensible*, *latent*, and *thermochemical* energy storage systems.

3.7.3 Flywheel Energy Storage System (FESS)

A FESS is actually electromechanical to transmit mechanical power into electrical power. An FESS is mainly used to store energy for a short or medium-term energy storage; an FESS involves a high initial cost with around \$5000/kWh; however, its maintenance cost is low at 19/kW annually. FESSs are characterized by low-energy density and high-power density, and short-term charging and discharging cycles (Arani et al. 2017). Due to their high efficiencies, FESSs are widely used in railways, hybrid vehicles, trains, wind turbines, and marine and space systems. Figure 3.15 shows three basic structures of an FESS, and its main components are *rotating mass*, *bearings*, *motor-generator*, and *frequency inverter*. To sustain high efficiency, magnetic bearings are often used to reduce power loss in transmission. An FESS can be integrated with different electric machines such as a *synchronous machine* (SM), *induction machine* (IM), and *switched reluctance machine* (SRM). SM and IM are commonly used in high-speed and high-power applications, while SM exhibits

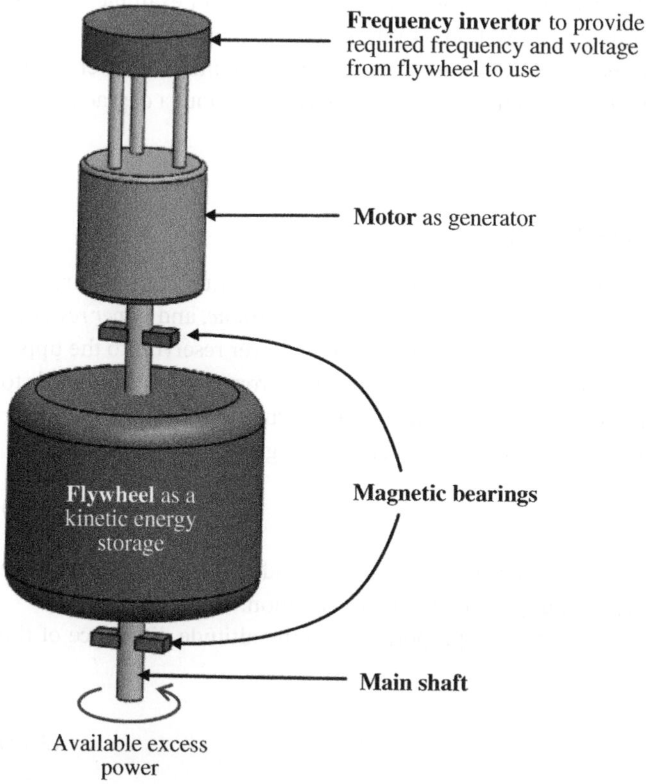

Figure 3.15 Structure of a flywheel energy storage system (FESS) (*Source:* Mahmoud et al. (2020)/with permission of Elsevier)

better performance in terms of its low inrush at the start. SRM is rarely used due to a high current ripple (Mahmoud et al. 2020). An FESS stores energy in terms of kinetic energy as,

$$E = \frac{1}{2} I \omega^2 \tag{3.14}$$

where E is the stored energy, I is the moment of inertia of the rotating mass, and ω is the angular velocity.

The specific energy per unit of mass is proportional to *the shape factor*(K) of the rotating mass as (Mahmoud et al. 2020),

$$\frac{E}{m} = K \frac{\sigma_{\max}}{\rho} \tag{3.15}$$

where E/m is the specific energy per unit of mass, $\sigma_{\max}$ is the maximum stress, and ρ is the density of the rotating mass. The shape factors K of some common flywheel shapes are 1, 0.606, 0.305, and 0.5 for *disc of Laval*, *solid disk*, *thick ring*, and *thin ring*, respectively.

When flywheels are made of steel materials and the rotations are supported by traditional bearings, axial angular velocities are limited to a few thousands of *revolutions per minute* (RPM). When the flywheels are made of carbon fiber materials, isolated in vacuums with reduced drag forces, and supported by the bearings with magnetic levitation, the axial angular velocities can reach up to

60,000 RPM. FEESs have been practically applied to store renewable energy. For example, the first FEES developed by Amber Kinetics offered 32 kW-hour (kWh) storage with a power rate of 8 kW for a 4-hour discharging cycle. More than 10 FEES units were used and accumulated over 38,000 operating hours for more than 880 full cycles of charging and discharging without a degradation of capability (Will et al. 2019).

3.7.4 Pumped Hydro Energy Storage (PHES)

A PHES is characterized by its flexibility, long lifespan, and low cost of maintenance. Figure 3.16 shows four main components of a PHES are *the pumping system*, *hydro turbine*, and *upper reservoir* and *lower reservoir*. The excess energy is used to pump water from the lower reservoir to the upper reservoir, and the stored energy is consumed when the water is released from the upper reservoir to the lower reservoir. The stored energy is gravitational due to the pressure difference of the lower and upper reservoirs. A PHES stores energy in terms of potential energy as,

$$E = \rho V g(h_U - h_L) \tag{3.16}$$

where E is the stored potential energy, ρ is the density of the fluid, h_U and h_L are the altitudes of the upper and lower reservoirs, respectively, and g is the gravity acceleration.

Therefore, the specific energy per unit of mass is proportional to the altitude difference of the upper and lower reservoirs as,

$$\frac{E}{m} = g(h_U - h_L) \tag{3.17}$$

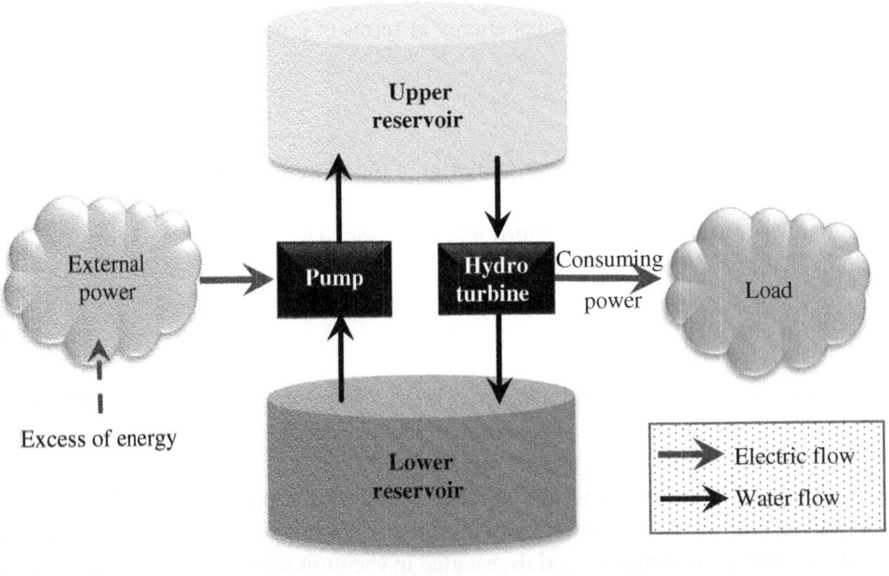

Figure 3.16 Structure of a pumped hydro energy storage (PHES) system

3.7.5 Compressed Air Energy Storage (CAES)

A CAES system uses compressed air to store energy for later use when the energy is needed. The energy from the source that exceeds the demand is stored, and it can be then released when the energy demand exceeds the supplied energy. CAES can be efficient in storing unused energy; however, a large-scale CAES application involves a significant heat loss since air compression generates heat; therefore, it is beneficial to store the generated heat in air compression and release the heat in air compression since the same amount of heat is needed to expand the air.

Figure 3.17 shows the basic structure of CAES. *A compressor* uses a motor as the prime mover to compress air; note that the motor is powered by the excess energy from an energy source such as a wind turbine or PVs. *A heat exchanger* is coupled with the compressor to collect and store thermal energy in air compression, and the heat is extracted at the air discharge phase. An expander uses an air turbine to do the work of the mechatronic system. The energy from the source is converted by CAES in five steps as shown in Figure 3.17: (1) the excess energy is used to run the motor and compress air; (2) the air is compressed to store energy; (3) the generated thermal energy is stored; (4) the stored air and heat are discharged to the generator; (5) the collected energy is consumed to do works in the application.

3.7.6 Gravity Energy Storage (GES)

GES is proposed to store an unprecedentedly large amount of energy for a time duration of 6–14 hours whenever the energy is needed (Energy Vault 2024). The working principle of GES is similar to PHES in the sense that the energy is stored in the form of potential energy; however, mobile custom-made composite blocks instead of water are used to store potential energy by hydraulic lifting. To store energy, an electrical pump is used to pump water beneath a moveable solid piston to lift a mass; to use energy, the water under high pressure is routed to a turbine (i.e., generator) to generate electricity for use.

Figure 3.18 shows the basic structure of a GES system (Berrada et al. 2021; Emrani et al. 2022; and Tong et al. 2022), and more varieties of GESs have been discussed in detail by Botha et al. (2021). The mobile mass is placed in *a piston* inside *the container*. In a charging mode, the excess energy is converted into potential energy by lifting the piston with an upward fluid flow by a pump. In a discharging mode, the electricity is generated by a descending movement of the piston, and the movement is regulated by a valve. In discharging, the piston forces the water and transfers work

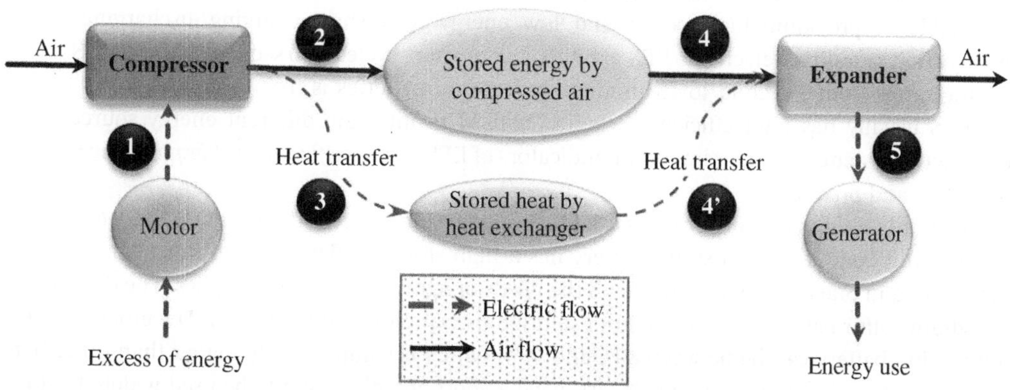

Figure 3.17 Structure of a compressed air energy storage (CAES) system

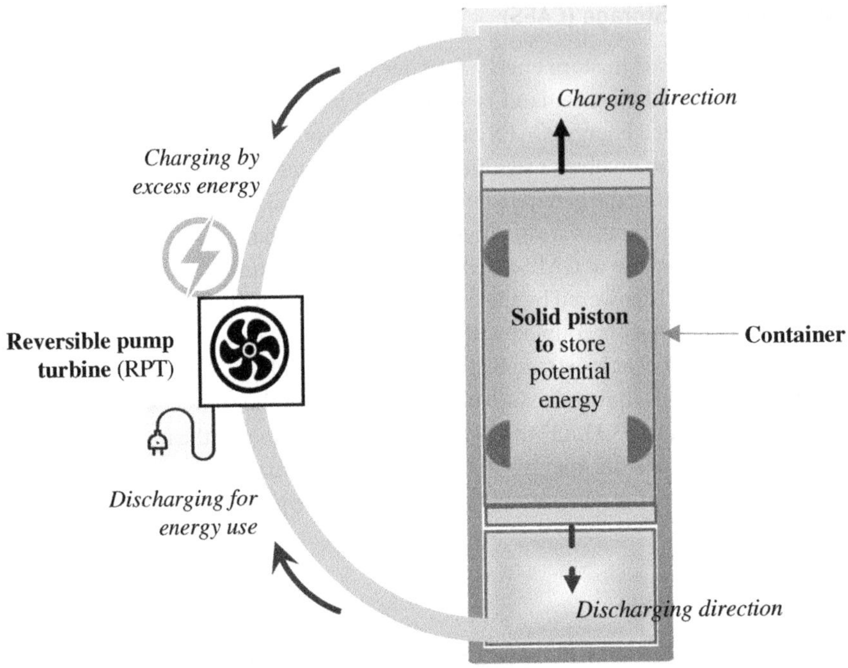

Figure 3.18 Structure of gravity energy storage (GES) system

in its translating motion, and the water flow with induced kinetic energy is reconverted into electrical energy by a turbine.

3.7.7 Electrical Energy Storage (EES)

An EES system can be one of *electrostatic and electromagnetic*, *direct electrical*, *combined engine and fuel*, and *electrochemical* systems. Electrostatic and electromagnetic EES systems are the mechatronic implementations of corresponding mechanical energy storage systems. Different EES technologies have their own advantages and disadvantages; for example, the batteries as electrochemical EES systems have the advantages of high *electrodialysis* (ED), low self-discharge, and installation cost. Superconducting magnetic energy storage (SMES) as an electromagnetic EES is capable of providing high power and low energy in a short charging/discharging cycle. A capacitor as a direct electrical EES exhibits a very short cycle (<10 seconds), and SMES has a cycling range from 1 second to 15 minutes, and that of batteries is from 5 minutes to 24 hours. An EES usually has high efficiency when it is used to integrate different energy sources into one electric system. The main important indicators of EESs are *specific energy*, *lifespan*, *power*, *independence*, and *protection*.

Commonly used materials for various battery energy storage are *sodium sulfur* (NaS), *lead-acid*, *lithium*, and *flow*. A sodium sulfur battery has a high storage efficiency, and it is adaptable and mostly used in practical applications with the needs of lightweight power supplies. The drawbacks of sodium sulfur batteries are the sodium solution preparation and high costs. The efficiency of a lithium-ion battery can be between 80% and 90%, which is comparatively higher than other batteries; therefore, lithium-ion batteries also have a short charging time to be used widely to store electrical energy; however, the performance of lithium-ion batteries varies with the operating temperature and the cost of lithium-ion batteries is relatively high. A large number of variations of lead-acid batteries are available in terms of dimensions and configurations, and such batteries are widely used in distributed power grids due to their cost-effectiveness. A flow battery stores electrochemical energy by dispersing ions in liquid electrolytes; it provides a high life cycle and offers the adaptability to different applications; however, it shows the limitations in implementation cost, energy density, and conversion efficiency (Rana et al. 2022). Figure 3.19 shows the scheme of a rechargeable battery energy storage system. It consists of a set of battery cells that can be connected in series or parallel; each cell consists of an anode, cathode, and electrolyte. Cells are discharged and charged subjected to a positive and negative voltage difference between the anode and the cathode, respectively.

Different fuels can be used in an EES to store electrical energy. Fuels in EES can be *hydrogen*, *synthetic natural gas* (SNG), *biofuel*, *fuel cell*, and *biogas*; in particular, fuel cell can be regenerative to utilize renewable energy sources in integrated mechatronic systems. A fuel cell stores chemical energy in the form of hydrogen or other fuels; it can be used to produce electrical power efficiently and cleanly. For example, when hydrogen is used, side products are water and heat besides electricity; in addition, a regenerative fuel cell's EES is scalable and applied in small systems such as computers to large systems such as automobiles and space systems. Fuel cells outperform other fuels in several aspects: (1) a fuel cell can operate at a higher efficiency than those of other fuels in the sense that more than 60% of chemical energy can be converted to electrical energy directly, fuel cells can operate at higher efficiencies than combustion engines and can convert the chemical energy in the fuel directly to electrical energy with efficiencies capable of exceeding 60%. (2) Fuel cell use only emits water; no carbon dioxide is emitted when compared to other combustion engines; using a fuel cell does not cause any air pollution. (3) A fuel cell's EES is involved in few moving components and it can operate quietly.

Figure 3.20 shows the structure of a regenerative fuel cell; the energy storage side of fuel cell consists of two electrodes (i.e., an *anode* and a *cathode*) that are sandwiched around an *electrolyte*. In a discharging cycle, hydrogen and oxygen are fed to the anode and cathode, respectively; it produces electricity and heat as long as the fuel is supplied. The *catalyst* at the anode converts hydrogen into protons and electrons, the electrons take a path of an external circuit to the cathode to form an

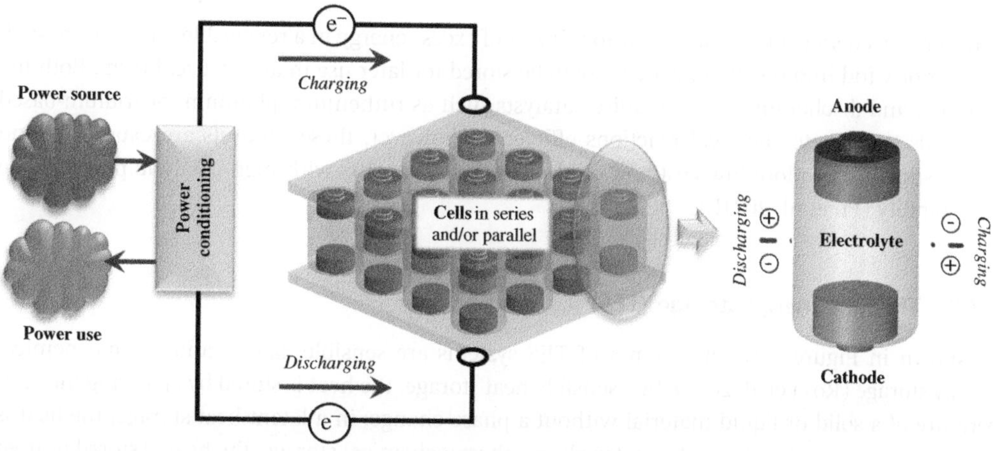

Figure 3.19 Battery scheme (*Source:* Azzuni and Breyer (2018)/with permission of Elsevier)

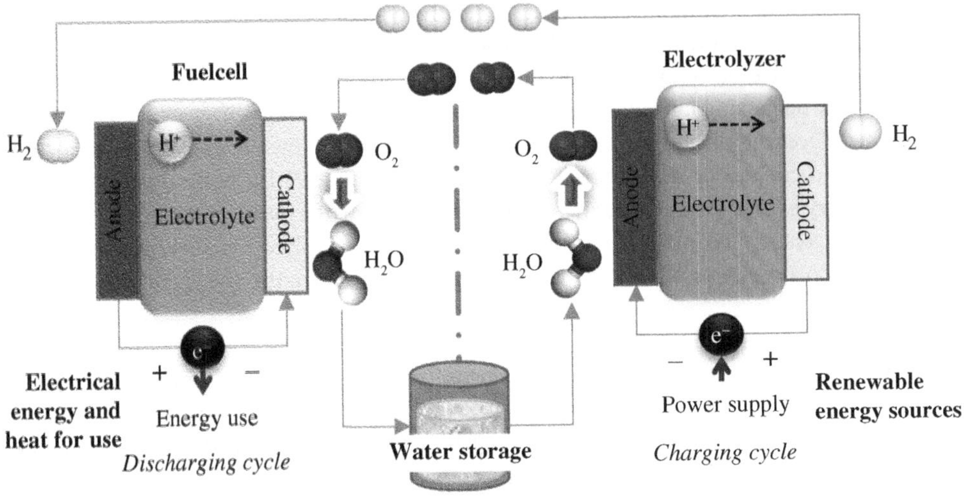

Figure 3.20 Structure of a regenerative fuel cell (*Source:* Pu et al. (2021b)/with permission of Elsevier)

electrical flow, and the protons take a path of electrolyte to the cathode, and have the chemical reaction with oxygen and electrons to produce heat and water as follows (Energy 2024):

$$\left.\begin{array}{ll}\text{Anode:} & 2H_2 \rightarrow 4H^+ + 4e^- \\ \text{Cathode:} & 4H^+ + O_2 + 4e^- \rightarrow 2H_2O \\ \text{Fuel cell:} & 2H_2 + O_2 \rightarrow 2H_2O\end{array}\right\} \quad (3.18)$$

In a charging cycle, the renewable energy is converted and applied a negative voltage difference of anode and cathode, and this leads to the reversible chemical reactions at the anode and cathode as follows:

$$\left.\begin{array}{ll}\text{Anode:} & 2H_2O \rightarrow 4H^+ + O_2 + 4e^- \\ \text{Cathode:} & 4H^+ + 4e^- \rightarrow 2H_2 \\ \text{Fuel cell:} & 2H_2O \rightarrow 2H_2 + O_2\end{array}\right\} \quad (3.19)$$

Reversible chemical reactions allow to transfer of excess energy in a renewable energy form such as solar or wind into the chemical energy to be stored for later use in an electrical form. Both the charging and discharging cycles require catalysts such as ruthenium, platinum, or iridium-based materials to activate chemical reactions efficiently; however, these catalysts are expensive, and it is desirable to explore low-cost and earth-abundant elements with high catalytic performance and stability (Pu et al. 2021b).

3.7.8 Thermal Energy Storage (TES)

As shown in Figure 3.21, three types of TES systems are sensible, latent, and thermochemical energy storage (Rosa et al. 2021). In a sensible heat storage, the heat is stored by changing the temperature of a solid or liquid material without a phase change; in a latent heat storage, the heat is stored by changing the phase of a material; in a thermochemical storage, the heat is stored or used by means of endothermic or exothermic chemical reactions of two components, respectively.

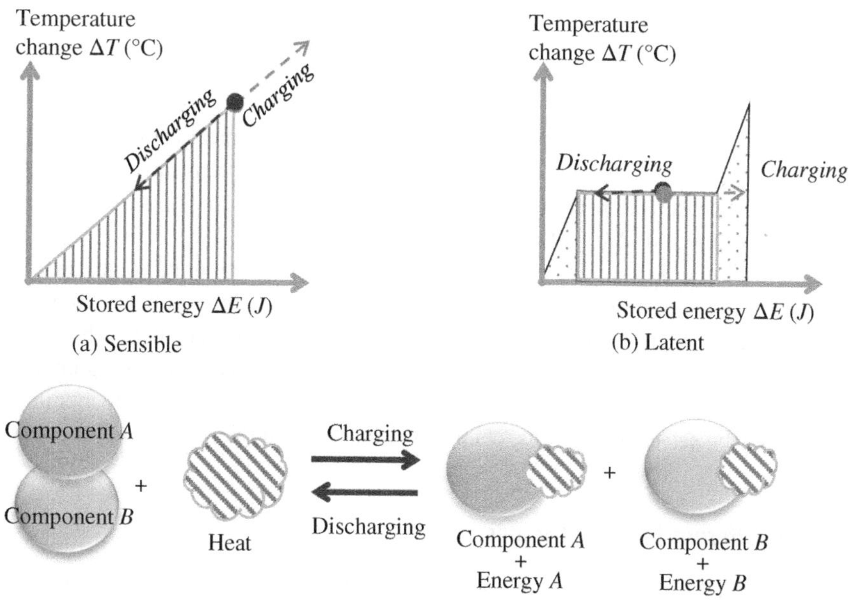

Figure 3.21 Three types of thermal energy storages (*Source:* Rosa et al. (2021)/with permission of Elsevier)

The amount of energy (Q) to be stored by a sensible heat storage is determined by,

$$Q = \rho V c_p \Delta T \tag{3.20}$$

where ρ, V, and c_p are the density, the volume, and the specific heat of the material at its given phase; ΔT is the change of temperature. Solid materials such as metals, rocks, sand, and concretes and fluids such as water, oils, and molten salts are commonly used in sensible energy storages. The materials such as concrete, sand, or rock remain in their solid phases in a wide range of temperatures; they are suitable to store energy at both low and high temperatures for large-scale applications. Selecting the right materials for sensible energy storage also considers other factors such as specific heat, stability, cost, environmental friendliness, and energy density.

The amount of energy (Q) to be stored by a latent energy storage is determined by,

$$Q = m \cdot \left[c_{ps}(T_m - T_i) + \Delta q + c_{pl}(T_f - T_m)\right] \tag{3.21}$$

where Q is the amount of the storage energy; T_m, T_i, and T_f are the melting, initial, and fluid temperatures, respectively; m is the mass; c_{ps} and c_{pl} are the specific heats of the material at its solid and fluid phases, respectively; Δq is the latent heat of fusion. Commonly used latent materials are classified into (1) *organic materials* such as paraffin and non-paraffin; (2) *inorganic materials* such as salt hydrates and metal alloys; and (3) eutectic materials such as organic eutectics, inorganic eutectics, and mixed organic and inorganic eutectics (Rosa et al. 2021). Latent energy storages exhibit the advantages of high energy density and isothermal heat storage and retrieval properties for certain applications; however, they show limitations due to poor thermal conductivity for high-level power systems (Yadav et al. 2021).

A thermochemical energy storage keeps or releases energy by reversible endothermic and exothermic chemical reactions, and its working principle can be described as,

$$\left.\begin{array}{l}(A \cdot B) + Q \underset{\text{exothermic}}{\overset{\text{endothermic}}{\rightleftarrows}} (A \cdot Q_A) + (B \cdot Q_B) \\ Q = n_{A \cdot B} \Delta H_r\end{array}\right\} \tag{3.22}$$

where A and B represent the two chemical components; $(A \cdot B)$ is the chemical compound composed of A and B; Q is the amount of heat to activate an endothermic chemical reaction; Q_A and Q_B are the amount of heat stored by components A and B, respectively; $n_{A \cdot B}$ is the number of moles of the chemical compound; ΔH_r is the reaction enthalpy in J/mol$_{A \cdot B}$.

Note that thermochemical energy storage is still at its preliminary developmental stage; further research effort is needed to make thermochemical energy storages practical since it offer 8–10 times energy density higher than that of sensible heat storage and two times energy density higher than that of a latent energy storage. Moreover, a thermochemical energy storage is capable of storing energy under ambient conditions with minimized energy loss. The overall performance of a thermochemical energy storage depends on the selected materials, chemical reactions, and operation conditions.

3.7.9 Comparison of Different Energy Storages

A common problem with a renewable energy source such as wind or solar power is intermittency; therefore, a self-powered mechatronic system is required to store a sufficient amount of energy for short- or long-term operations. Mechanical, electrical, and TESs can be used to store energy for mechatronic systems (Gutsch and Leker 2022). Energy storage systems should be selected based on a number of criteria in the applications such as power rating, energy density, energy capacity, efficiency, and lifespan. Table 3.9 shows a comparison example of different energy storage solutions by Olabi et al. (2021). Figure 3.22 shows a comparison example of different energy storage solutions in terms of lifespan and efficiency by Rana et al. (2022).

Electrical energy storages by fuel cells, supercapacitors, and lithium-ion batteries are widely used in computers, households, electric vehicles, wireless devices, industrial drives, and power generation. In comparison with supercapacitors, fuel cells and lithium-ion batteries have a high-energy

Table 3.9 Examples of comparison of different energy storages

	Power density (W/L)	Energy density (Wh/L)	Power rating (MW)	Energy capacity (MWh)	Efficiency (%)	Lifetime/year
Large-scale compressed air	(0.5, 2)	(1, 6)	(100, 1000)	<1000	(40, 70)	(20, 40)
Small-scale compressed air	>2	>6	(0.003, 10)	<0.1	65	>23
Pumped hydrogen	(0.45, 1.5)	(0.5, 2)	(100, 5000)	(500, 5000)	(70, 85)	(40, 60)
Lithium-ion battery	(1000, 10,000)	(100, 500)	(1, 100)	(0, 10)	(75, 97)	(4, 20)
Lead-acid battery	(1, 500)	(40, 90)	(0, 400	(1, 40)	(63, 90)	(5, 15)
Supercapacitor	>100,000	(10, 30)	(0, 0.3)	(0, 0.0005)	(84, 95)	(10, 30)
Fuel cell	>500	(500, 3000)	<50	0.312	(20, 66)	(5, 20)

Source: Olabi et al. (2021)/with permission of Elsevier.

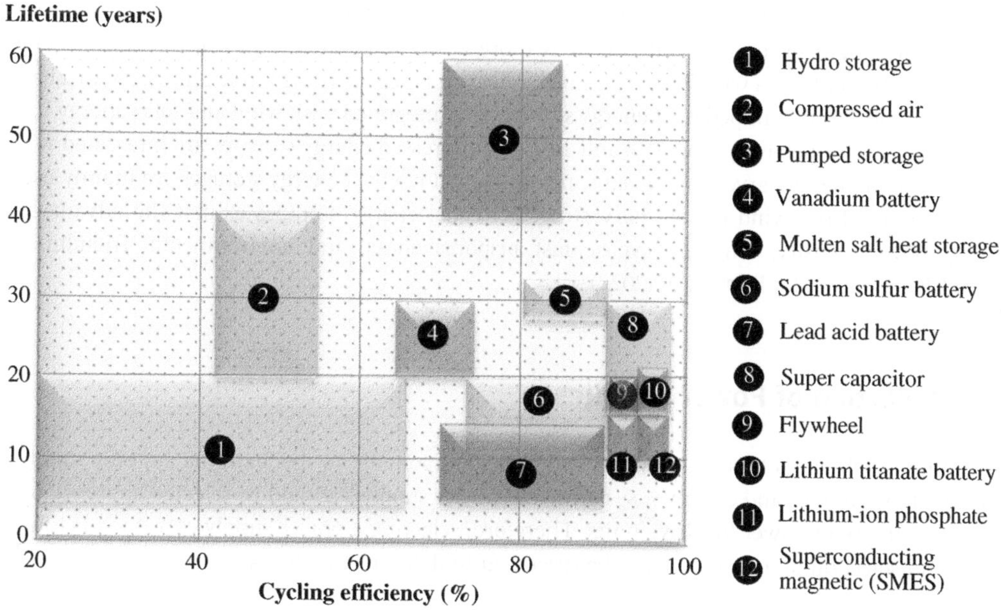

Figure 3.22 Comparison example of different energy storage solutions by lifespan and efficiency (*Source:* Rana et al. 2022/with permission of Elsevier)

density but limited power density; note that the latter is the most critical to enhance the dynamic performance of an energy storage system. From this aspect, supercapacitors dominate the applications with high-level power, quick charging cycles, and superior lifespan (Yadlapalli et al. 2022).

Mahmoud et al. (2020) discussed the integration of three mechanical energy storage systems with solar and wind renewable energy systems. As shown in Table 3.10, mechanical energy storage systems are evaluated based on storage capacity, initial and operating costs, lifespan, and

Table 3.10 Comparison of three mechanical energy storage systems

Mechanical EES	Advantage	Disadvantage
Flywheel (FEES)	• No pollution • Long lifetime • Discharges huge amount of power in minutes • Low cost per kWh	• Limited charge and discharge time • Cannot stand alone with a PV plant
Pumped heat (PHES)	• High efficiency • Stability • Low cost per kWh • Long discharge time	• High capital cost • Low energy density • Large footprint area
Compressed air energy storage (CAES) system	• Flexibility • Long discharge time • Fast startup • Low cost per Wh • Stability	• High capital cost • Low efficiency • When natural gas is used, reheat air emits CO_2.

Source: Mahmoud et al. (2020)/with permission of Elsevier.

environmental factors. Moreover, the type of energy source is a critical factor in selecting an energy storage system. For example, wind energy is more easily integrated with a mechanical energy source since power from a wind turbine is mechanical and is easily transmitted and stored mechanically. 78% and 22% of pumped heat energy storages are integrated with wind and solar energy harvesting systems, respectively, and 85% and 15% of compressed energy storages are integrated with wind and solar energy harvesting systems, respectively. Flywheel energy storages are mainly used to store energy from wind energy harvesting. Flywheel energy storages exhibit the advantages of high-power density, long life cycles, fast responsiveness, and environmental friendliness (Li and Palazzolo 2022; Rana et al. 2022).

3.8 Selection of Power Supplies

Drive units such as motors or solenoids are basic components to mechatronic systems such as machine tools, robots, and conveyor systems, and power supplies should be selected appropriately to provide the right power for drive units. Therefore, selecting a power supply corresponds to the sizing and selection of a motor. It will be discussed in detail in Chapter 4.

3.9 Summary

From the perspective of energy conversion, any mechatronic system can be modeled as a set of functional modules to transform power from one form to another. Moreover, since energy cannot be generated, the operation of a mechatronic system requires energy sources, and there is an emerging need to integrate power supply and storage systems together with advanced mechatronic systems for high-level independency, reliability, and prolonged lifespans in complex environments. It is critical to select the right energy source of a mechatronic system, and it is beneficial to integrate energy harvesting and storage systems with the mechatronic system; moreover, the FRs of a mechatronic module can be usually modeled in terms of its functions of energy conversion for energy harvesting, storing, transmitting, and consuming for actuations. To preserve the sustainable environment globally, more and more mechatronic systems have to rely on renewable energy, and numerous technologies have been developed to harvest the energy of the environment and convert it into the power in a right form to operate mechatronic systems. It is highly desired to integrate the energy harvesting systems with mechatronic systems.

Problems

3.1 Which of the following is not a unit of energy: (a) kilowatt · hour, (b) joule, (c) Newton · meter, (d) $\frac{\text{kilogram}\cdot\text{meter}^2}{\text{second}^2}$, and (e) $\frac{\text{kilogram}^2,\ \ \text{meter}}{\text{second}}$.

3.2 A mass moves on a horizontal plane from an initial speed $v_0 = 0$ to v in a time period of t, the power required to accelerate the mass in this time period is (a) $\frac{mv^2 t}{2}$, (b) $\frac{mv^2}{2}$, (c) $2mv^2$, (d) $v\sqrt{\frac{m}{2t}}$, or (e) $\frac{mv^2}{2t}$.

3.3 A 1-kg mass and a 4-kg mass have the same kinematic energy of 200 (J), which mass has a higher speed and what are their corresponding moving speeds?

3.4 An object with a mass of $m = 5$ kg is pulled up in a plane at an inclined angle of $\theta = 45°$ with the floor. The *coefficient of friction* (CoF) at contact surface is $f = 0.08$. and the pulling force is $T = 100$ N. Assume that the initial speed $v_0 = 0$ (m/s) and it is moved with a distance of $L = 10$ m over the plane; calculate (1) the works done by the pulling force, the friction force, and gravity; and (2) the speed of the object after it is moved for 10 m over the plane.

3.5 An elevator lifts a mass of 800 kg; it is capable of accelerating the mass from rest (i.e., $v_0 = 0$ m/s) up to $a = 7.5$ m/S^2 in one second, what is the minimum power the elevator should have for this capability?

3.6 A PV panel has an effective area of 8.0 m^2 and an efficiency of 14.0%; the annual average solar radiation H_y is estimated as 1500 kWh/m^2/year, estimate the amount of harvested energy annually.

3.7 A smart home is equipped with a refrigerator (1.0 kW) and a central air-conditioning system (1.5 kW) that are operated for 24-hours a day, a microware with (1.0 kW) with the operation for 1-hour daily, and a washing machine with (2.0 kW), which is operated for 3-hours weekly. All of them are powered by a solar energy system. Assume that the average daily solar radiation is 2000 kWh/m^2/year and the panel efficiency is 15%; determine the minimal size of the solar energy system to support the operations of these appliances.

3.8 Assume that the effective area of a wind turbine $A = 80.0$ m^2, air density ρ is 1.225 kg/m, the average wind speed is 4.5 m/s, and the coefficient of power c_p is 0.32; determine the maximum output power of the wind turbine.

3.9 Assume that the mass flow rate (m) of steam in a geothermal generator is 5 (kg/s), the specific heat (c) of the cooling water is 4180 (W/kg·°C), the temperatures of the condenser cooling water at entry and exit are 20 °C and 35 °C, respectively, and the efficiency (η) of a geothermal generator is 20%; determine the output power of the geothermal generator.

3.10 Assume that the daily energy demand of a mechatronic system is 0.05 (kWh), and it is expected to recharge its battery in 60 days; determine the minimal capacity of a 24 (V) battery.

3.11 Select a power source for a semi-automated boat lift that is capable of lifting of up to 15,000 pounds to a height of 2 m in 30 seconds; provide your rationale of your design.

References

Alami, A.H. (2019). *Introduction to Mechanical Energy Storage*. Cham: Springer https://doi.org/10.1007/978-3-030-33788-9_1.

An, J., Wang, Z., Jiang, T. et al. (2020). Reliable mechatronic indicator for self-powered liquid sensing toward smart manufacture and safe transportation. *Materials Today* 41 (2020): 10–20.

Arani, A.A.K., Karami, H., Gharehpetian, G.B., and Hejazi, M.S.A. (2017). Review of flywheel energy storage systems structures and applications in power systems and microgrgids. *Renewable and Sustainable Energy Reviews* 69 (2017): 9–18.

Aurora (2015). Review synthesis of alternative power supply. https://intrans.iastate.edu/app/uploads/2018/10/remote_RWIS_alt_power_supplies_w_cvr.pdf (accessed 12 September 2024).

Azzuni, A. and Breyer, C. (2018). Energy security and energy storage technologies. *Energy Procedia* 2018: 237–258.

Berrada, A., Emrani, A., and Ameur, A. (2021). Life-cycle assessment of gravity energy storage systems large-scale application. *Journal of Energy Storage* 49 (2021): 102825.

Botha, C.D., Kamper, M.J., and Wang, R.J. (2021). Design optimization and cost analysis of linear Vernier electric machine-based gravity energy storage systems. *Journal of Energy Storage* 44 (2021): 103397.

Department of Energy (2002). Using distributed energy resources: a how-to guide for federal facility managers. https://www.nrel.gov/docs/fy02osti/31570.pdf (accessed 12 September 2024).

Emrani, A., Berrada, A., Arechkik, A., and Bakhouya, M. (2022). Improved techno-economic optimization of an off-grid hybrid solar/wind/gravity energy storage system based on performance indicators. *Journal of Energy Storage* 49 (2022): 104163.

Energy (2024). Full cells. https://www.energy.gov/eere/fuelcells/fuel-cells (accessed 12 September 2024).

Energy Information Administration (EIA) (2007). Annual Energy Review 2007. https://www.osti.gov/servlets/purl/1212314 (accessed 12 September 2024).

Energy Vault (2024). Transforming utility scale gravity energy storage. https://www.energyvault.com/gravity (accessed 12 September 2024).

ExxonMobil (2024). Global fundamentals. https://corporate.exxonmobil.com/Energy-and-innovation/Outlook-for-Energy/Global-energy-fundamentals (accessed 12 September 2024).

Ghazanfarian, J., Mohammadi, M.M., and Uchino, K. (2021). Piezoelectric energy harvesting; a systematic review of reviews. *Actuators* 10 (2021): 312.

Gutsch, M. and Leker, J. (2022). Global warming potential of lithium-ion battery energy storage systems: a review. *Journal of Energy Storage* 52 (2022): 105030.

Isermann, R. (1999). Modelling, identification and simulation of mechatronic systems. *IFAC Proceedings* 32 (2): 879–890.

Isermann, R. (2000). Mechatronic systems: concepts and applications. *Transactions of the Institute of Measurement and Control* 22 (1): 29–55.

ITU and GeSI (2012) An energy-aware survey on ICT device power supplies. https://www.itu.int/dms_pub/itu-t/oth/4B/01/T4B010000070001PDFE.pdf (accessed 12 September 2024).

Li, X. and Palazzolo, A. (2022). A review of flywheel energy storage systems: state of the art and opportunities. *Journal of Energy Storage* 46 (2022): 103576.

Mahmoud, M., Ramadan, M., Olabi, A.G. et al. (2020). A review of mechanical energy storage systems combined with wind and solar applications. *Energy Conversion and Management* 210 (2020): 112670.

Maleque, M.A., Adebisi, A.A., and Shah, Q.H. (2012). Energy and cost analysis of weight reduction using composite brake rotor. *International Journal of Vehicle Structures and Systems* 4 (2): 69–73.

Mateu, L. and Moll, F. (2005). Review of energy harvesting techniques and applications for microelectronics (Keynote Address). *Proceedings of SPIE 5837, VLSI Circuits and Systems II* (30 June 2005). https://doi.org/10.1117/12.613046.

Necula, C., Gramescu, B., Comeaga, D.C. et al. (2014). A survey of energy harvesting as power supply for mechatronic systems. *Romanian Review Precision Mechanics, Optics, and Mechatronics* 45 (2014): 179–184.

Niccolai, L., Bassetto, M., Quarta, A.A., and Mengali, G. (2019). A review of smart dust architecture, dynamics, and mission applications. *Progress in Aerospace Sciences* 106 (2019): 1–14.

Olabi, A.G., Wilberforce, T., Ramadan, M. et al. (2021). Compressed air energy storage systems: components and operating parameters – a review. *Journal of Energy Storage* 34 (2021): 102000.

Oreilly (2022). Storage. https://www.oreilly.com/library/view/handbook-of-energy/9780080464053/xhtml/CHP022.html (accessed 12 September 2024).

Otter, M. and Elmqvist, H. (1997). Energy flow modeling of mechatronic systems via object diagrams. *Proceedings of the 2nd MATHMOD*, Vienna (1997), pp. 705–710.

Pandian, S.R. (2010) Intelligent mechatronic technologies for green energy systems. *Proceedings of the 2010 ASEE Gulf-Southwest Annual Conference*, McNeese State University. http://aseegsw.com/past%20Proceedings/FD4-3_Pandian.pdf (accessed 12 September 2024).

Pu, X., An, S., Tang, Q. et al. (2021a). Wearable triboelectric sensors for biomedical monitoring and human-machine interface. *IScience* 24 (1): 102027.

Pu, Z., Zhang, G., Hassanpour, A. et al. (2021b). Regenerative fuel cells: recent progress, challenges, perspectives and their applications for space energy system. *Applied Energy* 2021: 116376.

Rana, M.M., Uddin, M., Sarkar, M.R. et al. (2022). A review on hybrid photovoltaic – battery energy storage system: current status, challenges, and future directions. *Journal of Energy Storage* 51 (2022): 104597.

Rosa, M.D., Afanaseva, O., Fedyukhin, A.V., and Bianco, V. (2021). Prospects and characteristics of thermal and electrochemical energy storage systems. *Journal of Energy Storage* 44(2021): 103443.

Sun, H.H. (2017). Miniature wind energy harvesters. Doctoral thesis. Faculty of Physical ad Applied Sciences, University of Southampton. https://eprints.soton.ac.uk/416874/1/Final_submission_Huihui_Sun_290417.pdf (accessed 12 September 2024).

Tiefenbeck, V., Tasic, V., Schob, S., and Staake, T. (2018). Mechatronics to drive environmental sustainability: measuring, visualizing, and transforming consumer patterns on a large scale. https://cocoa.ethz.ch/downloads/2013/12/1603_EnergyInformaticsVienna_Tiefenbeck_2013submitted.pdf (accessed 12 September 2024).

Tong, W., Lu, Z., Chen, W. et al. (2022). Solid gravity energy storage: a review. *Journal of Energy Storage* 53 (2022): 105226.

Wang, Z.L. and Song, J.H. (2001). Piezoelectric nanogenerators based on zinc oxide nanowire arrays. *Science* 312 (5771): 242–256, 2006.

Wang, Z.L., Wang, X., Song, J.H. et al. (2008). Piezoelectric nanogenerators for self-powered nanodevices. *IEEE Pervasive Computing* 7 (1): 49–55.

Wavelength (2013). Power supply basics. https://www.teamwavelength.com/download/applicationtechnotes/an-ldtc3.pdf (accessed 12 September 2024).

Will, S., Senesky, M., Kwok, W.T. et al. (2019). Flywheel systems for utility scale energy storage. California Energy Commission. Publication Number: CEC-500-2019-012. https://www.energy.ca.gov/sites/default/files/2021-05/CEC-500-2019-012.pdf (accessed 12 September 2024).

Yadav, A., Verma, A., Kumar, A. et al. (2021). Review advances on enhanced thermal conduction in phase change materials using carbon nanomaterials. *Journal of Energy Storage* 43 (2021): 103173.

Yadlapalli, R.T., Alla, R.R., Kandipati, R., and Kotapati, A. (2022). Super capacitors for energy storage: progress, applications and challenges. *Journal of Energy Storage* 49 (2022): 104194.

Yuan, M., Li, C., Liu, H. et al. (2021). A 3D-printed acoustic triboelectric nanogenerator for quarter-wavelength acoustic energy harvesting and self-powered edge sensing. *Nano Energy* 85 (2021): 105962.

Yun, Y., Jang, S., Cho, S. et al. (2021). Exo-shoe triboelectric nanogenerator: toward high-performance wearable biomechanical energy harvester. *Nano Energy* 80 (2021): 105525.

Zhang, J., Chen, Z., Kankala, R.K. et al. (2021). Self-propelling micro-/nano-motors: mechanisms, applications, and challenges in drug delivery. *International Journal of Pharmaceutics* 596 (2021): 120275.

4

Actuating Systems

4.1 Introduction

Any machine including a mechatronic system is designed to fulfill specific functions. Typically, the purpose of a mechatronic system can be represented by a series of tasks such as changing the state of an object of interest, loading/uploading an object, transferring a material from one form to another, moving an object along a specific path, collaborating with humans, and empower humans as shown in Figure 4.1. It shows that actuators are essential components in a mechatronic system to convert the input energy of objects in one form into another form to perform expected tasks in applications.

Figure 4.2 generalizes the roles and interactions of an actuator in a mechatronic system (Janocha 2004). *An actuator* is to execute control commands and transfers the input energy in its original form into the output energy in its expected form used for taking specific action on targeted object(s). An operation of an actuator refers to *an actuation* that takes a physical action on an object or a process. For more examples other than those in Figure 4.1, an actuation can be to eject a part from a conveyor system and it can be triggered by a sensed event. An actuation can be to clamp a workpiece when it is loaded on the workstation in a machining or inspecting process.

An actuation can be enabled electrically, mechanically, pneumatically, or hydraulically, and the controller of an actuator is responsible to transform control commands into actions. An actuator can be generalized as an energy converter, and the form of energy can be represented by some relevant quantities such as *forces*, *displacements*, *accelerations*, and *power* as discussed in the previous chapter.

When an actuator is modeled as an energy converter, its roles in a mechatronic system can be specified clearly by its interfaces to *input energy* (source and form), *output energy* (form and target), and *control commands* as shown in Figure 4.3. The input energy is converted by two transducers: *the first transducer* converts input energy into output energy and act on the targeted object; *the second transducer* operates the first transducer based on control commands. Some researchers refer to the second transducer as *power amplification and modulation*. The second transducer converts a control signal at a low power into an acting signal that delivers the required input power to the energy conversion unit (Pace 2024).

According to the generic model of actuators in Figure 4.3, actuators in mechatronic systems can be classified based on the types of their constitutive elements such as (1) if the energy source is *active* or *passive*, (2) the forms of output energy, (3) types of objects to be acted, and (4) types of the transducers in implementations.

Mechatronics for Complex Products and Systems: Project-Based Design Approaches for Robots, Cyber-Physical Systems, Digital Twins, and Other Emerging Technologies, First Edition. Zhuming Bi.

Companion website: www.wiley.com/go/bi/mechatronics

Figure 4.1 Actuating examples in mechatronic systems (*Source:* (a) Boston Dynamics / https://www.pcmag.com/news/watch-boston-dynamics-robot-dog-open-a-door / Last accessed on July 19, 2024; (b–f) SANSEI AIR / https://airdanshin.co.jp/air-menshin/ / last accessed on August 26, 2024)

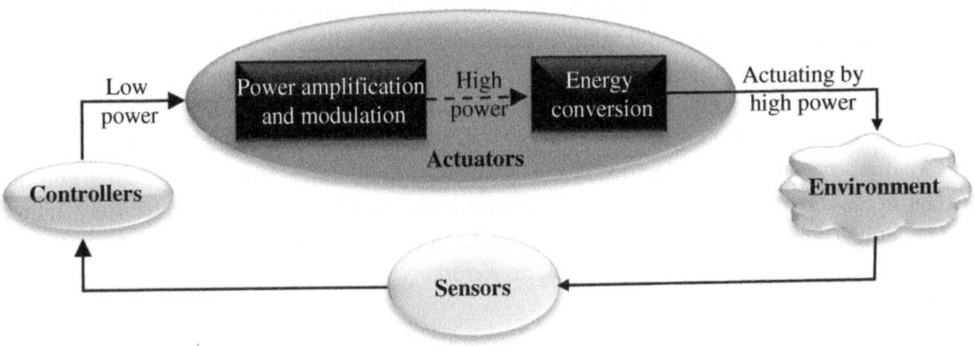

Figure 4.2 Generalized representation of actuator

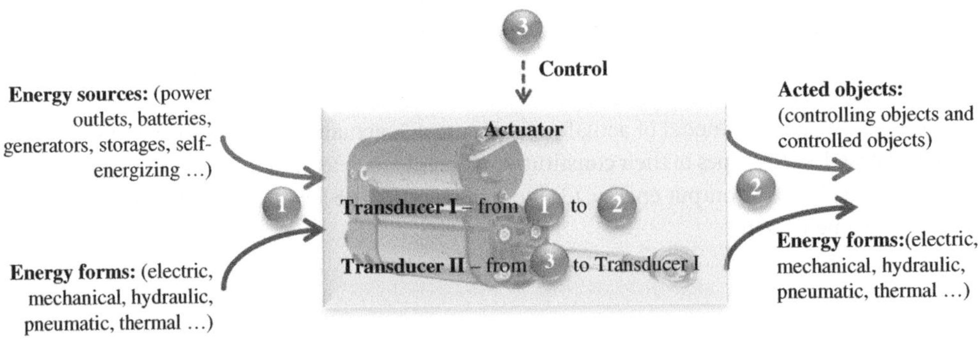

Figure 4.3 An actuator as an energy converter

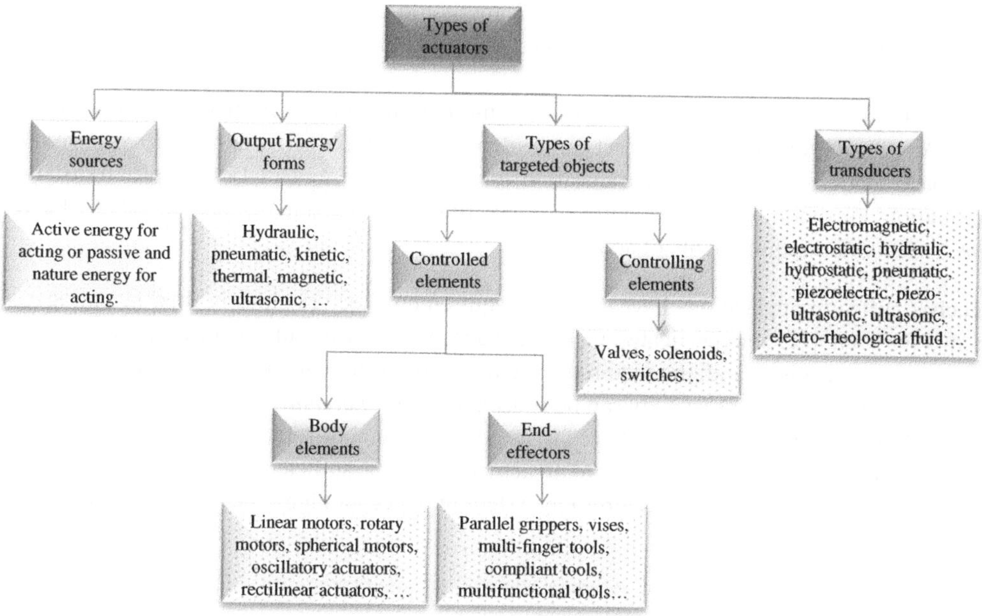

Figure 4.4 Classifying actuators by types of energy sources, output energy, targeted objects, and transducers

As shown in Figure 4.4, from the perspective of energy sources, an *active* actuator requires an external energy source; an electric energy is most commonly used to trigger actuations. A *passive* actuator utilizes the natural energy such as heat to trigger actuations. Some examples of passive actuators can be a spring element to store energy or thermometer that uses the phenomenon of thermal expansion of material in measurement. The energy output for actuations can be in a form of *hydraulic, pneumatic, kinetic, thermal, magnetic,* or *ultrasonic* energy. For the type of energy conversion, the conversion occurring in an actuator can be *electric, electromagnetic, electrostatic, hydraulic, hydrostatic, pneumatic, piezoelectric, piezo-ultrasonic, ultrasonic,* or an *electro-rheological fluid.*

Figure 4.4 also shows that actuators can be classified based on types of the targeted objects (Redekar et al. 2022). An actuator can act on either a *controlled element* or a *controlling element.* A controlling element can be a *solenoid, valve,* or *switch* used to control other components; a controlled element can be a *body element* to transfer motion or force internally or an end-effector to act on an external object in the environment. A body element can be *linear* or *rotary* motor, *mobile platform,* or *wrist,* and an end-effector can be one of a *parallel gripper,* a *vise,* a *multi-finger tool,* a *compliant tool,* or a *multifunctional tool.* In this chapter, the basic principles of energy conversion are discussed, and the commonly used types of actuators are introduced, respectively. In Chapter 13, end-effectors will be discussed in detail.

4.2 Functional Requirements (FRs)

Due to high diversifications of the classifications in Figure 4.4, it is difficult to generalize the basic *functional requirements* (FRs) of actuators. However, energy forms in actuators can be mechanical, pneumatic, electrical, hydraulic, or hybrid. Different actuators may share the similarity in transferring motions or energy from one form to another in performing their tasks (Redekar et al. 2022). As follows, basic FRs of actuators are discussed by taking an example of transferring motion and force.

Most actuating systems are designed to transfer motions and torques from a source to a targeted object. Any complex motion of an object can be decomposed into a series of one *degree-of-freedom* (DoF) basic motions, that is, a 1-DoF translation, a 1-DoF rotation, and a 1-DoF coupled translation and rotation. Furthermore, stroke lengths or angles of rotation need to be defined. Table 4.1 gives some basic motion requirements of actuators.

Velocity. Different applications have different requirements for velocities in their actuating systems. For example, a sophisticated pick-and-place system is usually required to have a high velocity. On the other hand, the required velocities of joints in a collaborative robot must be confined for safety reasons. A velocity can be *translational* or *rotational*, and the change of a translational or rotational velocity is referred to as a translational or rotational acceleration.

Operational load. The power supplied to an actuator must be sufficient for the actuator to fulfill given tasks. For example, a transportation system may need a higher power to move a heavier

Table 4.1 Common functional requirements (FRs) of motion actuators

Functional requirement (FR)	Description
Motion profile	The requirements of an expected motion are often primary, one *degree-of-freedom* (DoF) can be linear, rotary, or spiral; moreover, the stroke of a linear or spiral motion and the angular range of rotation is usually specified.
Cycle time	The cycle of actuation varies from one application to another; for example, a loading/unloading process in high-volume production usually has a short cycle time, while the operation speed of a collaborative robot must be confined to assure the safety of collaborative workers. Therefore, the cycle time of actuation must be specified in designing or selecting an actuator. Note that the cycle time is determined by allowable velocity and acceleration of the actuator.
Operational load	The power output by the actuator must be sufficient to carry the external load; the required power depends on both the operational speed and force or torque over the targeted object. Therefore, the power requirement can be very high when the mass or the operational speed of the targeted object is very high.
Repeatability, resolution, and accuracy	Repeatability is the closeness of targeted location, orientation, force, torque, or power when the actuator takes the same actuation repeatedly; resolution is the finest difference an actuator can make in changing its output; accuracy refers to the difference of the expected and actual outputs. All of them are required for an actuator to sustain consistent quality of operations.
Dynamic characteristics and responsiveness	Dynamic characteristics refer to the properties of an actuator in responding to the environmental change. Dynamic characteristics depend not only on the capability of actuating elements but also on the performance of sensing and controlling elements.

load; in contrast, an actuator on a camera is for the adjustment of focus, and this requires a limited amount of power. In real-life applications, the required power (P) is usually unknown, and the power supplies should be with varying loads.

Resolution, accuracy, and **repeatability.** These parameters define the precision with which the actuating system can achieve in operating the system. *Resolution* is defined as the smallest controllable change in output possible by the actuator. A typical example has the requirements on resolution to achieve the expected resolution, accuracy, and repeatability. When it is used to locate some components in a printed circuit board with a separation of the minimum size of 2 mm, in which case the positioning resolution of the robot should be less than 2 mm. *Accuracy* determines how close is the estimated value to the true value of a control variable. Accuracy is affected by numerous factors such as assembly tolerances, temperatures, and contact conditions. *Repeatability* reflects the closeness of a control variable in a number of repeated operations. Repeatability is affected by some factors such as the backlashes and hysteresis of actuators.

Dynamic behaviors/responsiveness. The dynamic response of an actuator is characterized by the velocity and acceleration of the actuator over time. Dynamic behaviors/responsiveness depend greatly on the effectiveness of feedback from the sensors and control systems. It is desirable to have mathematical models with first, second, third, or even higher order to analyze dynamic behaviors. Characterizing dynamic behaviors uses some techniques including step response analysis and frequency response analysis.

Compliance is the movement of a component in reaction to a force exerted on it. *Compliance* is usually defined as the stiffness of the system (low compliance), or sponginess/softness (high compliance) of the actuation system. Differing compliance requirements may be presented for different actuation systems. For example, actuation systems in machine tools would require high stiffness and therefore low compliance is used in order to ensure the required machine tool precision during machining. On the other hand, a positioning system, such as a system used to place a tool in a tool holder, would require a high compliance so as to allow the tool to be inserted with ease into its tool holder.

Operation load can be usually quantified as torque (τ) and power (P). Torque and power are the most important characteristics in selecting motors for actuators (GE 2024). Torque and power are not independent since they are associated with each other as,

$$P = \tau\,(2\pi N)/33{,}000 \tag{4.1}$$

where P is the power in horsepower (*HP*), N is the rotational speed in the number of *revolutions per minute* (RPM), and τ is the torque in lb-ft.

The power to a motion actuator is from the equipped motor, and Figure 4.5 shows the relation of motion and torque in an *alternating current* (AC) motor. An AC motor for the motion actuator can be characterized by locked torque, pull-up torque, pull-out torque, rated torque, locked current, and full-load current as explained in Table 4.2 (Redekar et al. 2022).

It is desirable to use standardized motors in actuators and commercial motors are mostly standardized. Figure 4.6 shows four prescribed motors by the *National Electrical Manufacturers' Association* (NEMA) that are developed to meet normal industrial needs in most economic ways (Agrawal 2007). Other than motions and torques, the constraints in selecting a motion actuator may also be affected by sizes, weights, auxiliary equipment, harsh environments, and needs of maintenance (Redekar et al. 2022).

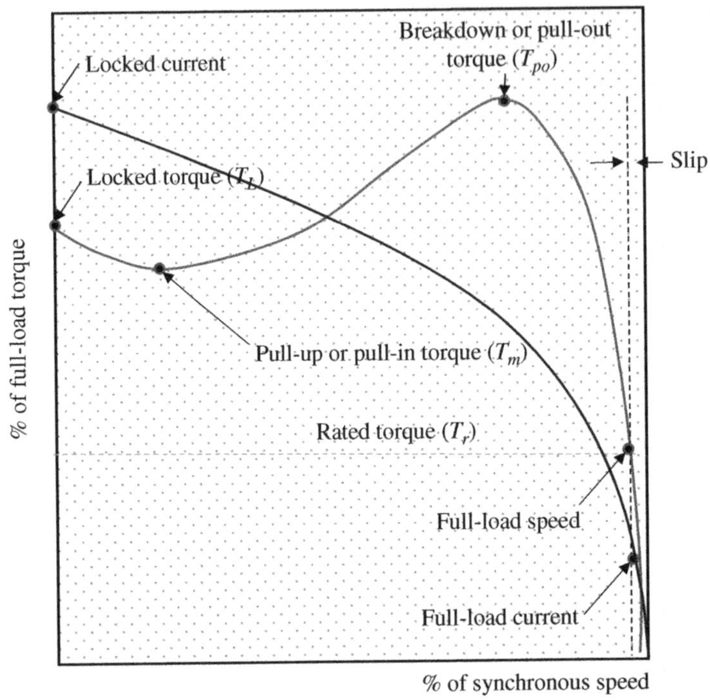

Figure 4.5 Torque and current with respect to the speed of an AC motor

Table 4.2 Characterizing an AC motor for motion actuator

Parameter	Description
Locked torque or starting torque	The starting torque a motor can produce at rest when the rated voltage is applied at the rated frequency. It is expressed as a percentage of the full-load torque.
Full-load torque	The torque required to produce the rated horsepower at the full-load speed.
Pull-up torque	The minimum torque that is produced in the acceleration period from rest to the speed corresponding to the breakdown torque. When no breakdown torque exists, it is the minimum torque from rest to the rated full-load speed. A pull-up torque is expressed as a percentage of full-load torque.
Breakdown torque	The maximum torque the motor is able to produce when the rated voltage is applied at the rated frequency without an abrupt speed drop. The breakdown torque is expressed as a percentage of full-load torque.
Full-load current	The steady current for an induction motor when it operates for the full-load torque by applying the rated voltage at the rated frequency.
Locked current	The steady-state current is when the rotor is locked by applying the rated voltage at the rated frequency.

4.3 Design Variables (DVs)

In this section, an electric motor is used as an example to illustrate its *design variables* (DVs) to meet FRs of a motion actuator in the previous section. In an electric motor, the power supply relates to its nameplate voltage and operation frequency; while the nameplate voltage is usually less than the nominal voltage of the power supply system.

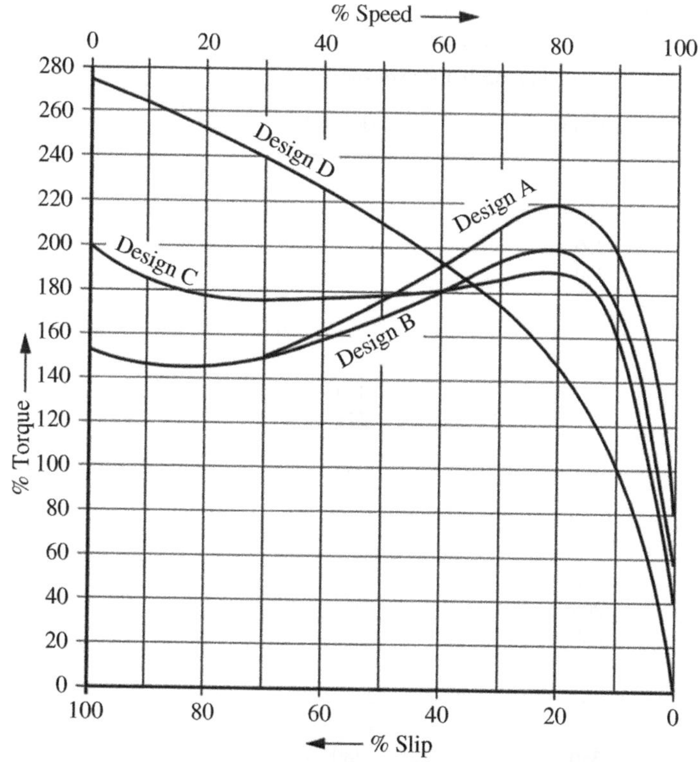

Figure 4.6 Speed–torque characteristics of four motors in NEMA standard

The nominal voltage of a power supply system can be specified in a large voltage range, and nominal commonly used frequencies are 50 and 60 Hz. Note that the nameplate voltage of the motor affects the produced power and it relates to the nominal voltage of the power supply system. For example, GE (2024) described some AC motors with a frequency of 50 or 60 Hz, nominal voltages of power supply, and corresponding nameplate voltages in Table 4.3.

Motors are typically *Mechatronic Functional Modules* (MFMs). Figure 4.7 classifies the DVs of an electric motor for an electric actuator. Besides the nameplate voltage of the motor, and the nominal frequency, and voltage of the power supply system, other main DVs are the choices associated with controls, power supplies, instrumentations, and safety assurance. *Control variables* affect the transfer function of an actuator. Control parameters typically include masses and inertia of constitutive components and load elements, motion profiles, control strategies, bandwidths, and frequencies. A *power supply* can be *stationary* or *mobile* depending on whether the application is fixed; a power supply can be electrical, using an internal combustion engine, or generated from steam or hydraulic power. An actuator typically uses a closed-loop control that is required to measure output variables such as displacements, velocities, accelerations, and forces or torques. One has to design or select the methods for measurements when subjected to the constraints on space, mounting, and cost; moreover, the performance of instruments is affected by various operating factors such as temperature, heat generation and dissipation, and vibration. Finally, it is a primary requirement to use an actuator safely. A number of safety factors must be taken into consideration; these include to power down automatically in an emergent situation, monitor fail modes, and take immediate actions

Table 4.3 Power versus operating voltages in GE motors with 60 Hz

	Motor nameplate voltage	
Power supply system voltage	**<125 HP**	**>125 HP**
Single-phase, 60 Hz		
120	115	—
208	200	—
240	230	—
Multiphase, 60 Hz		
208		—
240		—
480		460
600		575
2400		2300
4160		4000
Single-phase, 50 Hz		
	110	—
	200	—
	220	—
Multiphase, 50 Hz		
	200	—
	220	—
	380	380
	415	425
	440	440
	550	550
	3000	3000

Source: Adapted from GE (2024).

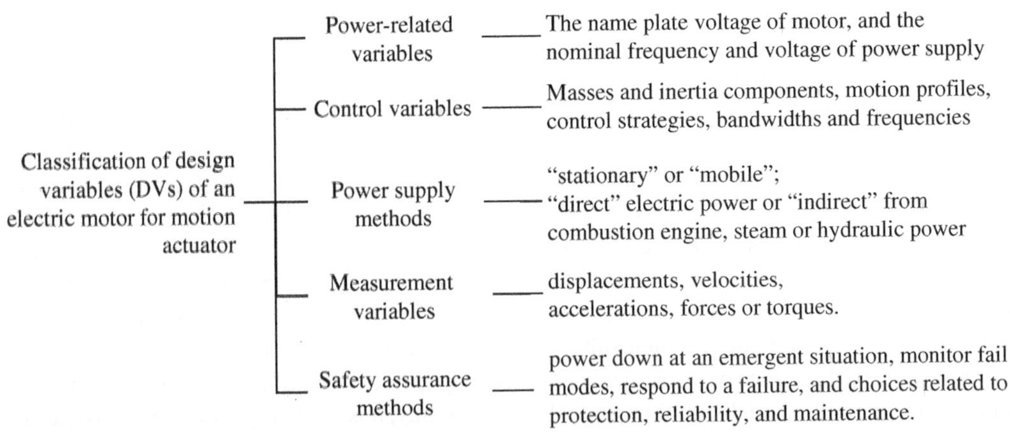

Figure 4.7 Main *design variables* (DVs) of electric motor for motion actuator

when a failure occurs. The safety variables also include those related to protection, reliability, and the need for routine maintenance.

4.4 Basics of Energy Conversion

Actuators are also types of transducers to transform energy from one form to another. Energy cannot be created or destroyed; however, energy can be transformed from one form to another. While an electrical form is mostly preferable in energy transmission and conversion, electrical power is rarely available from nature directly, it is not always the preferable way to consume energy in all applications. In many mechatronic systems, energy or power should be converted from one form to another so that energy can be consumed to do work naturally. Figure 4.8 shows seven types of energy conversion commonly used in various mechatronic systems.

4.4.1 Mechanical Energy Conversion

In an actuator with mechanical energy conversion, input energy is converted into output energy in mechanical forms such as strain energy, kinematic energy, and potential energy. Figure 4.9 shows an example of a solid-state actuator that is capable of converting electrically induced strain energy into output mechanical energy to do work when subjected to an external force *F*. In contrast to conventional hydraulic or pneumatic actuators that need separate power units with large electric

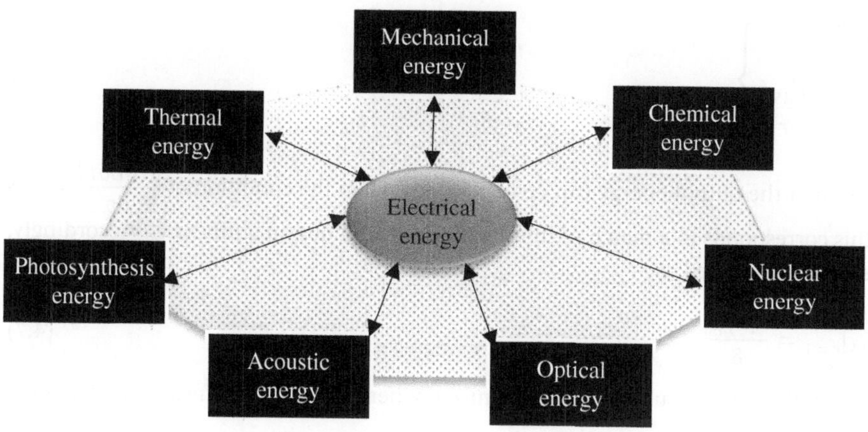

Figure 4.8 Common types of energy conversion in actuations

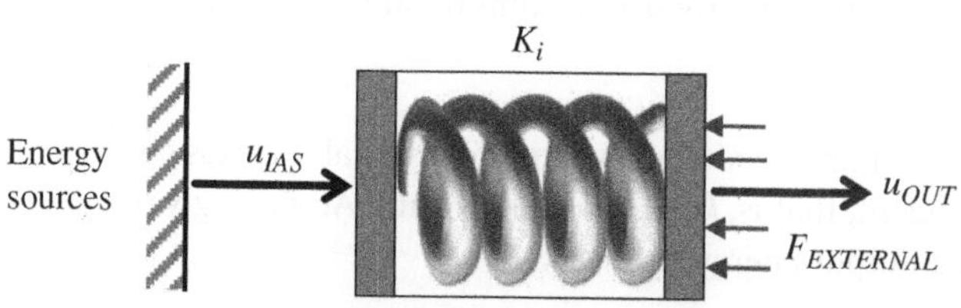

Figure 4.9 Example of a solid-state actuator

motors and pumps, a solid-state actuator directly converts electrical energy into mechanical energy; it is capable of producing a high force even with a limited motion range in MEMS applications (Giurgiutiu et al. 2000).

Let the input of the actuator be an initially actuated strain u_{IAS}, the outputs of the actuator are the displacement u_{OUT} and associated output energy E_{OUT}.

Assume that the equivalent stiffness coefficient of the actuator is k_i, the output strain u_{OUT} can be determined by u_{IAS} and k_i as,

$$u_{OUT} = u_{IAS} - \frac{F_{EXTERNAL}}{k_i} \tag{4.2}$$

Equation (4.2) shows that no output strain is produced when the external force reaches F_{BLOCK} as

$$F_{BLOCK} = k_i u_{IAS} \tag{4.3}$$

The output strain energy can also be found in Eq. (4.2) as,

$$E_{OUT} = \frac{1}{2} k_i u_{OUT}^2 \tag{4.4}$$

Once the output strain u_{OUT} is determined in Eq. (4.2), the external force $F_{EXTERNAL}$ can be converted by an external spring with an equivalent stiffness of k_e as,

$$k_e = \frac{F_{EXTERNAL}}{u_{OUT}} \tag{4.5}$$

Let $r = k_i/k_e$ and substituting it back to Eqs. (4.2) and (4.4) gives,

$$\left.\begin{aligned} u_{OUT} &= \frac{u_{IAS}}{1+r} \\ E_{OUT} &= \frac{r k_i u_{IAS}^2}{2(1+r)^2} \end{aligned}\right\} \tag{4.6}$$

Eq. (4.6) shows that the output energy can be maximized when the condition of $\frac{d(E_{OUT})}{dr} = 0$ is satisfied, and this corresponds to a match of input stiffness k_i and output stiffness k_e. Accordingly, the output energy is maximized as,

$$E_{OUT,MAX}|_{r=1} = \frac{k_i u_{IAS}^2}{8} \tag{4.7}$$

Figure 4.10 shows that the output energy is maximized when the output stiffness k_e matches the input stiffness k_i.

Example 4.1 A solid-state actuator produces a free input stroke of $u_{ISA} = 120\ \mu m$ with an internal stiffness $k_i = 400$ kN/mm, what is the condition to produce maximum output energy?

Solution

According to Eq. (4.7), the output energy will be maximized when the external stiffness coefficient (k_e) matches the internal stiffness coefficient, that is, $k_e = k_i = 400$ kN/mm. Accordingly, Eq. (4.7) can be used to calculate the maximized output energy as,

$$E_{OUT,MAX}|_{r=1} = \frac{k_i u_{IAS}^2}{8} = \frac{(400 \times 10^6)(120 \times 10^{-6})^2}{8} = 0.72$$

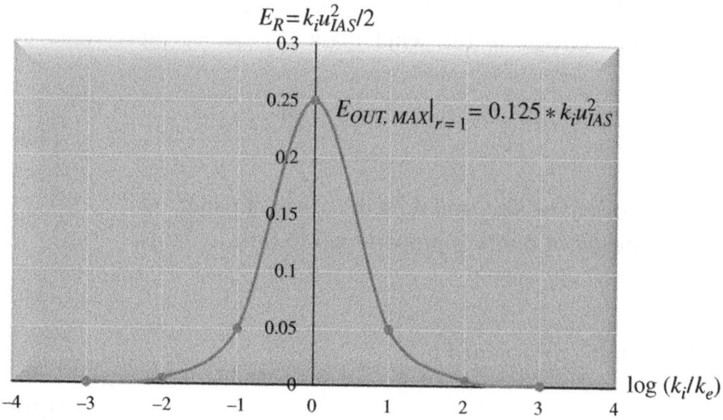

Figure 4.10 The maximized output energy occurs when the output stiffness k_e matches input stiffness k_i

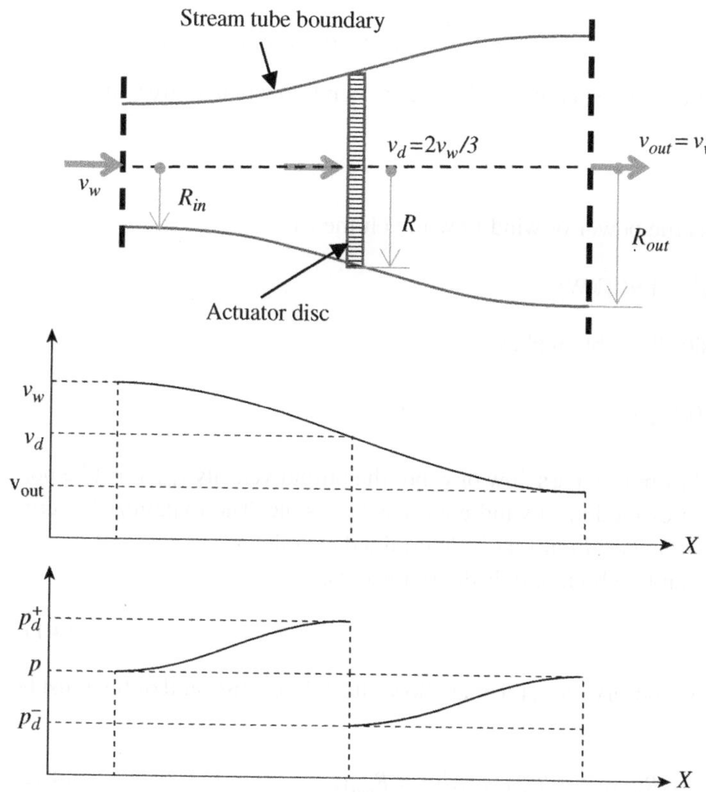

Figure 4.11 The velocity distribution for the Betz limit

With the growing popularity of utilizing renewable energy in mechanical systems, more and more actuators are MFMs that are integrated with energy harvesting systems. In such MFMs, mechanical energy is harvested and converted into electrical energy for use. Figure 4.11 shows a design by which wind energy is converted into electrical energy.

Assume that a vertical disc has a radius of R, and a wind flow passes the disc snugly without a mass loss perpendicularly. The kinetic energy of the wind mass (E) can be represented by (Alipranties 2014)

$$E = \frac{1}{2}mv_w^2 = \frac{1}{2}\rho\pi R^2 h v_w^2 \tag{4.8}$$

where ρ is the density of air, h is the height of disc, and v_w is the speed of wind.

Under a constant speed of v_w, the height of disc h is proportional to time t, that is,

$$h = v_w t \tag{4.9}$$

The total power of the wind flow can be determined by,

$$P = \frac{E}{t} = \frac{1}{2}\rho\pi R^2 v_w^3 \tag{4.10}$$

Example 4.2 Assume that the air density $\rho = 1.25\left(\frac{\text{kg}}{\text{m}^3}\right)$, the velocity of wind $v_w = 10\left(\frac{\text{m}}{\text{s}}\right)$, and the radius of disc as $R = 1$ (m); determine (1) the total power of wind flow in disc area and (2) the power density per area.

Solution

Applying Eq. (4.10), one obtains the power of wind flow in volume as

$$P = (0.5)(1.25)\pi(1)^2(10)^3 = 1.96\ (\text{kW})$$

Sequentially, the power density $\overline{P}$ is obtained as,

$$\overline{P} = \frac{P}{A} = \frac{1}{2}\rho v_w^3 = 0.625\ (\text{kW/m}^2)$$

When the energy is harvested from the mass flow of wind, the output velocity v_{out} must be positive, and this implies that a full extraction of wind energy is infeasible. The maximized energy extraction can be determined by *the Betz limit* that is determined as follows.

The total mass within the streamline boundary is determined as,

$$m = \rho V = \rho\pi R^2 v_w t \tag{4.11}$$

Following the principle of mass conservation, the mass flow rate at inlet, disc, and outlet must be constant, that is,

$$\dot{m} = \rho\pi R_{in}^2 v_w = \rho\pi R^2 v_d = \rho\pi R_{out}^2 v_{out} \text{ or } R_{in}^2 v_w = R^2 v_d = R_{out}^2 v_{out} \tag{4.12}$$

Assume that the atmosphere pressure at the inlet and the outlet be p, applying Bernoulli's principle one gets,

$$\left.\begin{aligned} p + \frac{1}{2}v_w^2 &= p_d^+ + \frac{1}{2}v_w^2 \\ p + \frac{1}{2}v_{out}^2 &= p_d^- + \frac{1}{2}v_w^2 \end{aligned}\right\} \tag{4.13}$$

The drag force (F) on the disc can be found as

$$F = \left(p_d^- - p_d^+\right)\pi R^2 \tag{4.14}$$

Substituting Eq. (4.13) into (4.14) yields

$$F = \frac{1}{2}\rho\pi R^2\left(v_d^2 - v_{out}^2\right) \tag{4.15}$$

Furthermore, the conservation of momentum is applied, and the drag force from actuator disc reduces the global momentum of the flow, and this reduction is related to the difference in input and output momentum rates as,

$$F = \dot{m}(v_w - v_{out}) \tag{4.16}$$

Since the mass flow rate at the location of the disc can be defined as,

$$\dot{m} = \rho\pi R^2 v_d \tag{4.17}$$

Using Eqs. (4.15)–(4.17), one gets,

$$v_d = \frac{v_w + v_{out}}{2} \tag{4.18}$$

Accordingly, the extracted power can be determined as,

$$P = Fv_d = \frac{1}{4}\rho\pi R^2(v_w - v_{out})(v_w + v_{out})^2 \tag{4.19}$$

Let $\frac{dP}{dv_{out}} = 0$, the extracted power can be maximized as,

$$\left.\begin{matrix} P_{max} = \left(\frac{16}{27}\right)\left(\frac{1}{2}\rho\pi R^2 v_w^3\right) \\ v_{out} = \frac{v_w}{3} \end{matrix}\right\} \tag{4.20}$$

Therefore, the factor 16/27 = 59% is the maximum power coefficient, and P_{max} is commonly referred to as the Betz limit (Alipranties 2014).

Example 4.3 Assume that the air density $\rho = 1.25 \left(\frac{\text{kg}}{\text{m}^3}\right)$, the radius of the disc of a horizontal-axis wind turbine is $R = 50$ m, and the wind velocity at its inlet is $v_w = 15$ m/s, determine the Betz limit for the maximized extracted energy.

Solution

Applying Eq. (4.20), the extracted energy is maximized when $v_{out} = \frac{v_w}{3} = 5$ m/s, and the corresponding energy limit is,

$$P_{max} = \left(\frac{16}{27}\right)\left(\frac{1}{2}\rho\pi R^2 v_w^3\right) = 9.817\ (\text{mW})$$

The Betz limit can be used to determine the maximum possible efficiency (η) of a wind turbine. The mass conservation of the wind flow in Eq. (4.12) gives the relation of the radii of the inlet and disc as,

$$R_{in}^2 = \frac{v_d}{v_w} R^2 \tag{4.21}$$

Therefore,

$$\eta_{max} = \frac{P_{out}}{P_{in}} = \left(\frac{16}{27}\right) \frac{\left(\frac{1}{2}\rho\pi R^2 v_w^3\right)}{\left(\frac{1}{2}\rho\pi R_{in}^2 v_w^3\right)} = \frac{\left(\frac{16}{27}\right)}{\left(\frac{2v_w}{3v_w}\right)} = 88.9\% \tag{4.22}$$

In practice, a wind turbine operates at a low efficiency in a range of (40–45%) and the extracted power (*P*) is estimated empirically as,

$$P = \left(\frac{1}{2}\right) C_p(\lambda, \theta)\rho\pi R^2 v_w^3 \tag{4.23}$$

where C_p is the performance coefficient determined by the tip-speed ratio (λ) and the pitch angle (θ) of the blades; the tip–speed ratio represents the ratio of the circumferential speed of a blade tip over the wind speed as,

$$\lambda = \frac{\omega_w R}{v_w} \tag{4.24}$$

where ω_w is the angular velocity of a blade in rad/s. C_p can be maximized when $\frac{\partial C_p}{\partial \theta} = 0$, and the dependence of the performance coefficient on a pitch angle is typically shown in Figure 4.12a; Figure 4.12b shows the extracted power accordingly when $R = 80$ m, and v_w is in the range of (3, 15) (m/s) (Alipranties 2014).

4.4.2 Electromechanical Energy Conversion

Electromechanical devices are used for energy conversion between mechanical and electrical forms. Electromechanical devices are mainly used as (1) *transducers* such as speakers and microphones to transform the signals from one form to another; (2) *actuators* such as relays and solenoids to produce the forces for certain motions; (3) continuous *energy conversions* such as generators to convert mechanical energy to electrical energy.

Despite the power in an electrical subsystem and a mechanical system being in different forms, the energy is conserved in conversion. Therefore, even power-relevant variables in different forms are quite different; they play equivalent roles in producing power, and these variables as well as associated system components show similarities in modeling an actuating system. Take an example of a linear or rotary actuator, Pelz (2003) discussed relevant variables and components analogous to those in electrical power sources in Table 4.4.

In any of the aforementioned applications, an electromechanical system usually consists of (1) *a mechanical subsystem* such as a plunger in a linear actuator or a rotor in a motor, (2) *magnetic subsystems* such as magnetic cores and air gaps, and (3) *an electrical subsystem* with electric circuits such as windings. Figure 4.13 shows main parameters and variables in an electromechanical system. The electrical subsystem is governed by *Ohm's Law*, *Kirchoff's Voltage Law* (KVL), and *Kirchoff's Current Law* (KCL) on currents, voltages, resistors, and power (kaliasgoldmedal 2024). The mechanical subsystem is governed by *Newton's Laws* on positions, velocities, and accelerations.

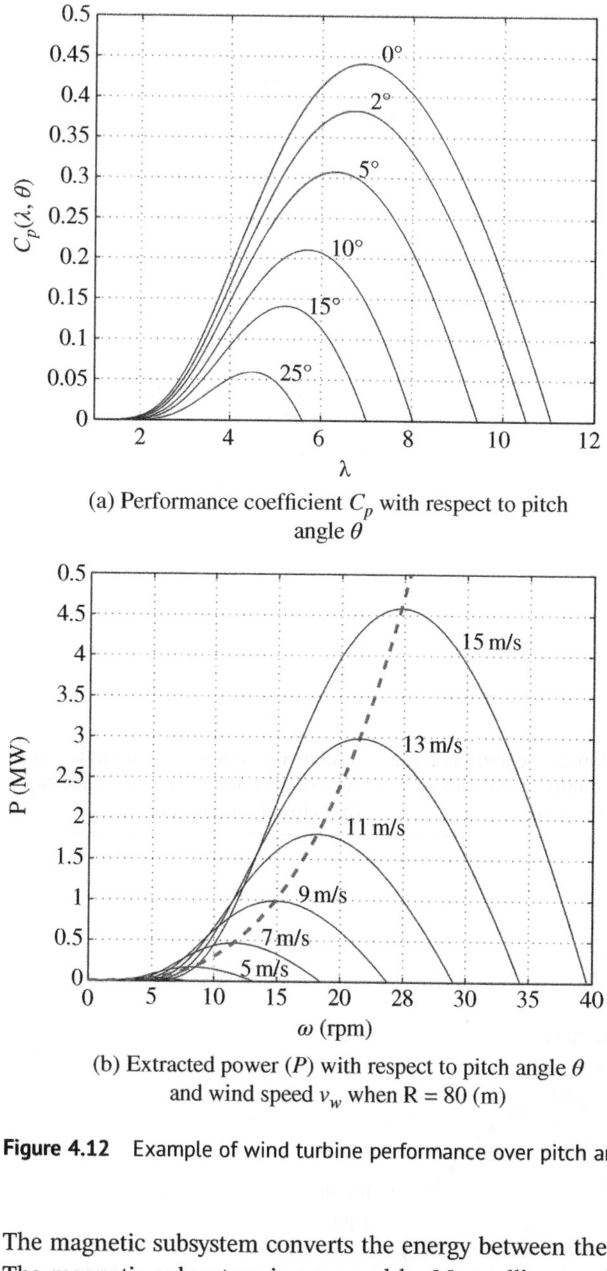

(a) Performance coefficient C_p with respect to pitch angle θ

(b) Extracted power (P) with respect to pitch angle θ and wind speed v_w when R = 80 (m)

Figure 4.12 Example of wind turbine performance over pitch angle θ and wind speed v_w

The magnetic subsystem converts the energy between the electrical and mechanical subsystems. The magnetic subsystem is governed by Maxwell's equations about magnetic flux, flux density, and field strength.

Between the electrical and magnetic subsystems, the magnetic flux is interacted with the current in the circuit to produce a force or torque that will be used in the mechanical subsystem. Between the mechanical and magnetic subsystems, the movement of the moving part causes the change in the magnetic flux and induces an electromotive force (emf) in the electrical subsystem. For energy conservation, the product of the torque and speed in the mechanical subsystem equals the product of the emf and current.

Table 4.4 Analogous variables and components of electrical power with a linear and rotary actuator

Electrical power	Linear actuator	Rotary actuator
Current (i)	Force (F)	Torque (M)
Voltage (u)	Velocity (v)	Angular velocity (ω)
Coil: $u(t) = L \cdot \frac{d}{dt} i(t)$	Spring: $v(t) = \frac{1}{k} \cdot \frac{d}{dt} F(t)$	Torsion spring: $\omega(t) = \frac{1}{k} \cdot \frac{d}{dt} M(t)$
Capacity: $i(t) = C \cdot \frac{d}{dt} u(t)$	Inertia: $F(t) = m \cdot \frac{d}{dt} v(t)$	Rotational inertia: $M(t) = J \cdot \frac{d}{dt} \omega(t)$
Resistor: $i(t) = \frac{1}{R} \cdot u(t)$	Damping: $F(t) = b \cdot v(t)$	Rotational damping: $M(t) = b \cdot \omega(t)$
Power dissipation by resistor: $P(t) = u(t) \cdot i(t)$	Power dissipation by damping: $P(t) = v(t) \cdot F(t)$	Power dissipation by damping: $P(t) = \omega(t) \cdot M(t)$
Magnetic energy: $T(t) = \frac{1}{2} \cdot L \cdot i^2(t)$	Potential energy: $T(t) = \frac{1}{2} \cdot \frac{1}{k} \cdot F^2(t)$	Potential energy: $T(t) = \frac{1}{2} \cdot \frac{1}{k} \cdot M^2(t)$
Electrostatic energy: $T(t) = \frac{1}{2} \cdot C \cdot u^2(t)$	Kinematic energy: $T(t) = \frac{1}{2} \cdot m \cdot v^2(t)$	Kinematic energy: $T(t) = \frac{1}{2} \cdot J \cdot \omega^2(t)$
Transformer: $i_1 \cdot u_1 = i_2 \cdot u_2$ $i_1 = \alpha \cdot i_2$ $u_1 = \frac{1}{\alpha} \cdot u_2$	Lever: $F_1 \cdot v_1 = F_2 \cdot v_2$ $F_1 = \alpha \cdot F_2$ $v_1 = \frac{1}{\alpha} \cdot v_2$	Lever: $M_1 \cdot \omega_1 = M_2 \cdot \omega_2$ $M_1 = \alpha \cdot M_2$ $\omega_1 = \frac{1}{\alpha} \cdot \omega_2$
Sum of all currents at a node is zero; sum of all voltages in a closed loop is zero	Sum of all forces at a point is zero; sum of all relative velocities in a closed loop is zero	Sum of all moments at a point is zero; sum of all relative angular velocities in a closed loop is zero

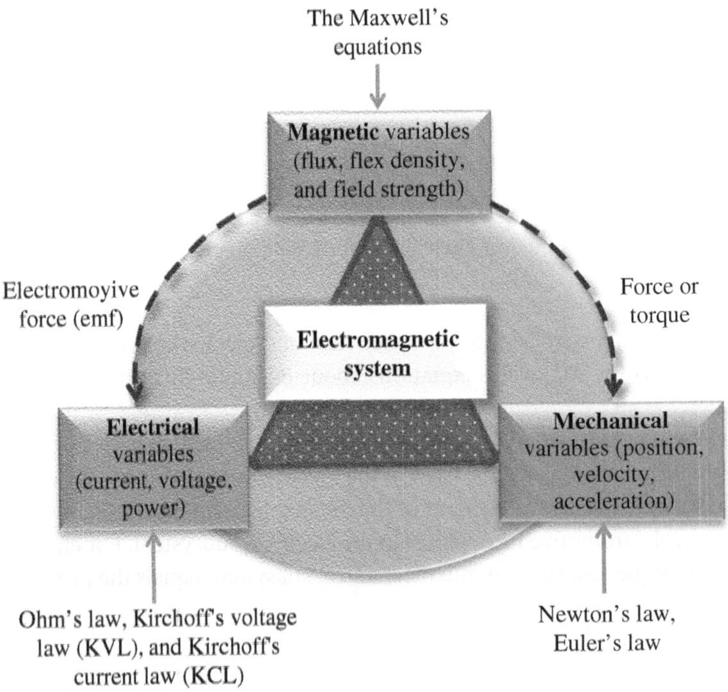

Figure 4.13 Structure of an electromechanical system

Assume that a conductor with a length of l is placed in a magnetic field with a uniform *magnetic flux density* of B, when the conductor moves at a speed of $\mathbf{v}$, the *electric and magnetic field (emf)* is induced in the conductor as,

$$e = lv \times B \tag{4.25}$$

where e is the electric/magnetic field, when the direction of the electric/magnetic field is determined by *the right-hand rule* based on the cross product of the velocity $\mathbf{v}$ and the magnetic field B. The unit of the magnetic field B is *Tesla*; one Tesla is equal to webers/m^2 since B relates to the magnetic flux ϕ as,

$$\phi = B \cdot A \cdot \sin(\theta) \tag{4.26}$$

where B is the strength of magnetic flux/field in Tesla (T) or webers/m^2; A is the area of the conductor through which the magnetic field lines project (m^2); θ is the angle of the magnetic field links and the plane of the conductor.

Example 4.4 Determine the voltage of an open circuit between two brushes that make contacts on a rotating Faraday's disc in the magnetic field of B as shown in Figure 4.14.

Solution

Figure 4.14 shows that two brushes make the contacts at the radii of r_1 and r_2, respectively. Taking a finite segment of length $\Delta l = dr$ in $r \in [r_1, r_2]$, Eq. (4.25) can be used to determine the corresponding electromagnetic field (emf) as,

$$de = d(lv \times B) = (dr)(\omega_r r) \times (B) = \omega_r rBdr \tag{4.27}$$

Integrating the emf in $r \in [r_1, r_2]$ one gets,

$$e = \int_{r_1}^{r_2} de = \int_{r_1}^{r_2} \omega_r rBdr = \frac{\omega_r B}{2}\left(r_2^2 - r_1^2\right) \tag{4.28}$$

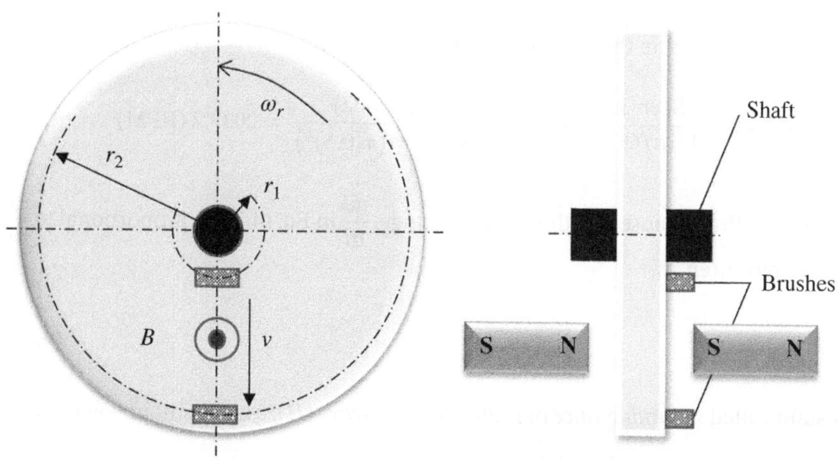

Figure 4.14 Voltage by a rotating Faraday's disc in a magnetic field

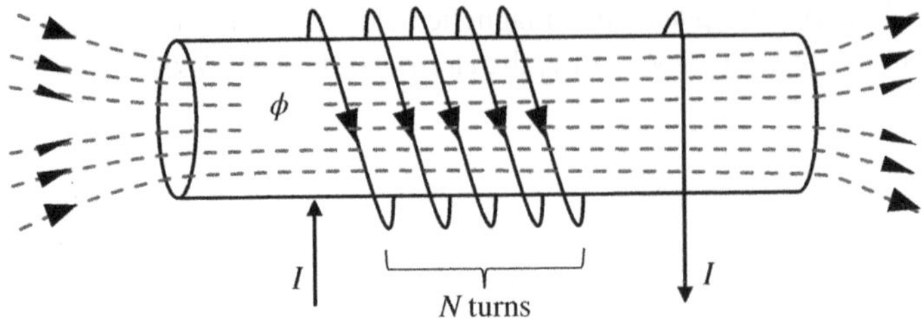

Figure 4.15 Self-induced electromagnetic field (emf)

A current-carrying conductor can produce a magnetic field. On the other hand, a magnetic field can be used to create an electric current in a conductor. When a magnetic flux linking a conductor is changed, an emf is induced, and a current will be produced if the conductor forms a complete loop or circuit.

A conductor can be *self-induced* in an application when the magnetic flux is changed due to the change of a current or a relative motion. Figure 4.15 shows the case where a magnetic field is established through the current in the coil. When the current in the coil changes, the flux linking the coil changes accordingly, and this self-induces an emf. According to Lenz's Law, the magnitude and direction of the induced emf are found as,

$$e = -N\frac{d\phi}{dt} \tag{4.29}$$

where e is the emf in voltage; N is the number of turns in windings; $\frac{d\phi}{dt}$ is the rate of the magnetic flux in webers/t.

Example 4.5 A fan blade has a length of 0.5 m. It rotates in the magnetic field of $B = 7.5 \times 10^{-5}$ Tesla. Assume that the rotating axis is perpendicular to the plane where the magnetic field is and the induced emf is $e = 0.002$ V, what is the rotating speed (n) of the fan blade in *RPM*?

Solution

Eq. (4.29) can be used to determine the rotating speed (n) as

$$e = N\frac{d\phi}{dt} \Rightarrow 0.02 = (1)\frac{B(\pi r^2)}{1/(n/60)} \Rightarrow n = \frac{(60)(1)(0.002)}{(7.5\times 10^{-5})(\pi(0.5)^2)} = 2037.2\ (\text{RPM})$$

When the magnitude of flux is caused by the current change, $\frac{d\phi}{dt}$ in Eq. (4.29) is proportional to $\frac{dI}{dt}$, and Eq. (4.29) can be rewritten as,

$$e = -L\frac{dI}{dt} \tag{4.30}$$

where L is the constant called *self-inductance* or *inductance* in *Henry* (*H*), e is the emf in voltage, and $\frac{dI}{dt}$ is the rate of current change in *A/second*.

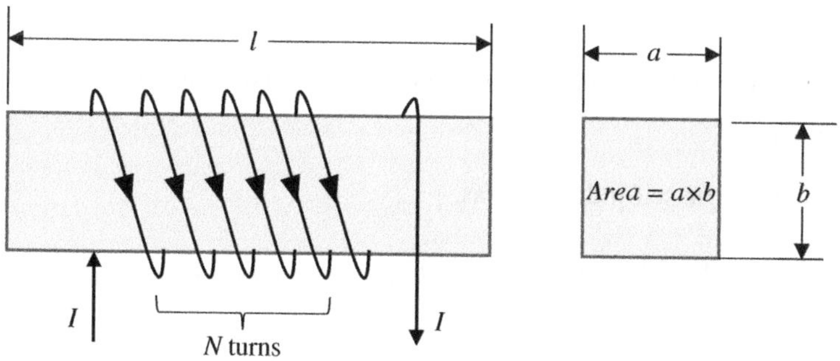

Figure 4.16 Inductance of an iron-cored solenoid

The inductance can be determined by three different ways. *The first method* is to measure the emf (e) when subjected to the specified $\frac{dI}{dt}$, and the inductance can be found as $L = e/\frac{dI}{dt}$. *The second method* is to measure the magnetic flux when subjected to the specified current I, and the inductance can be found as $L = N\phi/I$. *The third method* is to calculate the inductance based on the geometric dimensions of the conductor. Take an example of an iron-cored solenoid as shown in Figure 4.16, the inductance of the solenoid is calculated as (Slametumy 2018),

$$L = \frac{N^2(Area)\mu_0\mu_r}{l} = \frac{N^2}{S} \tag{4.31}$$

where N is the number of turns of windings; l and a are the length and the width of the solenoid, respectively; μ_0 is the permeability of free space (vacuum), that is, $\mu_0 = 4\pi \times 10^{-7}$ Henry/m; μ_r is the relative permeability of a magnetic material such as iron, that is, $\mu_r = \frac{\mu_{iron}}{\mu_0} = 6.3654 \times 10^4$; S is the *reluctance* that is calculated as the ratio of the magnetic path length (l) by the permeability times the cross-sectional area A as *relectance* $= \frac{l}{Area \cdot \mu_0\mu_r}$.

Example 4.6 Assume that a coil has 500 turns in windings, a magnetic flux of $\phi = 5 \times 10^{-6}$ webers is produced by a current of $I = 10$ A; determine the inductance of the coil.

Solution

The second method $L = N\phi/I$ is used to calculate the inductance as,

$$L = \frac{N\phi}{I} = \frac{(500)(5 \times 10^{-6})}{(10)} = 2.5 \times 10^{-4}\ \text{Henry}$$

Example 4.7 A wooden toroid has a diameter of 0.5 (m) with a cross-sectional area of 5.0×10^{-6} m^2; the relative permeability is $\mu_r = 1$. It is uniformly wound with a coil of 800 turns and it carries a current of $I = 2A$; determine (1) the self-inductance of the coil and (2) the induced emf when the current is uniformly reduced to zero in 10 ms.

Solution

The length (l) of the conductor (the wooden toroid) can be determined based on its diameter as,

$$l = \pi D = \pi(0.5) = 1.571\ (\text{m})$$

Eq. (4.31) is then used to calculate the self-inductance as,

$$L = \frac{N^2(Area)\mu_0\mu_r}{l} = \frac{(800)^2(5.0\times10^{-6})(1)(4\pi\times10^{-7})}{(1.571)} = 2.46\times10^{-6}.\ \text{Henry's}$$

Eq. (4.30) is used to calculate the voltage of emf as,

$$e = -L\frac{dI}{dt} = -\left(2.46\times10^{-6}\right)\frac{2}{(10\times10^{-3})} = -4.92\times10^{-4}\ (\text{V})$$

In converting energy from one form to another, energy is conserved in the actuating system. Corresponding to the three subsystems in Figure 4.13, energy in an electromechanical system presents in three forms, that is, *electrical energy*, *mechanical work*, and *magnetic energy*.

The instantaneous power of self-induced conductor L with a current I can be found as,

$$p = e\cdot I = LI\frac{dI}{dt} \tag{4.32}$$

The total magnetic energy can be found by integrating Eq. (4.32) along the current path as,

$$E_m = \int_0^l LI\frac{dI}{dt} = \frac{1}{2}LI^2 \tag{4.33}$$

Assume that a coil of N-turns wound and situated in a magnetic field; it has a uniform cross-sectional area of *Area* and a length of the current path of l. Eq. (4.34) can be used to calculate the magnetic energy as,

$$E_m = \frac{1}{2}LI^2 = \frac{1}{2}\left(\frac{N^2(Area)\mu_0\mu_r}{l}\right)I^2 = \frac{1}{2}(\mu_0\mu_r(Area\cdot l))H^2 \tag{4.34}$$

where E_m is the magnetic energy in Joule; $H = \dfrac{NI}{l}$ and $area\cdot l$ is the volume of uniform magnetic field in m^3.

Eq. (4.35) can be further used to find the energy in a unit volume ($E_{m,u}$) of magnetic field as,

$$E_{m,u} = E_m/(Area\cdot l) = \frac{1}{2}(\mu_0\mu_r(Area\cdot l))H^2/(Area\cdot l) = \frac{B^2}{2\mu_0\mu_r} \tag{4.35}$$

where $E_{m,u}$ is the unit magnetic energy in Joule/m^3 and $H = \dfrac{B}{\mu_0\mu_r}$.

Eqs. (4.34) and (4.35) can also be used to correspond the current I with the magnetic flux density as,

$$I = \frac{Bl}{N\mu_0\mu_r} \tag{4.36}$$

Example 4.8 An air-cored solenoid (i.e., $\mu_r = 1$) has a length of 0.5 m, a radius of 0.05 m, and it is wound with 500 turns; determine the stored energy when there is a current of $I = 4$ A.

Solution

Eq. (4.35) can be used to calculate the stored energy as,

$$E_m = \frac{1}{2}\left(\frac{N^2(Area)\mu_0\mu_r}{l}\right)I^2 = = \frac{1}{2}\left(\frac{(500)^2(\pi(0.05)^2)(1)(4\pi\times10^{-7})}{(0.5)}\right)(4)^2 = 0.0126 \text{ Joule}$$

Figure 4.17 shows two cases where the attraction forces are induced by the emf. Figure 4.17a shows the case where the load is attached to one of the magnetic poles (i.e., S or N). The attractive force on the object can be determined by the energy conservation that the work on load ($F \cdot dx$) comes from the change of the magnetic energy due to the change of relative distance of the two poles (i.e., dx) in Figure 4.17a as,

$$F \cdot dx = E_{m,u} \cdot (Area \cdot dx) \tag{4.37}$$

Eq. (4.35) can be substituted in Eq. (4.37) to find the induced attraction force as,

$$F = E_{m,u} \cdot (Area) = \frac{B^2(Area)}{2\mu_0\mu_r} \tag{4.38}$$

The attraction force on the load in Figure 4.17b is simply doubled due to the attraction areas on both of south and north poles.

Example 4.9 An actuator is set up as in Figure 4.14b as a lifting mechanism. The U-shaped conductor is made of iron with a length of 0.5 m, a cross-sectional area of 8.0×10^{-4} m^2; a coil is wound with 800 turns on each side. Assume that the permeability of the magnet is $\mu_r = 600$; calculate the current to produce a lifting force of 50 kg.

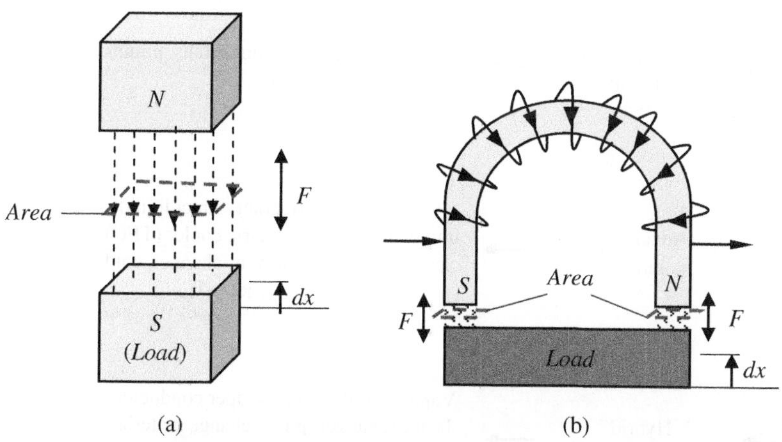

Figure 4.17 Induced forces in a magnetic field: (a) single attraction area and (b) double attraction area

Solution

Eq. (4.37) can be used to calculate the required strength of magnetic flux for an attraction force of 50 kg as,

$$B = \sqrt{\frac{2\mu_0\mu_r(F/2)}{Area}} = \sqrt{\frac{2(600)(4\pi \times 10^{-7})(50/2)}{8.0 \times 10^{-4}}} = 3.873 \text{ Tesla}$$

In Eqs. (4.34) and (4.44), one has $H = \frac{NI}{l} = \frac{B}{\mu_0\mu_r}$; therefore, the required current can be found as,

$$I = \frac{Bl}{N\mu_0\mu_r} = \frac{(3.873)(0.5)}{(800+800)(600)(4\pi \times 10^{-7})} = 1.605 \text{ A}$$

Electromagnetic forces are relatively weak at a micro-scale due to the scaling problem that magnifies the effect of surface force such as friction; on the other hand, electrostatic actuators show their advantages in (1) ease of minimization, (2) conventional materials can be micromachined and integrated in chip, (3) it can be easily controlled even though a high-level voltage circuit may be required, and (4) low power is required for activation, especially for the purpose of positioning (Ceyssens et al. 2019).

4.4.3 Thermomechanical Energy Conversion

Advanced mechatronic devices are required to respond to changes quickly in an ever-compact size. A fast action by an actuator in a compact size requires power with a high density. Mechatronic systems can be classified in terms of heat fluxes into *high*, *ultra-high*, and *extreme-high* heat fluxes in the ranges of (10^2, 10^3), (10^3, 10^4), and (10^4, ∞) W/cm^2, respectively. The performance of mechatronic components is affected by a temperature change significantly. Arshad et al. (2020) found that over 55% of failures occurring in mechatronic systems are caused by an elevated temperature. To manage an operating temperature, heat must be dissipated appropriately. There is a need to develop sophisticated actuators to dissipate the heat and control an operating temperature by thermomechanical energy conversion.

Figure 4.18 classifies the actuators for heat dissipation based on thermomechanical energy conversions. The heat can be dissipated either *actively* or *passively*. *An active heat dissipation* uses an

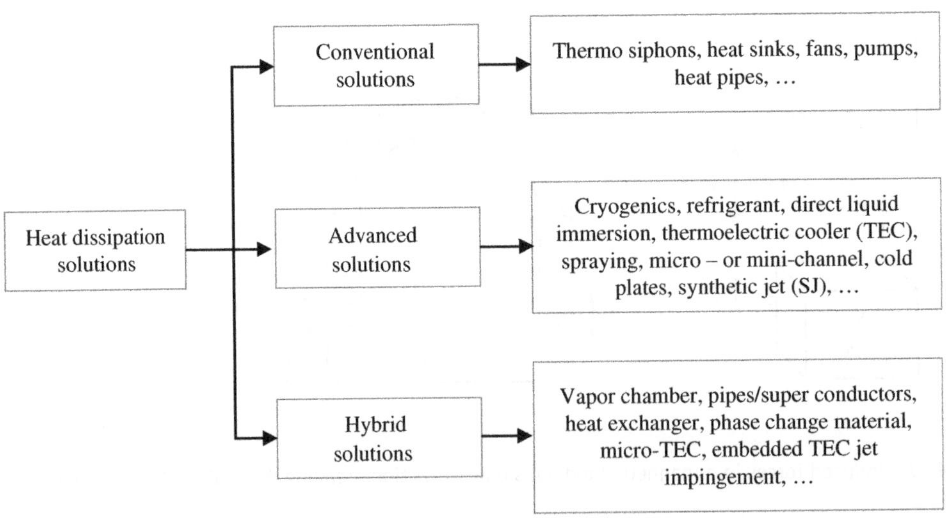

Figure 4.18 Classification of actuators for heat dissipation

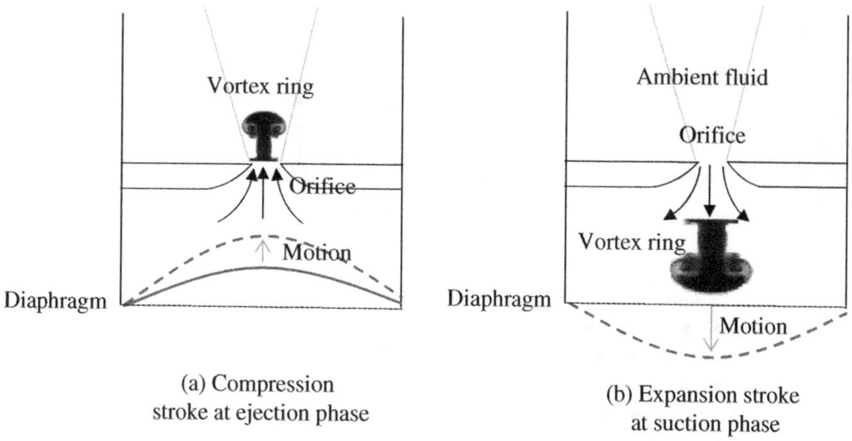

Figure 4.19 Working principle of a synthetic jet cooler

external device such as a pump, compressor, or fan to remove a large amount of heat at a high rate. *A passive heat dissipation* utilizes the gravitational buoyancy or capillary force to circulate the cooling fluid. Some common actuators for heat dissipation are *thermoelectric coolers*, *phase changing coolers*, and *synthetic jets*.

Taking an example of a synthetic jet in Figure 4.19, it is a fluidic device similar to a Helmholtz resonator. A thin surface such as a membrane or diaphragm is actuated by a fluctuating voltage, a cyclic suction and ejection of fluid is produced out of a narrow slot or orifice, and the fluid flows to dissipate heat (Arshad et al. 2020).

Each operating cycle consists of an ejection phase and a suction phase. At *an ejection phase*, the high-temperature fluid in the cavity is ejected from the orifice and formed as a vortex ring due to the inward movement, and the vortex moves away due to its initial velocity. At *a suction phase*, the diaphragm is deformed in an opposite way by a negative pressure, and the low-temperature fluid around is drawn back to the cavity, and this reduces the overall temperature in the cavity.

The heat can be dissipated as mechanical strain energy when an actuator uses some special materials. Some materials such as *shape memory alloys* (SMAs) respond to heat stimuli mechanically; such materials can be used as actuators in structural health-monitoring systems and self-healing systems. Abavisani et al. (2021) referred to SMAs as *multi-functional materials* and classified SMAs into *magnetic* SMAs, *polymer* SMAs, and *ceramic* SMAs. A shape memory material memorizes its original shape when it is deformed; it will return to its original shape when it is stimulated by heat. For example, SMA in Figure 4.20 exhibits different crystalline structures (i.e., martensite and austenite) subject to temperature changes. Accordingly, two distinguished features of SMA are (1) shape memory effect and (2) superelasticity. After the alloy undergoes a plastic deformation when subjected to a load, it will return to its original shape when it is heated and unloaded. Superelasticity happens when motion and force are generated due to the shape memory effect and energy is stored at a high temperature.

4.4.4 Electro-stimulated Materials

Electro-stimulated materials, such as electro-active polymers, shape memory polymers, and sand hydrogels, piezoelectric materials, are able to change their inherent mechanical and responsive

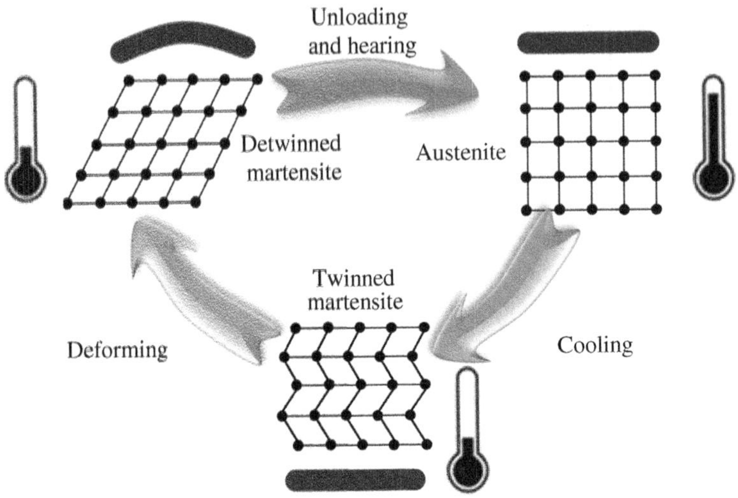

Figure 4.20 Thermomechanical efforts in *shape memory alloys* (SMAs)

attributes when subjected to low-energy stimuli. Electro-stimulated materials are initially in their transition states that exhibit viscoelastic properties of a solid and liquid; gels have three-dimensional cross-linked structures. Gels are made from natural materials such as collagens and dextrans and synthetic polymers, and produce good electric or mechanical responses to external stimuli. These materials have been widely used for wearable devices, actuators, sensors, and biomedical applications (Imdad et al. 2019).

Gao et al. (2023) introduced a supramolecular gel–elastomer system that can grasp an object by iontronic adhesion as shown in Figure 4.21. Some materials have the properties of electro-adhesion that can be used to produce continuous, astrictive, and reversible adhesion forces in robotic applications. A dynamically cross-linked geo-elastomer system was designed; it included an ionic organohydrogel as adhesive electrodes and a resilient polyurethane with a high density of electrostatic energy as dielectric layers. The system provided cohesive heterolayer bonding for adhesive prehension.

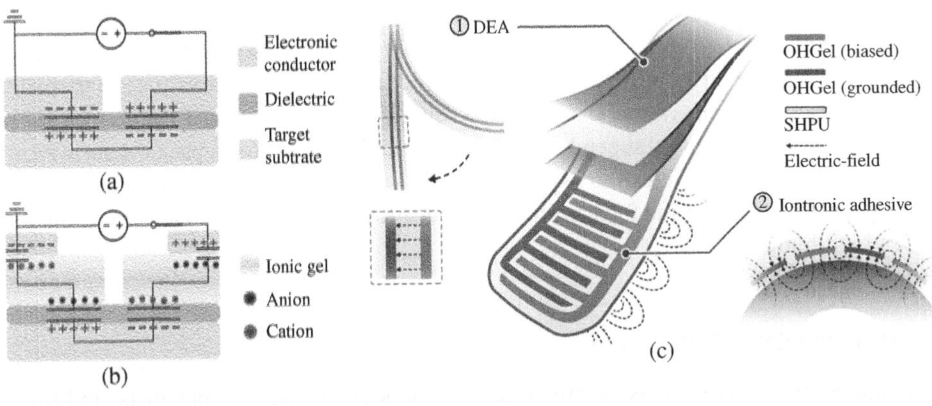

Figure 4.21 An electro-adhesion system (a) unactuated state, (b) actuated state and (c) actuator in product (*Source:* Gao et al. (2023)/Springer Nature/CC BY 4.0)

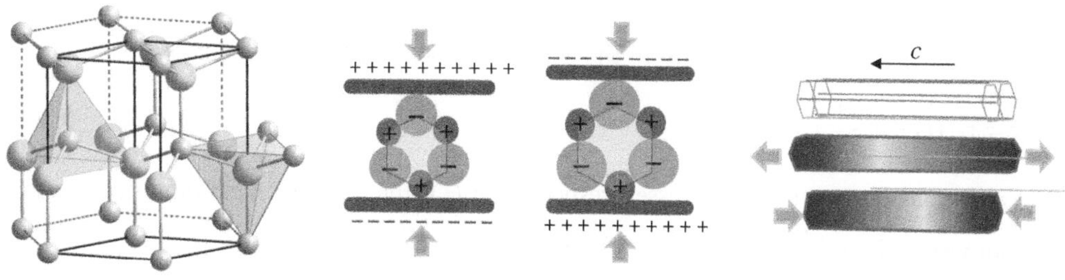

(a) Z_nO crystal structure (b) Piezoelectric effect under compression and tension (c) Axial strain distributions

Figure 4.22 Energy conversion by piezoelectric materials

As shown in Figure 4.22, when ZnO crystal is with a tetrahedral structure, Zn^{2+} and O^{2-} are arranged layer-by-layer along the length direction of a nano-tube. In an undeformed state, the charge centers of cations and anions are aligned with each other; in a deformed state by an external load, the charge centers are misaligned and thus form an electric dipole for a piezo-potential; and the free electrons are driven to flow through an external circuit. This corresponds to an energy conversion and reduces the piezo-potential to reach a new equilibrium state. Therefore, a nanogenerator is capable of generating a steady stream of pulse current in an external circuit when a fluctuating load is applied to the piezoelectric material (Fang et al. 2016).

4.4.5 Magneto-rheological Fluid Energy Conversion

Magneto-rheological fluid (MRF) is a type of smart material widely used as a functional material of actuators in brakes, automotive dampers, medical equipment, and vibration controls. MRF is capable of transforming its liquid state to a semi-solid state in a magnetic field. Such a phase of MRF can be changed in a very short time and it is reversible. MRF includes suspended magnetizable particles at the micro-size in the carrier fluid. These particles are aligned in a magnetic field, and this leads to a chain-like structure to resist a flow and change the rheological properties of the fluid (Bahiuddin et al. 2018). The direction and amplitude of the current between two electrode coils determine the behaviors of MRF materials, such as contraction, extension, and the generated force by the actuators (Kaluvan et al. 2016).

As shown in Figure 4.23, in the activated state ("ON"), MRF under the yield stress exhibits viscoelasticity, whose modulus depends on the magnetic field intensity. An MRF is also known to be subject to *shear thinning*, that is, the viscosity exceeding yield decreases with an increase of shear rate. In an unactivated state ("OFF"), the MR fluid is still non-Newtonian and temperature dependent; however, the change of viscosity is ignorable and it can be treated as a Bingham plastic to

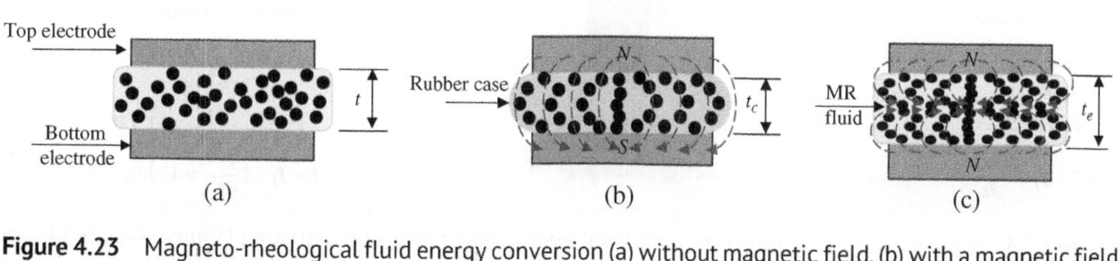

Figure 4.23 Magneto-rheological fluid energy conversion (a) without magnetic field, (b) with a magnetic field in rubber, and (c) with a magnetic field in fluid

simplify analysis, that is, $\tau = \tau_y + K\dot{\gamma}$, where τ and τ_y are shear and yield stresses and $\dot{\gamma}$ is the shear rate. Accordingly, the behavior of MRF in the activated mode can be represented by,

$$\tau = \tau_y(H) + \eta \frac{dv}{dz}, \tau > \tau_y \tag{4.39}$$

where τ and τ_y are shear and yield stresses, respectively; H is the magnetic field intensity; η is the Newtonian viscosity; and $\frac{dv}{dz}$ is the velocity gradient in the z-direction.

4.4.6 Nano-level Energy Conversion

Actuating mechanisms at nano-level are quite different from those at micro- or macro- levels. Nano-level actuations rely greatly on phenomena or forces on surfaces. The concept of self-power nanotechnology by Wang and Song (2006) has led to an emerging research area on energy harvesting and sensing. A nano-system can be self-powered by *piezoelectricity*, *pyroelectricity*, or *triboelectricity*. Accordingly, three energy harvesting technologies are called as *piezoelectric nanogenerators* (PENGs), *triboelectric nanogenerators* (TENGs), and *pyroelectric nanogenerators* (PyENGs). Both PENGs and TENGs obtain electric power from kinetic energy; while PyENGs obtain electric power from heat energy (Askari et al. 2022).

Figure 4.24 shows that triboelectric effect is used to convert small-scale mechanical energy into electrical current. The principles of nano- or micro-level energy conversion were tracked back to Maxwell's equation for displacement current. It was found that the second term $\frac{\partial P}{\partial t}$ in the Maxwell's equation was related directly to the output current of a *TENG* (Wang 2017). Therefore, the original Maxwell's equation was expanded to include the polarization density PS in the displacement vector; PS represented induced electrostatic charges on the surface by triboelectrification effect. TENGs were used to harvest small-scale mechanical energy for self-powered sensors and actuators without an external power supply.

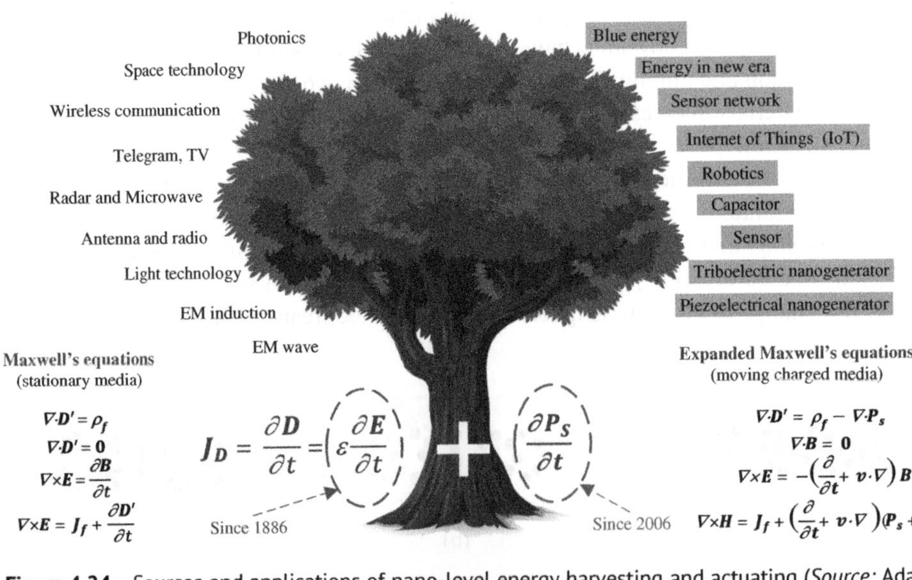

Figure 4.24 Sources and applications of nano-level energy harvesting and actuating (*Source:* Adapted from Cao et al. (2023))

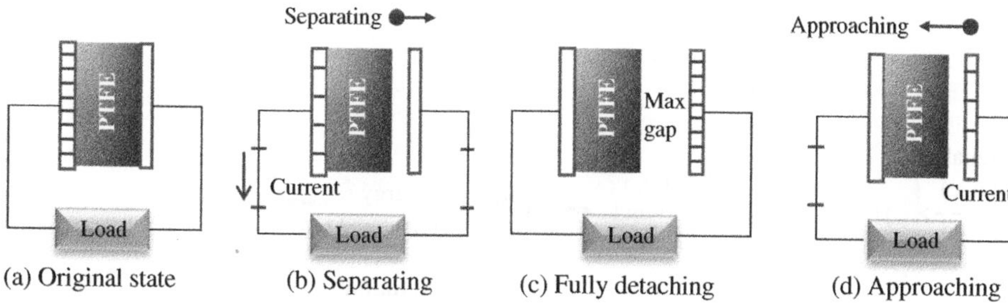

Figure 4.25 TENG operated in a *contact-separation mode*

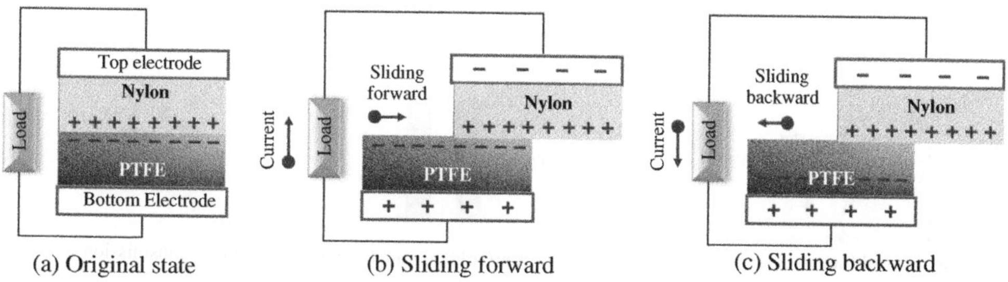

Figure 4.26 TENG operated in a *sliding mode*

A TENG is able to convert mechanical energy into electrical energy for use. The energy is converted based on electromagnetic, triboelectric, electrostatic, and piezoelectric mechanisms. TENGs have shown great potential to harvest and use energy for smart things from small scale to road infrastructure, precision agriculture, and massively networked sensors and actuators at large scale. Figures 4.25 and 4.26 show two basic operating modes of the operation of a TENG, which are (1) the *contact-separation mode* and (2) the *sliding mode* (Barkas et al. 2019).

4.5 Main Components

Actuators are an integral part of mechatronic systems that power the systems so as to interact with targeted objects in applications in expected ways. Actuators are indispensable components in various mechatronic systems such as smart cameras, home appliances, air conditioners, cars, and aircraft. Rao (2024) argued that the four basic function blocks of an actuating system are *energy source*, *controller*, *power converter*, and *load*, while others treated energy sources and loads as input and output of a mechatronic system, respectively.

From a system perspective, an actuator serves as an interface between the controller (i.e., a data processing unit) and an actuating system (i.e., a mechanical execution unit). The actuator converts low-power control signals with motion instructions into operating signals at a high energy level enough to take expected actions. Signal transducers are combined with amplifier elements to exploit the physical transformation principles governing the inter-relationships between various forms of energy (Reif 2015). Therefore, an actuator consists of two main components to (1) translate control signals (low-voltage) and (2) execute control signals (high-voltage), respectively. Figure 4.27

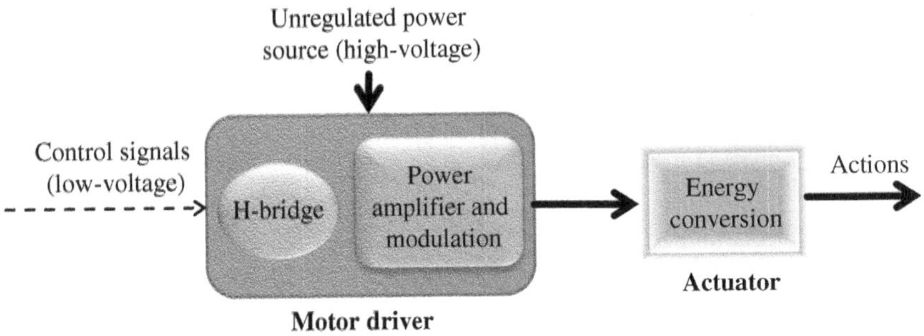

Figure 4.27 Main components of an actuating system

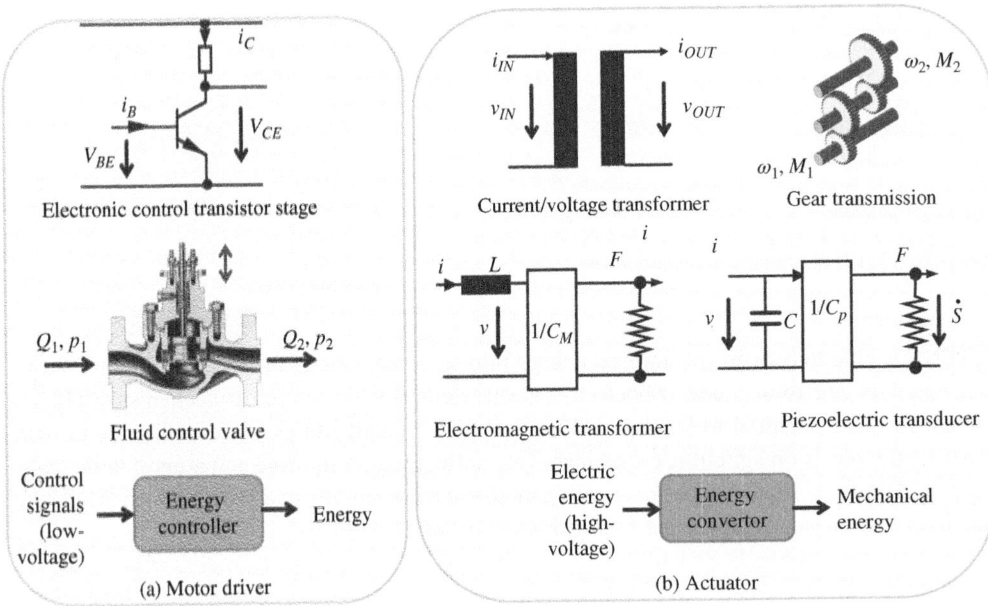

Figure 4.28 Examples of basic components in an actuating system (Emerson 2018)

shows an example of the main components in an actuator that is driven by a motor (Kanpur 2017). The actuating system consists of a motor driver and actuator. The motor driver takes low-voltage signals to modulate unregulated power; the actuator converts electrical energy into mechanical energy in execution. Figure 4.28 shows the examples of some basic elements that are used to implement motor drivers and actuators in actuating systems (Janocha 2004, Brown 2017, Emerson 2024).

4.6 Valve and Electric Actuators

Actuators used in mechatronic systems have been classified in different ways. As shown in Figures 4.4 and 4.29, actuators can be classified based on inputs, outputs, conversions of energy, and execution mechanisms that include (1) the forms of output energy such as heat, light, sound,

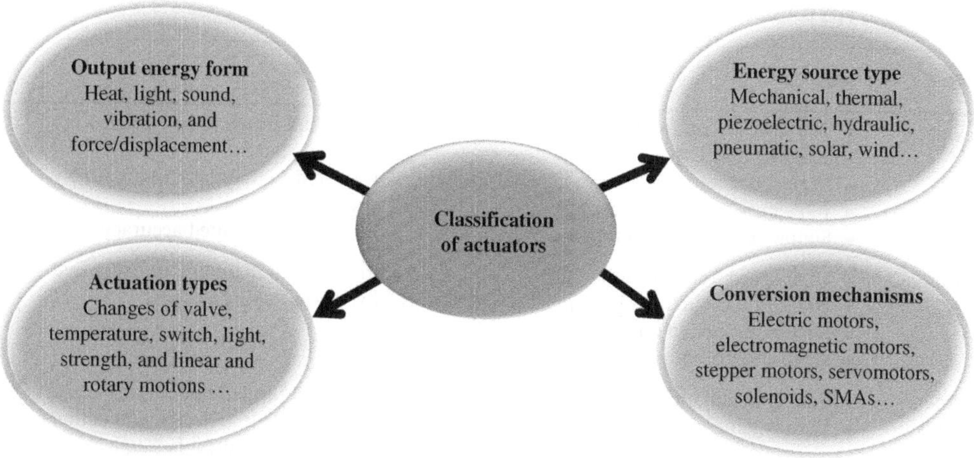

Figure 4.29 Classification of actuators

and mechanical actuators. (2) The types of energy sources such as *mechanical*, *thermal*, *electrical*, *piezoelectrical*, *hydraulic*, and *pneumatic* (Brown 2017). (3) The types of actuations such as the change of states such as a valve, temperature, switch, light, and strength and motions (i.e., *linear*, *rotary*, and coupled motion). (4) The types of energy conversion mechanisms such as electromagnetic motors, AC and DC servomotors, SMAs, stepper motors, and solenoid motors (Rao 2024).

4.6.1 Valve Actuators

Figure 4.30 shows that valves are very common parts used to dictate the amount of fluid being allowed to pass through energy sources. From this perspective, a *valve* can be an essential part of an actuator to convert energy from one form to another. On the other hand, an actuator might be an essential part of valve control as well when an encapsulated MFM is developed to control the

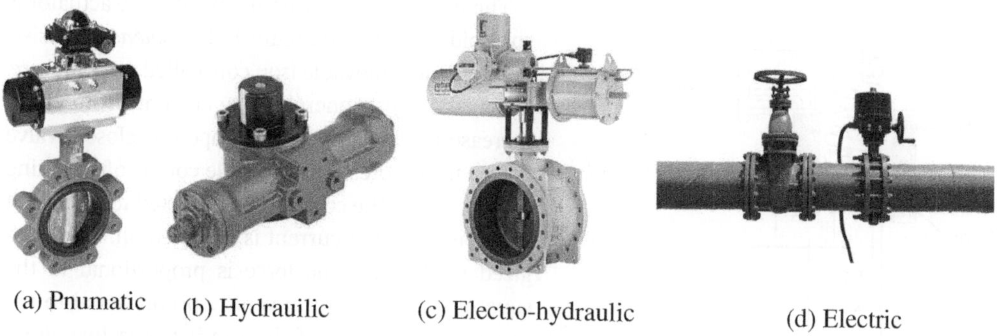

Figure 4.30 Four types of valve actuators (*Source:* (a) AliExpress.com/https://www.aliexpress.us/item/1005006022245897.html?gatewayAdapt=4itemAdapt#nav-description / Last accessed on July 19, 2024; (b) NYANGA MOVER / https://nyangamovervalveint.co.uk/ / Last accessed on July 19, 2024; (c) Rotork/https://www.rotork.com/en/media/latest-news/electro-hydraulic-actuators-deliver-improved-safety-at-indias-tank-farms / Last accessed on July 19, 2024; (d) SixAxis Company / https://www.saferack.com/glossary/motor-operated-valve-mov/ / Last accessed on July 19, 2024)

Table 4.5 Pros and cons of four different valve actuators

Types	Pros	Cons
Pneumatic	• The lowest cost and most cost-effective actuator type • High speed and fast-acting • 100% duty cycle, long-lasting • Simple to operate and repair	• Large and heavy to be installed in compact space • Required to maintain an air supply system • Control with limited accuracy
Hydraulic	• High speed and fast acting • 100% duty cycle, long lasting • Simple to operate and repair • Control with high-accuracy	• Required to maintain an expensive hydraulic supply system
Electro-hydraulic	• Simple installation • Data logging options available	• Slow and limited duty cycle (<30%) • Inapplicable to fail-safe applications • Difficult to repair; it is fragile
Electric	• High speed and fast acting • Supports fail-safe applications • Customized controls with high accuracy; 100% duty cycle • Remote diagnostics/data logging	• Most expensive option • Difficult to repair

valve. In such a case, it can be better referred to as a *valve actuator* to distinguish traditional valves since the MFM is sophisticated to control the openness of a fluid or air flow. A valve actuator is built into the valve housing, which dictates when the valve will open or close by a control system.

Common valve actuators are classified based on energy sources into pneumatic, hydraulic, electro-hydraulic, and electric actuators (see Figure 4.30). To select a valve actuator for an application appropriately, Table 4.5 provides *pros* and *cons* of these four types of actuators (Cowan Dynamics 2024).

Note that a duty cycle in Table 4.5 is determined by the times at "ON" and "OFF" states. For example, when a solenoid is actuated for 30 seconds, then OFF for 90 seconds, and then its duty cycle can be determined as $30\frac{\text{sec}}{30\text{ sec} + 90\text{ sec}} = 25\%$ duty cycle.

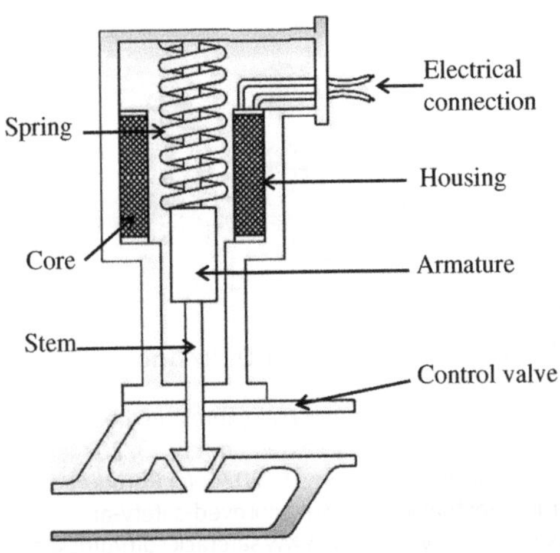

Figure 4.31 Solenoid as a valve actuator

The most commonly used type of valve actuator is solenoid as shown in Figure 4.31. *A solenoid* consists of a coil and a movable iron core called *an armature.* When the coil is energized by current, it moves to increase the flux linkage and open or close a valve between the cores. The movable core is often spring loaded so that the core can be retracted to its default position when the current is switched off. The energized electromagnetic force is proportional to the square of current and inversely proportional to the square of the width of the gap between two cores. Solenoids are cost-effective and primarily used to operate latches, locks, switches, and triggers.

Like the designs of other MFMs, it is desirable to use a standardized valve actuator as possible. Even though no standard component is available, it is

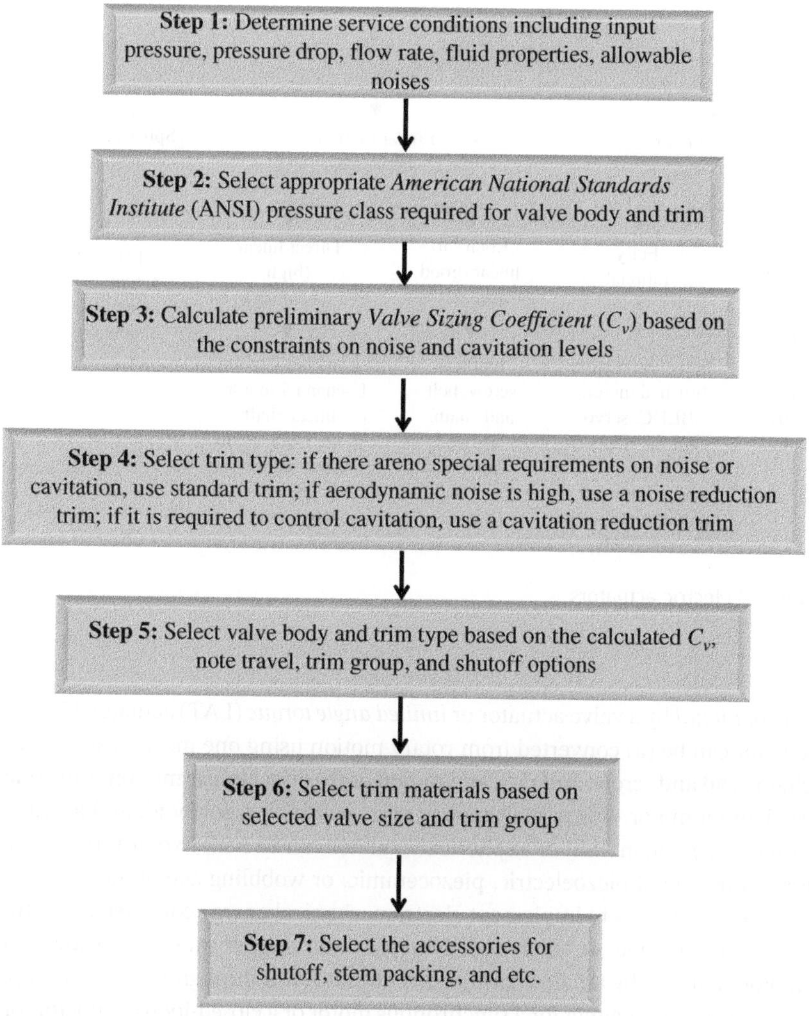

Figure 4.32 Flowchart of valve sizing (*Source:* Adapted from Emerson (2024))

beneficial to standardize the design process so that the satisfactory valve actuator can be designed in a smaller number of iterations. Figure 4.32 gives an example of a standardized procedure in determining the size of a fluid valve actuator (Emerson 2024); one can see that the design spaces at each step consist of standardized components for specific FRs.

4.6.2 Electric Actuators and Motors

Electric actuators are the most popular actuators in mechatronic systems. Various electric actuators are available, and they can be classified in different ways. For example, electric actuators can be classified based on types of targeted mechanical motions as shown in Figure 4.33 (Redekar et al. 2022). An electric actuator can be one of *rotary*, *linear*, *spherical*, and *seismic* motions.

Furthermore, (1) the motion in a rotary actuator can be *fully rotational* by an induction motor, brushed motor, brushless DC motor (BLDC), servo motor, stepped motor, synchronous motor, or

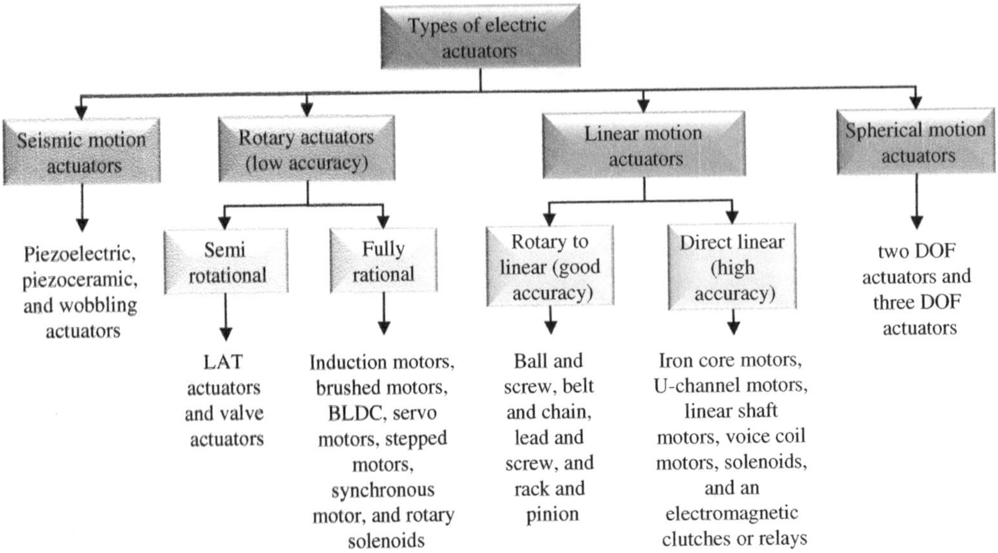

Figure 4.33 Classification of electric actuators

rotary solenoid or *semi-rotational* by a valve actuator or *limited angle torque* (LAT) actuator. (2) The motion in a linear actuator can be (a) converted from rotary motion using one mechanism of ball and screw, belt and chain, lead and screw, and rack and pinion or (b) direct linear motion using one of iron core motor, U-channel motor, linear shaft motor, voice coil motor, solenoid, and an electromagnetic clutch or relay. (3) The motion of a spherical actuator can be with two or three DoFs. (4) A seismic actuator can be one of piezoelectric, piezoceramic, or wobbling actuators.

Electric actuators use electric motors to implement electromechanical energy conversions. Common types of electric motors include *AC motors*, *DC motors*, and *stepper motors* as shown in Figure 4.34. An AC motor is driven by an alternating current that flows through stator windings. The rotor can be either a permanent magnet for a synchronous motor or a closed-loop conductor for an induced asynchronous motor. The speed of an AC motor is controlled based on the frequency of the input signal. A DC motor uses a direct current signal. It may use a permanent magnet as the stator magnetic pole or a stator winding to produce a magnetic field. Both the rotor armature and the stator winding are energized to drive the motor. When a permanent magnet are used, the speed of the DC motor is controlled by the voltage applied to the armature. When the stator windings is used, the speed of DC motor is controlled by the current to the stator.

A stepper motor is another type of electric motor that rotates by steps, it moves by a fixed number of angular resolutions. Different from an AC or DC motor, a stepper motor is capable of rotating a specific number of revolutions or fractions of a full revolution; it can control a specified angular displacement rather than the speed of a continuous rotation. Due to these features, stepper motors are widely used in robotic systems where the displacements of joints must be controlled accurately in path planning.

A stepper motor also consists of a fixed *stator* and a rotational *rotor*. A stator is designed with a number of teeth on which coils are wired; a rotor is made of a permanent magnet, a variable reluctance iron core, or a hybrid of two as shown in Figure 4.35. By energizing one or more of the stator phases, a magnetic field is generated by the current flowing in the coil, and this aligns the rotor with

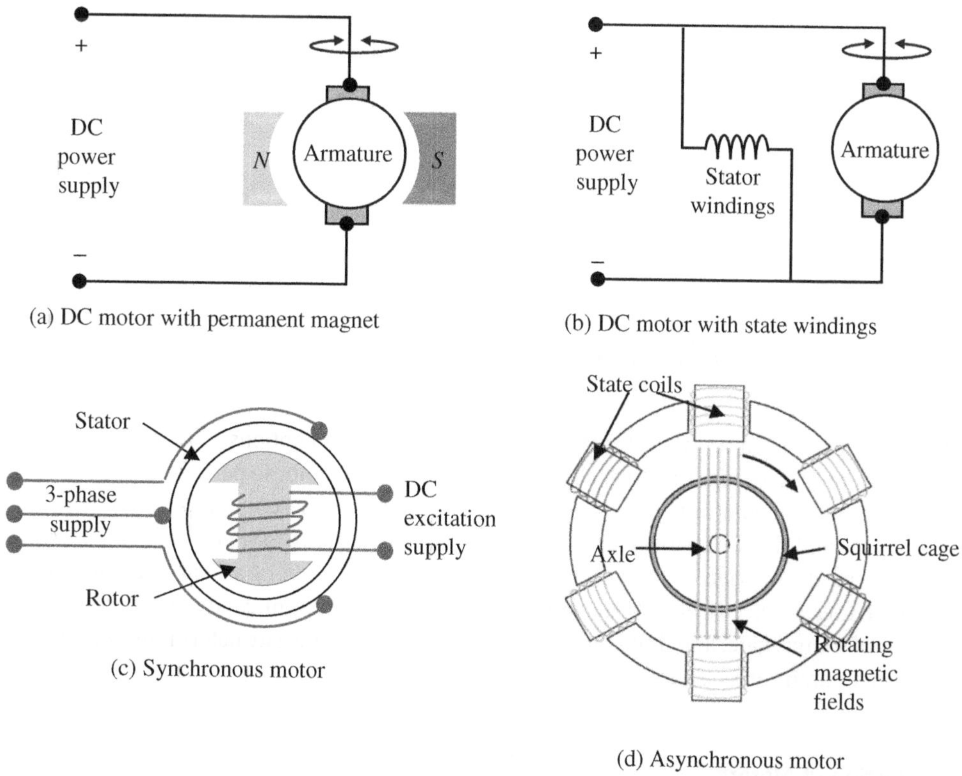

Figure 4.34 Types of DC and AC motors

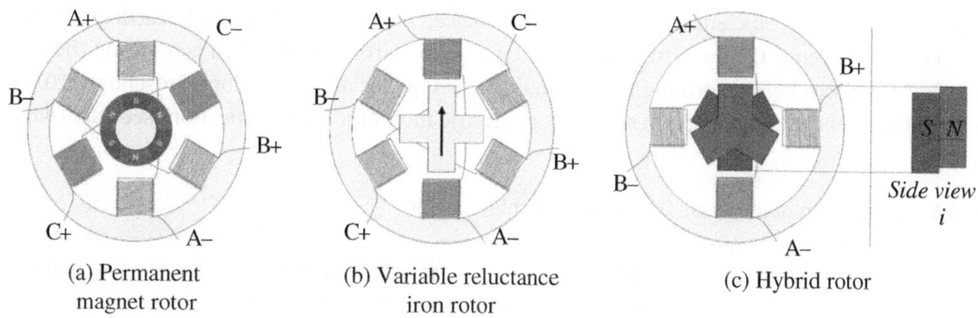

Figure 4.35 Three types of rotors in stepper motors

the field. By switching different phases on coils in sequence, the rotor is rotated continuously until the specified angular displacement is reached.

As shown in Figure 4.36, despite the varieties of stepper motors, stepper motors share the commonality of using a sequential switching to control a set number of coils wound on the stator. A sequential switching is used to energize appropriate drive circuitry, the switching frequency determines the rotational speed, and the number of switching actions determines the angular displacement corresponding to specific time (Fiore 2020).

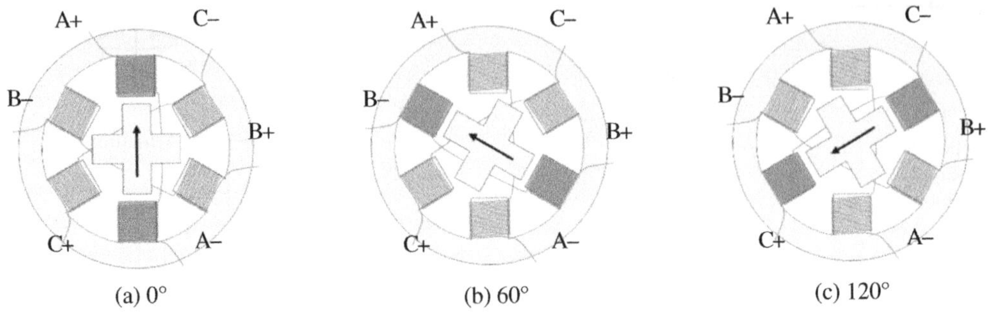

Figure 4.36 Sequential switching in stepper motors

Sequential switching can be implemented by one of three common options: (1) step/direction by sending a pulse on the *step* pin, the drive changes its output to the motor to perform a step, and the level of the *direction* pin determines the direction of movement. (2) Phase/enable by energizing the phase of a stator winding, the phase determines the direction of current. (3) *Pulse-width modulation* (PWM) by setting the voltage of the gate of the low-side or high-side of a field-effect transistor (FET); an FET uses an electrical field to control a current flow in a semiconductor. It consists of the terminals of source, gate, and drain, and the voltage applied on the gate determines if a current is from source to drain or vice versa.

4.6.3 Selection of Motors

When an electric actuator is used in an MEM, an electric motor should be selected appropriately by taking into consideration of FRs of energy conversion in the application and the specifications of available electric motors. In general, an electric motor is selected based on the requirements of output voltage, speed, power, efficiency, lifespan, maintenance, noise, regulation, and starting torque. Motors Quick Reference Guide (Groschopp 2022) is a good starting point when comparing the different motor types to help determine the optimal motor type for an application. Groschopp (2022) gave an example of the comparison in selecting an electric motor for their application (see Table 4.6).

Table 4.6 Comparison of universal, DC, AC, and brushless motors

	DC	AC	Universal AC and DC	Brushless AC and DC
Voltage type	DC	AC	AC, DC	AC, DC
Speed (RPM)	350–6000	1200–3400	8000–over 20 000	2300–3800
Power	Medium	Low–medium	Very high	high
Efficiency (%)	60–70	40–80	55–70	65–80
Lifespan	Medium	Very high	Low	Very high
Maintenance	Medium	Very low	High	Very low
Noise	Medium	Quiet	Noisy	Very quiet
Speed regulation	Fair	Good	Poor	Excellent
Starting torque	Very high	Low–medium	High	Very high

Source: Adapted from Groschopp (2022).

4.7 Summary

A mechatronic system takes its actions in applications by its actuating system. An actuating system consists of various actuators, and each actuator is capable of converting energy in one form such as electric, hydraulic, pneumatic, solar, and thermal energy into another form that can be controlled to fulfill a task. Actuators are a part of the mechatronic systems to do real work in the physical world.

The primary function of an actuating system is to produce output energy in a controllable manner to do work in applications. Therefore, actuators are highly diversified due to the differences in sources and target sources; energy can be converted in many different ways including mechanical, electromechanical, thermomechanical, electro-stimulated, MRF, and nano-level conversions. Actuating systems can be classified based on the forms of source and target energy, and the approach and the mechanism to implement energy conversions. DVs in designing an actuator are determined by the selection of energy conversion.

Implementing an actuating system requires that of a control subsystem and an energy conversion subsystem. Valve actuators and electric actuators are the most commonly used components of controlling and energy-converting mechanisms in actuating systems. Selecting valves or electric actuators for a mechatronic system must take into consideration of the requirements on output voltage, speed, power, efficiency, lifespan, maintenance, noise, regulation, and starting torque simultaneously.

Problems

4.1 A solid-state actuator produces a free input stroke of $u_{ISA} = 120\ \mu$m with an internal stiffness $k_i = 400 \frac{\text{kN}}{\text{mm}}$; what is the external stiffness coefficient when the output energy is 0.4 J?

4.2 A solid-state actuator produces a free input stroke of $u_{ISA} = 200\ \mu$m with an internal stiffness $k_i = 860 \frac{\text{kN}}{\text{mm}}$; what is the condition to produce maximum output energy?

4.3 Assume that the air density $\rho = 1.255 \left(\frac{kg}{m^3}\right)$, the velocity of wind $v_w = 6.5 \left(\frac{m}{s}\right)$, and the radius of disc as $R = 8.5$ (m) in Figure 4.11, determine (1) the total power of wind flow in disc area and (2) the power density per area.

4.4 Assume that the air density $\rho = 1.255 \left(\frac{\text{kg}}{\text{m}^3}\right)$, the radius of the disc of a horizontal-axis wind turbine is $R = 80\ m$, and the wind velocity at its inlet is $v_w = 12.5$ m/s, determine the Betz limit for the maximized extracted energy.

4.5 Determine the voltage of an open circuit between two brushes that make contacts on a rotating Faraday's disc in a magnetic field of B as shown in Figure 4.14 when $r_1 = 1$ cm, $r_2 = 5$ cm, $\omega_r = 628 \frac{\text{rad}}{\text{s}}$, and $B = 5$ Tesla.

4.6 As shown in Figure 4.37, a conductor has a length of 0.5 (m) situated at the right angle to a uniform magnetic flux/field of $B = 1.0$ Tesla, it moves at a velocity of 20 m/s, determine (*a*) the induced electromagnetic field (emf) in the conductor and (*b*) the electromotive force (emf) when the conductor moves at an angle of 60° with respect to the magnetic field.

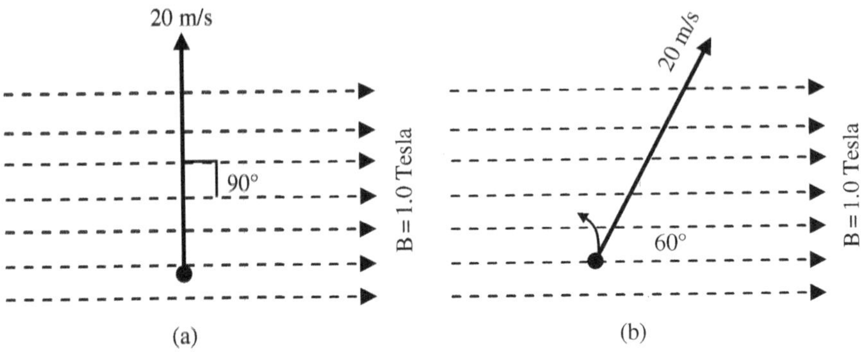

Figure 4.37 Calculation of emf in Problem 4.5 (a) 90 degrees and (b) 60 degrees

4.7 The wing space of an aircraft is 60.0 (m), and the aircraft is flying horizontally at a speed of 1200.0 km/h. Assume that the Earth' magnetic field is vertical with the strength of $B = 6.5 \times 10^{-5}$ Tesla, calculate the voltage difference between the wing tip and the fuselage of aircraft.

4.8 Assume that a coil has 600 turns in windings, a magnetic flux $\phi = 7.5 \times 10^{-6}$ webers is produced by a current of $I = 12$ A, determine the inductance of the coil.

4.9 A wooden toroid has a diameter of 0.25 (m) with a cross-sectional area of 4.0×10^{-6} m^2; the relative permeability is $\mu_r = 1$. It is uniformly wound with a coil of 1600 turns and it carries a current of $I = 5$ A; determine (1) the self-inductance of the coil and (2) the induced emf when the current is uniformly reduced to zero in 2 ms.

4.10 An air-cored solenoid (i.e., $\mu_r = 1$) has a length of 0.6 m, a diameter of 0.05 m, and it is wound with 1200 turns; determine the stored energy when there is a current of $I = 5$ A.

4.11 A current of $I_1 = 35$ mA passes through a self-inductance coil with $L = 500$ mH, determine the released magnetic energy when the current drops to $I_2 = 10$ mA.

4.12 An actuator is set up as in Figure 4.14b as a lifting mechanism. The U-shape conductor is made of iron with a length of 0.4 m, a cross-sectional area of 1.2×10^{-3} m^2, and a coil is wound with 600 turns on each side. Assume that the permeability of the magnet is $\mu_r = 600$, calculate the current to produce a lifting force of 60 kg.

References

Abavisani, I., Rezaifar, O., and Kheyroddin, A. (2021). Multifunctional properties of shape memory materials in civil engineering applications: a state of the art review. *Journal of Building Engineering* https://doi.org/10.1016/j.jobe.2021.102657.

Agrawal, K.C. (2007). *Electrical Power Engineering Reference & Applications Handbook*, 1e. CRC Press.

Alipranties, D.C. (2014). Fundamentals of wind energy conversion for electric engineers. https://engineering.purdue.edu/~dionysis/EE452/Lab9/Wind_Energy_Conversion.pdf (accessed 12 September 2024).

Arshad, A., Jabbal, M., and Yan, Y. (2020). Synthetic jet actuators for heat transfer enhancement – a critical review. *International Journal of Heat and Mass Transfer* 146: 118815.

Askari, H., Xu, N., Barbosa, B.H.G. et al. (2022). Intelligent systems using triboelectric, piezoelectric, and pyroelectric nanogenerators. *Materials Today* 52 (2022): 188–206.

Bahiuddin, I., Mazlan, S.A., Shapiai, M.I. et al. (2018). A new constitutive model of a magneto-rheological fluid actuator using an extreme learning machine method. *Sensors and Actuators A* 281 (2018): 209–221.

Barkas, D.A., Psomopoulos, C.S., Papageorgas, P. et al. (2019). Sustainable energy harvesting through triboelectric nanogenerators: a review of current status and applications. *Energy Procedia* 157 (2019): 999–1010.

Brown, P. (2017). *Sensors and Actuators: Technology and Applications*. New York: Library Press.

Cao X, Xiong Y, Sun J, Xie X, Sun Q and Wang ZL (2023) Multidiscipline applications of triboelectric Nanogenerators for the intelligent era of internet of things, *Nano-Micro Letter*https://doi.org/10.1007/s40820-022-00981-8.

Ceyssens, F., Sadeghpour, S., Fujita, H., and Puers, R. (2019). Actuators: accomplishments, opportunities and challenges. *Sensors and Actuators A* 295 (2019): 604–611.

Cowan Dynamics (2024). Valve actuator types for industrial process control. https://www.cowandynamics.com/blog/valve-actuator (accessed 12 September 2024).

Emerson (2018). Fisher™ Sliding-Stem Valve Selection Guide. https://www.emerson.com/documents/automation/product-bulletin-fisher-sliding-stem-valve-selection-guide-en-126266.pdf (accessed 12 September 2024).

Emerson (2024). Control valve handbook, the sixth version. https://www.emerson.com/documents/automation/control-valve-handbook-en-3661206.pdf (accessed 12 September 2024).

Fang, F.R., Tang, W., and Wang, Z.L. (2016). Flexible nanogenerators for energy harvesting and self-powered electronics. *Advanced Materials* https://doi.org/10.1002/adma.201504299.

Fiore, C. (2020). Stepper motors basics: types, uses, and working principles. https://www.monolithicpower.com/stepper-motors-basics-types-uses (accessed 12 September 2024).

Gao, D., Thangavel, G., Lee, J. et al. (2023). A supramolecular geo-elastomer system for soft iontronic adhesives. *Nature Communications* 14: 1990.

GE (2024). AC motor selection and application guide. https://www.apprep.com/Site/images/motor_resources/motor_selection_guide.pdf (accessed 12 September 2024).

Giurgiutiu, V., Pomirleanu, R., and Rogers, C.A. (2000). Energy-based comparison of solid-state actuators. https://citeseerx.ist.psu.edu/viewdoc/download?doi=10.1.1.194.6479&rep=rep1&type=pdf (accessed 12 September 2024).

Groschopp (2022). The basics of motor selections. https://www.groschopp.com/wp-content/uploads/motor-selection-whitepaper-groschopp-1.pdf (accessed 12 September 2024).

Imdad, A., Li, X., Chen, X. et al. (2019). A review of electro-stimulated gels and their applications: present state and future perspectives. *Materials Science & Engineering* 103: 109852.

Janocha, H. (2004). *Actuators Basics and Applications*. Berlin Heidelberg New York: Springer-Verlag.

Kaliasgoldmedal (2024). Principles of electromechanical energy vonversion. https://kaliasgoldmedal.yolasite.com/resources/MAEM/MAEM_U_1.pdf.

Kaluvan, S., Thirumavalan, V., Kim, S., and Choi, S.B. (2016). A new magneto-rheological fluid actuator with application to active motion control. *Sensors and Actuators A: Physical* 239 (2016): 166–173.

Kanpur, H. (2017). Sensors and actuators. http://robo.sntiitk.in/assets/docs/Sensors-and-Actuators-1.pdf.

Pace, C. (2024). MFE3004 Mechatronics I – mechatronic systems design. www.um.edu.mt/courses/studyunit/MFE3004.

Pelz, G. (2003). *Mechatronic Systems Modelling and Simulation with HDLs.* ChichesterEngland: Wiley.

Rao, R. (2024). What is an actuator? types, principles, and applications. https://www.wevolver.com/article/what-is-an-actuator-principles-classification-and-applications (accessed 12 September 2024).

Redekar, A., Deb, D., and Ozana, S. (2022). Functionality analysis of electric actuators in renewable energy systems—a review. *Sensors* 22 (2022): 4273. https://doi.org/10.3390/s22114273.

Reif, K. (2015). *Automotive Mechatronics – Automotive Networking Driving Stability Systems Electronics.* Springer Fachmedien Wiesbaden.

Slametumy (2018). Electromagnetic induction. https://slametumy.wordpress.com/wp-content/uploads/2018/11/bab-7-induktansi.pdf (accessed 12 September 2024).

Wang, Z.L. (2017). On Maxwell's displacement current for energy and sensors: the origin of nanogenerators. *Mater. Today* 20: 74–82. https://doi.org/10.1016/j.mattod.2016.12.001.

Wang, Z.L. and Song, J. (2006). Piezoelectric nanogenerators based on zinc oxide nanowire arrays. *Science* 312 (5771): 242–246.

5

Sensing Systems

5.1 Introduction

A mechatronic system is the same as any other artificial system in the sense that it is designed for specific purposes in its applications. Moreover, a mechatronic system, or any of *mechatronic functional modules* (MFMs) in the system, fulfills its functions by providing or changing one or a set of physical quantities such as a motion, a state of an object, a force, and an amount of heat. The system or MFM implements its functions by inputting and transferring one of a set of source physical quantities based on specified governing equations or models (Bi 2018).

To verify if a given function is fulfilled adequately, the corresponding physical quantities as inputs, outputs, or associated variables in a system should be measured to ensure that governing equations and constraints of the system are fully satisfied when the system transfers inputs into outputs. As shown in Figure 5.1, it is in particular true when the control of a physical quantity in the system is in a closed-loop mode. The state of this physical quantity must be measured, so that the impact of the change of this physical quantity can be re-evaluated in each control cycle to take corresponding action(s) in the next control cycle.

A system involves a set of parameters and variables where a part of the variables are controlled to yield expected system outcomes. Since inputs and outputs of a system depend on each other, the system inputs must be known or given correctly if the excepted system outcomes are to be produced. System outputs must be known if unknown system inputs are determined. Both inputs and outputs of system must be known if the system models are to be calibrated. In all the above three scenarios, it becomes essential to measure a certain physical property and compare the measurement with the target value of the property at a given time. *A sensor* is an instrument or device to measure, record, display, or respond to the physical properties of system. A sensor is a type of transducer, as shown in Figure 5.2. A sensor measures a physical property by converting one form of energy into another form that can be understood by an observer.

We have discussed in Chapter 2 that one of the key characteristics of a mechatronic system is its *adaptability* and *resilience* to deal with changes and uncertainties in a complex environment. The controls of and adaptable or resilient system are usually *data driven* or *event driven*. A set of critical system variables are monitored, measured, and analyzed to empower the mechatronic system to achieve high-level *diagnosability*, *predictability*, and *adaptability*. The higher the level of intelligence of a system is, the larger the amount of data about the system state is obtained, and more sensors will be deployed in measuring these physical quantities. Due to the importance of sensing

Mechatronics for Complex Products and Systems: Project-Based Design Approaches for Robots, Cyber-Physical Systems, Digital Twins, and Other Emerging Technologies, First Edition. Zhuming Bi.

Companion website: www.wiley.com/go/bi/mechatronics

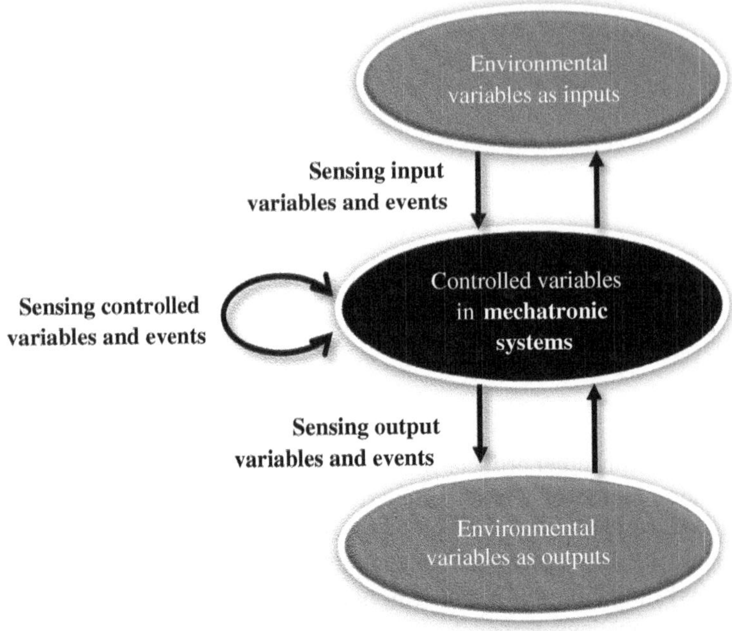

Figure 5.1 Variables to be measured in a mechatronic system

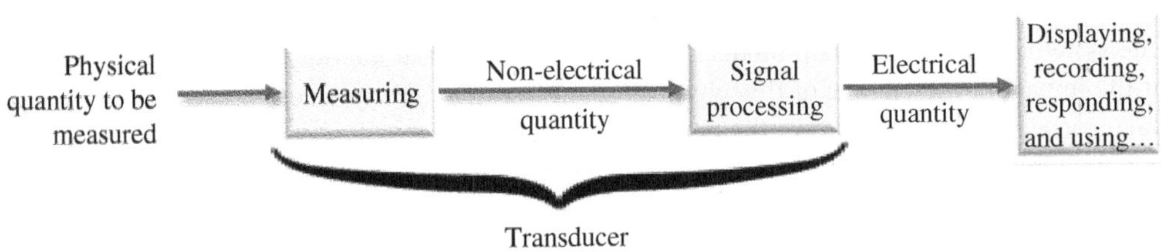

Figure 5.2 Sensor as a type of transducer

systems, more and more mechatronic systems are developed upon the infrastructure of *Internet of Things* (IoT). IoT makes sensors *omnipresent* since sensors are essential components of smart things such as cars, airplanes, cellular phones, machine tools, and home appliances.

The number of sensors reflects the advancement of a mechatronic system. Figure 5.3 shows an example of a group of the sensors in a passenger car where dozens of the sensors are equipped to measure system variables for the operations of all critical components. Figure 5.4 shows the evolution of smartphones over years. The functionalities and intelligence of a smartphone depended greatly on the types and numbers of the sensors that are embedded in the phone. It can be easily understood that more advanced functions rely on more feedbacks about environmental variables from sensors. Without the information about variables to be controlled and environmental factors to be affected, no human activities can be performed by machines automatically.

Figure 5.5 shows the conclusion by Precedence Research (2022) that the market size of sensors- was estimated as \$205.8 billion in 2022, and it would expand to reach \$508.64 billion in 2032 globally. It was expected at a *compound annual growth rate* (CAGR) of 8.4% from 2023 to 2033.

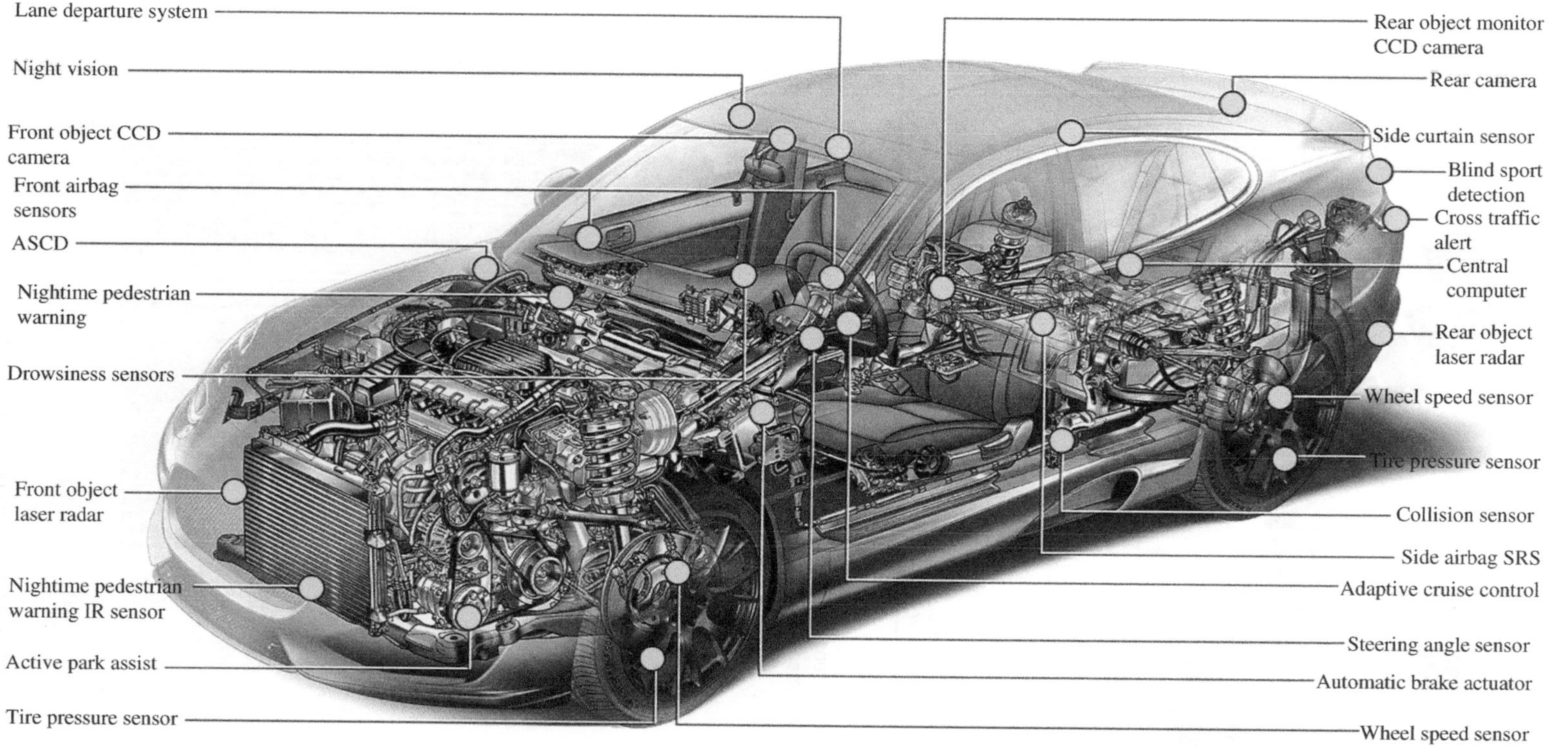

Figure 5.3 Commonly used sensors in a generic car (*Source:* Behance (2024)/Beaudaniels.com)

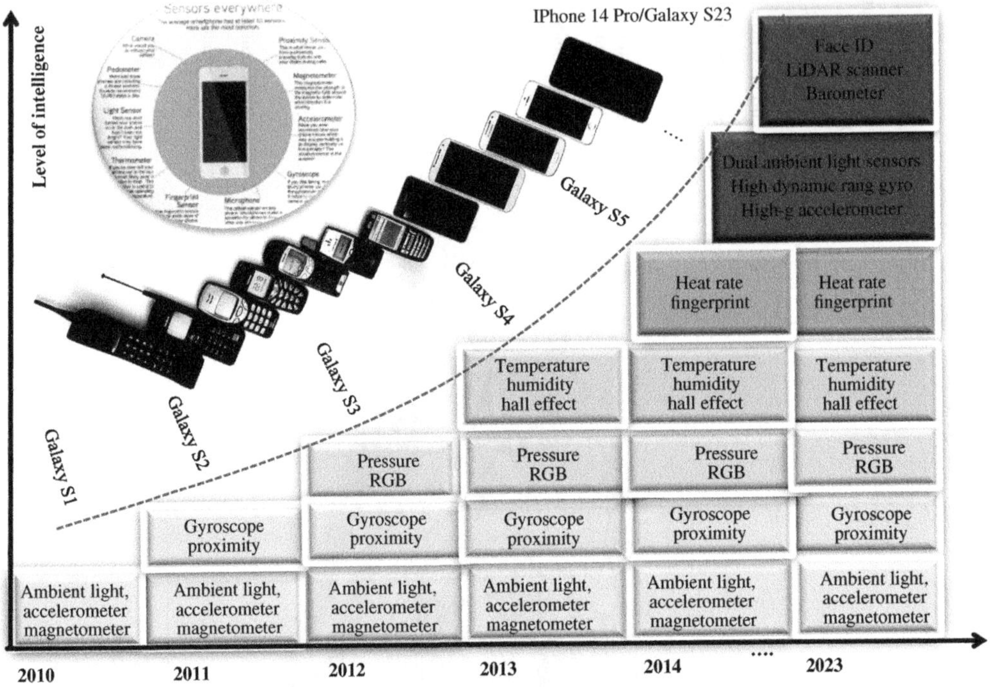

Figure 5.4 Number of sensors versus smartness of a cell phone (Qualcomm 2014; Apple 2024; Wiki 2024a) (*Source:* Jojhnjoy / Wikimedia Commons / Public domain)

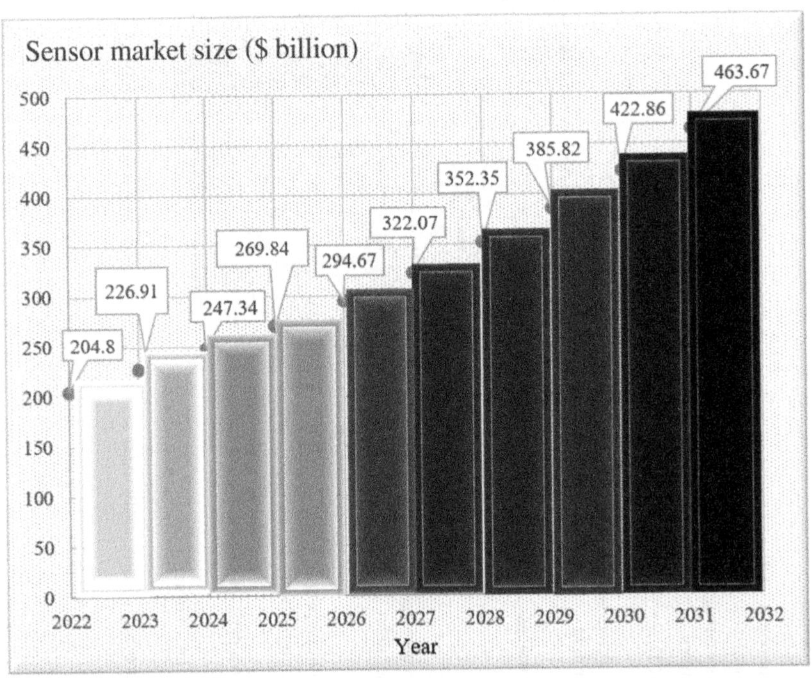

Figure 5.5 The growth of global sensor market predicted (*Source:* Adapted from Precedence Research (2022)

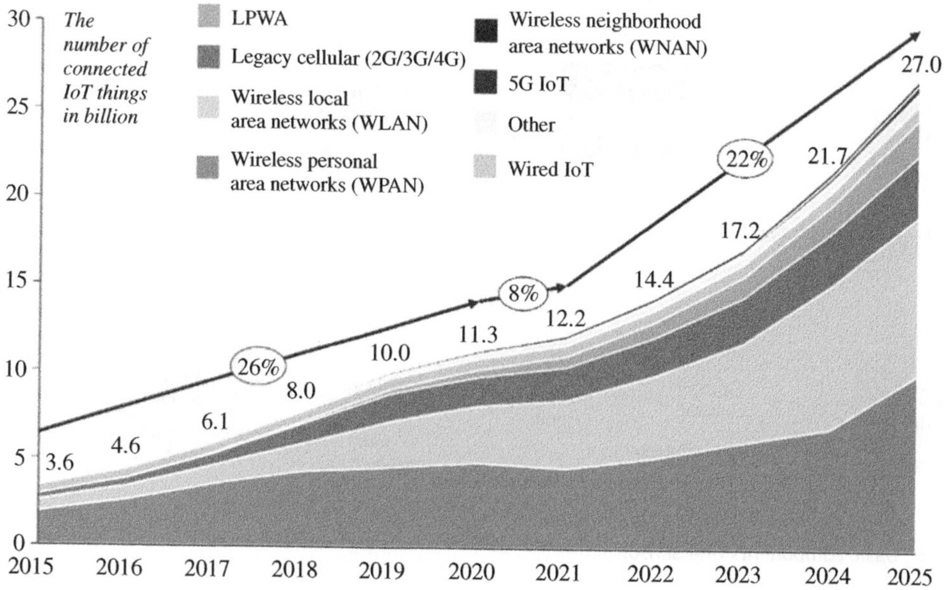

Figure 5.6 The growth of smart things in IoT over years (*Source:* Adapted from IoT Analytics (2024))

These growths were attributed to the ever-increasing demands for consumer appliances and electronics, surveillance systems, navigation, entertainment, sustainable systems, and IoT. For example, more and more *micro-electro-mechanical systems* (MEMSs) sensors equipped with bus capabilities and embedded processing have been widely applied in automotive, medical, manufacturing, defense, and military. These advanced sensors are becoming smart things in sensor networks.

In turn, the prevalent use of various sensors in various mechatronic applications has contributed to the growth of IoT significantly. Figure 5.6 shows the smart things that would be networked in IoT will reach 27 billion in 2025; as a comparison, that number was 3.6 billion in 2015 over the Internet.

5.2 Sensors, Actuators, and Transducers

Similar to an actuator, a sensor is also a transducer to convert energy in its primary form into an electric signal in another form that can be understood and used by an observer. From this perspective, both sensors and actuators are transducers as shown in Figure 5.7. Alternatively, a specific transducer can be either of *a sensor* or *an actuator.*

Figure 5.8 shows basic components of a sensor or a sensing system. It consists of (1) *a sensing element* that is used to measure a physical quantity in environment; (2) *a sign-processing unit* that is used to clean, process and, convert raw data into a readable format; and (3) *an A/D converter* that is used to translate an analogue signal to digital signal; a digital signal can be readily understood in a digital world.

In addition, a sensing element can be *active* where the power source such as light, audio, or electrical voltage is available to measure the physical quantity of interest. Moreover, when the desirable outcome is an electrical signal, the sensor actually converts raw data into a desirable electrical

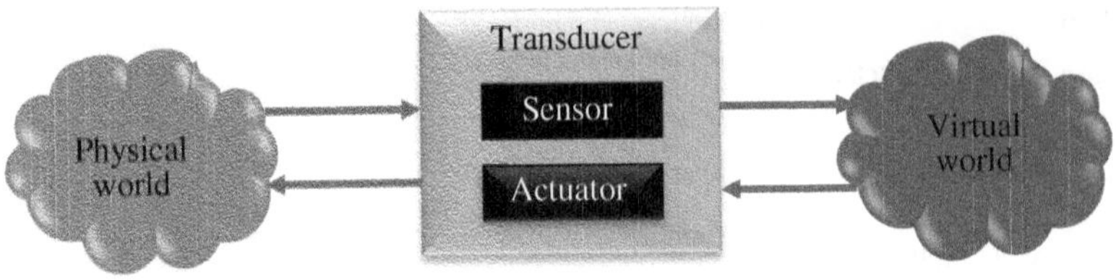

Figure 5.7 Both sensors and actuators are transducers

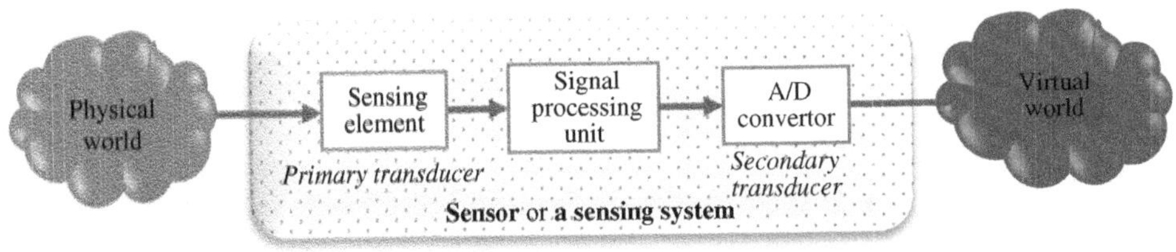

Figure 5.8 Basic components of a sensor

signal in two steps by *a primary transducer* performed by the sensing element and *a secondary transducer* performed by an A/D convertor, respectively.

Conventional transducers refer mainly to primary transducers that are reliable, relatively large-size, and based on mature technologies such as a thermocouple to measure temperature or a magnetic compass to show a direction. With the increasing demands to measure more physical quantities at *micro-* or even *nano-*levels, more and more micro-electronic transducers are developed. Some examples of commonly used micro-electronic transducers are photodiodes to detect lights, piezo-resistors to measure fluid pressures, and micro-accelerometers to detect accelerations of objects. The sensors at micro- and nano-levels are highly sensitive but less reliable or robust.

5.3 Classifications

Numerous sensors are used to measure billions of physical quantities. To design or select right sensors in designs of mechatronic systems, understanding the classifications of available sensors can be very beneficial. Sensors have been classified using many aspects. Sensors were also classified based on their applications, approaches to convert energy in different forms, types of output signals, and sensing principles (Bi and Wang 2010). Alternatively, Figure 5.9 shows four commonly used criteria in classifying sensors: (1) *signal types*, (2) *power supplies*, (3) *operation modes*, and (4) *types of physical quantities* to be measured.

5.3.1 Types of Quantities to be Measured

In a mechatronic system, as long as a relevant variable should be monitored, analyzed, or controlled, a corresponding sensor must be equipped to measure this variable. Moreover, a variable may represent one of the different quantities of the physical mechatronic system. Table 5.1 classifies commonly used physical quantities based on the domains of the representative physical behaviors.

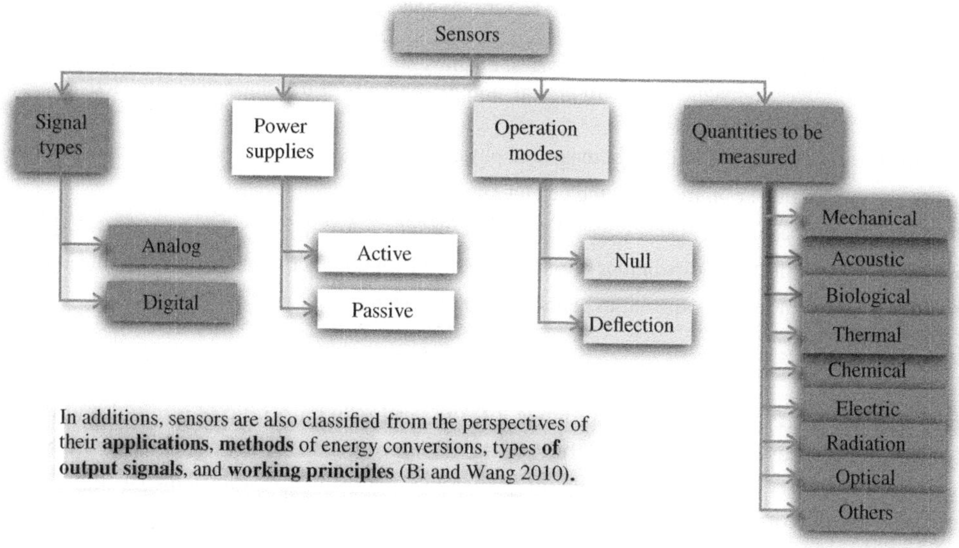

Figure 5.9 Classifications of sensors

Table 5.1 Classification of commonly used physical quantities in mechatronics

Physical behavior	Measured quantity
Mechanical	Displacement, position, velocity, acceleration, force, strain, stress, pressure, and torque
Electrical	Charge, voltage, current, electric field (i.e., amplitude, phase, and polarization), conductivity, and permittivity
Magnetic	Amplitude, phase, polarization, flux, and permeability
Thermal	Temperature, flux, specific heat, conductivity, convection, and radiation
Optical	Refractive index, reflectivity, and absorption
Acoustic	wave (amplitude, phase, and polarization), spectrum, and wave velocity
Biological* and *chemical	Fluid concentration, pressure, velocity, and contamination

5.3.2 Requirements Related to Measurement

Other than common *functional requirements* (FRs) for an MFM such as functions, lead-time, cost, reliability, and interfaces, a sensor should also satisfy some additional FRs relevant to the implementation of measurements. These additional FRs are commonly defined in terms of working principles, dimensions, formats of measurands, overall sizes, flexibility, and durability as shown in Table 5.2 (Mahdi 2014).

5.3.3 Specifications Related to Measurement

Basic *specifications* related to measurement include (1) *the accuracy* to reflect the measured error by comparing the measured and true values of a physical quantity, (2) *the resolution* to represent the

Table 5.2 Functional requirements (FRs) related to sensing operation

Working principles	are applied to measure a physical quantity, and commonly used working principles are electro-magnetic, electro-optical, and piezo-electrical transformation, and active or passive ultraviolet
Dimensions	refer to the number of measured physical variables for certain a physical phenomenon
Formats	can be *continuous* or *analog* and *analog* or *discrete*
Overall sizes	specifies allowable physical sizes of a sensor
Flexibility	reflects the capabilities of a sensor to deal with unanticipated changes and uncertainty
Durability	reflects the capabilities of a sensor for an extended lifespan

smallest possible increment a sensor can detect, (3) *the sensitivity* for a ratio of original and measured values of the measured quantity, (4) the *repeatability* or *precision* for the closeness of measured values in repeated tests, and (5) the *working range* in which a sensor is capable of measuring a physical quantity satisfactorily. Note that these specifications are different but also closely relevant, and Figure 5.10 illustrates their differences and relations.

To make an output signal readable and understandable, *a transfer function* is usually needed to transfer an input physical signal into an output electrical signal. In defining a transfer function, some basic specifications include (1) *the dynamic range* as the ratio of maximum and minimum amplitudes of measurand, (2) *the linearity* for the deviation of the output from a best-fit straight line of measurand, and (3) *the bandwidth* as a frequency range that is defined by specifying the lowest to the highest cutoff frequencies.

When a physical quantity is fluctuating over time, *hysteresis* is another specification that should be characterized. As shown in Figure 5.11, a hysteresis is a deviation of the measured output at a specified amplitude of an input signal when it is reached from two opposite directions.

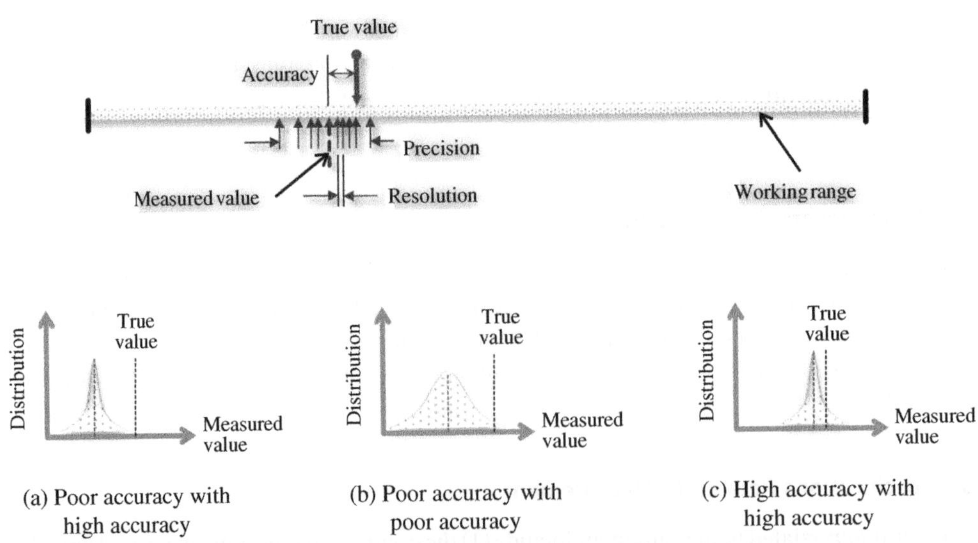

Figure 5.10 Basic measurement specifications of sensors

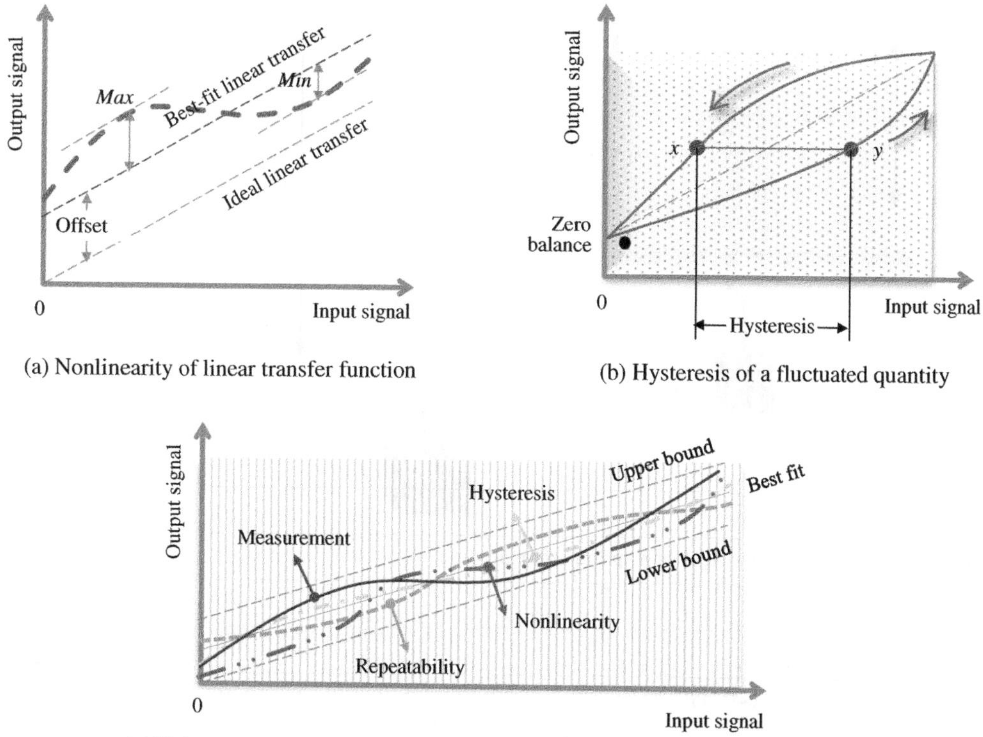

Figure 5.11 Hysteresis and nonlinearity of a fluctuating quantity

5.4 Working Principles

Depending on the nature of a physical quantity to be measured, an appropriate working principle should be selected to transfer the physical quantity into an understandable electrical signal accurately. To this end, a sensor includes one or more transducers. Transducers can be *mechanical*, *electrical*, *optical*, *magnetic*, or any other way that is capable of transferring an original signal into a readable signal. In this section, some commonly used working principles of transducers are introduced.

5.4.1 *Hooke's Law*

Most of the physical quantities, such as force, pressure, displacement, and velocity, are relevant to energy and power. In an elastic region, the response of an object to such a physical quantity is linear; the relations for forces and deflections are governed by *Hooke's law* of materials. Springs are the simplest models of elastic materials subjected to external loads. Springs have been widely used to measure force, displacement, deflection, and pressure.

Figure 5.12 shows the working principle of a spring. According to Hooke's law, the deflection (δ) of a spring element is proportional to the force (F) that is applied on the spring as,

$$F = k \cdot \delta \tag{5.1}$$

where k is the spring constant that is determined by the number of coils (N), spring index (C), wire diameter (d), and transverse elastic modulus (G) as (Budynas and Nisbett 2014),

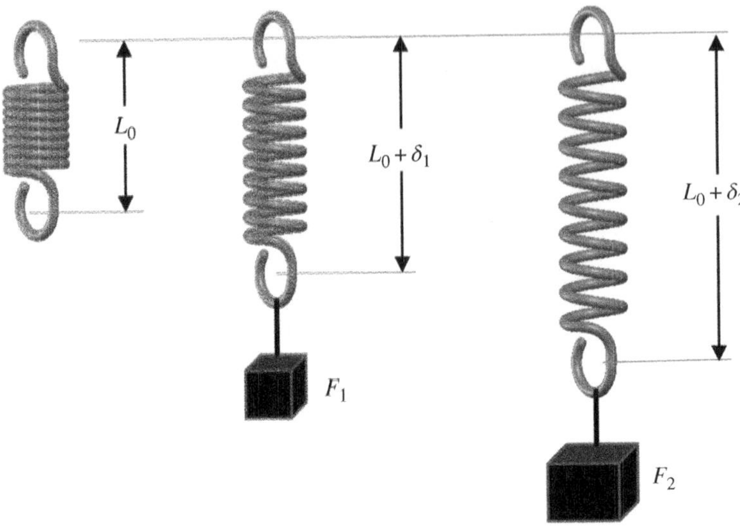

Figure 5.12 The working principle of a spring element

$$k = \frac{Gd}{8C^3N\left(1 + 1/\left(2C^2\right)\right)} \tag{5.2}$$

Example 5.1 A spring element is made of a hard steel wire with a transverse elastic modulus of $G = 7.85 \times 10^4$(MPa), the wire diameter is $d = 2.54 \times 10^{-3}$(m), and the spring index is $C = 6$, determine the number of coils (N) if the spring rate (k) is expected to be k =18.0 (kN/m).

Solution

Equation (5.2) can be used to determine the number of coils as,

$$N = \frac{Gd}{8C^3k\left(1 + 1/\left(2C^2\right)\right)} = \frac{(7.85 \times 10^{10})(2.54 \times 10^{-3})}{8(6)^3(18.0 \times 10^3)\left(1 + 1/\left(2(6)^2\right)\right)} = 6.3226 \text{ (coils)}$$

Many elastic solids can be simply represented by an equilibrium spring. As shown in Figure 5.13, a pressure sensor can be modeled as a spring element where the deflection of a thin film is proportional to the difference of the pressures on two sides.

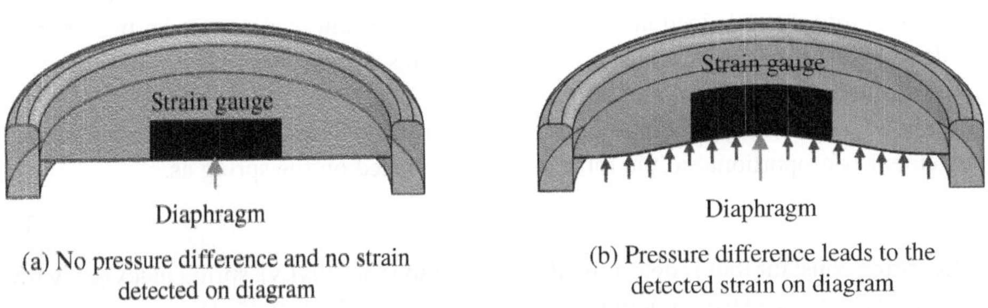

Figure 5.13 Schematic of a pressure sensor

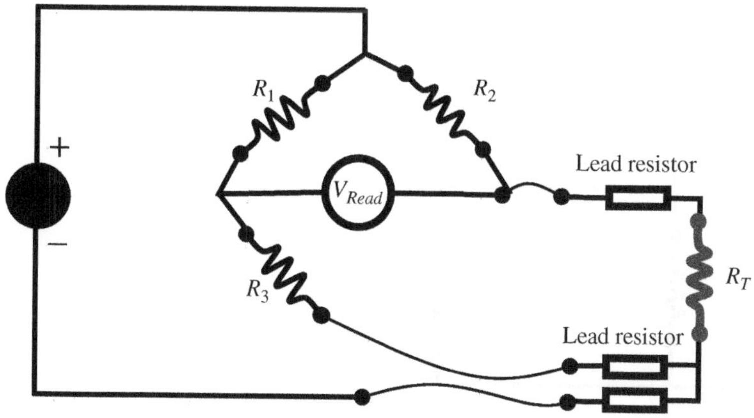

Figure 5.14 Ohm's law as working principle of a thermometer

5.4.2 Ohm's Law

Ohm's law is empirical, describing the relation of voltage (V), current (I), and resistance (R) in an electronic circuit as,

$$V = I \cdot R \tag{5.3}$$

Equation (5.3) involves in three variables. Assume that two of three variables in Eq. (5.3) are known, the equation can be used to determine the third one that is changed with a physical quantity to be measured. Figure 5.14 shows an example of using Ohm's law in measurement. It is a thermometer where ohm's law is applied to detect the change of temperature.

The change of temperature leads to the change of the resistance in a certain electrical path as,

$$R = R_0(1 + \alpha(T - T_0)) \tag{5.4}$$

where R_0 and T_0 are the reference values of resistance and temperature, respectively; R is the resistance when temperature is changed to be T; α is the coefficient of the thermal expansion of the material. Equation (5.4) can be rewritten to determine the temperature T based on the detected R as,

$$T = T_0 + \frac{R - R_0}{\alpha R_0} \tag{5.5}$$

Example 5.2 A sensing probe of a thermometer is made of platinum with a thermal expansion coefficient of $\alpha = 9.0 \times 10^{-3}$ (°C)). The reference resistance of the thermometer is $R_0 = 109.73\ \Omega$ at a reference temperature of $T_0 = 25$°C, determine the range of temperature change when the measured resistance is in the range of (80, 120)Ω.

Solution

Equation (5.5) can be used to determine the range of the temperature change as,

$$\begin{cases} T_{min} = T_0 + \dfrac{T_{min} - R_0}{\alpha R_0} = 25 + \dfrac{80 - 109.735}{(9.0 \times 10^{-3})(109.735)} = -5.108\ (°C) \\ T_{max} = T_0 + \dfrac{T_{max} - R_0}{\alpha R_0} = 25 + \dfrac{120 - 109.735}{(9.0 \times 10^{-3})(109.735)} = 35.394\ (°C) \end{cases}$$

Therefore, the range of the temperature is $[T_{min},\ T_{max}] = [-5.108, \quad 35.394](°C)$.

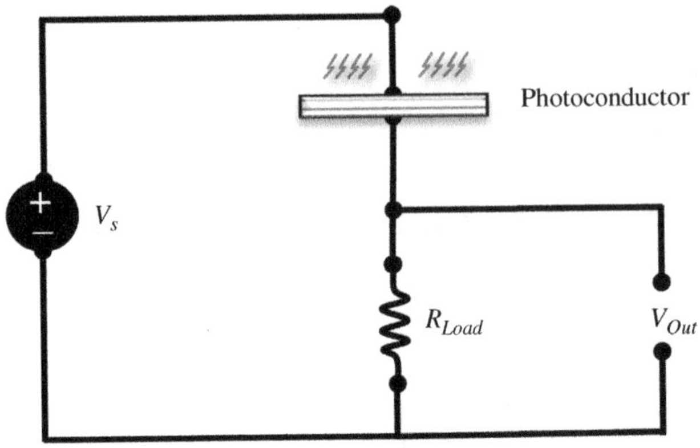

Figure 5.15 Photoconductivity subjected to optical stimulus

5.4.3 Photoconductivity

Photoconductivity refers to the dependence of optical and electrical behaviors where the electrical conductivity of a material is changed when it absorbs electromagnetic energy from visible, ultraviolet, or infrared (IR) light or from gamma radiation. To utilize the phenomenon of photoconductivity in measurement as shown in Figure 5.15, a photoconductor is made of a semiconductor material whose conductivity is changed with the number of free carriers generated by optical illumination.

Assume that a semiconductor has n and p numbers of electrons and hole concentrations, respectively, its conductivity (σ) can be determined by,

$$\sigma = e(\mu_e n + \mu_h p) \tag{5.6}$$

where σ is the conductivity ($/(\Omega \cdot \text{cm})$), e is the constant for electronic charge (1.6×10^{-19}C), and μ_e and μ_h are the mobilities of electrons and holes ($\text{cm}^2/(\text{second} \cdot \text{V})$), respectively.

In the absence of optical illumination, the numbers of electrons and hole concentrations are at an equilibrium state, and the corresponding conductivity of the material is called as *dark conductivity* (σ_0) as,

$$\sigma_0 = e(\mu_e n_0 + \mu_h p_0) \tag{5.7}$$

where n_0 and p_0 are the numbers of electrons and hole concentrations in an equilibrium state, respectively.

When the material is illuminated by a light with sufficient photon energy, carriers more than the numbers in the equilibrium state are generated, and the conductivity is changed as,

$$\Delta\sigma = e(\mu_e \Delta n + \mu_h \Delta p) \tag{5.8}$$

where $\Delta n = n - n_0$ and $\Delta p = p - p_0$ are the exceeded numbers of electrons and hole concentrations, respectively.

Note that the resistance (R_d) of a semiconductor depends on the conductivity (σ), thickness (l), and the effective area (A) as,

$$R_d = \frac{1}{\sigma}\frac{l}{A} \tag{5.9}$$

Moreover, the change of resistance (ΔR_d) corresponds to that of conductivity as,

$$\Delta R_d = -\frac{1}{\sigma^2}\frac{l}{A}\Delta\sigma \tag{5.10}$$

Substituting Eq. (5.9) into Eq. (5.10) yields,

$$\frac{\Delta R_d}{R_d} = -\frac{\Delta \sigma}{\sigma} \tag{5.11}$$

where $\frac{\Delta R_d}{R_d}$ and $\frac{\Delta \sigma}{\sigma}$ are the relative changes of resistance and conductivity, respectively.

Photoconductors are widely used to measure the responses to optical factors such as photo-resistors, photodiodes, and phototransistors; light detectors; and IR detectors.

5.4.4 Hall Effect

A Hall effect refers to a phenomenon of generating a potential difference when an electrical conductor is transverse to an electric charge (e) in a magnetic field (B) that is perpendicular to the current direction. As shown in Figure 5.16, the magnetic force (F_m) is perpendicular to the plane formed by the movement (v_d) and the magnetic field, and it is determined by,

$$F_m = e \cdot v_d \cdot B \cdot \sin(\theta) \tag{5.12}$$

where F_m is the magnetic force in *Newton*; e is the charge in Coulombs; v_d is the drift velocity of the charge in m/s; B is the magnetic field in Teslas $= N/(C \cdot m/s)$; θ is the angle from F_m to B in the plane of $v_d \times B$.

Example 5.3 Assume that a charge ($e = 50 \times 10^{-9}$ C) occurs to a rod by rubbing, it moves in the x-direction with a velocity of ($v_d = 10$ m/s) in the Earth's magnetic field ($B = 0.5$ Gauss $= 5.0 \times 10^{-5}$ Teslas) toward the y-direction as shown in Figure 5.17, determine the magnetic force on the rod.

Solution

Equation (5.12) can be used to determine the magnetic force on the rod as,

$$F_m = e \cdot v_d \cdot B \cdot \sin(\theta) = \left(50 \times 10^{-9}\right)(10)\left(5.0 \times 10^{-5}\right) \cdot \sin\left(90^{\circ}\right) = 2.5 \times 10^{-11}\ (N)$$

It can be seen that the magnetic force by the Earth's magnetic field is ignorable in many applications.

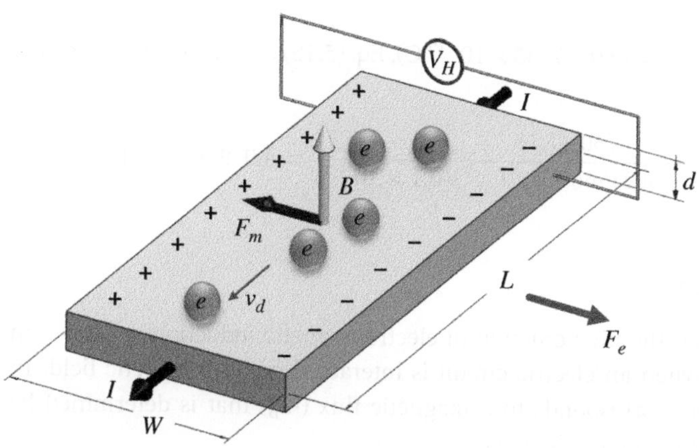

Figure 5.16 Hall effect and Hall voltage

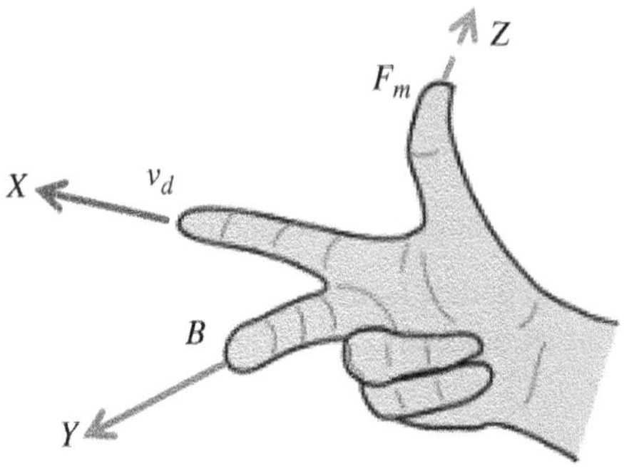

Figure 5.17 Magnetic force (F_m) on a moving charge (v_d) in a magnetic field (B)

A current (I) can be evaluated based on the drift velocity as,

$$I = n \cdot e \cdot A \cdot v_d \tag{5.13}$$

where n is the density of the charge carriers, $A = w \cdot d$ is the cross-sectional area of the wire, and w and d are the width and thickness of the wire, respectively.

Removing the drift velocity in Eqs (5.12) and (5.13) one gets,

$$F_m = \frac{B \cdot I}{\boldsymbol{n} \cdot A} \tag{5.14}$$

Moreover, the electric force by charge buildup can be found as,

$$F_e = V_H \frac{e}{w} \tag{5.15}$$

where V_H is the potential difference by the Hall effect (i.e., Hall voltage or Hall electromotive force (emf); w is the width of the cross section.

Based on the equilibrium condition where $F_m = F_e$, the potential difference, or the Hall voltage (V_H), can be calculated from Eqs (5.14), (5.15) with Eq. (5.13) as,

$$V_H = \frac{I \cdot B}{n \cdot e \cdot d} = \frac{\boldsymbol{n} \cdot e \cdot A \cdot v_d \cdot B}{n \cdot e \cdot d} = w \cdot v_d \cdot B \tag{5.16}$$

where V_H is the Hall emf or Hall voltage; I is the current in ampere; B is the magnetic field in Teslas; n is the number of coils; e is a charge; d and w are the thickness and the width of B.

Example 5.4 A Hall probe is made of a copper strip whose electronic density is $n = 8.5 \times 10^{28}$ elecotrons/m^3 and the charge for an electron is $e = 1.60217663 \times 10^{-19}$ C. The cross section of the strip is $w \times d = (2.0 \text{ cm} \times 0.1 \text{ cm})$; determine the Hall voltage when the current is $I = 50$ A and the magnetic field is $B = 1.1$ T.

Solution

Note that the charge of an electron is ($e = 1.60217663 \times 10^{-19}$ C), Eq. (5.16) can be used to determine the Hall voltage as,

$$V_H = \frac{I \cdot B}{n \cdot e \cdot d} = \frac{(50)(1.1)}{(8.5 \times 10^{28})(1.60217663 \times 10^{-19})(0.1 \times 10^{-2})} = 4.039 \times 10^{-6} \text{ (V)}$$

5.4.5 Faraday's Law of Induction

Faraday's law of induction describes the phenomenon of electromagnetic induction as shown in Figure 5.18. An emf is generated when an electric circuit is interacted with a magnetic field. In a magnetic field (B), a loop of wire corresponds to a magnetic flux (Φ_B) that is determined by an integral over a surface (Σ) bounded by the loop as,

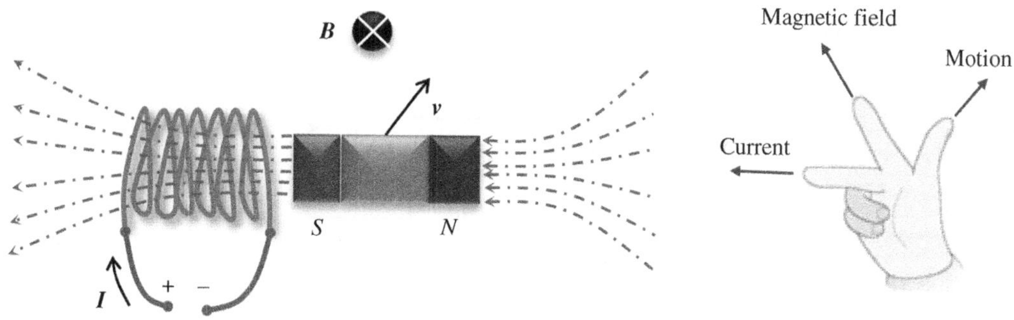

Figure 5.18 Faraday's law and right-hand rule

$$\Phi_B = \iint_{\Sigma(t)} B(t)\, dA \tag{5.17}$$

where dA is the finite element of surface area of a surface $\Sigma(t)$ that changes with time, $B \cdot dA$ represents the amount of flux through dA. The magnetic flux passing through the loop is proportional to the number of magnetic field lines through the loop. The change of magnetic flux in N number of loops induce an emf (ε) of,

$$\varepsilon = -N\frac{d\Phi_B}{dt} \tag{5.18}$$

where ε is the emf, and Φ_B is the magnetic flux. The right hand rule is used to determine the direction of the emf ε.

Example 5.5 Determine the induced emf when a coil has triple loops with a radius of 6.0 cm; the average $B(t)$ changed from 0.05 T to 0.25 T in 0.1 seconds.

Solution

Equation (5.18) can be used to determine the emf as,

$$\varepsilon = -N\frac{d\Phi_B}{dt} = -(3)\frac{(0.25-0.05) * \pi(6.0 \times 10^{-2})^2}{0.1} = 0.0678 \text{ (V)}$$

As shown in Figure 5.19, Faraday's law of induction also tells that a coil resists a change in a magnetic field by generating an opposite voltage or current, and the circuits for primary and secondary coils have the relations of,

$$\frac{V_S}{V_P} = \frac{N_S}{N_P} \tag{5.19}$$

where V_S, N_s and V_p, N_p are the voltages induced by the primary and secondary coils, respectively.

5.4.6 Curie–Weiss Law

Magnetism is the property of a material to respond to a magnetic field by attracting or repelling. Magnetic susceptibility (χ) measures how much a substance can get magnetized in a magnetic field. *The Curie–Weiss law* describes the magnetic susceptibility in a paramagnetic region when

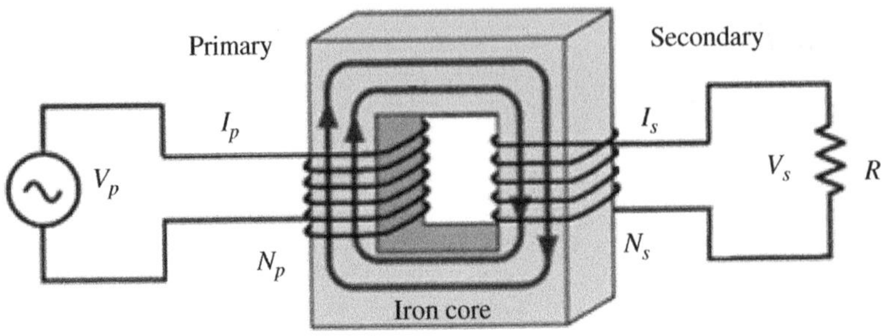

Figure 5.19 Faraday's law of induction for voltage transformation

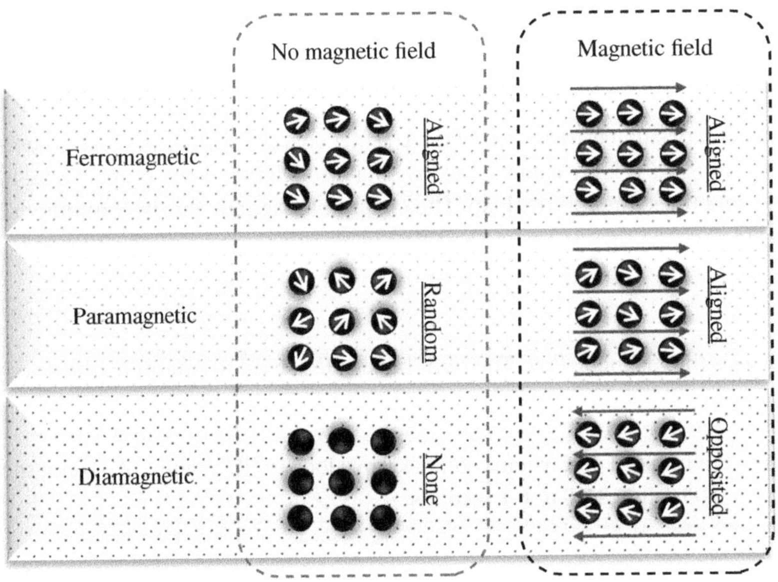

Figure 5.20 The Curie–Weiss law

temperature is above the Curie point. The magnetic susceptibility is dimensionless and inversely proportional to the ratio of the temperature change as,

$$\chi = \frac{C}{T - T_C} \tag{5.20}$$

where χ is the magnetic susceptibility (dimensionless); C is the material-specific Curie constant (Kelvin); T and T_c are the absolute temperature and material-specific Curie temperature in Kelvin, respectively. For example, the Curie temperatures for iron (Fe), gadolinium (Gd), and nickel (Ni) are 1043, 293, and 631 Kelvin, respectively. Assume that the singularity of susceptibility occurs when $T = T_c$, a ferromagnet is involved in spontaneous magnetization when $T < T_c$. as shown in Figure 5.20.

The magnitudes of the resulting magnetization (M) and the applied magnetic field (H) are corresponded with each other by the magnetic susceptibility (χ) as,

$$M = \chi H \tag{5.21}$$

where M is the resulting magnetization (A/m) and H is the applied magnetic field (A/m).

The Curie–Weiss law is very useful to detect the singularity in the susceptibility when the temperature equals the Curie temperature. The condition of singularity corresponds to a case when the law is applicable. Singularity is also referred as gravitational singularity or space–time singularity. The Curie–Weiss law also describes the magnetic susceptibility of a ferromagnet in the paramagnetic region above the Curie point. Note that the magnetic amplitude of a magnet determines the torque in an external magnetic field. No matter it is an electric current in a coil, a magnetic bar, a planet, a molecule, or an electron, it is involved in the corresponding magnetic moment.

5.4.7 Time of Flight (ToF)

Time of Flight (ToF) determines a displacement, speed, or relevant quantity by measuring the traveling time of a light, wave, or an acoustic or electromagnetic signal from an emitter to a receiver as shown in Figure 5.21. The time to receive a signal that is reflected back by the surface of a targeted object can be used to locate the targeted object relative to a sensor. ToF is used in electronics to implement *magnetic resonance angiography*, *mass spectrometry*, *near-IR spectroscopy*, and *planar Doppler velocimetry*.

ToF can also be used to model the relations of displacement, velocity, and acceleration of an object in kinematics, Figure 5.22 shows the trajectory of an object that is initially launched at a velocity of (v) at an angle θ. ToF (t) for the object to reach the same altitude can be determined by,

$$h(\theta, t) = v \cdot \sin(\theta) \cdot t - \frac{g}{2}t^2 \tag{5.22}$$

When $h(\theta, t) = 0$, the ToF can be found from Eq. (5.22) as,

$$t = \frac{2v \cdot \sin(\theta)}{g} \tag{5.23}$$

The maximum travel distance $s_{max}(\theta)$ is proportional to ToF as,

$$s_{max} = \frac{v^2}{g} \sin(2\theta) \tag{5.24}$$

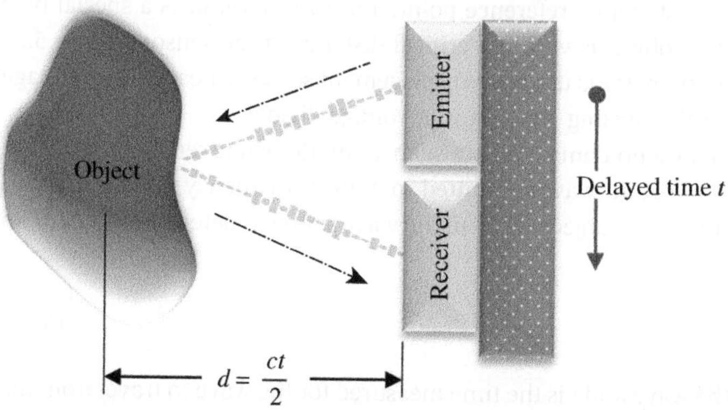

Figure 5.21 Time of flight (ToF) principle

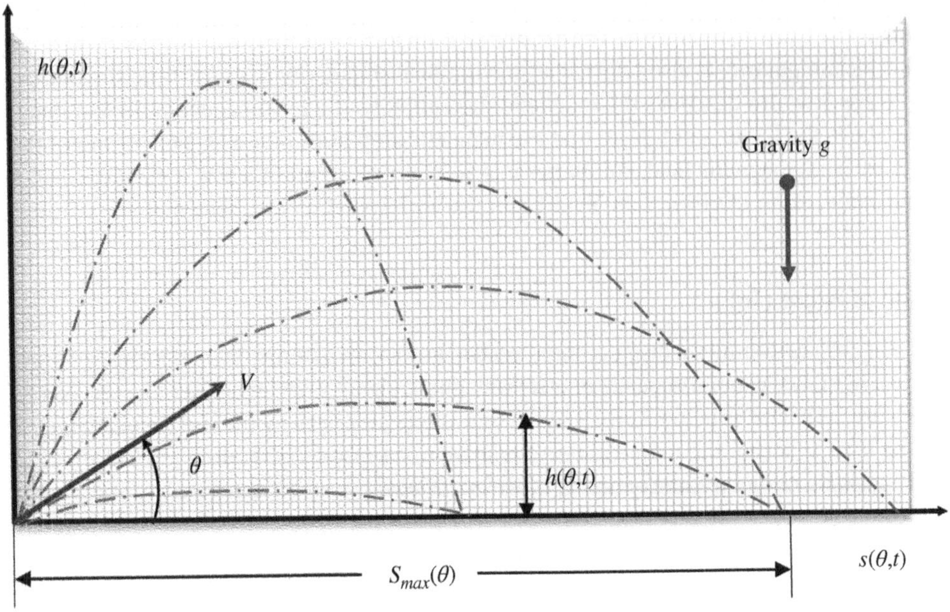

Figure 5.22 Flight angle versus flight distance

5.5 Types of Physical Quantities

The world consists of matter and energy; any physical quantity to be measured in a mechatronic system relates to matter and energy that are distributed through space and time. In this section, most common physical quantities in mechatronic systems are introduced and corresponding sensors to measure these quantities are discussed.

5.5.1 Displacement, Position, and Proximity

A displacement sensor measures a relative distance when an object moves. Displacement sensors differ from one to another depending on types of the motions to be measured. Four basic motion types are *translation*, *rotation*, *oscillation*, and *reciprocation*. For examples, *a position senor* measures an absolute distance of an object from a reference point; a *proximity sensor* is a special type of position sensor to determine if an object is within a critical distance of the sensor. Figure 5.23 shows an example in which a potentiometer is used as a displacement sensor whose output voltage is proportional to the resistance of the sensing element in a voltage divider.

A long distance can be measured by a no-contact sensor such as an ultrasonic sensor as shown in Figure 5.24. In an ultrasonic sensor, sound wave is emitted in a given frequency within a typical range of 2 ~ 13 MHz, and the distance of an object (d) from the wave source is determined by *ToF* in Figure 5.21 as,

$$d = \frac{v \cdot t}{2} \tag{5.25}$$

where v is the traveling speed of the wave and t is the time measured for the wave to travel from the emitter to the receiver. Both *sound navigation and ranging* (SONAR) and *radio detection and ranging* (RADAR) use the TOF principle to measure a distance but with different electromagnetic waves.

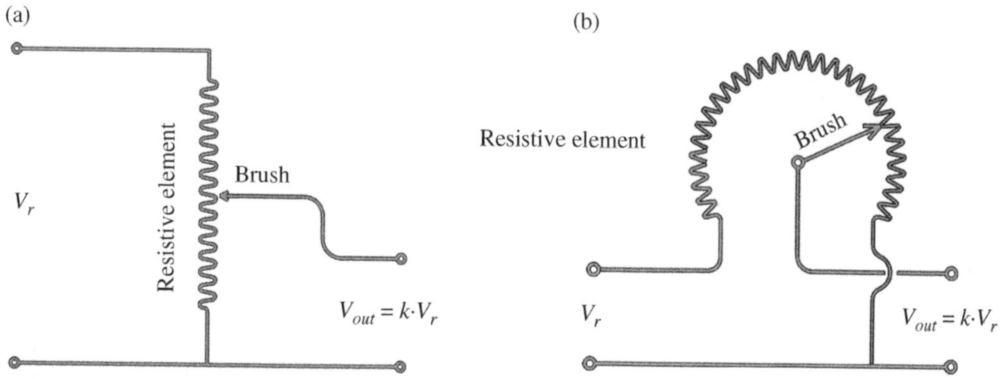

Figure 5.23 Voltage divider for position measurement

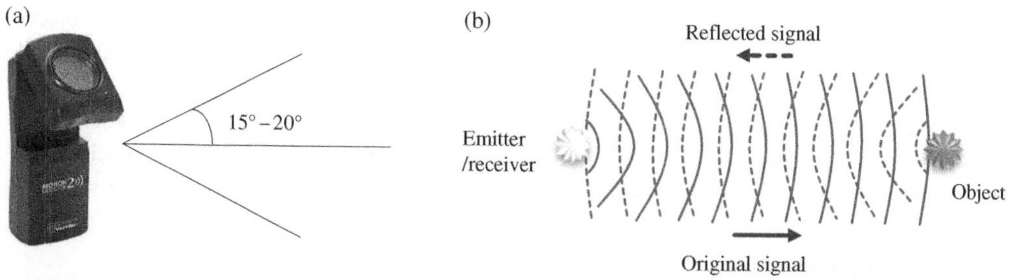

Figure 5.24 A sensor to measure long distance (*Source:* Flinn Scientific / https://www.flinnsci.com/motion-detector-2/tc2321/ / Last accessed on July 19, 2024)

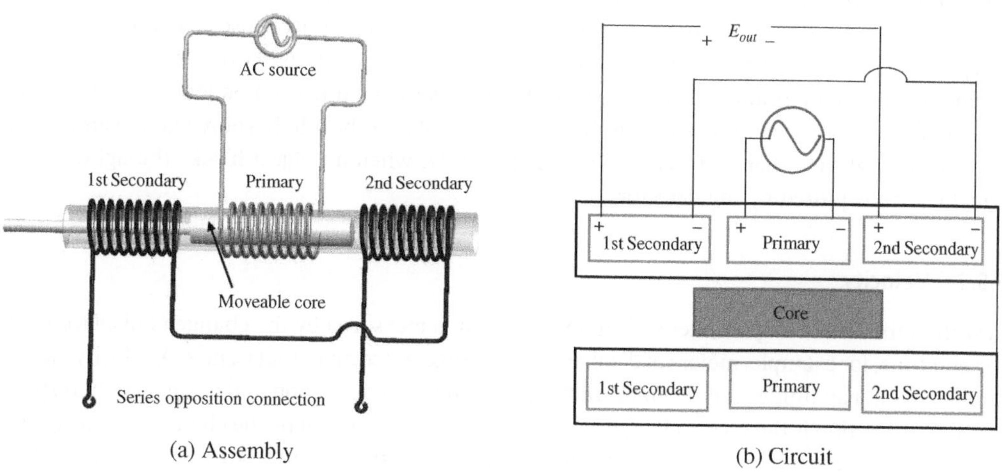

Figure 5.25 *Linear variable differential transformer* (LVDT)

A *linear variable differential transformer* (LVDT) is inductance-based. It is featured with "infinite" resolution. As shown in Figure 5.25, it has a ferromagnetic core that can move in three coils. No physical contact is made between the movable core and coil structure; therefore, it has a long life-span since no friction or wear is involved in operation. Either of *direct current* (DC) or *alternating*

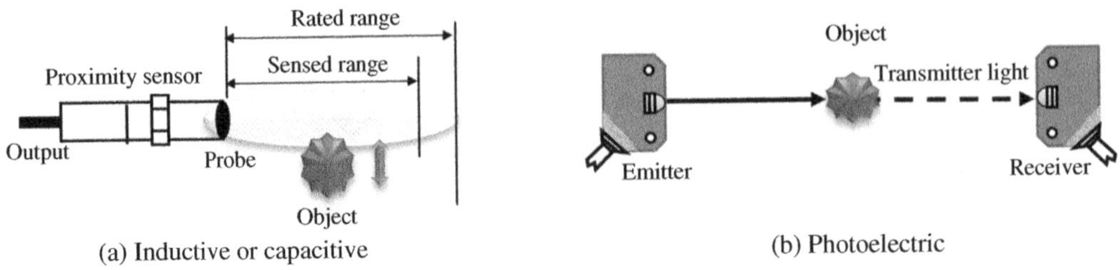

Figure 5.26 Examples of proximity sensors

current (AC) can be used in coils; typical frequency bandwidths are 250 Hz for DC-LVDT and AC-LVDT, respectively. LVDT has its linearity in operation and a typical range of accuracy is 0.1–1%. The measurement range is from units of millimeters up to 1 m.

A *proximity sensor* is of no-contact and used to detect the presence of an object in a specific range. A proximity sensor can be implemented by using light, sound, *IR* or an electromagnetic field in detecting an object. The output of a proximity sensor will be "0" or "1" for a state if an object is within the range of detection. Proximity sensors are widely used in vehicles, aircrafts, mobile devices, production lines, and security systems.

Most commonly used proximity sensors are either "*inductive*" or "*capacitive*." An inductive sensor uses an electromagnetic field, and it can only detect metal objects. In contrast, a capacitive sensor can detect objects with either of metal and nonmetal materials as long as the materials are able to carry electrical charges. Therefore, capacitive sensors are widely used to detect metal, wood, glass, water, or plastic objects.

Figure 5.26 shows that a proximity sensor can also be photoelectric. When an object is present in a light path between an emitter and a receiver, the receiver will not be able to receive transmitted light. The working principle of an ultrasonic proximity sensor is the same as a photoelectric proximity receiver except that an ultrasonic wave is used instead of light.

Kinney (2022) gives a comparison of four commonly used proximity sensors in Table 5.3.

A *photogate* is a special type of photoelectrical proximity sensor; it is used to detect the presence of an object at a certain location. It can also be used to measure the times or frequencies of the events that an object passes a specific location. It is also accomplished by a light source at one end and a light detector at the other end. As shown in Figure 5.27, when an object blocks the light, a step signal can be captured by the detector.

5.5.2 Velocity

Velocity is the rate of displacement of an object, and it is measured by the change of displacement per unit time, for example, meters of displacement change per second, that is, m/s. A velocity has its direction and magnitude, and it can be *angular* and linear for a rotation and a translation, respectively. Acceleration is the rate of change of velocity, and it is measured by the change of velocity per unit time, for example, meters of velocity change per second, that is, m/s^2. An acceleration also has its direction and magnitude.

Speed is the amplitude of a velocity in a specific direction, and tachogenerators are most commonly used device to measure a speed around a rotary axis. As shown in Figure 5.28, a *tachometer* or *tachogenerator* is electromechanical; it produces electrical power from mechanical energy when an axis rotates. The generated voltage (V_O) is proportional to the rotational speed. A tachogenerator determines the rotary direction by the polarity of its output voltage.

Table 5.3 A comparison of four commonly used proximity sensors

Type	Description	Sensing range	Sensing materials
Inductive	Noncontact to detect ferrous objects, ideally mild steel thicker than 1 mm. It consists of four major components: a *ferrite core* with *coils*, an *oscillator*, a *Schmitt trigger*, and an *output amplifier*.	4–40 mm	Iron, steel, aluminum, cooper, etc.
Capacitive	Noncontact to detect both metallic and nonmetallic objects in powder, granular, liquid, and solid forms. It is ideal for sight glass monitoring, tank liquid level detection, and hopper powder level recognition.	3–60 mm	Liquids, wood, granules, plastics, glass, etc.
Photoelectric	Noncontact to detect an object in a range by light; it consists of an emitter such as a light emitting diode (LED) or laser diode, a photodiode or phototransistor receiver, and supporting electronics for signal processing.	1–60 mm	Silicon, plastics, paper, metals, etc.
Ultrasonic	Noncontact to detect an object in a range by ultrasonics similar to a photoelectric proximity sensor.	3–30 mm	Cellophane, foam, glass, liquids, powders, etc.

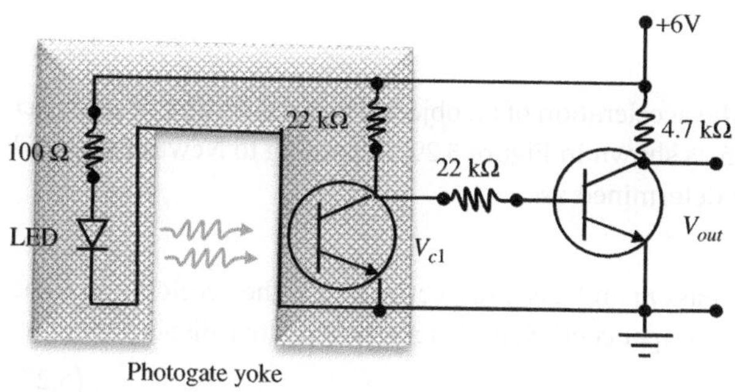

Figure 5.27 Circuit of a photogate sensor

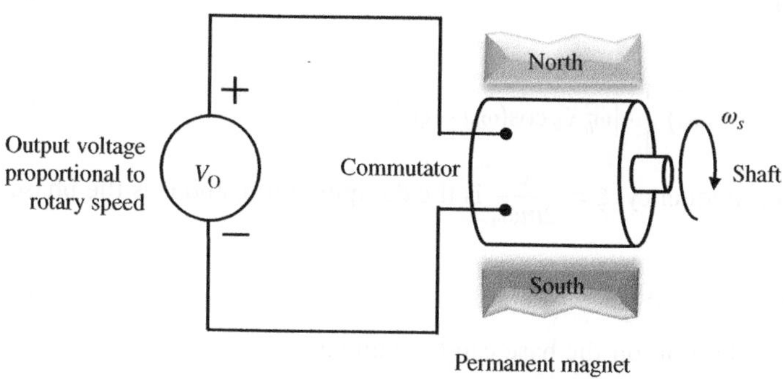

Figure 5.28 A tachogenerator

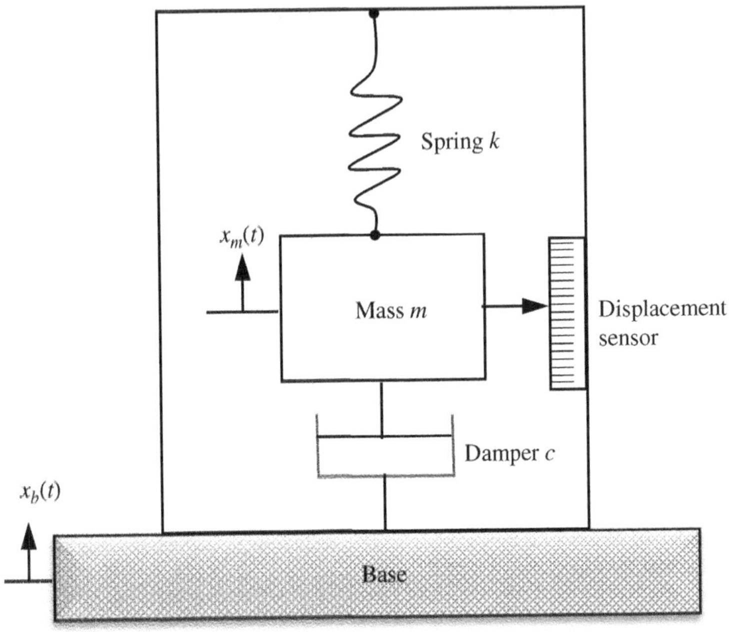

Figure 5.29 Structure of an accelerator

5.5.3 Acceleration

An *accelerometer* is designed to measure the acceleration of an object. The acceleration of an object can be measured by a mass–spring system as shown in Figure 5.29. According to Newton's second law, the force applied on mass m can be determined as,

$$ma = -k(x_m - x_b) - c(\dot{x}_m - \dot{x}_b) \quad (5.26)$$

where x_m and x_b are the displacement of mass m and base, respectively; a is the acceleration to be measured; k is the spring rate; and c is the damper coefficient. Let $z(t)$ be a relative displacement of,

$$z(t) = x_m(t) - x_b(t) \quad (5.27)$$

When the base is involved in a harmonic vibration of $x_b(t) = X_b \cos(\omega_b t)$ where X_b and ω_b are the amplitude and frequency of the vibration, respectively, Eq. (5.26) can be reformatted as,

$$m\ddot{z} + c\dot{z} + kz = m\omega_b^2 X_b \cos(\omega_b t) \quad (5.28)$$

Solving the above equations yields,

$$z(t) = \left[1/\sqrt{(\omega_n^2 - \omega_b^2)^2 + (2\xi\omega_n\omega_b)^2}\right] \cdot \left[\omega_b^2 X_b \cos(\omega_b t - \varphi)\right] \quad (5.29)$$

where $\omega_n = \sqrt{k/m}$ is the resonant frequency; $\xi = \dfrac{c}{2m\omega_n}$ is the damping ratio; and φ is the phase angle determined by,

$$\varphi = \tan^{-1}\left[2\xi(\omega_b/\omega_n)/\left(1 - (\omega_b/\omega_n)^2\right)\right] \quad (5.30)$$

The acceleration caused by the vibration on the base can be found as,

$$\ddot{x}_b(t) = -\omega_b^2 X_b \cos(\omega_b t) \quad (5.31)$$

When $\omega_b \ll \omega_n$, one has $\varphi \approx 0$ and,

$$\ddot{x}_b(t) = -\omega_n^2 z(t) \quad (5.32)$$

Equation (5.32) shows that the mechanical sensitivity of an accelerometer is approximately inverse of the square of the resonant frequency. The higher the resonant frequency ω_n is, the lower

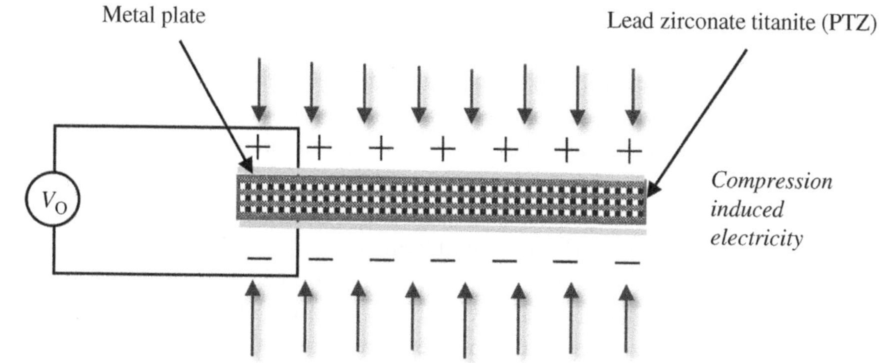

Figure 5.30 Piezoelectric effect in accelerometers

the mechanical sensitivity is. It is clear from Figure 5.29 that the most critical element in an accelerometer is the sensing element to measure the change of displacement over time.

The working principle of the majority of accelerometers is the piezoelectric effect due to their compactness as shown in Figure 5.30. The piezoelectric effect occurs due to a piezoelectric crystal when it is deformed when subjected to an external force. A force not only deforms the crystal but also causes its polarization; thus it forms an electric field inside the crystal. Therefore, an external force deforms the crystal and generates a voltage. The relation of the generated voltage and applied force can be modeled to convert the voltage output into acceleration.

Other ways to sense acceleration include capacitive effect, bubble effect, and photo effect. By all means, common accelerometers are classified based on the techniques to measure a displacement as *piezoelectric*, *piezoresistive*, *capacitive*, and *servo* accelerometers (Utmel Electronic 2024). Here, a servo accelerometer is introduced, while other three types will be covered in the next section since an acceleration is determined based on force measurements.

A servo acceleration sensor is a closed-loop sensing system to enhance dynamic performance and linearity and increase a dynamic range. As shown in Figure 5.31, the displacement of mass from its equilibrium position is sensed and amplified to generate a force for the mass to return to its home position. This improves the accuracy, expands ranges of measurement, and enhances anti-jamming capability. Therefore, servo accelerometers are widely used in inertial guidance and navigation systems; they are also used in high-precision calibration and vibration measurement.

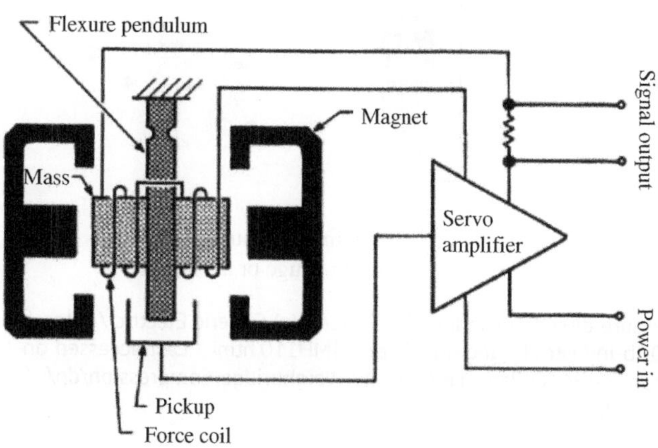

Figure 5.31 A servo accelerometer (*Source:* Fierceelectronics (2023)/Questex LLC)

Accelerometers can be mainly characterized by: (1) output types that can be a digital or analog voltage proportional to acceleration to be measured, (2) number of motion axes, (3) the amplitudes of measurands, (4) sensitivity, (5) bandwidth, (6) resistance/cache mechanism, and (6) accumulative errors. For example, a piezoelectric accelerometer has its advantages of easy fabrication, few components, and high resonant frequency; however, it has very high thermal transient sensitivity and high base strain sensitivity.

Every accelerometer has its characteristics that can be customized to a certain application. For example, a piezoelectric accelerometer has (1) a low cutoff frequency of 0.1–1 Hz for 5%; (2) a light and compact size that can be as light as 0.7 g for a miniaturized accelerometer; (3) a wide measurement range of up to +/− 500 g; (4) a range of sensitivity from 5 to 100 mv/g; (5) a broad bandwidth typically from 0.2 to 5 kHz; (6) the range of operating temperature from −70 °C to 150 °C. In comparison, a capacitive accelerometer has (1) a good performance when the frequency is low, and it can measure gravity; (2) heavier and bigger than a piezoelectric accelerometer with similar capability of measurements; (3) a measurement range of up to +/− 200 g; (4) a range of sensitivity from 10 to 1000 mv/g; (5) a broad bandwidth typically from 0 to 800 Hz; (6) a range of operating temperature from −65 °C to120 °C.

5.5.4 Force

A force can be measured *directly* or *indirectly*. Figure 5.32a and b shows the cases where the forces are measured directly by a scale and indirectly by a strain gauge, respectively. In applications, most of the forces are measured indirectly. An indirect force sensor measures a defect of the object instead of the applied force, and the measurement is not taken when the force is applied.

5.5.4.1 Direct Contact Sensors

A micro-switch detects a contact by a mechanical force directly. As shown in Figure 5.33, when a force is applied to a probe, the state of a mechanical contact is changed and detected.

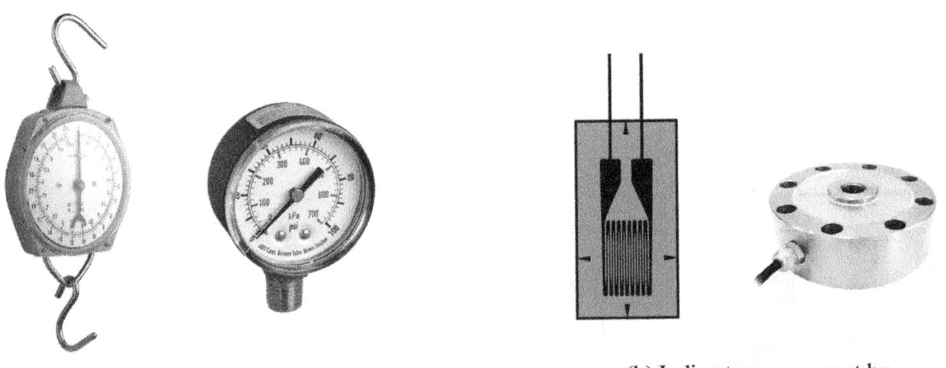

(a) Direct measurement by a scale or pressure gauge

(b) Indirect measurement by a strain gauge or a loadcell

Figure 5.32 Measurement of force or pressure directly or indirectly (*Sources:* (a) Generic Electric / https://www.webstaurantstore.com/avaweigh-110-lb-industrial-hanging-scale/334MH110.html / Last accessed on July 19, 2024; (b) ATO / https://www.amazon.com/ATO-Pancake-Capacity-Weighbridge-Compression/dp/B07VQGZMD8 / Last accessed on July 19, 2024)

(a) One directional (b) Multi-directional (c) Roller

Figure 5.33 Examples of micro switches: (a) one directional reed switch, (b) multidirectional reed switch, and (c) roller contact switch (*Source:* (a) Honeywell / https://www.amazon.com/Honeywell-BZ-2RW8244-A2-SWITCH/dp/B00MEIE9I0 / Last accessed on July 19, 2024; (b) Radwell International UK Ltd / https://www.radwell.com/Buy/SCHNEIDER%20ELECTRIC/SCHNEIDER%20ELECTRIC/XD5PA22 / Last accessed on July 19, 2024; (c) SUNS International / https://suns-international.myshopify.com/products/suns-international-z-15gq22-panel-mount-roller-plunger-15a-micro-switch-1 / Last accessed on July 19, 2024)

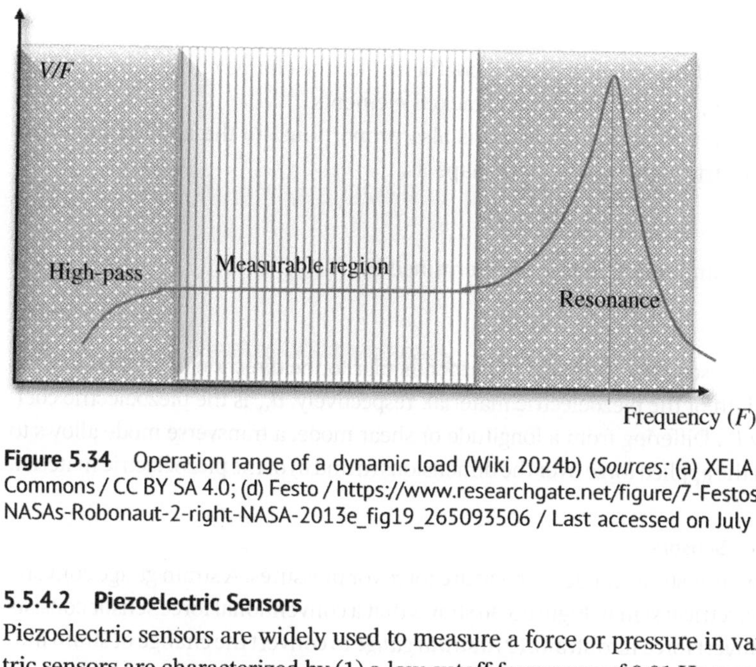

Figure 5.34 Operation range of a dynamic load (Wiki 2024b) (*Sources:* (a) XELA Robotics / Wikkimedia Commons / CC BY SA 4.0; (d) Festo / https://www.researchgate.net/figure/7-Festos-ExoHand-left-Festo-2012-NASAs-Robonaut-2-right-NASA-2013e_fig19_265093506 / Last accessed on July 19, 2024)

5.5.4.2 Piezoelectric Sensors

Piezoelectric sensors are widely used to measure a force or pressure in various MEMSs. Piezoelectric sensors are characterized by (1) a low cutoff frequency of 0.01 Hz even as they are inapplicable to measure static load; Figure 5.34 illustrates the operation range of frequency of a dynamic load (Wiki 2024b); (2) excellence in measuring a dynamic load in high frequency; (3) high resolutions; (4) compact size and light weight. However, piezoelectric sensors are affected by fluctuating operating temperatures and they become inapplicable when an operating temperature is too high.

A piezoelectric sensor is a type of direct force sensor that measures a force based on a piezoelectric effect in piezoelectric materials such as quartz and certain silicates. As shown in Figure 5.35, when a dynamic force is applied in a piezoelectric material, it causes a measurable flow of electrons along the surface of the crystal called as "nano-current" that is measurable. The measured current can be linearized and amplified to detect the applied force.

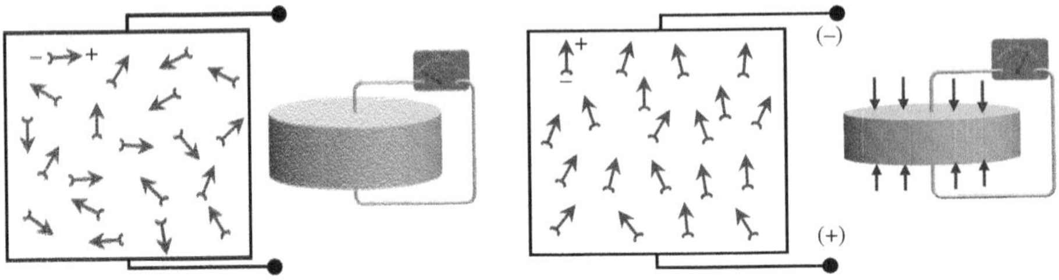

(a) Randomly distributed orientations in polarization domain

(b) Polarization in DC electric field when force in applied

Figure 5.35 Working principle of a piezoelectric force sensor

A piezoelectric material can be operated in one of the three modes, that is, *transverse*, *longitudinal*, and *shear*. In a longitudinal mode, the amount of charge displaced is strictly proportional to the applied force and independent of the piezoelectric element's size and shape. Therefore, the only way to change the charge output is to multiple elements (n) in series or in parallel as,

$$Q_x = d_{xx}F_x \cdot n \tag{5.33}$$

where Q_x is the charge output; d_{xx} is the piezoelectric coefficient for the charge released by the force (pC/N); F_x is the applied force; n is the number of stacked elements.

In a shear mode, the amount of charge displaced is also proportional to the applied force and independent of the piezoelectric element size and shape as,

$$Q_x = 2d_{xx}F_x \cdot n \tag{5.34}$$

In a transverse mode, the amount of charge is determined by,

$$Q_x = 2d_{xy}F_y \cdot b/a \tag{5.35}$$

where the force (F_y) is transverse; Q_x is generated in perpendicular to the direction of applied force; a and b are the depth and width of the piezoelectric material, respectively; d_{xy} is the piezoelectric coefficient for the charge Q_x by F_y. Differing from a longitude or shear mode, a transverse mode allows to customize the sensitivity of the applied force over the dimensions (a and b) of the piezoelectric material.

5.5.4.3 Conventional Force Sensors

A conventional force sensor uses strain gauges to measure forces or pressures. A strain gauge converts a force or pressure into an electrical signal. Figure 5.36 shows that a conventional force sensor consists of (1) *a spring element* to convert force into strain, (2) *a strain gauge* to convert the change of strain into that of resistance, and (3) *a bridge circuit* to convert resistance changes into voltage changes.

A strain gauge is a resistive device that changes its resistance when the shape of the attached object changes. In a bridge circuit, four strain gauges are connected as four resistors (i.e., R_1, R_2, R_3, and R_4) in a Wheatstone bridge. Under an equivalent condition when there is no load applied on the spring element, one has,

$$\frac{R_1}{R_2} = \frac{R_4}{R_3} \tag{5.36}$$

When a force is applied on a spring element, a strain occurs to each of strain gauges (R_i), which changes its resistance accordingly into $R_i + \Delta R_i$, where $i = 1, 2, 3,$ or 4.

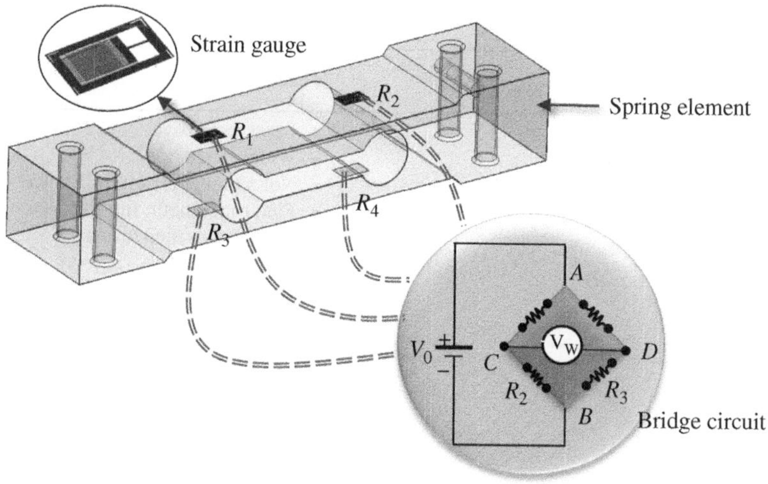

Figure 5.36 A strain gauge in a force sensor

$$V_W = \frac{R_1 R_2}{(R_1 + R_2)^2}\left(\frac{\Delta R_1}{R_1} - \frac{\Delta R_2}{R_2} + \frac{\Delta R_3}{R_3} - \frac{\Delta R_4}{R_4}\right)V_0 \tag{5.37}$$

where V_0 is the voltage of source power between A and B; V_W is the measured voltage between C and D.

Furthermore, when $R_1 = R_2 = R_3 = R_4 = R$, Eq. (5.37) can be simplified as,

$$V_W = \frac{V_0}{4R}(\Delta R_1 - \Delta R_2 + \Delta R_3 - \Delta R_4) \tag{5.38}$$

Strain gauges can be customized based on the forces to be measured (Choi et al. 2013). Table 5.4 compares three strain gauges that are designed to measure a force on bulk materials, a vibration on bending elements, and a strain subjected to large temperature changes, respectively.

5.5.5 Pressure

Pressure (p) is the force applied on a unit area of surface of an object; the force is distributed over the surface of the object. The International System of Units (SI) of pressure are N/m^2 and $kg/(m \cdot s^2)$.

A pressure sensor is designed to measure a pressure in applications. Pressure sensors can be classified by different aspects such as measurement modes or working principles. The mode of pressure measurement can be *absolute*, *gauge*, or *differential* as shown in Figure 5.37: (1) an *absolute pressure* is measured based on the reference of vacuum or zero point; one side of the senor measures the pressure of the medium and the other side is sealed as a vacuum; (2) a *gauge pressure* is measured by taking the reference of the atmospheric pressure; one side of the sensor measures the pressure of the medium and the other side is connected to atmosphere; (3) *a differential pressure* measures the difference between two sides of the pressure sensor.

Figure 5.38 shows that the pressure sensors can be implemented using different working principles including *resistive*, *capacitive*, *piezoelectric*, *optical*, or MEMSs. In these sensors, a mechanical displacement by pressure is turned into an electrical signal by (1) a *resistive sensor* that detects the change of electrical resistance of a strain gauge bonded to the diaphragm subjected to the pressure to be measured; (2) *a piezoresistive sensor* that corresponds to a resistive change of a semiconductor material due to the strain associated with the deflection of a diaphragm; it can

Table 5.4 Comparison of three strain gauge types

Type	Illustration	Features
Foil strain gauge	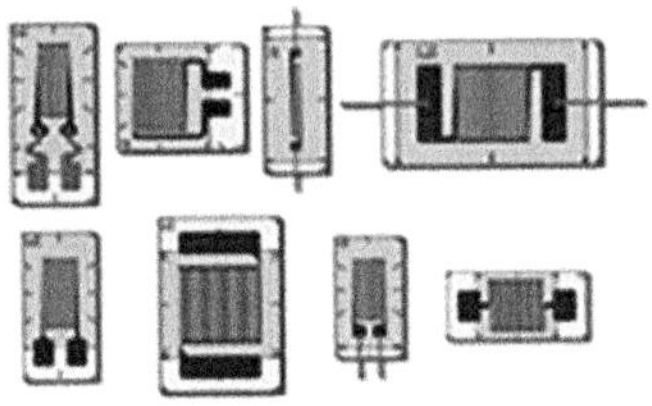	• Least expensive • Widely used • Not suitable for long distance • Electromagnetic Interference • Sensitive to moisture and humidity
Wire strain gauge	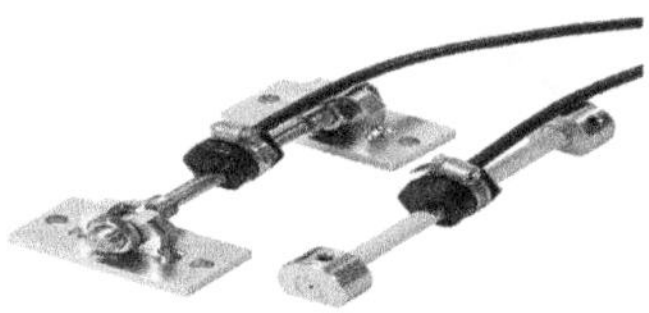	• Determines strain from frequency of an AC signal • Bulky
Fiber optic gauge	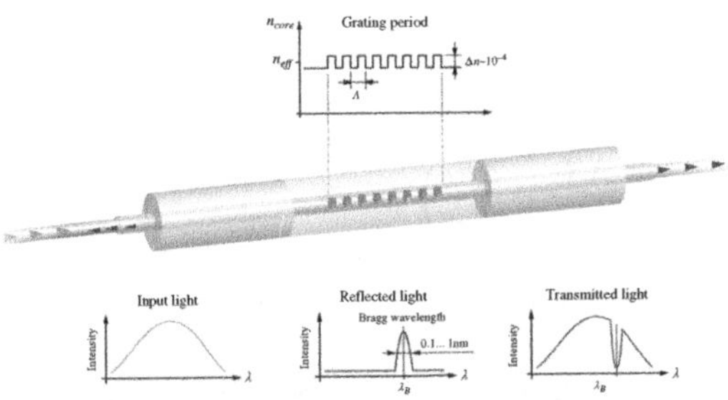	• Uses optical fibers • Is applicable to measure when subjected to large temperature changes • Compact size • High cost • Fragile

Source: Choi et al. (2013)/MDPI/CC BY 4.0.; Roctest Ltd / https://roctest.com/en/product/sm-5-series-surface-mount-vibrating-wire-strain-gauge-2/ / Last accessed on July 19, 2024.

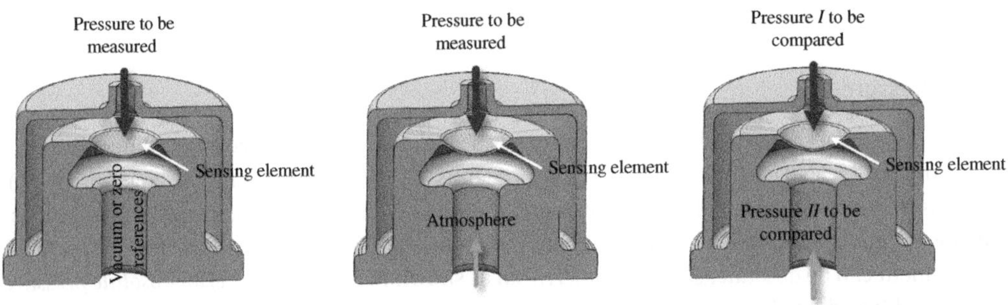

(a) Absolute mode (b) Gauge mode (c) Differential mode

Figure 5.37 Three modes of pressure measurement

measures pressure more sensitively since the magnitude of the change can be 100 times great than the resistance change in comparison with that in a resistive sensor; (3) *a capacitive sensor* that detects a change of a capacitance when a plate deflects subjected to pressure; it can measure a pressure as small as 10 mbar and carry a large overload in comparison with other types of sensors; (4) a*n optical sensor* that utilizes interferometry to detect the deflection of an optical fiber that is induced by pressure; due to its insensitivity to electromagnetic interference, an optical sensor can be used in a noisy environment such as in a radiography equipment; (5) *an MEMS*

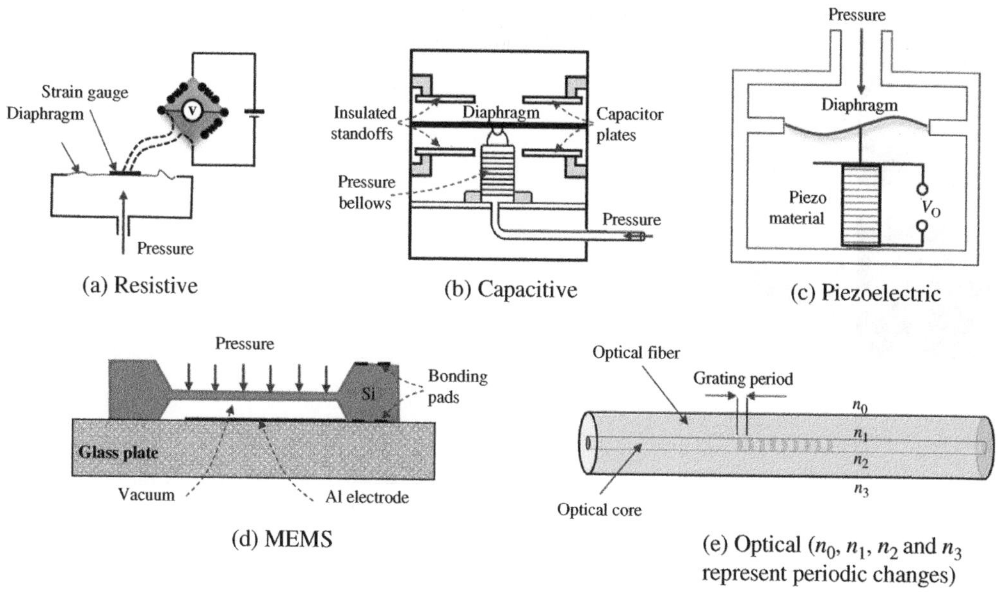

Figure 5.38 Five working principles of pressure measurement (*Source:* Adapted from Avnet (2023))

sensor that includes a micro-scale piezo or capacitive element to convert a pressure-induced deflection into an electrical output; an MEMS sensor is usually surface mounted with a small dimension of less than 2 mm by 2 mm.

5.5.6 Contacts

A contact reflects the state if a surface is touched by an external object or two objects are contacted; the contacts need to be detected in numerous applications such as typing on a keyboard or using a smartphone through its touch screen. *A touch sensor* is designed to detect and record a physical touch as shown in Figure 5.39. Touch sensors usually are simple, small, and low cost and are made to replace traditional mechanical switches. A contact can be sensed by the changes of *resistance, capacity, IR light,* or *acoustic wave* (Shawn 2019).

The location or amplitude of a contact should also be sensed. Such a touch sensor is commonly referred to as *a tactile senor.* A tactile sensor can be responsive to a force, pressure, or torque that makes a contact between two objects at certain locations. Tactile sensors are very popular in computer users' interfaces, home and security systems, and various robotic applications. Figure 5.40 shows some applications where touch sensors are used for force controls in robotic grasping.

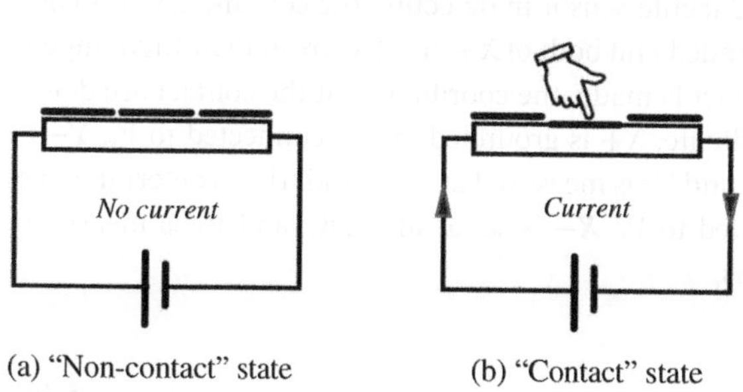

Figure 5.39 Structure of a touch sensor

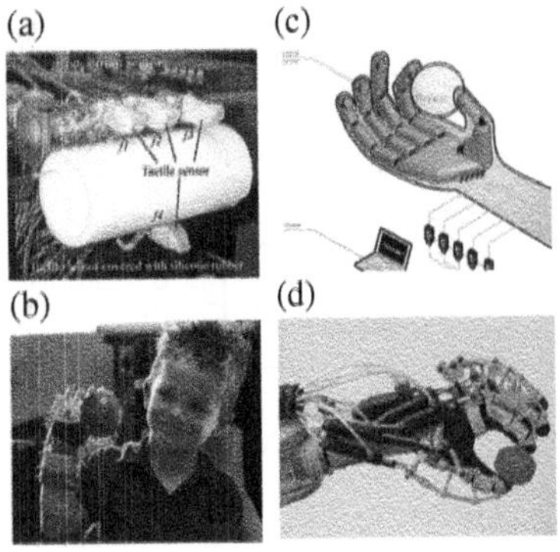

Figure 5.40 Examples of tactile sensors in robotic grasping (*Source:* (b) Intel Free Press/Flickr.com)

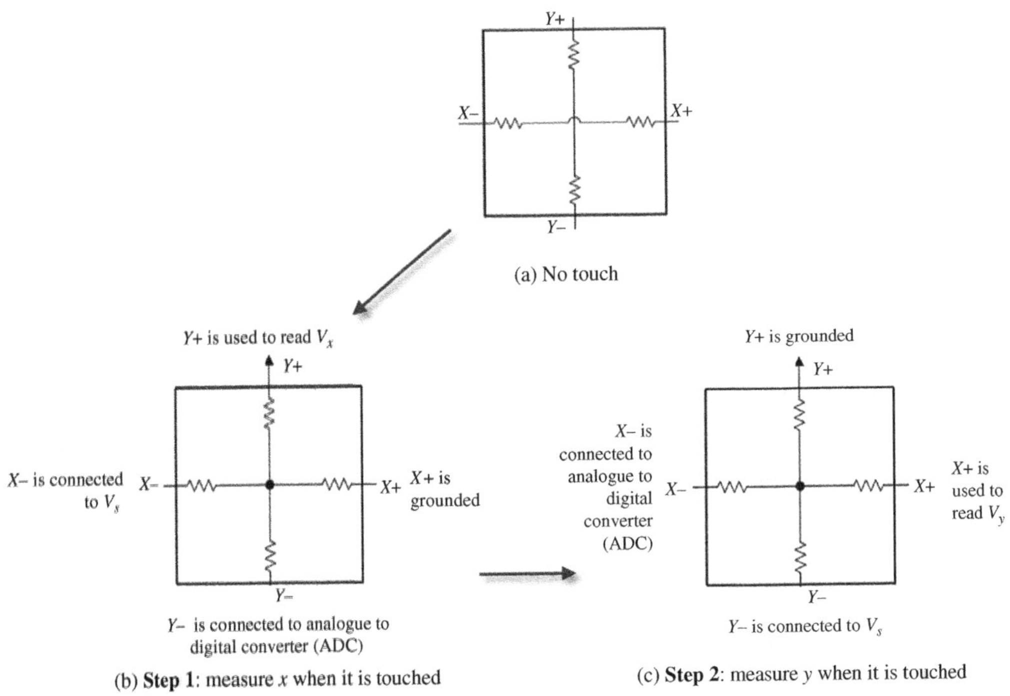

Figure 5.41 Determination of coordinates in two steps in a tactile sensor (*Source:* Alves et al. (2021)/ MDPI/CC BY 4.0)

Figure 5.41 shows the procedure of a 2 by 2 tactile sensor in detecting the coordinates (x, y) of a contact. When no contact is made, $X+$ is grounded and both of $X+$ and $Y+$ are at their high impedance states, and $Y-$ is a pull up. When a contact is made, the coordinates at the contact are determined in two steps: to determine the x-coordinate, $X+$ is grounded, $X-$ is connected to V_s, $\mathbf{Y-}$ is set as an analogue to digital converter (ADC), and Vx is measured at $Y+$. Similarly, to determine the y-coordinate, $Y+$ is grounded, $Y-$ is connected to V_S, $\mathbf{X-}$ is set as an ADC, and Vy is measured at $X+$. The coordinates are then calculated as,

$$\left.\begin{aligned} x &= \frac{V_x}{V_s}\cdot(\text{width of screen}) \\ y &= \frac{V_y}{V_s}\cdot(\text{height of screen}) \end{aligned}\right\} \tag{5.39}$$

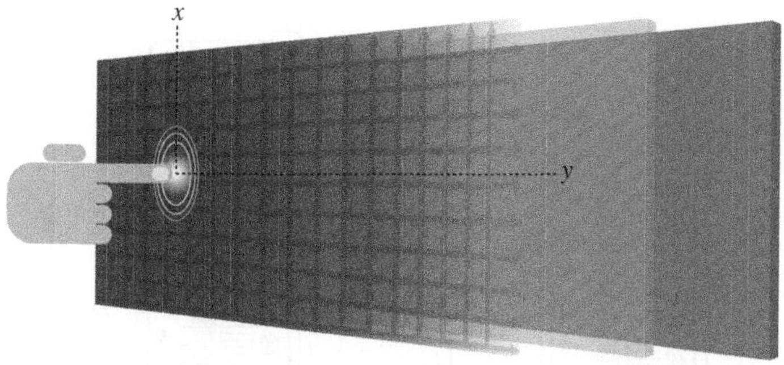

Figure 5.42 An infrared tactile sensor (*Source:* Alves et al. (2021)/MDPI/CC BY 4.0)

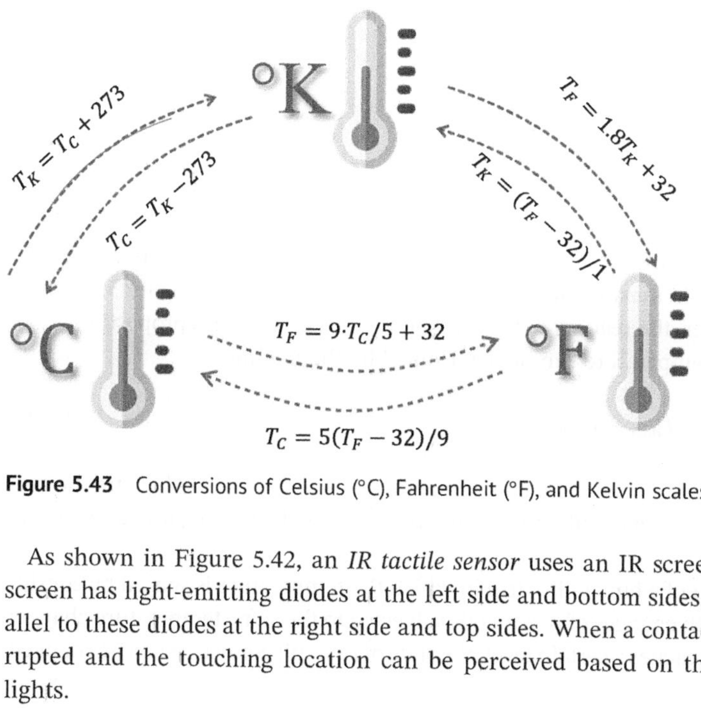

Figure 5.43 Conversions of Celsius (°C), Fahrenheit (°F), and Kelvin scales (°K)

As shown in Figure 5.42, an *IR tactile sensor* uses an IR screen to detect a contact. An IR screen has light-emitting diodes at the left side and bottom sides, and it has the receivers parallel to these diodes at the right side and top sides. When a contact occurs, the lights are interrupted and the touching location can be perceived based on the coordinates of interrupted lights.

5.5.7 Temperature

Temperature is a physical quantity of matter to expresses the perceptions of hotness and coldness quantitatively. The sensor to measure temperatures of matter is called as a thermometer. A thermometer can be calibrated in one of the three temperature scales, that is, (1) the Celsius scale (°C), (2) the Fahrenheit scale (°F), and (3) the Kelvin scale (°K). In particular, the Kelvin is used predominantly for scientific purposes as one of several base units in the SI (Wiki 2024c). Figure 5.43 shows the conversions among three temperature scales.

A temperature can be measured in numerous ways directly or indirectly since any matter will respond to temperature change in a form of dimension, resistance, stress, strain, or other physical quantities. Cottrell (2023) classified techniques for temperature measurements into *thermometers*, *probes*, and *noncontact measurements*.

A conventional thermometer uses mercury as the medium to detect a temperature change, mercury has the coefficient of volume expansion of 1.80×10^{-4}/°C, and the tube size of a thermometer

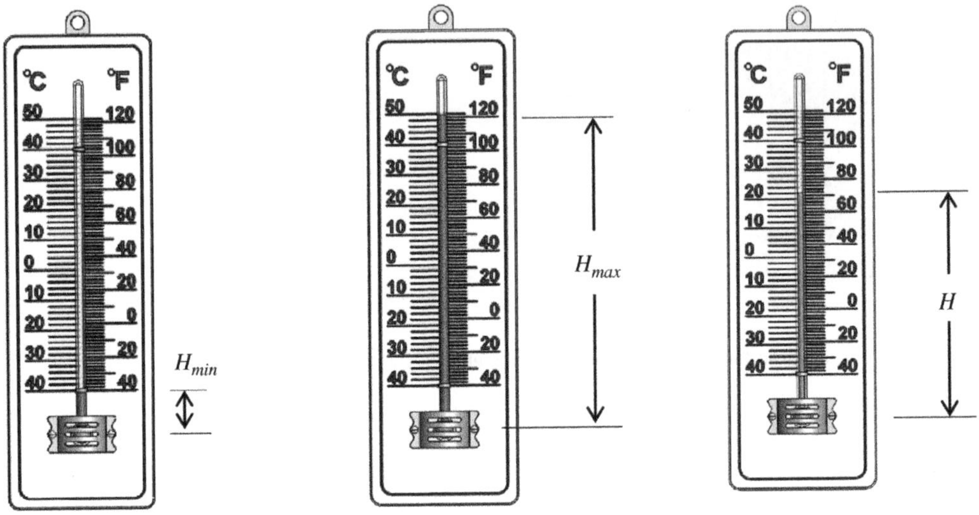

(a) Reading H_{min} at T_{min} (b) Reading H_{max} at T_{max} (c) Reading H at T

Figure 5.44 Measurement of temperature by a thermometer

can be designed so that the range of meter reading can be corresponded to that of the volumetric change between the minimal and maximal temperature readings. As shown in Figure 5.44, since the tube has a uniform diameter, the reading of the volumetric change becomes proportional to a temperature change, and the temperature can be determined by the reading as,

$$T = T_{min} + \frac{H - H_{min}}{H_{max} - H_{min}}(T_{max} - T_{min}) \tag{5.40}$$

where T_{min}, and T_{max} are the minimal and maximal temperatures and H_{min}, and H_{max} are the corresponding height readings at minimal and maximal temperatures; H is the height reading at the temperature T.

Figure 5.45. shows the design of a bimetallic strip in which the temperature is sensed by considering the difference of the length changes of two metals subjected to the same temperature change; when the temperature change reaches a certain level, the state of a circuit connector can be changed in a control system.

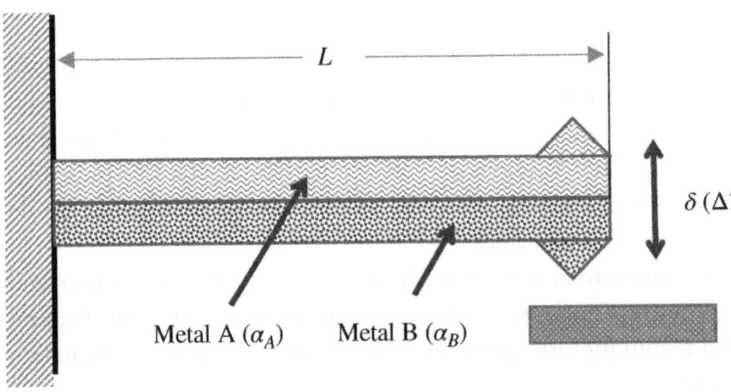

Figure 5.45 A bimetallic strip

$$\left.\begin{aligned} L_A &= L \cdot (1 + \alpha_A(T - T_o)) \\ L_B &= L \cdot (1 + \alpha_B(T - T_o)) \end{aligned}\right\} \tag{5.41}$$

where T_o is the reference temperature when the lengths of the plates of the two metals are at their neutral lengths of L; α_A and α_B are the thermal expansion coefficients of the two metals, respectively; T is the temperature to be measured; L_A and L_B are the lengths of the two plates subjected to a temperature change $(T - T_o)$. Since the two metal plates are bonded, the difference of L_A and L_B will cause a perpendicular deflection δ that will be proportional to the temperature change $(T - T_o)$; therefore, the temperature (T) is changed from the reference temperature T_o to a certain level, the deflection δ will trigger a state change of a circuit connector.

Early thermometers used liquid or gas as media to sense temperature; however, the invention of thermocouple created a new direction in advancing thermometry, and *resistance elements, thermopiles*, and *semiconductor materials* are widely used as probes or transducers in thermometers.

A thermocouple utilizes the thermoelectric effect for the conversion of the temperature change to the electric voltage. A thermoelectric effect can be one or a combination of three separate effects, that is, *Seebeck effect* associated with temperature and voltage, *Peltier effect* associated with voltage and heat, and *Thomson effect* emphasized on nonuniformity of the Seebeck effect.

These effects take effect when two or more different materials form a conductive junction as shown in Figure 5.46. Different pair of materials (A and B) can be used in a thermocouple and each material pair exhibits different characteristic of temperature range and voltage. Table 5.5 shows the material pairs of commonly used thermocouples and corresponding characteristics.

The voltage produced by a thermoelectrical effect is usually small in the millivolt range, and the relation of voltage and temperature can be defined as,

$$V = \int_{T_{ref}}^{T_{sense}} (S_1(T) - S_2(T))\, dT \tag{5.42}$$

where V is the voltage; T_{sense} and T_{ref} are the sensed and reference temperatures, respectively; $S_1(T)$ and $S_2(T)$ are the thermopower coefficients of the materials A and B, respectively.

Since the relation of voltage and temperature in Eq. (5.42) is nonlinear, a lookup table is required to correspond to the temperature to the voltage. HeyPete (2016) gives examples of the lookup tables for the thermocouples with different types in Figure 5.47.

5.5.8 Chemical Particles

Chemical materials are detected and measured mostly in chemical processes, mine explorations, and healthcare applications. However, more and more chemical sensors are used in mechatronic

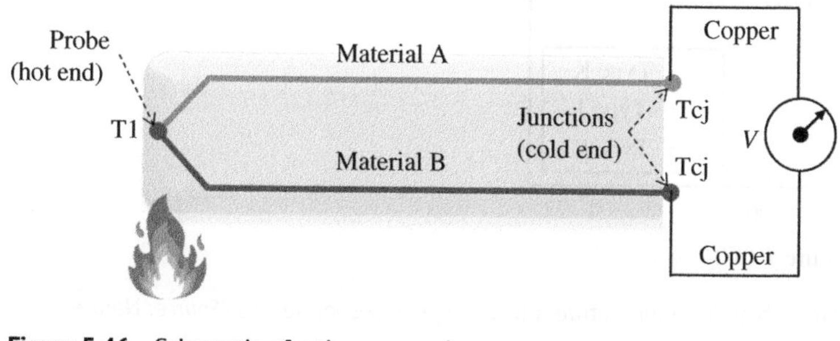

Figure 5.46 Schematic of a thermocouple

Table 5.5 Commonly used thermocouples and characteristics

Type	Materials	Minimal temperature (°C)	Maximum temperature (°C)	Minimal voltage (mV)	Maximum voltage (mV)
J	Iron constantan (Cu–Ni)	0	750	0	42.281
T	Copper constantan (Cu–Ni)	−250	350	−5.18	17.819
K	Cromel (Ni–Cr) Alumel (Ni–Al)	−200	1250	−5.891	50.644
E	Cromel (Ni–Cr) Constantan (Cu–Ni)	−200	900	−8.825	68.787
N	Nicrosil (Ni–Cr–Si) Nisi (Ni–Si–Mg)	−260	1300	−4.336	47.513
S	Platinum - 30% rhodium Platinum	−50	1768	−0.236	18.693
B	Platinum - 30% rhodium Platinum - 6% rhodium	0	1820	0	13.82
C	Tundsten - 5% rhenium Tungsten - 26% rhenium	0	2320	0	37.107

Source: Cottrell (2023)/Analog Devices, Inc.

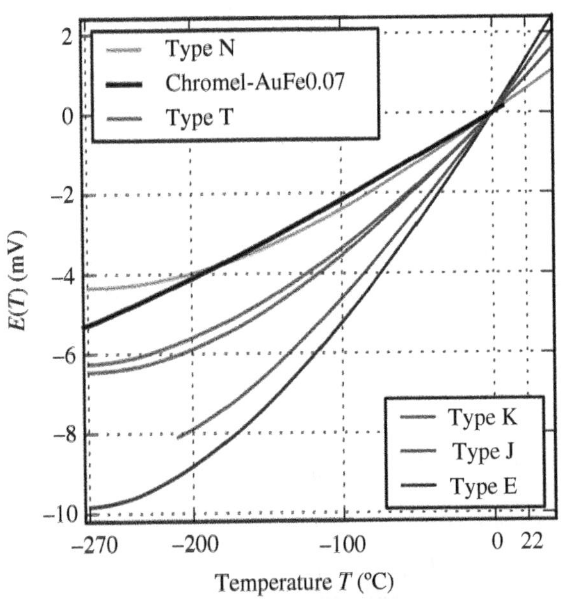

Figure 5.47 Examples of lookup tables for temperature and voltage correspondences (*Source:* Nanite/Wikimedia/CC BY 3.0)

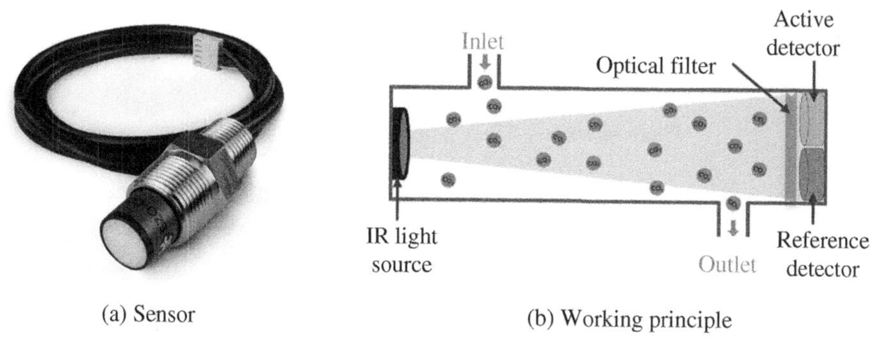

Figure 5.48 A CO_2 gas sensor (*Source:* Atlas (2024)/Atlas Scientific)

products due to the growing concern of sustainability of economy, society, and environment in sustainable manufacturing. Here, sensing *carbon dioxide* (CO_2) is used as an example to elaborate the needs of developing sensors to detect chemical particles.

Nondispersive IR (NDIR) is the most common type of sensors to measure carbon dioxide (CO_2). As shown in Figure 5.48, NDIR uses IR lamp to generate light in a tube that is filled with a sample of air. The air flows from inlet to outlet toward an optical filter in front of an IR light detector, and the IR light detector measures the amount of the IR light that passes through the optical filter.

5.6 Optical Encoders

An optical encoder is a type of motion sensor that emits light at one side of a coded disk and receives light at the other side to track the movement of an object. The encoder measures the motion based on the interruption of light. Figure 5.49 shows the basic construction of an optical encoder for a rotary encoder. As the code disc rotates, it shutters light from an LED, the receiver gets light and transmits it into an electronic signal in a sine or square waveform.

The measurement by an optical encoder can be either incremental or absolute as shown in Figures 5.49 and 5.50, respectively. An incremental encoder only has one row of slits to measure a relative angle; therefore, the code disc can be made at a low cost while the resolution is very limited. An absolute encoder has multiple rows of slits, the more rows it has, the better resolution the encoder can achieve; however, the cost to make a code disc is increased when the number of rows is increased. Bi and Zhang (2021) created a parametric model to represent an encoder family with different resolutions as shown in Figure 5.51. Moreover, an absolute encoder can detect the position of a shaft when a motor stops or restarts.

5.6.1 Resolutions

An absolute encoder represents an angular position of shaft as a digital word; for example, a 9-digital "001011011" for an encoder with 8 rows of slits. Commonly available encoders usually have up to 16 bytes to measure positional resolution. The resolution of an absolute encoder can be determined by,

$$\Phi_{abs} = 360^{\circ}/2^{n} \tag{5.43}$$

where n is the number of rows of the disc and Φ_{abs} is the resolution, that is, the minimized angle to be detected.

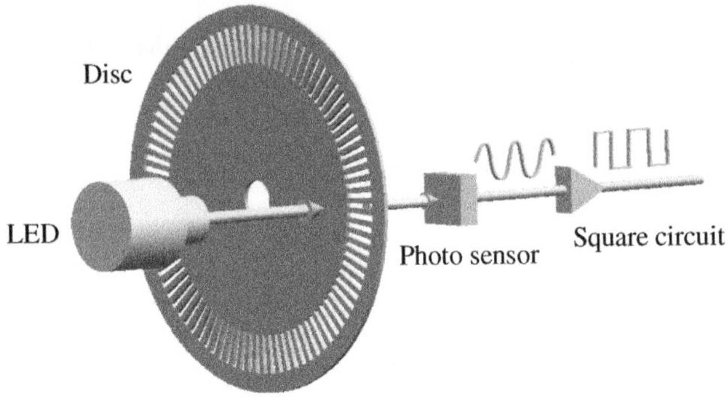

Figure 5.49 Diagram of an incremental encoder

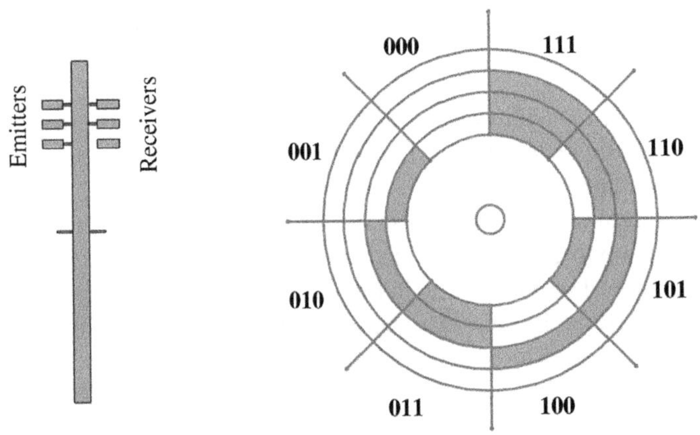

Figure 5.50 Diagram of an absolute encoder

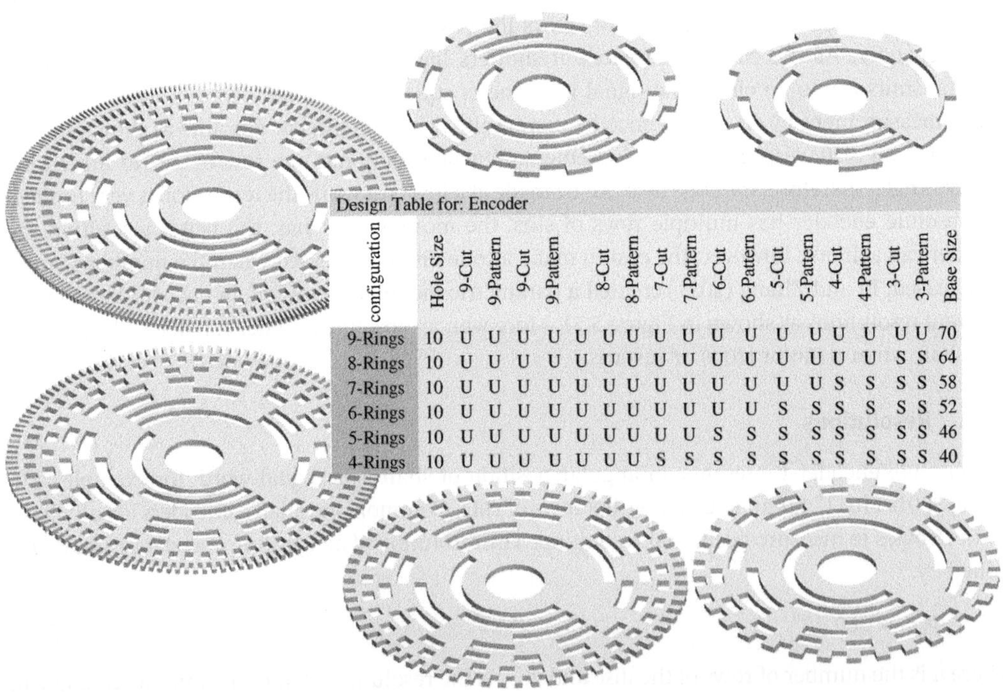

Design Table for: Encoder

configuration	Hole Size	9-Cut	9-Pattern	9-Cut	9-Pattern		8-Cut	8-Pattern	7-Cut	7-Pattern	6-Cut	6-Pattern	5-Cut	5-Pattern	4-Cut	4-Pattern	3-Cut	3-Pattern	Base Size
9-Rings	10	U	U	U	U	U	U	U	U	U	U	U	U	U	U	U	U	U	70
8-Rings	10	U	U	U	U	U	U	U	U	U	U	U	U	U	U	U	S	S	64
7-Rings	10	U	U	U	U	U	U	U	U	U	U	U	U	U	S	S	S	S	58
6-Rings	10	U	U	U	U	U	U	U	U	U	U	U	S	S	S	S	S	S	52
5-Rings	10	U	U	U	U	U	U	U	U	U	S	S	S	S	S	S	S	S	46
4-Rings	10	U	U	U	U	U	U	U	S	S	S	S	S	S	S	S	S	S	40

Figure 5.51 A parametric model of an encoder family with different resolutions (*Source:* Bi and Wang (2021)/ John Wiley & Sons)

Equation (5.43) shows that the resolution of an encoder can be increased by increasing the number (n) of rows of the disc. However, a large n might cause the difficulty in manufacturing thin features on the disc. Alternatively, the resolution of an encoder can be increased by integrating a gearing set at the coupler between the shaft and encoder as,

$$\Phi_{abs} = 360^{\circ}/(\rho \cdot 2^{n}) \tag{5.44}$$

where ρ is the gear ratio of the coupler.

Example 5.6 If an encoder disc has 16 rows, what is the resolution of the encoder?

Solution

Using the above equation and setting $n = 16$ gives,

$$\Phi_{abs} = 360^{\circ}/2^{16} = 0.005493^{\circ}$$

Example 5.7 Select an optical encoder to achieve a resolution of $\Phi_{abs} = 0.18°$ ∅ or better.

Solution

Using the above equation and setting $\Phi_{abs} = 0.18°$,

$$\Phi_{abs} = 360^{\circ}/2^{n} < \, = 0.18^{\circ}$$

$$n = \frac{\log\left(\frac{360^{\circ}}{0.18^{\circ}}\right)}{\log(2)} = 10{,}965 \Rightarrow 11 \text{ bits}$$

The resolution of an incremental encoder also depends on the number of the rows of slits. Moreover, a quadrature technique is often adopted to increase the resolution of an incremental encoder. As shown in Figure 5.52, the quadrature technique allows to divide each opaque or clear patch on a disc by 2 bits. Quadrature signals are offset by a half of phase, each patch is divided into 2^2 or 4 positions, and the resolution of an incremental encoder can be found as,

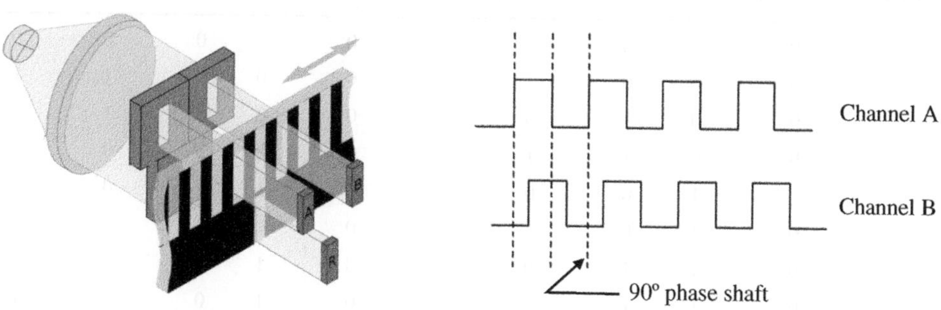

Figure 5.52 Using a quadrature technique in an incremental encoder (*Source:* MatthiasDD/Wikimedia commons/CC BY-SA.3.0)

$$\Phi_{inc} = \frac{360^{\circ}}{\left(C_{patch} \cdot 2^2\right)} = 90^{\circ}/C_{patch} \tag{5.45}$$

where C_{patch} is the number of the rows of the disc and Φ_{inc} is the resolution.

Example 5.8 For an incremental encoder with the "quadrature" technique, if an incremental encoder has 500 lines, what is its resolution?

Solution

Let $C_{patch} = 500$, applying Eq. (5.45) finds $\Phi_{inc} = 0.18°$

5.6.2 Decoding

When a reading from the rows of slits is converted into the position of shaft, either of *binary-coded decimal* (BCD) or *gray code* can be used. As shown in Table 5.6. BCD is a type of binary encodings of decimal numbers where each decimal digit is represented by a fixed number of bits, usually four or eight bits. Gray code was proposed by Frank Gray in 1947; it is an ordering of the binary numeral system such that two successive values differ in only one bit (binary digit), and it is less error prone but requires a converter to determine actual value of a shaft position.

Example 5.9 BCD and gray code are used in the disc designs in Figure 5.53a and b, respectively. Determine the shaft positions for the readings in the corresponding shadow masks, respectively.

Solution

Examine the redlines in Figure 5.53a and b, respectively, determine the values for 4-digits; and the shaft positions can be determined based on BCD and gray code accordingly.

Table 5.6 BCD and gray code encoding systems

	BCD				Gray code				
Decimal digit	2^3 (MSB)	2^2	2^1	2^0 (LSB)	4 (MSB)	3	2	1 (LSB)	
0	0	0	0	0	0	0	0	0	0
1	0	0	0	1	0	0	0	1	1
2	0	0	1	0	0	0	1	1	0
3	0	0	1	1	0	0	1	0	1
4	0	1	0	0	0	1	1	0	0
5	0	1	0	1	0	1	1	1	1
6	0	1	1	0	0	1	0	1	0
7	0	1	1	1	0	1	0	0	1
8	1	0	0	0	1	1	0	0	0
9	1	0	0	1	1	1	0	1	1

MSB, most significant bit; LSB, least significant bit.

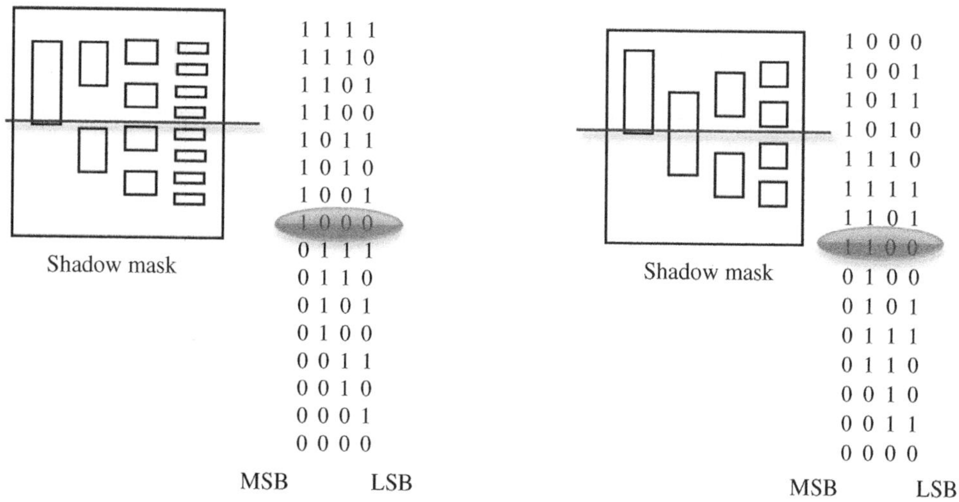

(a) BCD

(b) Gray code

Figure 5.53 Decoding examples using BCD and gray code

5.7 Sensors in MEMS

MEMS is the most critical to mechatronic systems; MEMS is one of the most promising technologies to combine silicon-based microelectronics with micromachining technologies for advanced mechatronic products. MEMS is an integrated technology to make micro-scale mechatronic devices that are capable of sensing, controlling, and operating micro-scale objects and generating effects on macro-scale objects (Prime 2022). MEMS has been developed based on early *integrated circuit* (IC) fabrication but extended widely to other non-IC techniques to reduce cost by batch fabrication and expand the capabilities of devices by integrating thousands of MEMS devices on a single silicon wafer.

As shown in Figure 5.54 for the examples of MEMS devices, micro-level mechanisms, sensors, actuators, and electronics are all integrated to fulfill the functions of an MEMS; especially, microsensors are used to detect states of system variables by measuring mechanical, thermal, magnetic, chemical, or electromagnetic quantities; microelectronics processes the acquired data to generate

(a) Compliant mechanism at scale of 0.2 μm

(b) HP MEMS chip for sensing

(c) MEMS structure

Figure 5.54 MEMS structure and examples (*Source:* (a) NASA / https://2019.spaceappschallenge.org/challenges/planets-near-and-far/memory-maker/teams/automata/project / Last accessed on July 19, 2024)

Table 5.7 Types of microsensors in MEMS

Domain	Quantities to be sensed
Mechanical	Force, pressure, velocity, acceleration, and position
Thermal	Temperature, entropy, heat, and heat flow
Chemical	Concentration, composition, and reaction rate
Radiant	Electromagnetic wave intensity, phase, wavelength, polarization reflectance, refractive index, and transmittance
Magnetic	Field intensity, flux density, magnetic moment, and permeability
Electrical	Voltage, current, charge, resistance, capacitance, and polarization

Source: Adapted from Prime (2002).

control commands for micro-actuators to drive micro-mechanisms to react or create some form of changes in applications.

Similar to macro-level sensors, microsensors can be classified by the physical quantities to be sensed as shown in Table 5.7 (Prime 2002).

However, fabrication of MEMS is very different from that of macro-scale mechatronic device in the sense that a macro-scale mechanical device is assembled from a set of functional components, while the components in an MEMS are fabricated through the same series of processes as shown in Figure 5.55 (Staals 2016).

It becomes clear that the design of microsensors in MEMSs is quite different from the sensors in macro-scale mechatronics. Taking an example of a micro-accelerometer in Figure 5.56, the typical specifications are as follows:

1) High precision dual axis accelerometer with signal-conditioned voltage outputs in a single IC
2) Sensitivity from 20 to 1000 mV/g
3) High accuracy and temperature stability
4) Low power (less than 700 μA typical)
5) 5 mm × 5 mm × 2 mm LCC package
6) Low cost ($0.58–2.06 (digikey))

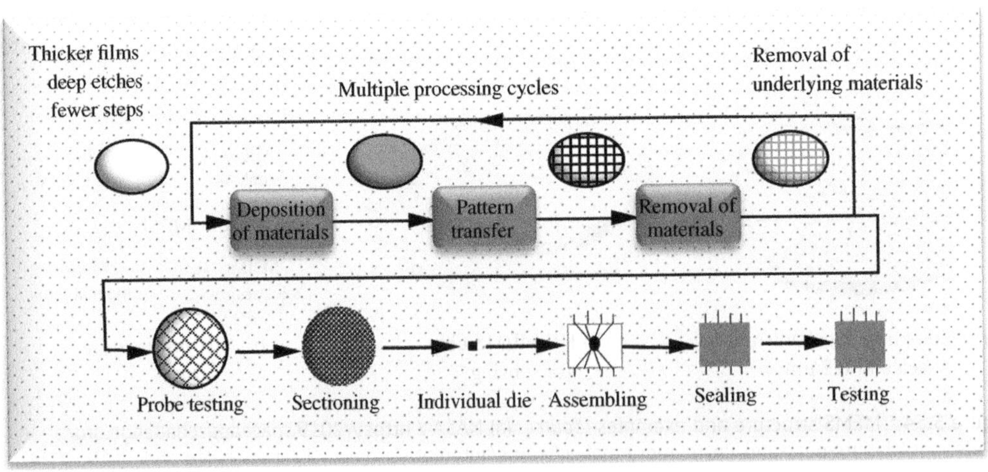

Figure 5.55 Fabrication of MEMS as similar to IC

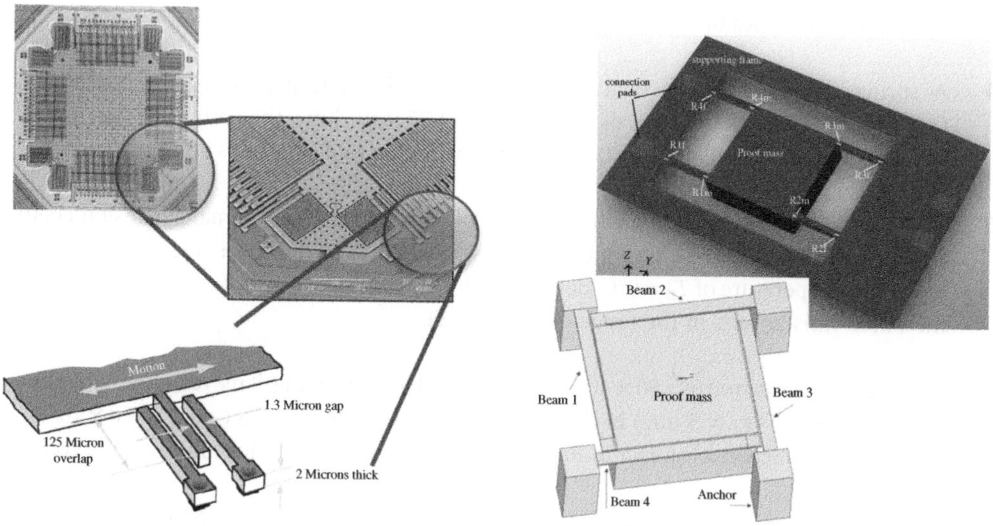

(a) Capacitive (b) Piezoresistive

Figure 5.56 Microsensor examples (*Source:* (a) Circuit Digest /https://circuitdigest.com/tutorial/what-is-mems-various-mems-devices-and-applications / Last accessed on July 19, 2024)

Figure 5.56 shows that a conventional mass–spring system will not be applicable at a micro-level, and all of machine elements such as springs and joints must be substituted by compliant mechanisms.

5.8 Summary

A mechatronic system has a set of system variables that should be controlled to achieve expected system outcomes, and expected outcomes are represented by a number of output variables. Both of system inputs and outputs should be measured to ensure that they are given or produced correctly. Therefore, sensors are essential MFMs in a mechatronic system to measure physical properties of objects and compare the measured results with targeted values of physical properties.

The more sensors a mechatronic system uses, the more variables can be monitored and controlled, and the higher level of intelligence of the mechatronic system is. A sensor is basically a transducer to convert a physical quantity from its original form into a readable digital signal, any working principle of a transducer can be used to design and implement a sensor. Commonly used working principles are the Hook's law, Ohm's law, photoconductivity, Hall effect, Faraday's law of induction, Curie–Weiss law, and ToF. Sensors should be designed and selected based on the types of physical quantities to be detected, environmental factors, operational ranges, accuracies, costs, and other characteristics.

Problems

5.1 What is the function of a transducer? Give one example of a primary transducer and its function.

5.2 Describe one example of a sensor that makes a direct measurement and the other one that makes an indirect measurement. Clarify your criterion of calling a measurement as "direct" or "indirect."

5.3 What are main types of transducers for measuring displacement?

5.4 A spring element is made of a hard steel wire with a transverse elastic modulus of $G = 7.85 \times 10^4$(MPa), the wire diameter is $d = 1.27 \times 10^{-3}$(m), the spring index is $C = 8$, and the number of coils is $N = 7$, determine the spring rate (k) of this spring element.

5.5 The sensing probe of a thermometer is made of platinum with a thermal expansion coefficient of $\alpha = 9.0 \times 10^{-3}$ (°C). The reference resistance if the thermometer is $R_0 = 99.996\ \Omega$ at a reference temperature of $T_0 = 0$°C, determine the range of the temperature change when the measured resistance is in the range of (50, 100) Ω.

5.6 Assume that an electron ($e = 80 \times 10^{-9}$ C) occurs to a ball by rubbing, it moves in the x-direction with a velocity of (v_d = 5 m/s) in the Earth's magnetic field (B = 0.5 Gauss = 5.0×10^{-5} T) toward the y-direction as shown in Figure 5.17, determine the magnetic force on the rod.

5.7 An electron ($e = 1.60217663 \times 10^{-19}$ C) is moving at 5.00×10^3 m/s in a 2.50 T magnetic field, the induced magnetic force is 1.50×10^{-15} N, determine the angle of the velocity of the electron with respect to the direction of the magnetic field.

5.8 A Hall probe is made of a copper strip whose electronic density is $n = 8.5 \times 10^{28}$ elecotrons/m^3. The cross section of the strip is $w \times d$ = (5.0 cm × 0.125 cm), determine the Hall voltage when the current is I = 20 A and the magnetic field is B = 1.5 T.

5.9 Determine the induced electromotive force (emf) when a coil has two loops with a radius of 5.0 cm, the average $B(t)$ changed from 0.01 T to 0.3 T in 0.2 seconds.

5.10 You are building a robot to move all-terrain with obstacles. Given your assumptions of robot size, onboard computation, power, and wireless communication, describe two to four types of sensors you would expect. Explain the role of each sensor and why you selected it.

5.11 Select an absolute optical encoder to achieve a resolution of $\emptyset_{abs}$ = 0.1 degree or better.

5.12 For an incremental encoder with the "quadrature" technique, if an incremental encoder has 1800 lines, what is its resolution?

5.13 What is the function of the Wheatstone bridge (Figure 5.57)? For the 1-element Wheatstone bridge as shown, the input voltage is 12 (V). What is the change in output voltage v_0 if R_3 changes from 1.0 to 1.1 kΩ? Assume that R_1 = 10.0 kΩ, R_2 = 1.2 kΩ, and R_4 = 10.0 kΩ and that only R_3 of the four resistors changes over time.

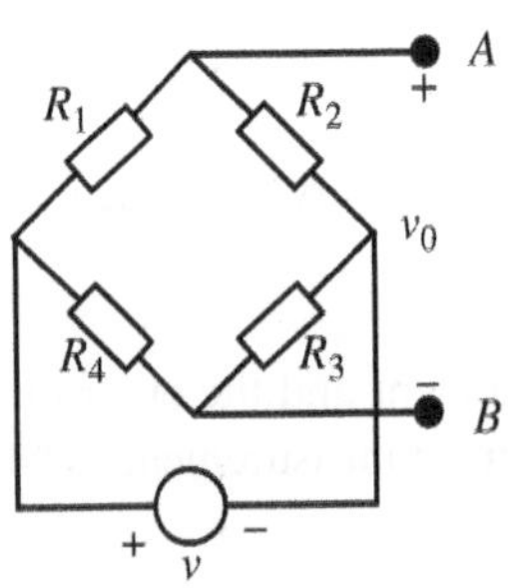

Figure 5.57 Wheatstone bridge

References

Alves, C., Custodio, T., Silva, P. et al. (2021). smartPlastic: innovative touch-based human-vehicle interface sensors for the automotive industry. *Electronics* 10 (11): https://doi.org/10.3390/electronics10111233.

Apple (2024). iPhone 14 pro Max – technical specifications. https://support.apple.com/kb/SP876?locale=en_US, last accessed on Sept. 12, 2024.

Atlas (2024). How does an NDIR CO_2 sensor work? https://atlas-scientific.com/blog/how-does-an-ndir-co2-sensor-work, last accessed on Sept. 12, 2024.

Avnet (2023). Types of pressure sensors: a comprehensive overview. https://www.avnet.com/wps/portal/abacus/solutions/technologies/sensors/pressure-sensors/types, last accessed on Sept. 12, 2024.

Behance (2024). Vehicle sensors. https://www.behance.net/gallery/56001209/Generic-car-update/modules/330040133, last accessed on Sept. 12, 2024.

Bi, Z.M. (2018). *Finite Element Analysis: A Systematic and Practical Approach*, 1e. Academic Press.

Bi, Z.M. and Wang, L.H. (2010). Advances in 3D data acquisition and processing for industrial applications. *Robotics and Computer-Integrated Manufacturing* 26 (5): 403–413.

Bi, Z.M. and Wang, X.Q. (2021). *Computer Aided Design and Manufacturing*, 1e. Wiley-ASME Series.

Budynas, R.G. and Nisbett, J.K. (2014). *Shigley's Mechanical Engineering Design*, 10e. McGraw Hill.

Choi, S.W., Kim, Y., Kim, J.M., and Park, H.S. (2013). Field monitoring of column shortenings in a high-rise building during construction. *Sensors* 13 (11): 14321–14338.

Cottrell, M.S. (2023). Temperature. https://web.mst.edu/~cottrell/me240/resources/temperature/temperature.pdf (accessed 12 September 2024).

Fierceelectronics (2023). A practical approach to vibration detection and measurement part 1: physical principles and detection techniques. https://www.fierceelectronics.com/sensors/stmicroelectronics-wins-best-sensors-2023-ultralow-power-lis2dux12-accelerometer-0 (accessed 12 September 2024).

HeyPete (2016). MAX31855 temperature linearization. https://blog.heypete.com/category/sensors/thermocouples (accessed 12 September 2024).

IOT Analytics (2024). State of IoT 2023: Number of connected IoT devices growing 16% to 16.7 billion globally. https://iot-analytics.com/number-connected-iot-devices (accessed 12 September 2024).

Kinney, T.A. (2022) An overview of proximity sensors. https://www.machinedesign.com/automation-iiot/sensors/article/21831577/baumer-electric-an-overview-of-proximity-sensors (accessed 12 September 2024).

Mahdi, I.A. (2014). Sensors and transducers. https://uotechnology.edu.iq/dep-MechanicsandEquipment/Lectures%20and%20Syllabus/Lectures/Same/Fourth%20Grade/Measurements%202.pdf (accessed 12 September 2024).

Precedence Research (2022). Sensor market size to hit around usd 508.64 billion by 2032. https://www.precedenceresearch.com/sensor-market (accessed 12 September 2024).

Prime (2002). *An Introduction to MEMS, Prime Faraday Partnership*. Loughborough University.

Qualcomm (2014). Behind the sixth sense of smartphones: the snapdragon processor sensor engine. https://www.qualcomm.com/news/onq/2014/04/behind-sixth-sense-smartphones-snapdragon-processor-sensor-engine (accessed 12 September 2024).

Shawn (2019). Touch sensors: what it is, how it works, Arduino guide. https://www.seeedstudio.com/blog/2019/12/31/what-is-touch-sensor-and-how-to-use-it-with-arduino (accessed 12 September 2024).

Staals, R.J.M. (2016). *Accelerated Wafer Z-Map for Layout Independent Leveling*, Eindhoven University of Technology, https://pure.tue.nl/ws/portalfiles/portal/46923242/840214-1.pdf (accessed 12 September 2024).

Utmel Electronic (2024). Introduction to acceleration sensors. https://www.utmel.com/blog/categories/sensors/introduction-to-acceleration-sensors (accessed 12 September 2024).
Wiki (2024a). Mobile phone. https://en.wikipedia.org/wiki/Mobile_phone (accessed 12 September 2024).
Wiki (2024b). Incremental encoder. https://en.wikipedia.org/wiki/Incremental_encoder (accessed 12 September 2024).
Wiki (2024c). Temperature. https://en.wikipedia.org/wiki/Temperature (accessed 12 September 2024).

6

Bridging Physical and Cyber Systems

6.1 Introduction

A mechatronic system is a convergence of physical and virtual things. The convergence requires to bridge the physical and virtual worlds for seamless interactions of *mechatronic functional modules* (MFMs). As shown in Figure 6.1, the signals in the virtual world are digitized, while the signals in the physical world are mostly continuous and analogue. In other words, the signals in the physical and virtual worlds have different formats and characteristics; they cannot be interacted directly without mutual interpretations. Therefore, a set of sophisticated MFMs is needed to bridge the physical and virtual worlds for the exchanges, storages, transmissions, and analyses of data. In this chapter, data and signals in the physical and virtual worlds are discussed, the needs for data conversions are elaborated, and the methods and basic functional elements for data conversions are presented.

6.2 Characteristics of Signals

The differences in signals in the physical and virtual world are discussed in this section. In general, signals can be classified into *analog signals* and *digital signals,* which will be discussed below.

6.2.1 Analog Signals

Most of data and signals in the physical world are analog signals. *An analog signal* is made up of a continuum of the values within a given range. An analog signal corresponds to a directly measurable quantity by a conversion from another physical quantity. Figure 6.2 shows three examples of analog signals: (1) the reading of the height of mercury level in a thermometer that is affected by surrounding temperature (see Figure 6.2a), (2) the reading in a speedometer that reflects an acceleration or deceleration when the car is in use (see Figure 6.2b), and (3) the volume increase in a stereo system when the volume knob is turned (see Figure 6.2c).

Figure 6.3 also shows two examples of analog signals in the physical world: (1) driving torques in a tripod robot and (2) the braking torques of a collaborative robotic platform when a robot was controlled to follow a specific task (Bi et al. 2008; Bi and Wang 2010).

Mechatronics for Complex Products and Systems: Project-Based Design Approaches for Robots, Cyber-Physical Systems, Digital Twins, and Other Emerging Technologies, First Edition. Zhuming Bi.

Companion website: www.wiley.com/go/bi/mechatronics

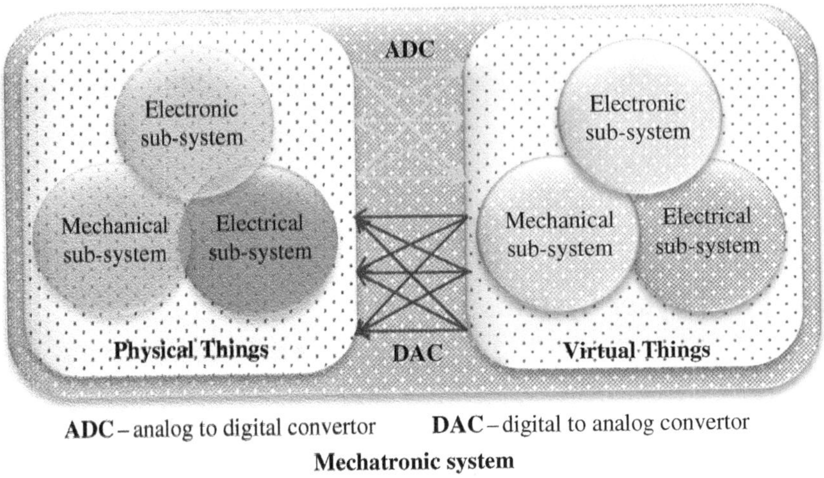

Figure 6.1 Bridging physical and virtual things in a mechatronic system

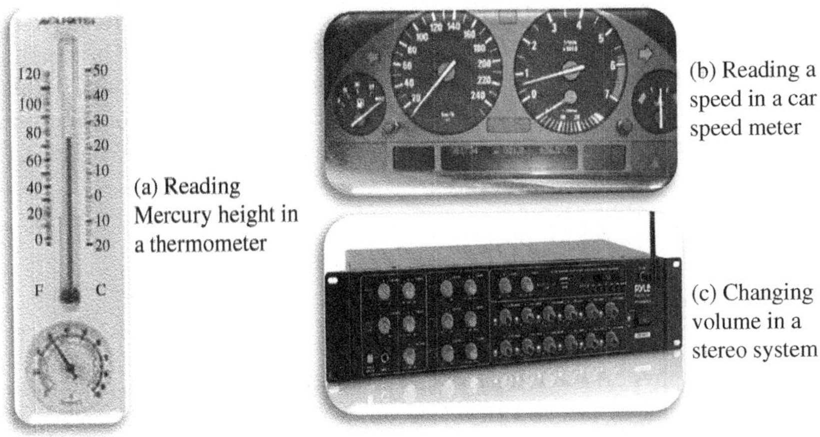

Figure 6.2 Examples of analog signals (*Sources:* (a) Chaney Instrument / https://www.amazon.com/Chaney-Instruments-Glass-Thermometer-Humidity/dp/B0007VE0QW / Last accessed on July 19, 2024; (b) craelectronics / https://www.youtube.com/watch?app=desktop&v=HS4N1nFojy0 / Last accessed on July 19, 2024; (c) Pyle / https://pyleusa.com/products/pt8050ch / Last accessed on July 19, 2024)

While analog signals are *continuous* and represent actual states of physical quantities, they show some disadvantages when they are collected, transmitted, and converted into electrical signals. Some main disadvantages of analog signals are,

1) when analog signals are converted and amplified, various noises embedded in original signals affect the true-worth of original signals;
2) since analog signals are continuous, the amount of original data is theoretically infinite, and this poses the practical challenge to store data with good accuracy considering the storage limits; and
3) implementing controls by using analog signals directly is usually more expensive than that through digital signals.

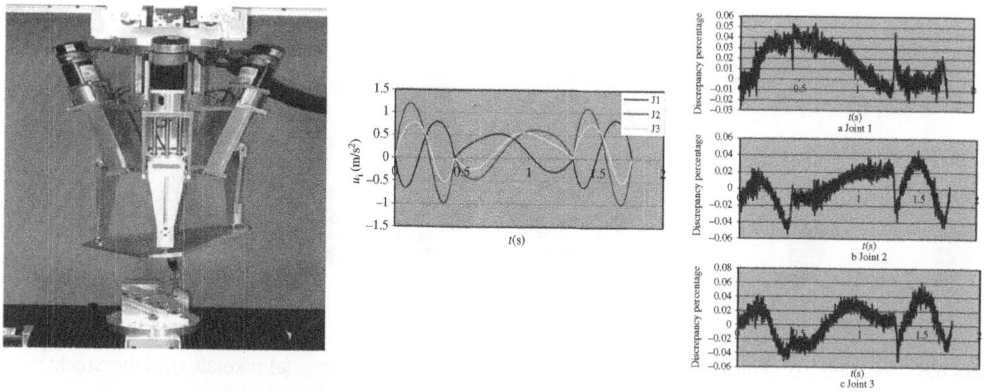

(a) Recorded driving torques in a tripod-base rod robot

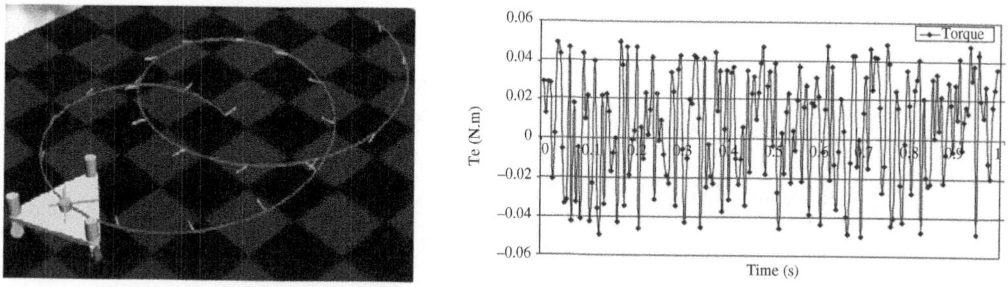

(b) Braking torques in a collaborative robotic platform

Figure 6.3 Examples of analog signals in robotic controls (*Sources:* (a) Bi et al. (2008)/Springer Nature; (b) Bi and Wang (2010)/with permission of Elsevier)

Based on the aforementioned reasons, analog signals should be usually converted into their digital forms in use. Most of physical quantities are continuous and analog; these quantities should be transferred into their corresponding digital forms.

6.2.2 Digital Signals

A digital signal represents data as a sequence of discrete values. A digital signal takes one value from a finite set of possible values at a given time. When a digital signal is represented by a set of binary bits, each bit only has two states that are commonly referred as the binary states of "*0*" or "*1*". Figure 6.4 shows three examples of digital signals where the digit "*0*" and "*1*" stand for the "OFF" and "ON" state of an object, respectively.

The advantages of using a digital signal or system are,

1) the noise caused by the physical circuit can be reduced after an analog signal is digitized and converted into a set of binary states with "0" and "1";
2) binary signals can be easily stored in computers with minimized storage;
3) the techniques for producing digital systems have been greatly advanced; digital systems can be made at an extremely low cost in comparison with conventional analog systems for equivalent functions.

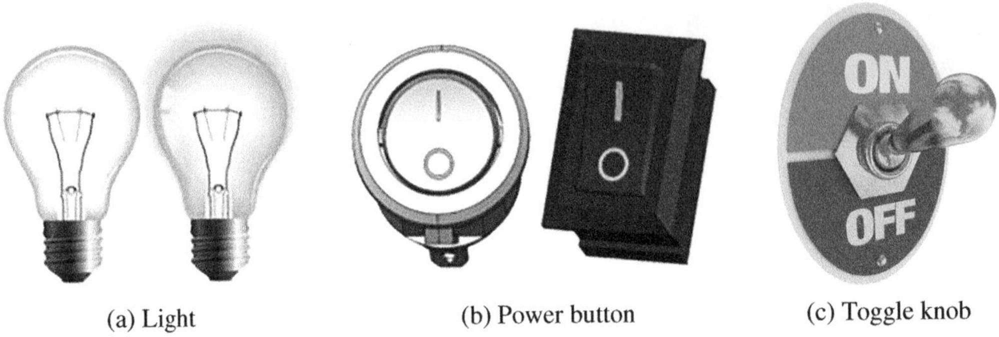

Figure 6.4 Examples of digital signals with "ON" and "OFF" states (*Sources:* (a) pixelalex/Adobe Stock; (c) iQoncept/Adobe Stock)

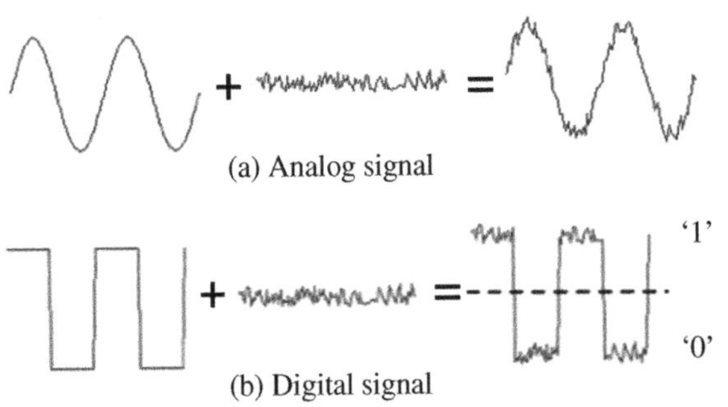

Figure 6.5 Comparison of noises in analog and digital signals

Note that no signal from a transducer is ideal without any noise. By saying this, digital signals also have some noises that are similar to analog signals. By comparing an analog and digital signals in Figure 6.5, a digital signal is less susceptible to noise than an analog signal. As shown in Figure 6.5b, the voltage value is unnecessary to determine exactly the state of a bit in a digital signal. This simplifies the data conversion and reduces the cost of achieving high accuracy of a variable in its digital form.

6.3 Conversions of Digital and Analog Signals

In the physical world, controlling an object begins with acquiring a continuous signal in an analog form. The analog signal is then converted into the digital form and analyzed to produce a control command. On the other hand, the control command is then converted back into its analog signal for execution. As shown in Figure 6.6, a *digital signal processing* (DSP) is responsible for the conversions of data between physical and digital worlds. DSP presents, converts, and allocates data or information in an appropriate bandwidth and capacity.

To control a physical quantity, the corresponding analog signal will be acquired and processed through the digital world in a series of activities in DSP as detailed in Figure 6.7.

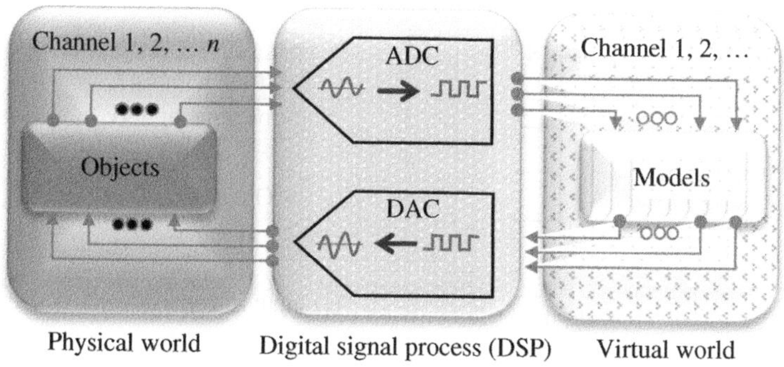

Figure 6.6 Digital signal processing (DSP) in data conversions

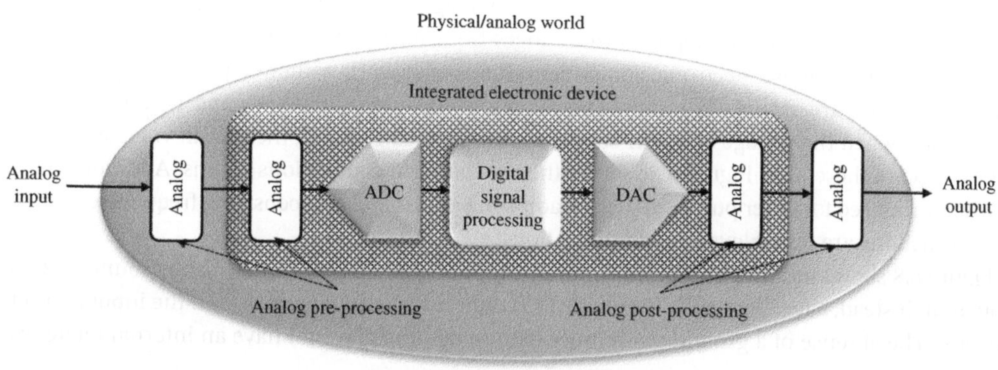

Figure 6.7 Data flow of a controlled variable in physical and digital worlds

6.4 Basic Electronic Elements for DSP

Data conversions in DSP require some basic electronic components to transform a signal in one form to another. Two types of primary electronic components are *operational amplifiers* (op-amps) and *comparators*. These two types of electronic components are discussed in the following sections, respectively.

6.4.1 *Operational Amplifiers (Op-Amps)*

The amplitudes of physical signals are usually very low. These signals must be amplified before they can be further processed. op-amps are essential building blocks to amplify physical signals in an analog electronic system.

Figure 6.8 shows the configuration of an op-amp. An op-amp is a *direct current* (DC) coupled amplifier for a high voltage gain with a differential input and a single-ended output. An op-amp can produce an output voltage that is more than 100,000 times than that of input voltage. As shown in Figure 6.8a, an op-amp has two inputs called *non-inverting input* (+) and *inverting input* (−), respectively, and the voltage at its output is,

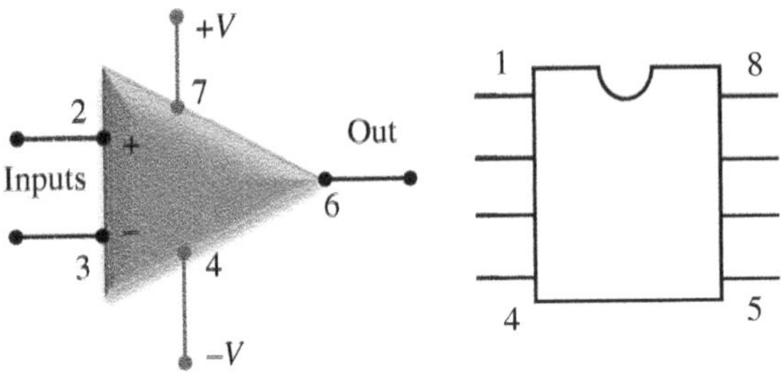

(a) Diagram (b) Top view of pins

Pin	Function
2	Inverting input (v_-)
3	Non-inverting input (v_+)
4	$-V$ supply
6	Output (v_{out})
7	$+V$ supply

(c) Input and output

Figure 6.8 Configuration of an op-amp

$$v_{out} = A_{OL}(v_+ - v_-) \tag{6.1}$$

where v_{out} is the output voltage; v_+ and v_- are the voltages at non-inverting and inverting inputs, respectively; A_{OL} is the open-loop gain that ranges from 10^5 to 10^7 at a low frequency.

An op-amp is an active electronic component. It does not work without a power supply. Therefore, pins 4 and 7 are connected to supply power at low and high voltages –V and + V, respectively. In practice, when the voltage of power supply is not specified at 4 or 7, the default voltage value of ±12 V is set at these pins. Figure 6.8c shows the standardized connections of pins. Additional pins can be introduced for other purposes such as adjusting offsets and compensating frequencies, while these functions are less relevant to amplifications.

Figure 6.8 shows negative and positive power supplies at pin 4 and 7; there is no ground or zero potential. Instead, a feedback path is provided to connect the output current with the input current directly. The absence of a ground pin implies that an op-amp does not have an internal reference point of voltage.

Modern op-amps are usually very durable as long as the power supply is set appropriately. Moreover, an op-amp must be wired in a circuit based on the configuration of pins before the power is actually switched on. Other than amplifying physical signals as essential steps to complete various tasks including buffering, inverting, non-inverting, differentiating, and summing, op-amps can be used to process and manipulate an analog signal in many other ways such as filtering (i.e., low-pass, high-pass, band-pass, integrator, and differentiator), and oscillating, comparing, sourcing (voltage, current), converting (voltage-to-current, current-to-voltage), and some nonlinear applications (Ultimate Electronics 2024).

As shown in Figure 6.9a, an ideal op-amp is primarily used as a voltage amplifier; it includes two inputs and one output. It corresponds to a subtraction of two inputs ($v_+ - v_-$) followed by a

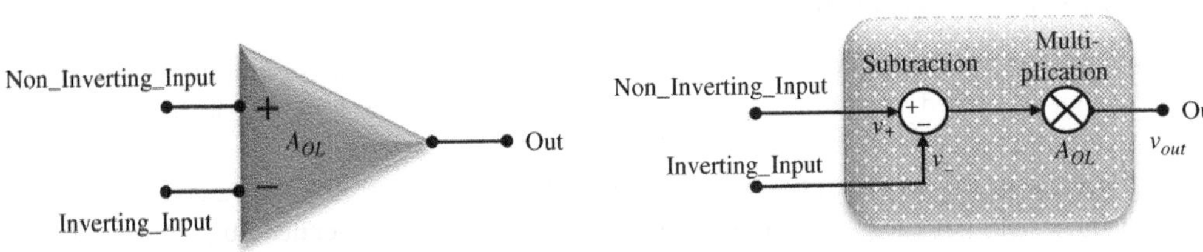

(a) Ideal op-amp (b) Subtraction and multiplication

Figure 6.9 Configuration of ideal op-amp and its applications multiplication of A_{OL} as shown in Figure 6.9b. Therefore, an ideal op-amp can be alternatively modeled as a v *controlled voltage source* (VCVS).

To model an op-amp in a simplified way, the following five assumptions are made regarding an ideal op-amp:

1) the input impedance is infinite; there is no current flow into or out of the input terminal. In other words, the state of the input terminal can only be measured by voltage;
2) the output impedance is zero; the output voltage is constant and holds as v_{out} for whatever the amount of current flows in or out;
3) the non-inverting and inverting inputs are fully balanced so that $v_{out} = A_{OL}(v_+ - v_-)$; when there is an offset voltage (v_{input_offest}) in the real-world, $v_{out} = A_{OL}(v_+ - v_- + v_{input\_offest})$;
4) an ideal op-amp omits the connection to the power supply and the power consumption is ignorable;
5) an ideal op-amp has a large gain approaching infinity; it has infinite bandwidth and the open-loop gain (A_{OL}) is constant and large for any frequency of interest.

As shown in Figure 6.10, op-amps in the real world cannot satisfy the above assumptions strictly. However, their input impedances are very high, even though not infinite. Output impedances are very low, even though not zero. The impacts of offsets must be taken into consideration. Amplification gains change with frequencies even though the bandwidths are very high. Since real op-amps are powered, the output voltages are limited by positive and negative voltages of the power supply.

In a real-world op-amp, a source signal can be connected to either the non-inverting input or the inverting input as shown in Figure 6.11a and b, respectively. Accordingly, the feedback path is connected to either of two inputs.

When the source signal is connected to the non-inverting input in Figure 6.11a, the feedback is connected to the inverting input. Since there is no current between the two inputs, the feedback voltage v_f is associated with v_i as $v_f \approx v_i$, and analyzing the feedback circuit leads to,

$$v_{out} = A_v v_i = \left(1 + \frac{R_f}{R_i}\right) v_i \tag{6.2}$$

where $A_v = \left(1 + \frac{R_f}{R_i}\right)$ is the gain, and R_f and R_i are the two resistors in the feedback circuit.

Equation (6.2) shows that the output voltage has the same sign as the input voltage. The ratio of two resistors determines the gain of the op-amp: the larger the ratio is, the higher the gain of

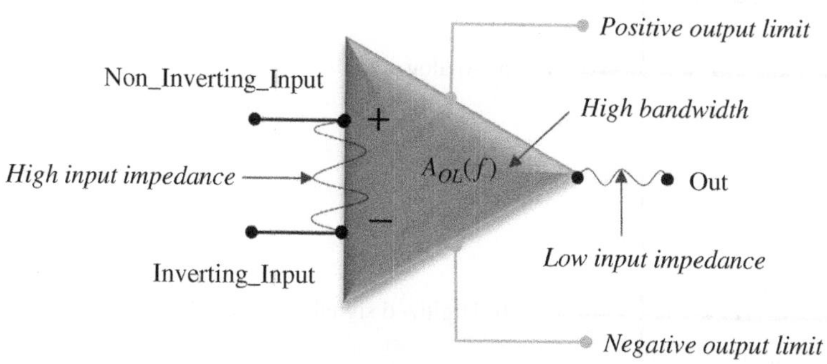

Figure 6.10 Configuration of a real-world op-amp

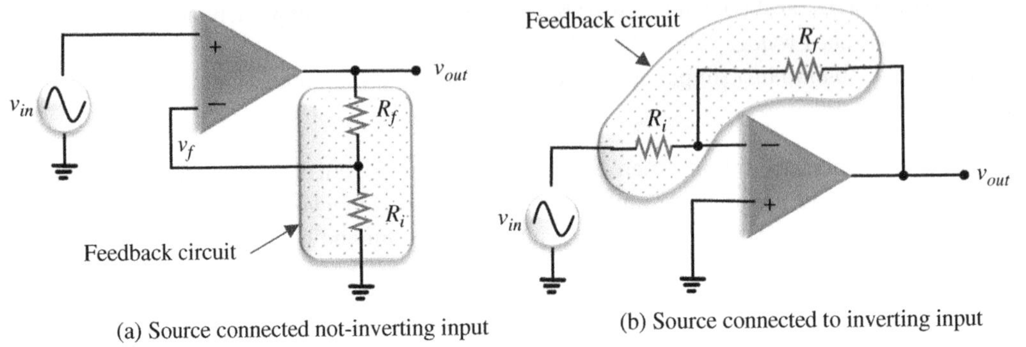

Figure 6.11 Two options of connecting source signal in a real-world op-amp

the op-amp is. One special case is when $R_i = R_f = 0$; the corresponding gain becomes $A_v = 1$, and the corresponding op-amp can be used as a buffer.

When both of the source signal and the feedback signal are connected to the inverting input in Figure 6.11b, the voltage v_f at the inverting input is approximately grounded at the non-inverting input, and analyzing the feedback circuit leads to,

$$v_{out} = A_v v_i = -\frac{R_f}{R_i} v_i \tag{6.3}$$

where $A_v = -\frac{R_f}{R_i}$ is the gain, and R_f and R_i are the two resistors in the feedback circuit.

Equation (6.3) shows that the output voltage has the opposite sign with the input voltage. The ratio of two resistors determines the gain of the op-amp: the larger the ratio is, the higher the gain of the op-amp is.

6.4.2 Comparators

To digitize an analog signal, a comparator should be used. As shown in Figure 6.12, a comparator is compared with a reference signal to evaluate which signal has a higher value. An op-amp can be used as a comparator when its two inputs are connected to reference and source voltages,

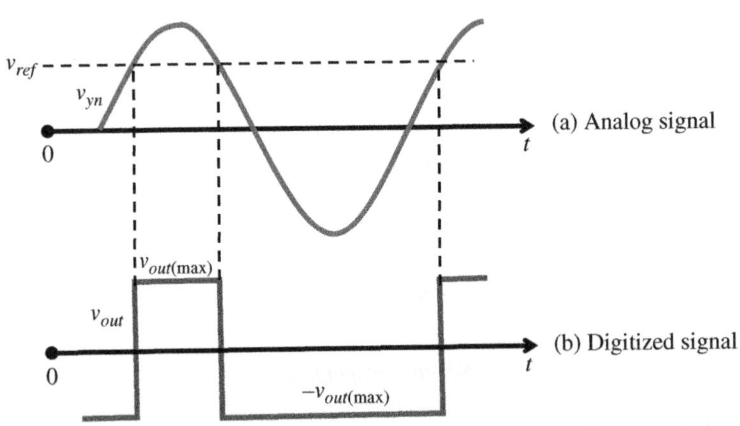

Figure 6.12 Digitizing an analog signal by a comparator

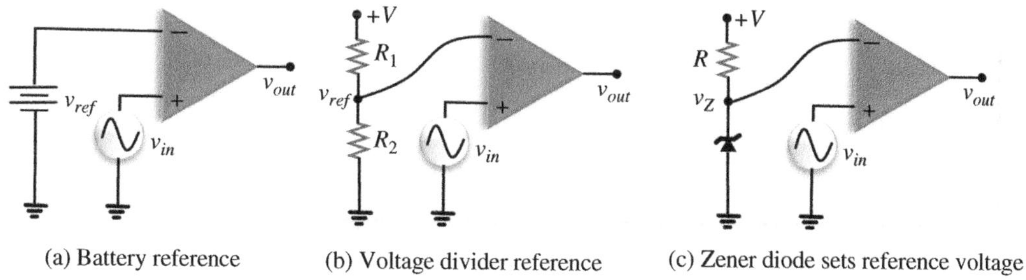

Figure 6.13 Three ways to set up a reference signal in a comparator

respectively. A comparator compares two signals and decides if the amplitude of one signal is larger than the other. Moreover, when an op-amp is used as a comparator, the compared result is then amplified, a small difference between the source and reference signals leads to a high output voltage attributed to a large gain of op-amp.

Figure 6.13 shows three common ways to set up a reference voltage in a comparator. The reference voltage and source signals are connected to the non-inverting and inverting inputs, respectively.

6.5 Digitization

When a continuous signal is digitized as a discretized signal, some information will be lost. On the other hand, when a digital signal is converted back into a continuous signal, some assumptions have to be made to convert a finite dataset into an infinite dataset. It is essential to understand how two types of signals are converted into each other, so that the most critical features of data can be preserved.

Digitization is implemented through two main processes, that is, *sampling* and *quantization*. Sampling and quantization processes are discussed in the following sections, respectively.

6.5.1 Sampling

A sampling process is to represent a signal in a continuous time domain into that in the discretized domain with uniform time intervals as in Figure 6.14. Sampling deals with the difference between an analog signal and a digital signal in the time domain.

The most critical parameter in a sampling process is *sampling frequency* (f_s). A sampling frequency is the number of sampling operations per second, and it should be selected so that all key features of interest in the original signal can be preserved in reconstructing a digital signal based on the sampled dataset. Figure 6.15 shows the effect of the sampling frequency on the performance of the sampling process. The higher the frequency of a sampling process is, the more data samples will be extracted and preserved, and a high cost will be involved due to computation, storage, and transmission. Therefore, the frequency of the sampling process should be selected based on the characteristics of the analog signal, the accuracy of digital data, and the limits of computing resources (Analog 2023).

A *sampling theorem* is applied in designing a sampling process. An appropriate sampling process means that after a sampling process, the original signal can be reconstructed from the sampled signal to capture its key attributes accurately. According to the sampling theorem, the sampling

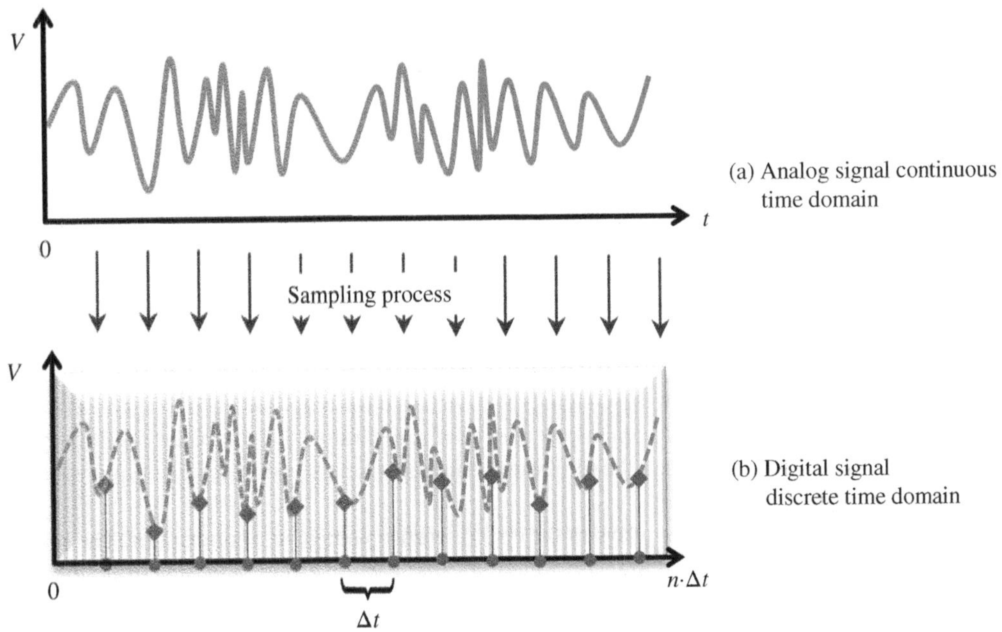

Figure 6.14 Sampling of analog signal in the time domain

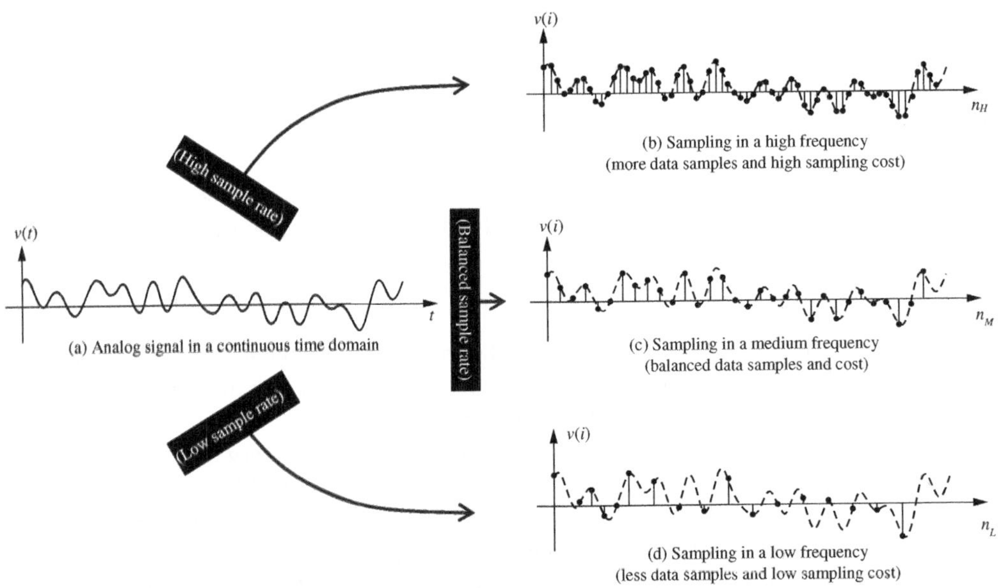

Figure 6.15 Frequency versus accuracy and storage in a sampling process

frequency (f_s) must be compared with the frequency of the source signal to be sampled to determine the appropriateness of a sampling rate.

A source signal may involve in multiple frequencies, while the sampling theorem applies to the highest frequency of the source signal with key attributes of the corresponding physical quantity.

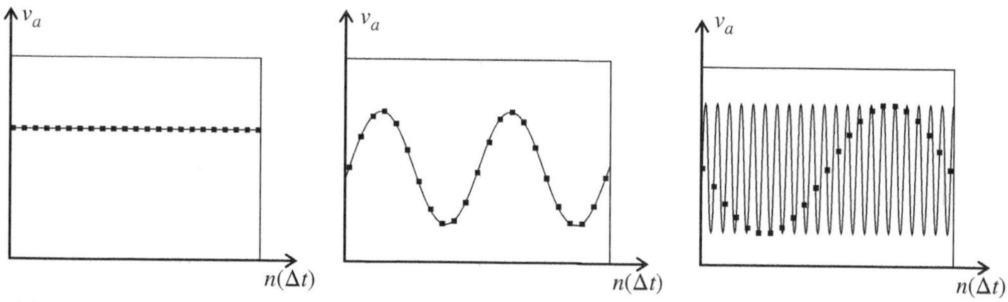

(a) A source signal of f_a=0 Hz and the sampling theorem is satisfied.

(b) A source signal of f_a=90 Hz and the sampling theorem is satisfied.

(c) A source signal of f_a=950 Hz, the sampling theorem is not satisfied and aliasing occurs.

Figure 6.16 Aliasing phenomenon in a sampling process

Let the highest frequency of a source signal be f_m, the sampling frequency (f_s) to preserve the key attributes of the source signal must satisfy the condition of,

$$f_m < \frac{f_s}{2} \tag{6.4}$$

where $f_s/2$ in Eq. (6.4) is called *the Nyquist frequency*, and the number of $(2 \cdot f_m)$ is called as *the Nyquist rate.*

Example 6.1 Assume that the sampling frequency is $f_s = 1000$ samples/s, determine if it is appropriate to digitize an analog signal with the frequency of $f_a = 0$, 90, and 950 cycles in Figure 6.16, respectively.

Solution

For the case in Figure 6.15a, $f_a = 0 < 500 = \frac{f_s}{2}$, the sample theorem is still satisfied. All of the key attributes of the source signal can be reconstructed from the sampled signal.

For the case in Figure 6.16(b), $f_a = 90 < 500 = \frac{f_s}{2}$, the sample theorem is still satisfied. All of the key attributes of the source signal can be reconstructed from the sampled signal. These samples represent accurately the sinusoid because there is no other sinusoid that can produce the same samples. The frequency of the source signal is 0.09 of the sampling frequency (1000). Equivalently, 1000/90 = 11.1 samples are taken over a complete cycle of a sinusoid cycle.

For the case in Figure 6.16(c), $f_a = 950 > 500 = \frac{f_s}{2}$, the sample theorem is violated. Not all of the key attributes of the source signal can be reconstructed from the sampled signal. This is an improper sampling of the source signal since another sine wave can produce the same samples. The source signal has been misrepresented by another signal, and this phenomenon is called *aliasing*. The true identity of the source signal has been hidden by the sampling process.

Example 6.2 Let a source signal be $x(t) = 50 \cdot \sin(105t)\cos(180t)$, determine the sampling frequency f_s so that the source signal can be reconstructed from the sampled signal appropriately.

Solution

It is clear that the sampling theorem should be applied in solving this problem. However, the source signal is involved in multiple components with different frequencies (i.e., $f_{a,1} = 105/(2\pi)$ (Hz) and

$f_{a,2} = 180/(2\pi)$ (Hz)), and these frequencies should be combined to determine the corresponding sampling frequency based on Eq. (6.4),

$$f_{a,12} = f_{a,2} + f_{a,2} = (105 + 180)/(2\pi) = 45.36 \text{ (Hz)}$$

Using Eq. (6.4) finds the sampling frequency must satisfy,

$$f_s > 2f_{a,12} = 2(45.36) \approx 91 \text{ (Hz)}$$

6.5.2 Quantizing

A signal represents the change of a specific physical quantity over time. It implies that a signal has two dimensions that should be digitized, respectively. In the previous section, a sampling process is performed to digitize a continual time domain into a discrete time domain. In this section, a *quantizing process* is performed to digitize a continual amplitude function into a finite set of discrete values that change over time. The quantizing process is to assign a digital word or number to each state at each discrete sampling time.

Figure 6.17 shows the information loss by quantization. The voltage at any specific time can be extracted from an analog signal in Figure 6.17a, and the extracted voltage is the true value in the source signal. In comparison, the number of data samples in Figure 6.17b are finite due to the sampling process; the voltage at a specific sampling time is digitized by the quantification process.

The extracted voltage is a digital value with a *quantizing error*. As shown in Figure 6.18 (Analog 2023), a quantizing error is the smallest possible change of an analog signal that can result in a change of a digital output. A quantizing error depends on the range of reference voltage and number of bits as,

$$\Delta V = V_R/2^N \tag{6.5}$$

where ΔV is the resolution; N is the number of bits in quantification; 2^N is the total number of states; and V_R is the range of the reference voltage. In the example illustrated in Figure 6.17, $N = 3$, $2^N = 8$, $V_R \in [0, 1]$, and $\Delta V = 1/8$ accordingly. Therefore, the quantizing error will be half of the resolution (1/16), and this error is inherent and unavoidable in the conversion of an analog signal into its digital form.

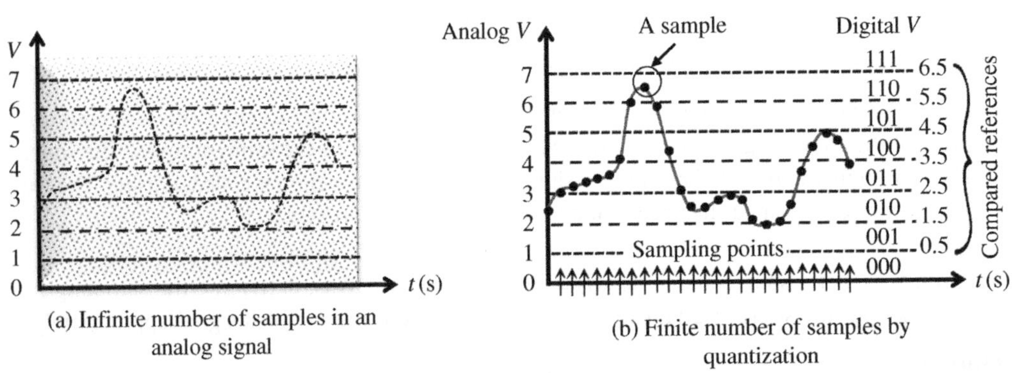

Figure 6.17 Quantization, range, and resolution

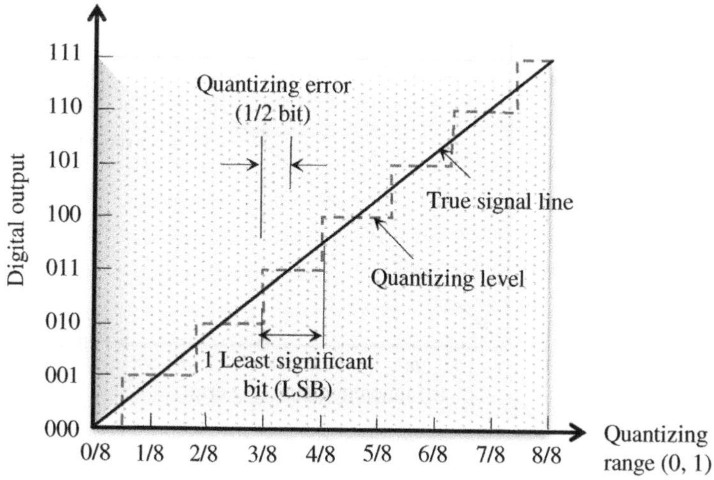

Figure 6.18 Parameters that affect quantizing errors

Example 6.3 Assume that the input analog voltage is changing between 0 and 5 V and use an 8-bit analog-to-digital conversion (ADC) to draw the output: (1) calculate the resolution of the converter; (2) determine the maximum voltage of the output signal; and (3) what is the ideal input when the output digital word is "10000000"?

Solution

(1) The resolution is determined by the changing range ($V_R = 5$ V) of voltage and the number of bits ($N = 8$) by Eq. (6.5); in this case $\Delta V = 5/(2^8) = 19.53$ (mV). (2) The maximum possible output is calculated as $V_{max} - \Delta V = 4.98047$ (V). (3) When the output digital word is "1000000," an ideal voltage without the consideration of an inherent error is $V_{min} + V_R \cdot (1/2^1 + 0/2^2 + 0/2^3 + 0/2^4 + 0/2^5 + 0/2^6 + 0/2^7) = 2.5$ (V).

Example 6.4 Assume that the input analog voltage is changing between −5 and 5 V, and a 10-bit ADC is used in quantization, (1) calculate the number of quantization levels; (2) calculate the resolution in digitization.

Solution

(1) The number of the quantization levels is completely determined by the number of bits ($N = 10$) in the converter; in this case, it is $2^{10} = 1024$. (2) The resolution is determined by the changing range ($V_R = 10$ V) of voltage and the number of bits ($N = 10$) by Eq. (6.5); in this case, $\Delta V = 10/(2^{10}) = 9.765625$ (mV).

6.5.3 Sampling and Quantizing in Analog-to-Digital Conversion (ADC)

While converting an analog signal into a digital signal, the information in the analog signal might not be preserved correctly, since the sampling process and quantizing process will introduce certain inherent errors in the time domain and the amplitude domain, respectively.

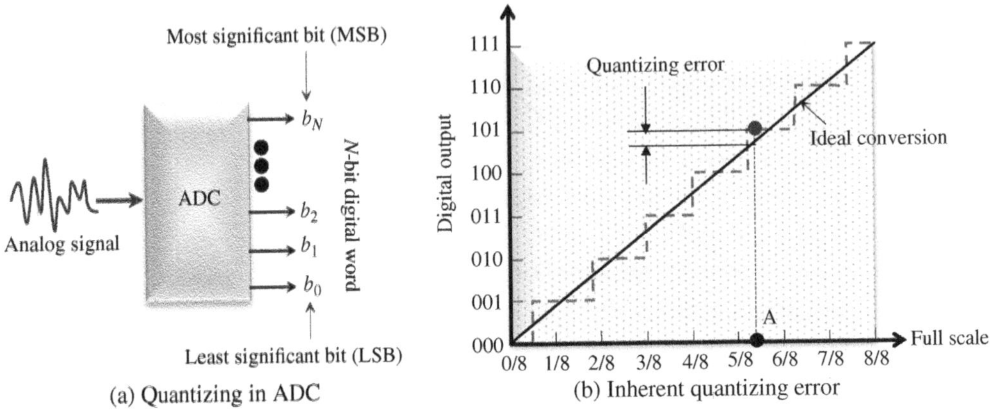

Figure 6.19 Quantizing error in ADC

In Figure 6.19, let the reference range of voltage be (0, 1)V, and its amplitude will be converted into a digital word with $N = 3$ bits; the finite set of possible digital words with 3 bits are eight binary combinations, i.e., {000, 001, 010, 011, 100, 101, 110, and 111}. Furthermore, the following observations can be made:

1) The maximum value in this quantization process is 7/8 (V) for a 1 (V) of the analog signal; it causes 1/8 (V) of inherent error.
2) *The least significant bit* (LSB) is the most right number of the digital word; in this example, it is '001' and it is equal to a voltage of 1/8 (V) of the analog signal.
3) *The most significant bit* (MSB) is the most left number of the digital word; in this example, it is "100" and it is equal to ½ of the voltage of the full-scale of the analog signal.
4) The resolution in quantization is determined by the number of finite states for the given range of the analog signal. When the number of bits in a digital word is given as N, the resolution can be expressed as $1/(2^N)$; in the case illustrated in Figure 6.19, it is 1/8.
5) The value of the largest digital number is "111" corresponds to the value of full-scale voltage minus one unit of LSB.
6) The quantizing error can be reduced by increasing the number of bits used in conversion.

Assume that (1) the range of an analog signal be $[V_{ref-}, V_{ref+}]$ where V_{ref-} and V_{ref+} the low and high reference voltages of the source analog signal, respectively; (2) the digital word consists of N bits, an analog input with V_{ref-} or V_{ref} is converted into a digital code of "0" and "$2^N - 1$," an analog input of V_k can be converted to its digital code (k) as,

$$k = \frac{(2^N - 1) \times (V_k - V_{ref-})}{(V_{ref+} - V_{ref-})} \tag{6.6}$$

where k is the digital code for the analog voltage of $\check{V}_k$; Eq. (6.6) can be used to determined

$$\check{V}_k = V_{ref-} + \frac{(V_{ref+} - V_{ref-})}{(2^N - 1)} k \tag{6.7}$$

where $\check{V}_k$ is the analog voltage with an inherent error $(V_k - \check{V}_k)$.

Example 6.5 The low and high reference voltages of an analog signal are $V_{ref-} = 1$ (V) and $V_{ref+} = 4$ (V), respectively, if the analog-to-digital conversion (ADC) uses 10 bits in the conversion, (1) what is the digital code for an analog 3.2 (V)? (2) What is the inherent error in quantizing a 3.2 (V)? (3) Find the corresponding voltages for the digital words of (i) "0000011001," (ii) "0001010000," (iii) "0111110000," (iv) "0111110100," (v) "1011010000," (vi) "1100100000," and (vii) "1110000100."

Solution

(1) Using Eq. (6.6) one gets,

$$k = \frac{(2^N - 1) \times (V_k - V_{ref-})}{(V_{ref+} - V_{ref-})} = \frac{(2^{10} - 1) \times (3.2 - 1)}{(4 - 1)} = 750.2$$

Decode 750.2 into 10-bit digital word as,

	MSB	b_8	b_7	b_6	b_5	b_4	b_3	b_2	b_1	LSB
Binary	2^9	2^8	2^7	2^6	2^5	2^4	2^3	2^2	2^1	2^0
Decimal	512	256	128	64	32	16	8	4	2	1
750.2	1	0	1	1	1	0	1	1	1	0

Therefore, the digital code for an analog 3.2 (V) is 1011101110.

(2) A digital code "1011101110" has its decimal value of 750; using Eq. (6.7) to convert the digital code "1011101110" into its analog form yield,

$$\breve{V}_k = V_{ref-} + \frac{(V_{ref+} - V_{ref-})}{(2^N - 1)} k = 1 + \frac{(4-1)}{(2^{10} - 1)} 750 = 3.1994 \text{ (V)}$$

Therefore, the inherent error in quantization is $V_k - \breve{V}_k = 0.006$ (V).

(3) The digital words in a binary form are translated into a decimal form at first, and Eq. (6.7) is further used to determine the corresponding voltages as below:

	(i)	(ii)	(iii)	(iv)	(v)	(vi)	(vii)
Binary	0000011001	0001010000	011110000	0111110100	1 011 010 000	1 100 100 000	1 110 000 100
Decimal	25	80	240	500	720	800	900
↓				Using Eq. (6.7)			
$\breve{V}_k$	1.07	1.23	1.70	2.47	3.11	3.35	3.64

Example 6.6 The low and high reference voltages of an analog signal are $V_{ref-} = 0$ (V) and $V_{ref+} = 10$ (V), respectively. if a distinguishable voltage difference 0.625 (V) is required in digitizing the analog signal, how many bits are required in a digital word? What are the finite set of digital voltage states?

Solution

The problem shows that the resolution is $\Delta V = 0.625$ (V) and $V_R = V_{ref+} - V_{ref-} = 10$ (V), using Eq. (6.5) one gets,

$$N = \log_2\left(\frac{V_R}{\Delta V}\right) \log_2\left(\frac{10}{0.625}\right) = 4$$

Therefore, a digital word with 4 bits should be used, and the output has a finite set of $2^4 = 16$ voltage states as below:

Output state	Voltage range (V)	Output state	Voltage range (V)
0	0.000–0.625	8	5.000–5.625
1	0.625–1.250	9	5.625–6.250
2	1.250–1.875	10	6.250–6.875
3	1.875–2.500	11	6.875–7.500
4	2.500–3.125	12	7.500–8.125
5	3.125–3.750	13	8.125–8.750
6	3.750–4.375	14	8.750–9.375
7	4.375–5.000	15	9.375–10.000

In practice, it is critical to ensure that an ADC has more bits than the application initially requires. For example, if an application requires 10 bits for the accuracy of a digital signal, then a 12-bit converter will be more suitable. In other words, the achievable accuracy must be always less than the total number of available bits (Microchip 2023).

Since an analog signal is two dimensional; one relates to the change of amplitude, and the other relates to the change of frequency. Therefore, the accuracy of an ADC depends on (1) the resolution of amplitude in quantization and (2) the resolution of frequency in sampling. A digitized signal with an inherent error can be expressed as,

$$\breve{v_k}(x) = \lim_{\Delta x \to 0} \frac{v(x) + v(x + \Delta x)}{2\Delta x} \tag{6.8}$$

where $v(x)$ and $v(x + \Delta x)$ are the voltages at two sides of the sampling region, Δx is the resolution determined by the number of bits in Eq. (6.5). The accuracy of increases with an increase of the number of bits in sampling or quantizing.

However, an increase of the number of bits in sampling or quantizing increases computing cost for data acquisition, transmission, and processing; therefore, the number of bits for a certain analog signal should be selected appropriately to make a trade-off of accuracy and cost of hardware and software. Table 6.1 provides the recommended numbers of bits for analog signals with different levels of accuracy.

Table 6.1 Number of bits versus accuracy level

Accuracy level	Number of bits
Low	8-bit or lower
Medium	10-bit and12-bit
High	16-bit
Extremely high	24-bit or higher

6.6 Analog-to-Digital Conversion (ADC)

An *analog-to-digital conversion* (ADC) is an electronic component to convert an analog signal into a digital signal with multi-level states without considering its essential content; ADC digitizes both of time and amplitude components of a signal.

There are three types of ADCs, that is, (1) integrating ADCs, (2) flash ADCs, and (3) successive approximation ADCs. However, any type of ADCs has three basic functional modules as shown in Figure 6.20. *The first type* is a comparator; every type of ADC has one or more comparators. Each comparator receives two analog voltage inputs and produces one digital output to a bit of ADC. If the voltage at the "+" input is higher than that at the "—" input, the comparator produces its output as "1"; otherwise, it produces its output as "0" at the bit. *The second type* is a linearity reference type that is used as the reference to compare the input signals in the converting process; it determines the differential and integral nonlinearities of ADC; an integrating ADC uses capacitors as linearity references, and a successive approximation ADC uses digital-to-analog conversions (DACs) as linearity references. The third type is a voltage reference to determine the full scale of conversion (Loewenstein 2000).

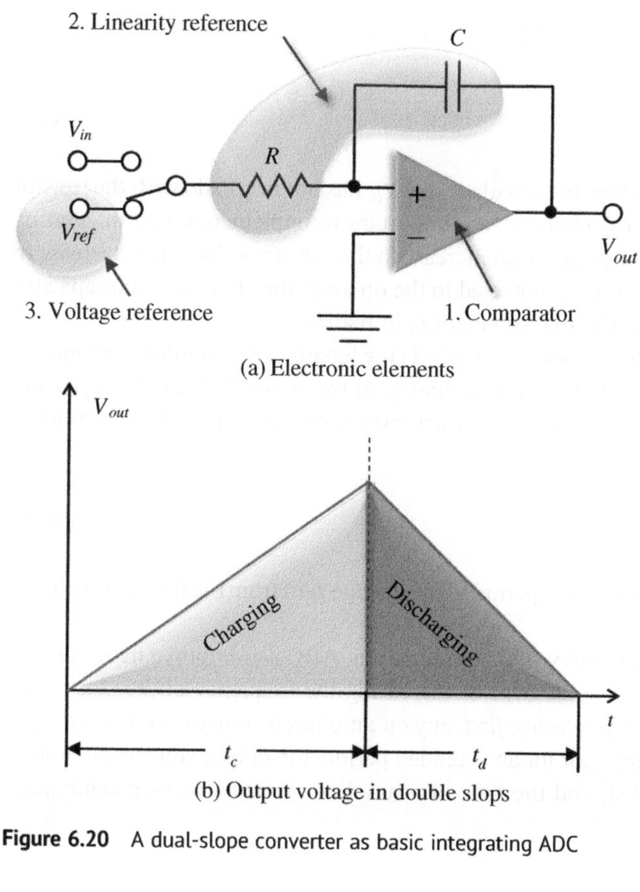

Figure 6.20 A dual-slope converter as basic integrating ADC

6.6.1 Integrating ADC

An integrating ADC converts an analog voltage into its digital form by an integrator. A dual-slope converter in Figure 6.20 shows the basic implementation of an integrating ADC. The analog signal is connected to the input of the integrator for a fixed run-up period; it is followed by connecting a known reference voltage with an opposite polarity to the integrator unit; and the output of the integrator returns to zero in a run-down period. The analog voltage can be computed from the given reference voltage, the fixed run-up period, and the measured run-down period. The run-down period is measured in units of the converter's clock for a higher resolution. However, pursuing a higher resolution may affect the converting speed.

For the electronic elements shown in Figure 6.20, the combination of the resistor (R) and the capacity (C) serves as a linearity reference as,

$$V_{out} = -\frac{V_{in}}{RC} t_{in} + V_{initial} \tag{6.9}$$

where V_{in} is the voltage at the input of the integrator; t_{in} is the connecting time; $V_{initial}$ is the initial voltage before the input is connected.

Assuming the initial voltage ($V_{initial}$) at each cycle of conversion to be zero, applying Eq. (6.9) in both of the charging and discharging phases in Figure 6.20b one gets,

$$\frac{V_{in}}{RC} t_c + \frac{V_{ref}}{RC} t_d = 0 \tag{6.10}$$

where t_c and t_d are the charging and discharging times by V_{in} and V_{ref}, respectively.

Rewriting Eq. (6.10), one gets,

$$V_{in} = -\frac{t_d}{t_c} V_{ref} \tag{6.11}$$

It is interesting to note that Eq. (6.11) for the calculation of V_{in} has nothing to do with the resistor (R) and the capacity (C). However, these elements are important in implementing an integrating ADC since the voltage in Figure 6.20b involves in an increase in the run-up period and a decrease in the run-down period. Since the integrator is connected to the op-amp, the charges to the capacitor by V_{in} and V_{ref} are different and affect the measurement t_d in reality.

The resolution of an integrating ADC is associated with (1) the length of the run-down period (t_d) and (2) the resolution of time measurement, or the frequency of the controller's clock (f_{clock}), and the required resolution (i.e., the number of bits N) is associated with the length of the run-down period for a full scale input $V_{in} = -V_{ref}$ as,

$$t_{d,full-scale} = -\frac{2^N}{f_{clock}} \tag{6.12}$$

where $t_{d,full-scale}$ is the length of the run-down period to the voltage zero from the full-scale voltage; N is the number of bits.

One should note that Eq. (6.12) does imply that an integrating ADC can improve its resolution arbitrarily by using a longer time of measurement or increasing the frequency of the clock. The resolution is limited due to the following reasons: *first*, any op-amp has its range of output voltage; when its input is connected to the integrator for an extended period, the output voltage will not be increased since its maximum is limited, and the time of a run-down time becomes meaningless.

Second, the accuracy of the comparator is used as the null detector for the zero voltage at the output, and it has its noise limits in its circuit that limits the resolution in detecting the zero voltage at the output. *Third*, the resolution is constrained by the quality of the capacity, the capacity is not perfectly linear and varies over time.

6.6.2 Flash Converter

A flash converter is also known as a parallel ADC; it is the fastest way to convert an analog signal into its digital form. Flash converters are widely applied in the applications with a large bandwidth; however, they consume more power than other types of ADC. Therefore, the resolution of a flash converter is generally limited to 8 bits. Since an *N*-bit digital word requires $2^N - 1$ number of comparators, the high-speed of a flash converter is a trade-off with its high cost of implementation.

As shown in Figure 6.21, a flash converter uses a set of parallel comparators for quantization. Op-amps without a feedback are used for comparators. All outputs of the comparators are converted into one digital work. For each comparator, a known signal is connected as the reference voltage, the analog signal is connected to the other input, and its output tells if the analog signal is greater than the reference voltage. A comparator yields a "1" when its analog input voltage is higher than the connected reference voltage; otherwise, the comparator output becomes "0."

When the voltage of input signal increases and exceeds the reference voltage of the first comparator, its output is switched to a high voltage level indicating a state of "00000001." When the voltage of input signal increases continuously, the sequential comparators are switched to a high voltage until the input signal reaches its maximum, and the state becomes "11111111"; therefore, the output state indicates the voltage of the input signal. For example, if the voltage of an analog input is between those of X_4 and X_5, all of the comparators from X_1 to X_4 have an output of "1," while the rest of the comparators from X_5 to X_{2^N-1} have an output of "0." The position where the output is

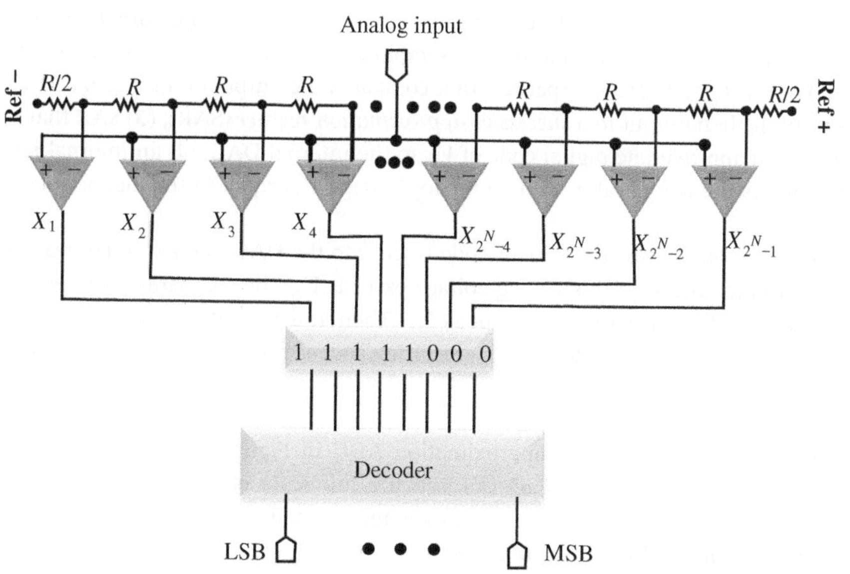

Figure 6.21 Configuration of a flash converter

Table 6.2 Comparators' outputs and states of ADC

Comparators' positions							→	Digital words		
X_1	X_2	X_3	X_4	X_5	X_6	X_7	→	D_2	D_1	D_0
0	0	0	0	0	0	0	→	0	0	0
1	0	0	0	0	0	0	→	0	0	1
1	1	0	0	0	0	0	→	0	1	0
1	1	1	0	0	0	0	→	0	1	1
1	1	1	1	0	0	0	→	1	0	0
1	1	1	1	1	0	0	→	1	0	1
1	1	1	1	1	1	0	→	1	1	0
1	1	1	1	1	1	1	→	1	1	1

switched from "1" to "0" corresponds to the case where the analog signal becomes smaller than the reference voltage at that position.

Table 6.2 shows the correspondences of the comparators' outputs with the states of ADC for a 3-bit flash converter. The 3-bit example required 7 converters, 6-bits would require 63, and 8-bits would need 255 comparators and equivalent number of resistors. Therefore, the cost of this type of converter is very high, since the number of comparators and resistors required increases rapidly.

6.6.3 Successive Approximation

A successive-approximation ADC converts an analog signal into its digital form by a binary search through all possible quantization levels. Figure 6.22 shows that a successive-approximation ADC consists of four main functional elements, that is, (1) a sampling and holding circuit to acquire the analog voltage (V_{in}) of the signal; (2) a comparator that compares the output of an internal DAC with V_{in}; and the result is then output to a *successive approximation register* (SAR); (3) SAR that is responsible to provide an approximate digital code of V_{in} to the internal DAC; (4) an internal reference DAC that provides the comparator with an analog voltage V_{ref} equal to the digital output of SAR_{in}.

SAR is initialized so that MSB is set as "1." This code is fed into the DAC to have an equivalent analog ($V_{ref}/2$) into the comparator. If this analog voltage exceeds V_{in}, the comparator causes the SAR to reset this bit to "0"; otherwise, this bit is kept as "1." Then, the next bit is set to "1" and the same test is done until every bit has been tested. The resulting code is finally output by the SAR at *the end of the conversion* (EOC).

Example 6.7 Assume that the successive-approximation ADC in Figure 6.23 has 8 bits, it is used to digitize analog signal with $V_{in} = 7.65$ (V) and the full scale of digitization is set as $V_{full\_scale} = 10$ (V). Illustrate the process of the successive-approximation ADC in digitizing V_{in}. Evaluate the inherent error in by this converting process.

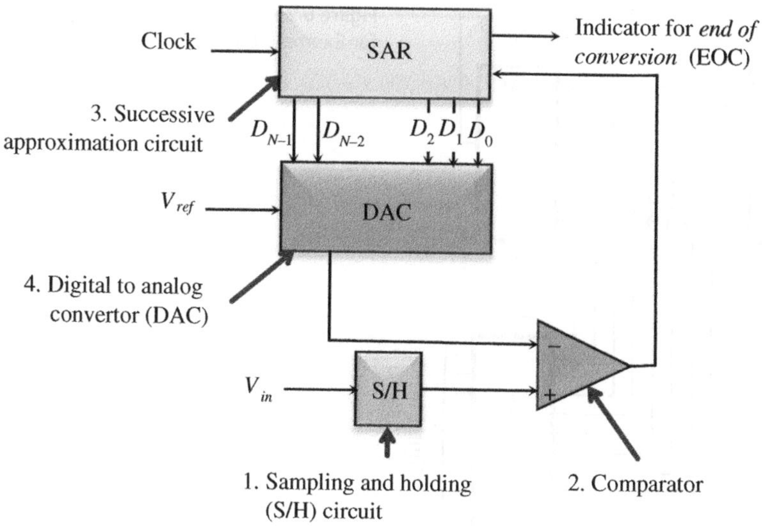

Figure 6.22 Main functional components of a successive-approximation ADC

Figure 6.23 Main functional components of successive-approximation ADC

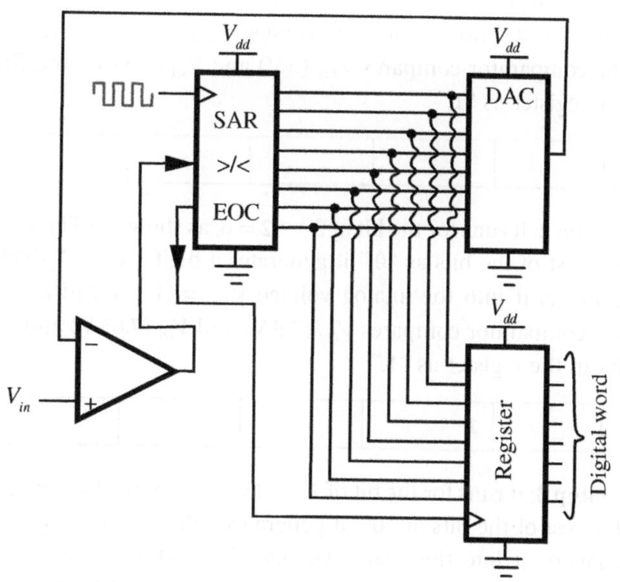

Solution

The ADC has 8 bits, therefore, the digitizing process by the successive-approximation ADC involves in eight steps from MSB to LSB sequentially as shown in Figures 6.24–6.31.

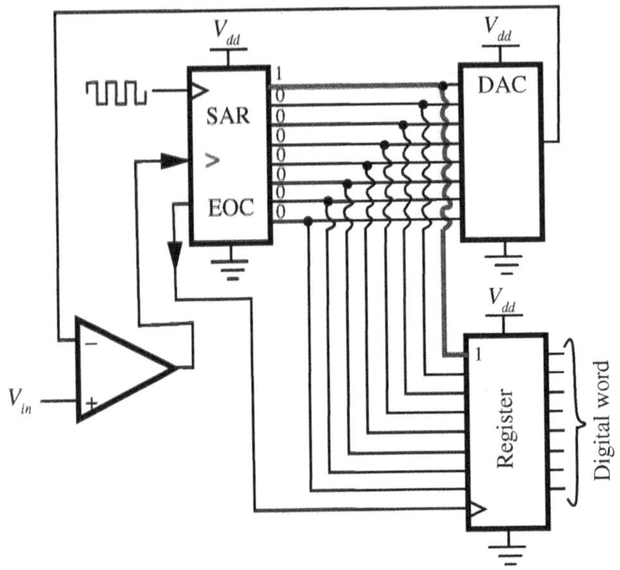

Figure 6.24 Step 1 of execution of ADC for Example 6.7

Step 1: it runs for MSB, that is, $2^N - 1 = 7$, as shown in Figure 6.24. The SAR sets the MSB as “1” and the rest of the bits as “,” it generates a digital word “10000000” as the input of DAC, the DAC converts it into the analog voltage $V_{ref} = (1 + 0) \cdot V_{full_{scale}}/2 = 5$ (V) to the inverting input, the comparator compares V_{ref} (5 V) and V_{in} (7.65 V) and finds $V_{in} > V_{ref}$, and it sets the MSB in the register as “1,”

1							

Step 2: it runs for the bit of $2^N - 2 = 6$, as shown in Figure 6.25. The SAR sets the sixth as “1” and the rest of the bits as “0,” it generates a digital word “**1**1000000” as the input of DAC, the DAC converts it into the analog voltage $V_{ref} = (1 + 1/2) \cdot V_{full_{scale}}/2 = 7.5$ (V) to the inverting input, the comparator compares V_{ref} (7.5 V) and V_{in} (7.65 V) and finds $V_{ref} > V_{in}$, and it sets the seventh bit in the register as “1,”

1	**1**						

Step 3: it runs for the bit of $2^N - 3 = 5$, as shown in Figure 6.26. The SAR sets the fifth as “1” and the rest of the bits as “0,” it generates a digital word “**11**100000” as the input of DAC, the DAC converts it into the analog voltage $V_{ref} = (1 + 1/2 + 1/4) \cdot V_{full_{scale}}/2 = 8.75$ (V) to the inverting input, the comparator compares V_{ref} (8.75 V) and V_{in} (7.65 V) and finds $V_{ref} < V_{in}$, and it sets the sixth bit in the register as “0,”

1	**1**	**0**					

Step 4: it runs for the bit of $2^N - 4 = 4$, as shown in Figure 6.27. The SAR sets the fourth as “1” and the rest of the bits as “0,” it generates a digital word “**110**10000” as the input of DAC, the DAC converts it into the analog voltage $V_{ref} = (1 + 1/2 + 1/8) \cdot V_{full_{scale}}/2 = 8.125$ (V) to the inverting input, the comparator compares V_{ref} (8.125 V) and V_{in} (7.65 V) and finds $V_{ref} < V_{in}$, and it sets the fifth bit in the register as “0,”

1	1	0	0				

Step 5: it runs for the bit of $2^N - 5 = 3$, as shown in Figure 6.28. The SAR sets the third bit as "1" and the rest of the bits as "0," it generates a digital word "**11001**000" as the input of DAC, the DAC converts it into the analog voltage $V_{ref} = (1 + 1/2 + 1/16) \cdot V_{full_{scale}}/2 = 7.8125\,(V)$ to the inverting input, the comparator compares V_{ref} (7.8125V) and V_{in} (7.65 V) and finds $V_{ref} < V_{in}$, and it sets the fourth bit in the register as "0,"

1	1	0	0	0			

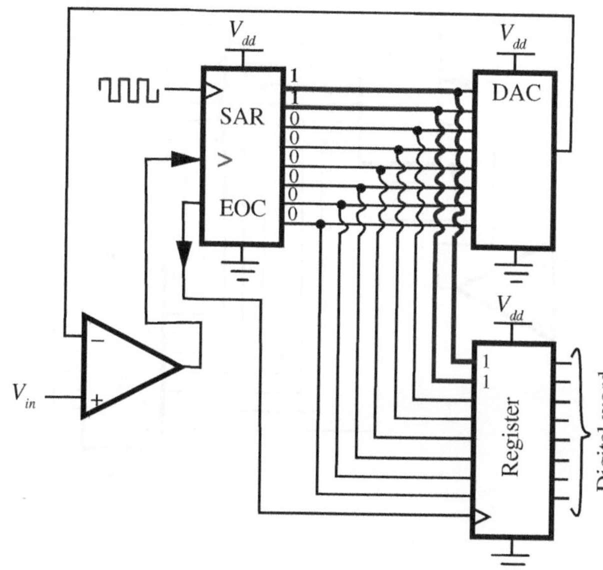

Figure 6.25 Step 2 of execution of ADC for Example 6.7

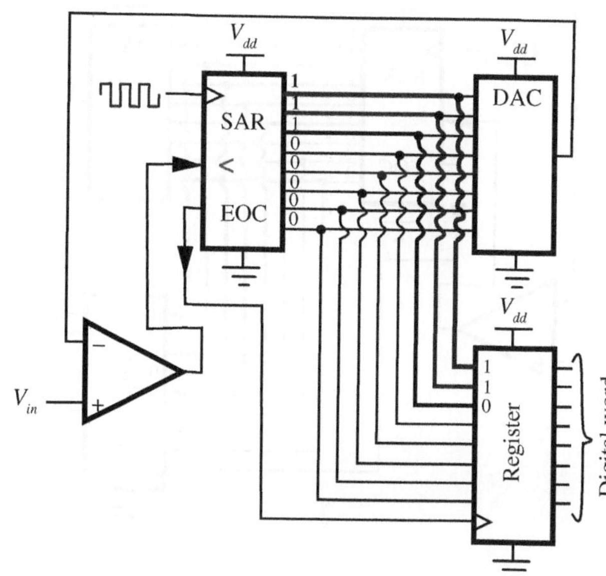

Figure 6.26 Step 3 of execution of ADC for Example 6.7

Step 6: it runs for the bit of $2^N - 6 = 2$, as shown in Figure 6.29. The SAR sets the second bit as "1" and the rest of the bits as "0," it generates a digital word "**11000**100" as the input of DAC, the DAC converts it into the analog voltage $V_{ref} = (1 + 1/2 + 1/32)\cdot V_{full_{scale}}/2 = 7.65625(\text{V})$ to the inverting input, the comparator compares V_{ref}(7.65625 V) and V_{in} (7.65 V) and finds $V_{ref} < V_{in}$, and it sets the third bit in the register as "0,"

1	1	0	0	0	0		

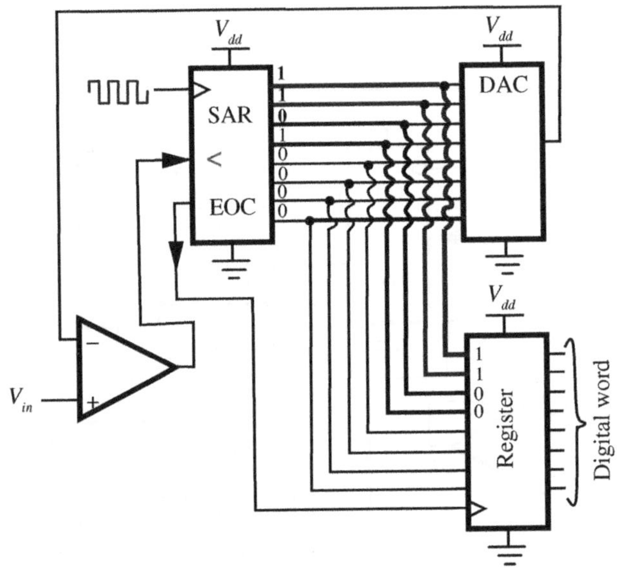

Figure 6.27 Step 4 of execution of ADC for Example 6.7

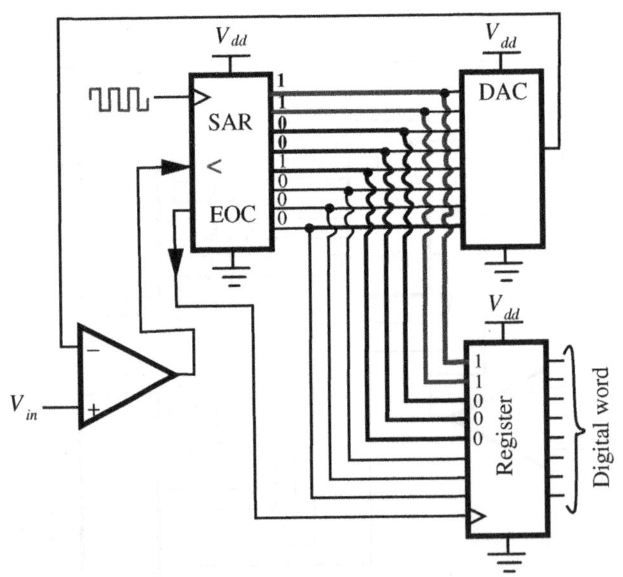

Figure 6.28 Step 5 of execution of ADC for Example 6.7

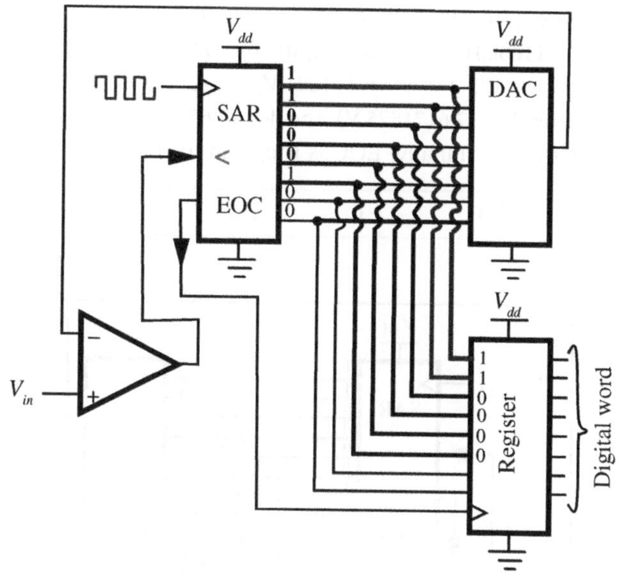

Figure 6.29 Step 6 of execution of ADC for Example 6.7

Step 7: it runs for the bit of $2^N - 7 = 1$, as shown in Figure 6.30. The SAR sets the first bit as "1" and the last bit as "0," it generates a digital word "**110000**10" as the input of DAC, the DAC converts it into the analog voltage $V_{ref} = (1 + 1/2 + 1/64) \cdot V_{full_{scale}}/2 = 7.578125\,(\text{V})$ to the inverting input, the comparator compares V_{ref} (7.578125 V) and V_{in} (7.65 V) and finds $V_{ref} > V_{in}$, and it sets the second bit in the register as "1,"

1	1	0	0	0	0	1	

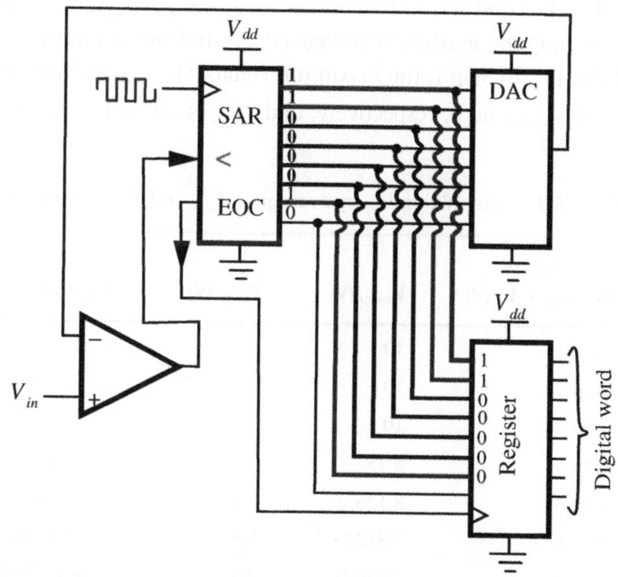

Figure 6.30 Step 7 of execution of ADC for Example 6.7

Step 8: it runs for the LSB, as shown in Figure 6.31. The SAR sets LSB as "1," it generates a digital word "**11000011**" as the input of DAC, the DAC converts it into the analog voltage $V_{ref} = (1 + 1/2 + 1/64 + 1/128) \cdot V_{full_{scale}}/2 = 7.6171875\,(\text{V})$ to the inverting input, the comparator compares V_{ref} (7.6171875 V) and V_{in} (7.65 V) and finds $V_{ref} > V_{in}$, and it sets the first bit in the register as "1." Finally, EOC is triggered to complete the conversion process.

1	1	0	0	0	0	1	1

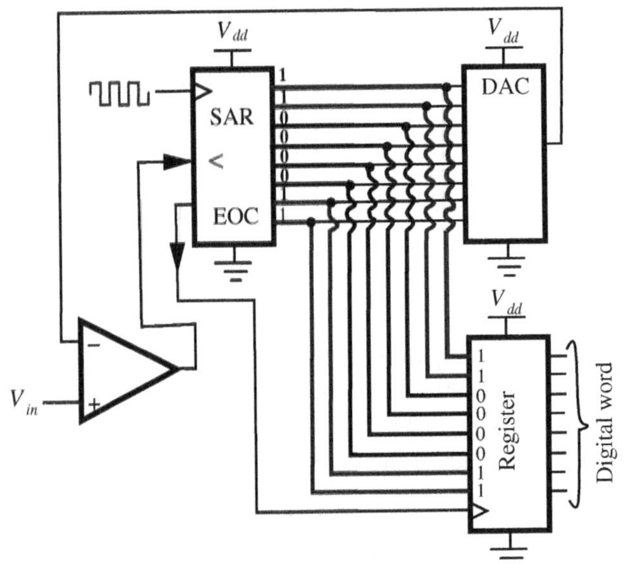

Figure 6.31 Step 8 of execution of ADC for Example 6.7

For the hand verification of reference voltage (V_{ref}) at each step sequentially, it can be calculated by averaging the maximum voltage (V_{max}) and the minimum voltage (V_{min}) at each step as shown in Table 6.3. In Step 1, the maximum voltage ($V_{max,0}$) and the minimum voltage ($V_{min,0}$) are initialized as $V_{full\text{-}scale}$ and 0, respectively, and $V_{ref,i}$ is set as $V_{ref,i} = \frac{V_{max,i} - V_{min,i}}{2}$; for a sequential step, when

Table 6.3 Determining V_{ref} by averaging V_{max} and V_{min} in successive approximation

V_{in} = 7.65 (V) V_{full_scale} = 10 (V)	$V_{max,i}$ (V)	$V_{min,i}$ (V)	$V_{ref} = (V_{max,i} - V_{min,i})/2$	$V_{in} > V_{ref,i}$?	Bit
Step 1	10	0	5	True	$D_7 = 1$
Step 2	10	5	7.5	True	$D_6 = 1$
Step 3	10	7.5	8.75	False	$D_5 = 0$
Step 4	8.75	7.5	8.125	False	$D_4 = 0$
Step 5	8.125	7.5	7.8125	False	$D_3 = 0$
Step 6	7.8125	7.5	7.65625	False	$D_2 = 0$
Step 7	7.65625	7.5	7.578125	True	$D_1 = 1$
Step 8	7.65625	7. 578125	7. 617875	True	$D_0 = 1$

$V_{in} > V_{ref,i}$, $V_{min,i+1}$ is set as $V_{ref,i}$ (i.e., $V_{min,i+1} = V_{ref,i}$) and $V_{max,i+1}$ remains unchanged (i.e., $V_{max,i+1} = V_{max,i}$); otherwise, $V_{min,i+1}$ remains unchanged (i.e., $V_{min,i+1} = V_{min,i}$), and $V_{max,i+1}$ is set as $V_{ref,i}$ (i.e., $V_{max,i+1} = V_{ref,i}$). The same step repeats until the LSB is processed.

To determine the inherent error in above converting process, one can see the digitized word for $V_{in} = 7.65$ (V)

is "**11100011**"; therefore, the corresponding analog voltage is,

$$\widetilde{V}_{in} = V_{full_{scale}} \frac{1 \times 2^8 + 1 \times 2^7 + 0 \times 2^6 + 0 \times 2^5 + 0 \times 2^4 + 0 \times 2^3 + 1 \times 2^2 + 1 \times 2^1}{2^8 - 1} = 7.617\,\text{V}$$

The error of conversion is $error = \left|V_{in} - \widetilde{V}_{in}\right| = 0.033$ (V).

Example 6.8 Find the digital form of an analog voltage $V_{in} = 5$ (V) by an 8-bit successive approximation ADC for $V_{full\_scale} = 15$ (V); determine the inherited error due to this ADC by hands.

Solution

The approach illustrated in Table 6.4 is used to solve this problem, and Table 6.4 shows the digitizing process for $V_{in} = 5$ (V).

Table 6.4 shows that the digitized word for $V_{in} = 5$ (V) is "**01010101**"; therefore, the corresponding analog voltage is,

$$\widetilde{V}_{in} = V_{full_{scale}} \frac{0 \times 2^8 + 1 \times 2^7 + 0 \times 2^6 + 1 \times 2^5 + 0 \times 2^4 + 1 \times 2^3 + 0 \times 2^2 + 1 \times 2^1}{2^8 - 1} = 5\,\text{V}$$

The error of conversion is $error = \left|V_{in} - \widetilde{V}_{in}\right| = 0$ (V).

Traditionally, the role of each bit corresponds to a bit weight in a successive approximation ADC. When a bit is set as "1," it adds its bit weight into the total analog value. Therefore, the working principle of a successive approximation ADC can be represented graphically. Figure 6.32a shows the bit weights in an 8-bit word, and Figure 6.32b illustrates the digitizing process for $V_{in} = 5.60$ (V) and $V_{full\_scale} = 10$ (V).

Table 6.4 Digitizing V_{in} in $V_{full\_scale} = 10$ (V) by 8-bit successive approximation

$V_{in} = 5$ (V) $V_{full\_scale} = 15$ (V)	$V_{max,i}$ (V)	$V_{min,i}$ (V)	$V_{ref} = (V_{max,i} - V_{min,i})/2$	$V_{in} > V_{ref,i}$?	Bit
Step 1	15	0	7.5	False	$D_7 = 0$
Step 2	7.5	0	3.73	True	$D_6 = 1$
Step 3	7.5	3.75	5.63	False	$D_5 = 0$
Step 4	5.63	3.75	4.69	True	$D_4 = 1$
Step 5	5.63	4.69	5.16	False	$D_3 = 0$
Step 6	5.16	4.69	4.93	True	$D_2 = 1$
Step 7	5.16	4.93	5.05	False	$D_1 = 0$
Step 8	5.05	4.93	4.99	True	$D_0 = 1$

Bit	Weight (w_i) (V)
7	$10/(2^1) = 5$
6	$10/(2^2) = 2.5$
5	$10/(2^3) = 1.25$
4	$10/(2^4) = 0.625$
3	$10/(2^5) = 0.3125$
2	$10/(2^6) = 0.15625$
1	$10/(2^7) = 0.07815$
0	$10/(2^8) = 0.039075$

(a) Table of bit weights

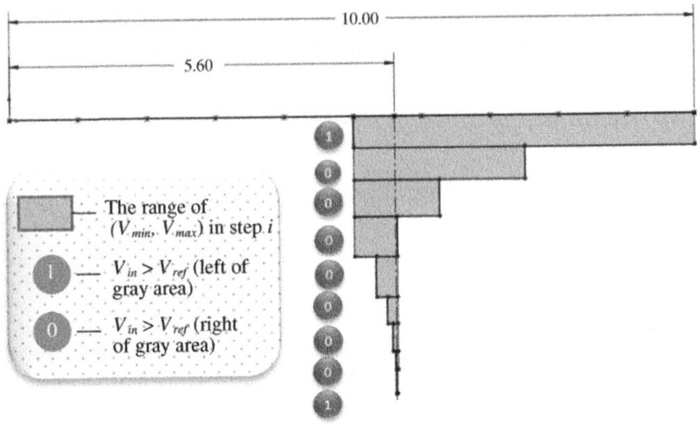

(b) Digitizing V_{in} = 7.65 (V) and V_{full_scale} = 10 (V)in 8 bits

Figure 6.32 Working principle of successive approximation elaborated by bit weights

6.7 Holding Process in Sampling

An analog signal generally varies over time, while a digitizing process by integrating ADC or successive approximation ADC takes time. If an input signal is directly connected to the ADC, the digitizing process is likely unreliable and unstable. Therefore, it is necessary for the ADC to acquire a sample at a time and hold the voltage for a certain period to convert into its digital form completely, and this is achieved by a sampling and holding (S/H) circuit.

Figure 6.33 shows the diagram of a simplified S/H circuit. When the switch is connected, it samples the input voltage; when the switch is open, it holds the sampled voltage since the capacitor is charged. Each cycle of conversion consists to two periods: *the first period* is the acquisition time to charge the capacitor when the switch is turned on; *the second period* is the conversion time from the start of a conversion (e.g., turning on the MSB in an SAR) until the EOC (e.g., turning ON/OFF bit-0 in an SAR).

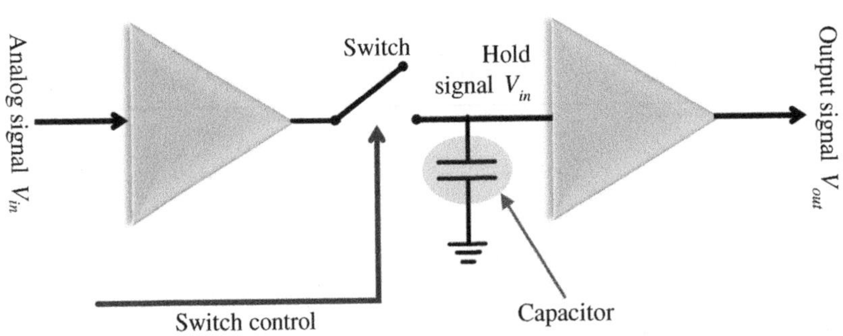

Figure 6.33 Diagram of simplified *sampling and holding* (S/H) circuit

6.8 Digital-to-Analog Conversion (DAC)

Digital-to-analog conversion (DAC) is to produce an analog signal from a digital signal; DACs are generally required when cyber systems in the virtual world are used to control objects in the physical world. In principle, a DAC can be implemented by adding the contributions of individual bits based on their positions as,

$$\widetilde{V}_{in} = \sum_{i=0}^{N-1} b_i w_i \tag{6.13}$$

Example 6.9 Assume that an 8-bit DAC is used to representing an analog signal with $V_{full\text{-}scale}$ = 10 (V), determine the corresponding voltages of analog signals for the digital words of "11100111," "00101010," and "10111101."

Solution

The bit weights of an 8-bit DAC for $V_{full\text{-}scale}$ = 10 (V) have been provided in Figure 6.32a. It is reused here in weighting the contributions of individual bits for three digital words as follows.

Bit	Weight (w_i) (V)	"11100111" → (b_i)	"00101010" → (b_i)	"10111101" → (b_i)
7	$10/(2^1) = 5$	1	0	1
6	$10/(2^2) = 2.5$	1	0	0
5	$10/(2^3) = 1.25$	1	1	1
4	$10/(2^4) = 0.625$	0	0	1
3	$10/(2^5) = 0.3125$	0	1	1
2	$10/(2^6) = 0.15625$	1	0	1
1	$10/(2^7) = 0.07815$	1	1	0
0	$10/(2^8) = 0.039075$	1	0	1
	$\sum_{i=0}^{N-1} b_i w_i$	9.023475 (V)	1.64065 (V)	7.382825 (V)

The weight of a bit in a digital word decreases from SMB to LSB; the weight of an LSB corresponds to the resolution of DAC. According to Eq. (6.5), the conversion resolution is determined by (1) the voltage range V_R and (2) the number of bits (N) in the digital word. Table 6.5 shows the change of the resolution with the number of bits when V_R is set as $V_R = 10$ (V).

In physical implementations, the contributions of individual bits to an analog signal can be added either as *voltages* or as *currents*; accordingly, two types of DACs are *weighted resistor* DACs and *R-2R Ladder* DACs.

6.8.1 Weighted Resistor DAC

A weighted resistor DAC uses a set of binary weighted resistors in the inverting adder circuit. Each binary resistor is controlled by a digital switch, the switch is turned "on" or "off" when the value of the corresponding bit is "1" and "0," respectively.

In Figure 6.34, the virtual short concept indicates that the voltage at the inverting input of the op-amp is the same to the voltage at the non-investing input; in other words, the node at the inverting input has a zero volt since the non-inverting input is ground (Tutorrisalpoint 2023),

Table 6.5 Resolution versus number of bits when $V_R = 10$ (V)

Number of bits (N)	Quantization levels (2^N)	Resolution ΔV (weight of LSB)	Relative resolution (%)	dB (20·log ($\Delta V/V_R$))
8	256	0.03906	0.0003906	−48.0
10	1,024	0.00977	0.0000977	−60.0
12	4,096	0.00244	0.0000244	−72.0
14	16,384	0.000610	0.00000610	−84.0
16	65,536	0.000153	0.00000153	−96.0
18	262,164	0.000038	0.00000038	−108.0
20	1,048,576	0.0000095	0.000000095	−120.0

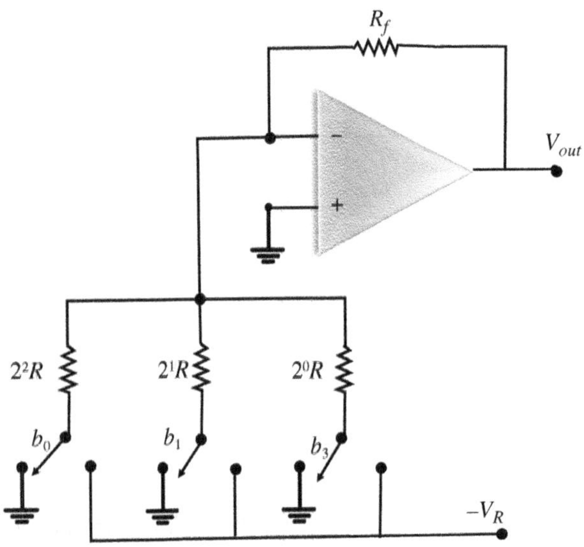

Figure 6.34 3-Bit weighted resistor DAC

$$\frac{0 + V_R b_2}{2^0 R} + \frac{0 + V_R b_1}{2^1 R} + \frac{0 + V_R b_0}{2^2 R} + \frac{0 - V_0}{R_f} = 0 \tag{6.14}$$

Rewriting Eq. (6.14) finds V_0 as,

$$V_0 = \frac{V_R R_f}{R}\left(\frac{b_2}{2^0} + \frac{b_1}{2^1} + \frac{b_0}{2^2}\right) \tag{6.15}$$

Furthermore, let $R = 2R_f$, V_0 in Eq. (6.15) becomes,

$$V_0 = \frac{V_R}{2}\left(\frac{b_{N-1}}{2^0} + \frac{b_{N-2}}{2^1} + \cdots + \frac{b_0}{2^{N-1}}\right) \tag{6.16}$$

Since 3 bits in a digital word have a set of eight combinations from "000" to "111," a 3-bit DAC has eight possible values of the output voltages.

Generically, when the number of bits is set as N, the voltage in a weighted resistor DAC can be found as,

$$V_0 = \frac{V_R}{2}\left(\frac{b_2}{2^0} + \frac{b_1}{2^1} + \frac{b_0}{2^2}\right) \tag{6.17}$$

A weighed resistor DAC is very fast and the gate slew rate can be -3 V/ns, and the resistors can be scaled to any expected output impedances; however, two main disadvantages of a weighted resistor DAC are (1) the difference of the resistors corresponding to MSB and LSB is increased with the number of bits in a digital word increase; (2) when the number of bits increases, it is challenging to produce different resistors accurately.

6.8.2 R-2R Ladder DAC

An R-2R ladder DAC provides an inexpensive and simple way to convert a digital signal into its analog form. It differs from a weighted resistor DAC in sense that the weights of individual bits are summed in currents rather than voltages. In an R-2R ladder, two precision resistors are converting a digital binary number into an analog signal that is proportional to the value of the digital number. The word "ladder" refers to a ladder-like configuration of the resistors in the network., and Figure 6.35 shows an R-2R ladder for a N-bit DAC.

There are only two types of resistors (i.e., R and 2R) in the ladder configuration. The switches are analog switches that are controlled by digital signals. In different levels of the ladder, a series of currents, that is, $I/2$, $I/4$, ... , $I/(2^N - 1)$ is generated.

Figure 6.36a shows the bottom level of the ladder, two $2R$ resistors in parallel means that two units of current I_0 will be split equally into 2 of I_0 in each branch; Figure 6.36b shows the second

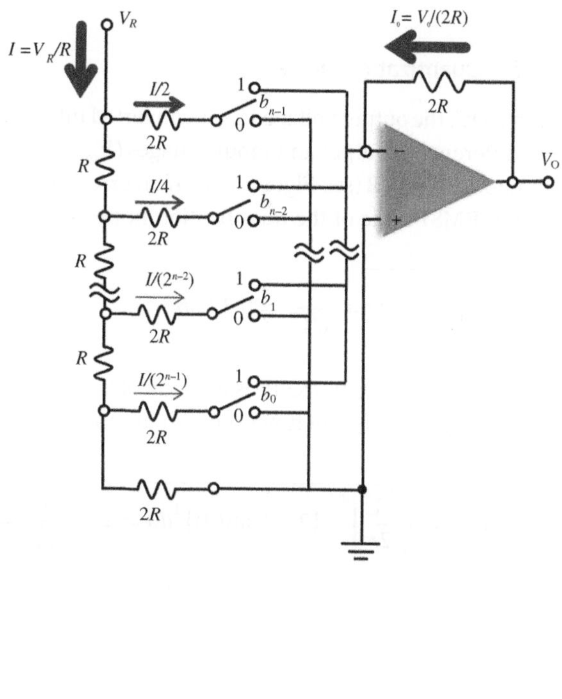

Figure 6.35 Configuration of an R-2R ladder DAC with N-bit

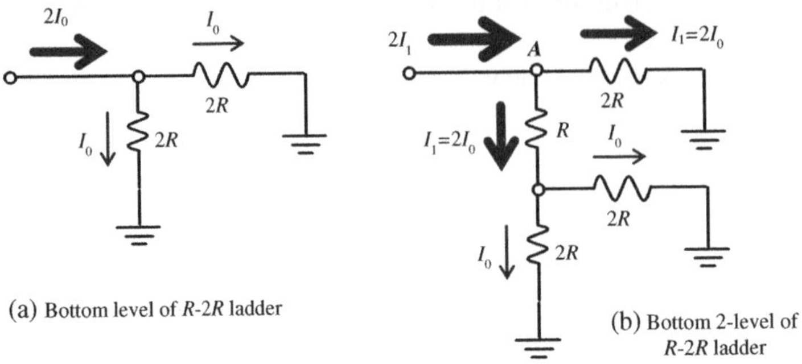

Figure 6.36 Bottom levels of an R-2R ladder

to the bottom level of the ladder; the Thévenin resistances of the two branches split at A both equal 2*R*, so the current into this node will split evenly, that is, $I_1 = 2I_0$.

More levels can be added continuously in the ladder under the required number of bits are met. For an R-2R ladder with *N*-bit, the current of the whole serious will be $I_N = 2^N \cdot I_0$, For the voltage drop across the horizontal resistors, $V_N = 2R \cdot I_N = 2^{(N+1)}R \cdot I_0$. Since the input voltage is $V_{in} = (2^N \cdot R)/I_0$, I_0 can be found as $I_0 = V_{in}/(2^N \cdot R)$. Assume that the total current at the summing junction is $V_d \times I_0$, then the output voltage can be found as,

$$V_{out} = (-R_f)\cdot(V_{in} \times I_0)/R_T \tag{6.18}$$

where V_{out} are the output analog voltage, R_T is the equivalence resistor of the R-2R ladder, R_f is the feedback resistor, and V_{in} is the voltage is digital form.

Since voltage switches are replaced by current switches, all nodes remain at a constant voltage. There is no need to charge/discharge node capacitances. *Complementary metal–oxide–semiconductor* (CMOS) gates can be used as switches, and the output terminals of all switches are grounded as 0 V.

6.8.3 Quantization Noise

In an ADC, the output voltage V_{out} is converted into discrete levels that does not follow V_{in} precisely. The difference of output and input voltages ($V_{out} - V_{in}$) is called as quantization noise, and it has an amplitude of ± LSB (see Figure 6.37). If all error values are with an equal probability, the *root mean square* (RMS) value of the quantization noise is,

$$\sqrt{\int_{-1/2}^{1/2} x^2 dx} = \frac{1}{\sqrt{12}} = 0.3\,LSB \tag{6.19}$$

A *signal-to-noise ratio* (SNR) is the ratio of the amplitude of the source signal to the noise level. Let a source signal be a sine function with an amplitude of 2^{N-1}, the corresponding RMS is,

$$V_{rms} = \sqrt{\frac{1}{2\pi}\int_{-\pi}^{\pi} \{2^{n-1}\sin(\theta)\}^2 d\theta} = 2^{n-1}\frac{1}{\sqrt{2}} = 0.35 \times 2^n$$

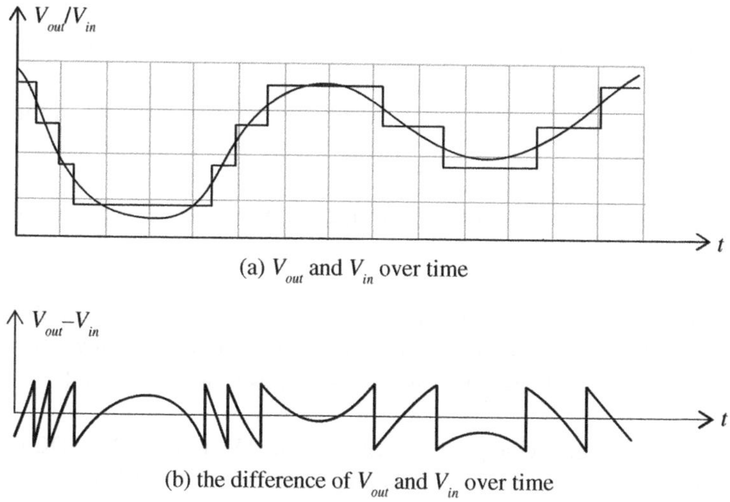

Figure 6.37 Quantization noise in ADC

The SNR of the quantization is determined as,

$$20\log_{10}\left(\frac{0.35\times 2^N}{0.3}\right) = 1.8 + 6N\ (\text{dB})$$

6.9 Summary

A mechatronic system is an integration of a set of physical mechatronic components together with their digital models in the virtual world. It is essential to transmit and exchange data between the physical and virtual worlds so that the mechatronic system and its application environment can understand each other mutually. Since the signals in the physical and virtual worlds are in different formats, that is, analog and digital formats, ADCs and DACs are developed to transfer signals from one way to the other. Op-amps and comparators are two basic types of electronic elements to implement an ADC or DAC. ADCs are classified into three basic types, that is, integrating ADCs, flash converters, and successive approximation ADCs. DACs are classified into weighted resistor DACs and R-2R ladder DACs. In applications, an ADC is characterized by the sampling and quantizing processes that determine the accuracies of time digitization and amplitude digitization.

Problems

6.1 For $x(t) = 5\sin(5\pi t) + \sin(20\pi t)$, can you reconstruct $x(t)$ from the samples of $x(t)$ at a rate of 21 Hz? Explain how you obtain your answer.

6.2 Let the sampling frequency is $f_s = 500$ samples/s, determine if it is appropriate to digitize an analog signal with the frequency of $f_a = 5$, 90, and 1200 cycles, respectively.

6.3 Assume that the input analog voltage is changing between 0 and 10 V and use an 8-bit analog-to-digital conversion (ADC) draw the output, (1) calculate the resolution of the converter;
(2) determine the maximum voltage of output signal; (3) what is the ideal input when the output digital word is "10001001"?

6.4 What is the minimal sampling rate to reconstruct an input signal $x\ (t) = 5\ \sin^2\ (5\pi t) + \cos\ (20\pi t)$ correctly?

6.5 Suppose that there is a 10-bit A/D converter with $V_{ref-} = 0$ (V) and $V_{ref+} = 5$ (V): (1) what is the digital code for an analog 2.2 (V)? What is the error from analog-to-digital conversion (ADC)? (2) Find the corresponding voltage values for the ADC result of "0000011001."

6.6 Assume that the input analog voltage is changing between −2 and 2 V, and an 8-bit ADC is used in quantization, (1) calculate the number of quantization levels; (2) calculate the resolution in digitization.

6.7 The low and high reference voltages of an analog signal are $V_{ref-} = 2$ (V) and V_{ref}, = 5 (V), respectively; if the ADC uses 8 bits in the conversion, (1) what is the digital code for an analog 4.2 (V)? (2) What is the inherent error in quantizing a 3.2 (V)? (3) Find the corresponding voltages for the digital words of (i) "00001001," (ii) "01010000," (iii) "01110000," (iv) "01111100," (v) "10110000," (vi) "11001000," and (vii) "11100100."

6.8 The low and high reference voltages of an analog signal are $V_{ref-} = 0$ (V) and V_{ref}, = 5 (V), respectively; if a distinguishable voltage difference 0.125 (V) is required in digitizing the analog signal, how many bits are required in a digital word? What are the finite set of digital voltage states?

6.9 Find the digital form of an analog voltage $V_{in} = 4$ (V) by an 8-bit successive approximation ADC for $V_{full\_scale} = 10$ (V); determine the inherited error due to this ADC by hands.

6.10 Find the digital value of V_{in} using a 4-bit ADC for $V_{in} = 6.28$ (V) and $V_{full\_scale} = 10$ (V), and determine the inherited error due to the analog-to-digital conversion.

6.11 Assume that an 8-bit DAC is used to representing an analog signal with $V_{full\text{-}scale} = 8$ (V), determine the corresponding voltages of analog signals for the digital words of "11100111," "00101010," and "10111101."

6.12 Discuss one approach to implement DAC.

References

Analog (2023). Fundamental of sampled data systems. https://www.analog.com/media/en/training-seminars/design-handbooks/data-conversion-handbook/chapter2.pdf (accessed 12 September 2024).

Bi, Z.M. and Wang, L. (2010). Dynamic control model of a cobot with three omni-wheels. *Robotics and Computer-Integrated Manufacturing* 26: 558–563.

Bi, Z.M., Lang, S.Y.T., and Verner, M. (2008). Dynamic modeling and validation of a tripod-based machine tool. *International Journal of Advanced Manufacturing Technology* 37 (3–4): 410–421.

Loewenstein, E.B. (2000). Analog-to-digital converters. http://dsp-book.narod.ru/MISH/CH85.PDF (accessed 12 September 2024).

Microchip (2023). Analog and interface product solutions. https://www.microchip.com/downloads/en/devicedoc/21841a.pdf (accessed 12 September 2024).

Tutorrisalpoint (2023). Digital to analog converters. https://www.tutorialspoint.com/linear_integrated_circuits_applications/linear_integrated_circuits_applications_digital_to_analog_converters.htm (accessed 12 September 2024).

Ultimate Electronics (2024). The ideal op-amp (operational amplifier). https://ultimateelectronicsbook.com/ideal-op-amp (accessed 12 September 2024).

7

Signal Conditioning and Processing

7.1 Introduction

Figure 7.1 shows the procedure in which a physical quantity in a mechatronic system is sensed, processed, conditioned, transmitted, processed, and then utilized in system controls. Due to the need to use collected data in closed-loop control of the mechatronic system, an input signal to the system must be in its right form, and it is usually different from its original form. The conversion from its original form to a new form will be achieved by *signal conditioning and processing* (SCP).

Depending on the expected form of the designated signal, more than one signal-conditioning function, such as amplification and filtering, should be performed on the signal. Figure 7.2 exemplifies the SCP in a weight measurement system. It is clear from this example that many factors may prevent the original signal collected from a source from being used in its application directly. To use the collected signal, it requires some intermediate circuitries to prepare the signal in its right form. These intermediate circuitries are referred to as electronic elements for SCP that will be discussed in depth. In this chapter, *first*, the features of basic electronic elements are reviewed. *Second*, common SCP operations are classified into three categories, that is, signal cleaning, signal conditioning, and signal transmission; some typical tasks and corresponding implementations in each category are discussed in Sections 7.3, 7.4, and 7.5, respectively. *Third*, available SCP tools are overviewed. *Finally*, the introduction and discussions on SCP are summarized in Section 7.6.

7.2 Basic Concepts in Electronic Circuits

In a mechatronic system, MFMs themselves are modularized and built from the modules at low levels. For example, commonly used electronic elements such as op-amps and comparators in Chapter 6 are built upon low-level electronic elements such as resistors, capacitors, and inductors. Therefore, the concepts and features of these basic electronic elements are reviewed in this section before we discuss the implementations of electronic elements for SCP.

7.2.1 Charge, Current, Voltage, and Power

An electronic circuit is for the movement of electrons; each electron carries a charge of $1.602176634 \times 10 \times 10^{-19}$ C. Electrons move and carry *charges* (Q) that affect *current* (I) and *voltage* (V) in the circuit. Therefore, the change of energy or work in a circuit relates to the movement of electronics fundamentally.

Mechatronics for Complex Products and Systems: Project-Based Design Approaches for Robots, Cyber-Physical Systems, Digital Twins, and Other Emerging Technologies, First Edition. Zhuming Bi.

Companion website: www.wiley.com/go/bi/mechatronics

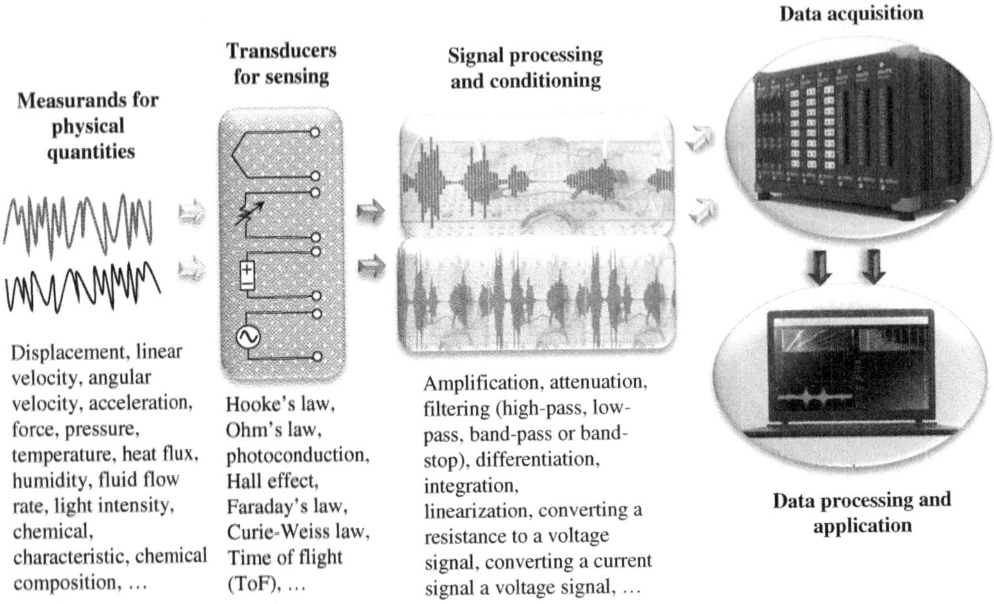

Figure 7.1 Signal conditioning and processing (SCP) in data flow from source to system control (*Source:* (a) Xtemp (Pty) Ltd / https://www.xtemp.co.za/product/iolite-data-acquisition/ / Last accessed on July 19, 2024; (b) Dewesoft / https://downloads.dewesoft.com/brochures/dewesoft-dsa-nvh-brochure-en.pdf / Last accessed on July 19, 2024)

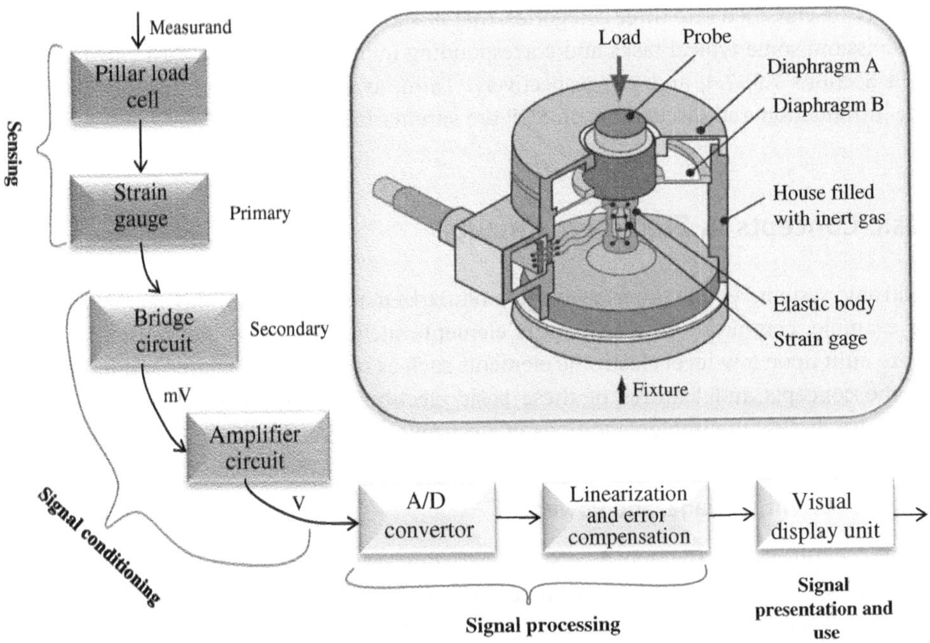

Figure 7.2 Signal conditioning and processing (SCP) in a weight measurement system

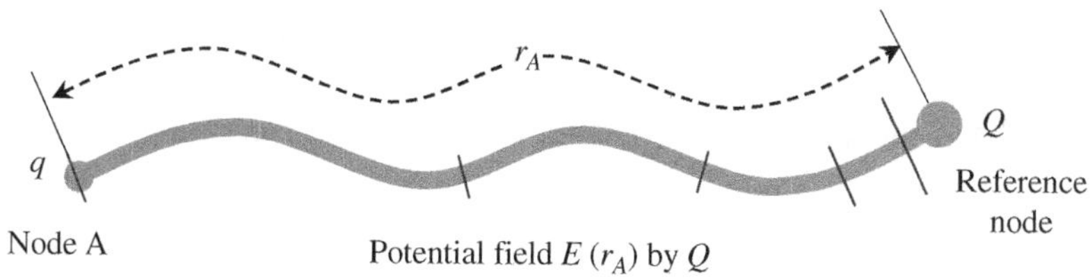

Figure 7.3 Coulomb's law applied two nodes in a circuit

Coulomb's Law governs the behaviors of static charges in a circuit. When the electrons move in the circuit, they work in an electric field that is described by *electric potential* (or voltage), and it is associated with *electric potential energy*. Note that the electric potential by Q is *directional*, and it is along a radical line pointing away from Q. Figure 7.3 illustrates an electric field between an arbitrary node A by taking a reference node in the circuit. Let the amount of charge at the reference node be Q and the amount of charge at node A be q.

The potential attractive force (F) of q by Q can be determined by Coulomb's law as (Khanacademy 2024),

$$F = q \cdot E = q \cdot \frac{Q}{4\pi\varepsilon_0 r_A^2} \tag{7.1}$$

where F is Coulomb's force between q and Q; ε_0 is a constant; r_A is the distance of node A from the reference node; $E = \dfrac{Q}{4\pi\varepsilon_0 r_A^2}$ is referred to as the electric field at node A (d_A) by Q.

The force (F) drives q to move in the circuit, and this involves an amount of work (W) since the distance to the reference node is changed. The work W by F can be found as,

$$dW = F \cdot dr = q \cdot E \cdot dr = q \cdot \frac{Q}{4\pi\varepsilon_0 r^2} dr \tag{7.2}$$

where dr is the infinite step of distance change when q moves in the circuit; dW is the corresponding amount of work by the movement of q.

When the charge q moves from node A to B, Eq. (7.2) can be used to determine the change of the work (W_{AB}) as,

$$W_{AB} = \int_{r_A}^{r_B} dW = \int_{r_A}^{r_B} q \cdot \frac{Q}{4\pi\varepsilon_0 r^2} dr = q \cdot \left(\frac{Q}{4\pi\varepsilon_0 r_B} - \frac{Q}{4\pi\varepsilon_0 r_A}\right) = q \cdot (U_B - U_A) \tag{7.3}$$

where $U_A = \dfrac{Q}{4\pi\varepsilon_0 r_A}$ and $U_B = \dfrac{Q}{4\pi\varepsilon_0 r_B}$ represent the electric potential energy stored in the charge q when it is distanced r_A and r_B from Q, respectively.

U_A and U_B in Eq. (7.2) suggest that a property about the location of q from Q can be generalized as,

$$U_r = \frac{Q}{4\pi\varepsilon_0 r} \tag{7.4}$$

where U_r represents the amplitude of an electric potential field in the circuit charged by Q at the reference node.

The difference in the amplitudes of an electric potential field at two locations (i.e., A and B) is commonly referred to as the voltage difference (V_{AB}) between nodes A and B. The voltage difference can be determined by Eq. (7.3) as (Spinning Number 2023),

$$V_{AB} = \frac{W_{AB}}{q} = U_B - U_A = \frac{Q}{4\pi\varepsilon_0 r_B} - \frac{Q}{4\pi\varepsilon_0 r_A} \tag{7.5}$$

Let $r_A = \propto$ and $r_B = r$ in Eq. (7.5), the corresponding voltage difference is called *an absolute voltage* at r as,

$$V = \left.\frac{Q}{4\pi\varepsilon_0 r_B}\right|_{r_B = r} - \left.\frac{Q}{4\pi\varepsilon_0 r_B}\right|_{r_A = \propto} = \frac{Q}{4\pi\varepsilon_0 r} = \frac{kQ}{r} \tag{7.6}$$

where $k = \frac{1}{4\pi\varepsilon_0}$ is a constant equal to 9.0×10^9 $(\mathrm{N} \cdot \mathrm{m}^2/\mathrm{C}^2)$; r is the distance of a test charge (q) from Q at $r_A = \propto$.

Equation (7.3) can also be used to determine the amount of the transferred power (W) by the charge (q) subjected to a potential field with a voltage of (V),

$$W = q \times V \tag{7.7}$$

where the power (W) is measured in the unit of joules (J); the units of q and V are *Coulomb* and *Voltage*, respectively.

The movement of electrons forms a flow of electric charge called *Current* (I), and it is the amount of moved charge in a time unit as,

$$I = \frac{q}{t} \tag{7.8}$$

where t is the amount of time to move charge q. The units of t and I are *second* (s) and *Ampere* (A), respectively.

The rate of the transferred power (P) in a unit of time can be derived from Eqs. (7.7) and (7.8) as,

$$P = \frac{W}{t} = \frac{q}{t} \times V = I \cdot V \tag{7.9}$$

where the unit of power is *Joules per second* or *Watts* in the *International System of Units* (SI) system.

Aforementioned Coulomb's law alone cannot represent the behaviors of electrons in a circuit completely since the circuit involves two variables (e.g., q and Q) from Eq. (7.1) to Eq. (7.8). In this case, *Ohm's law* is introduced to model the relation of I and V. Ohm's law states that a current (I) between two nodes is proportional to the voltage difference (V) of these nodes as,

$$I = \frac{V}{R} \tag{7.10}$$

where R is called an *impendence* in a circuit.

7.2.2 Resistor, Capacitor, and Inductor

Figure 7.4 shows the properties of an impendence in three basic electronic elements, that is, resistors, capacitors, and inductors.

Figure 7.4a shows a resistor whose resistance (R) is independent of current I or voltage V. The resistance R is determined by the size and material of the resistor as,

$$R = \rho\frac{L}{A} \tag{7.11}$$

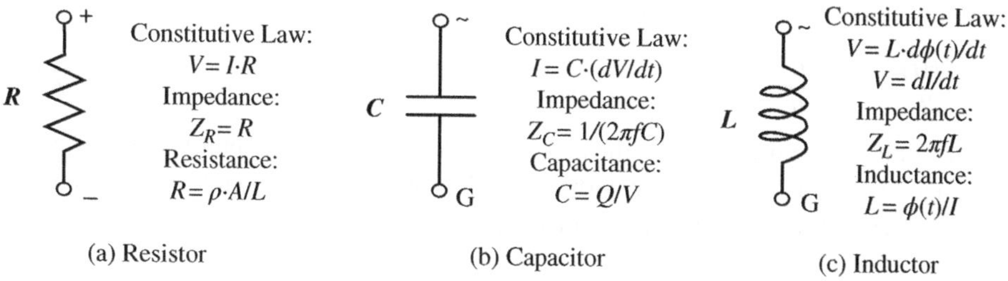

Figure 7.4 Resistor, capacitor, and inductor

where L and A are the length and the cross-sectional area of the resistor, respectively; ρ is the resistivity in *Ohm-meters.* The unit of R is *Ohms* in the SI unit system. The impedance (Z_R) of a resistor is constant as R.

Figure 7.4b shows a capacitor that stores the electrical energy in an electric field by accumulating electric charges on two spaced surfaces. The pair of two surfaces are insulated with each other. A capacitor is characterized by its capacitance (C) as,

$$C = \frac{Q}{V} \tag{7.12}$$

where Q and V are the amount of charge and the applied voltage in the capacitor, respectively. The unit of C is *farad* in the SI unit system. One farad means that one Coulomb of charge Q on a conductor causes one voltage across the capacity.

The impedance (Z_C) of a capacitor is,

$$Z_C = \frac{1}{2\pi fC} \tag{7.13}$$

where f is the frequency of the current or voltage change.

Figure 7.4c shows an inductor that stores the electrical energy in a magnetic field when an electric current flows through it. An inductor is usually made of coils. When a current flows into the coil, the magnetic field changes, and this induces an *electromotive force (emf)* voltage in the conductor, this phenomenon is described as *Faraday's law of induction.* Therefore, an inductor is characterized by its inductance (L) as,

$$L = \frac{\phi(t)}{I} \tag{7.14}$$

where $\phi(t)$ is a magnetic flux generated by current I. The inductance (L) of an electric element is determined by the geometry of a current path and the magnetic permeability. The unit of L is *Henry.*

According to Faraday's law of induction, when the current flows through the circuit, the inductor creates a changing flux and includes a voltage in the inductor. The voltage (ε) induced by a change of magnetic flux can be found as,

$$\varepsilon = -\frac{d\phi(t)}{dt} \tag{7.15}$$

Moreover, Eq. (7.14) can be reformatted as,

$$\phi(t) = L \cdot I \tag{7.16}$$

Substituting Eq. (7.16) into (7.15) yields,

$$\varepsilon = -\frac{d\phi(t)}{dt} = -L\frac{dI}{dt} \tag{7.17}$$

An inductance can also be measured by the amount of *electromotive force* (emf) generated for a given rate change of current. An inductor with $L = 1$ (*Henry*) produces an emf of 1 V when the current flows through the inductor at the rate of 1 A per second.

The impedance (Z_L) of a capacitor is,

$$Z_L = 2\pi fL \tag{7.18}$$

where f is the frequency of the current or voltage change.

7.2.3 Input Loading and Output Loading

Section 7.2.2 shows that the impedance of an electronic element affects the behaviors of the circuit. When an electronic element is used in SCP, it can be either at the input side or the output side of the circuit, as shown in Figure 7.5. The impedance of the electronic element must be carefully selected to minimize its impact on a source or an output signal.

Figure 7.5a shows the case of an input loading (R_s). The use of the input loading affects the amplitude of the source signal (V_{IN_L}) from V_S as,

$$V_{IN\_L} = \frac{R_{IN\_L}V_S}{R_S + R_{IN\_L}} \tag{7.19}$$

where R_{IN_L} is the equivalent resistance at that part of the loading circuit.

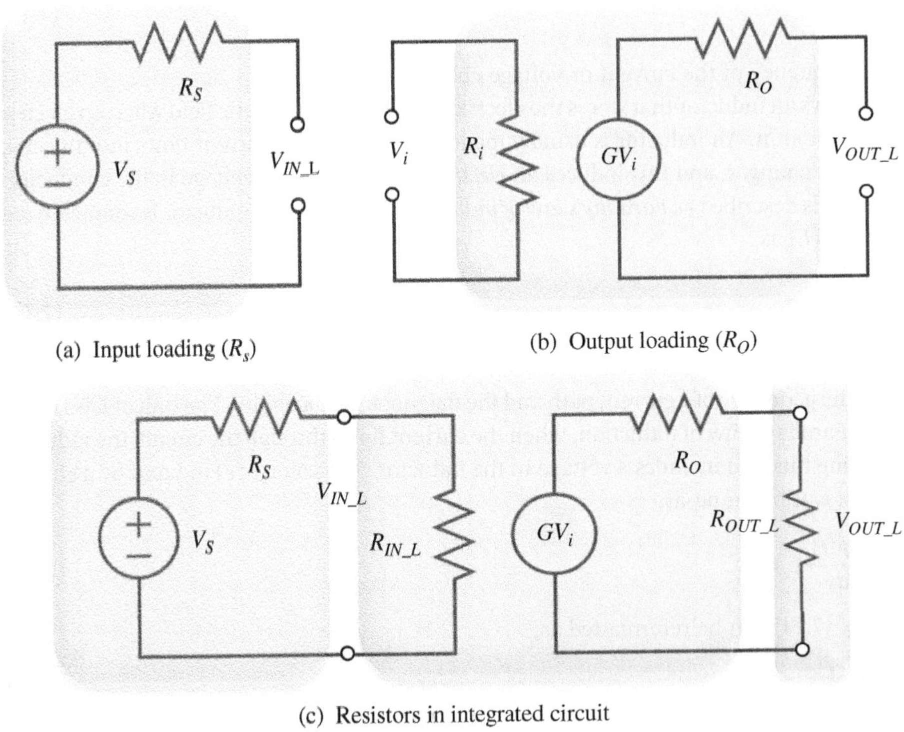

Figure 7.5 Input loading versus output loading

Equation (7.19) tells that the input loading R_S should be minimized to reduce its effect on V_{IN_L} from V_S.

Figure 7.5b shows the case of an output loading (R_O). The use of the output loading affects the amplitude of the loading signal (V_{OUT_L}) from GV_i as,

$$V_{OUT\_L} = \frac{R_L \cdot GV_i}{R_O + R_{OUT\_L}} \tag{7.20}$$

where G is the gain of the op-amp from input to output; R_{OUT_L} is the equivalent resistance at that part of the loading circuit.

Equation (7.20) tells that the output loading R_O should be maximized to reduce its effect on V_{OUT_L} from GV_i.

Figure 7.5c shows the ideal case when a signal processing unit is used to amplify a source signal with a gain of G; in other words, the resistors should be selected to be $R_{OUT\_L} >> G \cdot R_{IN\_L}$ and $R_{IN\_L} >> R_S$.

7.2.4 Basic Types of Signals

To extract certain features from a source signal, one or more basic signals are often applied in SCP. Figure 7.6 shows five types of basic signals that are commonly used in SCP.

Table 7.1 gives corresponding mathematic models of five basic signals.

Different basic signals have different effects in extracting the features from a source signal. Therefore, one has to choose the right signal in processing or conditioning a source signal.

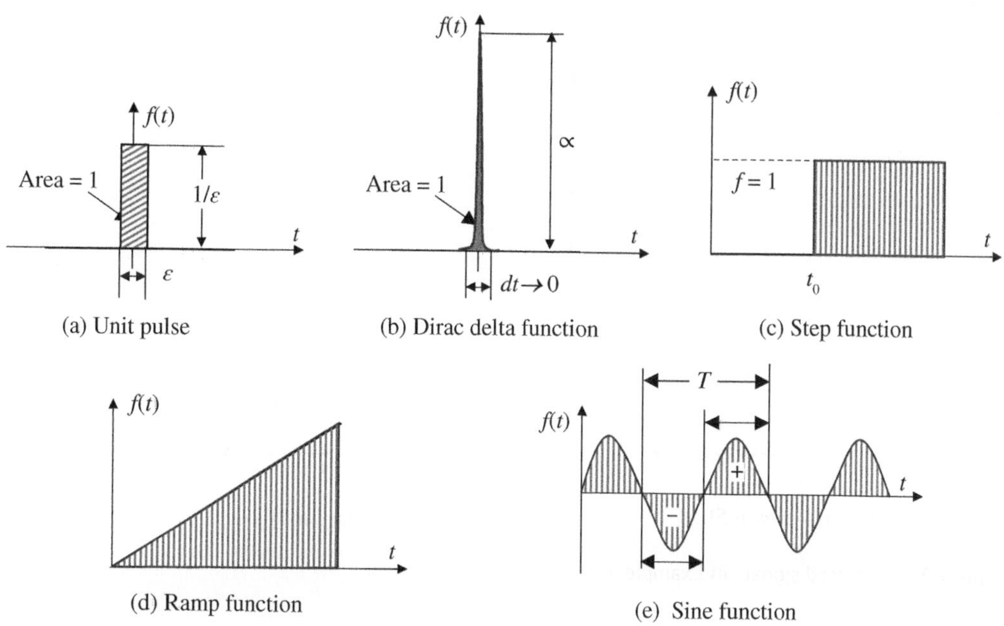

Figure 7.6 Five types of basic signals

Table 7.1 The mathematic models of five basic signals

Type	Function	Explanation
a) Unit pulse	$f(t) = \begin{cases} \frac{1}{\varepsilon} & -\frac{\varepsilon}{2} < t < \frac{\varepsilon}{2} \\ 0 & t \le -\frac{\varepsilon}{2} \text{ or } t \ge \frac{\varepsilon}{2} \end{cases}$	ε is very small value; effective signal area = $\varepsilon\left(\frac{1}{\varepsilon}\right) = 1$.
b) Dirac delta function	$f(t) = \lim_{\Delta t \to 0} f(t) = \propto$ $area = \int_{-\propto}^{+\propto} f(t)dt = 1$	The Dirac delta function can be treated as an extreme case of the unit pulse when the signal occurs suddenly.
c) Step function	$f(t) = \begin{cases} 1 & t \ge t_0 \\ 0 & t < t_0 \end{cases}$	It applies a threshold to pick a signal from a specific time.
d) Ramp function	$f(t) = \begin{cases} kt & t \ge 0 \\ 0 & t < 0 \end{cases}$	k is the slope of the ram function.
e) Sine function	$f(t) = \sin(\omega t + \phi)$	A sine function is periodical with a frequency of ω and a phase angle of ϕ.

Example 7.1 Assume that a step signal step$[t - x]$ is used to digitize an analog signal $f(t) = t$ at the sampling times of ($n = 0, 1, 2, 3, 5, 6, \ldots$), plot the digitized signals when the analog signals are processed by the following equations of (a) $f(t)*(step[t] - step[t-5])$ and (b) $step[t-1] + f(t)*step[t-4]$, respectively.

Solution

As a step function step$[t - x]$, we have $step(t-x) = \begin{cases} 1 & t \ge x \\ 0 & t < x \end{cases}$; as a ramp function, we have $f(t) = 0, 1, 2, 3, \ldots, 6$ when $n = 1, 2, 3, \ldots, 6$, respectively. Therefore, the results of the two operations can be evaluated *step by step* from $n = 1$ to $n = 6$ and plotted in Figure 7.7a and b, respectively.

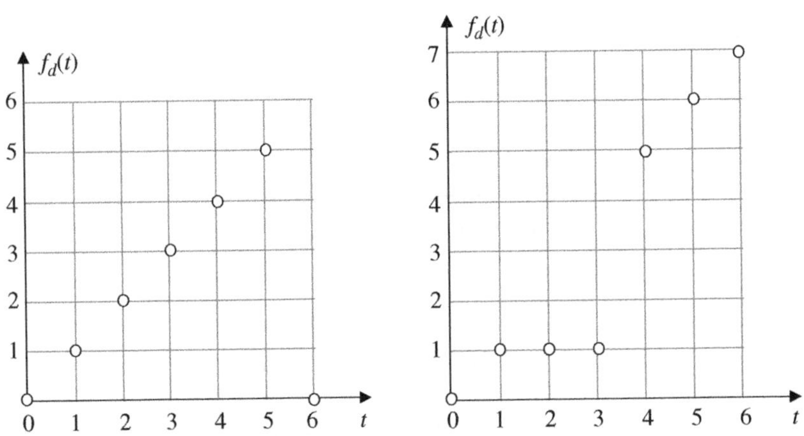

(a) $f(t)*(step[t] - step[t-5])$ (b) $step[t-1] + f(t)*step[t-4]$

Figure 7.7 Processed signals in Example 7.1

Technically, SCP is to combine a source signal with unknown variables with one or more signals with known variables, so that the features of interest can be extracted from the source signal. To design an electronic component for SCP, one has to understand what an output signal will be when several input signals are combined.

Example 7.2 Determine an output signal when it is combined from the input signals given as (1) $3\angle 150° + 5\angle -60° + 4\angle 120°$, (2) $\frac{(-1+j)^5}{1+j}$, and (3) $\frac{5}{2j}(\angle 60°) + \frac{\sqrt{2}e^{jx}}{2-j}$, respectively.

Solution

1) $3\angle 150° + 5\angle -60° + 4\angle 120°$

$$= [3\cdot \sin(150°) + 5\cdot \sin(-60°) + 4\cdot \sin(120°)]$$
$$+ [3\cdot \cos(150°) + 5\cdot \cos(-60°) + 4\cdot \cos(120°)]\cdot j$$
$$= \frac{-3\sqrt{3}+1}{2} + \frac{3-\sqrt{3}}{2}\cdot j = 2.1918e^{2.8481\cdot j}$$

2) $$\frac{(-1+j)^5}{1+j} = \frac{(-1+j)^5}{(1+j)}\frac{(1-j)}{(1-j)} = \frac{-(-1+j)^6}{2} = -\frac{\left(\sqrt{2}e^{-\frac{\pi}{4}\cdot j}\right)^6}{2} = -4e^{-\frac{3\pi}{2}\cdot j} = 4j$$

3) $$\frac{5}{2j}(\angle 60°) + \frac{\sqrt{2}e^{jx}}{2-j} = -2.5e^{j150°} + \frac{2e^{j(45°+x)}}{5}$$

Example 7.3 For the amplifier configuration in Figure 7.8, determine the output voltage v_O based on the given variables for input three voltages (v_1, v_2, v_3) and all resistors (R_1, R_2, R_3 R_f).

Solution

Since the non-inverting input of the op-amp is grounded, the voltage at node A is zero, and it is connected to the inverting input of the op-amp. Therefore, the currents go through the resistors (R_1, R_2, R_3, R_f) can be determined directly based on (v_1, v_2, v_3, v_O). Moreover, the currents flow into node A must be balanced as,

$$I_1 + I_2 + I_3 + I_f = \frac{v_1 - 0}{R_1} + \frac{v_2 - 0}{R_2} + \frac{v_3 - 0}{R_3} + \frac{v_O - 0}{R_f} = 0 \tag{7.21}$$

v_O can be derived from Eq. (7.21) as,

$$v_O = -R_f\left(\frac{v_1}{R_1} + \frac{v_2}{R_2} + \frac{v_3}{R_3}\right)$$

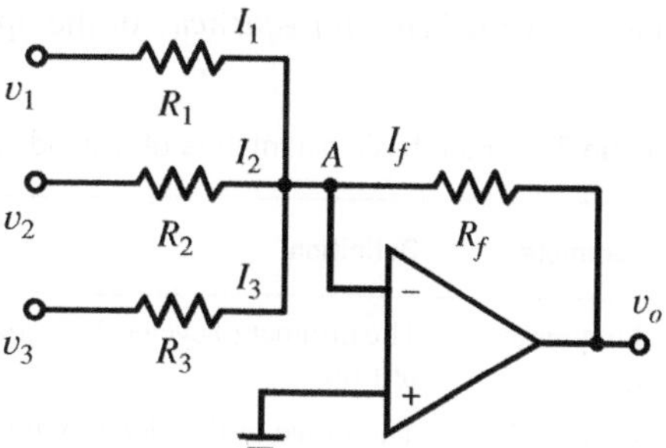

Figure 7.8 Signal conditioning and processing in Example 7.3

7.2.5 Main Parameters of Periodical Signals

The features of a signal are represented by a number of parameters, and Table 7.2 shows commonly used parameters of a steady periodic signal.

7.2.6 Amplitude and Phase Changes

SCP surely changes the frequency, amplitude, phase, or offset of a source signal with its purpose. One has to understand how these changes are represented or modeled correctly.

An amplitude change is commonly measured by *a gain*. A gain (G) is defined in its logarithmic scale as the ratio of the amplitudes of pre-processed and post-processed signals as,

$$G = \log_{10} \frac{v_{OUT}}{v_{IN}} \tag{7.22}$$

Moreover, *decibels* (dB) is often used as the standardized unit of a gain (G_{dB}) as,

$$G_{dB} = 20 \cdot G = 20 \cdot \log_{10} \frac{v_{OUT}}{v_{IN}} \tag{7.23}$$

Example 7.4 When the amplitude of a source signal (v_{IN}) is amplified to an output signal (v_{OUT}) by (a) 10 times and (b) 1000 times, respectively, determine the corresponding gain (G and G_{dB}) (a) and (b).

Solution

(a) Equations (7.22) and (7.23) are used to calculate G and G_{dB} when $\frac{v_{OUT}}{v_{IN}} = 10$.$G$ is found as $G = 1$ and G_{dB} is found as $G_{dB} = 20$ (dB). (b) Equations (7.22) and (7.23) are used to calculate G and G_{dB} when $\frac{v_{OUT}}{v_{IN}} = 1000$, G is found as $G = 3$ and G_{dB} is found as $G_{dB} = 60$ (dB).

It is worth noting that when an op-amp is used to amplify the magnitude of a signal, the gain from an input signal to the output signal can be constant only when the working frequency is within *the bandwidth of frequency*. In other words, the amplification of the magnitude of a signal may be affected by the range of working frequencies of op-amp.

As shown in Figure 7.9, if the frequency of a signal is above the upper frequency or below the lower frequency of the op-amp, the gain will be reduced significantly. The upper and lower frequencies are called *cut-off frequencies* of the op-amp.

Table 7.2 Four basic parameters of a steady periodic signal ($v(t) = a_0 + a \cdot \sin(2\pi f + \phi)$)

Parameter	Definition	Illustration
Frequency (f)	The number of cycles the signal repeats in a unit period of time	
Amplitude (a)	The range of the signal magnitude in a single period	
Phase angle (ϕ)	The angular component of the periodic signal	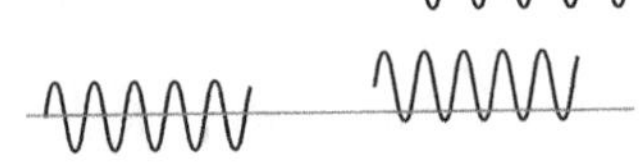
Offset (a_0)	The eccentrical component of magnitude of the periodical signal	

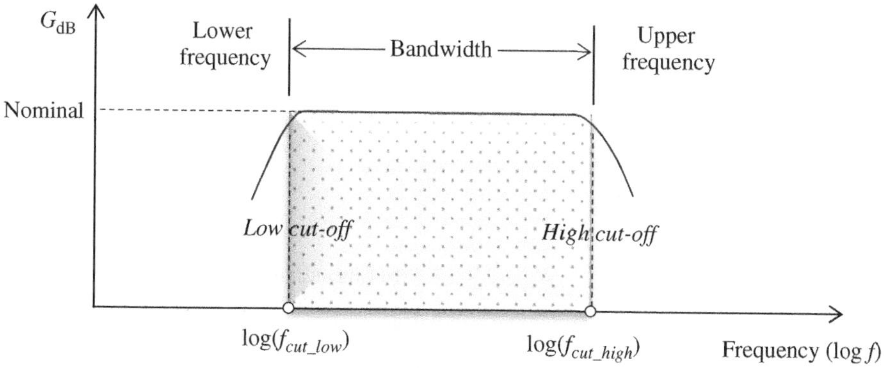

Figure 7.9 *Cut-off frequencies* of op-amp for constant gain

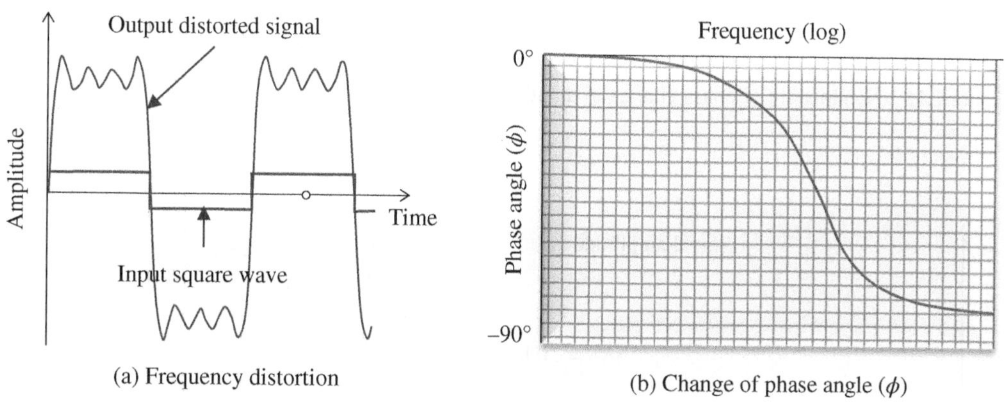

Figure 7.10 The effects of amplification on frequency distortion and phase angle (ϕ)

When the bandwidth of an op-amp is narrow, the shape of the source signal may be distorted over time, and this is referred to as *frequency distortion* as shown in Figure 7.10a. Amplifying a signal also affects the phase angle; this occurs even to a frequency within the bandwidth of working frequencies as shown in Figure 7.10b. Note that a phase change in an op-amp is unnecessarily linear as shown in Figure 7.11b; nonlinearity occurs due to the change of phase angle in amplification.

Another indicator to measure the performance of SCP for an amplitude change is *signal-to-noise ratio* (SNR) or (S/N). As shown in Figure 7.12, SNR measures the impact of background noise on the amplitude of a processed signal. The unit of SNR is dB. The larger the SNR is, the less the SNR is affected by background noise; otherwise, the noise suppresses the expected signal if the SNR is negative. Alternatively, SNR is sometimes defined as the logarithmic ratio of the power of processed signal and background noise.

When a number of tasks are involved in SCP, it becomes important to take into consideration the sequence of performing these tasks. In particular, a task to remove background noise must be performed as early as possible. Ideally, amplifying the source signals before any new noises are introduced.

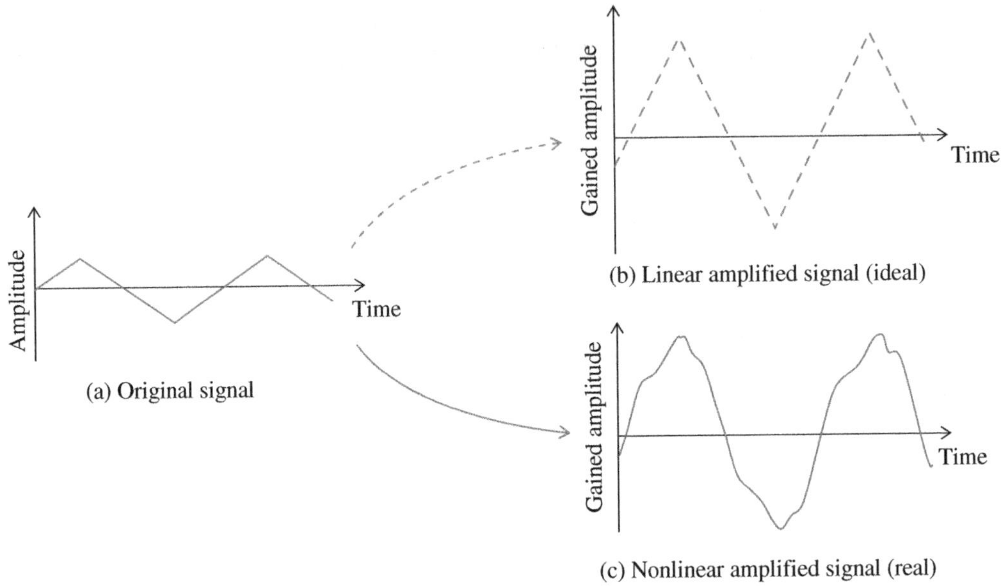

Figure 7.11 Nonlinearity of phase change in amplification

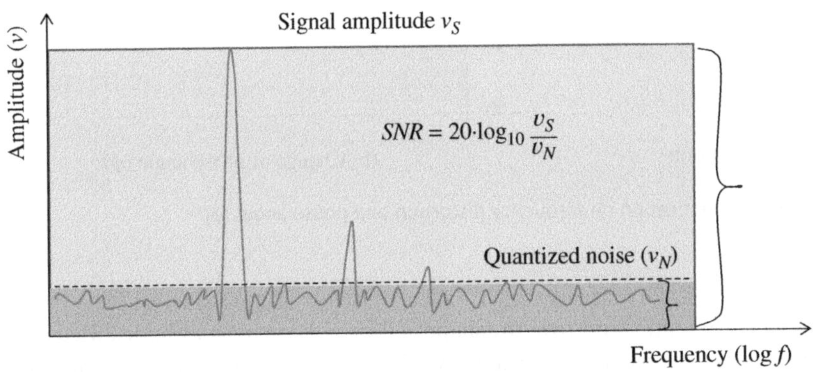

Figure 7.12 Signal-to-noise ratio (SNR)

The output of a sensor is usually at a millivolt or microvolt level. Such a weak signal is very easily affected by any noise that is introduced at different SCP stages, and this tends to cause erroneous data. The example in Figure 7.13 provides the rationale for why the noise should be removed before performing other tasks.

Example 7.5 In Figure 7.13, assume that the output of a J-type thermocouple travels a 10-m wire to arrive at a data acquisition system (DAQ). The output voltage at the probe is around 50 μV/°C, and the noise by the 10-m wire is around 200 μV. Determine the SNR for (a) an amplification of 500 times is applied after the wire's noise is introduced (i.e., at the input of DAQ) and (b) an

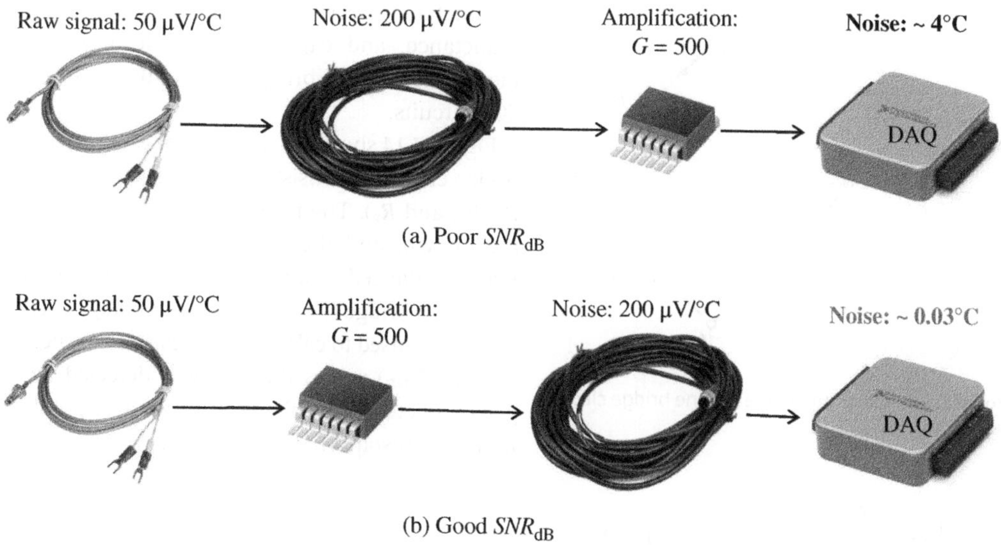

Figure 7.13 Remove noise at source to improve *signal-to-noise ratio* (SNR)

amplification of 500 times is applied before the wire's noise is introduced (i.e., at the output of J-type) (NI 2003).

Solution

a) For a temperature change of one degree (1°C), one has $V_S = 50\,\mu V$ and $V_N = 200\,\mu V$ before the amplification of $G = 500$ is applied. Therefore, using Eq. (7.23) yields,

$$SNR_{dB} = 20 \cdot \log_{10} \frac{(50 \times 10^{-6}) \times 500}{(200 \times 10^{-6}) \times 500} = -12\,(dB)$$

This implies that the raw data has been completely suppressed by noise; since the amplification is applied to both raw signal and noise.

b) For a temperature change of one degree (1°C), one has $V_S = 50\,\mu V$ with an amplification of $G = 500$; $V_N = 200\,\mu V$ is introduced after the raw signal is amplified. Therefore, using Eq. (7.23) yields,

$$SNR_{dB} = 20 \cdot \log_{10} \frac{(50 \times 10^{-6}) \times 500}{(200 \times 10^{-6})} = +42\,(dB)$$

This implies that SNR_{dB} has been greatly improved even with the same noise level since the amplification is only applied to raw signals.

7.2.7 Wheatstone Bridges

Using a Wheatstone bridge is an effective way to detect a small derivation of resistance from a balanced condition. Therefore, Wheatstone bridge circuits are used to measure a large variety of physical parameters such as strains, light intensity, and temperatures. The changes to these physical parameters affect the resistance, impedance, inductance, and capacitance of electronic elements. Wheatstone bridge is one of the most used SCP circuits.

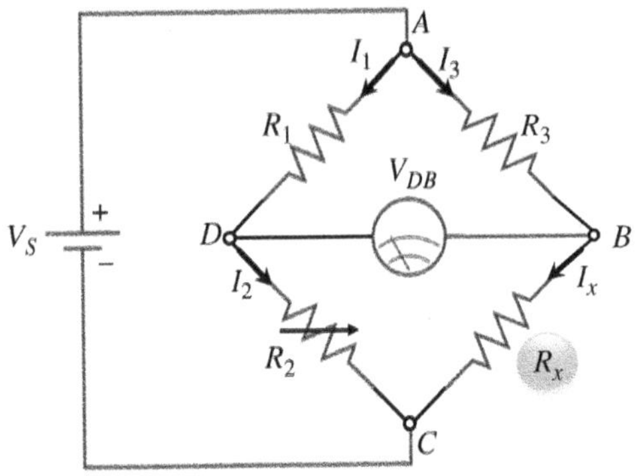

Figure 7.14 Diagram of Wheatstone bridge circuit

Figure 7.14 shows the diagram of a Wheatstone bridge circuit. It consists of four resistors (i.e., R_1, R_2, R_3, and R_x). The resistances of R_1, R_2, and R_3 are known, and the resistance is adjustable to reach a balanced condition when no current flows across the bridge (i.e., $V_{DB} = 0$). *Kirchhoff's circuit law* can be used to calculate the resistance when a voltage (V_{DB}) across the bridge is detected by a galvanometer. The voltage (V_{DB}) is corresponded to these resistances as,

$$V_{DB} = \left(\frac{R_2}{R_1 + R_2} - \frac{R_x}{R_x + R_3}\right) \cdot V_S \tag{7.24}$$

where V_S is the voltage of the power supply and Eq. (7.24) can be rewritten to obtain R_x from measured V_{DB} as,

$$R_x = \frac{R_2 \cdot V_S - (R_1 + R_2) \cdot V_{DB}}{R_1 \cdot V_S + (R_1 + R_2) \cdot V_{DB}} R_3 \tag{7.25}$$

Note that at a balanced condition, the resistance of R_2 is adjusted so that $V_{DB} = 0$, Eq. (7.25) gives the value of R_x at its initial stage as,

$$R_{x,0} = \frac{R_2}{R_1} R_3 \tag{7.26}$$

Let δR_x be a very small change from its original value $R_{x,\,0}$, that is, $\delta R_x << R_{x,\,0}$, Eq. (7.24) can be rewritten as,

$$V_{DB} = \left(\frac{R_2}{R_1 + R_2} - \frac{R_{x,0} + \delta R_x}{R_{x,0} + \delta R_x + R_3}\right) \cdot V_S = \left(\frac{R_{x,0}}{R_{x,0} + R_3} - \frac{R_{x,0} + \delta R_x}{R_{x,0} + \delta R_x + R_3}\right) \cdot V_S \doteq -\frac{\delta R_x}{R_x + R_3} \cdot V_S \tag{7.27}$$

The change of resistance of R_x can be obtained from Eq. (7.27) from V_{DB} as,

$$\delta R_x = -\frac{R_x + R_3}{V_S} \cdot V_{DB} \tag{7.28}$$

Example 7.6 A temperature sensor has a resistance of 100 Ω at 0°C is placed as R_x in the Wheatstone bridge with each of the other arms also being 100 Ω in Figure 7.13. The voltage of the power supply is $V_S = 6.0$ (V). If the resistance temperature coefficient of the sensing materials is $\alpha = 0.0039/$ °C, find the output voltage from the bridge per degree change in temp. Note that the variation of the resistance of the platinum with temperature can be represented as $R_{T+\Delta T} = R_T\,(1+\alpha\cdot\Delta T)$ or $\Delta R_T = \alpha\cdot\Delta T\cdot R_T$.

Solution

Let $R_1 = R_2 = R_3 = R_{x,0} = 100\ \Omega$ and $\Delta T = 1°C$, one has, $\Delta R_T = 0.39\ \Omega$, and applying Eq. (7.27) finds

$$V_{DB} \doteq -\frac{\delta R_x}{R_x + R_3}\cdot V_S = -\frac{0.39}{(100) + (100)}\cdot(6) = -0.013\ \text{V/°C}$$

7.3 Signal Cleaning

Signal cleaning aims to remove noise from a source signal that will be digitized for the detected states. The noise included in the source signal may cause the challenge of detecting the correct state of the signal at a certain time. A *Schmitt trigger* is an electronic gate that uses hysteresis to remove the noise from a signal. As shown in Figure 7.15b, a Schmitt trigger is different from an ordinary gate in sense that the state of the output voltage can only be switched when a threshold is reached.

Figure 7.15b shows that a Schmitter trigger has two levels of output voltage as,

$$\left.\begin{aligned} V_{ON} &= -\frac{R_2}{R_1}V_{IN,+} \\ V_{OFF} &= -\frac{R_2}{R_1}V_{IN,-} \end{aligned}\right\} \tag{7.29}$$

The working principle of a Schmitter trigger is its hysteresis characteristic. When the output is at its low state, the input has to go above V_{ON} before the state of the output voltage is triggered to its high state. Otherwise, when the output is at its high state, the input has to go below V_{OFF} before the output voltage is triggered to its low state. As shown in Figure 7.16a, this hysteresis phenomenon suppresses the effect of noise on the voltage stage. The farther apart V_{ON} and V_{OFF} are, the better immunity to noise can be achieved. Note that normal data such as a *transistor–transistor logic* (TTL) does not involve any hysteresis.

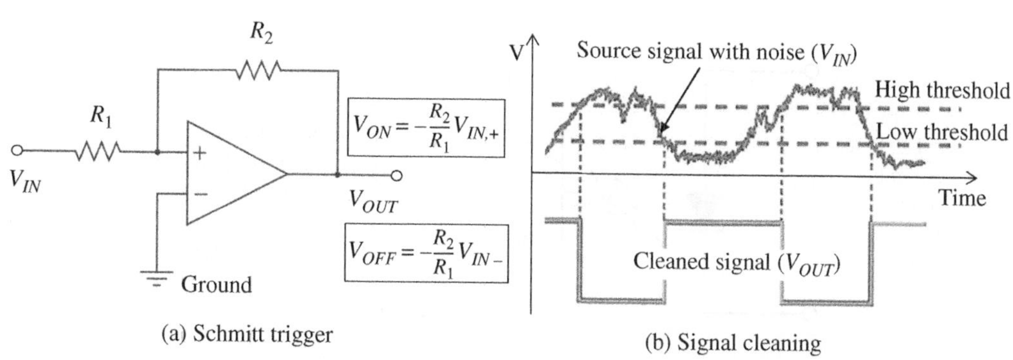

Figure 7.15 Schmitt trigger for signal cleaning

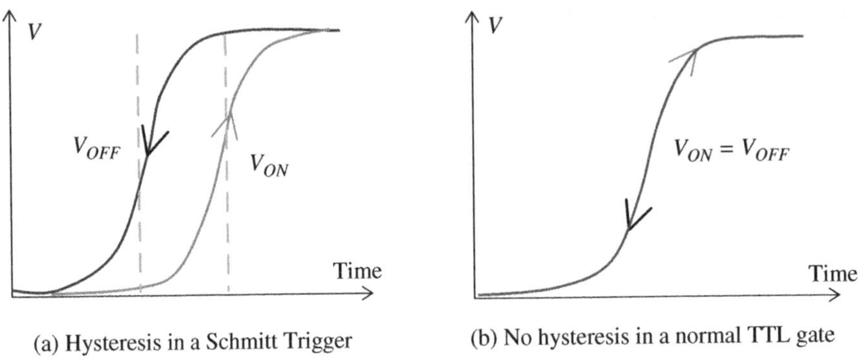

Figure 7.16 Hysteresis characteristic of Schmitt Trigger

7.4 Signal Isolation

Signal isolation aims to protect valuable instruments and measurements from dangerous and adverse effects caused by impulse noises, transient power surges, internal ground loops, or other hazard factors presented in a harsh industrial environment. An isolation is usually implemented by transmitting a signal to an isolation barrier without any direct electrical contact so that no undesired components in the signal can pass the barrier.

Three commonly used isolators are *light-emitting diodes* (LEDs), *capacitors*, and *inductors*.

7.4.1 Optical Isolation by Light-Emitting Diodes (LEDs)

A *light-emitting diode* (LED) utilizes *the electro-optical coupling principle* to isolate undesired components from a source signal. As shown in Figure 7.17, an LED is used together with a photodetector to transmit a signal across an isolation barrier, and light serves as the media to translate data. Optical isolation has the advantage of its immunity to electric and magnetic noise; however, it suffers from a slow transmission rate since it is restricted by the switching speed of an LED; moreover, an LED has a high dissipation of power and wear.

Since both an LED and phototransistor are current devices, input and output resistors are added into an optical isolator in Figure 7.18; this allows to processing of voltages instead of currents

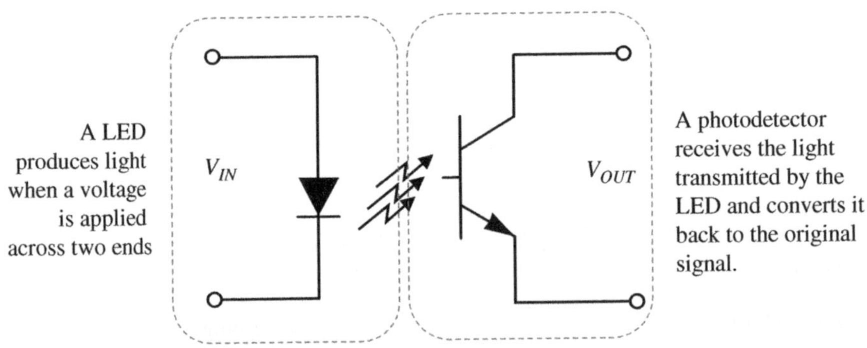

Figure 7.17 Optical isolation by *light-emitting diodes* (LEDs)

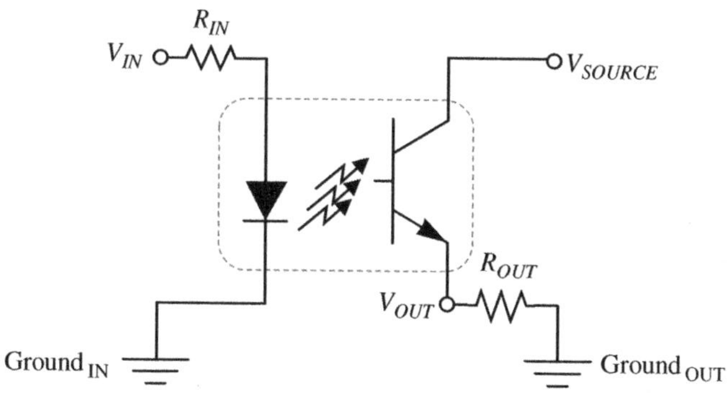

Figure 7.18 Input and output resistors in an optical isolator

directly. The values of resistors are consistent with the specifications of input and output currents. The amount of DC isolation provided by an optical isolator is usually in the range of kV. When V_{IN} is too large, it exposes a risk of producing an electrical arc and breaking isolation.

7.4.2 Capacitive Isolation by Capacitor

Capacitive isolation uses an electric field to change the level of charge between two plates of a capacitor. As shown in Figure 7.19, the charge across an isolation barrier can be detected, and it is proportional to the level of the measured signal. Capacitive isolation is immunity to magnetic noise. It provides a fast rate of data transmission. However, it uses an electric field to transmit data, and it is susceptible to interference by external electric fields.

7.4.3 Inductive Isolation by Inductor

As shown in Figure 7.20, some care has to be taken when an input voltage may involve a spark. An inductive isolator is used to isolate a *direct current* (DC) since an inductor is incapable of transmitting a signal with steady and constant amplitude (i.e., DC) as a power transmitter. An inductive isolator has the advantage of a high transmission rate but it becomes susceptible to interference from another magnetic field.

In summary, Table 7.3 compares the features of three basic types of isolators.

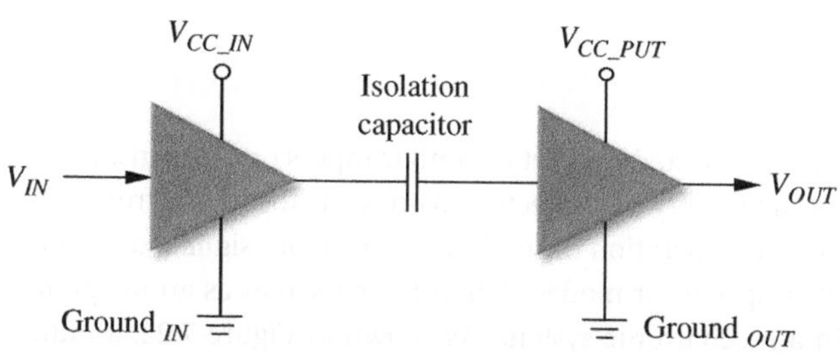

Figure 7.19 Capacitive isolation

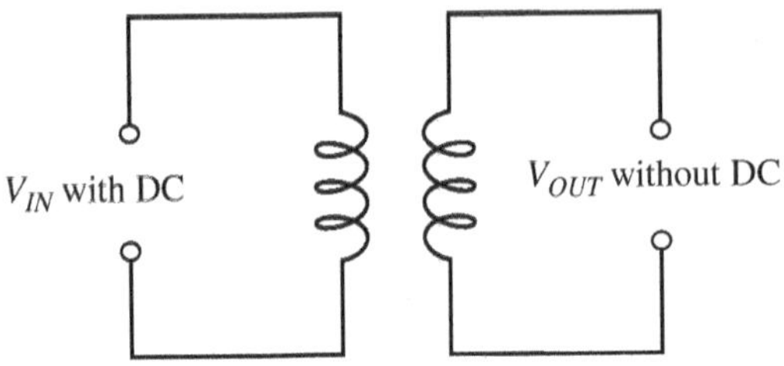

Figure 7.20 Inductive isolation

Table 7.3 Comparison of three types of isolators

Type	Advantages	Disadvantages
Optical	Immunity to electric field Immunity to magnetic field	Slow rate of data transmission high dissipation of power
Capacitive	Fast rate of data transmission Immunity to magnetic field	Susceptible to electric field interference
Inductive	Fast rate of data transmission Immunity to electric fields	Susceptible to magnetic field interference

7.5 Signal Transmission

During the course of the signal transmission, electronic elements in the circuit should be capable of minimizing noise and reducing distortion of signals. In this section, basic electronic elements used in transmitting signals are introduced individually.

7.5.1 Switches

A switch determines which signal or when the signal is transmitted at a terminal. Taking an example of an *analog switch* in Figure 7.21, it controls the "On" and "Off" of the signal to flow between two nodes in a circuit.

As shown in Figure 7.21, a switch is usually controlled by a digital signal, and common switch types include *signal-pole signal throw* (SPST), *signal-pole double throw* (SPDT), and *double-pole double throw* (DPDT). The resistor (R_{ON}) in a switch is to indicate a finite resistance of flow, its resistance is one basic specification of the switch.

7.5.2 Multiplexer

In many applications, a system output is determined by a set of similar inputs or the transmission path can be shared by multiple signals; it can be very cost-effective to integrate the inputs from these similar terminals. *Multiplexing* represents the operation of sending one or more signals over common transmission lines at different times, speeds, or modes. A *multiplexer* serves as an integrated interface to implement multiplexing in a mechatronic system. As shown in Figure 7.22, a multiplexer is usually a combinational logic circuit to switch one of multiple inputs and connect to a shared output terminal.

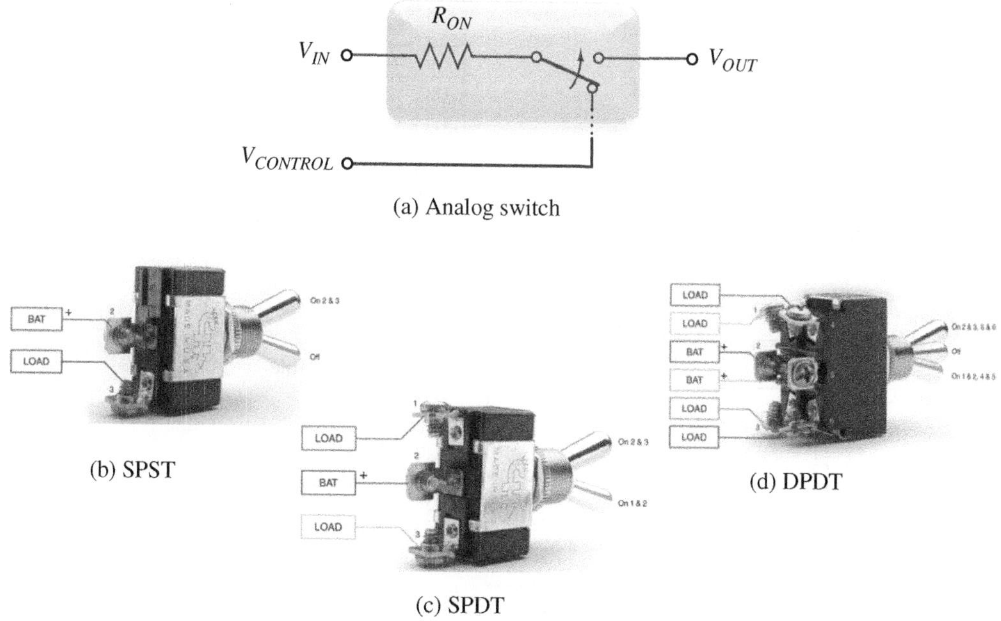

Figure 7.21 Analog switches and common types (*Source:* (b–d) Hi-Line Inc / https://hi-line.com/understanding-pole-switches / Last accessed on July 19, 2024)

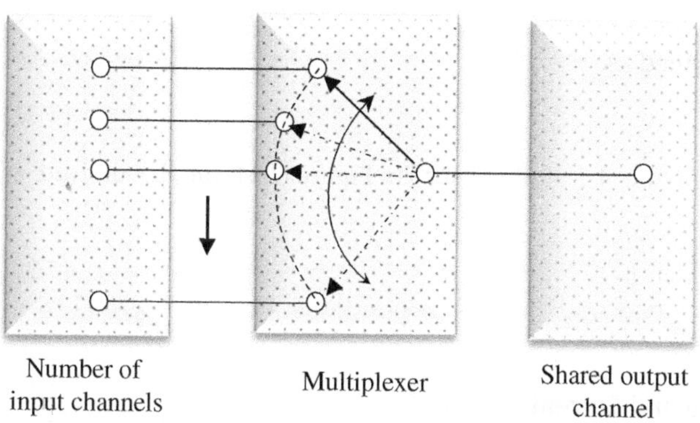

Figure 7.22 Illustration of basic multiplexer

The core component is a fast-acting rotary switch that connects one of the input channels to the output terminal, and it can be implemented either by high-speed logic gates or analog transistors. Figure 7.23 shows a design of a 2-to-1 multiplexer using basic logic NAND gates where two input currents are I_0 and I_1, and the output is Q. It is seen from the true table (Figure 7.23b) that when switch A is connected to I_0, Q is the same to I_0. Otherwise, when switch A is connected to I_1, Q is the same as I_1.

The Boolean expression for Figure 7.23c can be found as,

$$Q = \overline{A \cdot \overline{I_0}} \cdot I_1 + \overline{A} \cdot I_0 \cdot I_1 + A \cdot I_0 \overline{\overline{I_1}} \cdot + A \cdot I_0 \cdot I_1 \tag{7.30}$$

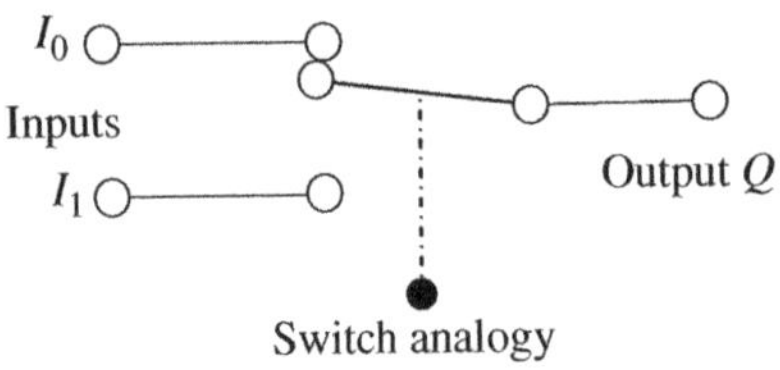

(a) SPDT

	Input		Output
A	I_0	I_1	Q
0	0	0	0
0	1	0	1
0	0	1	0
0	1	1	1
1	0	0	0
1	1	0	0
1	0	1	1
1	1	1	1

(b) True table

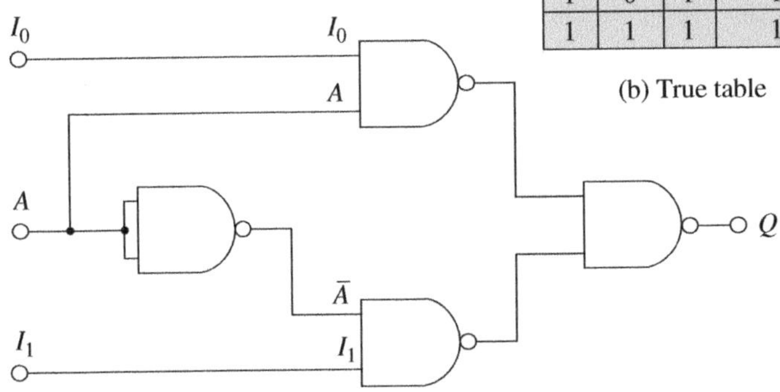

(c) NAND Gates in 2–1 multiplexer

Figure 7.23 2-1 Multiplexer

Equation (7.30) can be further simplified as,

$$Q = \overline{A} \cdot I_1 \cdot + A \cdot I_0 \tag{7.31}$$

Similarly, the Boolean expression of the 4-1 multiplexer in Figure 7.24 can be derived as (Electronics-tutorials 2024),

$$Q = \overline{ab} \cdot I_0 + a\overline{b} \cdot I_1 \cdot + \overline{a} \cdot b \cdot I_2 + a \cdot b \cdot I_3 \tag{7.32}$$

7.5.3 Protection from High Voltage and Current

When an input current or voltage is too high, an electronic circuit may be damaged. Therefore, it becomes necessary to protect a critical component by introducing one or several protection elements such as *resistors*, *fuses*, *Zener diodes,* and *optoisolators*.

For protection from a high current, a resistor connected in a circuit serially can limit the amplitude of the current. When a fuse is connected in two nodes, it breaks the connection when the current exceeds its threshold.

For protection from a high voltage, an optoisolator introduced in Figure 7.25 can decouple two circuits completely; a Zener diode can protect against a high voltage or prevent damage from a wrong polarity. As shown in Figure 7.25, a Zener diode has a specified voltage drop when it is used in a reverse bias; therefore, it can be used to regulate a voltage in a reverse bias. In other words, a Zener diode operates in the breakdown region, and it is capable of maintaining a nearly constant voltage under the conditions of widely varying currents.

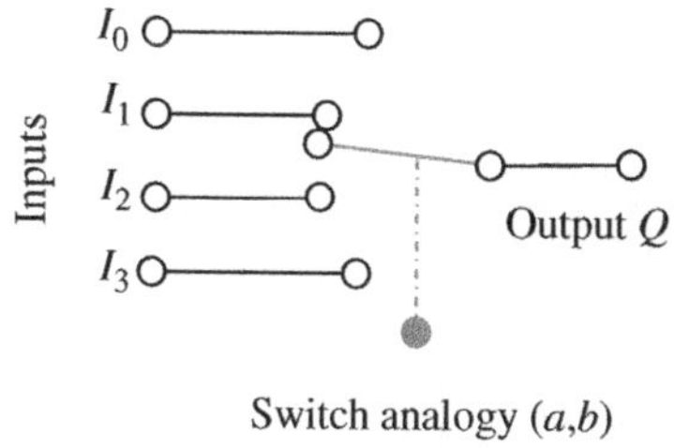

(a) 4PST

Input						Output
a	b	I_0	I_1	I_2	I_3	Q
0	0	x	x	x	1	0
0	1	x	x	1	x	1
1	0	x	1	x	x	0
1	1	1	x	x	x	1

(b) True table

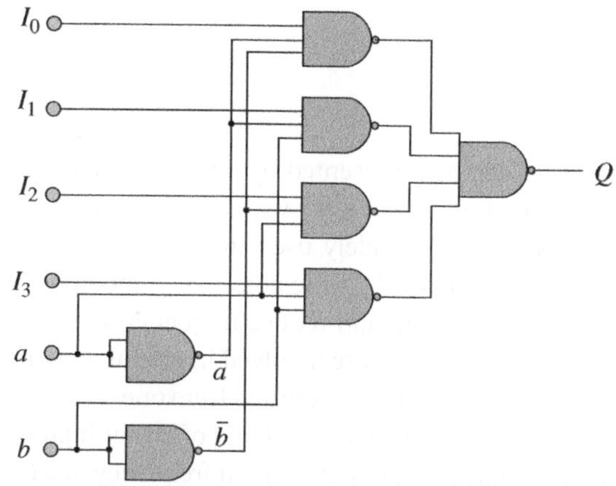

(c) NAND Gates in 4–1 multiplexer

Figure 7.24 4-1 Multiplexer

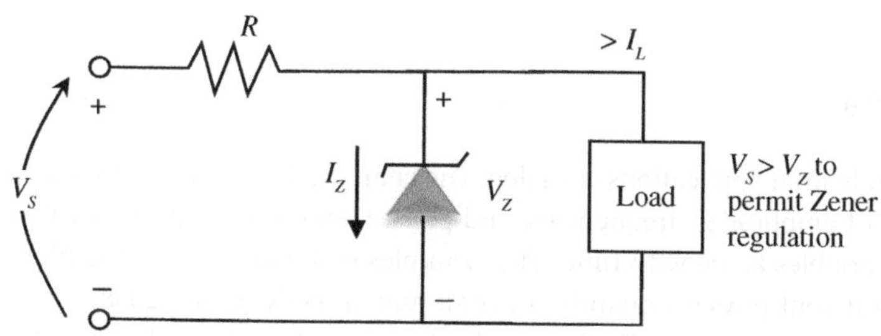

Figure 7.25 Zener diode for protection of high current

7.5.4 Modulation/Demodulation

The emerging Internet of Things (IoT) requires the ability to transmit or exchange information, data, and messages from one location to another in the shortest time delay.

Naturally, this follows a question that how fast the data transmission can be? Since transmission speed is closely associated with frequency, a signal can be modulated to enhance the range of signals; *modulation* is one of the basic electronic elements that are widely used in communication

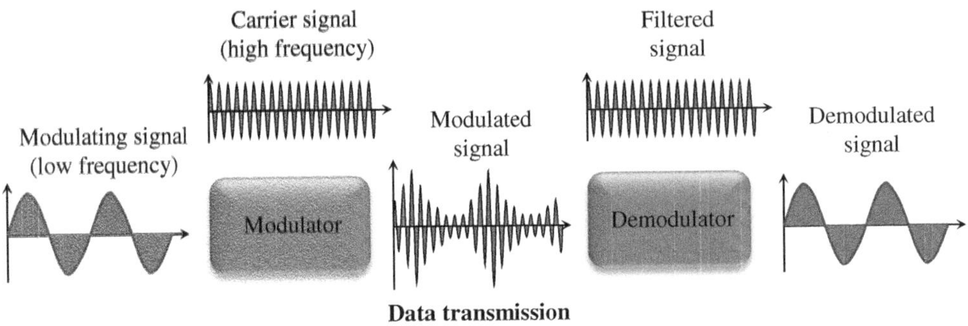

Figure 7.26 Modulation/demodulation

systems. Most signals in daily life can be represented by a number of sinusoidal waveforms, and modulation is the superimposition of the signal wave with a high-frequency carrier signal to speed up the signal transmission. Modulation is widely used in wireless communication to transport information through *mobile radio control* (MRC); the objective of modulation is to improve transmission quality, reduce power consumption, and reduce frequency spectrum.

Figure 7.26 shows that three types of signals are involved in modulation at different stages: (1) *a modulating signal* called a message signal to be transmitted from one location to another; a message signal is often referred to as a baseband signal with the band of low or high frequency. (2) *A carrier signal* has a high-frequency sinusoidal wave; due to its high frequency, it can travel much faster in comparison with the baseband signals; note that the frequency, amplitude, and phase of a carrier signal should be known. (3) *A modulated signal* is the outcome of modulation; when the modulation is accomplished, the output message is the mix of the baseband and carrier signals. In demodulating, an individual signal in the modulated signal can be extracted based on the amplitude, the frequency, or the phase of this constitutive signal.

7.6 Signal Conditioning

The raw data from real-world industrial applications is analog. The characteristics of analog signals are highly diversified in terms of amplitudes, frequencies, and phases, and most importantly the dynamics and noises of these variables alone with time. The examples in Table 7.4 show the high diversity of measurement for different physical quantities over a human body (Olson 2008).

Therefore, when the raw data from sensors is difficult to read or transmit, it needs conditioning to transform it into an easily read format. *A signal conditioner* performs a series of operations on a source signal to make it suitable for further processing. Figure 7.27 classifies the types of signal conditioning at a high level, and Figures 7.28 and 7.29 further classify and illustrate the types of signal conditioning for analog signals, respectively.

7.6.1 Amplification

Measurements from most sensors produce weak electric signals with low voltages. A voltage of typical measurement is in the range of millivoltage (i.e., 10^{-3} V), and a voltage of measurement in a microelectromechanical systems (MEMS) can be lower and in the range of microvolt (i.e., 10^{-6} V).

Table 7.4 Examples of diversified measurements of physical quantities over a human body

Physical Quantity	Sensing Method	Amplitude	Sampling Rate (Hz)
Blood flow	Electromagnetic	[1, 300] mL/s	[0, 20]
Blood pressure	Strain gage	[0, 400] mmHg	[0, 50]
Temperature	Thermistor	[32, 40] °C	[0, 0.1]
Respiratory rate	Impedance	[2, 50] times/min	[0.1, 10]
pO_2	pO_2 electrode	[30, 100] mmHg	[0, 2]
pCO_2	pCO_2 electrode	[40, 100] mmHg	[0, 2]
pH	pH electrode	[3, 13] pH units	[0, 1]
Cardiac output	dye dilution	[4, 25] L/min	[0, 20]
Electrocardiography	skin electrode	[0.5, 4] mV	[0.05, 150]
Electroencephalography	scalp electrode	[5, 300] μV	[0.5, 150]
Electromyography	needle electrode	[0.1, 5] mV	[0.5, 150]
Electroretinography	contact lens electrode	[0, 900] μV	[0, 50]

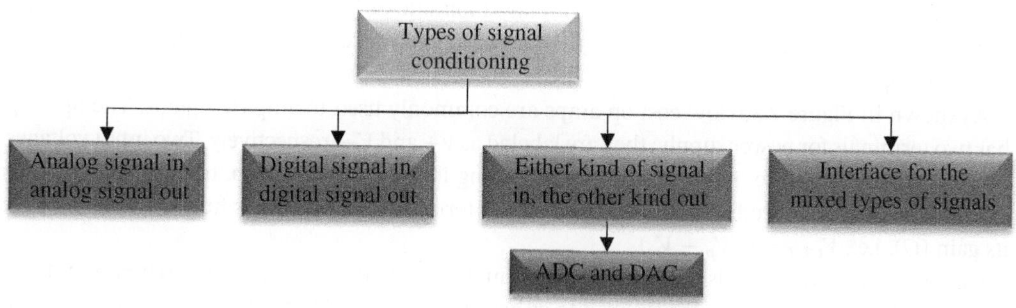

Figure 7.27 Types of signal conditioning at high-level

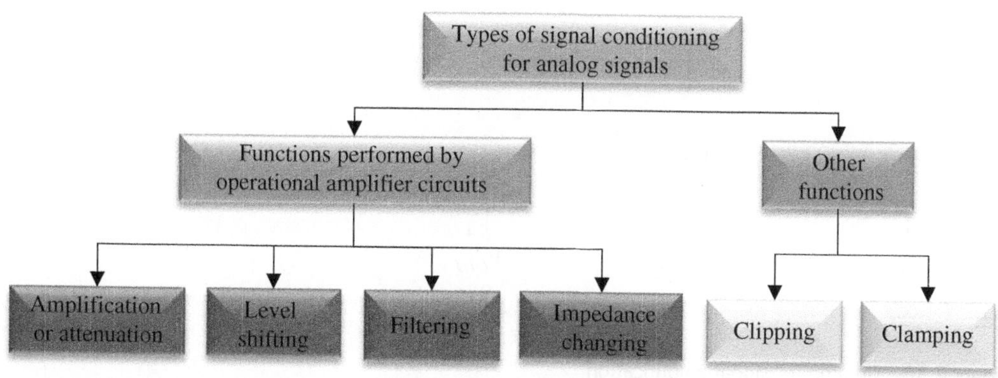

Figure 7.28 Types of signal conditioning for analog signals

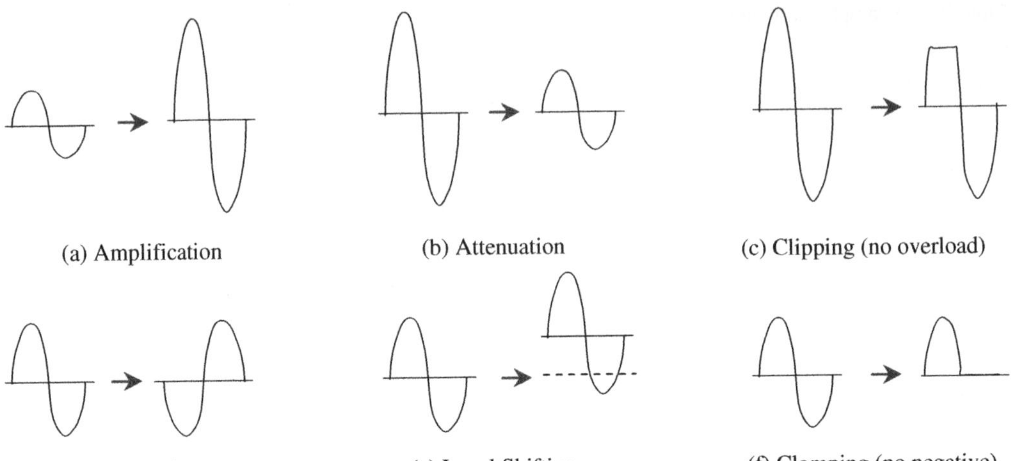

Figure 7.29 Illustration of types of signal conditioning for analog signals

Therefore, it is difficult to transmit such weak voltages over a wire with a long distance. Many signal processing units require an input voltage with an amplitude of 1–10 V. Signal amplification is performed by *an amplifier* to boost the strength of a signal and make it ready for further transmission or processing.

As shown in Figure 7.30, low-cost op-amps are commonly used to amplify signals. The op-amp has two terminals for power supply; they are labeled as $V+$ and $V-$, respectively. Two input voltages (V_n, V_p) are connected to inverting and non-inverting inputs of the op-amp, respectively, and the output voltage (V_O) is produced at the signal output terminal. An op-amp is fully characterized by its gain (G), i.e., $V_{OUT} = G(V_p - V_n)$.

It is also a good time to revisit the variety of op-amps. Note that an op-amp has two inputs and one output, and the output can be feedbacked to either of two inputs. Therefore, an op-amp has many configurations, which can be used to serve different functions. Figure 7.31 shows eight configurations and their corresponding functions.

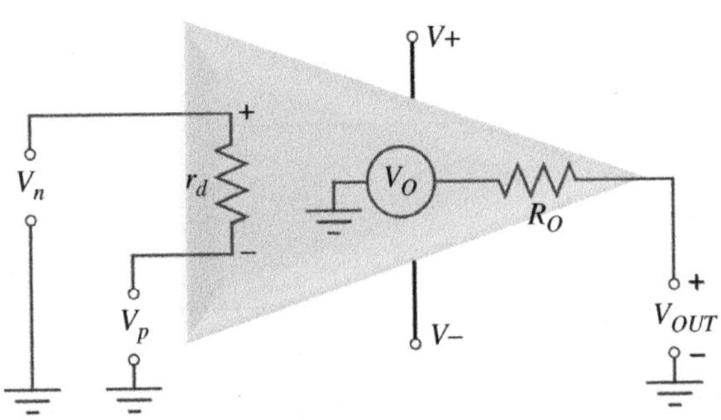

Figure 7.30 Op-amp for signal amplification

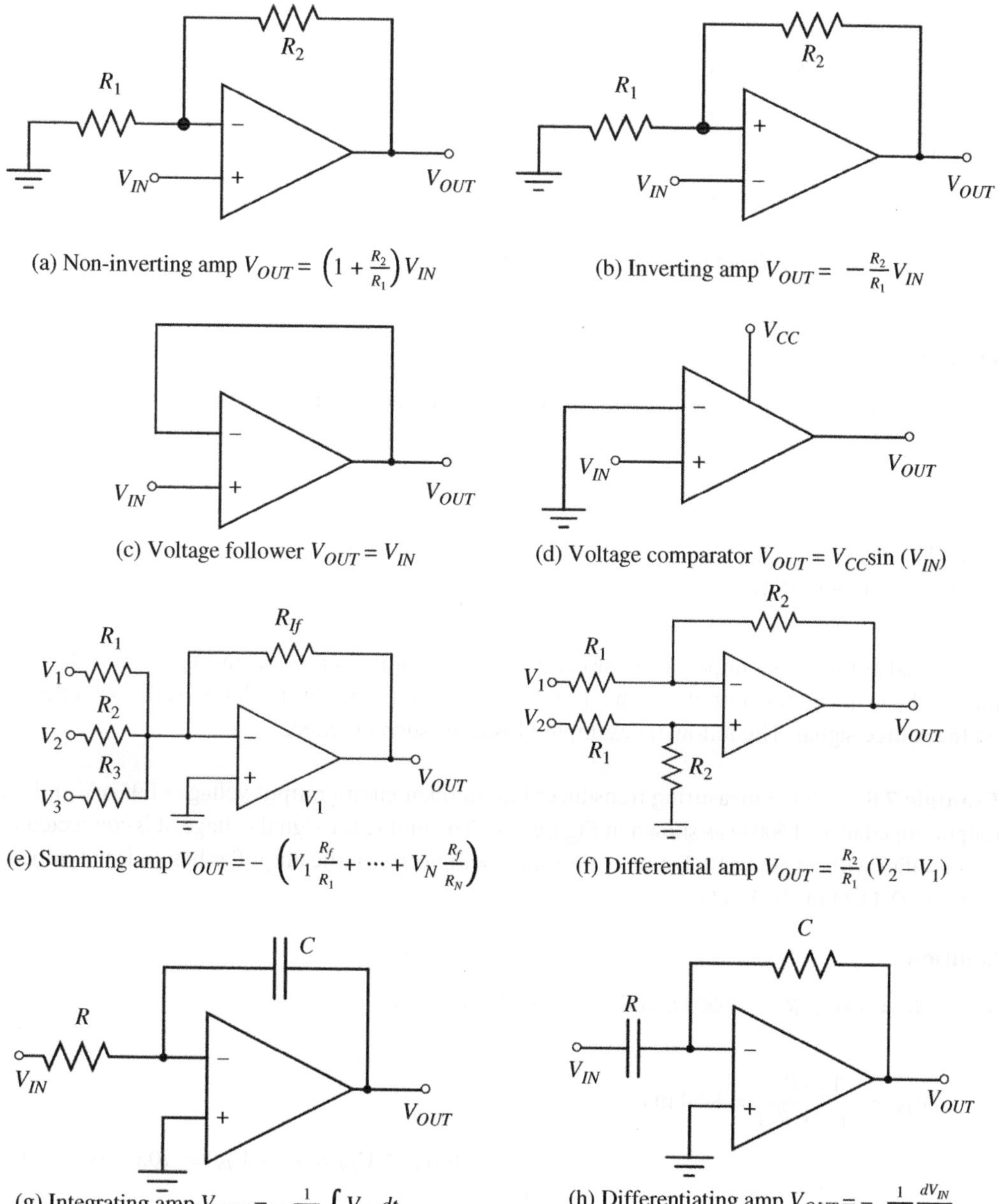

(a) Non-inverting amp $V_{OUT} = \left(1 + \frac{R_2}{R_1}\right) V_{IN}$

(b) Inverting amp $V_{OUT} = -\frac{R_2}{R_1} V_{IN}$

(c) Voltage follower $V_{OUT} = V_{IN}$

(d) Voltage comparator $V_{OUT} = V_{CC}\sin(V_{IN})$

(e) Summing amp $V_{OUT} = -\left(V_1 \frac{R_f}{R_1} + \cdots + V_N \frac{R_f}{R_N}\right)$

(f) Differential amp $V_{OUT} = \frac{R_2}{R_1}(V_2 - V_1)$

(g) Integrating amp $V_{OUT} = -\frac{1}{RC}\int V_{IN} dt$

(h) Differentiating amp $V_{OUT} = -\frac{1}{RC}\frac{dV_{IN}}{dt}$

Figure 7.31 Eight configurations of op-amps

Example 7.7 Figure 7.32 shows that there is a temperature change of 10°C at the thermocouple junction, and it produces an emf of 530 μV between V_1 and V_2. Let $R_1 = 10$ (kΩ), how to select R_2 if one expects the circuit to yield an output of $V_O = 10$ mV?

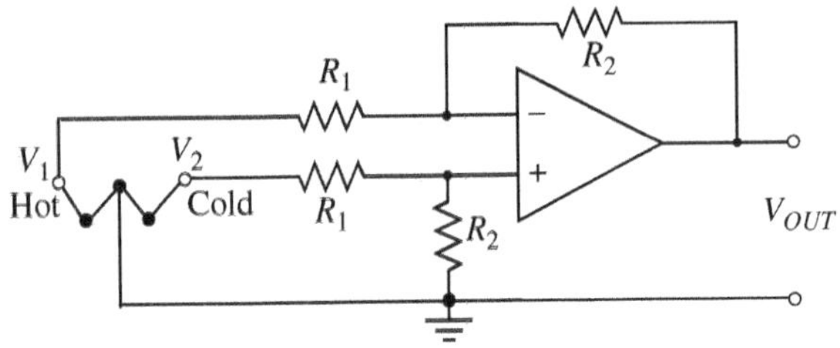

Figure 7.32 Example of determining R_2 in a differential amp

Solution

The following relation of input and output voltages in Figure 7.31f is used:

$$V_{OUT} = \frac{R_2}{R_1}(V_2 - V_1) \tag{7.33}$$

Let $(V_2 - V_1) = 5.3 \times 10^{-6}$ V, $R_1 = 10 \times 10^3\ \Omega$, and $V_{OUT} = 10 \times 10^{-3}$ V, Eq. (7.33) yields the solution of $R_2 = 1.89 \times 10^5\ \Omega$.

Note that as the output impedance, the connection to an amplifier with a given gain will affect the input voltage due to the output impendence; the higher an output impendence is, the less effect is on the source signal. The following example illustrates such an effect.

Example 7.8 A force-measuring transducer has an open-circuit output voltage of 95 mV and an output impedance of 500 Ω as shown in Figure 7.33. To amplify the signal voltage, it is connected to an amplifier with a gain of 10. Estimate *the input loading error* if the amplifier has an input impedance of (a) 4 k Ω or (b) 1 M Ω.

Solution

a) Let $R_S = 500\ \Omega$, $R_L = 4000\ \Omega$, and $V_S = 95$ mV, one has,

$$V_{IN} = \frac{V_S \cdot R_L}{(R_L + R_S)} = 84.4 \text{ mV}$$

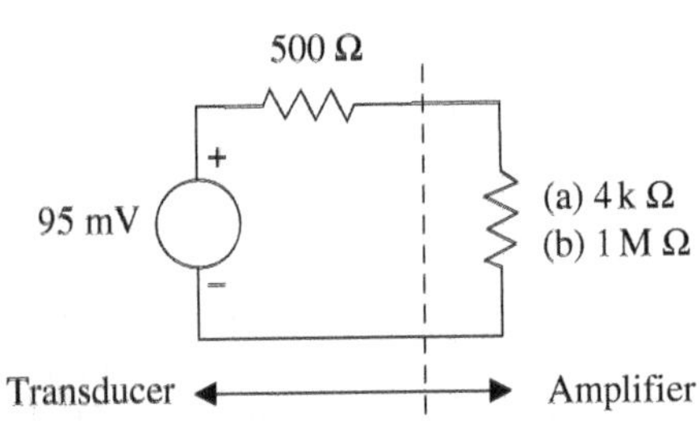

Figure 7.33 Example of determining input loading error

Therefore, $\Delta V_{IN} = V_S - V_{IN} = 10.6$ mV or 11% of V_S.

b) Let $R_S = 500\ \Omega$, $R_L = 1\ \text{M}\Omega$, and $V_S = 95$ mV, one has,

$$V_{IN} = \frac{V_S \cdot R_L}{(R_L + R_S)} = 94.95 \text{ mV}$$

Therefore, $\Delta V_{IN} = V_S - V_{IN} = 0.05$ mV or 0.05% of V_S.

An op-amp is very suitable to amplify the voltage of a source signal; however, its current output is very limited. To amplify the current further, a *bipolar junction transistor* (BJT) in Figure 7.34 can be used as a current amplifier.

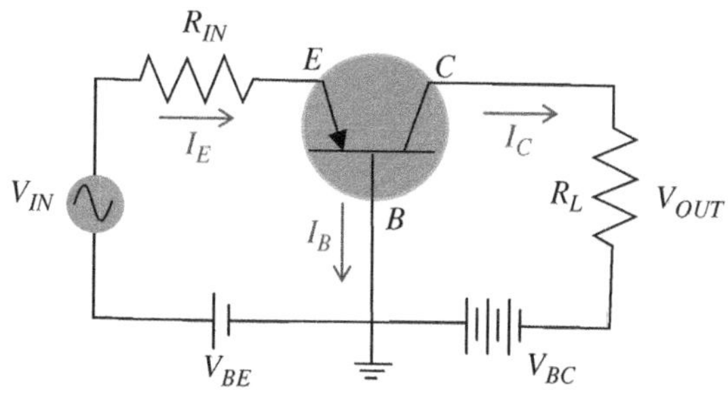

Figure 7.34 Bipolar junction transistor (BJT) as current amplifier

7.6.2 Attenuation

Since the working range of most data acquisition systems is either within the range of [0, 5] V or within the range of [0, 10] V, an input voltage with an exceeded voltage must be attenuated.

Attenuation refers to any reduction in the strength of an analog or digital signal. As shown in Figure 7.35, using a voltage divider can be the simplest attenuation circuit, and the output voltage can be reduced by a division as,

$$V_{OUT} = \frac{R_2}{R_1 + R_2} V_{IN} \tag{7.34}$$

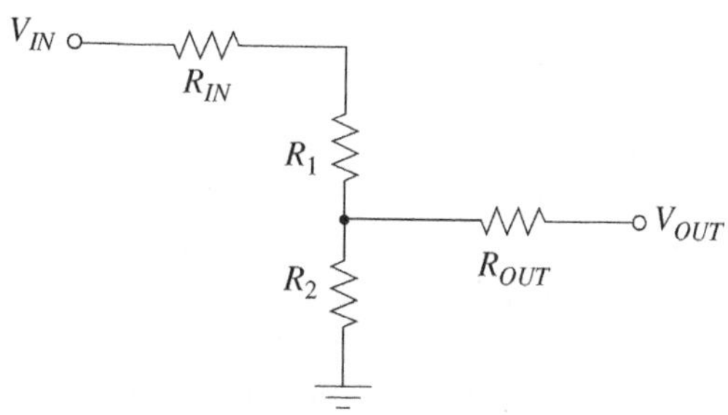

Figure 7.35 Diagram of a simple attenuator by a voltage divider

Furthermore, to make an attenuator functional appropriately, the output resistor (i.e., the impedance connected to the output) must be higher than the divider, and the input resistor (i.e., the impedance connected to source signal) must have an impedance of at most 1/10 that of the attenuator.

Therefore, the simplest attenuator becomes impractical when it is used for a multiplexer since the output impedance of a voltage divider is too high for the inputs of a multiplexer. Taking a 10:1 divider for a reading signal of 50 V as an example, when $R_1 = 900$ kΩ and $R_2 = 100$ kΩ are selected, the output voltage becomes 5 V. The impedance of a divider is too low in comparison with a source impedance of 1 MΩ.

7.6.3 Filtering

Figure 7.36 shows that *signal filtering* suppresses undesirable features while retaining desirable features of the signal.

The functional requirements (FRs) of a signal filter are defined by using the concepts of *signal-to-noise ratio* (SNR or S/N), *bandwidth*, *cutoff frequency*, and *break-point*. SNR is illustrated in Figure 7.12 and reintroduced here as,

$$\frac{S}{N} = 10 \cdot \log\left(\frac{P_S}{P_N}\right) \text{dB} = 20 \cdot \log\left(\frac{V_S}{V_N}\right) \text{dB} \tag{7.35}$$

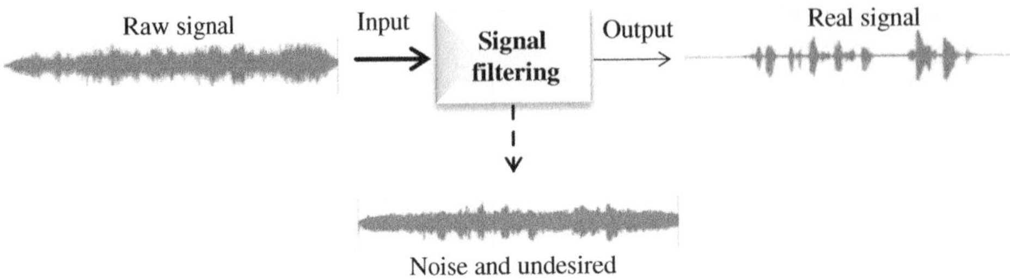

Figure 7.36 Suppressing undesirable features by filtering

where P_s and P_N are the power of source and noise signals; V_s and V_N are the voltages of source and noise signals, respectively.

Bandwidth refers to the frequency range in which an amplitude gain meets the condition of $|G(jw)| > 0.707$. *A break-point* refers to a point with a gain of −3 dB for power amplitude or an attenuation gain of −6 dB for voltage amplitude. *A cutoff frequency* is the frequency at which the filter breaks the source signal based on the break-point. To characterize a signal filter shown in Figure 7.37, the other four concepts are *pass band, transition band, stop band*, and *ripple*. *A pass band* is the frequency range that is allowed to pass the filter at the output. *A stop band* is the frequency range where the signal is attenuated. *A transition band* is the frequency range transiting from a pass band to a stop band or vice versa. *A ripple* is the residual periodical variation of a signal.

Figure 7.38 classifies the filters based on the types of attenuated frequency ranges into *low-pass, high-pass, band-pass, band-stop* (or *low-high-pass*), *low-band-pass, band-high-pass*, and *low-band-high-pass*.

Other than a classification based on applicable frequency bands, the filters can be also classified based on the criteria such as whether or not (a) a filtered signal is digital or analog and (b) the filtering process is active or passive, as illustrated in Figure 7.39.

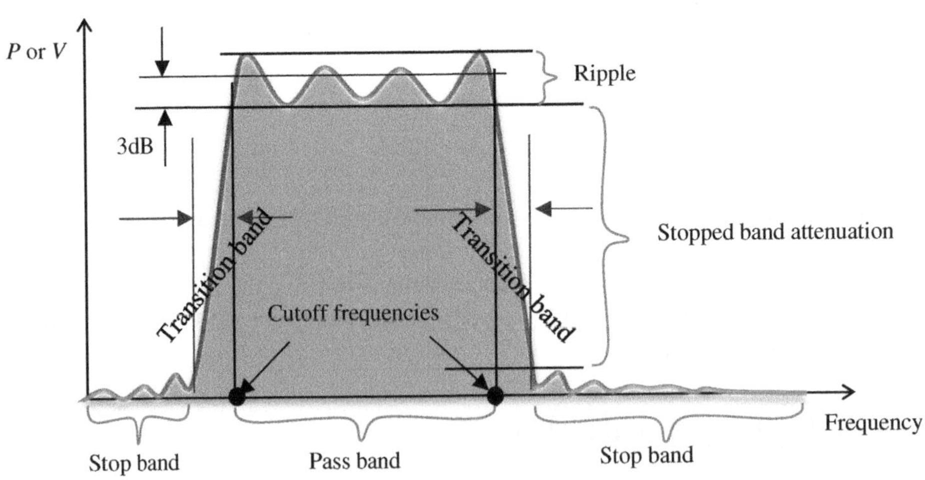

Figure 7.37 Concepts in characterizing a filtering process

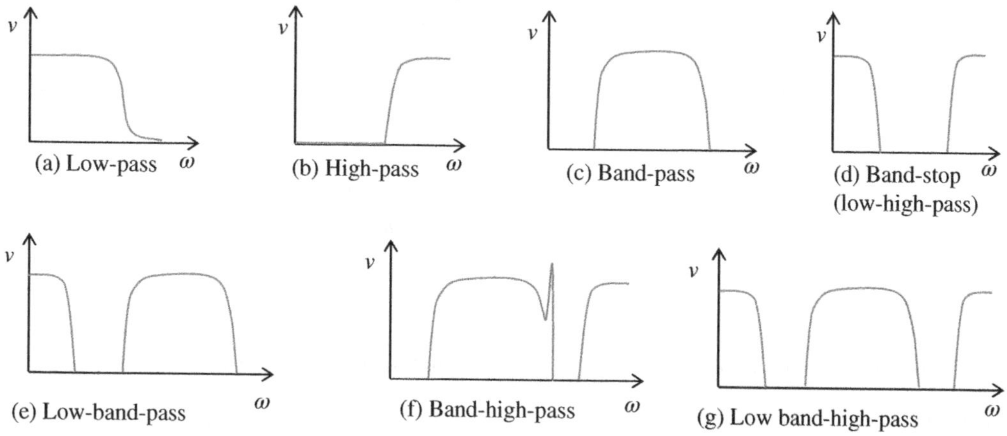

Figure 7.38 Classification of filters

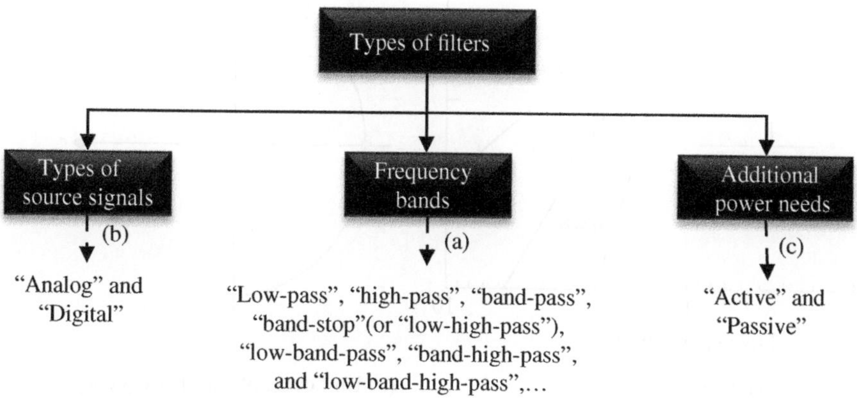

Figure 7.39 Three criteria in classifying filters

Figure 7.40 shows the difference between an active filter and a passive filter; a passive filter in Figure 7.40a does not require an additional power supply to filter a signal; filtering is performed by a set of passive electronic elements such as resistors, capacitors, and inductors. An active filter in Figure 7.40b requires its own source of power to filter a source's signal.

Three common low-pass filters are *Butterworth filters*, *Chebyshev filters*, and *Bessel filters*, which are shown in Figures 7.41, 7.42, and 7.43, respectively. Butterworth filters are characterized by the maximized flatness in the passband; as shown in Figure 7.41, this is no ripple in the passband. Moreover, a gain of a lowpass Butterworth filter is a function of the polynomial order and frequency.

A Chebyshev filter in Figure 7.42 has the fastest crisper change in slope to minimize the transient band. On the other hand, this increases the ripple in the passband. A Chebyshev filter without a ripple in passband has the same frequency response as a Butterworth filter, while a high-order Chebyshev filter is more satisfactory than a Butterworth filter as a notch filter.

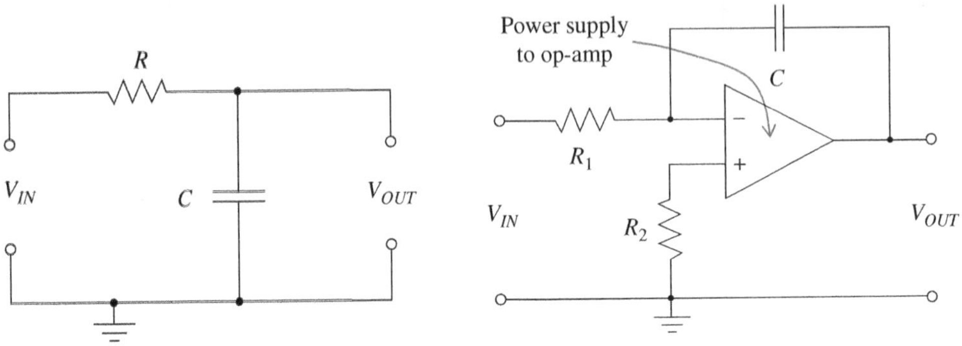

(a) No power in a passive filter (b) Power supply in an active filter

Figure 7.40 Active filter versus passive filter

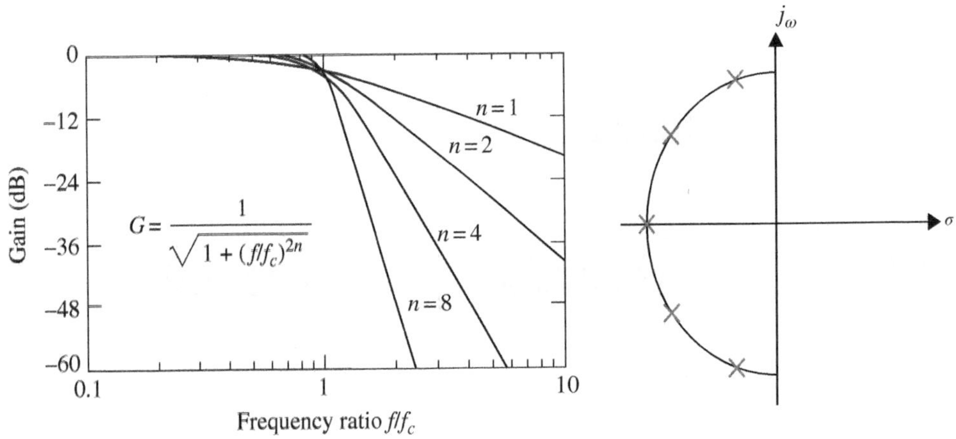

(a) Gain versus frequency (b) Unit cycle of cut-off frequency

Figure 7.41 Butterworth lowpass filter

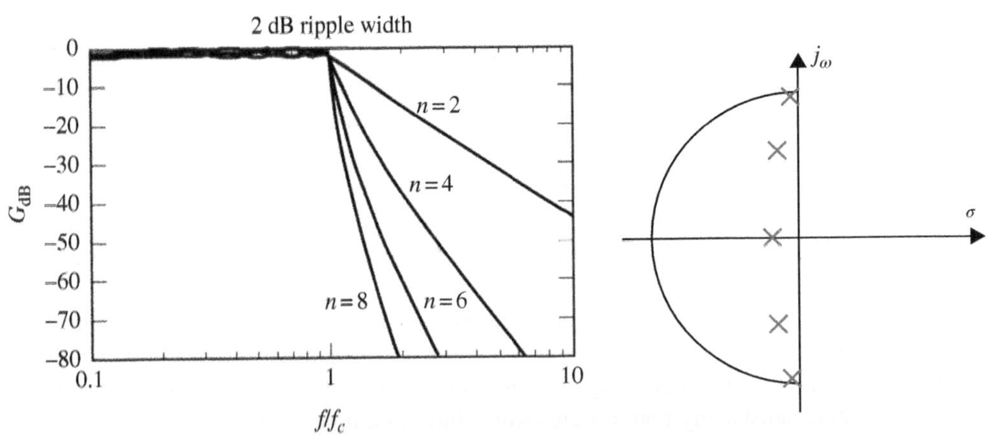

(a) Gain versus frequency (b) Unit cycle of cut-off frequency

Figure 7.42 Chebyshev lowpass filter

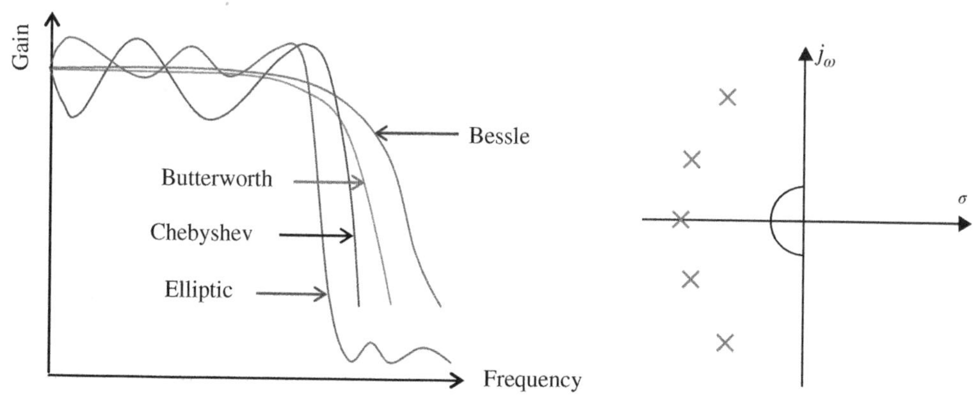

(a) Comparison of gains by different filters (b) Unit cycle of cut-off frequency

Figure 7.43 Bessel lowpass filter

Neither of Chebyshev filter nor the Butterworth filter sustains the phase of the source signal well. When an application (such as an audio signal) requires a proper phase shift, a Bessel filter in Figure 7.43 is ideal to preserve the wave shape in the filtered signal in the pass band. However, a Bessel filter has a much slower roll-off even more than a Butterworth filter. Therefore, Bessel filters are adopted when they are necessary. The roots of Bessel functions have the poles that lead to a maximally flat group delay in the pass band.

7.6.4 Linearization

An output of a steady variable is commonly linear; therefore, the measurement of such a steady variable requires linearization to characterize its state. In other cases, the nonlinearity in the measurement is caused by the sensing device. For example, when a thermocouple is used to measure a temperature, the output voltage is not proportional to the temperature change.

In general, *linearization* is a process of mapping the output from a sensor with the stimulus to correct measurement, and it can be achieved by either *software linearization* or *hardware linearization*. Figure 7.44 shows an example where the software linearization is adopted to interpret an output Seebeck coefficient (μV/°C) into the corresponding temperature for three types of thermocouples (i.e., Type-J, Type-K, and Type-S); the slop of a voltage curve is utilized to linearize the output temperature.

A thermocouple should be selected based on the range of temperature to be measured; it would be ideal that the Seebeck coefficient is varied as small as possible within the measured temperature range. An ideal linear thermocouple would have a contact Seebeck coefficient.

7.6.5 Conditioning Digital Signals

Most of the aforementioned signal conditioning techniques are used to process analog signals. After an analog signal is conditioned and digitized, its digital form might be further conditioned for sequential data processing, and digital signal conditioning and processing are software-controlled. Digital signal processes show advantages in the following aspects:

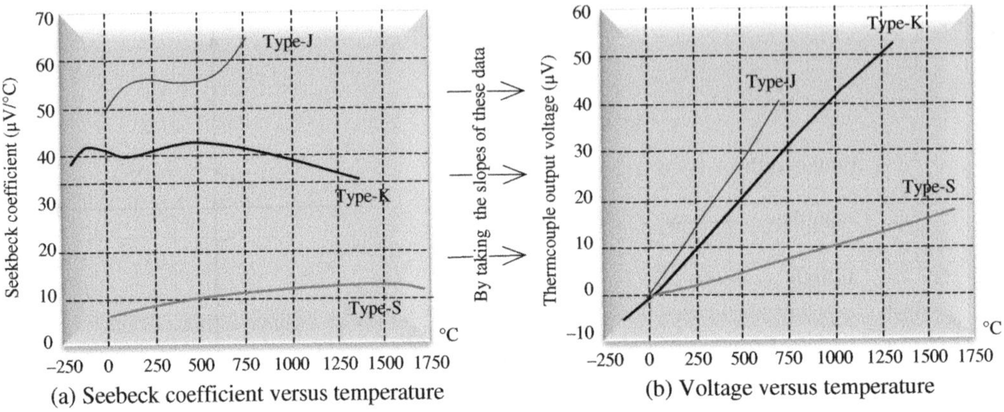

Figure 7.44 Software linearization of three thermocouple types

Versatility:	it is possible to implement some special functions that turn extremely difficult, impractical, or impossible to be achieved by hardware including *fast Fourier transform* (FFT) or "perfect" filters.
Stability:	a digital signal is immune to environmental factors such as temperature changes, power supply drifts, or interference.
Noiseless:	when the calculation is sufficiently precise, a digital signal is not affected by any noise in processing.
Linearity:	it sustains the linearity of input and output signals perfectly.

A digital signal often requires performing *pulse shaping* at a number of applications such as (1) optimizing the signal-to-noise ratio (SNR), (2) optimizing throughput, (3) minimizing ballistic deficit, (4) conditioning an input signal of ADC, and (5) providing a compromise solution to a set of conflicting requirements. Pulse shaping has three basic types: *the first one* is digital signal filtering to modify the bandwidth of the signal in its frequency domain. *The second one* is signal shaping to change the shape of the signal in its time domain. *The third one* is the Fourier transform to convert the signal from a frequency domain to a time domain or vice versa.

It is not a rare case that the duration of a certain state in a digital signal should be extended so that it can be detected by a microprocessor. This can be accomplished by taking a shoot at the moment of interest. As shown in Figure 7.45, when a trigger pulse is received by a shoot, the signal processing

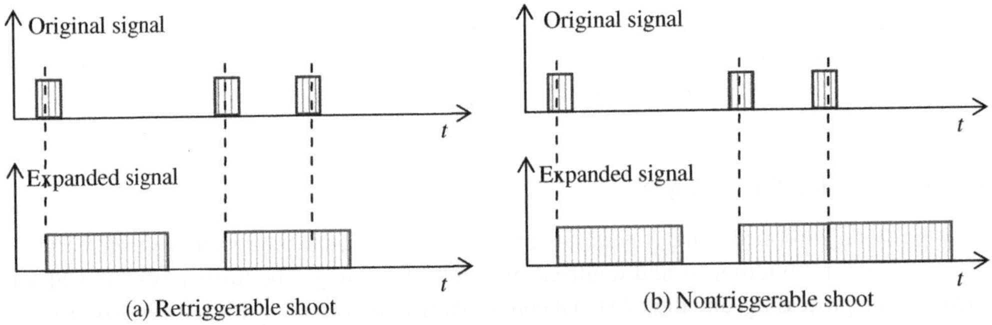

Figure 7.45 Retriggerable and non-retriggerable shoots in expanding pulse length by shooting

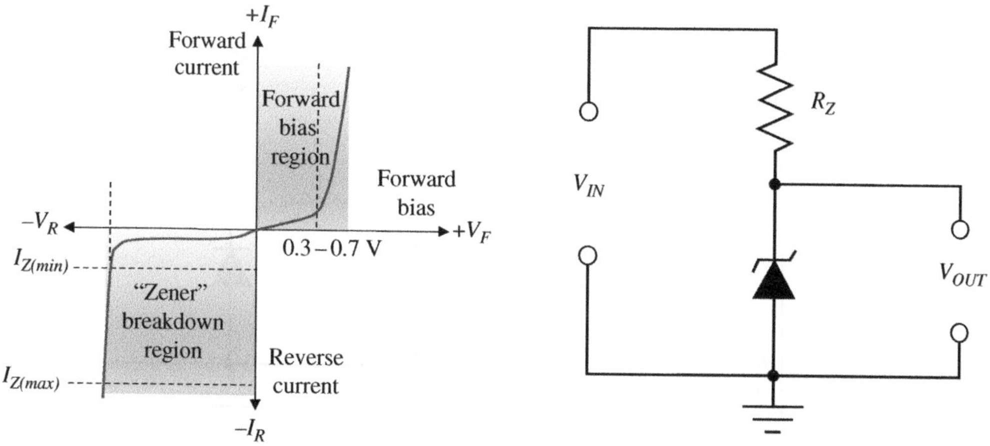

(a) Characteristic curve of Zener diode (b) Clipping circuit

Figure 7.46 Zener diode for signal clipping

unit will produce a pulse with a fixed length. Moreover, a shoot can be ***retriggerable*** or ***not retriggerable***. With a retriggerable shot, an active output will be extended when another trigger pulse occurs before the fixed length is terminated. With a non-retriggerable shoot, an active output pulse is always terminated with the same length.

7.6.6 Signal Clipping

Signal clipping has a similar purpose with signal attenuation to prevent circuitry damage due to an exceeded voltage. Taking an example, a sensor mounted on a car engine may occasionally pick up a noise with hundreds of volts and this will destroy a microprocessor. To protect a circuit from an exceeded noise, the signal can be chipped so that the amplitude of a signal never goes above a tolerated voltage. A Zener diode in Figure 7.46a can be used to clip a signal, and Figure 7.46b shows the corresponding circuit. The Zener diode serves as an isolator when the input voltage is less than the Zener voltage V_Z. It will conduct (i.e., let a current pass) once the voltage exceeds V_Z. Therefore, the output voltage will be the same as the input voltage until the latter exceeds V_Z. Once it occurs, the output voltage will remain constant as V_Z when the input voltage increases. The resistor is chosen in such a way that the maximum current through the Zener diode is below its allowable limit.

7.7 Signal Clamping

In some cases, the signal should be conditioned to ensure that it does not reverse its direction. Taking a car sensor as an example, if a signal becomes negative due to noise, it will destroy a microprocessor when the microprocessor is powered by *direct current* (DC). To avoid this, a signal should be clamped so that it never goes below zero, and a regular diode can be applied to clamp a signal. As shown in Figure 7.47a, a regular diode will conduct when it is forward biased, the output voltage will follow the input until the input goes below ~ −0.7 V. From then, the output will decrease. Note that this slight negative voltage will not affect the performance of most electronic elements. Similar

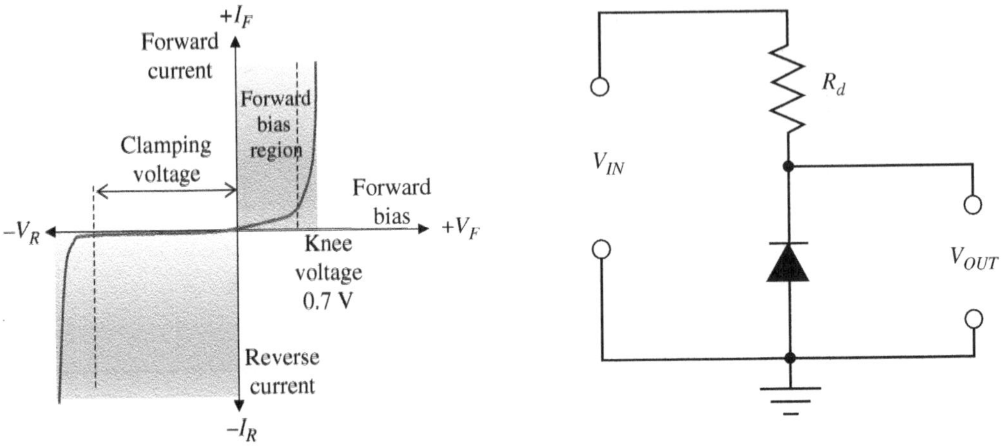

(a) Characteristic curve of regular diode

(b) Clamping circuit

Figure 7.47 Regular diode for signal clamping

to the circuit for signal clipping, the resistor should be chosen in such a way that the maximum current through the diode is within the allowable limit of the regular diode.

7.8 Summary

Mechatronic systems are often used in an environment with various disturbances, dynamic factors, and uncertainties. Measured data from sensors and transducers usually include randomness and noise that are prone to problematic in use. Moreover, the full-scale outputs of measurements are relatively weak, for example, mill- or micro-level voltages or currents. Therefore, signal conditioners become essential circuity to prepare a source signal for sequential processing in sequential steps.

Each conditioning circuit must ensure proper conditioning before the signal can be further processed adequately. For each source signal, a signal conditioner acts as the interface to scale up, filter, condition, or buffer a real-world signal. SCP covers a broad scope of tasks including cleaning, isolating, transmitting, and conditioning. Designers should be aware of the need for electronic circuits for signal conditioning and processing, they should be capable of defining the FRs of SCP and selecting appropriate electronic circuits to meet these FRs in designing a mechatronic system.

Problems

7.1 Assume that a step signal step $[t-x]$ is used to digitize an analog signal $f(t)=t$ at the sampling times of $n = 0, 1, 2, 3, 5, 6,...$), plot the digitized signals when the analog signals are processed by the following equations of (a). $f(t)*(step[t-2]-step[t-3])$ and (b). $step[t]+f(t)*step[t-2]$, respectively.

7.2 Determine an output signal when it is combined from the input signals given as (1) $3\angle 120° + 5\angle 60° + 4\angle -90°$, (2) $\frac{(1+j)^3}{1-j}$, and (3) $\frac{5}{2-j}(\angle 30°) + \frac{\sqrt{2}e^{jx}}{j}$, respectively.

7.3 For the amplifier configuration in Figure 7.8, determine the output voltage v_O based on the given variables for input three voltages $(v_1, v_2, v_3) = (2, 5, 8)$ V and all resistors $(R_1, R_2, R_3$ $R_f) = (10, 5, 20, 100)$ Ω.

7.4 When the amplitude of a source signal (v_{IN}) is amplified to an output signal (v_{OUT}) by (a) 20 times and (b) 500 times, respectively, determine the corresponding gain (G and G_{dB}) (a) and (b).

7.5 In Figure 7.12, assume that the output of a J-type thermocouple travels a 5-m wire to arrive at a data acquisition system (DAQ), the output voltage at the probe is around 25 μV/°C, and the noise by the wire is around 20 μV/m. Determine the SNR for (a) an amplification of 200 times is applied after the wire's noise is introduced (i.e., at the input of DAQ) and (b) an amplification of 200 times is applied before the wire's noise is introduced.

7.6 A temperature sensor has a resistance of 200 Ω at 0°C is placed as R_x in the Wheatstone bridge with each of the other arms also being 200 Ω in Figure 7.13. The voltage of power supply is $V_S = 12.0$ (V). If the resistance temperature coefficient of the sensing materials is $\alpha = 0.0043/$°C, find the output voltage from the bridge when the temperature is increased by 5°C.

7.7 As shown in Figure 7.31, there is a temperature change of 15°C at the thermocouple junction, and it produces an emf of 800 μV between V_1 and V_2. Let $R_1 = 20$ (kΩ), how to select R_2 if one expects a circuit yields an output of $V_O = 20$ mV?

7.8 A force-measuring transducer has an open-circuit output voltage of 125 mV and an output impedance of 750 Ω. To amplify the signal voltage, it is connected to an amplifier with a gain of 20. Estimate *the input loading error* if the amplifier has an input impedance of (a) 5 k Ω or (b) 1 M Ω.

References

Electronics-tutorials (2024). The Multiplexer. https://www.electronics-tutorials.ws/combination/comb_2.html (accessed 12 September 2024).

Khanacademy (2024). Electric potential, voltage. https://www.khanacademy.org/science/electrical-engineering/ee-electrostatics/ee-fields-potential-voltage/a/ee-electric-potential-voltage (accessed 12 September 2024).

NI (2003). Data acquisition and signal conditioning course manual. http://www.n4iqt.com/dallasis4/labview_7.0_daq_course.pdf (accessed 12 September 2024).

Olson, W.H. (2008). Basic concepts of medical instrumentation. http://ijlalhaider.pbworks.com/w/file/fetch/63677907/Basic%20Concepts%20in%20Medical%20Instrumentation%20by%20WH%20Olson_C01_Webster.pdf (accessed 12 September 2024).

Spinning Number (2023). Charge, current, voltage. https://spinningnumbers.org/a/basic-quantities.html (accessed 12 September 2024).

8

System Controls

8.1 Basics of Control Systems

A control system is the "brain" of a whole mechatronic system. It interprets sensed signals or humans' inputs into the commands that are sent to actuators to operate systems. Figure 8.1 shows the role of a control system in a mechatronic system. It is not overstated that all of the sub-systems we introduced in previous chapters are developed to implement and run the control system.

For example, the mechanical system is to take actions by following the commands from the control system. The sensing system is to acquire the system's state so that the control system can produce control commands correctly. The actuating system is to translate weak control commands into mechanical power that is sufficient enough to drive the mechanical system. The rest of the sub-systems such as *signal conditioning and processing* (SCP) are used to prepare data in the right forms for the control system to interact with other sub-systems. In fact, when a mechatronic system is relatively simple, all of the sub-systems, except for the mechanical system, can be merged into the control system as a whole mechatronic system.

The control system determines the intelligence level of the mechatronic system. The higher level of intelligence a mechatronic system has, the more advanced and complicated its control system will be. Figure 8.2 shows some good examples of mechatronic systems whose performance depends greatly on the powerfulness and functionalities of their control systems.

8.1.1 Complexity of Control Problem

Similar to any other system, the complexity of a control system depends on (1) the numbers and types of system parameters and control variables, (2) the relations of these parameters and variables to the system's behaviors, and (3) the changes and uncertainties of these parameters and variables over time. Due to complexity, critical variables must be monitored so that the discrepancies of these variables to the targets can be evaluated in a closed-loop control.

From this perspective, the complexity of a control system can be partially reflected by the number and types of sensors used to collect data from MFMs in a mechatronic system. In a data-driven or closed-loop control, the reliability of a control decision relies on the availability, accessibility, accuracy, and sufficiency of data acquired from system elements. Therefore, the reliability and abundance of sensed data are the foundation to design and implement a control system. Take an example of a smart car, Figure 8.3 shows some commonly used sensors (Guerrero-Ibáñez et al. 2018). Moreover, a smart car is typically equipped with 30–100 microprocessors to control various

Mechatronics for Complex Products and Systems: Project Based Design Approaches for Robots, Cyber-Physical Systems, Digital Twins, and Other Emerging Technologies, First Edition. Zhuming Bi.

Companion website: www.wiley.com/go/bi/mechatronics

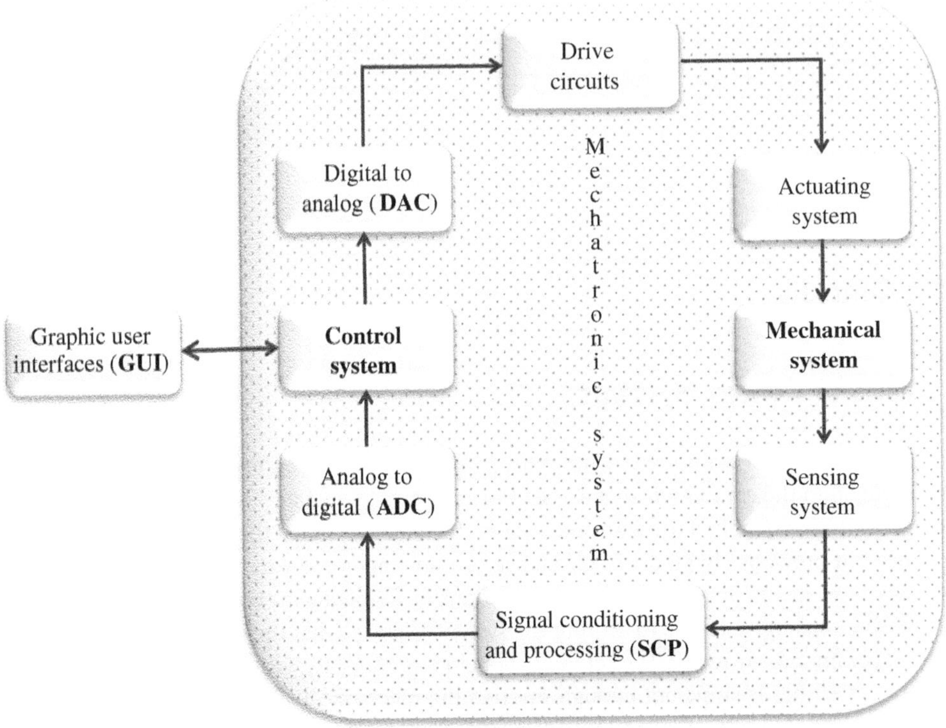

Figure 8.1 The role of control system in a mechatronic system

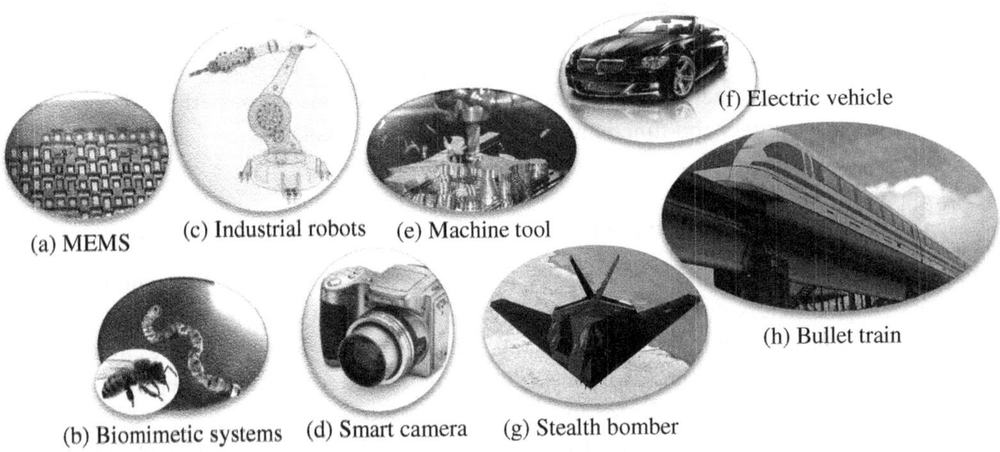

Figure 8.2 Mechatronic systems from microscale to macroscale (*Sources:* (a) Confovis / https://www.confovis.com/en/solutions/surface-metrology/ / Last accessed on July 19, 2024; (b1) Alekss/Adobe Stock; (b2) U.S. National Science Foundation / https://www.nsf.gov/news/mmg/mmg_disp.jsp?med_id=185000&from= / Last accessed on July 19, 2024; (d) Better Digital Photo / https://www.better-digital-photo-tips.com/camera-review-sites.html / Last accessed on July 19, 2024; (e) Weerg / https://www.weerg.com/guides/cnc-milling-turning / Last accessed on July 19, 2024; (f) BMW / https://www.collinsauto.ca/blog/is-bmw-the-best-in-luxury-sedans- / Last accessed on July 19, 2024; (g) US Air Force / Wikimedia Commons / Public Domain; (h) Stahlkocher / Wikimedia Commons / CC BY SA 3.0)

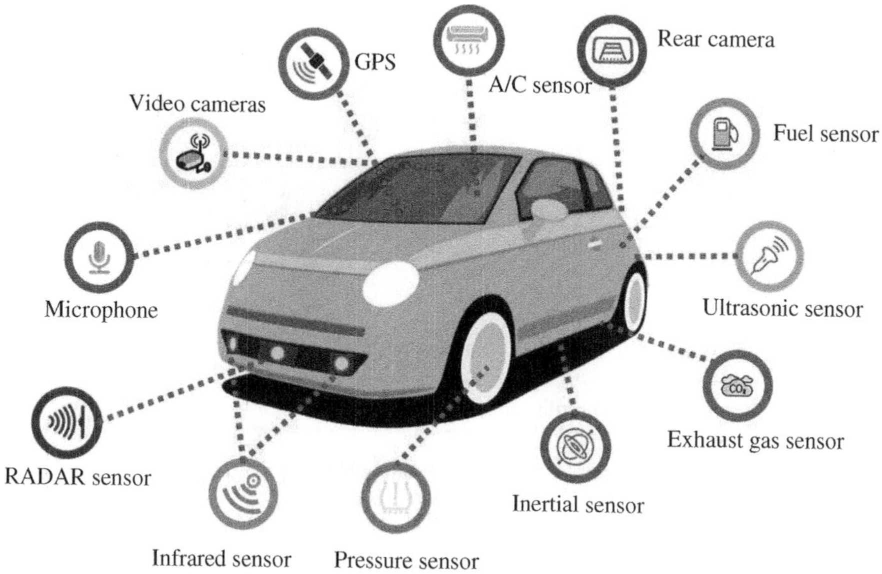

Figure 8.3 Commonly used sensors in a smart car (*Source:* Guerrero-Ibáñez et al. (2018), open access/ MDPI/CC BY 4.0)

sub-systems while driving. The more advanced a mechatronic system is, the more sensors will be needed to collect real-time data for precise system control.

8.1.2 Types of Control Problems

A control problem is typically formulated by specifying three components, that is, *input*, *output*, and *the system* to transform input to output. Accordingly, control problems may be diversified but generally classified into three types, as shown in Figure 8.4. The first type is *system identification,* which

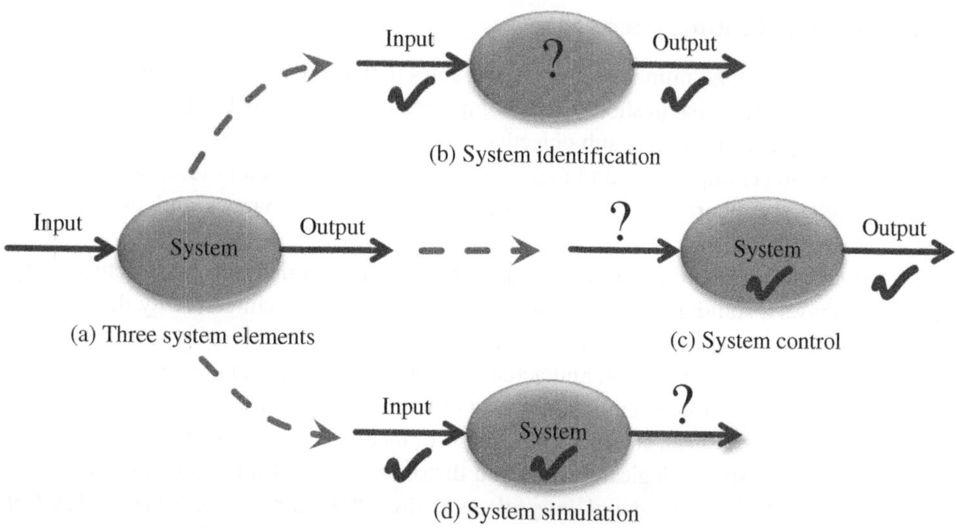

Figure 8.4 Three types of control problems

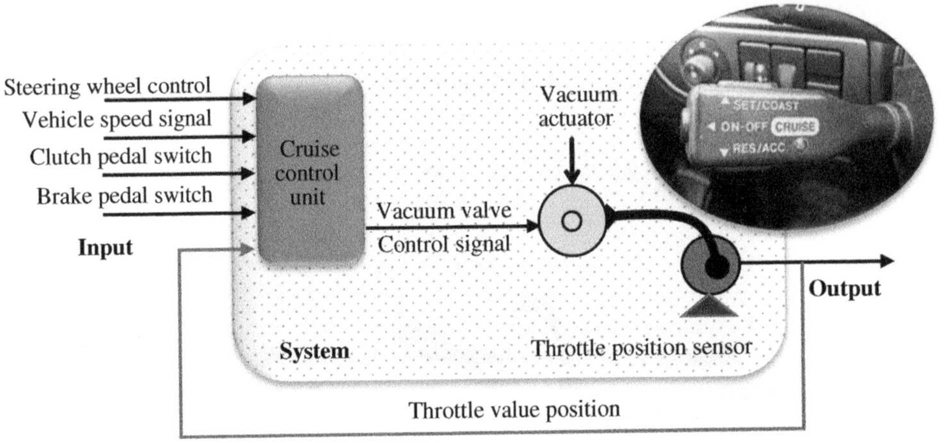

Figure 8.5 Example of cruise control in the car (*Source:* OSHA Safety Manuals / https://www.safetymanualosha.com/cruise-control-driving/ / Last accessed on July 19, 2024)

aims to understand the system by specifying the input and observing the output. The second type is *system control,* which aims to determine the right input when both the system and the expected output are known. The third type is *system simulation,* which aims to predict the output when both the system and the input are known.

The simplest form of system control is to (1) determine the desired output and (2) find a way to control the system's behavior to generate the expected output. Take an example of a cruise control in a car in Figure 8.5, the desired speed is first set up, and the control unit determines the openness of the throttle value based on the desired speed and other inputs. It produces and sends control signals to the vacuum actuator. Meanwhile, the position of the throttle value is monitored and feed-backed to the control unit, so that the correction can be made when the actual speed is different from the desired speed. Bear in mind that most mechatronic control systems are pretty complex and far beyond on/off systems; system variables are continuously controlled by closed-loop feedback.

8.1.3 Architecture of Control Systems

When a system involves a high number of system variables that are controlled and updated at different time scales, its control system should be decomposed so that control objectives at the higher system level can be replaced by a set of sub-objectives at lower component or device levels. Control decompositions help to (1) implement and maintain control units at different levels and domains independently and (2) support system reconfiguration to adapt to changes and uncertainties over time.

A control system can be decomposed by modularization, by using a hierarchical structure, or by both. Figure 8.6 shows a standardized five-layer hierarchical control architecture by the *International Society of Automation* (ISA). The lowest layer of the control pyramid is referred to as the field level. It consists of sensors, instruments, and actuators that are used to (1) measure process parameters such as displacement, force, temperature, and pressure and (2) control process variables by hydraulic, pneumatic, mechanical, electrical, electronic, or mechatronic actuators. The second one is the control level. It consists of logical devices and distributed control units such as a *programmable logic controller* (PLC), *proportional–integral–derivative* (PID), and *distributed control system* (DCS) controller. The devices at this level control or regulate the devices at the field level in closed-loop form. The feedback is received from the field level, and the decisions are made and sent back to

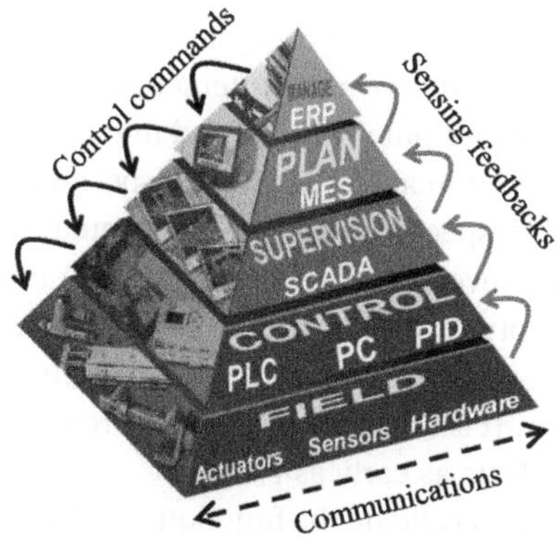

(a) Five-layer architecture

Layer	Functions	Tools
Manage	Enterprise Resource Planning (ERP)	Sage®, Oracle®, SAP®, Microsoft Dynamics 365®
Plan	Manufacturing Execution Systems	Hydra®, Bluestreak®, IQMS®, CIMAG®, Epicor®
Supervision	Statistical Control and Data Acquisition system	Ignition®, Indusoft®, Wonderware®, Kepware®, Factory Talk®
Control	distributed controls and programmable logical controls.	A-B®, Omron®, Mitsubishi®, Siemens®, Modicon®, ABB®
Field	The arms and legs of the manufacturing system. Traditional sensors lean on the computing power of the system it is attached to.	Field devices from simple switches to IIoT programmable edge computing sensors.

(b) Functions and tools at different layers

Figure 8.6 Five-layer enterprise control architecture (*Source:* SMCworld (2024) / SMC Pneumatics Ltd)

the devices at the field level so that the right actions are taken for these devices to meet expected goals. The controls from the third to the fifth layer serve as the roles for supervision, planning, and management, respectively (Rahman et al. 2021).

8.1.4 Design of Control Systems

In IEEE standards, a control system is defined as a collection of components organized to accomplish a specific function or a set of functions. The generic V-model procedure in Figure 8.7 can be

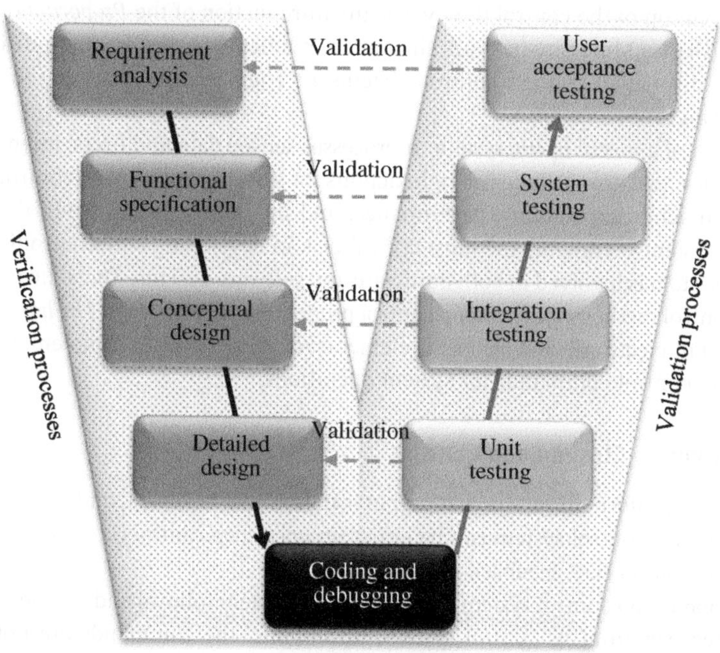

Figure 8.7 Design of control system by V-model

followed to design a control system as an MFM at a specified level or domain (Koziorek et al. 2018). *First*, it begins with the definitions of the specifications of the expected control system from the users' perspective; it focuses on system inputs and outputs based on the system's interactions with others in the environment. *Second*, the requirements for input and output are translated into those from the conversion of the inputs to the outputs; these are referred to as the functional specifications of the system. *Third*, the solutions to given functional specifications are conceptually designed. *Fourth*, the high-level functional specifications at high levels can be decomposed hierarchically, and detailed designs should be performed over conceptual designs at high levels. *Fifth*, detailed designs are coded and converted into programs on the software side, and corresponding hardware systems are configured and built on the hardware side. It is desirable to verify the results obtained at each step rigorously to minimize hidden design defects as early as possible.

On the other hand, the tasks at the right side of the V-model are to validate the functional units of the control system from the below level to the top level to ensure each and every functional unit achieves the expected functions. If there is a discrepancy in either a verification or validation process, previous steps are repeated until a resolution is found to resume the sequential design process.

8.2 Control Theory

Control theory is a combined field of control engineering and applied mathematics to control a dynamic system such as a robot, machine tool, an engineered process, or appliance. The objective of using the control theory is to model a dynamic system appropriately and predict and control the system's behaviors in response to environmental changes to produce desired system outcomes with minimized delay, overshoot, and steady-state errors (Wiki 2024). The development of the control theory can be traced back to the 19th century, when James Maxwell established the theoretical basis for operating governors. It was advanced in the late 1800s to develop the criteria for control stability. Another milestone in the evolution of the control theory was the introduction of the *Proportional Integral Derivative* (PID) by Nicolas Minorsky in 1992. Since then, mathematics has been applied heavily in modeling highly diversified process control systems in engineering, sociology, life sciences, and operations research.

A control system is the heart of a mechatronic system. It processes sensed data and users' intentions to generate commands for actuators to take the right actions for system operation. A control system plays a central role in achieving stability, performance, and the desired responses of the system. Designing a control system is arguably the most critical task in developing a mechatronic system. Moreover, a control system is involved in both hardware and software designs. *The software design* is to select the right control methods and develop control programs and interfaces. *The hardware design* is to select the right electronic and electrical elements; develop, select, or implement control methods; and execute control commands for system operations.

8.2.1 Open-Loop Control Versus Closed-Loop Control

Depending on whether or not the information of actual output is monitored, feedback, and used in determining control commands, a control system can be either *an open-loop control* (i.e., feedforward control) *or a closed-loop control* (i.e., feedback control).

Figure 8.8a illustrates an open-loop control system. The controlled system follows and responds to whatever commands it receives from its control system; meanwhile, the commands are not

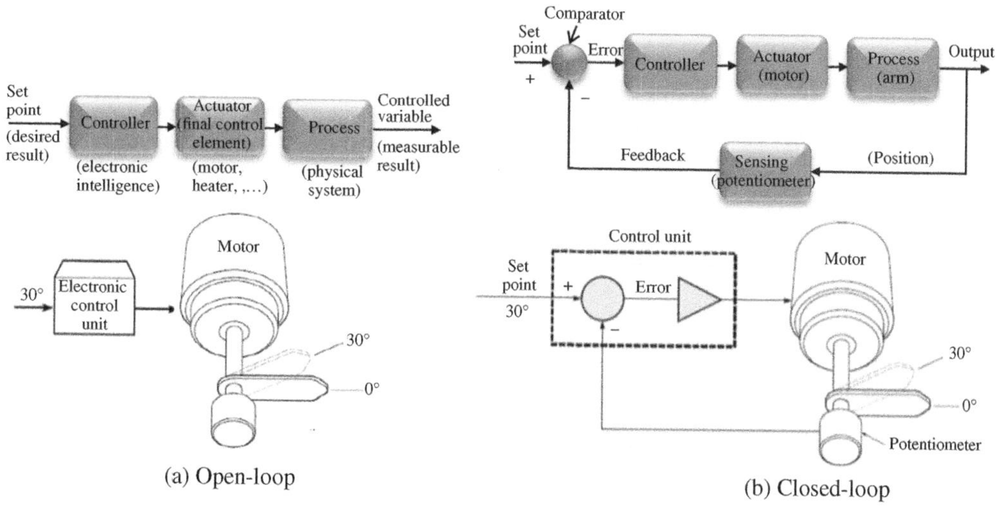

Figure 8.8 Open-loop control versus closed-loop control

affected by the system output, even if the system output is not expected. When the controller sends out a command to a motor to rotate a 30°, the motor will respond accordingly. However, the issued command will not be updated whether or not the motor actually rotates 30°.

Figure 8.8b illustrates a closed-loop control system. The controlled system follows and responds to the commands; however, the system output is sensed and feedbacked to the controller. The controller then compares the actual and desired signals to determine the discrepancy between two signals. If the discrepancy is larger than the given tolerance, the issued command is updated to refine the rotation of the motor. The feedback control will be continued until the desired position is reached within the given tolerance.

8.2.2 Process Control Versus Motion Control

A process control system in Figure 8.9a controls physical variables in an industrial or manufacturing process. Common process variables include the quantities relevant to pressure, temperature, chemical composition, and flow. Process control systems are widely used in continuous production lines for food, beverage, pharmaceuticals, oil, gas, *heating, ventilation, and air conditioning* (HVAC), and engineering materials. Process variables to be controlled must be sensed, monitored, and adjusted in a real-time manner.

A *motion control* system in Figure 8.9b controls positions, directions, speeds, and accelerations of motions in a machine, robot, or any other automated device. Motion control systems are commonly deployed in industrial and manufacturing systems such as *computer numerical controls* (CNCs), materials-handling systems, transportation, and assembly lines. Controllable variables are sensed to monitor system states and adjust the motions of MFMs based on sensed discrepancies of actual and expected measures such as displacements, velocities, or accelerations.

In comparison, a motion control system is generally more precise and faster, while a process control system is more suited for applications where control variables must be controlled in discretized

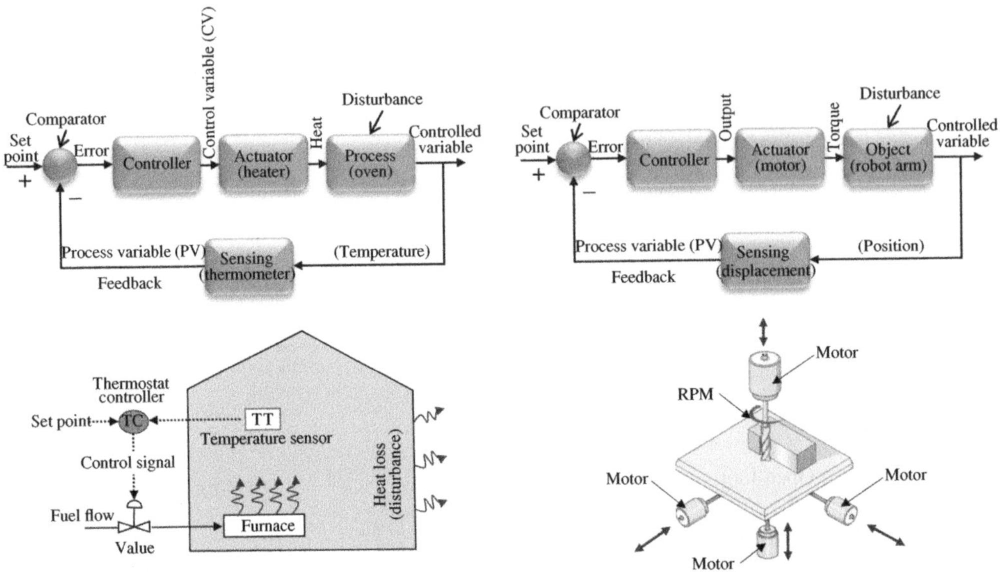

(a) **Process control** refers to an industrial process being electronically controlled for the purpose of maintaining a uniform correct output; commonly controlled variables are on/off, logic, events, sequences, …

(b) **Motion control** refers to control a system with motion. It provides the control for a device or machine such as a robot arm or a radar antenna); commonly controlled variables are displacement, speed, torque, path, …

Figure 8.9 Process control versus motion control

states, and commonly controlled variables are on/off states, event triggers, and the sequence of operations.

8.2.3 Steady Response Versus Transient Response

To characterize a control system, simple and standard signals are used as test signals, and the output signals are analyzed to characterize steady and transient responses of systems.

As shown in Figure 8.10, *a steady-state response* refers to a steady function of an input signal when the system is settled and the system starts working normally. A steady-state response is also called a force response. *A transient response* of a control system refers to the changes that occur to the transient of two conditions, that is, the first condition is at the time when an input signal is just applied to the system, and the second condition is the time when the output signal becomes steady. Figure 8.10b and c lists major characterized parameters of transient and steady-state responses, respectively.

8.2.4 Transfer Functions

A control system for *n* number of controllable variables can be viewed as an assembly of *n* number of *single-input and single-output* (SISO) systems. To facilitate our discussion, this section limits the scope of discussion on SISO, which is the simplest control system to transfer a single input into a single output. Moreover, SISO especially refers to a "1" and "0" digital system in control engineering. To implement a complex system with a large number of controllable variables, one needs to (1) control individual variables freely and (2) coordinate the behaviors of individual controllers at

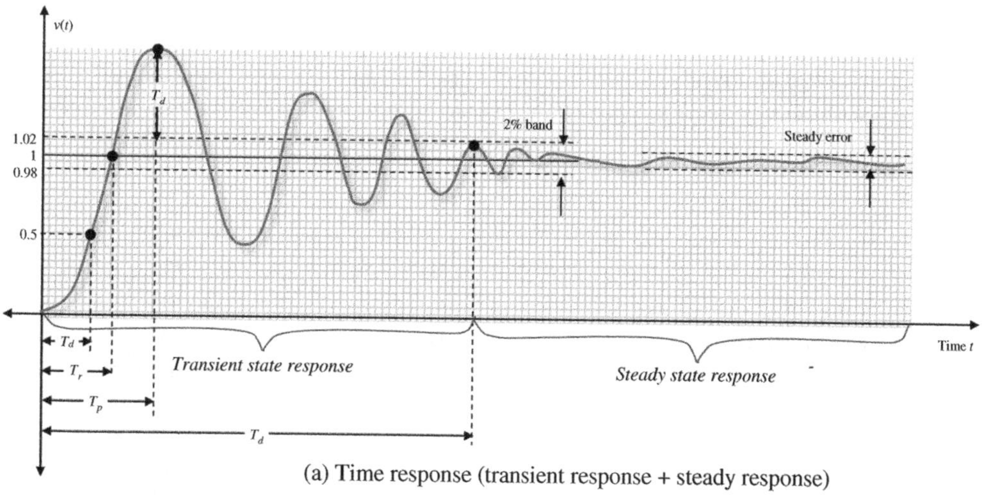

(a) Time response (transient response + steady response)

Characterization of transient response

Variable	Meaning
T_d	Delay time for 50% of amplitude
T_r	Rise time for 100% amplitude
T_p	Peak time for maximum output
M_p	Peak overshoot with %
T_s	Settling time with the output becomes steady

(b) Characterized parameters of transient response

Characterization of steady response

Variable	Meaning
τ	time reaches 63% of nominal voltage
f_c	cut-off frequency to stop/transfer signal
G	the ratio of output and input voltage
v_n	noise included in steady signal
f_n	nature frequency of the control system

(c) Characterized parameters of steady response

Figure 8.10 Characterizing time responses of signals

component or system levels. Essentially, one must understand how SISO works on individual variables.

In the fields of signal processing, information communication, and system controls, *transfer functions* are commonly used to model a control system. Transfer functions are extremely useful to model *linear time-invariant* (LTI) systems with the properties of *linearity* and *time invariance*:

1) *Linearity* means that the input and the output of the system are linearly corresponded. When a control system is linear, the system input and output consist of n components, that is, ($x_i(t)$, $i = 1, 2, \ldots, n$), and ($y_i(t)$, $i = 1, 2, \ldots, n$) for the input and the output, respectively. The summed input $\sum_{i=1}^{n} a_i x_i(t)$ produces a scaled and summed system output of $\sum_{i=1}^{n} a_i y_i(t)$.
2) *Time invariance* means that whenever a signal is input into the system, the system responds and produces an identical output signal except for a time delay (T), that is, when an input $x(t)$ is applied at time t, the system produces an output of $y(t)$. When the same input is applied at a time delay of T, the system produces an output of $y(t + T)$.

A transfer function is defined as a ratio of the input and output signals. It can be defined in either the time or frequency domain. Two transfer functions in the time and frequency domains can be transferred with each other by *Laplace transform* (LT) or *inverse Laplace transform* (ILT). Table 8.1 shows the conversions of common transfer functions in time and frequency domains, where s is a variable in a complex plane ($\sigma + i \cdot \omega$) that represents the output.

Table 8.1 Conversions of transfer functions in time and frequency domains

$f(t)$	$\mathcal{L}[f(t)] = F(s)$	Eq.	$f(t)$	$\mathcal{L}[f(t)] = F(s)$	Eq.
1	$\frac{1}{s}$	(1)	$\frac{ae^{at} - be^{bt}}{a-b}$	$\frac{s}{(s-a)(s-b)}$	(10)
$e^{at}f(t)$	$F(s-a)$	(2)	te^{at}	$\frac{1}{(s-a)^2}$	(11)
$u(t-a)$	$\frac{e^{-as}}{s}$	(3)	$t^n e^{at}$	$\frac{n!}{(s-a)^{n+1}}$	(12)
$f(t-a)u(t-a)$	$e^{-as}F(s)$	(4)	$e^{at}\sin(kt)$	$\frac{k}{(s-a)^2+k^2}$	(13)
$\delta(t)$	1	(5)	$e^{at}\cos(kt)$	$\frac{s-a}{(s-a)^2+k^2}$	(14)
$\delta(t-t_0)$	e^{-st_0}	(6)	$e^{at}\sinh(kt)$	$\frac{k}{(s-a)^2-k^2}$	(15)
$t^n f(t)$	$(-1)^n \frac{d^n F(s)}{ds^n}$	(7)	$e^{at}\cosh(kt)$	$\frac{s-a}{(s-a)^2-k^2}$	(16)
$f'(t)$	$sF(s) - f(0)$	(8)	$t\sin(kt)$	$\frac{2ks}{(s^2+k^2)^2}$	(17)
$f^n(t)$	$s^n F(s) - s^{(n-1)}f(0) - \ldots f^{(n-1)}(0)$	(9)	t	$1/s^2$	(18)
			t^2	$1/s^3$	(19)

Example 8.1 Determine the transfer function in the frequency domain for the relation of input voltage and output angular displacement (θ) in the circuit in Figure 8.11.

Solution

Let the current of the circuit be $i(t)$; the voltages at all elements meet the constraints of

$$v_{in} - R \cdot i - L\frac{di}{dt} - v_{emf} = 0 \tag{8.1}$$

where R is the resistance of the resistor, L is the inductance of the inductor, i is the current in the circuit, and v_{in} and v_{emf} are the input and output voltages, respectively.

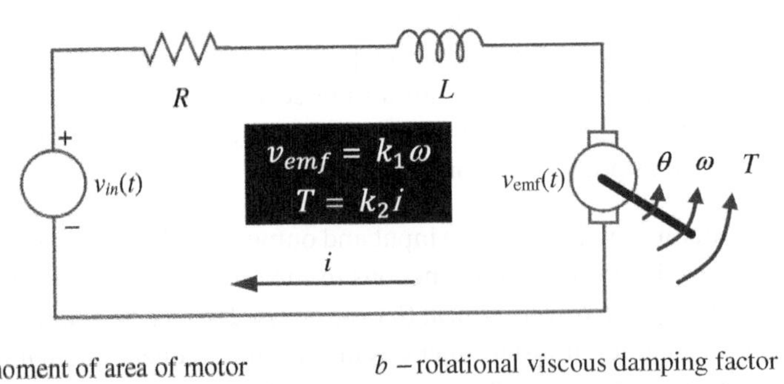

Figure 8.11 Circuit for transfer function example

To determine the transfer function of the circuit, i.e., the relation of v_{in} and θ, the relation of i with v_{in}, and v_{emf} must be defined, and the following governing equations of the motor are used:

$$\left.\begin{aligned} T &= J\frac{d\omega}{dt} + b\omega \\ \omega &= \frac{d\theta}{dt} \\ v_{emf} &= k_1\frac{d\theta}{dt} \\ T &= k_2 i \end{aligned}\right\} \tag{8.2}$$

where T is the torque, J is the rotational moment of inertia, b is the rotational damping, and k_1 and k_2 are two constants.

Substituting all relations in Eq. (8.2) into Eq. (8.1) yields the relation of v_{in} and θ as,

$$v_{in} - \frac{R}{k_2}\cdot\left(J\frac{d^2\theta}{dt^2} + b\frac{d\theta}{dt}\right) - \frac{L}{k_2}\left(J\frac{d^3\theta}{dt^3} + b\frac{d^2\theta}{dt^2}\right) - k_1\frac{d\theta}{dt} = 0 \tag{8.3}$$

Applying the Laplace transform equations in Table 8.1 and ignoring the transient terms in Eq. (8.3) get

$$V(s) - \left(\frac{R}{k_2}\cdot(Js^2 + bs) + \frac{L}{k_2}(Js^3 + bs^2) + k_1 s\right)\theta(s) = 0 \tag{8.4}$$

Equation (8.4) is rewritten as the transform function in the frequency domain as

$$\frac{\theta(s)}{V(s)} = \frac{1}{\frac{LJ}{k_2}s^3 + \frac{(RJ + Lb)}{k_2}s^2 + \frac{Rb + k_1k_2}{k_2}s} \tag{8.5}$$

A control system can be modeled by *block diagrams*. Every element in a control system is represented by a block; a block is a symbolic requirement of the transfer function for the corresponding element, and the control system is modeled by a set of interconnected blocks. The transform function of the control system can be determined by analyzing its block diagram.

Example 8.2 Determine the transfer function of the control system that is represented by the block diagram in Figure 8.12.

Solution

The input and the output of the block diagram are $R(s)$ and $C(s)$, respectively. It includes three feedbacks at nodes A, B, and C, respectively, and each of them is followed by an amplifier (i.e., G_1, G_2, and G_3). The transforms at nodes A, B, and C can be determined as:

	Input		
Node	**(+)**	**(−)**	**Output**
A	$R(s)$	$H_5\cdot C(s)$	$R(s) - H_5\cdot C(s)$
B	$(R(s) - H_5\cdot C(s))\cdot G_1$	$H_4\cdot C(s)$	$(R(s) - H_5\cdot C(s))\cdot G_1 - H_4\cdot C(s)$
C	$((R(s) - H_5\cdot C(s))\ G_1 - H_4\cdot C(s))\cdot G_2$	$H_3\cdot C(s)$	$(((R(s) - H_5\cdot C(s))\cdot G_1 - H_4\cdot C(s))\cdot G_2) - H_3\cdot C(s)$

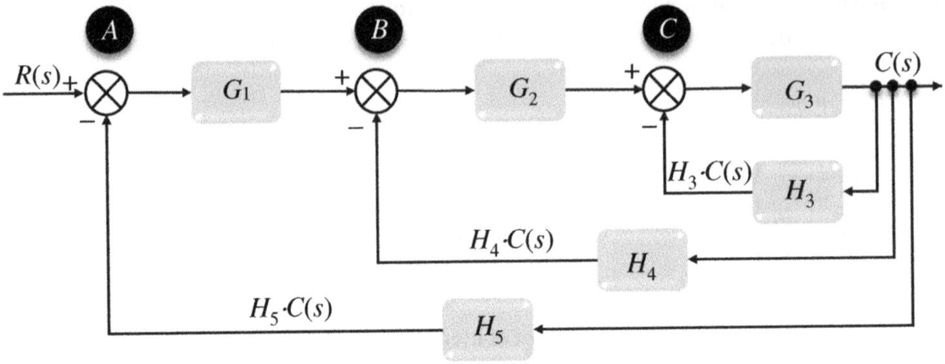

Figure 8.12 Block diagram for Example 8.2

Finally, the output from node C is amplified by G_3 as the output of the block diagram; therefore, one has,

$$((((R(s) - H_5{\cdot}C(s)){\cdot}G_1 - H_4{\cdot}C(s)){\cdot}G_2) - H_3{\cdot}C(s)){\cdot}G_3 = C(s) \tag{8.6}$$

Equation (8.6) can be rewritten to find the transfer function of the block diagram as:

$$\frac{C(s)}{R(s)} = \frac{G_1G_2G_3}{1 + H_3G_3 + H_4G_2G_3 + H_5G_1G_2G_3}$$

8.2.5 Orders of Control Systems

The order of a control system is defined as the highest power of s that is presented in the denominator of the closed-loop transform function $G(s)$ of *a unity feedback system*.

The higher the order of a control system, the more complicated the control system is. The order of the polynomial equation in the denominator of $G(s)$ relates to the number of energy storage elements in the system.

Figure 8.13 shows the block diagram of the first-order control system. The transfer function $G(s)$ and output signal $Y(s)$ can be found as:

$$G(s) = \frac{Y(s)}{R(s)} = \frac{1}{Ts + 1} \tag{8.7}$$

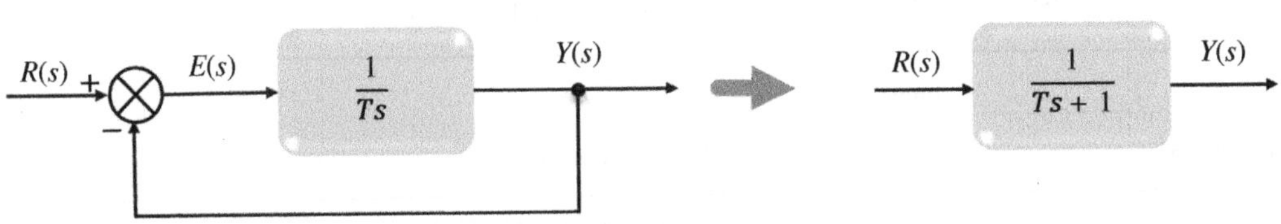

Figure 8.13 The first-order control system

For a unit input $f(t) = 1$, Table 8.1 shows that the corresponding Laplace transfer function is $R(s) = \frac{1}{s}$; therefore, on has:

$$Y(s) = G(s) \cdot R(s) = \frac{1}{(Ts+1)s} = \frac{1}{s} - \frac{1}{s+1/T} \tag{8.8}$$

Using Table 8.1 again for an inverse Laplace transform function of Eq. (8.8) finds the output signal in its time domain is

$$y(t) = 1 - e^{-t/T} \quad \text{for } t \geq 0 \tag{8.9}$$

Figure 8.14 characterizes the response of the first-order control system to the unit input by time constant T

Figure 8.15 shows the block diagram of the second-order control system. It is characterized by (1) the damping ratio ζ and (2) the natural frequency ω_n of the system.

The transfer function $G(s)$ and output signal $Y(s)$ can be found as:

$$G(s) = \frac{Y(s)}{R(s)} = \frac{\omega_n^2}{\omega_n^2 + s(s + 2\zeta\omega_n)} \tag{8.10}$$

For a unit input $f(t) = 1$, Table 8.1 shows that the corresponding Laplace transfer function is $R(s) = \frac{1}{s}$; therefore,

$$Y(s) = G(s) \cdot R(s) = \frac{\omega_n^2}{s\left(\omega_n^2 + s(s + 2\zeta\omega_n)\right)} \tag{8.11}$$

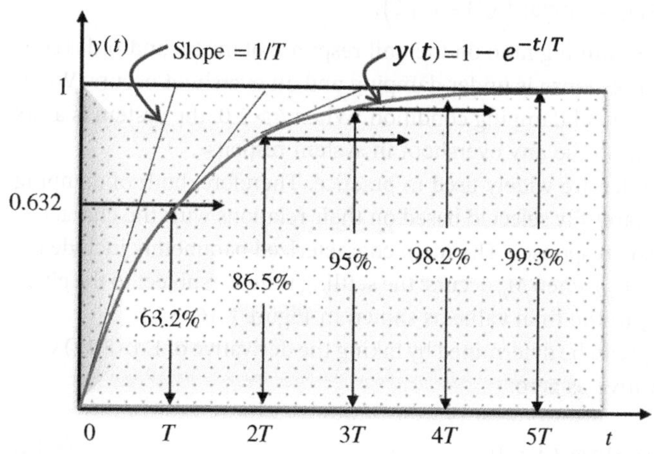

Figure 8.14 Response $y(t)$ of the first order control system to unit input ($f(t) = 1$)

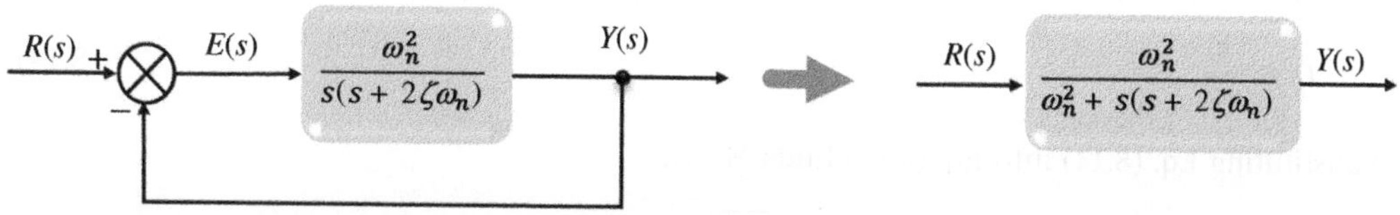

Figure 8.15 Block diagram of the second-order control system

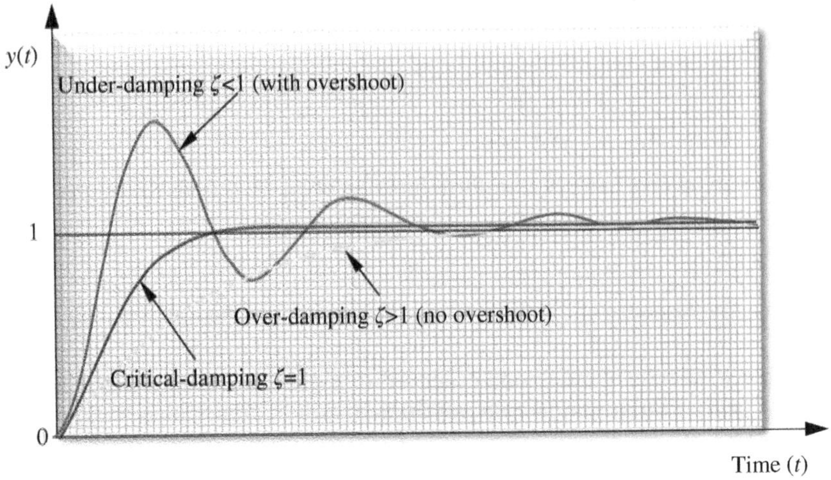

Figure 8.16 Impact of damping ratio on system response

Taking inverse Laplace transform of Eq. (8.11) finds the output signal $y(t)$ in its time domain as (Trumper 2003),

$$y(t) = 1 - e^{-\sigma t}\left(\cos \omega_d t + \frac{\sigma}{\omega_d}\sin \omega_d t\right) = 1 - \frac{\omega_n}{\omega_d} e^{-\sigma t}\sin(\omega_d t + \phi) \tag{8.12}$$

where $\sigma = \zeta\omega_n$, $\omega_d = \omega_n\sqrt{1-\zeta^2}$, and $\phi = \arctan\left(\sqrt{1-\zeta^2}/\zeta\right)$.

Figure 8.16 shows the impact of the damping ratio on the unit response of the second-order control system. When $\zeta < 1.0$, the system response is under damping and an overshoot occurs. When $\zeta = 1.0$, the system response is at its critical damping condition. When $\zeta > 1.0$, the system is at its overdamping condition and there will not be any overshoot in system response.

Control systems with the second order are widely used in practice. Therefore, both of damping ratio and natural frequency must be carefully selected based on their relations with the characterized parameters in transient and steady responses. The main characterized parameters include *the maximum overshoot* (M_p) and the time (t_p) when M_p occurs; the settling time (t_s) subject to the given tolerance band; and the rise time (t_r) to the final value as shown in Figure 8.16.

The overshoot (M_p) and the time (t_p) can be determined by taking the derivative of Eq. (8.12) with respect to t and setting its first derivative as zero:

$$\frac{d(y(t))}{dt} = e^{-\sigma t}\left(\frac{\sigma^2}{\omega_d}\sin \omega_d t + \omega_d \sin \omega_d t\right) = 0 \tag{8.13}$$

Equation (8.13) gives the condition for the first derivative being zero as $\omega_d t = (1 \pm 2n)\pi$ $n = 0, 1, 2, \ldots.$ For the first peak when $n = 0$, the first peak time (t_p) is calculated as:

$$t_p = \frac{\pi}{\omega_d} \tag{8.14}$$

Substituting Eq. (8.14) into Eq. (8.12) finds M_p as:

$$M_p = y(t) - 1 = e^{-\sigma t} = e^{-\frac{\sigma\pi}{\omega_d}} = e^{-\pi\zeta/\sqrt{1-\zeta^2}} \tag{8.15}$$

It is interesting to note that Eq. (8.15) only makes sense when $\zeta < 1$. Therefore, an overshoot exists only when the system is in the under-damping condition.

To find the rise time (t_r), let $y(t) = 1$ in Eq. (8.12), and t_r can be found as:

$$t_r = \arctan\left(-\frac{\omega_d^2}{\sigma}\right) = \arctan\left(-\frac{\omega_n\sqrt{1-\zeta^2}}{\zeta}\right) \tag{8.16}$$

To find the setting time (t_s), let the amplitude of the transient component of $y(t)$ be within the tolerance band of (Δ) as:

$$e^{-\sigma t} \leq \Delta \tag{8.17}$$

Equation (8.17) gives the setting time (t_s) for the tolerance band Δ as:

$$t_s = -\frac{\ln(\Delta)}{\sigma} = -\frac{\ln(\Delta)}{\zeta\omega_n} \tag{8.18}$$

8.2.6 Stability Analysis

The stability of a control system is very important to ensure the safety of control systems. A stable control system is capable of achieving stable outputs subjected to anticipated changes in disturbance of system inputs. Accordingly, control systems can be classified based on their stabilities into *absolutely stable* systems, *conditionally stable* systems, and *marginally stable* systems.

Transfer functions can be used to analyze the stability of a control system. By determining the poles and zeros of the transfer function, the stability of a control system can be determined. A system is *stable* if the real parts of all poles are negative (i.e., $\sigma_i < 0$), a system is marginally stable if the real-parts of all poles are not positive (i.e., $\sigma_i <= 0$), and a system is unstable if the real part of any pole is positive (i.e., $\sigma_i > 0$).

Assume a transfer function $H(s)$ about complex variable ($s = \sigma + j\omega$) be

$$H(s) = \frac{b_m s^m + b_{m-1}s^{m-1} + \ldots + b_1 s + b_0}{a_n s^n + a_{n-1}s^{n-1} + \ldots + a_1 s + a_0} \tag{8.19}$$

where n and m are the highest orders of polynomial terms, and a_i ($i = 1, 2, \ldots, n$) and b ($i = 1, 2, \ldots, m$) are the coefficients in the denominator and numerator, respectively.

Factoring the polynomials in the numerator and denominator converts Eq. (8.19) into

$$H(s) = \frac{N(s)}{D(s)} = K\frac{(s-z_1)(s-z_2)\ldots(s-z_{m-1})(s-z_m)}{(s-p_1)(s-p_2)\ldots(s-p_{n-1})(s-p_n)} \tag{8.20}$$

where $K = b_m/a_n$ is the gain factor of the system, and zeros (z_i ($i = 1, 2, \ldots, m-1, m$) and *poles* (p_i ($i = 1, 2, \ldots, n-1, n$) are the solutions of $N(s) = 0$ and $D(s) = 0$, respectively.

Giving $s = z_i$ in Eq. (8.20), $N(s) = 0$, and $\lim_{s\to z_i} H(s) = 0$; in a similar way, giving $s = p_i$ in Eq. (8.20), $D(s) = 0$, and $\lim_{s\to p_i} H(s) = \infty$.

Since the coefficients a_i ($i = 1, 2, \ldots, n$) and b_i ($i = 1, 2, \ldots, m$) in $D(s)$ and $N(s)$ are real, the poles and zeros must be either real values or in complex conjugate pairs, that is, ($p_i = \sigma_i$ and $p_{i+1} = \sigma_{i+1}$) or ($p_i, p_{i+1} = \sigma_i \pm j\omega_i$).

Example 8.3 Find the poles and zeros of a linear system represented by its differential equation of

$$\frac{d^2y}{dt^2} + 7\frac{dy}{dt} + 6y = \frac{du}{dt} + 1$$

Solution

The transfer function can be found from the differential equation as:

$$H(s) = \frac{N(s)}{D(s)} = \frac{s+1}{s^2 + 7s + 6}$$

Let $N(s) = 0$, there is only one zero, that is, $z_i = -1$.
Let $D(s) = 0$, there are two poles, that is, $p_1 = -1$ and $p_2 = -6$.

The poles, zeros, and gain factor K are all the properties of a transfer function; once these properties are given, the corresponding differential equation for the input–output system dynamics can be determined completely.

Example 8.4 Assume a system has a pair of conjugate poles $p_1, p_2 = -2 \pm j$, one zero $z_1 = -2$, and a gain factor of 5. Find the equation for input–output dynamic system dynamics.

Solution

Equation (8.20) can be used directly to determine the transfer function as

$$H(s) = \frac{N(s)}{D(s)} = (5)\frac{(s-(-2))}{(s+2-j)(s+2+j)} = \frac{5s+10}{s^2+4s+5}$$

Therefore, the above transfer function can be converted into the input–output system equation as

$$\frac{d^2y}{dt^2} + 4\frac{dy}{dt} + 5y = 5\frac{du}{dt} + 10$$

For a linear SISO system, its unforced response subject to given initial conditions is

$$y(t) = \sum_{i=1}^{n} C_i e^{\lambda_i t} = \sum_{i=1}^{n} C_i e^{p_i t} \tag{8.21}$$

where the constants C_i ($i = 1, 2, ..., n$) are determined based on initial conditions, the exponents λ_i ($i = 1, 2, ..., n$) are the roots of the characteristic equations that are also referred to as eigenvalues of the system. The characteristic equation in a control problem corresponds to the expression of denominator $D(s)$. Therefore, $\lambda_i = p_i$. The condition for the stability of a linear system can be determined directly by its transfer function as:

$$\lim_{t\to\infty} \sum_{i=1}^{n} C_i e^{p_i t} = 0 \tag{8.22}$$

As shown in Figure 8.17, when any solution of p_i is located in the positive size of the s-plane, the control system will be unstable.

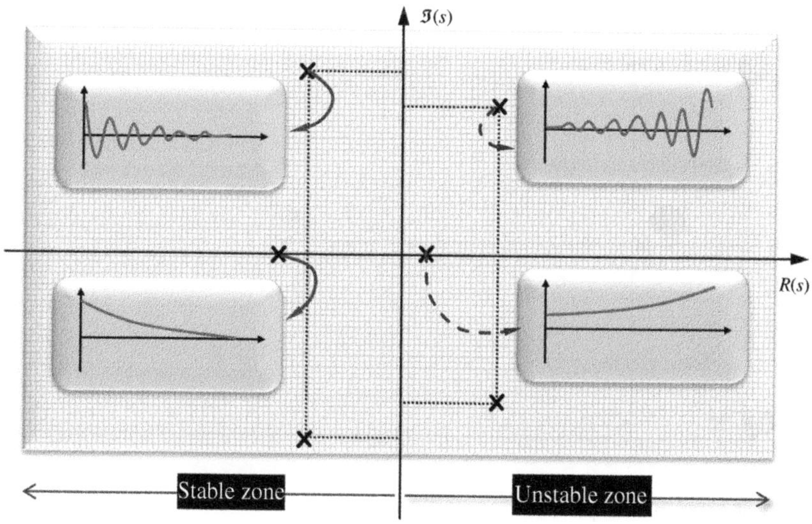

Figure 8.17 Impact of damping ratio on system response

The location of a pole in the s-plane determines the corresponding response of this component in system output, as below:

1) a real pole $p_i = -\sigma$ in the left half of the s-plane produces an output ($C_i e^{-\sigma t}$) with an exponential decay. The larger the distance of this pole from the origin of the s-plane, the faster rate the decay is; the closer this pole is to the origin, the slower rate the decay is.
2) A pole at the origin $p_i = 0$ produces an output that is constant in amplitude, determined by initial conditions.
3) A real pole in the right-half of the s-plane produces an output ($C_i e^{\sigma t}$) with an exponential increase; this indicates the system is unstable and uncontrollable.
4) A complex conjugate pole pair $p_i, p_{i+1} = \sigma \pm j\omega$ in the left half of the s-plane produces an output jointly as a decaying sinusoid of the form $C_i e^{-\sigma t} \sin(\omega t + \phi)$, where C_i and ϕ are determined by initial conditions. The rate of decay is determined by σ and the output is oscillated with a frequency of ω.
5) A pair of imaginary poles located on the imaginary axis ($\pm j\omega$) produce an oscillatory output with a constant amplitude that is determined by initial conditions.
6) A complex conjugate pole pair $p_i, p_{i+1} = \sigma \pm j\omega$ in the right half of the s-plane produces an output jointly as an increasing sinusoid of the form $C_i e^{\sigma t} \sin(\omega t + \phi)$, where C_i and ϕ are determined by initial conditions. The rate of increase is determined by σ and the output is oscillated with a frequency of ω; this indicates the system is unstable and uncontrollable.

Example 8.5 Discuss the stability of a control system with four poles in the s-plane as illustrated in Figure 8.18 (MIT 2024).

Solution

Figure 8.18 shows that the control system has four poles and no zero; all of the poles are in the left-half of the s-plane; therefore, the control system will be stable. The first and third poles produce an output with an exponentially decaying $C_1 e^{-3t}$ and $C_3 e^{-0.1t}$, respectively. The pair of the complex

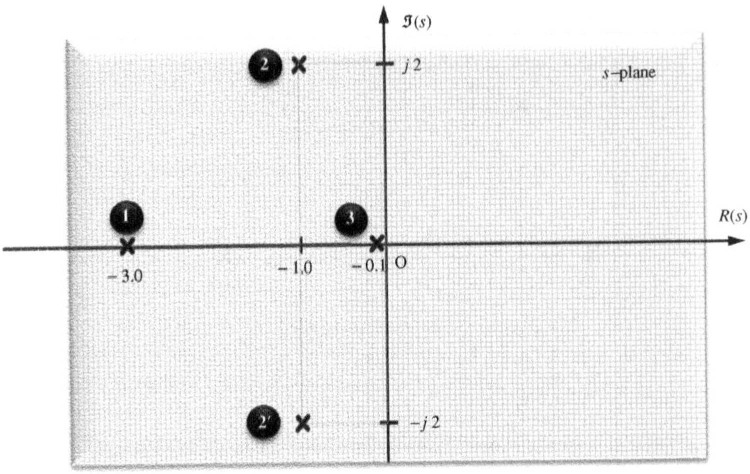

Figure 8.18 The solutions of poles in Example 8.5

conjugate poles produces an oscillatory output with an exponentially decaying $C_2e^{-t}\sin(2t+\phi)$, and the summed system output becomes,

$$y(t) = C_1e^{-3t} + C_3e^{-0.1t} + C_2e^{-t}\sin(2t+\phi) \tag{8.23}$$

Examining three components of system output in Eq. (8.23) finds that (1) the time constant τ of C_1e^{-3t} in Eq. (8.23) is 0.33 seconds; it will decay rapidly and it affects the output significantly in less than 4τ, that is, 1.33 seconds. (2) The time constant τ of $C_1e^{-0.1t}$ in Eq. (8.23) is 10 seconds; it will decay most slowly and it affects the output significantly for approximately 4τ, that is, 40 seconds, and it will dominate the transient response in the long term. (3) The output by the pair on conjugate poles $C_2e^{-t}\sin(2t+\phi)$ exhibits an overshoot; the time constant for decay is 1 second so the oscillation will decay in approximately 4 seconds.

The locations of poles also relate to the natural frequency and damping ratio, as shown in Figure 8.19; it represents the solutions of a transfer function for the classic second-order homogeneous system as:

$$\frac{d^2y}{dt^2} + 2\zeta\omega_n\frac{dy}{dt} + \omega_n^2y = 0 \tag{8.24}$$

The solutions to Eq. (8.24) can be derived as:

$$p_1, p_2 = -\zeta\omega_n \pm \omega_n\sqrt{\zeta^2-1} \tag{8.25}$$

When $\zeta \geq 1$, the system is overdamped since two poles are real and both lie in the left-half plane. When $0 \leq \zeta < 1$, the poles form a complex conjugate pair, that is,

$$p_1, p_2 = -\zeta\omega_n \pm j\omega_n\sqrt{\zeta^2-1} \tag{8.26}$$

Figure 8.19 shows the pair still on the left-half of the s-plane, and the poles lie on the circle with a radio of ω_n from the origin, and they are at an angle of $\pm\cos^{-1}(\zeta)$ from the negative real axis.

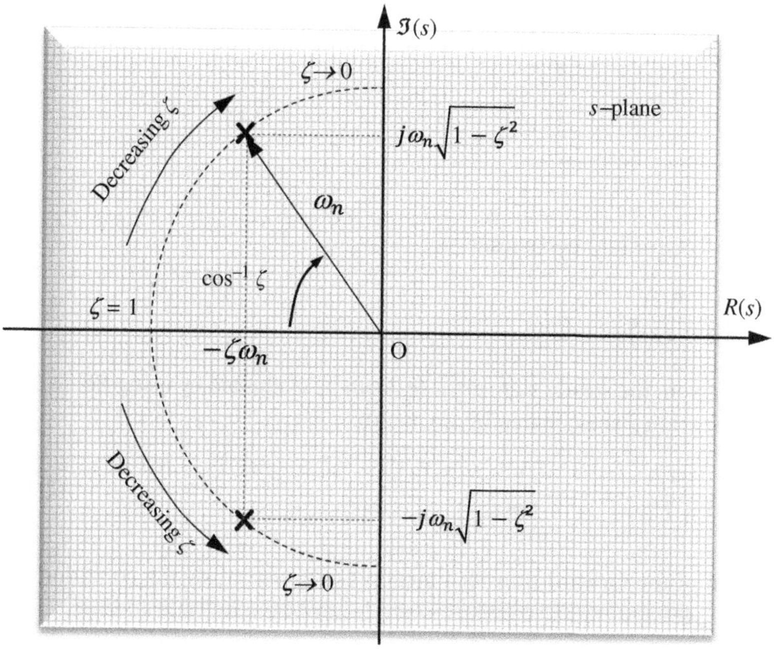

Figure 8.19 Locations of poles, nature frequency, and damping ratio

Therefore, the poles from an underdamped system lie on a semicircle with a radius of ω_n at an angle determined by the damping ratio ζ.

8.2.7 Accuracy of Control Systems

Accuracy reflects how well a control system follows the desired motion. The other relevant concept related to accuracy is repeatability; *repeatability* reflects how well the control system can repeat its moves. A *steady-state error* is a measure of the accuracy of a control system in tracking a command input; it refers to the deviation of the output of a control system from the desired response during a steady state. *The final value theorem* can be used to determine steady-state error as:

$$e_{ss} = \lim_{t \to \infty} e(t) = \lim_{s \to 0} sE(s) \tag{8.27}$$

where $E(s)$ is the Laplace transform of the error signal $e(t)$.

Taking into consideration the block diagram of a closed-loop control system with unity feedback in Figure 8.20, where $R(s)$ and $Y(s)$ are the Laplace transforms of input and output signals $r(t)$ and $y(t)$, respectively, the transfer function of the control system can be determined as:

$$\frac{Y(s)}{R(s)} = \frac{G(s)}{1 + G(s)} \tag{8.28}$$

Using Eq. (8.28), we find the Laplace transform of the error signal as:

$$E(s) = R(s) - Y(s) = R(s) - \frac{G(s)R(s)}{1 + G(s)} = \frac{R(s)}{1 + G(s)} \tag{8.29}$$

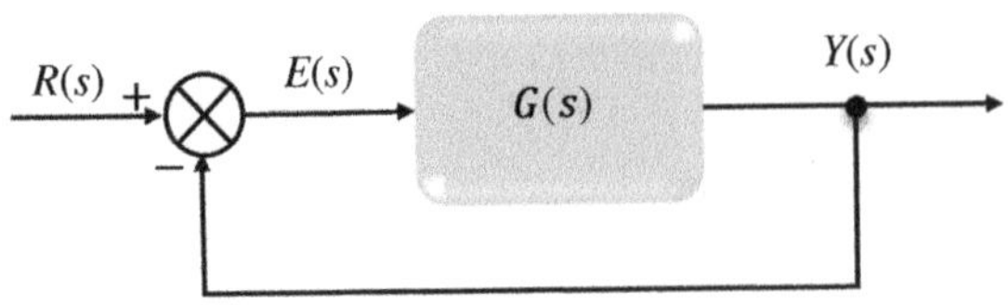

Figure 8.20 Closed-loop control system with unity feedback

Accordingly, the steady-state error (e_{ss}) becomes:

$$e_{ss} = \lim_{s\to 0} sE(s) = \lim_{s\to 0} \frac{s{\cdot}R(s)}{1+G(s)} \tag{8.30}$$

when both $R(s)$ and $G(s)$ are given, the steady-state error (e_{ss}) can be evaluated by using Eq. (8.30).

For a *step input*, $R(s) = 1/s$, and e_{ss} is evaluated by Eq. (8.30),

$$e_{ss} = \lim_{s\to 0} sE(s) = \lim_{s\to 0} \frac{s{\cdot}1/s}{1+G(s)} = \frac{1}{1+\lim_{s\to 0} G(s)} \tag{8.31}$$

To achieve $e_{ss} = 0$, the gain of the forward transfer function $G(s)$ must satisfy the condition of,

$$\lim_{s\to 0} G(s) = \infty \tag{8.32}$$

Equation (8.32) requires that the gain of the forward transfer function $G(s)$ is in the form of,

$$G(s) = \frac{(s-z_1)(s-z_2)\dots(s-z_{m-1})(s-z_m)}{s^k(s-p_1)(s-p_2)\dots(s-p_{n-1})(s-p_n)} \quad \text{and } k \geq 1 \tag{8.33}$$

Equation (8.33) shows that $G(s)$ must have one pole at the origin of the s-plane. This implies that at least one pure integration must be included in the forward path to ensure zero steady-state error, shown as "Output 1" in Figure 8.21. If no integration is present in the forward path, that is, $k = 0$, Eq. (8.32) will yield a finite error shown as "Output 2" in Figure 8.21.

For *a ramp input*, $R(s) = 1/s^2$, and e_{ss} is evaluated by Eq. (8.30) as:

$$e_{ss} = \lim_{s\to 0} sE(s) = \lim_{s\to 0} \frac{s{\cdot}1/s^2}{1+G(s)} = \frac{1}{\lim_{s\to 0} sG(s)} \tag{8.34}$$

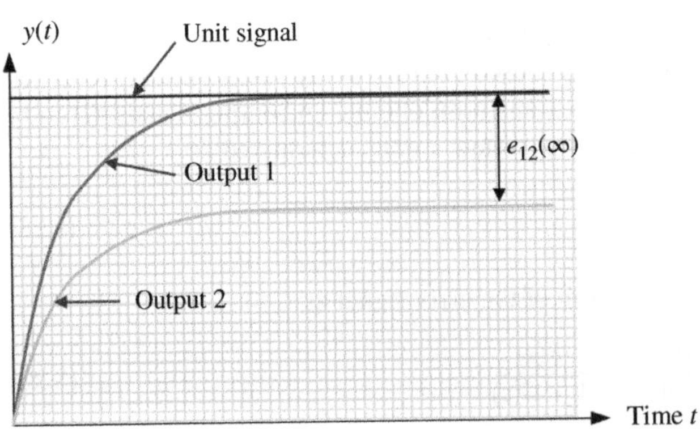

Figure 8.21 Steady-state error of step input

To achieve $e_{ss} = 0$, the gain of the forward transfer function $G(s)$ must satisfy the condition of,

$$\lim_{s \to 0} sG(s) = \infty \tag{8.35}$$

Equation (8.35) requires that the gain of the forward transfer function $G(s)$ is the form of,

$$G(s) = \frac{(s - z_1)(s - z_2)\ldots(s - z_{m-1})(s - z_m)}{s^k(s - p_1)(s - p_2)\ldots(s - p_{n-1})(s - p_n)} \quad \text{and } k \geq 2 \tag{8.36}$$

Equation (8.36) shows that $G(s)$ must have two poles at the origin of the s-plane; it implies that at least two pure integrations must be included in the forward path to ensure zero steady-state error, shown as "Output 1" in Figure 8.22. If there is only one integration present in the forward path, i.e., $k = 1$, Eq. (8.36) will yield a finite error shown as "Output 2" in Figure 8.22. If no integration is present in the forward path, it will yield an infinite error shown as "Output 3" in Figure 8.22.

For *a parabolic input*, $R(s) = 1/s^3$, and e_{ss} is evaluated by Eq. (8.30) as,

$$e_{ss} = \lim_{s \to 0} sE(s) = \lim_{s \to 0} \frac{s \cdot 1/s^3}{1 + G(s)} = \frac{1}{\lim_{s \to 0} s^2 G(s)} \tag{8.37}$$

To achieve $e_{ss} = 0$, the gain of the forward transfer function $G(s)$ must satisfy the condition of,

$$\lim_{s \to 0} s^2 G(s) = \infty \tag{8.38}$$

Equation (8.38) requires that the gain of the forward transfer function $G(s)$ is the form of,

$$G(s) = \frac{(s - z_1)(s - z_2)\ldots(s - z_{m-1})(s - z_m)}{s^k(s - p_1)(s - p_2)\ldots(s - p_{n-1})(s - p_n)} \quad \text{and } k \geq 3 \tag{8.39}$$

Equation (8.39) shows that $G(s)$ must have three poles at the origin of the s-plane; it implies that at least three integrations must be included in the forward path to assure zero steady state. If there

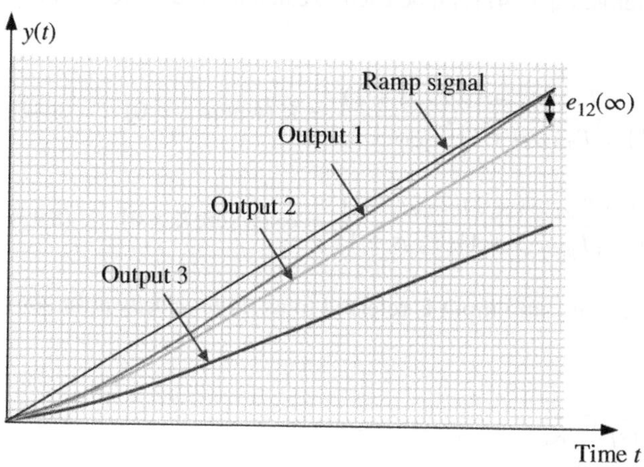

Figure 8.22 Steady-state error of ramp input

are only two integrations present in the forward path, that is, $k = 2$, Eq. (8.39) will yield a finite error. If there is only one or no integration present in the forward path, it will yield an infinite error.

8.2.8 Classification of Control Systems

Control systems can be classified based on their abilities to follow step inputs, ramp inputs, or parabolic inputs. The rationale is that system inputs are usually composed of basic signals. Here, a unity-feedback control system is taken into consideration, and its open-loop transfer function of $G(s)$ is

$$G(s) = \frac{K(T_{z,1}s + 1)(T_{z,2}s + 1)\ldots(T_{z,m-1}s + 1)(T_{z,m}s + 1)}{s^k\left(T_{p,1}s + 1\right)\left(T_{p,2}s + 1\right)\ldots\left(T_{p,n-1}s + 1\right)\left(T_{p,n}s + 1\right)} \tag{8.40}$$

where K is the open-loop gain, $T_{z,i}$ $(i = 1, 2, \ldots, m)$ and $T_{p,i}$ $(i = 1, 2, \ldots, n)$ are the time constants for m-zeros and n-poles, respectively.

The denominator has the term s^k that corresponds to a pole of multiplicity n at the origin of the s-plane; therefore, control systems are classified as type 0, 1, 2, ... when $k = 0, 1, 2, \ldots$, respectively. The higher the k is, the better the accuracy of the control system is; however, the system may raise a stability issue when k is too high.

Except for the terms s^k and K, other terms in $G(s)$ become "1" when s approaches 0 infinitely; this implies that the open-loop gain K relates directly to the steady-state error. The larger the K, the smaller the steady-state error is.

The steady-state error of a system for a unit-step input is referred to as *static position error constant* K_p; it can be determined from,

$$e_{ss} = \lim_{s\to 0} \frac{s}{1 + G(s)} \frac{1}{s} = \frac{1}{1 + G(0)} = \frac{1}{1 + K_p} \tag{8.41}$$

where $K_p = \lim_{s\to 0} G(s)$ and $G(s)$ is expressed in Eq. (8.40).

For type 0 and 1 or higher control systems, Eq. (8.41) can be used to calculate the corresponding K_p as:

$$K_p = \lim_{s\to 0} \frac{K(T_{z,1}s + 1)(T_{z,2}s + 1)\ldots(T_{z,m-1}s + 1)(T_{z,m}s + 1)}{s^k\left(T_{p,1}s + 1\right)\left(T_{p,2}s + 1\right)\ldots\left(T_{p,n-1}s + 1\right)\left(T_{p,n}s + 1\right)} = K \text{ (where } k = 0) \tag{8.42}$$

$$K_p = \lim_{s\to 0} \frac{K(T_{z,1}s + 1)(T_{z,2}s + 1)\ldots(T_{z,m-1}s + 1)(T_{z,m}s + 1)}{s^k\left(T_{p,1}s + 1\right)\left(T_{p,2}s + 1\right)\ldots\left(T_{p,n-1}s + 1\right)\left(T_{p,n}s + 1\right)} = \infty \text{ (where } k > 0) \tag{8.43}$$

The steady-state error of a system for a unit-ramp input is referred to as *static velocity error constant* K_v; it can be determined from,

$$e_{ss} = \lim_{s\to 0} \frac{s}{1 + G(s)} \frac{1}{s^2} = \frac{1}{\lim_{s\to 0} s\,G(s)} = \frac{1}{K_v} \tag{8.44}$$

where $K_v = \lim_{s\to 0} s\,G(s)$ and $G(s)$ is expressed in Eq. (8.40).

For type 0, 1, and 2 or higher control systems, Eq. (8.44) can be used to calculate corresponding K_p as:

$$K_v = \lim_{s\to 0} \frac{K(T_{z,1}s+1)(T_{z,2}s+1)...(T_{z,m-1}s+1)(T_{z,m}s+1)}{s^{k-1}\left(T_{p,1}s+1\right)\left(T_{p,2}s+1\right)...\left(T_{p,n-1}s+1\right)\left(T_{p,n}s+1\right)} = 0 \text{ (where } k=0) \quad (8.45)$$

$$K_v = \lim_{s\to 0} \frac{K(T_{z,1}s+1)(T_{z,2}s+1)...(T_{z,m-1}s+1)(T_{z,m}s+1)}{s^{k-1}\left(T_{p,1}s+1\right)\left(T_{p,2}s+1\right)...\left(T_{p,n-1}s+1\right)\left(T_{p,n}s+1\right)} = K \text{ (where } k=1) \quad (8.46)$$

$$K_v = \lim_{s\to 0} \frac{K(T_{z,1}s+1)(T_{z,2}s+1)...(T_{z,m-1}s+1)(T_{z,m}s+1)}{s^{k-1}\left(T_{p,1}s+1\right)\left(T_{p,2}s+1\right)...\left(T_{p,n-1}s+1\right)\left(T_{p,n}s+1\right)} = \infty \text{ (where } k>1) \quad (8.47)$$

The steady-state error of a system for a unit-parabolic input is referred to as *static acceleration error constant* K_a, it can be determined from,

$$e_{ss} = \lim_{s\to 0} \frac{s}{1+G(s)} \frac{1}{s^3} = \frac{1}{\lim\limits_{s\to 0} s^2 G(s)} = \frac{1}{K_a} \quad (8.48)$$

where $K_a = \lim\limits_{s\to 0} s^2 G(s)$ and $G(s)$ is expressed in Eq. (8.40).

For type 0, 1, 2, and 3 or higher control systems, Eq. (8.48) can be used to calculate the corresponding K_a as,

$$K_v = \lim_{s\to 0} \frac{K(T_{z,1}s+1)(T_{z,2}s+1)...(T_{z,m-1}s+1)(T_{z,m}s+1)}{s^{k-2}\left(T_{p,1}s+1\right)\left(T_{p,2}s+1\right)...\left(T_{p,n-1}s+1\right)\left(T_{p,n}s+1\right)} = 0 \text{ (where } k=0) \quad (8.49)$$

$$K_v = \lim_{s\to 0} \frac{K(T_{z,1}s+1)(T_{z,2}s+1)...(T_{z,m-1}s+1)(T_{z,m}s+1)}{s^{k-2}\left(T_{p,1}s+1\right)\left(T_{p,2}s+1\right)...\left(T_{p,n-1}s+1\right)\left(T_{p,n}s+1\right)} = 0 \text{ (where } k=1) \quad (8.50)$$

$$K_v = \lim_{s\to 0} \frac{K(T_{z,1}s+1)(T_{z,2}s+1)...(T_{z,m-1}s+1)(T_{z,m}s+1)}{s^{k-2}\left(T_{p,1}s+1\right)\left(T_{p,2}s+1\right)...\left(T_{p,n-1}s+1\right)\left(T_{p,n}s+1\right)} = K \text{ (where } k=2) \quad (8.51)$$

$$K_v = \lim_{s\to 0} \frac{K(T_{z,1}s+1)(T_{z,2}s+1)...(T_{z,m-1}s+1)(T_{z,m}s+1)}{s^{k-2}\left(T_{p,1}s+1\right)\left(T_{p,2}s+1\right)...\left(T_{p,n-1}s+1\right)\left(T_{p,n}s+1\right)} = \infty \text{ (where } k>2) \quad (8.52)$$

Table 8.2 summarizes the relations of inputs, system types, and static-steady errors based on the above analyses.

8.2.9 Frequency Responses

The variable in a transform function $G(s)$ is frequency. The response of a control system to the change in the frequency of an input signal can be analyzed by bode plots. *Bode plots* describe

Table 8.2 Steady-state errors related to inputs and types of systems

		Type 0 (k = 0)		Type 1 (k = 1)		Type 2 (k = 2)	
Input	**Steady-state error**	**Constant**	**Error**	**Constant**	**Error**	**Constant**	**Error**
Step $u(t)$	$\frac{1}{1+K_p}$	$K_p = K$	$\frac{1}{1+K}$	$K_p = \infty$	0	$K_p = \infty$	0
Ramp $t \cdot u(t)$	$\frac{1}{K_v}$	$K_v = 0$	∞	$K_v = K$	$\frac{1}{K_v}$	$K_v = \infty$	0
Parabola $\frac{t^2 u(t)}{2}$	$\frac{1}{K_a}$	$K_a = 0$	∞	$K_a = 0$	∞	$K_a = K$	$\frac{1}{K_a}$

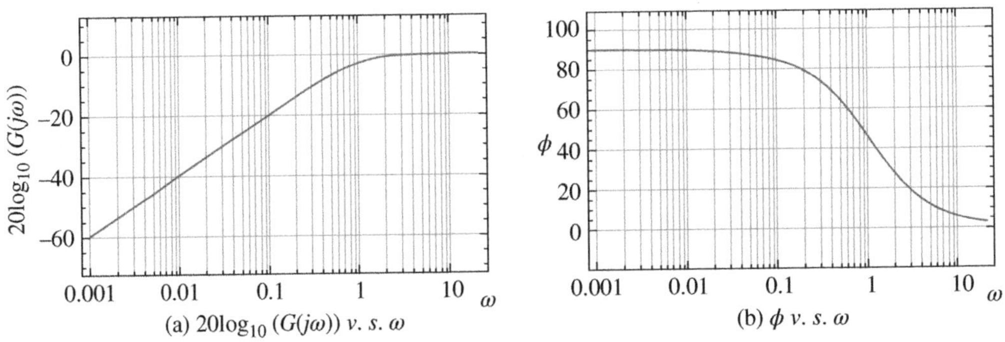

Figure 8.23 Bode plot for $G(s) = \frac{s}{s+1}$ (Wolframalpha 2024)

the frequency response of linear time-invariant systems, and the system response is reflected by the changes in magnitude and phase of an output signal.

As shown in Figure 8.23, a pair of bode plots consists of *a frequency–magnitude plot* and *a frequency–phase* plot. In a frequency–magnitude plot, the units for frequency and magnitude are "$\log_{10}$" and "$20 \cdot \log_{10}$," respectively. In a frequency–phase plot, the units for frequency and phase are "long" and "degree," respectively. To draw bode plots for a given transfer function $G(s)$, s is replaced by $j\omega$ so that $G(s)$ is expressed as:

$$G(s) = G(j\omega) = x(\omega) + jy(\omega) \tag{8.53}$$

Equation (8.53) gives for the functions of the magnitude $G(jw)$ and phase, along with ω as:

$$\left.\begin{aligned} 20\log_{10} G(s) &= 20\log_{10}\sqrt{x^2(\omega) + y^2(\omega)} \\ \phi &= \tan^{-1}\left(\frac{y(\omega)}{x(\omega)}\right) \end{aligned}\right\} \tag{8.54}$$

Many online tools are available to generate bode plots based on a given transfer function (Wolframalpha 2024). Conventionally, bode plots can also be roughly drawn by decomposing the transfer function into the components of gain, poles, and zeros and summing their contributions to magnitude and phase, as illustrated in many references (Swarthmore 2024).

Example 8.6 Derive the expressions of magnitude and phase with respect to frequency for the transfer function of $G(s) = \frac{s}{s+1}$ and plot its bode plots.

Solution

Step (1). Let $s = j\omega$ in $G(s)$, and it can be expressed as:

$$G(s) = G(j\omega) = \frac{j\omega}{(j\omega + 1)} \cdot \frac{(-j\omega + 1)}{(-j\omega + 1)} = \frac{\omega(1 - j\omega)}{(\omega^2 + 1)} \tag{8.55}$$

Step (2). By checking the real and imaginary components of $G(s)$ in Eq. (8.55), $x(\omega)$ and $y(\omega)$ are derived as:

$$x(\omega) = \frac{\omega}{(\omega^2 + 1)}; y(\omega) = -\frac{\omega^2}{(\omega^2 + 1)} \tag{8.56}$$

Step (3). Eq. (8.56) is used to find the magnitude and phase function of $G(s)$ as:

$$\left.\begin{aligned} 20\log_{10} G(s) = 20\log_{10}\sqrt{x^2(\omega) + y^2(\omega)} = 20\log_{10}\frac{\omega}{\sqrt{\omega^2 + 1}} \\ \phi = -\tan^{-1}(\omega) \end{aligned}\right\} \tag{8.57}$$

Step (4). Draw the Eq. (8.57) for a range of $\omega \subset [0.001, 10]$ get the bode plots of $G(s)$ as shown in Figure 8.23.

8.3 Proportional–Integral–Derivative (PID) Controls

As shown in Figure 8.24, a system is controlled to produce desired responses. Different controllers can be used to achieve the same objectives. Many control methods are available, and they can be generally classified into *classic controls*, *advanced controls*, and *intelligent controls*.

Proportional–integral–derivative (PID) controls are mostly used as classic control methods in industrial control systems and a variety of other systems that require modulated control

Figure 8.24 Description of control system

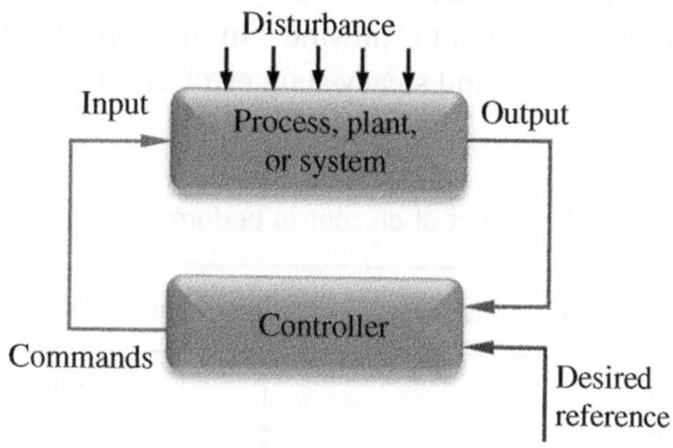

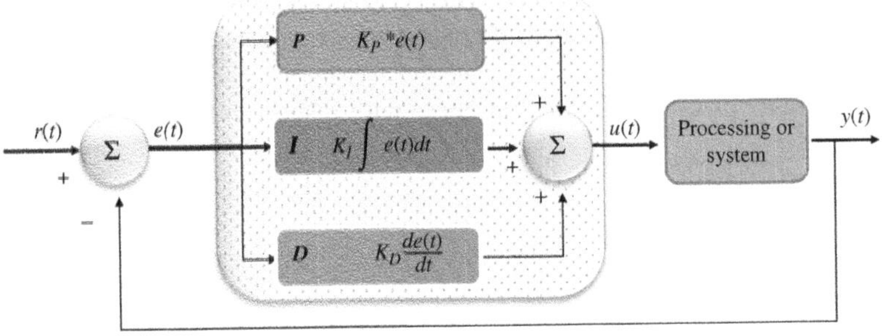

Figure 8.25 Proportional–Integral–Derivative (PID)

continuously. As shown in Figure 8.25, PIC calculates the error value $e(t)$ by comparing desire *set-point* (*SP*) and measured *process variable* (PV), and the output signals are corrected by *proportional* (*P*), *integral* (I), and *derivative* (*D*).

With the known $e(t)$, the output of the PID controller can be found as:

$$u(t) = K_P\, e(t) + K_I \int e(t)dt + K_D \frac{de(t)}{dt} \tag{8.58}$$

The transfer function of PID from $e(t)$ to $u(t)$ can be obtained by taking the Laplace transform of Eq. (8.58) as:

$$C(s) = \frac{U(e)}{E(s)} = K_D{\cdot}s + \frac{K_I}{s} + K_P \tag{8.59}$$

The proportional gain (K_P) adds a proportion of the control signal for the same level of error; introducing a given level of error in a controller tends to cause a closed-loop system to react more quickly; it also overshoots more. Meaning that increasing K_P tends to reduce steady-state error.

The derivative term (K_D) allows the controller to anticipate errors. If only K_P is used, the way to increase the control signal is if the error increases. However, a derivative component allows to increase control signal when the error begins sloping upward when the magnitude of the error is still small. The anticipation tends to introduce a damping in a system and decrease overshoot. Note that K_D has nothing to do with a steady-state error.

The integral term (K_I) tends to reduce steady-state error. If there is a persistent and steady error, the integrator accumulates and thus increases the control signal to reduce error. However, an integral term can make a control system oscillatory since when the error changes its sign, it takes time for the integrator to "unwind." In summary, the functions of K_P, K_I, and K_D on rise time, overshoot, settling time, and steady-state error are listed in Table 8.3.

Table 8.3 Impact of on control performance

Response	Rise time	Overshoot	Setting time	Steady-state errors
K_p	↓	↑	Small change	↓
K_I	↓	↑	↑	↓
K_D	Small change	↓	↓	No change

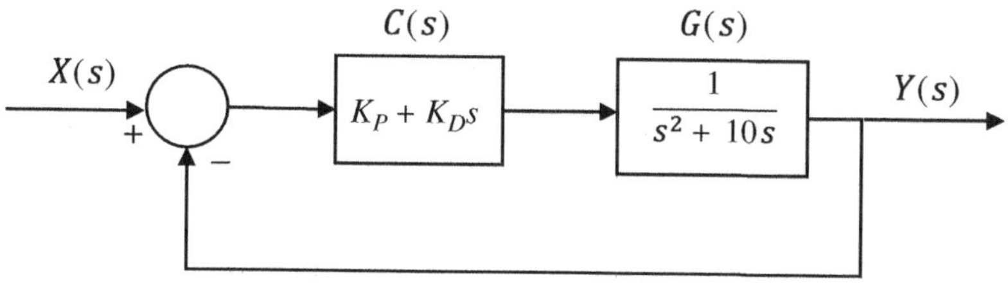

Figure 8.26 Block diagram of Example 8.7

Example 8.7 Consider the open-loop system $G(s) = 10/(s(s+2))$, design a PD controller for a unit step signal with the specifications of a rise time < 0.3 seconds and overshoot <10%.

Solution

First, the block diagram of the control system is drawn as shown in Figure 8.26, where a PD controller ($C(s)$) is introduced.

Second, the transfer function $H(s)$ of the control system can be found from Figure 8.26 as,

$$H(s) = \frac{C(s)G(s)}{1 + C(s)G(s)} = \frac{K_P + K_D s}{s^2 + (10 + K_D)s + K_P} \tag{8.60}$$

Write the polynomial expression in the denominator in the form of $s^2 + 2\zeta\omega_n s + \omega_n^2$, we have

$$\omega_n = \sqrt{K_P} \text{ and } \zeta = \frac{10 + K_d}{2\sqrt{K_P}} \tag{8.61}$$

Moreover, for a second-order system with a given natural frequency ω_n and damping ratio ζ, we have

$$\left.\begin{aligned} \text{Overshoot } \% &= \exp\left\{\frac{-\zeta\pi}{\sqrt{1-\zeta^2}}\right\} \quad \text{for } 0 < \zeta < 1 \\ t_r &\approx \frac{1}{\omega_n}(2.3\zeta^2 - 0.078\zeta + 1.12) \quad \text{for } 0 < \zeta < 1 \\ t_p &= \frac{\pi}{\omega_n\sqrt{1-\zeta^2}} \end{aligned}\right\} \tag{8.62}$$

Equation (8.61) and Eq. (8.62) can be used together to determine K_P and K_D in the PD controller.

8.4 Analog and Digital Implementation of PID Controllers

A PID controller can be readily constructed using basic electronic elements such as operational amplifiers, resistors, and capacitors. Analog PID controllers have been mostly replaced by digital electronic controllers. However, they have two inherent advantages, that is, greater reliability and fast responsiveness.

Figure 8.27 shows a construction of an analog PID with four basic components; three of them are used to generate proportional, integral, and derivative operations of error signal $e(t)$; the last one is

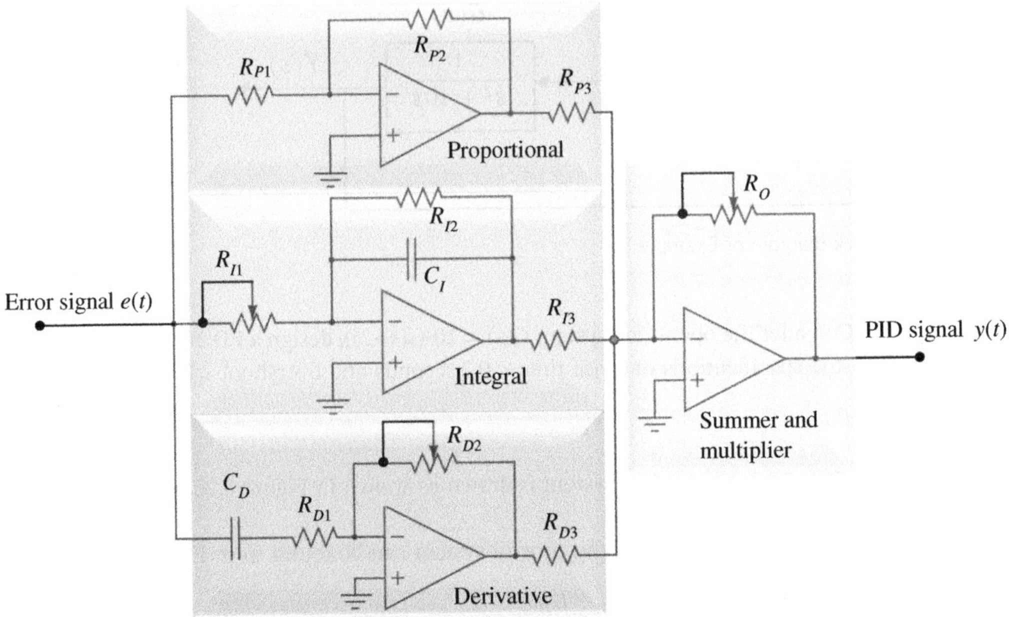

Figure 8.27 Analog PID implementation

used to sum and then amplify proportional, integral, and derivative signals as the output of the PID controller.

A PID controller can be alternatively implemented by a program that supports the operations, including multiplication, addition, and time delays. The equation for a PID controller is modeled in a continuous time domain, and it cannot be implemented directly by software. A transfer function in the s-domain must be first digitized and transferred into a representation in a discretized time domain (Bhandari and Csurcsia 2022).

Revisiting the control signal in Eq. (8.58),

$$u(t) = K\left(e(t) + \frac{1}{T_I}\int_0^t e(t)dt + T_D\frac{de(t)}{dt}\right) = K_P e(t) + K_I\int_0^t e(t)dt + K_D\frac{de(t)}{dt} \tag{8.63}$$

where $K = K_P$ is the proportional gain, $T_I = \dfrac{K}{K_I}$ is the integral time, and $T_D = \dfrac{K_D}{K}$ is the derivative time.

Taking a z-transform of Eq. (8.63) gets,

$$U(z) = \left[K_P + \frac{K_I}{1-z^{-1}} + K_D\left(1-z^{-1}\right)\right]E(z) \tag{8.64}$$

Equation (8.65) can be rewritten as:

$$U(z) = \left[\frac{(K_P + K_I + K_D) - (K_P + 2K_D)z^{-1} + K_d z^{-2}}{1-z^{-1}}\right]E(z) = \left[\frac{K_1 + K_2 z^{-1} + K_3 z^{-2}}{1-z^{-1}}\right]E(z) \tag{8.65}$$

where $K_1 = K_P + K_I + K_D$, $K_2 = -(K_P + 2K_D)$ and $K_3 = K_d$

Equation (8.65) can be rewritten as:

$$U(z) - z^{-1}U(z) = \left[K_1 + K_2 z^{-1} + K_3 z^{-2}\right]E(z) \tag{8.66}$$

Equation (8.66) can be converted back to its difference equation as (Toochinda 2011):

$$u[k] = u[k-1] + K_1 \cdot e[k] + K_2 \cdot e[k-1] + K_3 \cdot e[k-2] \tag{8.67}$$

Let the system output be obtained by a function readADC(), and the control signal u be output by a function writeDA(), and the control signal must be bounded based on the resolution of the DAC; for example, set $U_{min} = -2048$ and $U_{max} = 2047$ for a 12-bit DAC. The following *c* code can be used to implement PID control (Toochinda 2011).

```
double e, e1, e2, u, delta_u;
k1 = kp + ki + kd;
k2 = - (kp + 2*kd)
k3 = kd;
voidpid()
{
        e2 = e1;                          //update error variables
        e1 = e;

        y = readADC();                    //read variable from sensor
        e = setpoint - y;                 //evaluate error signal
        delta_u = k1*e + k2*e1 + k3*e2;   //determine the change of output signal
        u = u + delta_u;                  //update control signal
        if (u > Umax) u = Umax;           //check if the control signal exceeds upper limit
        if (u < Umin) u = Umin;           //check if the control signal exceeds lower limit
        writeDA(u)                        //send control signal
}
```

8.5 Advanced Controls

When the governing equations of a mechatronic system are known, a controller can be customized to generate a control signal based on the state and objective of the system. For example, *adaptive control* modifies the control law to cope with time-varying parameters. Taking the example of aircraft control, since the mass of an aircraft will decrease gradually due to fuel consumption, the control law must be adaptable to mass reduction. *Robust control* deals with uncertain or random factors. A robust control ensures that any possible change must be within an expected range in which the control laws are applicable. *Optimal control* optimizes the solution to a set of governing differential equations optimally, and two commonly used optimal controls are *model predictive control* (MPC) and *linear quadratic Gaussian* (LQG) control.

8.6 Intelligent Controls

A classic controller is applicable only when (1) the governing equations of the plant are known and (2) system parameters are known and constant. Otherwise, a controller must be flexible to deal with uncertainty or changes and capable of generating the commands by taking into consideration control objectives and the changes in system states and the environment simultaneously.

Intelligent control can be implemented by treating a physical system as a *block-box* in which the governing equations from inputs to outputs of the system are unknown. An intelligent controller produces control commands based on sensed data about inputs, outputs, and/or system states.

Intelligent controls are default options when mathematical models of a plant are not available, too complex, or system parameters are too hard to identify or vary in system operations. Two common intelligent controls that treat a physical system as a block-box are *fuzzy logic* and *artificial neural networks* (ANN).

8.6.1 Fuzzy Logic

Fuzzy logic is a computational paradigm based on how humans make decisions. A human is capable of reasoning through situations with uncertain, incomplete, or vague information. In such ways, she or he can respond to vague information such as "a temperature is rising" or "a car is decelerating" with precision actions such as "turn on an air-conditioner" or "press a gas pedal." An intelligent fuzzy control uses *the fuzzy set theory* as the mathematical foundation to reason decision-making activities when the information about the system is incomplete, imprecise, ambiguous, uncertain, or random. An intelligent fuzzy control manages a complex problem by a set of (*if...then*) rules heuristically, and the corresponding mathematical model is built upon fuzzy logic instead of differential equations. Intelligent fuzzy control is particularly effective when the control problem is highly nonlinear or developing an explicit mathematical model becomes infeasible.

Figure 8.28 shows an example in which an intelligent fuzzy control is used to control an "inverted pendulum" (Reyes 2024). A pole is mounted on the top of a cart with a pivot joint, and a reciprocating motion of the cart is controlled to keep the pole as straight as possible. In an implementation, the motion of the inverted pendulum is modeled by the position, speed, and acceleration of the cart; the fuzzy set for these variables is determined as shown in Figure 8.28b, and these variables are sensed and feedbacked to the fuzzy logic controller. The controller generates control commands based on fuzzy logic rules in Figure 8.28c.

8.6.2 Artificial Neural Network (ANN)

ANN is used as a nonlinear mathematical model that is capable of mimicking reasoning processes by biological neurons of a human's brain. ANN can be used to train a black-box model to map unknown functions and then transform inputs into outputs of the system.

An ANN model consists of (1) a set of neurons and associated weights (see Figure 8.29a) and (2) a constituted neuron network (NN) (see Figure 8.29b). ANN can be used in two alternative ways, that is, (1) identifying the parameters of a system and (2) controlling a system.

A mechatronic system usually involved in decision-making supports at multiple levels; the higher the level of a control unit is, the more uncertain and random factors are involved. In practice, it is often desirable to integrate intelligent controls at upper levels and classic controls at lower levels of mechatronic systems. For example, Hernandez-Barragan et al. (2020) integrated ANN in PID to implement the adaptive control of an omnidirectional platform, as shown in Figure 8.30.

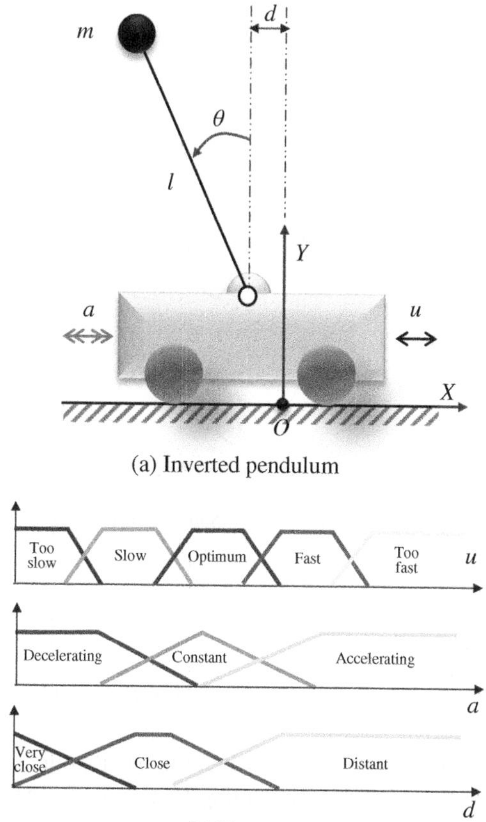

1. IF cart is on the left AND cart is going left THEN largely push cart to the right
2. IF cart is on the left AND cart is not moving THEN slightly push cart to the right
3. IF cart is on the left AND cart is going right THEN don't push cart
4. IF cart is centered AND cart is going left THEN slightly push cart to the right
5. IF cart is centered AND cart is not moving THEN don't push cart
6. IF cart is centered AND cart is going right THEN slightly push cart to the left
7. IF cart is on the right AND cart is going left THEN don't push cart
8. IF cart is on the right AND cart is not moving THEN push cart to the left
9. IF cart is on the right AND cart is going right THEN largely push cart to the left

(c) Fuzzy logic by position and velocity

Figure 8.28 Example of intelligent fuzzy control

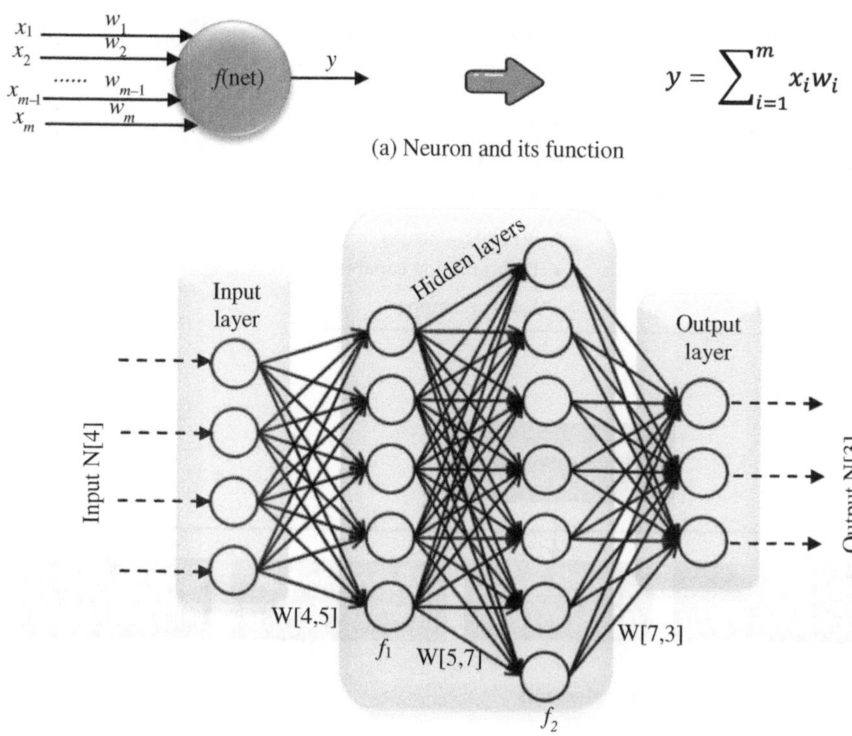

Figure 8.29 Neurons and neuron network (NN)

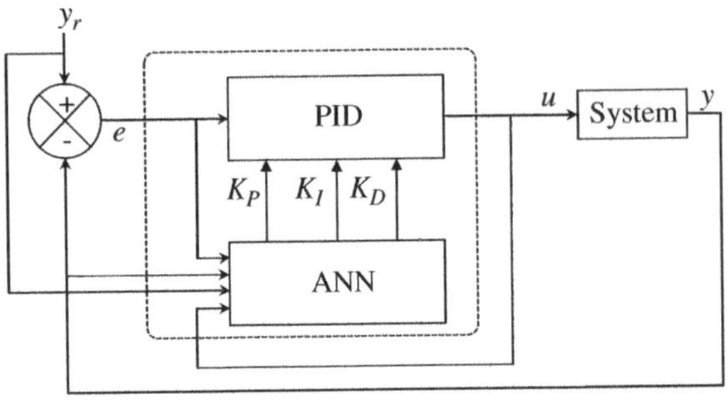

(a) Hybrid block diagram

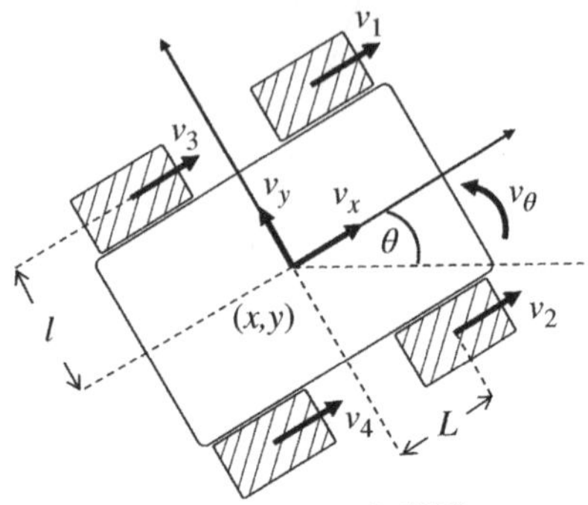

(b) Neuron network (NN)

Figure 8.30 Example of integrated ANN and PID controls in a mechatronic system (*Source:* Hernandez-Barragan et al. (2020)/MDPI/CC BY 4.0)

8.7 Design of Control System

Designing or selecting a control system for a mechatronic system must consider multiple criteria, such as those in Table 8.4 simultaneously.

It is desirable to use as many standardized control elements as possible when these components are available to meet control requirements. This helps to avoid reinventing wheels in system designs. Commonly used control elements are classified into five categories in Figure 8.31: microcontrollers, *programmable logic controllers* (PLCs), *digital signal processors* (DSPs), *field programmable gate arrays* (FPGA), and *personable computers* (PCs) with *data acquisition* (DAQ).

Table 8.4 Commonly used criteria in designing or selecting a control system.

• Price	• Required interrupt
• Size and weight	• Required hardware
• Number of digital inputs and outputs	• Communication interfaces
• Number of analog inputs and outputs	• Reliability
• Speed	• Memory
	• Programming capability
	• Software support

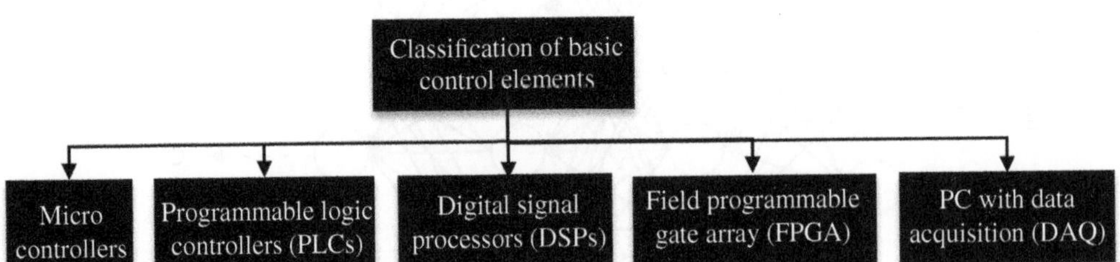

Figure 8.31 Classification of basic control elements

8.7.1 Microcontrollers

A microcontroller is a type of minicomputer specialized to fulfill certain tasks. It is also referred to as a *computer-on-chip*. A microcontroller consists of a *central processing unit* (CPU), *random-access memory* (RAM), *flash memory*, *serial bus interface*, *Input/Output (I/O) ports*, and occasionally, *electrical erasable programmable read-only-memory* (EEPROM), *analog-to-digital converter* (ADC), and *digital-to-analog converter* (DAC), as shown in Figure 8.32. A microcontroller can be programmed using different languages, such as assembly language, C, or C++.

When a microcontroller is used in a manufacturing line, it should be integrated with hardware such as power supplies for actuators. Microcontrollers are widely used in electrical appliances such as washing machines, microwaves, refrigerators, and air conditioners.

The key component in a microcontroller is the *CPU*, which is also referred to as a *microprocessor*. For industrial applications, two widely used microprocessor brands with 32-bit are Intel ×86 and Motorola 68000.

For prototyping a mechatronic system, Arduino boards in Figure 8.33 are the most widely used (Javapoint 2023). *Arduino* is an open-source, modular, and easy-to-use platform with its own microcontroller. Earlier Arduino boards were equipped with an 8-bit Atmel AVR microcontroller, and new Arduino models were equipped with a 32-bit Atmel ARM microcontroller.

8.7.2 Digital Signal Processing (DSP)

DSP is to perform arithmetic operations of signals at discrete times. DSP is performed by a microprocessor with sophisticated architecture such as *multiple buses* and *hardware multipliers* to support *parallel processing* and *fast sampling rates*. Accordingly, a *DSP* is specialized to reduce the number of instructions and operations for efficient arithmetic operations (see Figure 8.34). In contrast to microcontrollers, DSPs are more preferable when complex and iterative algorithms are used in mechatronic controls.

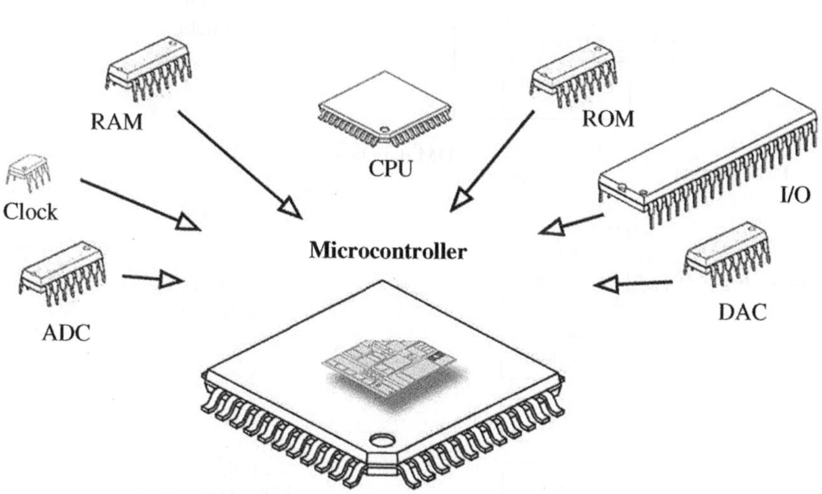

Figure 8.32 Basic components in a microcontroller (*Source:* Adapted from Arrow (2018))

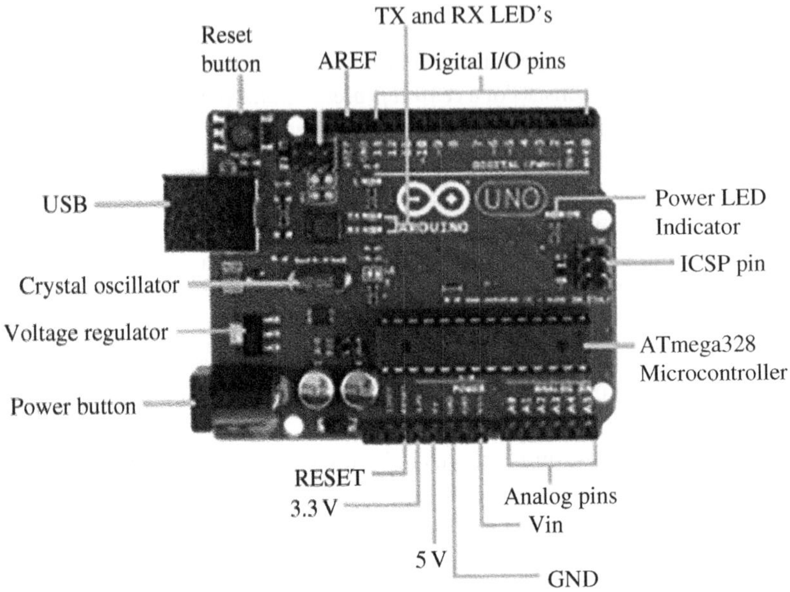

Figure 8.33 Structure of an Arduino board (*Source:* Javapoint (2023)/Javapoint)

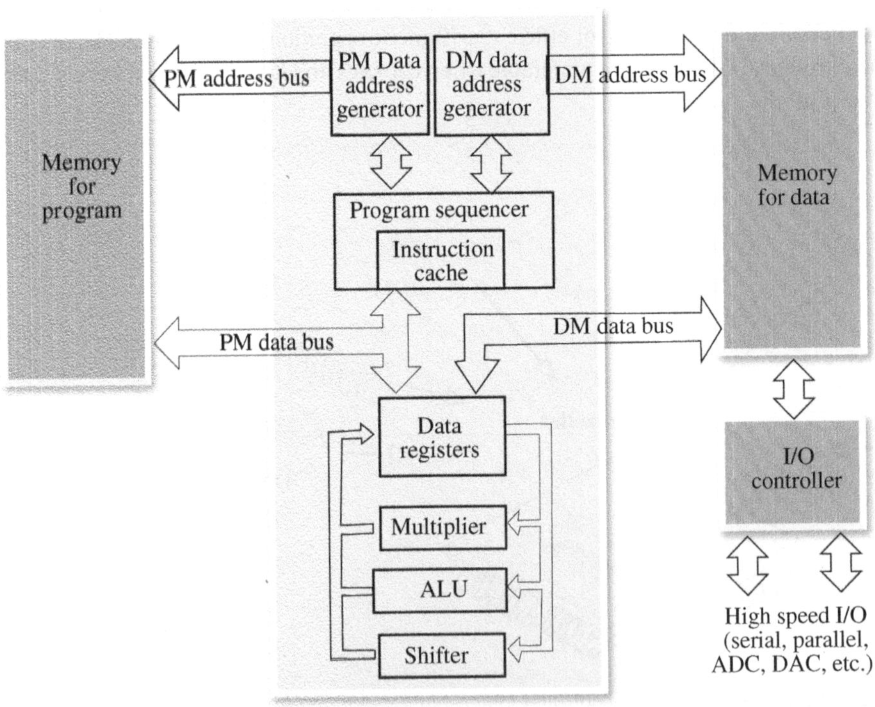

Figure 8.34 Architecture of digital signal processor (DSP) (*Source:* Smith (2023)/California Technical Publishing)

8.7.3 Field Programmable Gate Arrays (FPGA)

FPGA is reprogrammable after it is fabricated. In contrast to predetermined microcontrollers with fixed functions, FPGAs allow the programming of new features and functions to fulfill new tasks or adapt new standards. Reprogramming is implemented by reconfiguring hardware after it is installed in the field. FPGA can be used to perform any logical function that is traditionally performed by an *application-specific integrated circuit* (ASIC).

As shown in Figure 8.35, FPGA consists of an array of programmable logic blocks and interconnects. The array and interconnects can be configured to perform expected digital functions. FPGA is usually configured by using a *hardware description language* (HDL) that is similar to ASIC. It is extremely suitable for some applications such as telecommunication, automotive, and aerospace engineering since these applications require high flexibility and speedy programming.

The most significant feature of FPGA is its ability to update functions after it is purchased and installed in the field. By comparing a microcontroller and FPGA, (1) using a microcontroller can shorten the development time; (2) the peripherals of a microcontroller can be tested by suppliers, while FPGA has open-source soft peripherals and it is needed to be embedded and tested. On the other hand, a microcontroller is sophisticated and usually has better power efficiency; a microcontroller has a lower cost than an FPGA with similar functions; a microcontroller is especially useful when a control problem is simple but the solution should be developed in a large quantity.

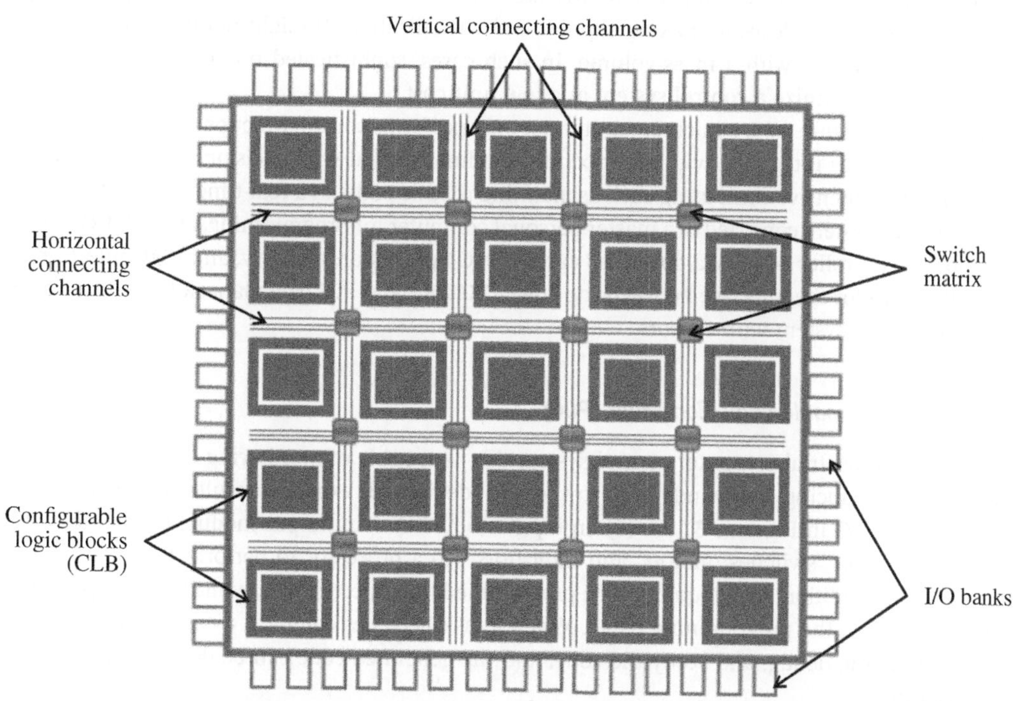

Figure 8.35 Structure of *field programmable gate arrays* (FPGA) (*Source:* Allabortfpga (2024)/Invent Logics)

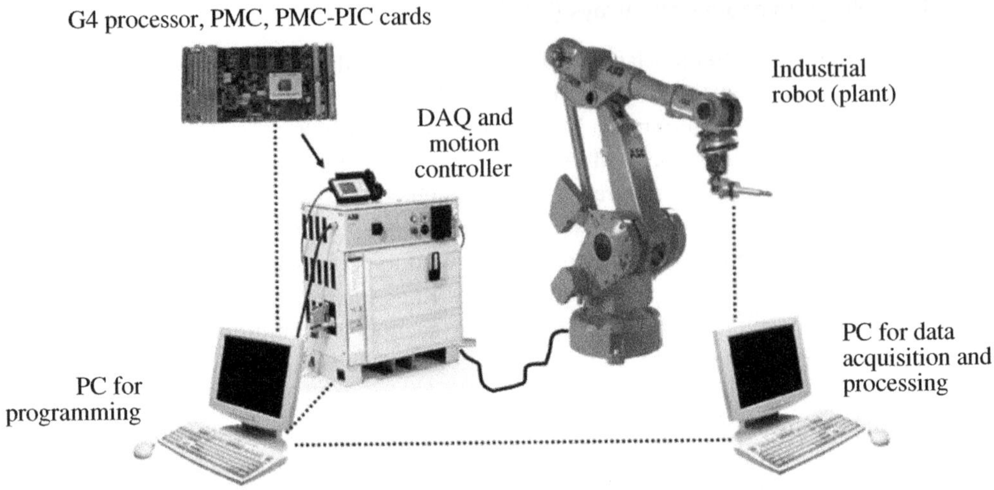

Figure 8.36 Example of using a minicomputer in mechatronic controls (*Source:* Soreling (2005)/Linköping University Electronic Press)

8.7.4 Microcomputers

When the control of a mechatronic system becomes very complex, a microcomputer can be used to process and analyze signals from multiple sources in depth. A microcomputer can be connected by a *data acquisition* (DAQ) system to transmit power and signals through I/O ports with the mechatronic system. In contrast, using a computer has the great advantage of superior computing capabilities for real-time decision-making support and virtualization. It might not be a cost-effective solution for products with a mass volume. In such cases, sophisticated microcontrollers should be designed to optimize a control system at a low unit cost.

As shown in Figure 8.36, a microcomputer not only performs control functions in a mechatronic system but also be integrated with other control modules to deal with decision-making activities, which require intensive computing such as planning robotic paths in a real-time mode. The example integrated system consisted of a master personal computer for programming, an ABB IRB-4400, the motion controller S4-plus, a power PC card with G4 processor, memory, and PMC interface, a PMC-PIC card for the interface between the power PC card and PCI backplane, a flash disc, and a data acquisition system with sensors.

8.7.5 *Programmable Logic Controller* (*PLC*)

A *PLC* is an industrial computer control system that is capable of monitoring the states of input devices and producing commands to control output devices for expected results. Two main components of a PLC in Figure 8.37 are (1) an input module and an output module to interact with sensors, actuators, switches, motors, and indicators; (2) a processor to perform logic operations in controls. In addition, PLC requires to be programmed for a specific task, while a simple program can be coded via the programming panel. When a task becomes complex, a computer should be used to program, debug, and verify the code, and the generated code can be downloaded to PLC for execution.

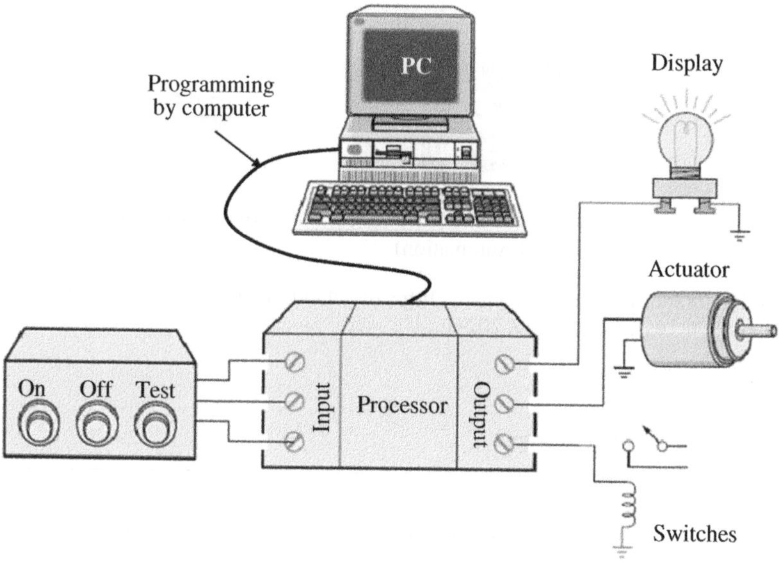

Figure 8.37 Example programmable logic controller

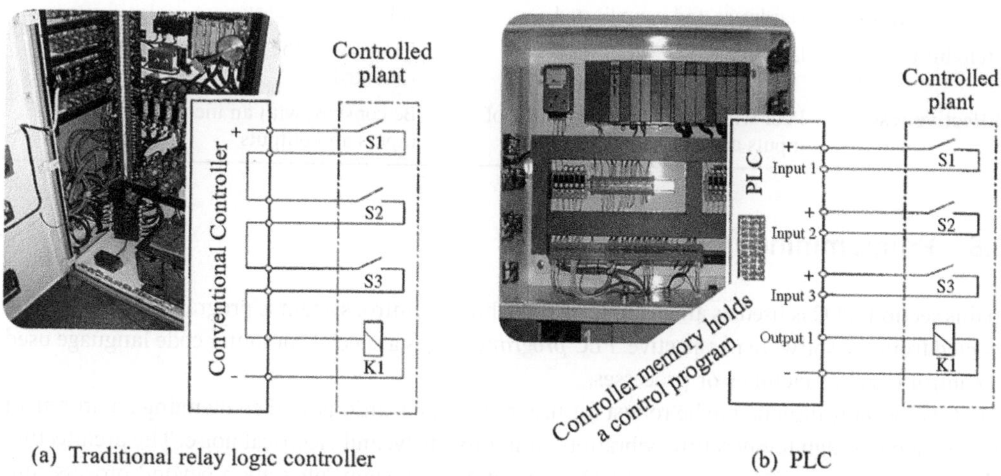

Figure 8.38 Comparison of relay logic controller and PLC (*Source:* (b) Shah Automation / https://www.facebook.com/ShahAutomation/ / Last accessed on July 19, 2024)

PLCs are developed to replace traditional relay logic controls. As shown in Figure 8.38a, a traditional relay logic control transmits the commands by wiring, while a PLC produces commands by issuing instructions, and commands are stored in memory as programs. PLC is the way to represent logical relations of inputs and outputs.

Figure 8.39 shows three popular brands of PLCs in markets. In general, using a PLC instead of a relay logic control brings multiple benefits. Table 8.5 lists the comparison of a relay logic controller and PLC.

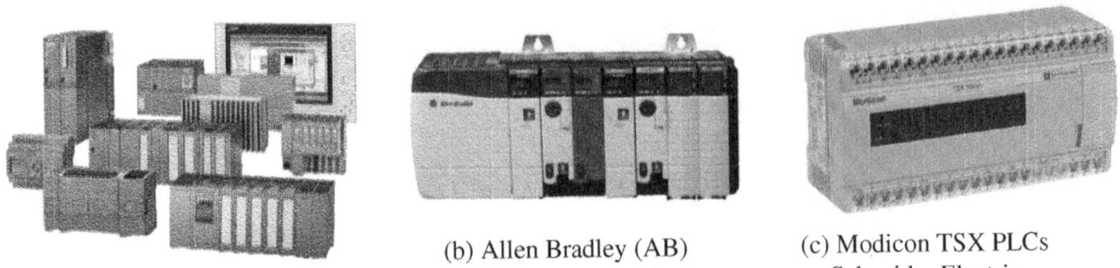

(a) Siemens simatic (b) Allen Bradley (AB) (Rockwell Automation) (c) Modicon TSX PLCs Schneider Electric

Figure 8.39 Three popular PLC brands (*Source:* (a) Siemens /https://mall.industry.siemens.com/mall/en/uk/Catalog/Products/5000005 / Last accessed on July 19, 2024)

Table 8.5 Comparison of PLC with relay logic controls

	Relay Logic Control	**PLC**
Source of commands	Wiring arrangements	Instructions in programs
Complexity	Increase the number of inputs and outputs	Be constant with the number of inputs and outputs
Implementation	Hardware and complicated	Software and easy
Reliability	Low	Fault-free if the program is appropriately validated
Effectiveness	Poor with an increased number of inputs and outputs	Be constant with an increased number of inputs and outputs

8.8 Programming in PLC

In this section, PLC is used as an example to show how a control system is programmed and implemented from the software perspective. *PLC programming* is a special computer code language used to control factory machines or processes.

PLC is specially designed to be robust for use in harsh, hazardous, and challenging environment settings such as high temperature, vibration, humidity, dirty, and electrical noise. The architecture of PLC supports an easy extension of I/O capabilities for system integration beyond process controls. PLCs are widely used in industrial processes, power supplies, transportation, packaging, glass industries, paper production, and automatic drainage monitoring. PLCs have shown their popularity and importance across many industries, with an amazing market value of around $14.6 billion in 2022 (Trainings 2024).

8.8.1 Data Structure and Flow

Figure 8.37 shows that a PLC consists of a processor, power supply, I/O modules, and a programming device. Here, the data structure and flow in a PLC are further discussed and shown in Figure 8.40.

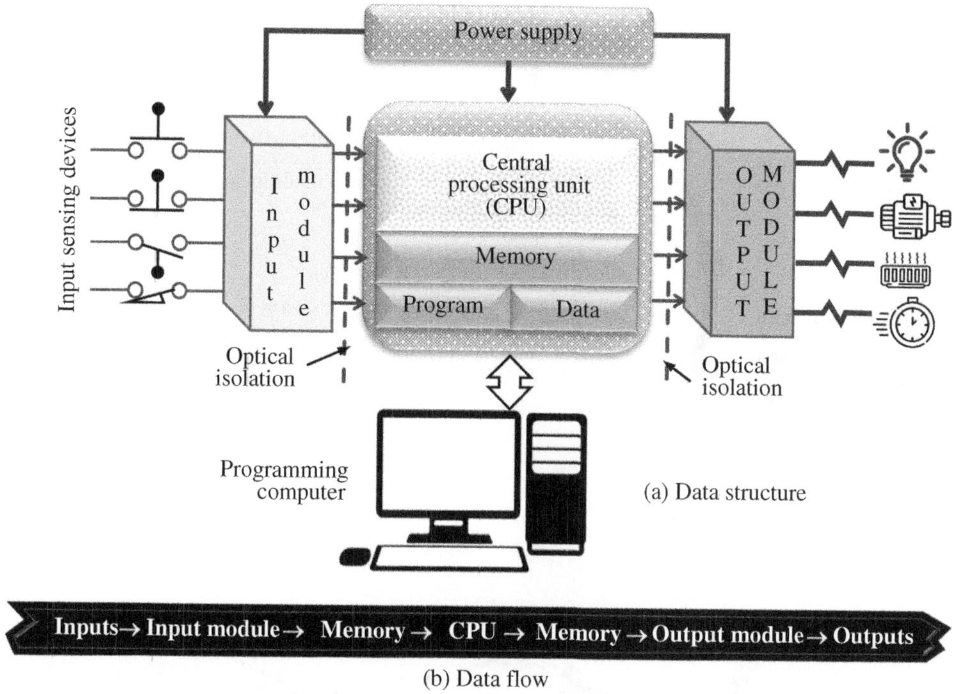

Figure 8.40 Structure and data flow of PLC

8.8.2 Operating Cycle

A PLC program consists of a series of operation cycles; each operating cycle consists of four steps: *step*-1 is an *input scan* to update the statuses of inputs; *step*-2 is a *program scan* to process the program logic based on given inputs; *step-3* is an *output scan* to activate or deactivate outputs based on program logic; *step-4* is *housekeeping* for communications and diagnosis.

As shown in Figure 8.41, the loop with the aforementioned steps is continually repeated and processed until it is terminated by the program. The program is stored in "nonvolatile" memory, and the program is stored even when the power is removed.

8.8.3 I/O Modules and Addresses

I/O is the connection of the outside world to the brain of PLC. A PLC system is usually equipped with dedicated modules for inputs and outputs separately. An input module monitors input signals from external devices such as push-buttons, switches, and sensors, as exemplified in Figure 8.42. An output module produces commands to control external devices such as relays, motor starters, and lights, as exemplified in Figure 8.43.

In PLC, every I/O port is uniquely addressed in an input or output data file. Table 8.6 shows an example of an address identifier to specify an I/O port uniquely.

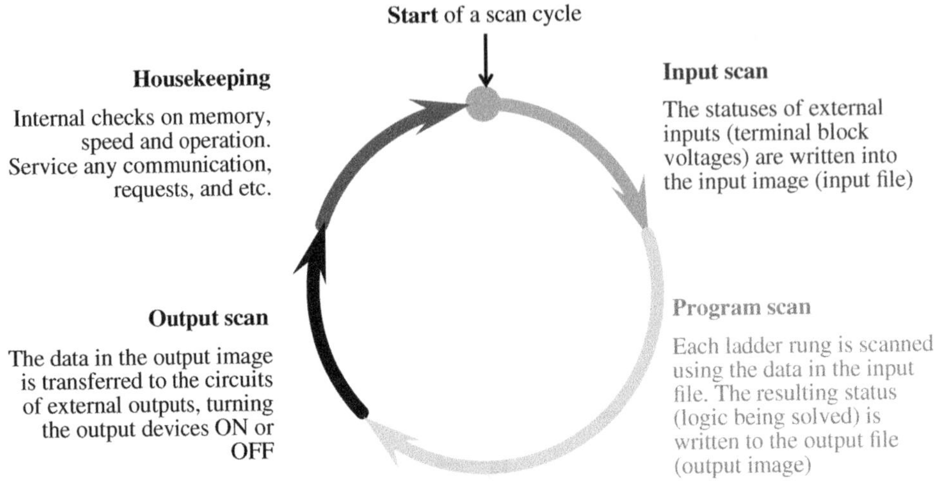

Figure 8.41 An operation cycle in PLC

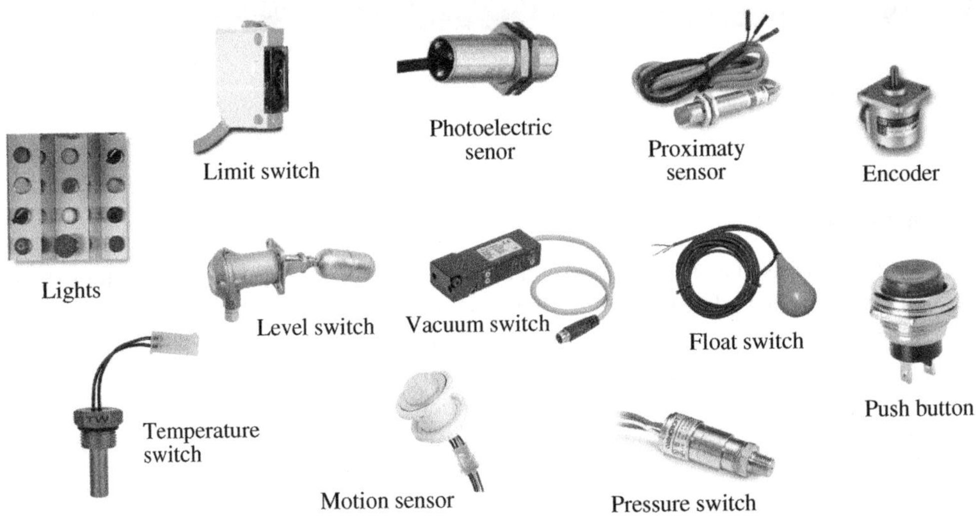

Figure 8.42 Examples of input devices (*Sources:* (c) Velleman Group / https://www.jameco.com/z/WPSE476-Velleman-Velleman-Inductive-Proximity-Sensor-Detection-Switch-by-Whadda_2354855.html / Last accessed on July 19, 2024; (e) Scribd Inc / https://www.scribd.com/document/405985229/9001CT0403-pdf / Last accessed on July 19, 2024; (f) Trimod Besta / https://www.trimodbesta.com/products/trimodbesta-levelswitches/standard-range/electric/?lang=en / Last accessed on July 19, 2024; (g) Millibar, Inc / https://millibar.com/vacuum-switches/ / Last accessed on July 19, 2024; (h) TRITON CONTROLS / https://www.ebay.com/itm/266453843623 / Last accessed on July 19, 2024; (i) Parts Express / https://www.parts-express.com/?utm_source=google&utm_medium=cpc&utm_campaign=1041101405&utm_content=53866483129&utm_term=parts%20express&gadid=495894331915&gad_source=1&gclid=Cj0KCQjwwO20BhCJARIsAAnTIVR8Mm8iJXEXQcSul0YDxjyl23_mdmUQcx5stP6ofbYHT5RWXlR-KCEaAqhPEALw_wcB / Last accessed on July 19, 2024; (j) Nason / https://www.nasonptc.com/products/switches/temperature/tw/ / Last accessed on July 19, 2024; (k) Quentacy / https://www.amazon.com/Quentacy-110-220V-Detection-Adjustable-Induction/dp/B093715WR9 / Last accessed on July 19, 2024; (l) Ashcroft, Inc / https://www.bricebarclay.com/product/ashcroft-a-series-miniature-watertight-pressure-switches/ / Last accessed on July 19, 2024)

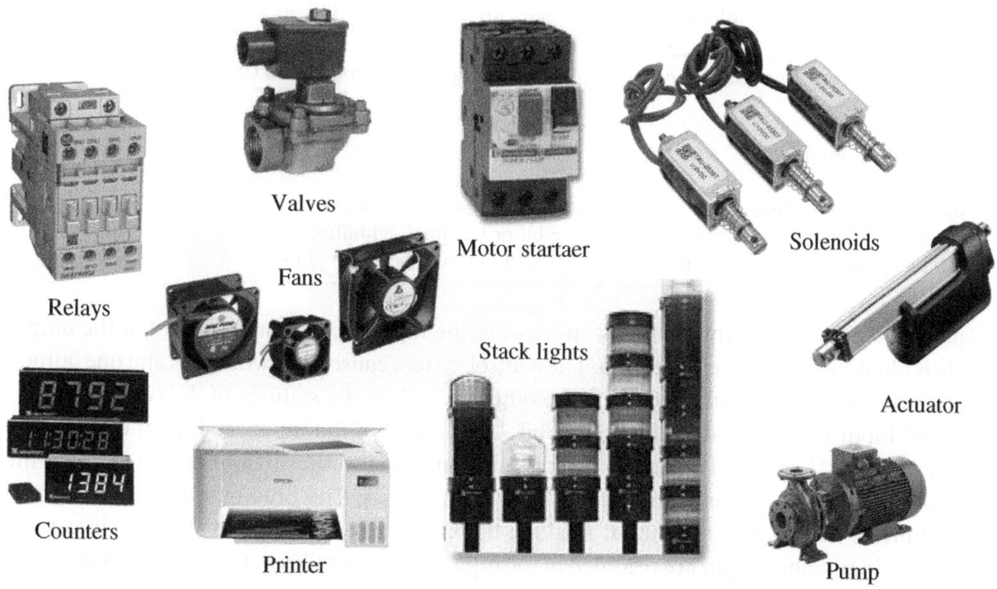

Figure 8.43 Examples of output devices (*Sources:* (a) Radwell International UK Ltd / https://www.radwell.co.uk/Buy/PARKER/FLUID%20POWER%20DIVISION/U322G2501?redirect=true / Last accessed on July 19, 2024; (b) Schneider Electric / https://www.amazon.com/SCHNEIDER-ELECTRIC-600-VAC-GV2ME10-Starter/dp/B00PSMWW8I / Last accessed on July 19, 2024; (c) Solenoids / https://www.aliexpress.com/item/32616755231.html / Last accessed on July 19, 2024; (d) Rockwell Automation / https://www.northcoast.com/product/detail/2054527/allen-bradley-rockwell-300-tod9310 / Last accessed on July 19, 2024; (e) Gurgaon / https://www.indiamart.com/proddetail/aub0912l-c-delta-cooling-fan-25346606933.html / Last accessed on July 19, 2024; (f) LINAK / https://www.linak.com/products/linear-actuators/ / Last accessed on July 19, 2024; (h) Epson / https://www.itubia.com/product/epson-ecotank-l3256-a4-color-3-in-1-printer/ / Last accessed on July 19, 2024; (i) Reckers Automation India Pvt Ltd / https://www.indiamart.com/proddetail/telemecanique-tower-light-13195018797.html?vcdimgform=1 c/ Last accessed on July 19, 2024; (j) North Ridge Pumps / https://www.northridgepumps.com/articles?id=219&slug=centrifugal-pump-troubleshooting-guide / Last accessed on July 19, 2024)

Table 8.6 Example of address identifiers

System-Defined Files			User-Defined File		
File Type	**Identifier**	**File Number**	**File Type**	**Identifier**	**File Number**
Output	O	0	Bit	B	10-255
Input	I	1	Timer	T	
Status	S	2	Counter	C	
Bit	B	3	Control	R	
Timer	T	4	Integer	N	
Counter	C	5			
Control	R	6			
Integer	N	7			
Float	F	8			
Ordinary/network		9			

The rules to specify the address of an I/O port are shown as follows:

An input address is formatted as *I*: *e*/*b*	An output address is formatted as *O*: *e*/*b*
where	where
I = identifier as input, : = slot delimiter, *e* = slot number of module, / = bit or terminal delimiter, and *b* = terminal number of the module.	*O* = identifier as input, : = slot delimiter, *e* = slot number of module, / = bit or terminal delimiter, and *b* = terminal number of the module.

Figure 8.44 shows an example of how an input at the input module or an output at the output module is identified in a control program. The control system consists of two inputs and one output, and there is one control logic to generate the output based on the statuses of the two inputs.

The first input is connected to the second bit "1" at slot 2 of the input module; therefore, it is addressed as "I:2/1" in case B. The second input is connected to the sixth bit "5" at slot 2 of the input module; therefore, it is addressed as "I: 2/5" in case A. The only output is connected to the third bit "2" at slot 4 of the output module; therefore, it is addressed as "O: 4/2" in case C. The states of inputs and output are recorded in the input and output files in the middle, respectively.

8.8.4 Elements of Logic Control

Table 8.7 shows five basic logic gates, corresponding operations, and symbols, and Figure 8.45 shows the truth tables of the logic operations for these gates.

In designing a control system, any complex logic can be constructed from the above basic logic gates.

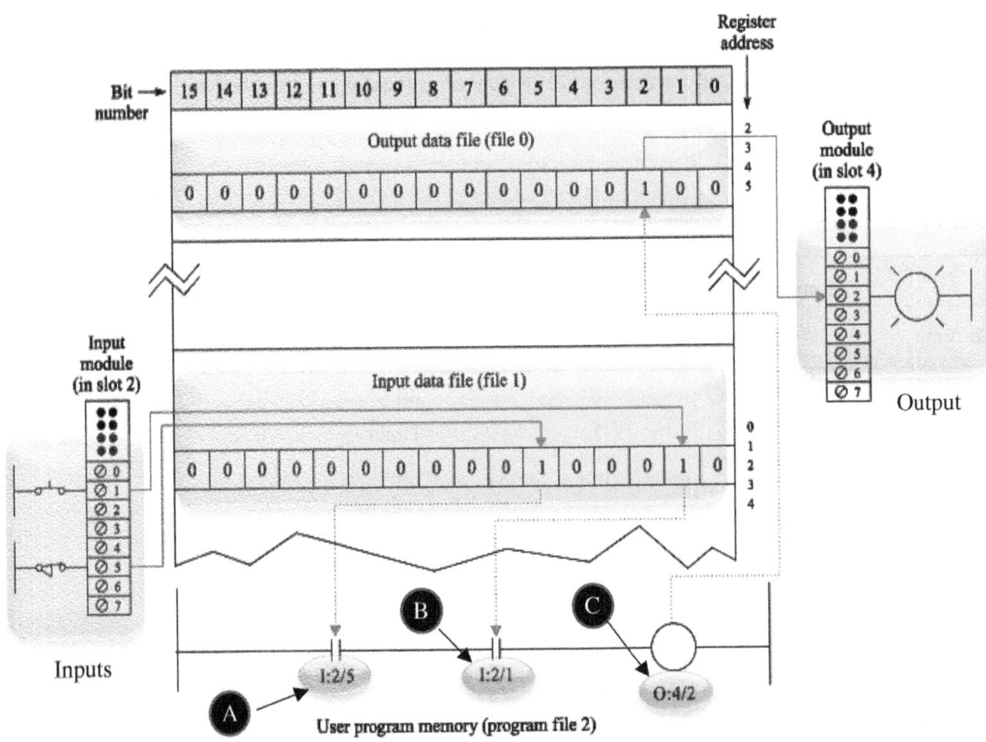

Figure 8.44 Identify and use I/O addresses in the program

Table 8.7 Five basic logic gates

Type	Operation	Symbol
AND	If both of the two inputs are "1," it outputs "1"; otherwise, it outputs "0."	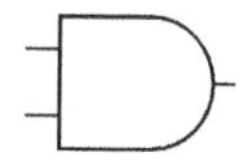
OR	If either of two inputs is "1," it outputs "1"; otherwise, it outputs "0."	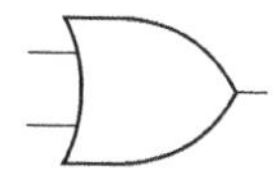
NOT	If the single input is "0," it outputs "1"; otherwise, it outputs "0."	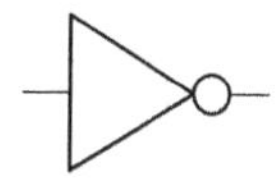
NAND	It is a combination of **AND** and **NOT**	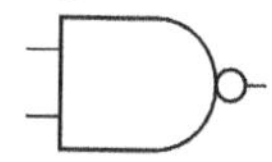
NOR	It is a combination of **OR** and **NOT**	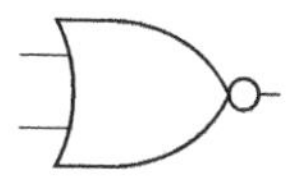

Example 8.8 Write out the truth table for the pushbutton system shown in Figure 8.46.

Solution

The pushbutton system in Figure 8.46 consists of three input signals (X1, X2, and Y1) and two output signals (Y2 and Y1), where the Y1 signal will be feedbacked to determine Y2 in the next scan. Y2 is determined by three logic gates as:

$$Y2 = (\overline{X2}) \cap (X1 \cup Y1) \tag{8.68}$$

Input		Output
X1	X2	Y=X1∩X2
0	0	0
0	1	0
1	0	0
1	1	1

(a) **AND** function Y=X1∩X2

Input		Output
X1	X2	Y=$\overline{\text{X1}\cup\text{X2}}$
0	0	1
0	1	0
1	0	0
1	1	0

(b) **NOR** function Y=$\overline{\text{X1}\cup\text{X2}}$

Input	Output
X1	Y=$\overline{\text{X1}}$
0	1
0	1
1	1
1	0

(c) **NOT** function Y=$\overline{\text{X1}}$

Input		Output
X1	X2	Y=$\overline{\text{X1}\cap\text{X2}}$
0	0	1
0	1	1
1	0	1
1	1	0

(d) **NAND** function Y=$\overline{\text{X1}\cap\text{X2}}$

Input		Output
X1	X2	Y=X1∪X2
0	0	0
0	1	1
1	0	1
1	1	1

(e) **OR** function Y=X1∪X2

Figure 8.45 Truth tables of five basic logic gates

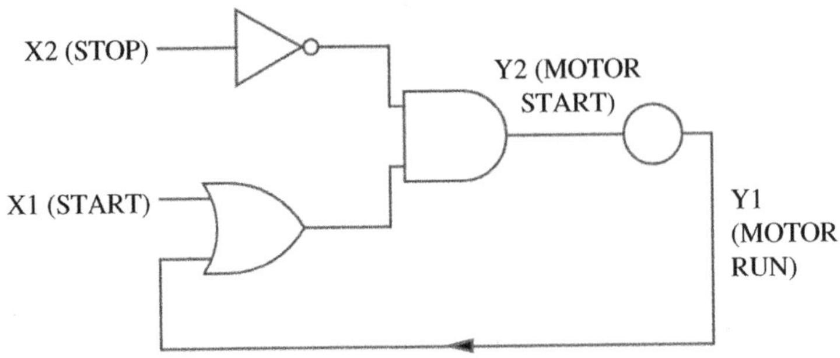

Figure 8.46 Control diagram of Example 8.8

Table 8.8 Truth table of the pushbutton system in Example 8.8

Input			Output
Y1	**X1**	**X2**	$Y2 = \overline{(X2)} \cap (X1 \cup Y1)$
0	0	0	0
0	0	1	0
0	1	0	1
0	1	1	0
1	0	0	1
1	0	1	0
1	1	0	1
1	1	1	0

Equation (8.68) is further used to determine the truth table of the pushbutton system, as shown in Table 8.8.

A truth table is used to determine the output based on given inputs at a time when the inputs are changed over time, a *combined logic* can be used to look into the change of output logic when the inputs change over time rather than remain constant.

Example 8.9 Determine the combined logic (Y) of an AND gate when the input signals (X1 and X2) are given in Figure 8.47a and b, respectively.

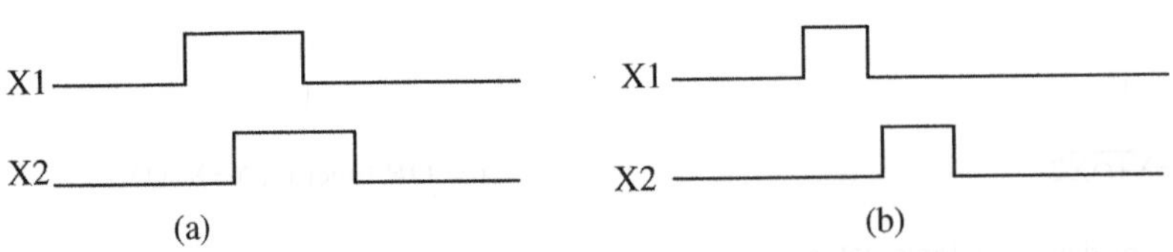

Figure 8.47 Two inputs changed over time in Example 8.9

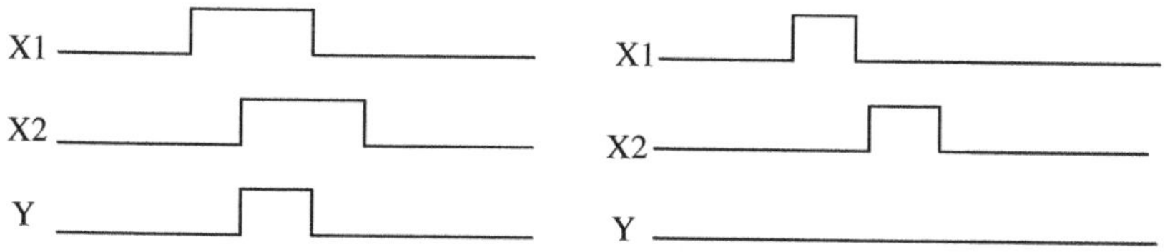

(a) (b)

Figure 8.48 Combined logic of two inputs in Example 8.9

Solution

For an AND gate, the corresponding logic operation is Y = X1∩X2; by applying this operation to the input signals shown in Figure 8.47, the corresponding combined logic can be determined and shown in Figure 8.48.

8.8.5 Ladder Logic Diagrams

There are a number of programming languages for PLCs, popular programming tools are *ladder logic diagram* (LLD), *function block diagram* (FBD), *statement list* (SL), *instruction list* (IL), *continuous function chart* (CFC), *structured control language* (SCL), and *sequential control and state graph* (graph).

LLD shows how logic elements are connected along horizontal rungs. As shown in Figure 8.49, the most commonly used logic elements are contacts, loads, timers, and counters. *Contacts*

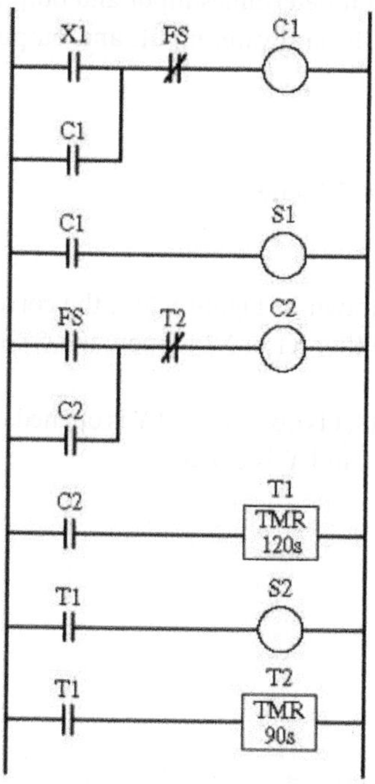

(a) Example of ladder logic

Ladder symbol	Hardware component
(a) —| |—	Normally open contacts (switch, relay, other ON/OFF devices)
(b) —|/|—	Normally closed contacts (switch, relay, etc.)
(c) —( )—	Output loads (motor, lamp, solenoid, alarm, etc.)
(d) TMR 3s	Timer
(e) CTR	Counter

(b) Common ladder symbol

Figure 8.49 Examples of logic elements in *ladder logic diagram* (LLD)

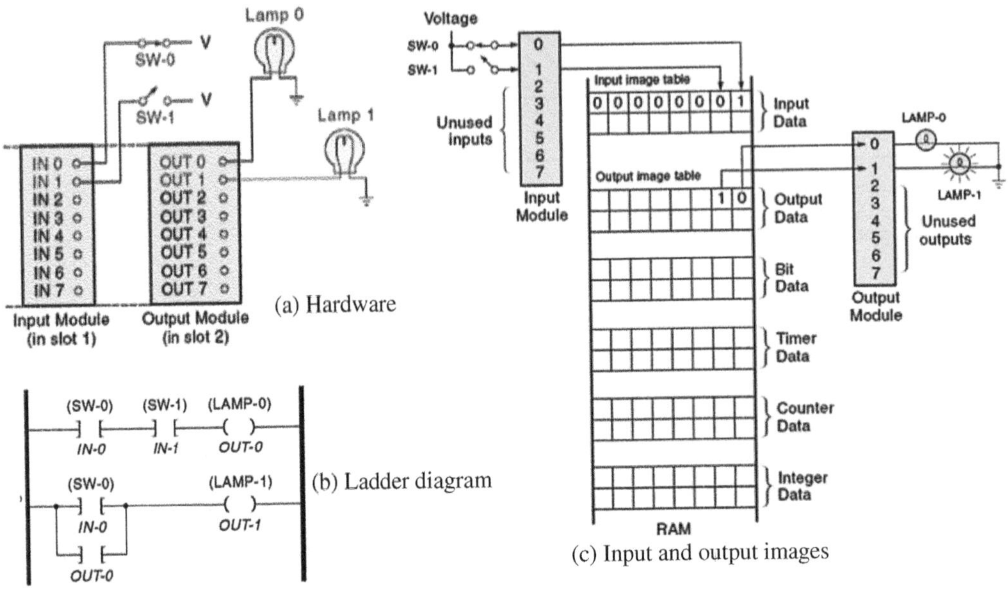

Figure 8.50 Examples of ladder logic diagram (LLD)

represent logical inputs such as limit switches and photodetectors; *loads* represent some output devices such as motors, lights, alarms, and solenoids; *timers* are the electronic devices to specify a length of delay; *counters* are the devices used to count pulses received.

LLD is closely related to the connected inputs and outputs, and it determines input and output images. Figure 8.50 shows an example of its correspondence with hardware, input, and output images.

Example 8.10 Construct a ladder logic diagram (LLD) for a NAND gate.

Solution

The truth table of a NAND gate is given in Figure 8.45d and redrawn in Figure 8.51a, the corresponding LLD can be readily determined in Figure 8.51b. When either X1 or X2 is open, the C coil is unexcited, the C contact remains closed, and Y is closed.

When both X1 and X2 are closed, the C coil is excited, the C contact is opened, and Y is opened. If X1 and X2 are closed, then C coil is excited, C contact is opened, and Y is closed.

Input		Output
X1	X2	$Y=\overline{X1 \cap X2}$
0	0	1
0	1	1
1	0	1
1	1	0

(a) Truth table of NAND gate

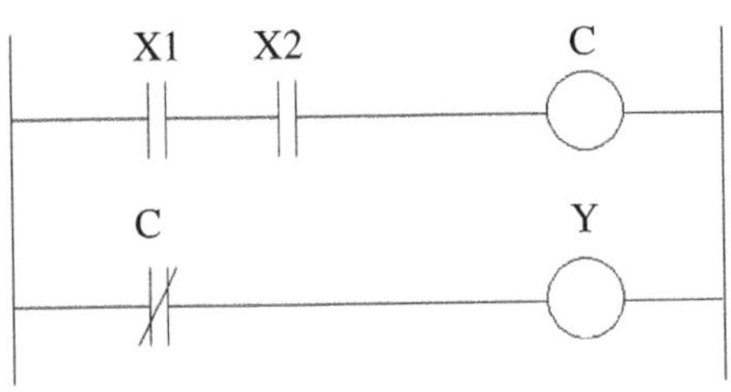

(b) Ladder logic diagram

Figure 8.51 Ladder logic diagram of a NAND gate

Input		Output
X1	X2	$Y = \overline{X1 \cup X2}$
0	0	1
0	1	0
1	0	0
1	1	0

(a) Truth table of NOR gate

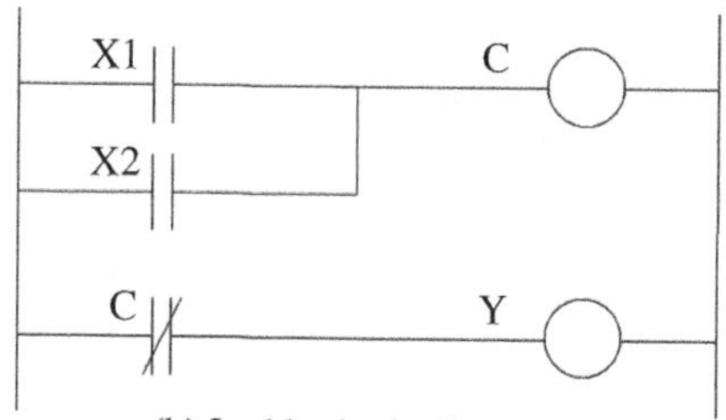

(b) Ladder logic diagram

Figure 8.52 Ladder logic diagram of a nor gate

Example 8.11 Construct a ladder logic diagram for an NOR gate.

Solution

The truth table of a NOR gate is given in Figure 8.45b and redrawn in Figure 8.52a; the corresponding ladder logic diagram can be readily determined in Figure 8.52b. If both of X1 and X2 are open, the C coil is unexcited and the C contact remains closed, therefore Y is on. Otherwise, if either or both of X1 and X2 are closed, the C coil is excited, the C contact is closed, and Y is closed.

8.8.6 Timers and Counters

A timer in programming is to record a time or set up a delay for an instruction. A timer operates as a relay timer, but it does not require wires to connect and use. Three main types of timers of PLC are *on-delay timers*, *off-delay timers*, and *retentive on-delay timers*, as shown in Figure 8.53.

When an on-delay timer is energized (i.e., "*1*" or "*True*"), it delays the output "*ON*" signal until the preset time of the timer is reached; the output bit is often called the timer's "*DONE*" bit. It implies that the timer has attained its preset time. When an off-delay timer is energized (i.e., "1" or "True"), it immediately outputs "ON" at its output bit (i.e., "DONE" bit). The timer then delays to turn the output "OFF" after the timer reaches the preset time. In practice, off-delay timers are less common than on-delay timers. A retentive on-delay timer is to delay the ON time, if the input goes off in between; it will resume the accumulator value, and if the input turns ON, it will start running for the resumed value.

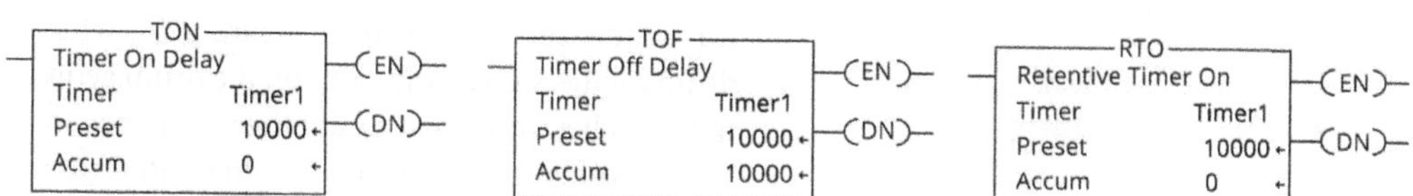

(a) On-delay (b) Off-delay (c) Retentive on-delay

Figure 8.53 Three types of PLC timers

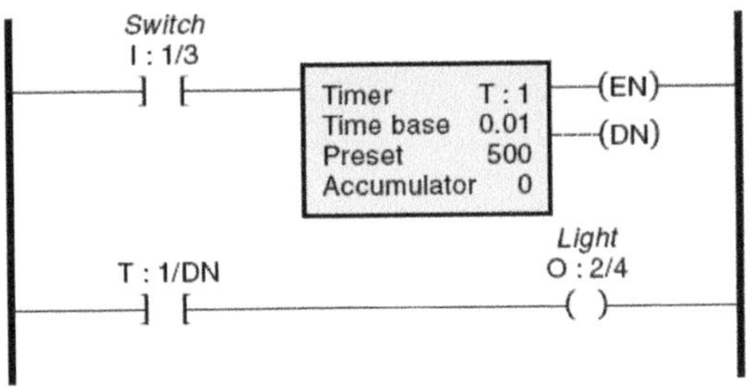

(a) Use a timer in control

Time base	The value "0.10" means that each count corresponds to 0.1 seconds
Preset	The value "500" means that the delay will last 500 counts, in other words, a total of 5 seconds (0.01 × 500 = 5 seconds)
Accumulator	holds the value of the current count
EN	A bit is TRUE as long as the timer rung is TRUE
DN	goes TRUE when the count meets or exceeds the preset value.

(b) Define a timer in control

Figure 8.54 Example of setting up and using a timer

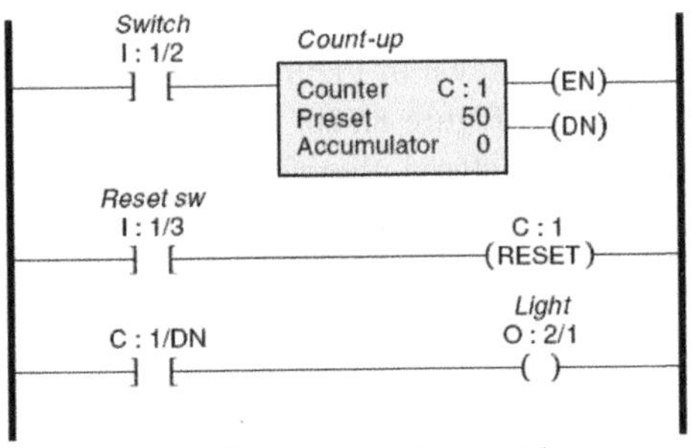

(a) Use a counter in control

Preset	When the counter gets to the preset value, it sets the DN bit TRUE and keeps on counting.
Accumulator	holds the value of the current count
EN	A bit is TRUE as long as the counter rung is TRUE
DN	goes TRUE when the count meets or exceeds the preset value.
RESET	A separate instruction with the same counter address; when this bit goest TRUE, the accumulator resets to zero (resets the counter)

(b) Define a counter in control

Figure 8.55 Example of setting up and using a timer

Figure 8.54 shows an example to set up and use a timer in a control system.

A counter in PLC is a functional block that counts up or down until it reaches a limit. When the limit is reached, the output is set. Figure 8.55 shows an example of how to set up and use a counter in a control system.

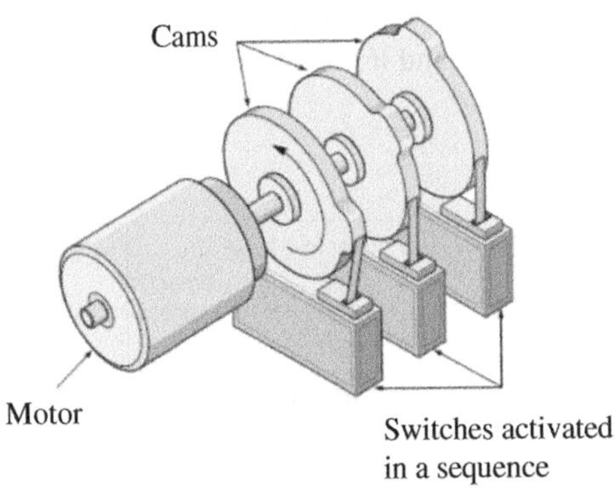

Figure 8.56 Traditional electromechanical sequencer

8.8.7 Sequencers

A *sequencer* is used when a repeating sequence of outputs is required. A repeating sequence of outputs is traditionally specified by an electromechanical sequencer, as shown in Figure 8.56; when a drum rotates slowly, the cams on the drum activate the switches from one to another in a sequence. The instructions by a sequencer allow a PLC to take a sequence of actions in order.

Figure 8.57 shows an example of defining a sequencer in a control system, and Figure 8.58 shows how a sequencer is used in a control system.

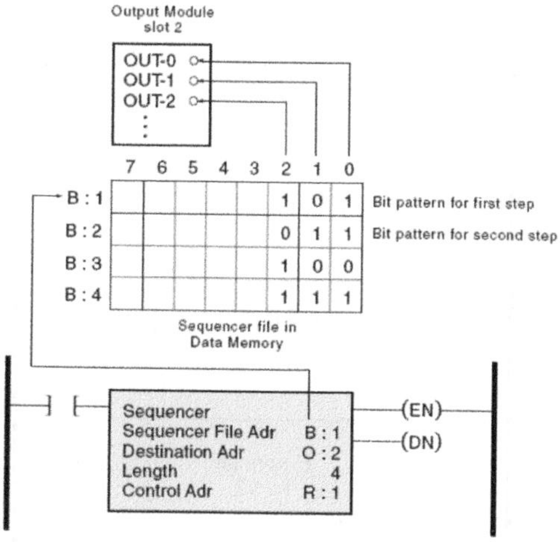

(a) Use a sequencer in control

Sequencer File Address	The first address of the sequence file that stores the sets of outputs
Destination Address	The output address (or slot) to which the sequence file words are transferred with each step.
Length	The length of the sequence file, or the number of the steps in the sequence
EN	goes TRUE when the sequencer rung is TRUE.
DN	goes TRUE when it has operated on the last word in the sequence file.
Control Address	Stores the control bits (EN, DN) and words (length) of the sequencer

(b) Define a sequencer in control

Figure 8.57 Example of defining a sequencer

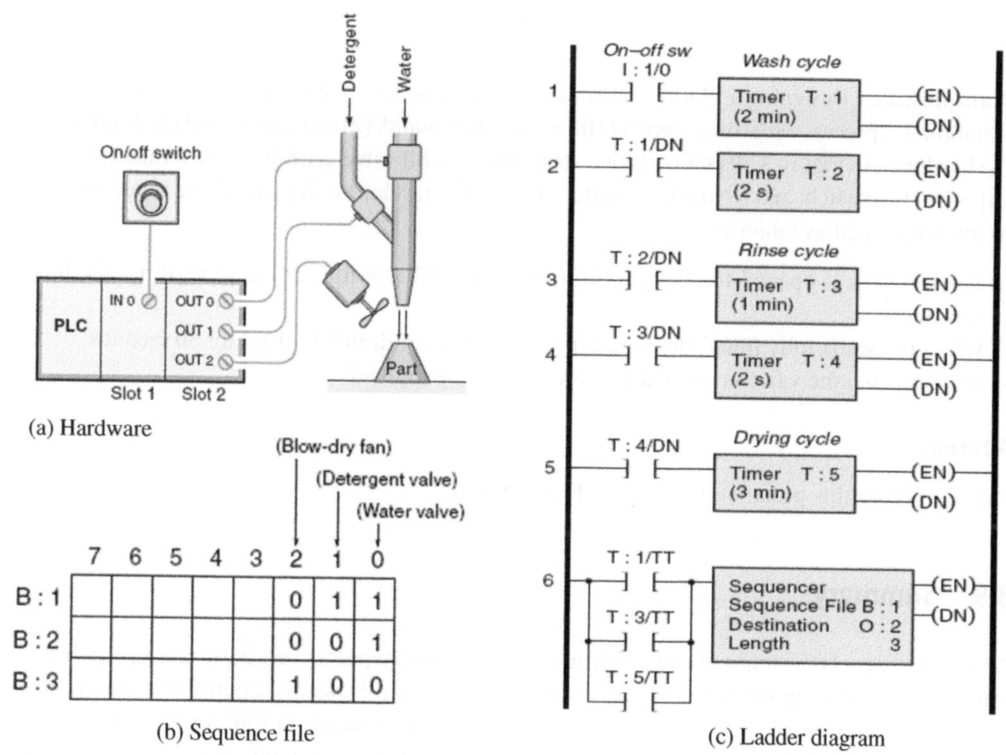

Figure 8.58 Example of defining and using a sequencer

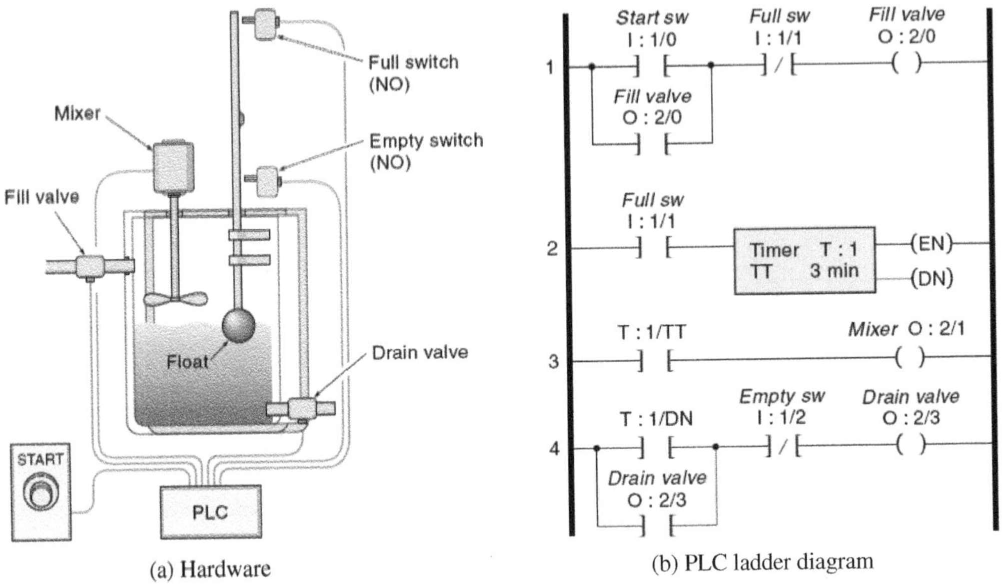

Figure 8.59 Example of PLC programming

Example 8.12 Designing a PLC to control a batch process is shown in Figure 8.59a. It is designed to automate a process with three steps: (1) fill a vat with a liquid, (2) mix liquid, and (3) drain the vat. The hardware in Figure 8.59a consists of a start button, a fill valve, a drain valve, a mixer, a floater with a NO full switch, and a No empty switch. When the start button is pushed, the sequence of the events is specified as follows:

1) The fill valve is opened to fill liquid into the vat until it is full; at the same time, the liquid is mixed.
2) When the vat is fully filled, the liquid in the vat will continually mixed for 30 seconds.
3) After mixing, the valve is opened and the vat is fully drained.

Solution

The solution to this problem is shown in Figure 8.59b.

8.9 Summary

A control system is the brain of a mechatronic system. This chapter covers basic concepts, theories, components, and programming methods of mechatronic control systems. Control systems are characterized in terms of openness, discrete or continuous controls, steady or transient controls, orders, and other criteria. PID control is used as an example of the implementation of certain control strategies. Moreover, it emphasizes the importance of using standardized MFMs in designing and implementing mechatronic control systems; common hardware components include microprocessors, Arduinos, and personal computers. In addition, mechatronic engineers should be able to program control systems in applications.

Problems

8.1 Determine the transfer function of the capacity voltage and input voltage of the RLC circuity in Figure 8.60.

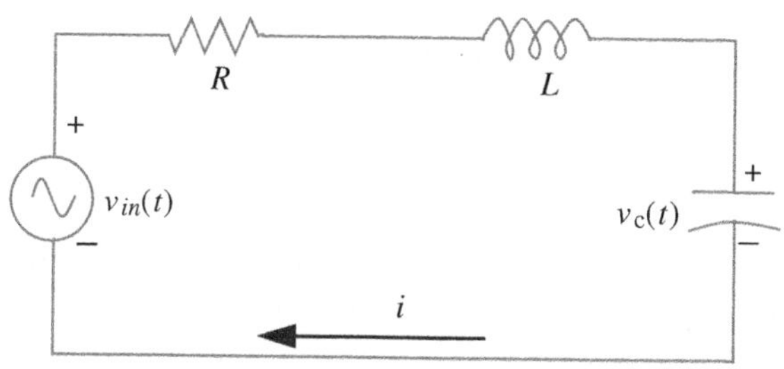

Figure 8.60 Transfer function in RLC circuit

8.2 Determine the transfer function of the input voltage ($v_{in}(t)$) and output voltage ($v_{out}(t)$) of the circuit in Figure 8.61.

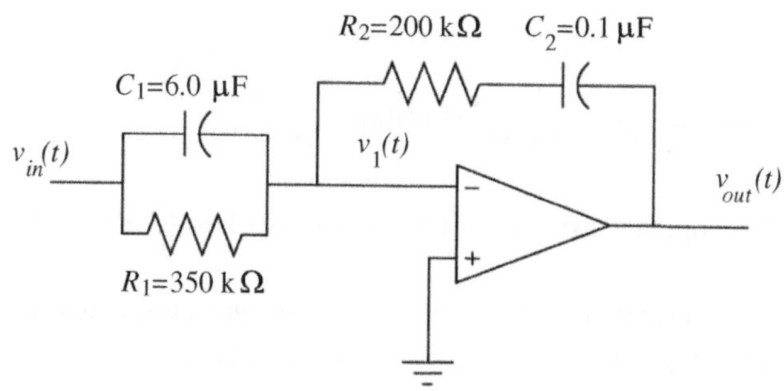

Figure 8.61 The transfer function for the circuit in Problem 8.2

8.3 Determine the transfer function of the control system that is represented by the block diagram in Figure 8.62.

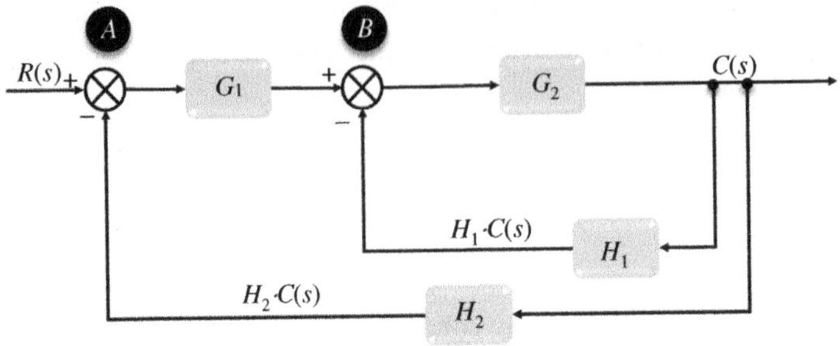

Figure 8.62 Block diagram for Problem 8.3

8.4 Find the poles and zeros of a linear system represented by its differential equation of

$$\frac{d^2y}{dt^2} + 15\frac{dy}{dt} + 8y = 2\frac{du}{dt} + 3$$

8.5 Assume a system has a pair of conjugate poles $p_1, p_2 = -3 \pm j$, one zero $z_1 = -5$, and the gain factor of 10. Find the equation for input–output dynamic system dynamics.

8.6 Derive the expressions of magnitude and phase with respect to frequency for the transfer function of $G(s) = \dfrac{s}{s^2+1}$ and plot its bode plots.

8.7 Consider the open-loop system $G(s) = 5/(s(s + 1))$, design a PD controller for a unit step signal with the specifications of a rise time < 0.2 second and overshoot <10%.

8.8 Determine the stability of the control systems with the following transfer functions:
(a) $G(s) = \dfrac{1}{(s+5)(s+2)}$; (b) $G(s) = \dfrac{1}{(s-5)(s+2)}$; (c) $G(s) = \dfrac{1}{(s-5)(s-2)}$;
(d) $G(s) = \dfrac{1}{(s^2+s+1)}$; (e) $G(s) = \dfrac{1}{(s^2-2s+3)}$; (f) $G(s) = \dfrac{1}{(s^2-2s)}$,

8.9 Write out the truth table of the operations: (1) $Y = (A \cdot B) + C$; (2) $Y = \left(A \cdot \overline{B}\right) + (A \cdot B)$

8.10 Determine the combined logic (Y1) and (Y2) of an "OR" and "NOR" gate, respectively, when the input signals (X1 and X2) are given in Figure 8.63a and b, respectively.

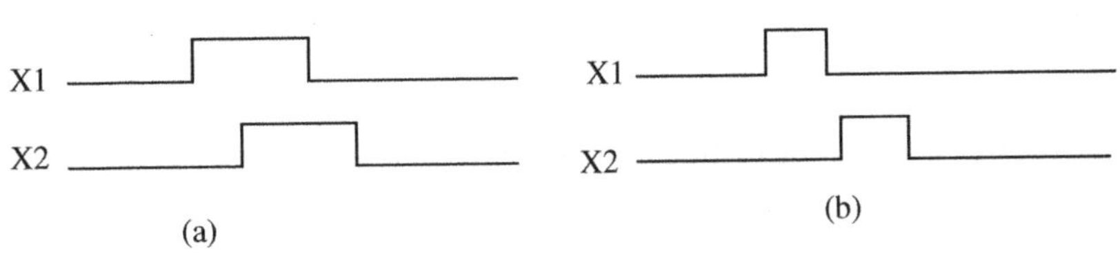

Figure 8.63 Two inputs changed over time in Problem 8.9

8.11 Construct a ladder logic diagram (LLD) for an AND and OR gate, respectively.

References

Allabortfpga (2024). Field programmable gate array. https://allaboutfpga.com/fpga-architecture (accessed 12 September 2024).

Arrow (2018). Introduction to microcontrollers. https://www.arrow.com/en/research-and-events/articles/engineering-basics-what-is-a-microcontroller (accessed 12 September 2024).

Bhandari, P. and Csurcsia, P.Z. (2022). Digital implementation of the PID controller. *Software Impacts* 13 (2022): 100306.

Guerrero-Ibanez, J., Zeadally, S., and Contreras-Castillo, J. (2018). Sensor technologies for intelligent transportation systems. *Sensors* 18 (4): 1212.

Hernandez-Barragan, J., Rios, J.D., Alanis, A.Y. et al. (2020). Adaptive single neuron aniti-windup PID controller based on the extended Kalman filter algorithm. *Electronics* 9: 636. https://doi.org/10.3390/electronics9040636.

Javapoint (2023). Arduino UNO. https://www.javatpoint.com/arduino-uno (accessed 12 September 2024).

Koziorek, J., Konecny, J., Gavlas, A. et al. (2018). Analysis of dataflows within industrial control system design. *MATEC Web of Conferences* 210: 02030. https://doi.org/10.1051/matecconf/201821002030.

MIT (2024). Understanding poles and zeros. https://web.mit.edu/2.14/www/Handouts/PoleZero.pdf (accessed 12 September 2024).

Rahman, M., Fentaye, A., Zaccaria, V. and Aslanidou, I. (2021). A framework for learning system for complex industrial processes. https://www.researchgate.net/publication/349412387, https://doi.org/10.5772/intechopen.92899 (accessed 12 September 2024).

Reyes, N.H. (2024). Artificial intelligence. https://www.massey.ac.nz/study/courses/artificial-intelligence-159302/ (accessed 13 December 2024).

SMCworld (2024). SMC Support Guide for Users–Europe & Africa. https://www.smcworld.com/upfiles/etc/network/Euroen.pdf (accessed 12 September 2024).

Smith, S.W. (2023). The scientist and engineer's guide to digital signal processing. https://www.dspguide.com/ch28/3.htm (accessed 12 September 2024).

Soreling, T. (2005). New interface for rapid feedback control on ABB-robots. http://www.ep.liu.se/exjobb/ikp/mtk/2005/2210 (accessed 12 September 2024).

Swarthmore (2024). Rules for drawing bode diagrams. https://lpsa.swarthmore.edu/Bode/BodeReviewRules.html (accessed 12 September 2024).

Toochinda, V. (2011). Digital PID controllers. https://www.scilab.org/discrete-time-pid-controller-implementation (accessed 12 September 2024).

Trainings (2024). PLC applications in various sectors. https://trainings.internshala.com/blog/plc-applications (accessed 12 September 2024).

Trumper, D.L. (2003). Second order step response. http://web.mit.edu/2.737/www/extra_files/unused%20files/trans.pdf (accessed 12 September 2024).

Wiki (2024). Control theory. https://en.wikipedia.org/wiki/Control_theory (accessed 12 September 2024).

Wolframalpha (2024). Bode plots. https://www.wolframalpha.com/input? (accessed 12 September 2024).

9

Digital Twins (DT-I), Digital Triads (DT-II), and Internet of Digital Triads Things (IoDTT)

9.1 Introduction

As discussed in Chapter 2, a mechatronic system is distinguished from traditional systems in individual disciplines by its special characteristics in modularization, integrability, multi-, inter-, cross- , and trans-disciplines, decentralization, adaptability, predictability, resilience, and sustainability.

This brings a significant challenge in designing a mechatronic system due to numerous factors such as: (1) design variables, constraints, and objectives in multiple domains and stages must be taken into consideration simultaneously; (2) system variables and states are changed with respect to time and their changes are in different time scales; (3) the design spaces for mechatronic functional modules (MFMs) in a mechatronic system for given sets of *functional requirements* (FRs) are becoming larger and larger due to ever-increasing numbers of functionalities, advanced technologies, and available off-the-shelf MFMs; (4) more and more mechatronic systems are required to be reconfigurable to adapt themselves to changes and uncertainties in applications.

As shown in Figure 9.1, an engineering design begins with the interpretation of *customers' requirements* (CRs) into FRs, it follows with conceptual and detailed designs of the system, and then designs and implementations of the system components (Wagg et al. 2020). Shetty and Kolk (2010) provided a detailed view of the designing process of a mechatronic system as shown in Figure 9.2. An engineering design can be viewed as *a process of augmenting information and knowledge step by step*. At each step, design concepts should be verified to see if they can be proceeded to the next stage when all of the design constraints are satisfied. From this perspective, an engineering design is *iterative* in nature, since the information at early design phases is insufficient to verify if all the constraints occurring to later phases are satisfied.

It is seen clearly from Figure 9.1 that *verification and validation* (V and V) is essential to move forward a mechatronic design. V and V should be performed virtually as much as possible to (1) reduce verification cost; (2) accelerate a design process; (3) analyze, evaluate, and compare more design alternatives for design optimization; and (4) incorporate knowledge-based engineering (KBE) for sustainable mechatronic designs.

As shown in Figure 9.3, the concept of *digital twins* (DT-I) was introduced to bridge the virtual and physical worlds (PWs). DT-I was initially introduced to correspond a physical entity to its virtual model, so that the former can be modeled, visualized, monitored, and simulated in the virtual world (VW). The virtual and physical entities were interacted using sensors and controllers.

Mechatronics for Complex Products and Systems: Project-Based Design Approaches for Robots, Cyber-Physical Systems, Digital Twins, and Other Emerging Technologies, First Edition. Zhuming Bi.

Companion website: www.wiley.com/go/bi/mechatronics

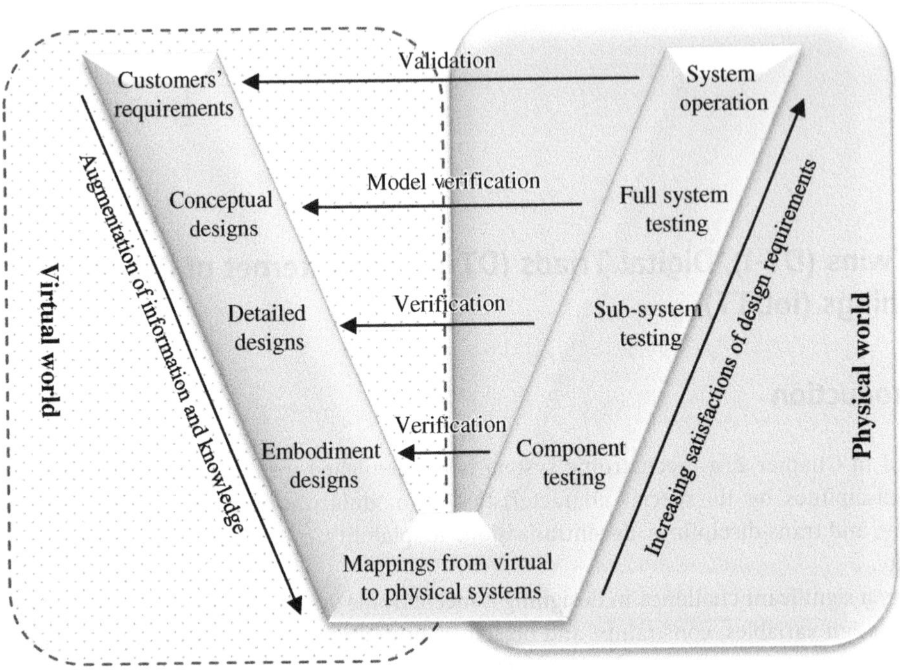

Figure 9.1 Traditional mechatronic design process

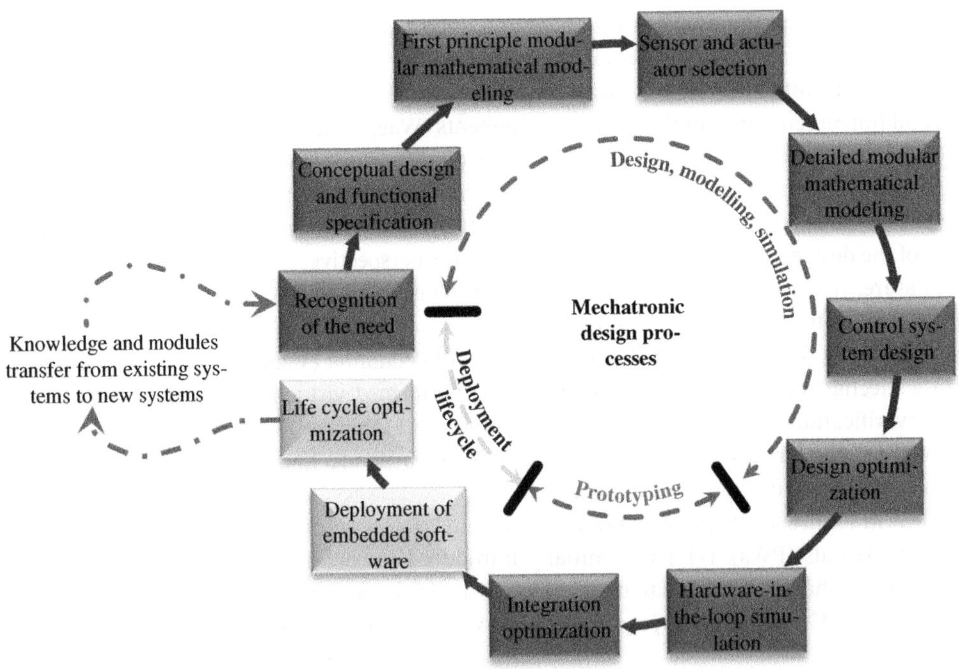

Figure 9.2 Mechatronic design process

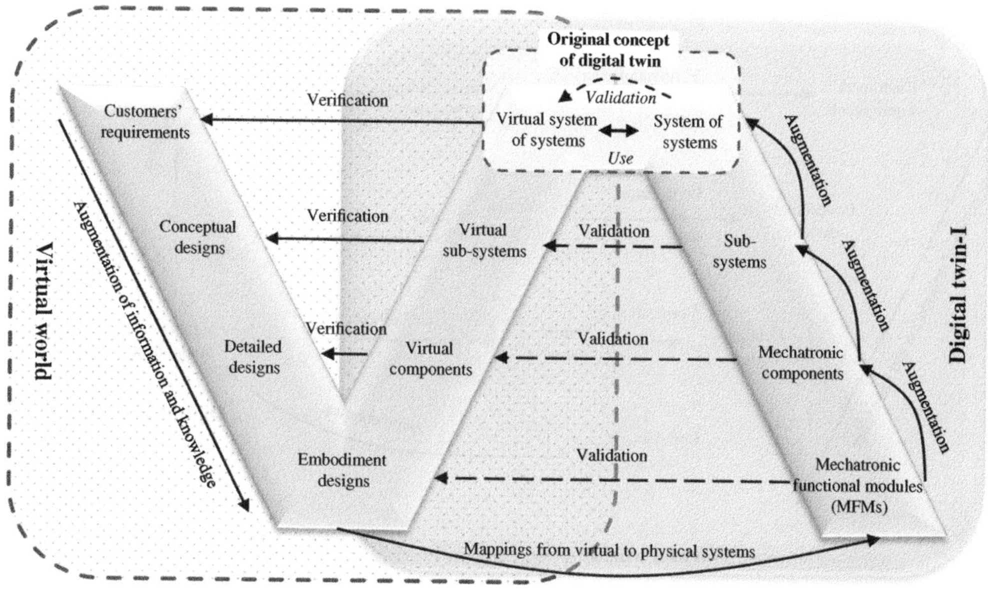

Figure 9.3 Digital twins (DT-I) in the mechatronic design

The concept of DT-I was later expanded so that one physical entity may correspond to multiple virtual entities, and this allowed one to model and analyze the system behaviors at multiple aspects of interests. Moreover, it becomes unnecessary to transform every virtual entity in the VW into its corresponding physical entity in the PW. A virtual entity would not be converted until it is verified against all design constraints and it was proven to be an optimal solution by comparing it with other solutions. This supported the *first-time-right* (FTR) practice from the VW to the PW (Bi and Wang 2020).

It is a continuous challenge to develop more and more complex mechatronic systems with advanced capabilities over time. *Knowledge-based system engineering* (KBSE) should be incorporated to reuse knowledge in all domains and phases to a great extent. As shown in Figure 9.4, KBSE helps to (1) minimize the effort on "reinventing a wheel," (2) reduce non-value-addition activities on repetitive tasks and verification, (3) full use of existing hardware and software resources, and (4) reduce ramp-up to configurate or re-configurate a mechatronic system for a set of specific FRs.

The above discussion shows that DT-I are vital enabling technologies to design and operate sustainable mechatronic systems. This chapter aims to gain a thorough understanding of DT-I-related concepts, existing technologies, the limitations and potential solutions in adopting DT-I for the development of sustainable mechatronic systems. The rest of the chapter is organized as follows: Section 9.2 gives an overview of the development of digital technologies. Section 9.3 discusses the roles of DT-I in mechatronic systems in detail. Section 9.4 identifies the limitations of existing digital technologies in meeting the design needs of advanced mechatronic systems. Sections 9.5 and 9.6 present new concepts of *digital-triads* (DT-II) and *the Internet of digital triad things* (IoDDT) to meet the needs of system adaptability, resilience, and sustainability at the component level and system level, respectively. Section 9.7 summarizes the applications of digital technologies in the sustainable development of mechatronic systems.

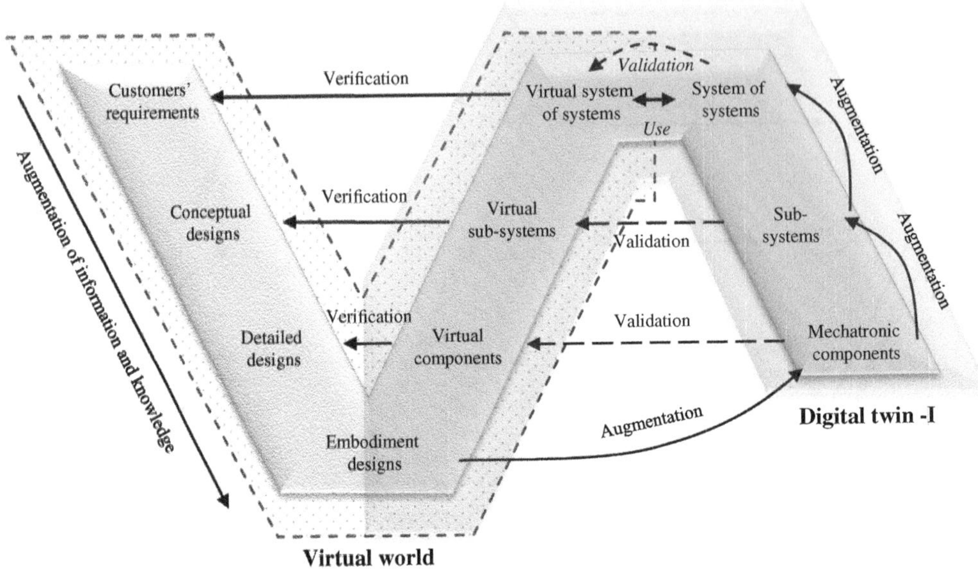

Figure 9.4 Incorporating MBSE and DT-I in the mechatronic design

9.2 Digital Twins (DT-I)

The era of digital transformation is characterized by numerous digital technologies such as Industry 5.0, Big Data Analytics (BDA), IoT, blockchain technologies (BCTs), artificial intelligence (AI), and metaverse. When digital technologies are applied in engineering, DT-Is become essential to represent physical entities in the digital world (DW) (Wagg et al. 2020). The complexity of an engineering process depends on that of the system to be designed. Due to the complexity of a mechatronic system, the N model in Figure 9.3 illustrates the importance of virtual verification and validation (V and V) by DT-Is in designing a mechatronic system. Since the introduction of DT-I in early 2000, the concept of DT-I has evolved over the years, and many similar concepts such as *digital twinning*, *digital counter parts*, *virtual engines*, and *intelligent prognostics* have been developed in different domains. The roles of DT-I-relevant technologies in engineering have been enhanced greatly when DT-Is are incorporated with more and more advanced digital tools such as *computer-aided design and manufacturing* (CADM), AI, communication and network, ontologies, middleware, BDA, BCT, and sensing technologies.

DT-I is essential to evaluate, verify, and validate cyber-physical systems (CPSs) in Industry 4.0 Uhlemann et al. (2017). DT-I allows engineers to optimize and implement corresponding physical systems with a minimized delay and cost. While it is ideal to use DT-I at all phases of a product life cycle, most existing DT-Is are developed and applied at an individual phase. For example, the DT-I by Classens et al. (2021) was used to diagnose faults in a precision mechanism for predictive maintenance. Rosen et al. (2019) argued that DT-Is should be used as value-chain models of next-generation mechatronic systems, and DT-I were used to verify and validate the interactions occurring to all levels and domains (i.e., from lower *connection*, to *conversion*, *cyber*, *cognition*, and to upper *configuration*) of physical mechatronic systems as shown in Figure 9.5. Moreover, the next-generation DT-Is would be expanded for their capabilities in (1) using semantic

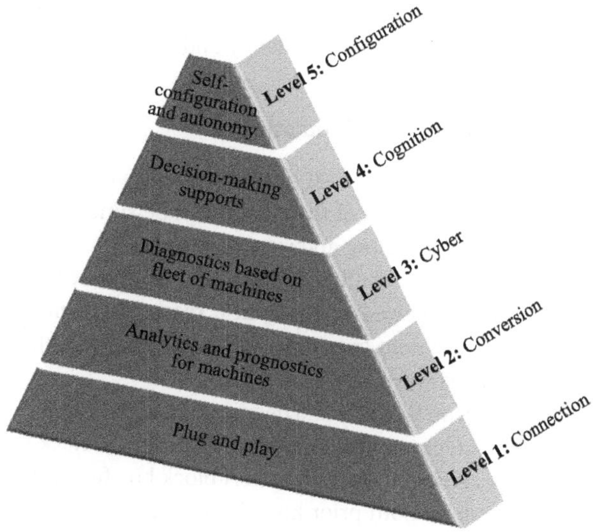

Figure 9.5 DT-Is for V and V of system interactions (5C)

technologies such as ontologies to add new artifacts at any time, (2) extending its coverage to pre-planning of required contexts, (3) closing the feedback loops of the development in early life cycle phases, (4) synchronizing their connection with physical twins in content of big data, and (5) configuring system of systems (SoS) for scalability and adaptability.

9.3 Enabling Technologies

Capabilities of DT-I models depend greatly on that of adopted enabling technologies. Desirable functional requirements (FRs) of a DT-I model include synchronized measurements, accurate predictions, supports for real-time communication, and computing power for analyses and processing of big data. Therefore, enabling technologies for DT-I models can be classified into those for data acquisition, modeling and simulation, communication, artificial intelligence (AI) and computing, and data processing capabilities (Song et al. 2023; Botín-Sanabria et al. 2022).

9.3.1 Data Acquisition

Once a physical twin is made by manufacturing and assembly processes, it sustains its interactions with the corresponding digital twin (DT) by sensing the application environment and providing the DT with live data about system states. The interface of the DT for data acquisition should be stable and the transmission rate should be high enough to synchronize the actions by digital and physical twins. Note that different applications may have different tolerances on transmission latency. *Supervisory control and data acquisition* (SCADA) supports data exchange between a control unit with its physical system. SCADA allows a microprocessor to collect the measurement data from a remote unit remotely. However, SCADA has a low rate of data updating. In comparison with SCADA, a *phase measurement unit* (PMU) is a multifunctional signal acquisition technique with a higher rate for updating data; PMU can be used to control and monitor a complex physical system in a real-time mode. In addition, SCADA and PMU can be combined in a DT-I model when digital and physical twins are interacted at different domains and levels with different requirements of updating rates of data exchanges. Slow tasks in order of *minutes* are communicated by SCADA, and the tasks that require lower latency and faster executions are communicated by PMU.

9.3.2 Modeling and Simulation

Modeling and simulation (M and S) creates a virtual model of a physical system, so that the behaviors and performance of the system can be simulated, analyzed, and predicted. Even though the techniques for modeling and simulation have been studied in multiple disciplines for a long time, there still are many challenges in modeling large-scale, closely coupled, and dynamic mechatronic systems with expected accuracy at a reasonable computing cost. For example, to cope with the change of the system over time, virtual models must be updatable to remove and add new mechatronic models when the tasks vary.

Depending on the level and degree of understanding of a physical system, its virtual model can be one of three types, that is, *white box*, *gray box*, or *black box*. A white box model fully relies on prior knowledge of the physical system. A block box fully relies on the abundant data collected from the physical system. Both prior knowledge and collected data are used to create a gray box model.

Note that a mechatronic system consists of MFMs at different levels; a mechatronic system can be modeled as a *system of systems* (SoS). There are challenges to model various interactions of sub-systems. While modeling the interactions among sub-systems requires reasonable simplifications and assumptions, over-simplification or inappropriate assumption may lead to deteriorated model behaviors even though respective sub-systems are modeled satisfactorily. In addition, there are the challenges to manage scales of system models: when a system is scaled-up, the numbers and types of system components and their interactions increase, and the demand for computational efforts will be increased exponentially. The techniques for model reduction should be advanced to simulate complex DT-Is with satisfactory speed and fidelity.

9.3.3 Communication Technologies

To synchronize the interactions of digital and physical twins, communication technologies must ensure global data access to the right place at the right time. Since the amount of data collected from a physical system keeps increasing, there are challenges in sustaining sufficient capacities and fast transmission rates of mechatronic systems locally and globally. The demand for a high bandwidth and reliable communication with minimized latency is identified as a bottleneck in implementing large-scale DT-Is. Wireless communications can support cloud-platforms and overcome geographical limits. Nowadays, the *fifth-generation* (5G) mobile network can fully support the integration of data-intensive *machine learning* (ML) applications in DT-Is; the delay by 5G can be reduced to milliseconds with a bandwidth of up to 20 gigabytes (GB)/second. Moreover, the sixth generation (6G) will further expand communication capabilities to support the interactions of decentralized DT-Is with a bandwidth of up to 1 terabytes (TB)/second.

9.3.4 Cloud Technologies

A mechatronic system can be built from a set of distributed and modular building blocks. The system is scalable by modifying, adding, or upgrading constitutive building blocks over time (Bradley et al. 2018). Since most of the building blocks, such as sensors, actuators, and controllers, are smart things with local autonomous decision-making capabilities, combining *cloud computing* (CC) and *edge computing* (EC) provides an ultimate solution (see Figure 9.6).

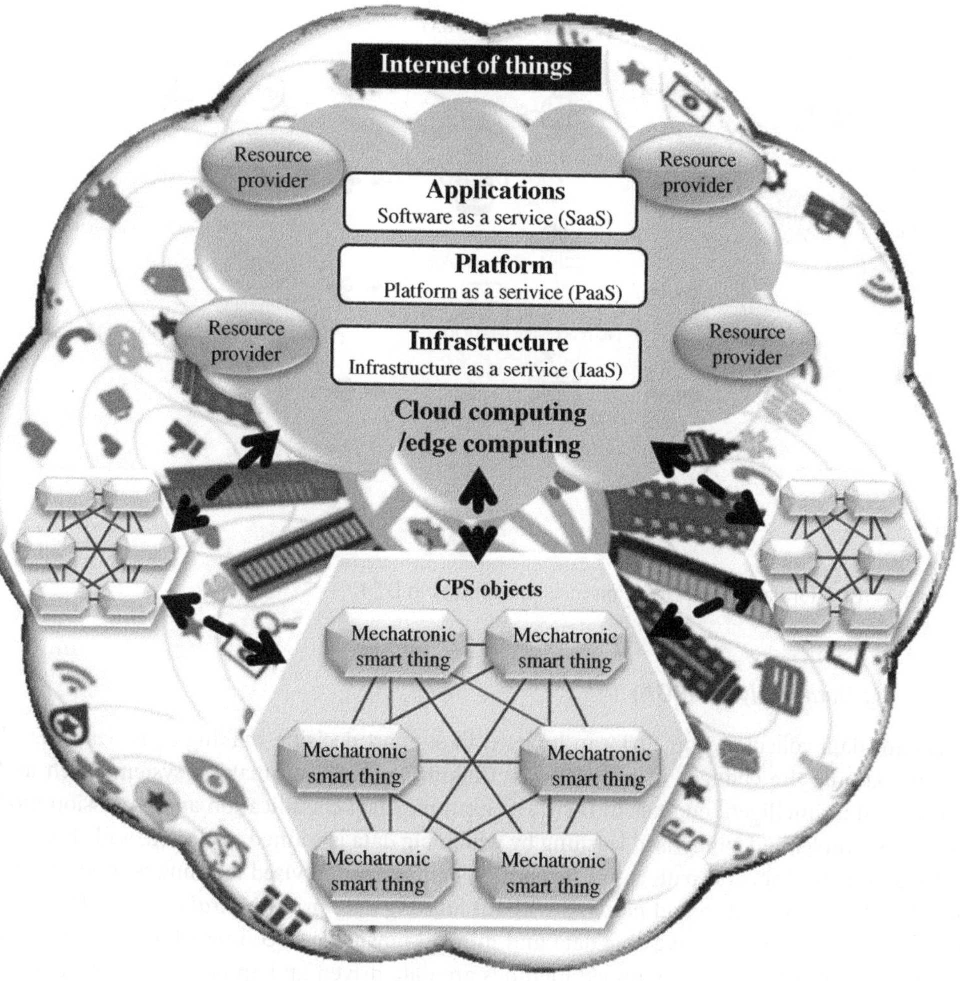

Figure 9.6 From mechatronics to the cloud

CC offers computing resources and services to smart things with limited hardware and software locally. The techniques for smart things to access cloud resources include *infrastructure-as-a-service* (IaaS), *platform-as-a-service* (PaaS), *software-as-a-service* (SaaS), and *EC*. IaaS is a CC model to provide computing power and storage as online services by networking resources with the Internet on an on-demand basis. PaaS is a CC model for a third-party service provider to provide hardware and software tools to users; the PaaS provider hosts hardware and software tools on its own infrastructures. SaaS is an application hosted on the cloud; it is accessible by users via a web browser, mobile app, or a thin client service.

CC provides a cost-effective solution through on-demand use or outsourcing of hardware and software in a centralized resource center. In contrast to CC, EC is the paradigm of distributed computing, and the computing and data storage are brought closer to end users to reduce the latency in time-critical processes such as local onsite decisions relevant to physical assets. In EC, bandwidth can be saved by pre-processing data before sending it to a cloud server; however, having more devices on the edge of network brings additional new security concerns.

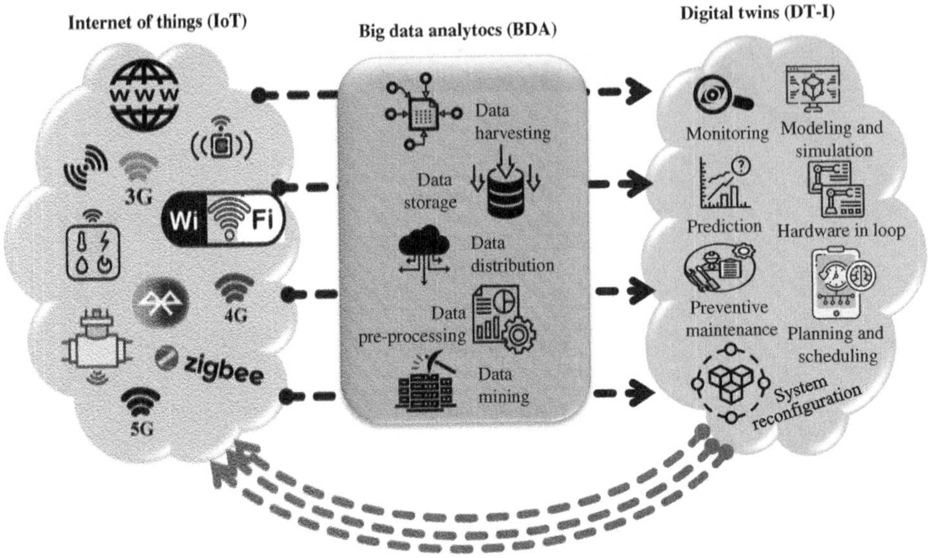

Figure 9.7 BDA for data-driven decision-making supports in DT-I

9.3.5 *Big Data Analytics* (BDA)

Abundant data collected from IoT can be used to support decision-making activities in DT-I in accommodating the changes in applications. Data-driven decision-making systems such as ML can be used as intelligent engines in DT-I. For example, data classification and regression models are used to implement supervised learning by which big data is trained, collected, and classified to understand behaviors and patterns of physical systems. Unsupervised learning is used to classify and identify data structure and patterns without *labeling* data prior. *Neural networks* (NN)-based deep learning is used to analyze big data and mined knowledge of complex DT-I in a reasonable time frame. All of the aforementioned methods are data driven and applicable to both black-box and gray-box models. Figure 9.7 shows the irreplaceable roles of BDA in using big data from IoT to implement data-driven decision-making supports in DTs (Botín-Sanabria et al. 2022).

9.4 From Digital to Physical Twins by Manufacturing

It is manufacturing to bridge the VW with the PW. Manufacturing is the most critical to implement *first-time-right* (FTR) practice of the conversion from a DT-I to its physical twin (Bi and Zhang 2021). On the other hand, a manufacturing system can benefit greatly from enabling technologies of DT-I due to the needs to deal with ever-increasing complexity and dynamics of a system.

A manufacturing system is a unified entity that is formed by a group of interacting and interrelated manufacturing assets to achieve its business goals in an enterprise. Due to the multidisciplinary nature of manufacturing activities, more and more mechatronic systems such as industrial robots, *computer numerical controls* (CNCs), machining centers, and *autonomous guided vehicles* (AGVs) are adopted to enhance manufacturing capabilities. In contrast to traditional systems, mechatronic systems have advanced functions and are able to cope with the high system complexity effectively. Note that the system complexity depends on four main factors, that is, (1) number and types of system elements, (2) number and types of the interactions of elements, (3) uncertainties and dynamics of surrounding environments, and (4) the changes of system states over time. Accordingly, the complexity of a manufacturing system has increased exponentially due to the increases of diversified products and applicable processes, and extensions of products' life cycles and operations in multiple disciplines.

More and more technologies advanced in multiple disciplines including Material Science, Manufacturing Engineering, Industrial Engineering, Management and Decisions, and Information and Communication Technologies (ICTs) have been adopted in mechatronic solutions to empower manufacturing systems. For example, *virtual manufacturing* (VM) allows an enterprise to access virtual assets over the Internet to expand system's capabilities through the Internet. On the other hand, adopting more multidisciplinary technologies complicates a manufacturing system at multiple dimensions since (1) both internal and external manufacturing resources are accessible, and a system becomes vulnerable to external attacks and privacy infringement; (2) the boundary of a manufacturing system with its application environment becomes more and more ambitious, and the enterprise faces fierce competitions with global competitors; (3) products' users are gaining more privileges for increased expectations on personalization, attractiveness, and environmental friendliness other than conventional demands such as functions, quality, costs, and lead times (Bi and Zhang 2021).

In designing, implementing, and operating a manufacturing system, the complexity must be manageable. Managing the system complexity poses an increasing challenge since more and more system elements are introduced to perform manufacturing businesses. Traditionally, *enterprise architecture* (EA) was introduced to assess the effectiveness of technological solutions in dealing with the system complexity. In an EA, system complexity is modeled and measured by (1) classifying and structuring system assets; (2) regularizing their interactions; and (3) making rules, policies, and guides for updating systems in their life cycles in the time domain.

In the digital era, newly developed technologies in sensing, networking, AI, CC, and BDA become available for enterprises to design, analyze, plan, schedule, and manage manufacturing resources and businesses effectively, and DT-I becomes one of the most effective digitizing tools to model constitutive elements, events, and relations of physical systems. However, EAs of most DT-Is are still structurally layered, gridded, or hieratical; such EAs actually affect system flexibility, adaptability, and resilience.

In summary, in designing, implementing, and using DT-Is, conventional EAs are exposed to some limitations such as limited capabilities in managing complexity, a high number of iterations for fixing defects exposed in later design phases, high development costs, and extended lead times. In the following sections, the concept of DT-I is re-examined from the perspective of the need to sustain flexibility and adaptability in a prolonged system life cycle. Its critical roles in mechatronic system developments are discussed to identify the limitations of DT-Is. To overcome the identified limitations of DT-Is, the new concepts of *DT-II* and *IoDDT* are introduced to represent sustainable mechatronic systems.

9.5 DT-Is in Manufacturing

A manufacturing system transforms raw materials or parts into finished products by a series of manufacturing processes and associated activities. The evolution of manufacturing systems can be examined by many aspects such as the changes of inputs, outputs, system elements, and available enabling technologies. In this section, the changes of system boundaries, business scopes, complexity and scales, and enabling technologies are discussed.

Traditionally, a manufacturing system was closed with a clear system boundary of internal manufacturing resources and external business environment. Nowadays, a manufacturing system become open with an ambiguous boundary since virtual manufacturing (VM) assets are accessible and more and more internal operations are closely interacted with external businesses. In addition, both the scale and the complexity of the system keep continuously increasing with the increases of functionalities and varieties of products, varieties and capabilities of machines and processes, and customers' expectations on products. Bi et al. (2008) and Bi (2011) summarized the evolution of system paradigms along with the increase of system complexity over time. Historically, the main system paradigms include *craft production*, *English system*, *mass production*, *lean production*, *mass customization*, and *sustainable manufacturing*.

As shown in Figure 9.8, in *a craft production* at the earlier time before 1850, manual tools were used to make simple products; the starting materials were directly from nature, and products were crude with basic functions. *English system* was introduced in the 1770s when James Watt industrialized steam engines with separate condensers, manufacturing processes were mechanized, power sources became essential to operate machines, and system performances were mainly valued by profits. *Mass production* was introduced in the 1800s when identical parts could be made in a production line, and they were exchangeable in assembled products. System inputs were expanded to include parts and components, and system performances emphasized on quality of products and productivity. Mass production led to the saturation of global manufacturing capacities for customers' needs, *lean production* brought a new way to increase system competitiveness by eliminating wastes of no-value-added activities, and the leanness was introduced to measure system performance. *Flexible manufacturing* was developed to make different products from the same system, system performances were evaluated by the effectiveness and responsiveness in producing personalized products. The trend of the environmental degradation has brought people's awareness of the sustainability of products and manufacturing processes; *sustainable manufacturing* was to optimize a manufacturing system from the perspective of system sustainability such as carbon

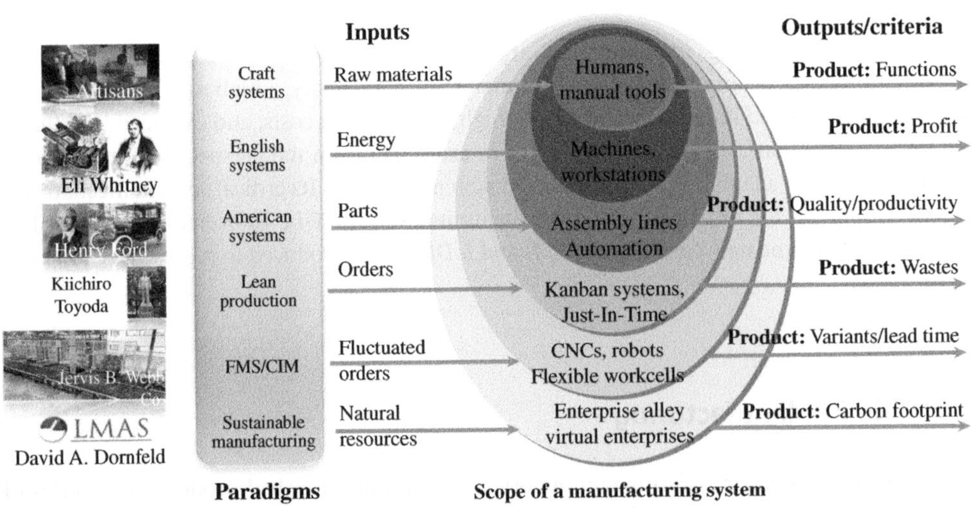

Figure 9.8 Increasing scales and complexity of manufacturing systems (Bi et al. 2008; Bi 2011)

footprints and consumption of natural resources. As a summary, with the evolution of manufacturing paradigms, the scales and complexity of manufacturing systems have been continuously increasing due to the increases of system elements, the interactions of system elements, and the changes and dynamic characteristics of environmental factors over time.

System paradigms were enabled by available technologies, and technology developments were fundamental drivers of advanced system paradigms (Jiang et al. 2014). A new manufacturing paradigm became practical when corresponding manufacturing technologies were applicable. As shown in Figure 9.9, manufacturing processes were automated when sensors and actuators were used in closed-loop controls. Newly developed information technologies (ITs) such as BCT, CPSs, *industrial IoT* (IIoT), and *human CPSs* (HCPSs) made digital manufacturing (DM) paradigm implementable (Bi and Zhang 2021). On the other hand, the demands for green, miniaturized, highly valued products have stimulated the studies in many emerging fields such as *nanoscience*, *environmental science*, *BDA*, *ML*, and *cybersecurity*.

A manufacturing system consists of *a material flow* and *an information flow*. In spite of the fact that final products are made from raw materials through a series of manufacturing processes in the material flow of *the physical world* (PW), advanced manufacturing technologies in Figure 9.9 are mostly applied in business activities at different phases and domains of the information flow in the *digital world* (DW). Figure 9.10 shows that manufacturing businesses can be classified into three groups based on the levels and phases of system life cycle where these manufacturing businesses are performed; a manufacturing business falls into one of the three groups: *digital world I* (DW-I), *digital world II* (DW-II), and PW. Due to ever-increasing system complexity, computer-aided techniques become more and more important to expand system capacities and improve the productivity and efficiency of manufacturing systems.

Using advanced manufacturing technologies automates (1) value-added processes in the material flow and (2) decision-making processes in the information flow, and the impact of automation on the performance of manufacturing systems was investigated by numerous researchers. Figure 9.10 shows the importance of automation from the perspective of the interactions between information

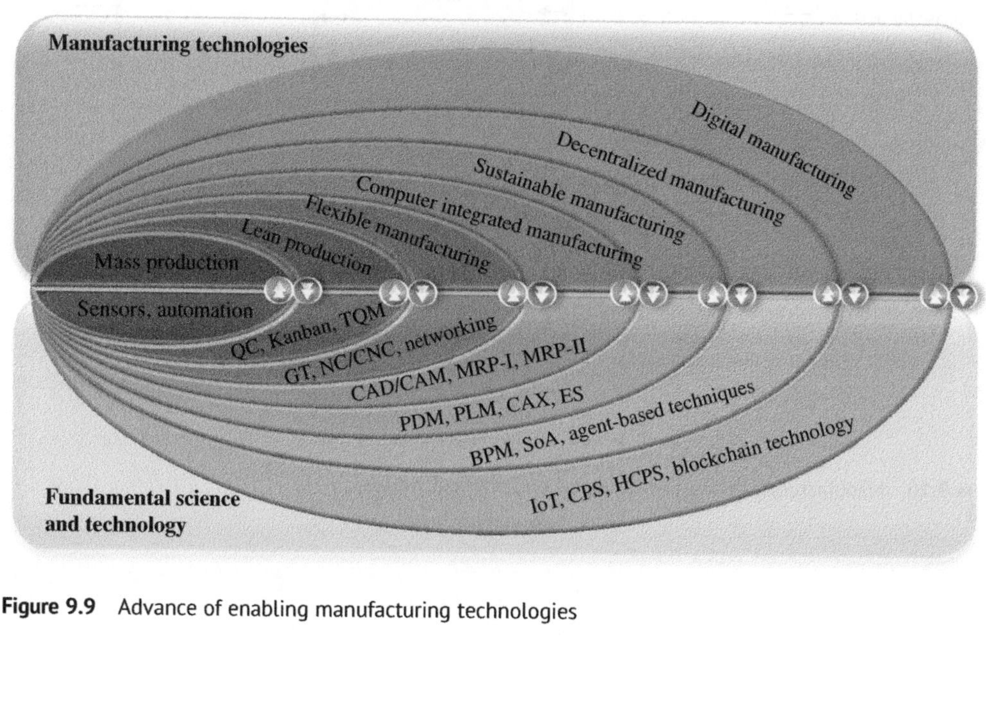

Figure 9.9 Advance of enabling manufacturing technologies

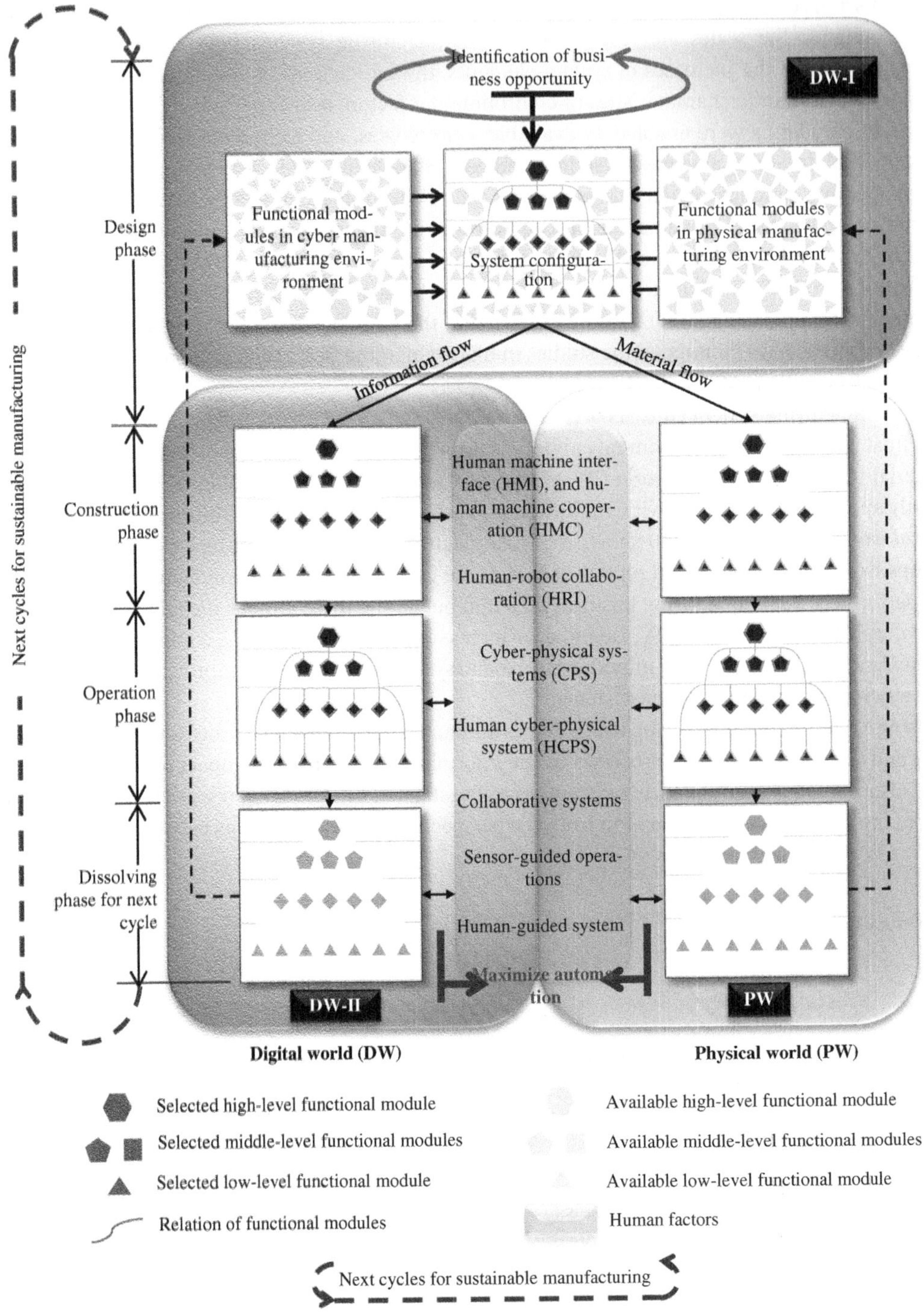

Figure 9.10 Manufacturing businesses in digital and physical worlds

and material flows. In the material flow, computerized devices such as *computer numerical controls* (CNCs) and *automated guided vehicles* (AGVs) were used to replace humans in various manufacturing processes. In the information flow, software tools were used to make decisions for plan, schedule, and control of manufacturing businesses at all phases and domains of system life cycle. While humans played their critical roles in achieving adaptability and agility of systems, humans must collaborate with computers seamlessly to deal with high-level complexity of manufacturing systems.

In the next section, system elements are classified based on their functions for manufacturing businesses in DW-I, DW-II, PW, and the interactions between two, existing EA concepts for the representations of system elements, digital technologies, and their relations are discussed to identify their limitations.

9.5.1 System Digitization

Figure 9.8 shows that system paradigms were evolved to deal with ever-increasing business scopes and complexities of manufacturing systems, Figure 9.9 shows that certain system paradigms became applicable when corresponding enabling technologies, and Figure 9.10 shows that enabling technologies allow more and more automations for (1) manufacturing processes in the PW and (2) decision-making supports at all domains and levels of businesses in the DW; moreover, system elements in physical and DWs must be interacted to smoothen materials and information flows in a unified manufacturing system. Therefore, a system paradigm can be characterized by the types of digitalized technologies as well as the interactions among system elements in the PW and DW.

The trend of digitizing manufacturing technologies was discussed by numerous researchers. As shown in Figure 9.11, *in the 1960s*, the digitization of manufacturing technologies begun when the first generation of computer-aided software tools was used to model parts, and more and more

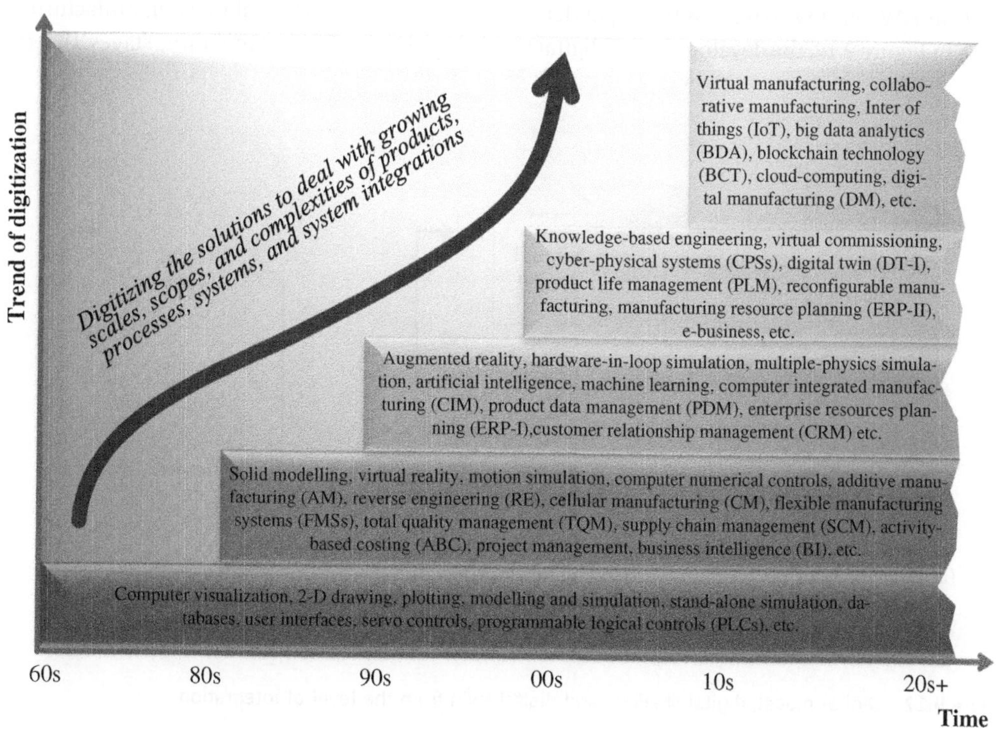

Figure 9.11 Trend of digitization in manufacturing

manual operations were mechanized in mass production. *In 1980s,* with the advance of computer hardware, solid modeling, CNCs, and reverse engineering (RE) were rapidly developed, and the digitized technologies such as cellular manufacturing (CM), flexible manufacturing systems (FMSs), total quality management (TQM), supply chain management (SCM), activity-based costing (ABC), and business intelligence (BI) were used to support decision-making activities at the levels of machine, work-cell, and the system. *In the 1990s*, more digital models were developed for designing, planning, scheduling, and controlling manufacturing businesses at the system level; typical digital tools developed in the 1990s were augmented reality (AR), hardware-in-loop simulation, multiple-physics simulation, AI, ML, computer integrated manufacturing (CIM), product data management (PDM), enterprise resources planning-I (ERP-I), and customer relationship management (CRM). *In the 2000s*, the newly developed digitized technologies were characterized by their capabilities in dealing with environmental changes and dynamic interactions; typical technologies were KBE, virtual commissioning, CPS, DT-Is, product life management (PLM), reconfigurable manufacturing systems (RMSs), manufacturing resource planning (ERP-II), and E-business. *In the 2010s and beyond*, digital technologies were advanced to achieve sustainability, resilience, and adaptability of manufacturing systems in the digital economy. Since the Internet can network everything and decision-making activities at system level were facing the challenge of big data in terms of volume, velocity, and variety (3V), the prevalent technologies included IoT, BDA, BCT, and CC.

9.5.2 Interactions of Physical and Digital Worlds

A manufacturing system is involved with numerous activities at different domains and levels of manufacturing businesses in both materials and information flows, and system elements must be integrated and interacted to achieve system-level goals. Therefore, enabling technologies are needed to support the interactions of system elements, in particular, the interactions of system elements in DW and PW. Based on the dependences of computer models on physical manufacturing assets in Figure 9.12, the development of digital technologies was classified into three stages, that is,

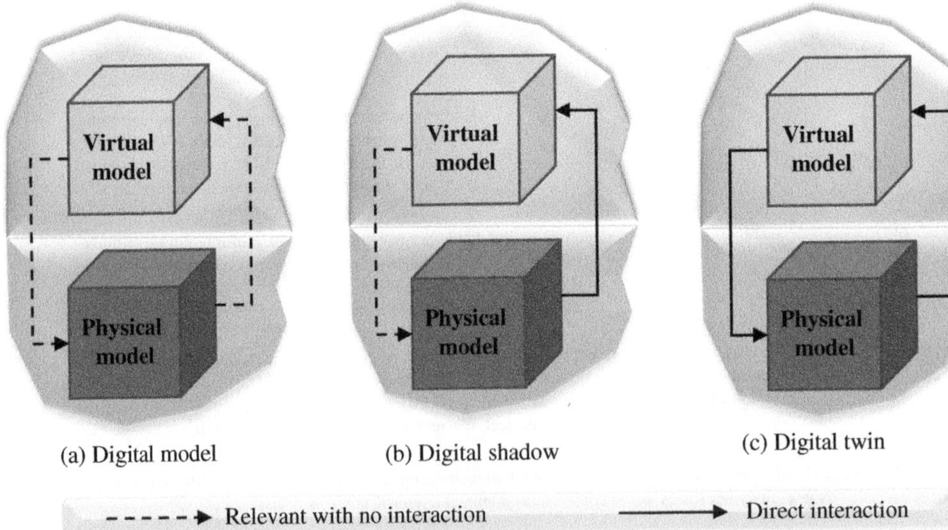

Figure 9.12 Digital model, digital shadow, and digital twin from the level of integration

the first stage for *digital models* that had no direct interaction with corresponding physical objects, the second stage for *digital shadows* that received unidirectional feedbacks from physical objects, and the third stage for *digital twins* that supported bidirectional interactions with corresponding physical objects. Similar to digital model and digital shadow, a DT was a part of digital transformation; it supported the integration of smart things and synchronization of physical and DT in shop floors for high productivity and adaptability of manufacturing systems.

In the digital era, the digitization of physical manufacturing assets such as products, machines, processes, and accessory tools becomes more and more prevalent, and the digitization of decision-making supports becomes ever critical at high level of business management. This trend of digitization is clearly evidenced by the rapid growths of (1) computer models and tools of factories, processes, peoples, and other smart things for physical counterparts and (2) bitcoins, virtual businesses, virtual enterprises, cloud services, digital economy, Blockchain-based applications. ITs such as CPSs have made it possible to implement the concepts of DT. CPSs are most critical to DT. Lu et al. (2020) surveyed the impact of digitization on manufacturing in terms of applications, connotations, reference models, and technical challenges. Due to their importance, many developed and developing countries have developed their strategical national programs with relevant DM technologies such as *Made in China 2025* by China, *Make in India* in India, *Advanced Manufacturing Partnership* in the United States, *Industrie 4.0* in Germany, and *Horizon 2020* in the European Commission.

9.5.3 Historical Development of DT-I

It has been a long history to use a mockup model or a high-fidelity computer model to support machine operations. The concept of DT-I was introduced in 1998 based on the idea of its digital counterpart by Alan Alda; DT-I was used to predict fleet maintenance. The earliest attempt was tracked back to the pairing approach proposed by NASA to reduce potential damages of the Apollo 13 spacecraft in its space exploration in 1970; engineers were able to simulate potential strategies in a mocked module before the mission of the spacecraft was actually executed; the pairing approach was advanced to use digital models that became critical to manage any space missions in the US space agency.

The concept of DT-I was inspired by the successful applications of digital models in aerospace and defense sectors. Negri et al. (2017) discussed the origin, history, and definitions of DT-I and found that there was no definition of DT-I that was unanimously accepted by researchers in the community. Existing definitions of DT were general to cover various systems with highly diversified system behaviors and properties. Two of the most cited ones are: (1) a DT was a mirrored model of a specific physical twin, and it was an integration of a high-fidelity simulation with an on-board monitoring system for health management, maintenance, and records of historical data. (2) A DT was a multi-physical and multiscale model to simulate a complex product or system using sensed data and physical models. Dynamic systems were modeled by differential equations in two-time scales; a low time was used to separate the evolution of the system properties from that of the instantaneous times. By all means, the core idea of DT-I remains the same in many definitions, that is, a DT refers to a digital model for synthesizing, real-time and multiscale monitoring, and inspecting of physical models, and it is mainly used to improve the resilience of artifact systems in their life cycle.

In numerous recent publications, the concept of DT-I becomes very comprehensive, and it is closely associated with BDA, DM, CPSs, and Industry 4.0. Qi and Tao (2018) related DT-I to big data, and they discussed their relations on the differences and similarities from the perspective of data in information flow. They found that big data and DT-I were complementary and should be integrated into DM. Note that DM was a concept for comprehensive manufacturing innovation in which various ITs were adopted and integrated to make better products with more functions and better quality in shortened times and at reduced costs. The driving forces, enabling technologies, and applications of DM were discussed extensively. DM represented an integrated suite of PLM tools, especially DM tools, for designs of products, processes, layouts, and systems.

DT-I was a replica of CPS for the integration of cyber and physical systems. Barnabas and Raj (2020) argued that the concept of CPS was inspired by human systems that consisted of smart things with networked sensors, processors, and actuators and interacted with PWs seamlessly; while DT-I was a collection of physics-based methods and analytical tools to represent present states of assets in DM. Therefore, DT-I should include the models for all aspects of physical assets including mechanical, thermal, electrical, fluid dynamics, chemical statistics, and economic aspects, DT-I used analytical tools and sensed data to predicate system outcomes in terms of flexibility, adaptability, maintainability, reliability, and availability.

DT-I is a comprehensive concept which closely relates to many DT-I. For example, Bi and Zhang (2021) discussed some main enablers of DT-I, the dependencies of other digital technologies, and potential applications. From the perspective of applications, DT can be implemented at different layers and domains, respectively, and DT were developed to model information exchanges across value chain, surveillance, real-time simulation, efficient commissioning, planning and scheduling systems, and machining systems (see Figure 9.13). Xiang et al. (2019) used DT-I in sustainable manufacturing to optimize the selection of materials to minimize environmental impacts; high-fidelity digital models were developed, and the data from both cyber and physical models was collected and fused to assess the environmental impacts of different materials. Kunath and Winkler (2018) proposed a DT framework for transparent order managements from placing orders to delivering final products. Sun et al. (2020) used DT for the assembly commissioning of high-precision products to improve assembly efficiency and achieve quality consistency. Tao et al. (2018) used DT-I for prognostics and health management to improve the efficiency and accuracy of complex products in ill-structured environments. Terkaj et al. (2019) proposed a digital platform for the design and operation of roll shop plants; this accelerated to respond the requests for quotations in effective ways by assessing design options using DT. It supported the interoperability of system elements, knowledge

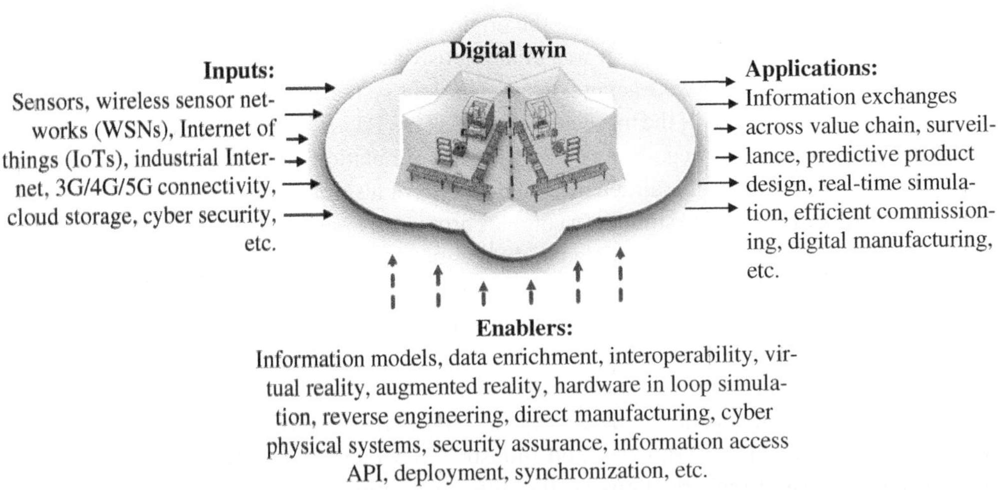

Figure 9.13 Topics relevant to DT-I

reuse by semantic technologies, and the implementation of ontologies based on common data models following technical standards.

DT was multidisciplinary and it involved engineering, information science, data science, controls, and computer sciences; and critical functions of DT were modeling and simulation, data fusion, collaboration, and services. Therefore, for the same physical entities, different DT can be created to add values to products, processes, systems, and organizations based on different measures. For example, the DT models by GE (2024) included *anomaly*, *thermal*, and *transient* DT to add the values by describing, analyzing, diagnosing, and predicting collectively.

DT-I has brought huge economic potential to enterprises since it allowed plugging and monitoring manufacturing resources at high level without the high degree of integration; however, vertical networking itself was insufficient to increase productivity effectively; the next step was to promote horizontal integration along value chains (Curry et al. 2022). D-I has attracted a great deal of attention for its applications in manufacturing, IIoT, healthcare, smart cities, automobiles, and retails. DT technology was ranked as one of the top 10 strategical technologies (AI, intelligent apps and analytics, intelligent things, DT, cloud to the edge, conversational platform, immersive experience, Blockchain, event-driven model, and continuous adaptive risk and trust) (Augustine et al. 2022). The relevant market size was estimated to be $6 billion and grow at 40% compound annual growth rate (CAGR) to $16 billion in all industry sectors by 2023, and the digital technologies were expected to reduce 50% of product development costs in 2015–2020 (Deuter and Pethig 2019; World Bank Group 2024). DT-I revolutionized the manufacturing industry in utilizing assets, reducing downtimes of equipment and production costs.

9.5.4 Communication and Integration

Few definitions of DT-I have clarified the interactions of its system elements but the emphasis is on the interactions of digital and physical twins. Here, a system element in DT-I can either in the digital or PW; an interaction in DT-I can be (1) *internal* in either of DW or PW or (2) *external* between the DW and PW.

DT-I can be classified into four groups in terms of types of system elements and interactions in Figure 9.14. (1) *One-to-one interaction*: only one digital model and one physical system are involved, and DT-I specifically referred to as cyber-physical integration; it fits well with the origin of DT concept. (2) *One-to-many interactions*: the PW is involved with multiple physical entities that interact with a single model in the DW directly. The relations of physical entities are modeled in the digital model rather than their direct interoperations in the PW. An example of one-to-many case is to use DT-I for planning and scheduling a work cell. (3) *Many-to-one interaction*: the PW is involved with one physical system that consists of interacted physical objects; while the digital model consists of multiple DT that represent the behaviors of the physical models at various aspects. Figure 9.15 shows an example of many-to-one case where one physical machining system corresponded to multiple digital models in the DW (Bi and Kang 2010; Bi et al. 2007a, 2007b). Moving bodies were sensed and real-time data was collected from the physical system and sent to the corresponding digital models for visualization, diagnosis, and closed-loop controls. (4) *Many-to-many interactions*: both the PW and DW involve in many entities; however, the entities in two worlds are heterogeneous in organization and an interaction can be internal in either of the PW or DW or external that occurs to two entities in different worlds. The GE DT model was a good example of DT-I with many-to-many interactions (GE 2021).

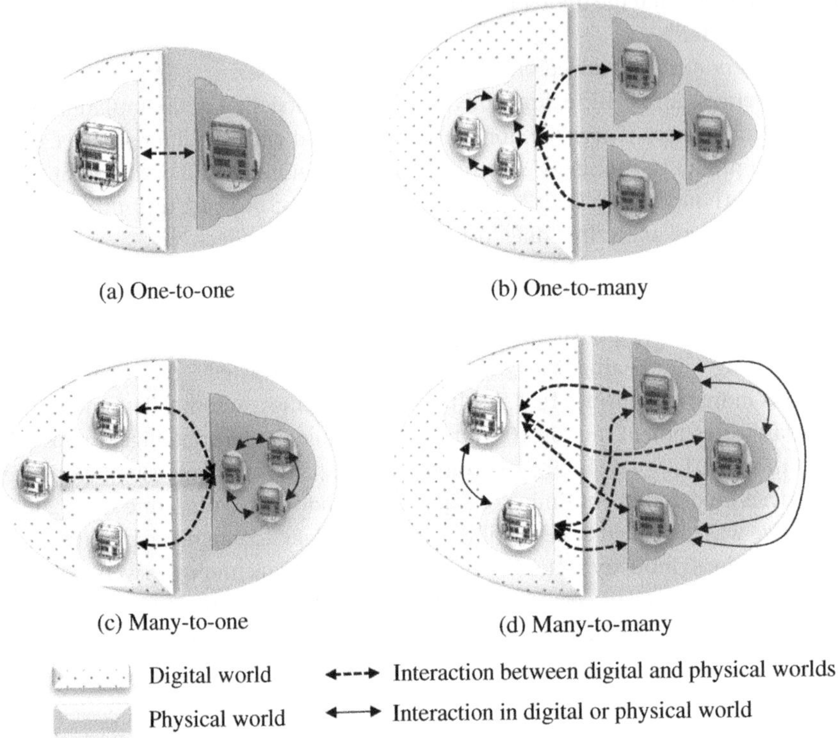

Figure 9.14 Interactions of digital and physical worlds

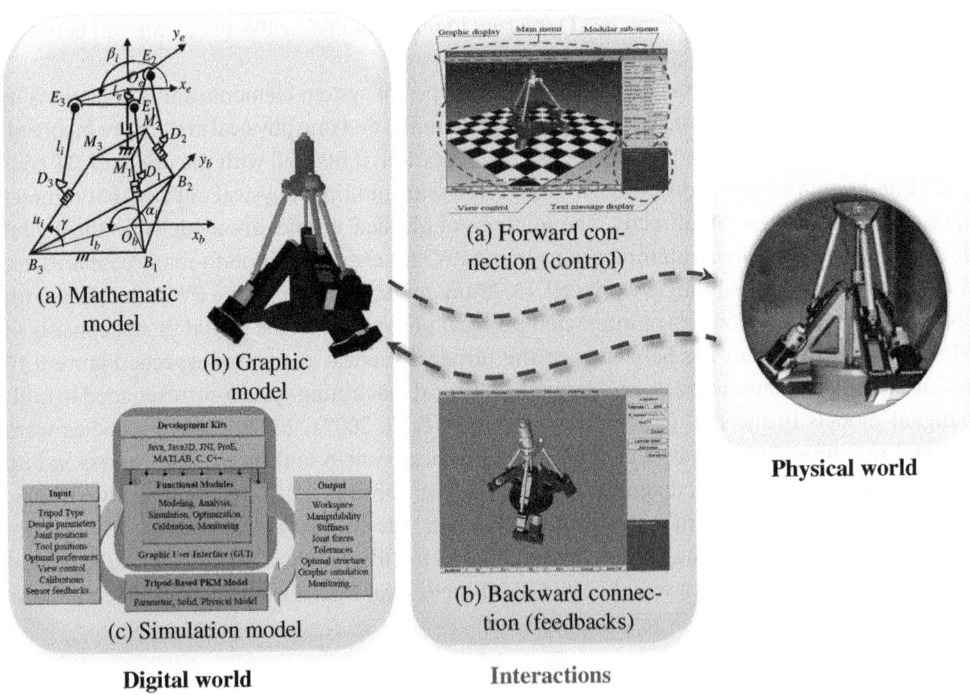

Figure 9.15 Example of many-to-one interaction (Bi et al. 2007a; Bi and Kang 2010)

9.5.5 System Architecture

An interaction occurs between two system elements; therefore, the number and types of interactions in DM depend on the organization of system elements. As an information system, many researchers used traditional hierarchal architecture to model the interactions in the DT of DT-I. Taking an example of Figure 9.16, *a five-layer architecture* was used (AMRC 2018; Altran 2019): the real-time data was collected from the application environment at *the device layer*; the data is cleaned and fused as useful information at *the network layer*, the information is further consolidated into consistent model at *the cyber layer*, the information model is used to make intelligent decisions based on dynamic data at *the cognition layer*, a number of digital models are selected, configured, and coordinated as a coherent system for optimized system performances at *the configuration layer.*

Manufacturing systems become more and more complex, and manufacturing assets tend to be distributed and decentralized. Therefore, some special functional units are needed to select suitable resources and compose them as workflows for specific enterprise missions (Viriyasitavat et al. 2019a, 2019b, 2019c, 2020a, 2020b). Tao et al. (2018a, 2018b) distinguished such functional units from digital and physical objects as *services* in a five-dimension DT model in Figure 9.17a, and the interactions were defined for the paired communications among *physical models*, *digital models*, and *services*. Nowadays, few manufacturing systems are fully automated, and humans still play irreplaceable roles to achieve high-level agility, flexibility, and adaptability in cost-effective ways. To model humans' roles in DT-I, Zhou et al. (2018, 2019) modeled humans in DT-I as shown in Figure 9.17b, and the interactions were classified into three categories, that is, *cyber–physical interactions*, *human–physical interactions*, and *human–cyber interactions*.

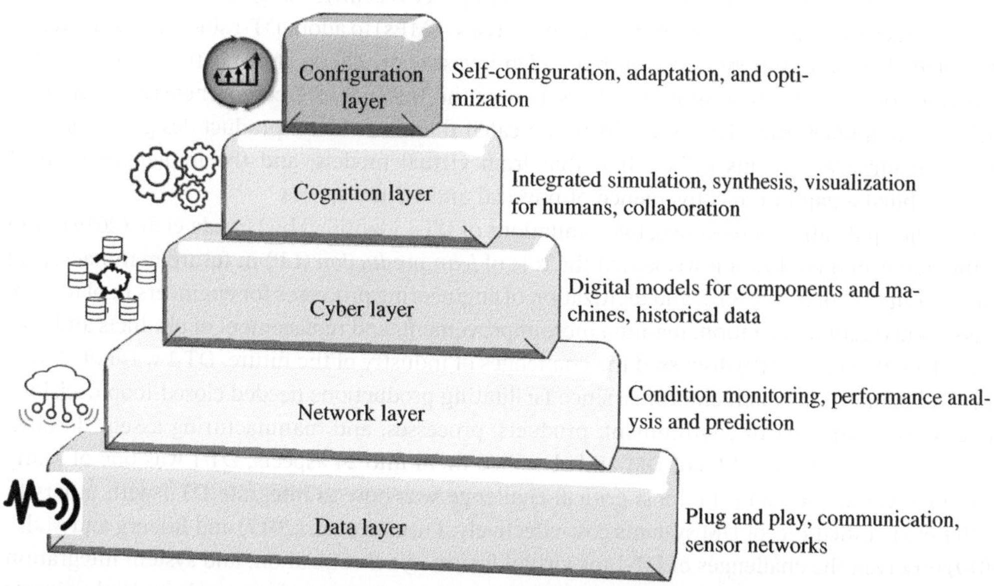

Figure 9.16 Traditional five-layer architecture of DT-I (AMRC 2018; Altran 2019)

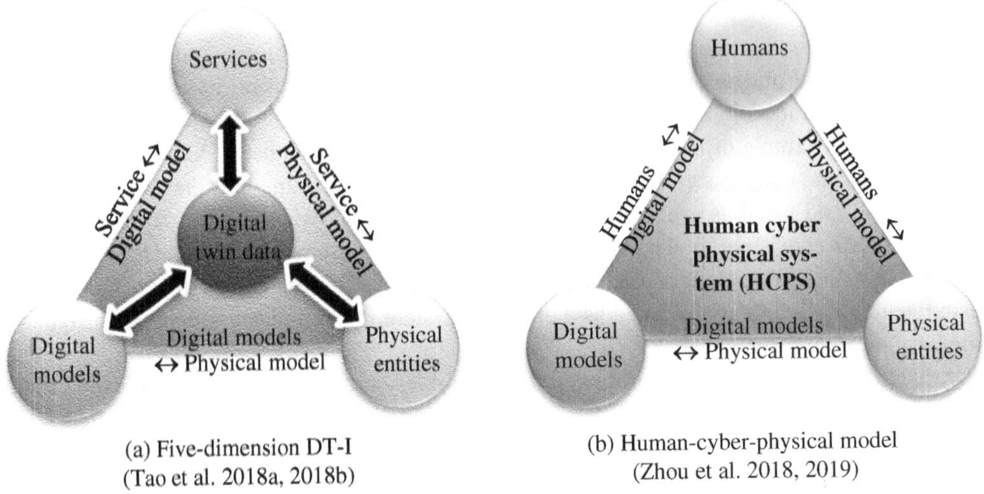

Figure 9.17 Two variants of DT-I system architecture

9.6 Limitations of DT-Is

Despite popularity of DT-Is, many researchers expressed their concerns on the limitations of DT-I. Wagner et al. (2019) discussed the challenges of adopting DT in designs and productions of high-performance products; they found that relevant technologies were confined within their respective disciplines, and there was a need to line DT with different domains for innovations in complex products. Tilson et al. (2010) argued that enterprise systems (ESs) were moved far beyond administrative and isolated tools and digital infrastructure should categorize all types of information artifacts; however, DT-I was insufficient since it was process centric. Singh et al. (2019) discussed the challenges for small- and medium-sized enterprises (SMEs) to adopt DT-I since it was treated as a high-tech hardware system that required a high investment. Zhang et al. (2019) looked into the limitations of DT-I in integrating digital assets over the Internet as a comprehensive solution to intelligent manufacturing. Tao et al. (2019) indicated that data-driven product design focused on data from physical systems rather than that from virtual models, and the implementation of DT-I exhibited a gap in the convergence of physical and virtual spaces.

From the application perspective, two limitations of DT-I identified by Umeda et al. (2019) were (1) the trends of digitalization weakened the role of *lean production* (LP) in future manufacturing and (2) limited supports of KBE and digitization of engineering processes for engineers in planning, design, construction, operation, maintenance, improvement, and replacement of products and processes. Panetto et al. (2019) discussed the challenges of industry of the future. DT-I was not able to model all connections in an enterprise since facilitating productions needed closed-loop decisions based on the dynamics of environment, products, processes, and manufacturing assets. da Silva et al. (2019) discussed the FRs of DM and classified them into 24 aspects; DT-I was one of many enabling technologies, and the most critical challenge was how to integrate DT-I with others to fulfill the identified FRs of DM systems cost-effectively. Putman et al. (2017) and Bilberg and Malik (2019) observed the challenges of DT-I for virtual fusion, synchronization, and system integration when multiple DT were involved; they proposed a system framework to track multiple objects simultaneously in both DW and PW. DHL (2021) suggested that DT-I must be extended to associate with product life cycle management (PLM) closely; while the adoption of DT-I was facing significant challenges in matching complex assets and their behaviors digitally with precision, promptness, and cost-effectiveness. DT-I faced the difficulty in dealing with rising complexity, need of agility, incomplete view of the systems, the identification of data and codifying relationships, updating information, and mindset (Altran 2019).

DT-I must support both the horizontal and vertical integration of manufacturing systems (Stark et al. 2019). DT-I was applied to provide quality of services for monitoring, maintenance, safety, and management of manufacturing processes; however, it was difficult to integrate a DT-I with the control system and there were very few successful cases to use DT to control physical systems directly; in addition, DT-I applications were mostly standalone, they were not scalable (Cimino et al. 2019). Last but not least, few studies were available on performance assessment of DT-I (Cognizant 2019). The DT-I concept was investigated by researchers in multiple disciplines from the perspectives of methods, technologies, operations, and businesses, and the understanding of DT-I was too diversified (Singh et al. 2019).

9.7 Advanced Attributes of Digital Entities in Manufacturing

We are interested in developing a generic model for system elements, which is sustainable for mechatronic systems. Therefore, the core values of a sustainable system in its new definition are treated as *functional requirements* (FRs). Consequently, a new representation of a generic system element must meet the following FRs:

FR1 – Fulfilling expected functions: a manufacturing system transforms raw materials into final products through a series of manufacturing processes in the material flows of systems. Manufacturing processes are performed at physical manufacturing assets. Therefore, making products (FR1) is the primary FR of a manufacturing system, and it should be defined based on the variants and volumes of products that customers need. In addition, a complex product is involved in many parts and components that are made through different manufacturing processes; physical manufacturing assets can be defined for products, components, parts, and processes, respectively. For the purpose of generality, one physical entity is involved with one system element; note such a physical entity can be at any level of product from a specific process, part, component, product, and to product family or product series over its life cycle.

FR2 – Dealing with task changes over time: a smart manufacturing system is sustainable; it is capable of reconfiguring itself to deal with the changes of customers' needs as well as the disturbance and uncertainties in manufacturing environments. Customers' needs correspond to the functionalities, variations, volumes, delivery times, and other expectations of products that are changing in the markets over time. The solutions of a system model to adapt changes are (1) the software flexibility of adjustable assets, (2) the hardware modularity which is capable of configuring systems by selecting different modules and assembling them in different ways, and (3) the combination of software and hardware flexibility.

FR3 – Supporting virtual analyses of system resources: to perform a manufacturing process successfully, engineers need to define, plan, program, verify, validate, control, and monitor the process to ensure the first-time right. With an ever-increasing complexity of manufacturing processes, performing these complex tasks is far beyond manual efforts. In Sustainable Manufacturing (SM), digital models are developed as DT-I of corresponding physical assets; DT are utilized for the optimization, simulation, and verification of manufacturing processes, and for controlling, monitoring, and diagnosing manufacturing processes in actual operations.

FR4 – Acquiring, processing, and mining data for use of digital models: running a manufacturing system is involved with numerous decision-making activities at all levels, domains, and aspects of businesses; on the other hand, SM emphasizes the responsiveness, adaptability, and resilience to the changes and uncertainties in dynamic environment. Therefore, decision-making activities are closed-loop and data-driven that rely on reliable and abundant data of everything in the system. Physical assets in the SM are sensed to acquire real-time data, these assets are networked to collect and share data, and the collected data is processed, mined, and utilized by digital models for decision-making supports.

FR5 – Making decisions for enhanced system smartness: the decisions for manufacturing operations are made based on the data collected from machines, operators, sensors, suppliers, markets, and users. With a rapid growth of networked things in a system, decision-making activities about manufacturing operations usually involve in exceptionally large data sets that are characterized as big data in terms of variety, volume, velocity, veracity, and value (5V). Big data is analyzed and mined to make smarter decisions on manufacturing operations and achieve better system performances at any aspects of interests such as agility, robustness, adaptability, flexibility, and resilience.

FR6 – Accessing virtual resources: the scope of manufacturing businesses has been continuously increased due to the growing complexity of products and the needs of extending the businesses over product life cycles. Manufacturing systems are highly pressured to be reconfiguring themselves to meet the changes of customers' needs over time. VM resources become more and more important for host enterprises to cover the increasing scope of manufacturing businesses. The information infrastructure of an ES should be capable of accessing virtual resources and supporting the interactions and interoperations of internal and external resources seamlessly.

FR7 – Supporting decision-making for incorporation-level businesses: in the digital era, manufacturing systems become more and more distributed and decentralized, and manufacturing businesses within a system become more closely related to the stakeholders such as suppliers, service providers, logistic systems, and users across boundaries of the system. Therefore, sustainable manufacturing operations require numerous decision-making processes at the incorporation level such as selection of suppliers or service providers, composition of workflows for emerging business opportunities, and reconfiguration of systems or virtual enterprise alliances for the changes of customers' needs. Decision-making support systems should be able to deal with the incorporation-level big data to support interactions and interoperations of enterprises with assured security, privacy, and responsiveness (Viriyasitavat et al. 2019a, 2019b, 2019c, 2020a, 2020b).

9.8 Concept of Digital Triad (DT-II)

A DT-I maps a physical object from the PW to digital model, so that the physical object can be modeled, monitored, and controlled with the appropriate integration of data and AI (Biz 2021). However, DT-I showed its limitations in supporting products or process designs in its life cycle since it experiences a continuous change over time and a DT-I works at a given time frame (Dickopf et al. 2019). Some researchers addressed this issue by system integration where DT-I becomes an essential component of an integrated system. For example, Augustine (2020) developed a DT-I-based system for the entire life cycle of products to identify and analyze problems in businesses; DT-Is were introduced at all stages of product life cycles from design, simulation, manufacturing, assembly, maintenance, disassembly, quality control, training, customization, finance, sales and marketing, and stakeholder engagement. Landahl et al. (2018) proposed a digital platform to simplify and merge multiple digital models into low-fidelity functional models; this allowed engineers to retrieving functions of existing assets, inserting new functions, and enabling technologies in new design concepts. From the perspective of DM, DT-I related only to its FRs of digitization, interoperation, and interaction (Siedler et al. 2019; Sierla et al. 2021). Liao et al. (2020) proposed an airframe DT-I where the probabilistic and prognostic method was adopted to quantify and update uncertainties in fatigue life assessment of aircraft in the US Air Force. Qi et al. (2021) proposed a five-dimensional DT model where all types of interactions among digital models, physical models, and services were classified into five dimensions.

Umeda et al. (2019) proposed a new concept the so-called *digital triplet* as the extension of DT-I. A digital triplet added ***the intelligent world*** together with cyber and PW for engineers to deal with engineering design processes. A digital triplet integrates three worlds to realize the integrated support of engineering activities throughout a product life cycle including design, manufacturing, use, maintenance, remanufacturing, and circulation of resources. However, the digital triplet emphasized two aspects, that is, (1) designers' interventions on DT and (2) support to transfer data, information, and knowledge from one stage to another stage in a product's life cycle.

To overcome the limitations of DT-I discussed in Section 9.6, a new concept the so-called digital triad is proposed here. As shown in Figure 9.18, a *digital triad* (DT-II) is a coalition of *life models*,

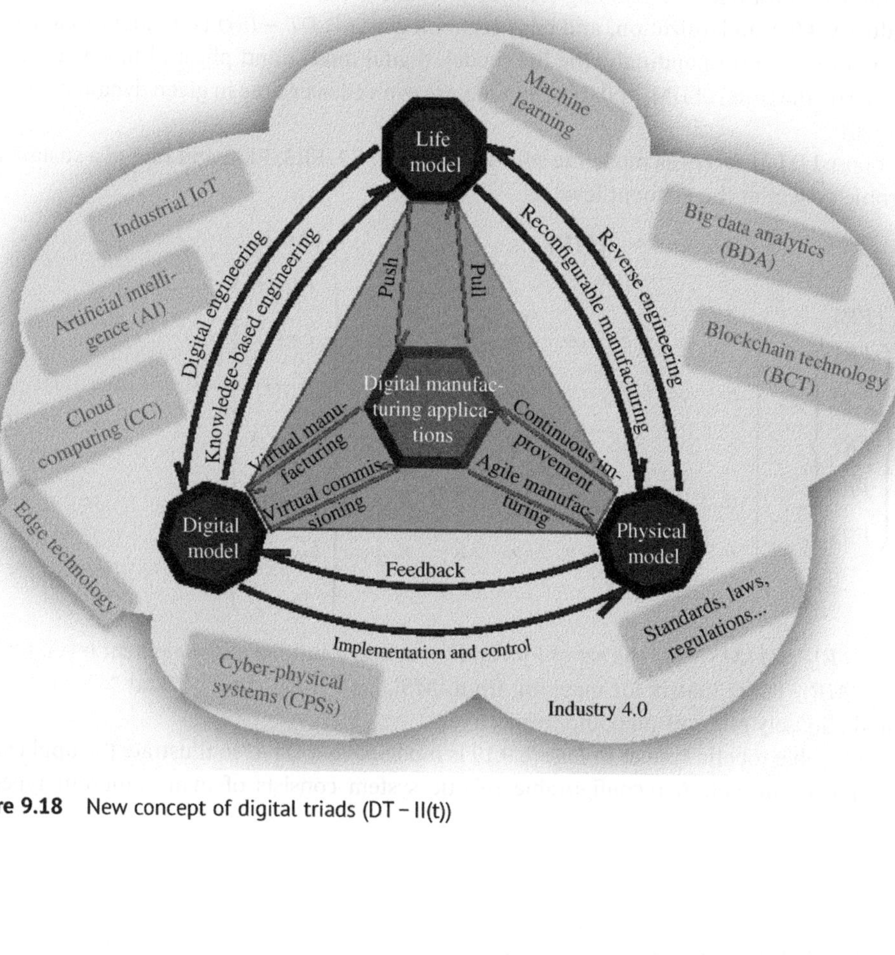

Figure 9.18 New concept of digital triads (DT – II(t))

digital models, and corresponding *physical models* and *interactions* among three models. Inspired by the idea for the representation of DT-I (Qi et al. 2021), a DT-II can be described as,

$$DT-II(t)=\{L_M(t),D_M(t),P_M(t),C_N(L_M(t),D_M(t),P_M(t),t),E_T(t)\}\quad t=t_1,t_2,\ldots,t_n,\ldots \tag{9.1}$$

where *DT-II* (t) is the system state of a DT-II at the given time t, $L_M(t)$, $D_M(t)$, and $P_M(t)$ are the life models, digital models, and physical models at t, respectively; $C_N(L_M(t), D_M(t), P_M(t), t)$ are the set of three interactions among the life, digital, and physical models at t; $E_T(t)$ is the set of enabling technologies to instantiate the DT-II; t_1, t_2, t_3, ... are the time frames to update a DT-II in a dynamic business environment.

Similar to the definition of DT-I, *a physical model* can be defined for one or a group of physical entities of products, processes, systems, or other functional units in the PW, and *a digital model* is one computer representation of the corresponding physical model in the DW; note that one physical model may involve in multiple digital models from different aspects. Differing from the definition of DT-I, (1) *a life model* $L_M(t)$ is introduced to sustain historical data, templates, components, and engineering design knowledge for optimization, upgradation, reconfiguration, and evolution of physical models over time. (2) A digital triad $DT-II(t)$ includes all types of interactions, $C_N(L_M(t), D_M(t), P_M(t), t)$ in integrating three models; other than the interactions of physical and digital models $C_N(D_M(t), P_M(t))$ in DT-I; the interactions $C_N(L_M(t), D_M(t), t)$ occur between life models and digital models and $C_N(L_M(t), P_M(t), t)$ between digital models and physical models. (3) $DT-II(t)$ is coined as a concept of the system whose implementation relies on various enabling technologies for creating, sustaining, and operating these models and facilitating the interactions among three types of models. Most relevant enabling technologies, $E_T(t)$, include IIoT, AI, ML, CPS, BDA, CC, BCT, Industry 4.0, standardization, and edge technologies. (4) $DT-II(t)$ be tailored to a specific application of DM; correspondingly, the life model, digital model, and physical model are customized based on the (FRs) of DM, and these models are evolved over time in given dynamic business environment.

The proposed DT-II serves as modeling solutions to FR1, R2, FR3, FR4, and FR5 of a sustainable mechatronic system at the element level as,

$$\underset{\{\boldsymbol{FR}\}_E}{\begin{Bmatrix} FR1\\ FR2\\ FR3\\ FR4\\ FR5 \end{Bmatrix}} = \underset{\{\boldsymbol{MR}\}_E}{\begin{bmatrix} \times & \circ & \circ & \circ & \circ & \circ & \times\\ \circ & \times & \circ & \circ & \circ & \circ & \times\\ \circ & \circ & \times & \circ & \circ & \circ & \times\\ \circ & \circ & \circ & \times & \circ & \circ & \times\\ \circ & \circ & \circ & \times & \times & \times & \times \end{bmatrix}} \underset{\{\boldsymbol{MS}\}_E}{\begin{Bmatrix} P_M(t)\\ L_M(t)\\ D_M(t)\\ C_{N,1}(t)\\ C_{N,2}(t)\\ C_{N,3}(t)\\ E_T(t) \end{Bmatrix}} \tag{9.2}$$

where the $\{\boldsymbol{FR}\}_E$ and $\{\boldsymbol{MS}\}_E$ are the set of FRs and modeling solutions at the element level, respectively; the $\{\boldsymbol{MR}\}_E$ is the matrix for mapping from $\{\boldsymbol{MS}\}_E$ and $\{\boldsymbol{FR}\}_E$; the "×" and "o" represent a closely and a loosely relevant mapping, respectively.

A reconfigurable robotic system in Figure 9.19 is modeled as a DT-II to illustrate the application of the proposed concept. A reconfigurable robotic system consists of many different types of

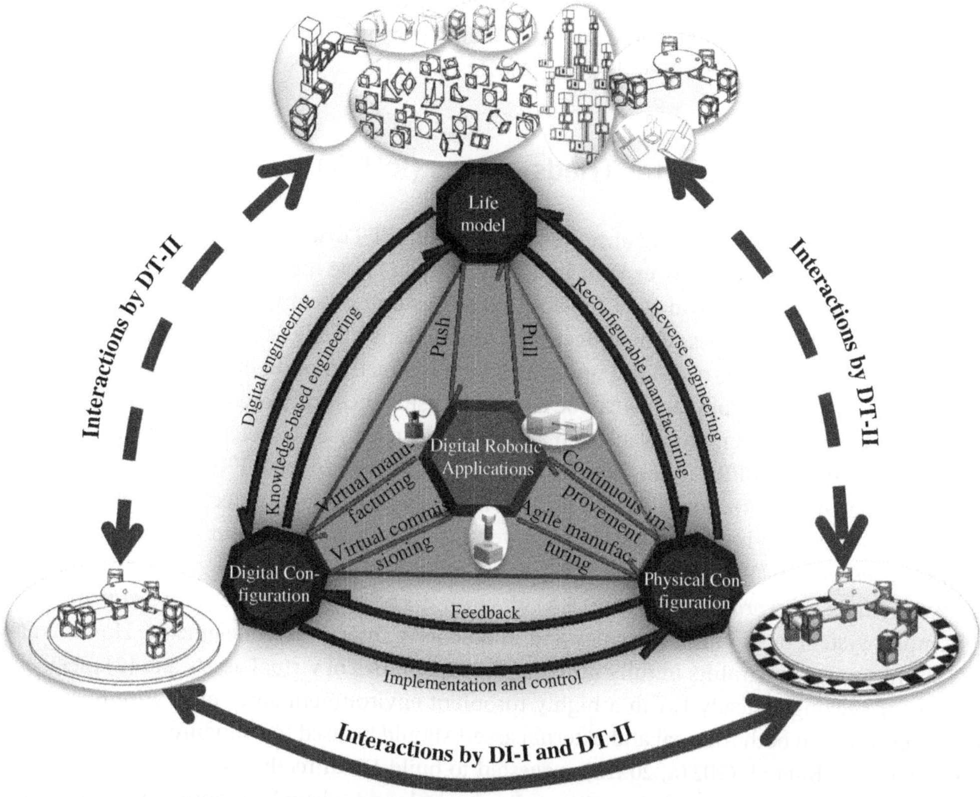

Figure 9.19 Reconfigurable robotic system as an example of DT – II(t)

modules that can be selected and assembled into different robotic configurations for specific applications. The DT-II consists of three types of models, that is, (1) a physical model for a robotic configuration that is built to perform a task for a specific application (such as loading/unloading, transporting, and inspecting) at time *t*; (2) a digital model for the corresponding physical robot at time *t*; and (3) a life model that keeps all data about standardized and customized modules, historical design cases, and computer tools for task-oriented designs.

Traditionally, DT-I was able to model the interactions between the digital and physical models (Bi et al. 2007a, 2007b; Bi and Kang 2010; Bi and Zhang 2021): the digital model was directly utilized for off-line programming and controlling a physical robot when a program was appropriately verified, and the physical robot was instrumented with the feedbacks to the digital model for on-line surveillance of actual robotic operations.

In comparison with DT-I, the DT-II(*t*) in Figure 9.19 provides an information system with a complete model for both product life cycle and KBE tools that are essential to task-orientated configuration designs. It supports (1) the interactions between the life model and digital model, so that a new digital model can be designed when a change occurs to robotic application and (2) the interactions between the life model and physical model, so that the data about robotic operations can be fully utilized to enrich the life model for sustainable manufacturing. Note that the DT-II in Figure 9.19 is generic and applicable to any reconfigurable or modular system such as flexible fixturing systems (Bi and Zhang 2001a) or modular machining tools (Bi et al. 2008); such systems have their capabilities of reconfiguring themselves for different tasks in a dynamic environment. High-level modularity and configurability were fundamental for a manufacturing system to achieve adaptability and sustainability (Bi and Zhang 2001b; Bi 2011).

9.9 The Internet of Digital Triads Things (IoDTT)

By examining the proposed DT-II against the limitations of DT-I discussed in Section 9.6, it is clear that DT-II shows its advantages over DT-I in the sense that (1) the life model is introduced to represent knowledge, methods, and data for both processes and outcomes of engineering designs. (2) DT-II covers all three types of heterogeneous interactions among life models, digital models, and physical models; in contrast, DT-I only covers the interaction of digital and physical objects. (3) Systematic assessment approaches can be developed using the formulated DT-II in eq. (1); when FRs of a mechatronic system are given, such approaches can be used to determine if current DT-II (t) can meet the specified FRs; if not, how to configure the system to meet FRs optimally. (4) DT-II has been proposed as an information system that includes all of enabling technologies $E_T(t)$ with specified roles in system implementation. However, DT-II itself is insufficient to model a complex *system of systems* (SoS) (Wang and Luo 2021; Wang et al. 2021a, 2021b).

Conventionally, *enterprise architecture* (EA) was introduced to manage the complexity and scale of a complex system by defining the types and interactions among system elements. However, an EA (1) brings some constraints in fully utilizing the capabilities of virtual assets and (2) meets the challenge of keeping a steady EA in a highly turbulent environment since enterprise businesses evolve over time and both internal and external assets should be used to configure systems dynamically. Therefore, Bi et al. (2021a, 2021b) suggested to build ESs directly upon the Internet-based information infrastructure to fully utilize all physical and virtual resources without extra constraints.

As shown in Figure 9.20, the concept of IoDTT is proposed to model DM systems. In the digital era, any DM applications are built on the information infrastructure ($\{E_T(t)\}$) that consists of all enabling technologies for data acquisition, processing, mining, and decision-making supports at

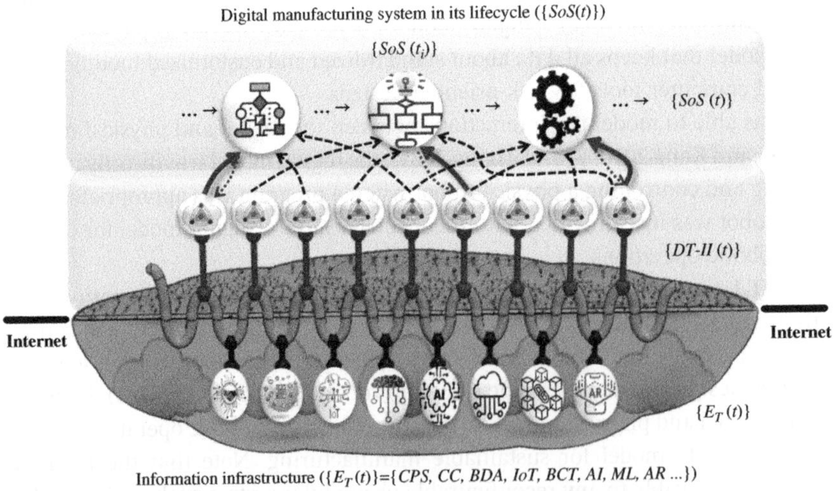

Figure 9.20 Internet of digital triad things (IoDTT) as system architecture

all levels and domains of manufacturing processes. These technologies become accessible to companies through the Internet. All smart things including {DT-II (t)} are networked; note that each DT-II has a life model to support its reconfiguration over time, respectively. The life cycle of DM is represented as a series of the system applications $SoS(t)$ at specific time. Each application will be initialized by one of its DT-II so-called the host DT-II$_h$ (t$_i$). The life model $L_{m,h}$ (t_i) of this DT-II is responsible to select other DT-II and compose a workflow as an SoS (ti)) to meet the FRs of the given applications. In return, the digital model $D_{m,h}$ (t_i) and physical model $P_{m,h}$ (t_i) of the host DT-II must be upgraded to reflect the changes of its interactions with others.

The proposed IoDTT serves as the modeling solutions to FR6 and FR7 of a smart manufacturing system at the SoS level as,

$$\overset{\{\boldsymbol{FR}\}_S}{\left\{\begin{matrix}\{\boldsymbol{FR}\}_E\\ \{\boldsymbol{FR}\}_N\end{matrix}\right\}} = \overset{\{\boldsymbol{FR}\}_S}{\left\{\begin{matrix}\{\boldsymbol{FR}\}_E\\ FR6\\ FR7\end{matrix}\right\}} = \overset{\{\boldsymbol{MR}\}_S}{\begin{bmatrix}\{\boldsymbol{MR}\}_E & \circ & \times\\ \circ & \times & \times\\ \circ & \circ & \times\end{bmatrix}} \overset{\{\boldsymbol{MS}\}_S}{\left\{\begin{matrix}\{\boldsymbol{MS}\}_E\\ \{SoS(t)\}\\ \{E_T(t)\}\end{matrix}\right\}} \tag{9.3}$$

where the $\{\boldsymbol{FR}\}_E$, $\{\boldsymbol{MR}\}_E$, and $\{\boldsymbol{MS}\}_E$ are the FRs, mapping relations, and modeling solutions determined in Eq (1) at the element level; the $\{\boldsymbol{FR}\}_N = \{FR6, FR7\}^T$ are the FRs at the system level; the {SoS(t)} is the set of system configurations in the system life cycle; the $\{E_T(t)\}$ is the information infrastructure with all of accessible enabling technologies. "×" and "o" represent a closely and a loosely relevant mapping, respectively.

To deal with the ever-growing complexity and scale of manufacturing systems, manufacturing assets can be of either physical or virtual, boundaries of DT-II are vaguer and vaguer, a system should actually consist of SoS for complex goals, and it must be configurable to adapt high-level changes in dynamic environments. IoDTT is proposed as system architecture of DM systems. It allows to unify all mainstream ITs as information infrastructure to model system assets and support heterogeneous interactions among system assets in the form of DT-II. IoDTT addresses the need to model configurability, scalability, and sustainability of DM.

DT-II and IoDTT provide a systematic way to model, analyze, design, and implement smart systems: the life model in DT-II provides all supports to engineering design processes within a DT-II, and the mechanism in an IoDTT for selecting and composing workflows for given applications closes the gap of the transaction from one SoS to another. To make DT-II and IoDTT practical in DM, future research efforts are needed at many aspects, and they are outlined herein. (1) *Designing a life model and integrating it with digital and physical models* for various applications such as a reconfigurable robotic system, as shown in Figure 9.18. Note that a life model should include data, knowledge, methods, and tools for the whole space of system configuration. (2) *Developing effective methods to configure an* SoS *from DT-II for specific applications.* This demands innovative approaches for various purposes such as project-oriented service selection, workflow composition, and security assurance (Viriyasitavat et al. 2019a, 2019b, 2019c, 2020a, 2020b). (3) *Developing quantitative system measures for the system properties such as functionality, stability, and sustainability, as well as reliability, robustness, and resilience in terms of these properties.* These measures are essential to assess and optimize manufacturing assets at the levels of DT-II and SoS (Bi et al. 2021a, 2021b). (4) *Standardizing and testing the interfaces of system elements.* System adaptability and reconfigurability are mainly achieved by selecting different system assets and collaborating them effectively (Bi et al. 2021a). (5) *Customizing cost-effective solutions of information technologies to specific applications* of data acquisition, processing, data mining, and decision-making supports for continuous improvement of the systems. Every enterprise has its uniqueness in terms of its strengths, assets, and associated partnerships; the solution of an ES must be tailored to enterprise and applications.

9.10 DT-Is and DT-IIs in Sustainable Mechatronic Systems

This section discusses some typical applications of DT-Is and DT-IIs in sustainable mechatronic systems. DT-Is and DT-IIs have the potential to add values to a sustainable mechatronic system by fulfilling some critical tasks as elaborated as shown below (Rasheed et al. 2019).

9.10.1 Monitoring and Controlling

Generally, the control model of a complex system is involved with certain simplifications and assumptions; it is unrealistic to (1) assume that these simplifications and assumptions are always valid or (2) expect predicted behaviors of the controlled physical system without discrepancies. DT provide cost-effective ways to deal with uncertainties, dynamics, and changes in using a mechatronic system. The DT of a physical system is developed to manage the complexity of a control model by monitoring critical variables of the physical system, and taking into consideration of the changes of these variables in closed-loop system controls. Bi and Wang (2008) gave an example where a DT of a tripod-based machine tool was developed and used to monitor and control a physical machine tool.

In order to monitor and control the physical machine tool in a real-time mode, the DT was modeled in Java 3D and it was used to monitor the system states based on the feedback data of displacements, velocities, and torques measured at the actuators of the machine tool. Figure 9.21 shows the scene-graph digital model of the machine tool that was mounted upside-down on a gantry system. The virtual universe object laid the foundation in creating this entire scene graph. All scene graphs were associated with corresponding virtual objects. A scene graph started with a BranchGroup node, which served as the root of a sub-graph, or branch graph, of the scene graph. A TransformGroup node in a branch graph specified the position, the orientation, and the scale of an object in the virtual universe. An object could be with the attributes for geometry, appearance, or both. A geometric object modeled the shape of the object. An appearance object modeled the appearance of the object such as color, texture, and material reflection. The physical system was controlled by its behavior nodes. Behavior nodes contained user-defined control codes, kinematic models, and state variables. The algorithms for signal processing were introduced into the code. In the illustration, the movable platform was controlled by one control node for on-line monitoring. As the digital model was connected with its physical counterpart through the control node, it could be used to control the machine tool remotely over the Internet.

The real-time data collected from the machine tool was transmitted to the digital model by a server-side servlet over the transmission control protocol (TCP) connection, and one data packet included 12 floating numbers and one long integer as,

1	2	3	4	5	6	7	8	9	10	11	12	13
x	y	z_e	θ_x	θ_y	x	y	u_1	u_2	u_3	FR	SS	CW

where the x and y are the coordinates of an independent x–y table attached to the gantry system; z_e, θ_x, θ_y are the desired motions of the end-effector platform; $u_{i\,=\,1,\,2,\,3}$ is the displacement of the ith leg along its guide-way; FR and SS denote the feed rate and the spindle speed of the spindle mounted on

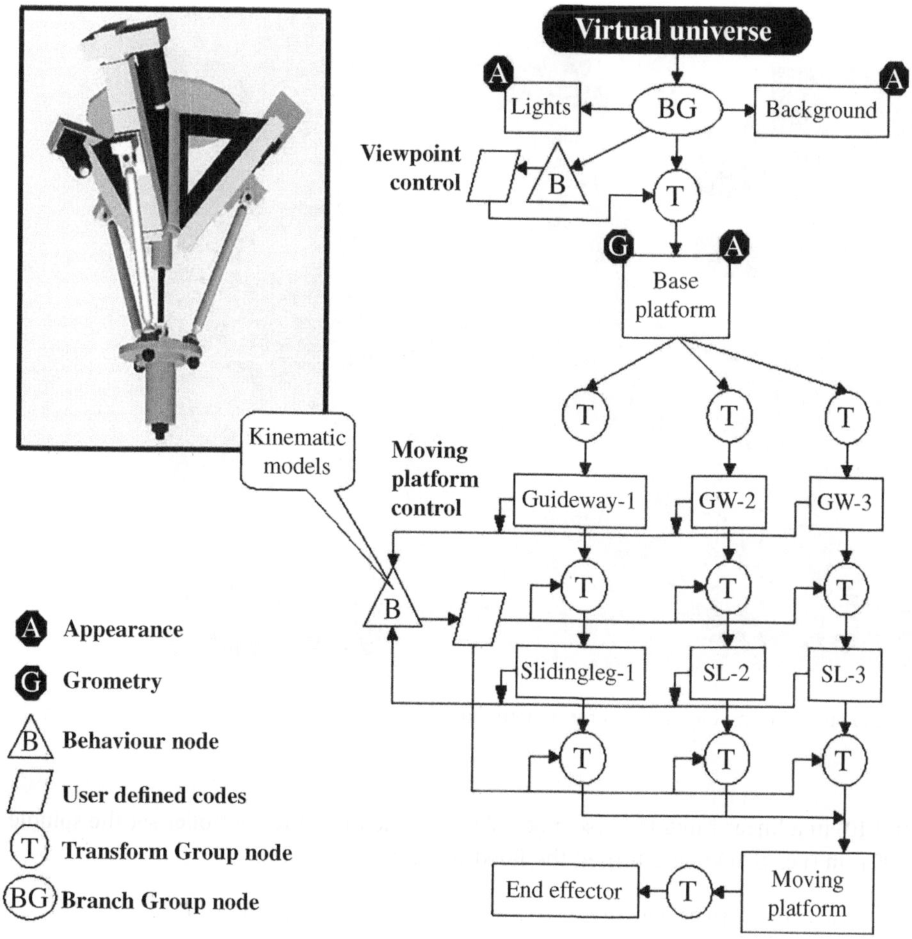

Figure 9.21 Digital twin of a tripod-based machine tool

the end-effector. CW is the control word indicating the system state including the operation mode such as the *manual*, *auto*, or *jogging* and the specified coordinate system. Since the digital model updated the kinematic model locally, only the first five floating numbers that represented the Cartesian working coordinates were used. The second set of five floating numbers represented joint coordinates; they were used by the digital model in the closed-loop control of the physical system.

The machine tool was controlled by the G code transmitted via the applet–servlet communication. The motion of machine tool could be controlled in either of *jogging control* or *NC control*. Jogging control controlled motion axes via a web user interface, respectively. Each button at the interaction controlled the motion along an axis. While *jogging control* was straightforward and easy to implement, it was applied only to the control based on forward-kinematic model. Since task-oriented control was based on an inverse-kinematic model, *numeric control* (NC) should be used to move multiple motion axes simultaneously. By using a DT-I, an experienced engineer could create and send G-code line by line to the machine-tool controller; or, a pre-generated G-code could be used to implement web-based machining. For example, the following G-code was sent to move the machine tool from the current position to the next with the increments along x-, y-, and z axes by

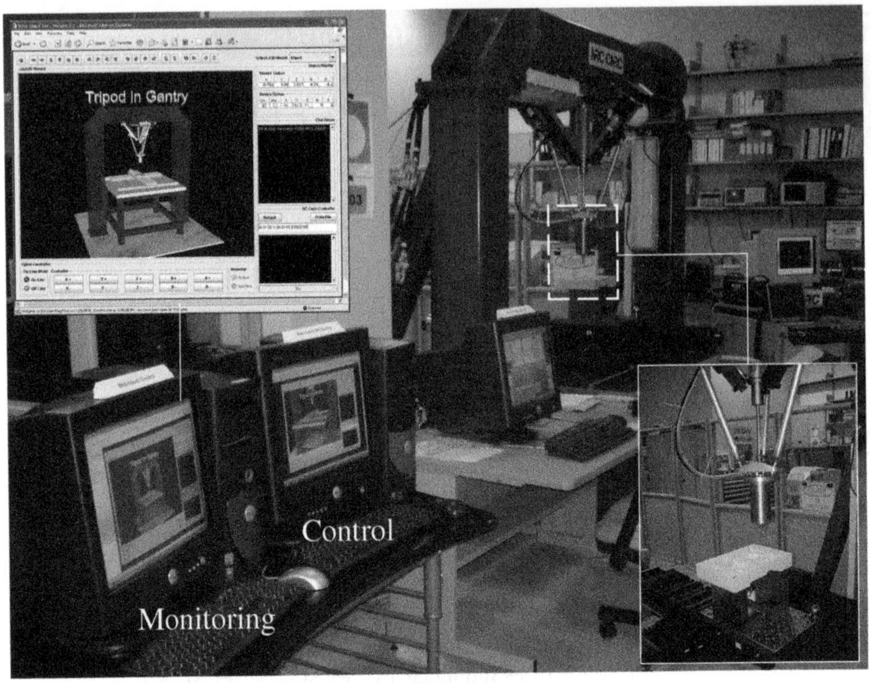

Figure 9.22 Example of using digital twins for remote monitoring and controlling

(20, −30, and 10) in a linear rapid traverse mode. At the same time, the controller set the spindle speed as 3000 rpm (i.e., S3000) and turned the flood coolant on (i.e. M8):

$$\mathrm{G0}\, x + 20\, y - 30\, z + 10\, \mathrm{S3000\ M8}$$

Figure 9.22 shows the scenario where the DT of the machine tool was used to monitor and control the physical system. Note that the DT was interacted with its physical system through the Internet; this implied that the web browser for the interactions of DT-I could be anywhere to monitor or control the physical system remotely as long as the network was accessible.

9.10.2 Data-Driven Decision-Making

An advanced mechatronic system is often required to deal with changes and uncertainties in a dynamic environment. The changes can be tackled by reconfiguring either its hardware or software system. In any case, the decisions on system reconfigurations must be made based on sensed data collected from the physical system and application environment. Since decisions should be evaluated for their effectiveness thoroughly before they can be executed in the PW, developing DT become mandatory in supporting data-driven decision-making activities for advanced mechatronic systems.

Bi and Lang (2007) developed a DT for an automated robotic coating system. The DT was used to process scanned data, determine tag points that cover all surfaces to be coated, and plan paths for robotic operations. The DT was implemented by a *data processing system* and an *automatic programming system*. The data processing system acquired and processed data related to products, robots, coating tools, and coating environment. In particular, it was able to process points clouds scanned from the surfaces of the product that was loaded into the physical coating workstation. The automatic programming system was to calculate tag points, generate robotic paths, and predict the coating performance of the generated paths.

Figure 9.23 shows that the inputs to the data processing system were (1) a points-cloud acquired by scanning an as-built part and (2) the original computer-aided design (CAD) model of as-designed part. The outputs of the data processing system were an accurate surface model of the part. It was a very complex process to generate a surface model from points-cloud and original CAD model and it involved the following tasks. *First*, the acquired data was filtered to eliminate invalid data. *Second*, the data from different views was registered and merged under the same coordinate system. *Third*, surfaces were reconstructed to generate accurate surfaces from points-cloud to update surfaces on

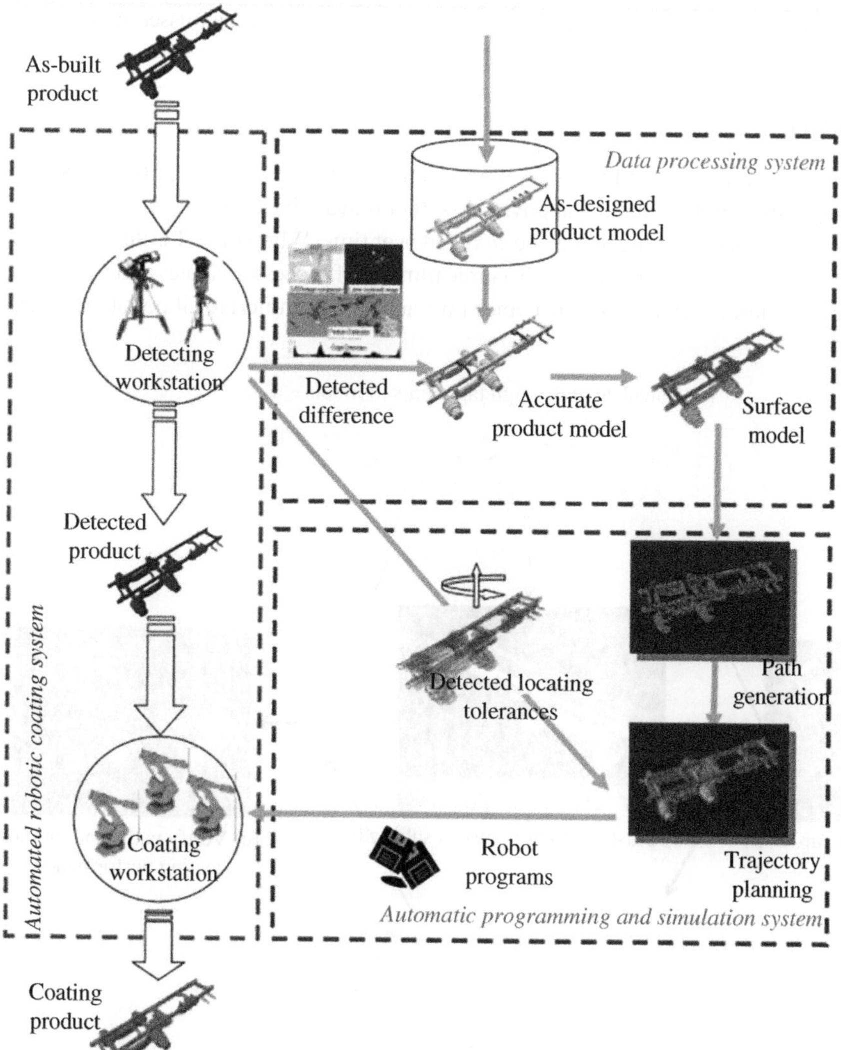

Figure 9.23 Digital twins of sensor-driven robotic coating system

original CAD models. Many software tools such as PolyWorks and Insight Toolkit (ITK) could be used to reconstruct surfaces from points clouds. The input of the automatic programming and simulation system was a surface model from the data processing system, and its outputs were executable robot programs to control robotic coating operations. The system also included a simulation module that was used to evaluate and validate robotic programs and coating quality. The system included DT for all motion devices including coating robots, a railway, and a fixturing system.

Figure 9.24 shows the sensor-driven programming and verification process of in the DT of the robotic coating system. One primary advantage of using a DT was its capability in predicting the outcomes from the expected actions on the system. The developed DT simulated robotic motions along the generated paths and assessed the quality of the coating processes in terms of overall coverages and thickness distributions. More specifically, system verifications included to (1) evaluate the performance of generated paths and trajectories; (2) validate system responses to changed surfaces; (3) detect potential physical collisions in coating; (4) predict the quality, productivity, and efficiency of the planned coating processes; (5) alert areas to be missed when 100% coverage became infeasible.

9.10.3 Fault Detections

A DT can be a write-box, gray-box, or black-box with given assumptions and simplifications; this implies that (1) not all information about the physical system is available and (2) even though certain information is known at beginning, it may be changed over time. When the DT is developed for a complex industrial process that is involved with a large number of process variables and is affected by many dynamic environmental factors, it becomes impractical to enumerate all possible defects

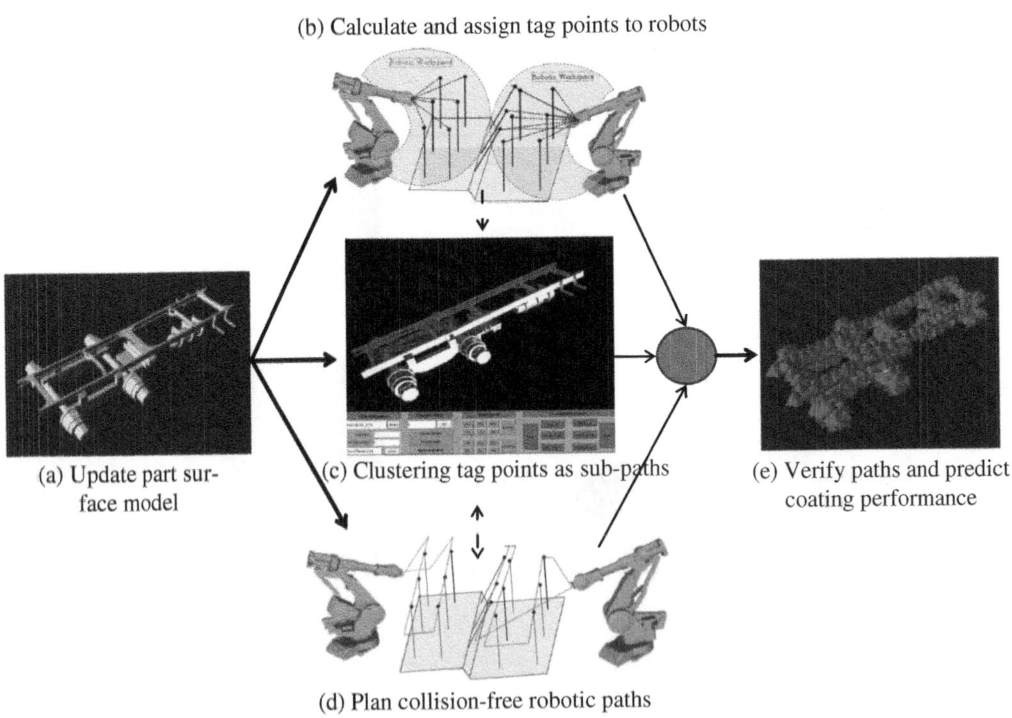

Figure 9.24 Sensor-driven programming and verification in a DT-I model

and provide the resolutions to these defects in a control system *in prior*. In practice, we rely on a DT of the industrial process to collect data from the physical system, and the DT is equipped with data-driven ML algorithms to monitor system states, predict the trend of system changes, detect faults, and identify the main causes of faults.

Several DTs were developed for complex chemical reactions in a blast furnace as shown in Figure 9.25. Different ML algorithms were incorporated into these digital models to extract features from multisource datasets, detect faults, and identify the causes of faults (Jiao et al. 2022; Fu et al. 2021; Ma et al. 2023; Wang et al. 2024).

Various ML algorithms were developed to classify data, extract features, and detect faults as shown in Figure 9.26. In a DT, the datasets were collected from multiple sources; the amount of data associated with abnormal working conditions was usually much less than the amount of data for normal working conditions. There was an imbalance between two datasets for abnormal and normal conditions, respectively. To achieve real-time performance in fault detection, it was proposed to preprocess existing and historical data under different working conditions in training ML models. The processed data was then converted into the same type for subsequent data processing. Matrix regression was used in a constructed ML model to cluster labeled features. This allowed

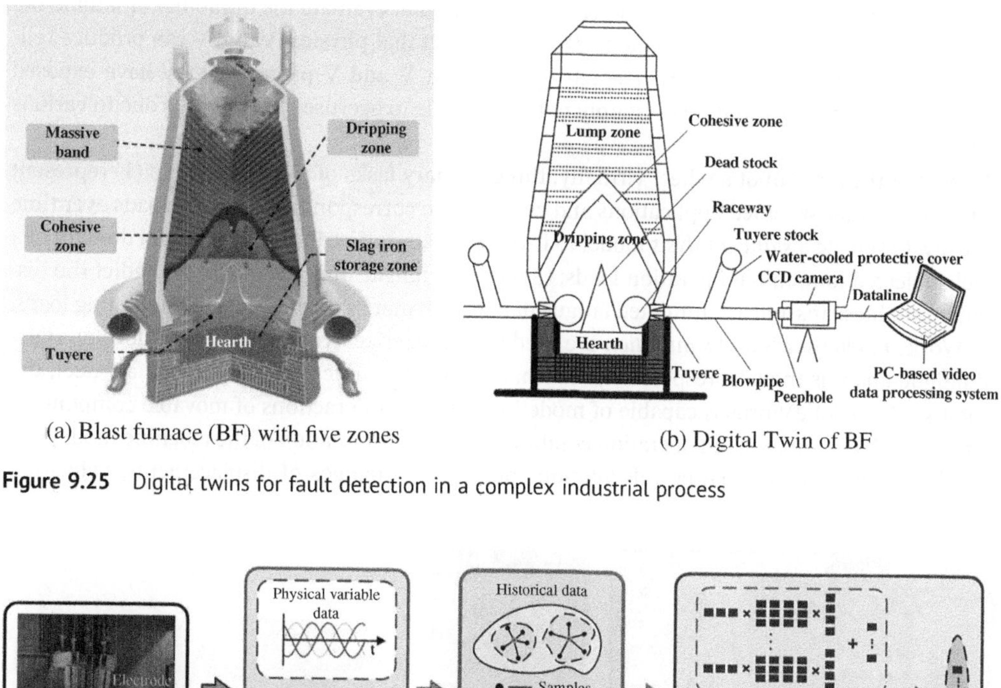

Figure 9.25 Digital twins for fault detection in a complex industrial process

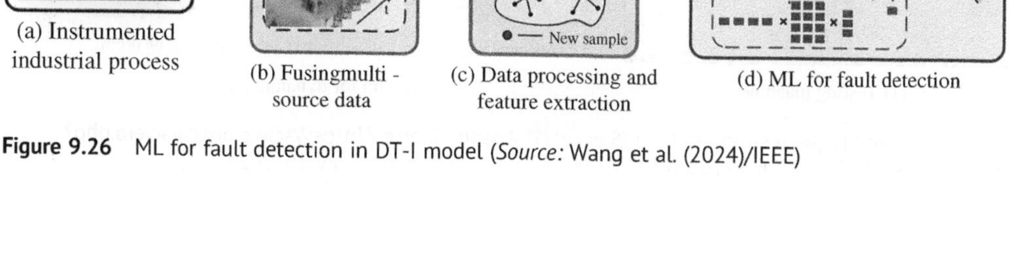

Figure 9.26 ML for fault detection in DT-I model (*Source:* Wang et al. (2024)/IEEE)

to correspond the datasets from different sources based on the similarities of system states when the data was collected.

9.10.4 Predication of Fatigue Life

Most of critical components in a mechatronic system fail due to fatigue; DT can be used to simulate the behaviors of these components and predict their fatigue lives accordingly. Bi et al. (2024) created a DT of a tire testing machine. It can be used to enhance the sustainability of the testing system by minimizing non-value-added businesses.

The client company makes a large variety of plastic wheels that are used in various small vehicles. Their wheels are usually customized to meet some specific requirements in different applications such as healthcare devices, home appliances, mobile devices, and accessories for agriculture or mining machinery. In these applications, the wheels of mobile devices are important components to sustain the stability in their stationary positions or movements. The wheels make direct contacts with these devices to corresponding environments as the supports to the weights and additional loads. Customers of the client company usually provide specific requirements for the wheels they request, and it is mandatory for the client company to verify and validate its products subject to specified operational conditions thoroughly. Conventionally designed wheels have to be prototyped, and V and V are performed by tedious experiments physically on testing machines. Figure 9.27 shows a setup of a testing machine that is used to evaluate the durability or fatigue life of a wheel subject to given cyclic motions. Despite the fact that physical V and V can produce reliable results and have been widely adopted by SMEs for V and V purposes; they have exposed numerous disadvantages, and it is certainly not sustainable to overuse physical tests due to various wastes.

To predict the lifespan of a wheel virtually, three primary tasks of modeling are to (1) represent road conditions for specified applications and characterize corresponding dynamic loads over time cyclically; (2) create a *static analysis* model to investigate stress distributed in the solid domain of a wheel subjected to nominal operation loads; (3) create a *fatigue analysis* model to predict the distributions of lifespans and accumulated fatigue damages in materials subjected to fluctuating loads. SolidWorks *motion study* and *simulation* are used to characterize dynamic loads and develop static and fatigue analysis models, respectively. To characterize dynamic loads applied on a wheel, the motion study in SolidWorks is capable of modeling dynamic interactions of movable components in a machine based on its given operation conditions, and the simulation of the machine model will yield all kinematic and dynamic information such as the changes of displacements, velocities,

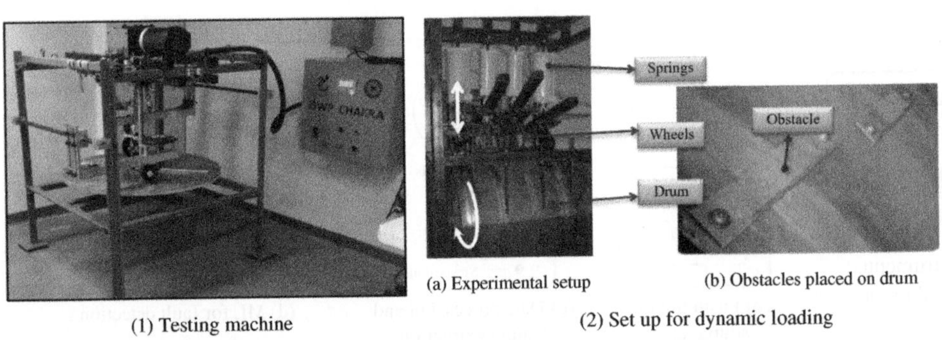

Figure 9.27 Tire fatigue testing machine (*Source:* (1) Anand Mhatre / https://scielo.org.za/scielo.php?script=sci_arttext&pid=S2226-72202017000100033 / Last accessed on July 19, 2024)

accelerations, and forces at joints and rigid bodies over time. The motion study module of SolidWorks is supported by powerful *automated dynamic analysis of mechanical systems* (ADAMS) in simulating kinematic and dynamic behaviors of a machine where ADAMS is a versatile tool that shows outstanding performance in simulating large-scale multi-body systems.

The DT is parametrized in the sense that (1) the same parametric model can be used to analyze and evaluate a large number of design options to optimize a product; (2) the knowledge people learn from one simulation model can be partially transferred to the knowledge of another simulation model based on some identified similarities such as dimensions, loads, materials, assemblies, contact conditions, and operational conditions. *A parametric study* on simulation models is enabled by the advanced tools in SolidWorks for *parametric modeling* and *design libraries*. For the motion simulation of a wheel assembly, Figure 9.28 illustrates an example of parametric models of wheels where all main dimensions of the wheel are parametrized, and their values can be customized to optimize specific evaluation criteria in a given application.

Using the DT to predict the fatigue life of a tire is involved with three steps. *At the first step,* the motion study model is created in SolidWorks to investigate kinematic and dynamic responses of the wheel to the changes of (1) surface conditions at contacts and operation variables, dynamic responses are characterized in terms of the fluctuated loads from the drum to the wheel. The motion study model is fully parametrized to include main features and dimensions, as illustrated in Figure 9.28 as well as the number, heights, widths, and distributed patterns of bumps on drum surface. From the simulation, the forces at all joints of the wheel are evaluated and they are further extracted as the inputs and then used to define corresponding loads in the sequential static and

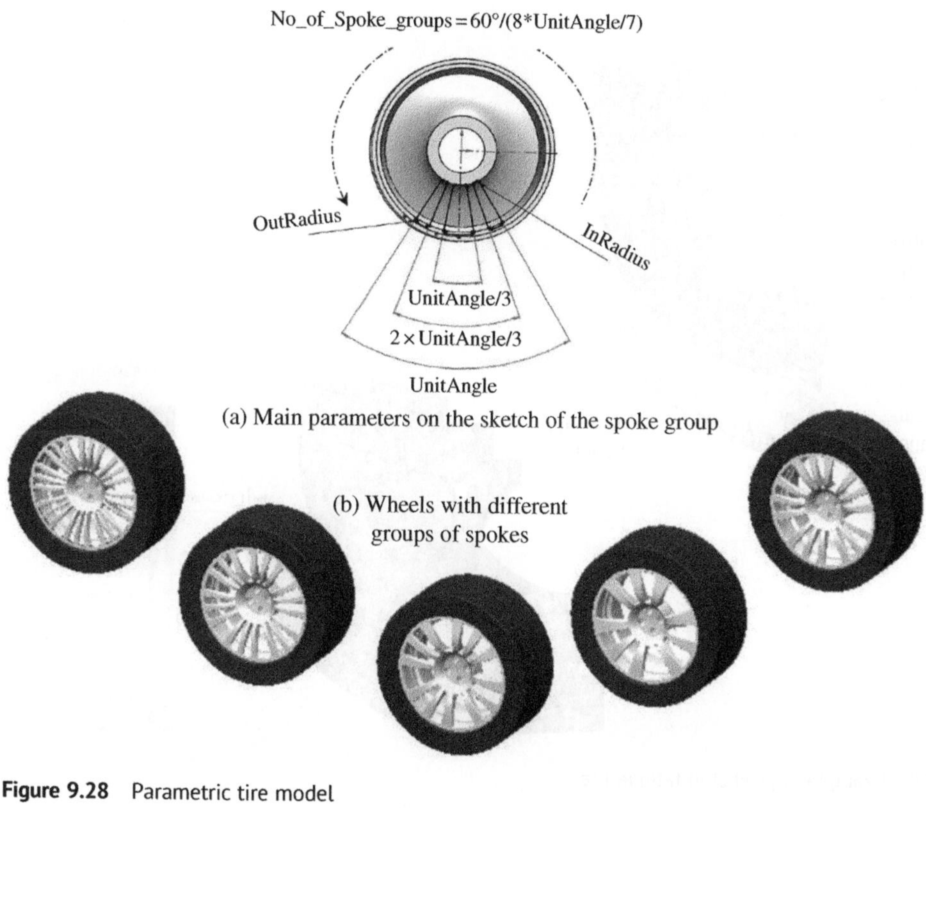

Figure 9.28 Parametric tire model

fatigue analyses. *At the second step*, a static analysis model is created for stress analysis of the wheel assembly under *nominal* external loads from two ends of its motion axes where the wheel assembly is supported. The results from the above static analysis are used to evaluate the *mean* and *alternating* components of a stress at any location of a solid domain in fatigue analysis. The criteria of static failure such as *maximum shear stress theory* (MSST) and *distorted energy theory* (DET) are selected to determine how the stress components in different directions are synthesized as an equivalent stress that is comparable to the yield or ultimate strength of the material. *At the third step*, the outcomes of dynamic loads to the wheel from a motion study model are used to determine fluctuating loads, and the criteria of fatigue failure are selected to weigh the effects of static and fatigue damages on overall strengths of the material. A fatigue analysis is then conducted to find the distributions of a lifespan and accumulated damages in the solid domain of the wheel specified in the fatigue analysis to determine the distribution of the failure damages and fatigue life over the wheel. Figure 9.29 shows an example of the distribution of the fatigue life subjected to given fluctuating loads.

9.10.5 *Virtual Verification and Validation* (V and V)

Generally, the control model of a complex system is involved with certain simplifications and assumptions; it is unrealistic to (1) assume that these simplifications and assumptions are always valid or (2) expect predicted behaviors of the controlled physical system without discrepancies. DT provide cost-effective ways to deal with uncertainties, dynamics, and changes in using a mechatronic system. The DT of a physical system is developed to manage the complexity of a control model by monitoring critical variables of the physical system, and taking into consideration of the changes of these variables in closed-loop system controls.

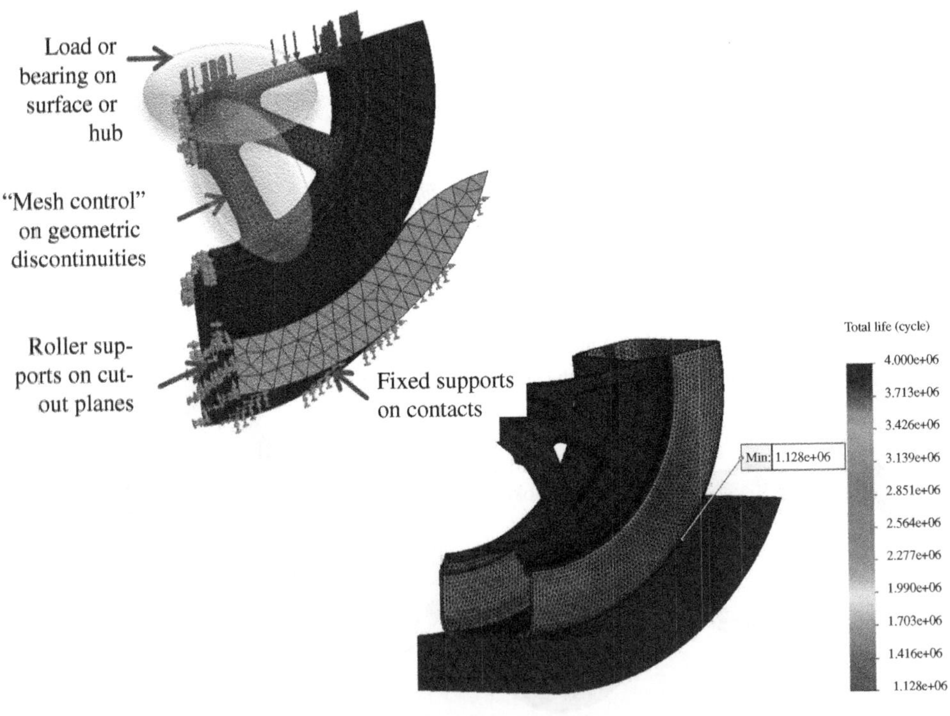

Figure 9.29 Example of predicted fatigue life

9.11 Summary

Design of mechatronic system and product is more complex than traditional mechanical systems, electrical systems, or controllers in their respective domains. To design mechatronic systems, DT-I can be used to (1) model, analyze, evaluate, and compare various design alternatives at the design phase and (2) monitor the systems and evaluate and predict the system performance at the application phase of mechatronic systems.

The concepts relevant to DT-I are overviewed, their importance in designing and developing mechatronic systems is discussed, existing DT-I technologies are reviewed to identify their limitations in supporting sustainable developments of mechatronic systems, and new concepts of DT-II and Internet of triad things (IoDDT) are introduced to expand the capabilities of digital technologies for sustainable mechatronic developments.

The conventional DT-I concept should be extended to overcome its limitations in representing (1) engineering design processes, (2) heterogeneous interactions of system elements other than cyber-physical interactions, (3) all business activities that determine the system performances, (4) the integration of other critical enabling technologies, (5) and the needs of system configurability, scalability, and sustainability. This chapter introduces a new concept of a DT-II by integrating a life model with digital and physical models as a triad, and DT-II has shown its significance to overcome the aforementioned limitations.

The proposed DT-II and IoDTT can be used to deal with highly diversified systems by customizing prioritized business goals such as sustainability when subjected to affordable digital technologies in achieving system goals. Moreover, the performance metrics for the system sustainability must be quantifiable so that different design solutions can be analyzed, evaluated, and compared to optimize system solutions. Future research efforts will be needed in many areas such as (1) developing quantifiable performance metrics and synergizing multiple performance metrics when they are considered simultaneously; (2) establishing design libraries that include commonly used mechatronic functional modules such as digital solutions (DSs) for FRs; (3) developing some design templates that correspond to DSs and FRs with consideration of the sustainability at both component and system levels; (4) using BDA and AI in dealing with the combinatory complexity in tailoring digital solutions to specific applications; (5) trading-off between system reconfigurations and the utilization of virtual assets in the system life cycle; (6) developing systematic approaches to verify and validate a system before it will be actually implemented in the PW; and (7) standardizing the design procedures for sustainable systems.

Problems

9.1 What is a DT-I? What are advantages of using a DT-I in mechatronic system? What are the limitations of DT-I? How can a DT-II overcome these limitations?

9.2 What is IoT? What are the limitations when DT-I is integrated into IoT? How can IoDDT overcome these limitations?

9.3 Create the DT for four racks that are illustrated in Figures 9.30–9.33, respectively. Predict max weights that these racks can support subjected to a *factor of safety* (FoS) of 2. The materials of all structural members are A36 steel.

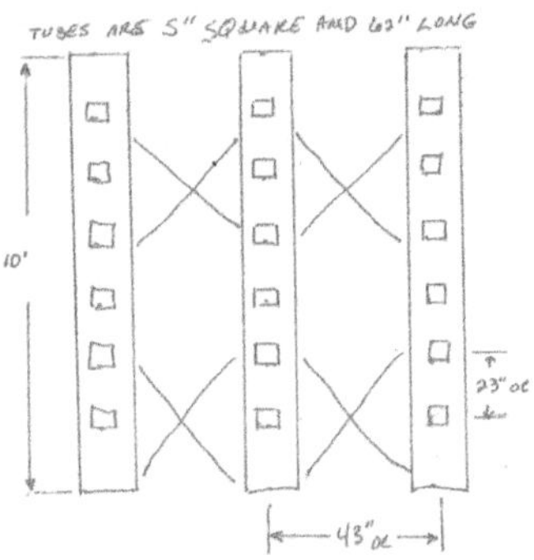

(a) Front view

(b) Real view

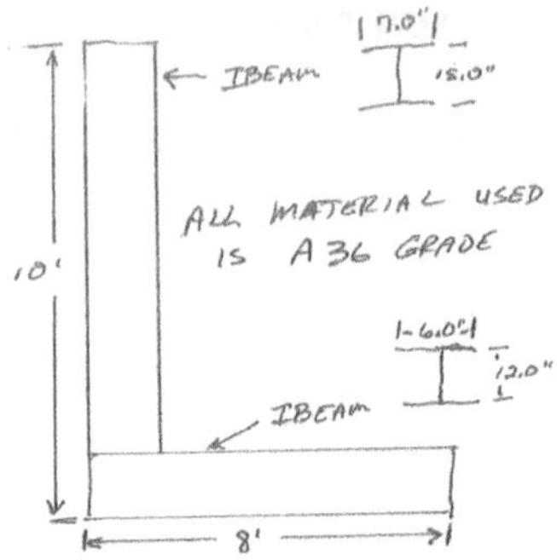

(c) Side view

(d) Four detail views (local)

Figure 9.30 Dimensions of rack 1

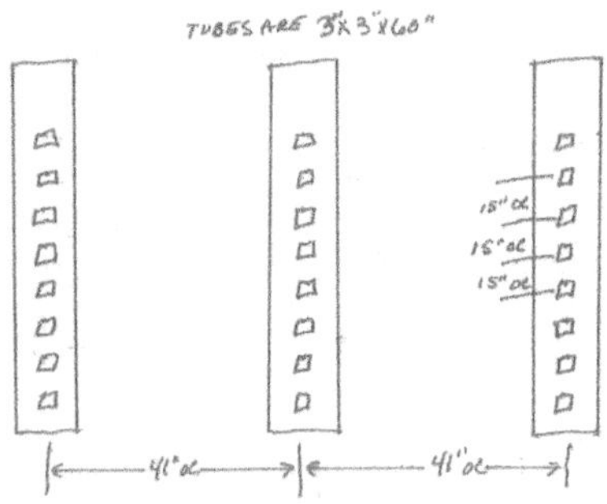

(a) Front view

(b) Real view

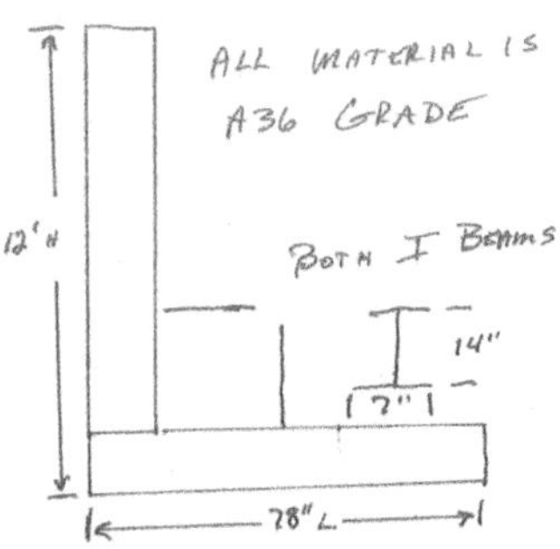

(c) Side view

(d) Five detail views (local)

Figure 9.31 Dimensions of rack 2

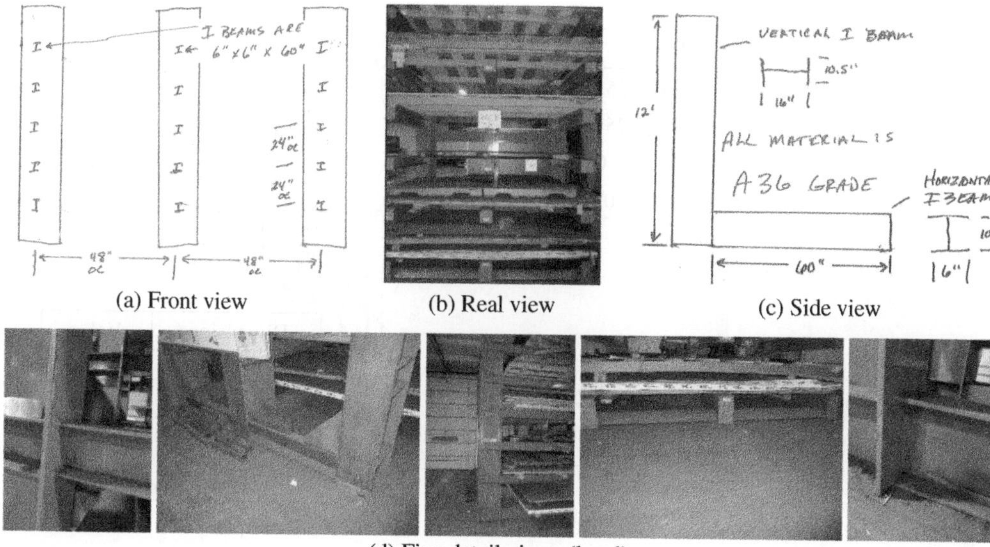

(a) Front view (b) Real view (c) Side view

(d) Five detail views (local)

Figure 9.32 Dimensions of rack 3

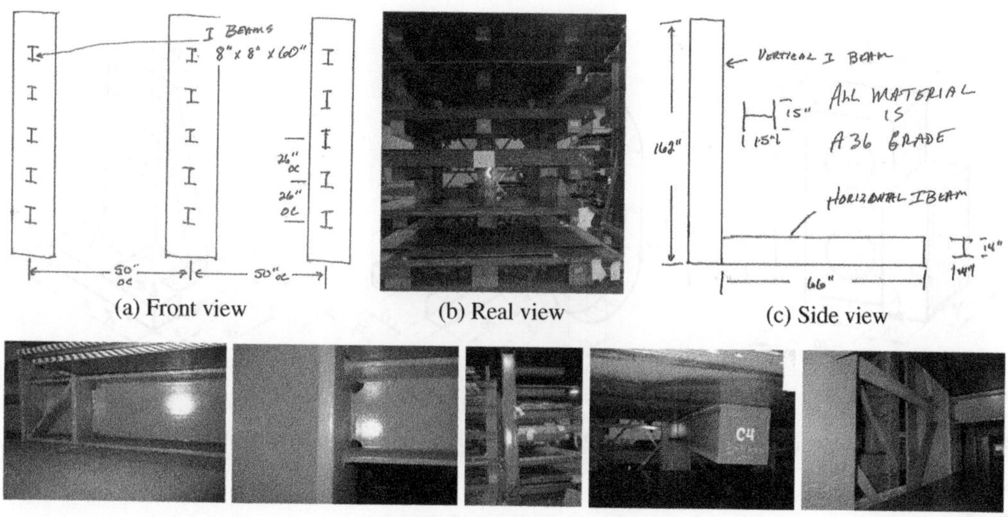

(a) Front view (b) Real view (c) Side view

(d) Four detail views (local)

Figure 9.33 Dimensions of rack 4

9.4 Choose a robot from the ABB web shop (ABB 2024). Download its CAD model (e.g., a SCARA robot in Figure 9.34), create a DT, define joints with the technical specifications in data sheet, and simulate the DT model to verify the workspace of the robot.

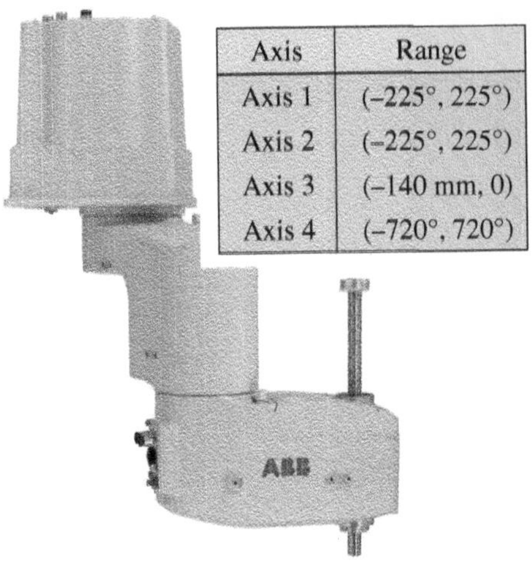

Axis	Range
Axis 1	(–225°, 225°)
Axis 2	(–225°, 225°)
Axis 3	(–140 mm, 0)
Axis 4	(–720°, 720°)

(a) ABB-SCARA

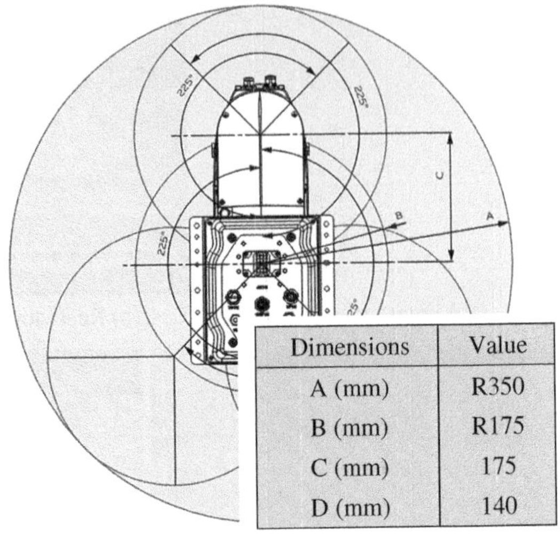

Dimensions	Value
A (mm)	R350
B (mm)	R175
C (mm)	175
D (mm)	140

(b) Task space

Figure 9.34 ABB SCARA model (*Source:* Adapted from ABB (2024))

9.5 Create digital models of the objects whose dimensions are shown in Figure 9.35a–c, respectively; use the digital models to determine their volumetric and mass properties.

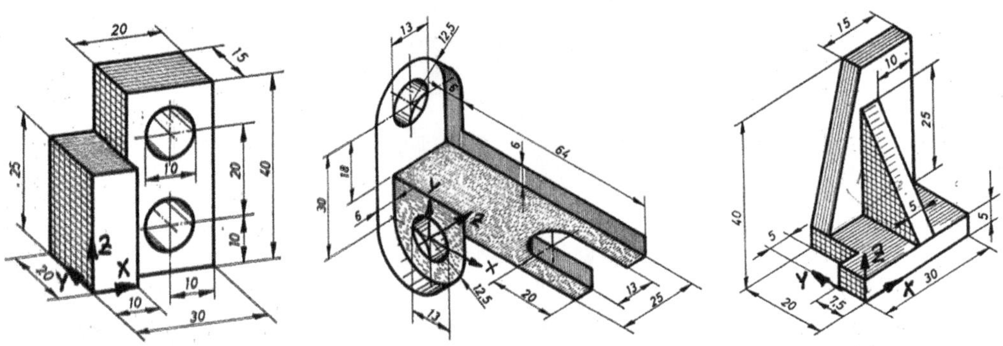

(a) (b) (c)

Figure 9.35 Dimensions of three objects in Problem 9.3 (unit: mm, material: case alloy steel)

References

ABB (2024). https://webshop.robotics.abb.com/us/scara-robot-irb-910inv-series.html (accessed 12 September 2024).

Altran (2019). Digital twins creating digital operations today to deliver business value tomorrow. https://capgemini-engineering.com/us/en/insight/the-digital-twin-how-to-create-value-now-and-prepare-for-digital-operations-for-tomorrow (accessed 12 September 2024).

AMRC (2018). Feasibility of an immersive digital twin: the definition of a digital twin and discussions around the benefit of immersion. www.amrc.co.uk/files/document/219/1536919984_HVM_CATAPULT_DIGITAL_TWIN_DL.pdf, (accessed 12 September 2024).

Augustine, P. (2020). Chapter four: the industry use cases for the digital twin idea. *Advances in Computers* 117 (1): 79–105.

Augustine, P., Raj, P., and Munirathinam (2022). *Enterprise Digital Transformation*, 1e. Auerbach Publications.

Barnabas, J. and Raj, P. (2020). Chapter nine – the human body: a digital twin of the cyber physical systems. *Advances in Computers* 117 (1): 219–246.

Bi, Z.M. (2011). Revisiting system paradigms from the viewpoint of manufacturing sustainability. *Journal of Sustainability* 3 (9): 1323–1340.

Bi, Z.M. and Kang, B. (2010). Enhancement of adaptability of parallel kinematic machines with an adjustable platform. *ASME Journal of Manufacturing Science and Engineering* 132 (9): 4003120.

Bi, Z.M. and Lang, S.Y.T. (2007). A framework for CAD- and sensor-based robotic coating automation. *IEEE Transactions on Industrial Informatics* 3 (1): 84–91.

Bi, Z.M. and Wang, L. (2008). *Design of a Reconfigurable Tripod Machine System and its Application in Web-Based Machining in Smart Devices and Machines for Advanced Manufacturing*, 189–218. Springer.

Bi, Z.M. and Wang, X.Q. (2020). *Computer Aided Design and Manufacturing*, 1e. Wiley-ASME Press Series.

Bi, Z.M. and Zhang, W.J. (2001a). Flexible fixture design and automation: review, issues and future directions. *International Journal of Production Research* 39 (13): 2867–2894.

Bi, Z.M. and Zhang, W.J. (2001b). Modularity technology in manufacturing: taxonomy and issues. *International Journal of Advanced Manufacturing Technology* 18 (5): 381–390.

Bi, Z.M. and Zhang, W.J. (2021). *Practical Guide to Digital Manufacturing - First-Time-Right from Digital Twin to Physical Twin*. Springer International Publishing.

Bi, Z.M., Lang, S.Y.T., Orban, P., and Verner, M. (2007a). An integrated design toolbox for tripod-based parallel kinematic machines. *ASME Journal of Mechanical Design* 129 (8): 799–807.

Bi, Z.M., Wang, L., and SYT, L. (2007b). Current status of reconfigurable assembly systems. *International Journal of Manufacturing Research* 2: 303–327.

Bi, Z.M., Lang, S.Y.T., and Shen, W.M. (2008). Reconfigurable manufacturing systems: the state of the art. *International Journal of Production Research* 46 (4): 967–992.

Bi, Z.M., Zhang, W.J., Wu, C. et al. (2021a). Generic design methodology for smart manufacturing systems from a practical perspective, part I – digital triad concept and its application as a system reference model. *Machines* 9 (10): 207.

Bi, Z.M., Zhang, W.J., Wu, C. et al. (2021b). Generic design methodology for smart manufacturing systems from a practical perspective, part II – systematic designs of smart manufacturing systems. *Machines* 9 (10): 208.

Bi, Z.M., Mikkola, A., Ip, A.W.P. et al. (2024). Virtual verification and validation to enhance sustainability of manufacturing systems. *IEEE Transactions on Automation Science and Engineering* https://doi.org/10.1109/TASE.2024.3370053.

Bilberg, A. and Malik, A.A. (2019). Digital twin driven human-robot collaborative assembly. *CIRP Annals - Manufacturing Technology* 68 (2019): 499–502.

Biz (2021). Industrial IoT: rise of digital twin in manufacturing sector. https://www.biz4intellia.com/blog/rise-of-digital-twin-in-manufacturing-sector (accessed 12 September 2024).

Botín-Sanabria, M., Mihaita, A., Peimbert-García, R. et al. (2022). Digital twin technology challenges and applications: a comprehensive review. *Remote Sensing* 14: 1335.

Bradley, D., Russell, D., Hegenberger, P. et al. (2018). Mechatronics & The Cloud, In X.-T. Chen E, Zhang Y, Zheng J and Bi ZM (2024) Multiple Collaboration Preserving Projection for Monitoring of Complex Industrial Process. *IEEE Transactions on Instrumentation and Measurement* 73: 1–9, Art no. 3502509 https://doi.org/10.1109/TIM.2023.3330211.

Cimino, C., Negri, E., and Fumagalli, L. (2019). Review of digital twin applications in manufacturing. *Computers in Industry* 113 (2019) 103130.

Classens, K., Heemels, W.P.M.H., and Oomen, T. (2021) Digital twins in mechatronics: from model-based control to predictive maintenance. *2021 IEEE International Conferences on Digital Twins and Parallel Intelligence (DTPI)*, Beijing China. https://www.koenclassens.nl/pdf/ClassensHeeOom2021d.pdf (accessed 12 September 2024).

Cognizant (2019). Nurturing digital twins: how to build virtual instances of physical assets to boost performance. https://www.cognizant.com/whitepapers/nurturing-digital-twins-codex4653.pdf (accessed 12 September 2024).

Curry, E., Scerri, S., and Tuikka, T. (2022). *Data Spaces: Design, Deployment and Future Directions*. Springer. https://link.springer.com/book/10.1007/978-3-030-98636-0 (accessed 12 September 2024).

Deuter, A. and Pethig, F. (2019). The digital twin theory. https://www.plmportal.org/en/the-digital-twin-theory-a-new-view-on-a-buzzword.html (accessed 12 September 2024).

DHL (2021). Digital twins in logistics: a DHL perspective on the impact of digital twins on the logistics industry. https://www.dhl.com/content/dam/dhl/global/core/documents/pdf/glo-core-digital-twins-in-logistics.pdf (accessed 12 September 2024).

Dickopf, T., Apostolov, H., Muller, P. et al. (2019). A holistic system lifecycle engineering approach – closing the loop between system architecture and digital twins. *Procedia CIRP* 84 (2019): 538–544.

Fu, Y., Luo, C., and Bi, Z.M. (2021). Low-rank joint embedding and its application for robust process monitoring. *IEEE Transactions on Instrumentation and Measurement* 70: 1–13, Art no. 3515313, https://doi.org/10.1109/TIM.2021.3075017.

GE (2021). GE digital twin: analytic engine for the digital power plant. https://www.ge.com/digital/sites/default/files/download_assets/Digital-Twin-for-the-digital-power-plant-.pdf (accessed 12 September 2024).

GE (2024). Digital twin model. https://www.ge.com/digital/applications/digital-twin (accessed 12 September 2024).

Jiang, S., Ong, Y.S., Zhang, J., and Feng, L. (2014). Consistencies and contradictions of performance metrics in multi-objective optimization. *IEEE Transactions on Cybernetics* 2014 (44): 2391–2404.

Jiao, H., Zhang, Y., Luo, C., and Bi, Z.M. (2022). Collaborative multiple rank regression for temperature prediction of blast furnace. *IEEE Transactions on Instrumentation and Measurement* 71: 1–10, Art no. 3515110, https://doi.org/10.1109/TIM.2022.3180408.

Kunath, M. and Winkler, H. (2018). Integrating the digital twin of the manufacturing system into a decision support systems for improving the order management process. *Procedia CIRP* 72 (2018): 225–231.

Landahl, J., Panarotto, M., Johannesson, H. et al. (2018). Towards adopting digital twins to support design reuse during platform concept development. *NordDesign*, August 14–17, 2018.

Liao, M., Renaud, G., and Bombardier, Y. (2020). Airframe digital twin technology adaptability assessment and technology demonstration. *Engineering Fracture Mechanics* 225 (2020): 106793.

Lu, Y., Witherell, P., and Jones, A. (2020). Standard connections for IIoT empowered smart manufacturing. *Manufacturing Letters* 26 (2020): 17–20.

Ma, C., Zhang, Y., and Bi, Z.M. (2023). Abnormal condition monitoring based on multiview data collaborative feature regression. *IEEE Transactions on Instrumentation and Measurement* 72: 1–10, Art no. 3517710, https://doi.org/10.1109/TIM.2023.3277987.

Negri, E., Fumagalli, L., and Macchi, M. (2017). A review on the roles of digital twin in CPS-based production systems. *Procedia Manufacturing* 11 (2017): 939–948.

Panetto, H., Lung, B., Ivanov, D. et al. (2019). Challenges for the cyber-physical manufacturing enterprises of the future. *Annual Reviews in Control* 47 (2019): 200–213.

Putman, N.M., Maturana, F., Barton, K., and Tilbury, D.M. (2017). Virtual fusion: a hybrid environment for improved commissioning in manufacturing systems. *International Journal of Production Research* 55 (21): 6254–6265.

Qi, Q. and Tao, F. (2018). Digital twin and big data towards smart manufacturing and industry 4.0: 360 degree comparison. *IEEE Access* 6: 3585–3593.

Qi, Q., Tao, F., Hu, T. et al. (2021). Enabling technologies and tools for digital twin. *Journal of Manufacturing Systems* 58 (B): 3–21.

Rasheed, A., San, O., and Kvamsdal, T. (2019). Digital twin: values, challenges and enablers. https://www.semanticscholar.org/reader/d062b576d80982078fddc7c090efd1df753a3413 (accessed 12 September 2024).

Rosen, R., Fischer, J., and Boschert, S. (2019). Next generation digital twin: an ecosystem for mechatronic systems? *IFAC PapersOnLine* 52 (15): 265–270.

Shetty, D. and Kolk, R.A. (2010). *Mechatronic System Design*. Stamford, CT: Cengage Learning.

Siedler, C., Langlotz, P., and Aurich, J.C. (2019). Identification of interactions between digital technologies in manufacturing systems. *Procedia CIRP* 81 (2019): 115–120.

Sierla, S., Azangoo, M., Rainio, K. et al. (2021). Roadmap to semi-automatic generation of digital twins for brownfield process plants. *Journal of Industrial Information Integration* https://doi.org/10.1016/j.jii.2021.100282.

da Silva, E.H.D.R., Anfelis, J., and de Lima, E.P. (2019). In pursuit of digital manufacturing. *Procedia Manufacturing* 28 (2019): 63–69.

Singh, S., Barde, A., Mahanty, B., and Tiwari, M.K. (2019). Digital twin driven inclusive manufacturing using emerging technologies. *IFAC PapersOnLine* 52 (13): 2225–2230.

Song et al. (2023). Digital twins for the future power system: an overview and a future perspective. *Sustainability* 15: 5259. https://doi.org/10.3390/su15065259.

Stark, R., Fresmann, C., and Lindow, K. (2019). Development and operation of digital twins for technical systems and services. *CIRP Annals - Manufacturing Technology* 68 (2019): 129–132.

Sun, X., Bao, J., Li, J., and Zhang, Y. (2020). Liu S and Zhou B (2019) a digital twin-driven approach for the assembly-commissioning of high precision products. *Robotics and Computer-Integrated Manufacturing* 61: 101839.

Tao, F., Qi, Q., Liu, A., and Kusiak, A. (2018). Data-driven smart manufacturing. *Journal of Manufacturing Systems* 2018 (48): 157–169.

Tao, F., Zhang, M., Liu, Y., and Nee, A.Y.C. (2018a). Digital twin driven prognostics and health management for complex equipment. *CIRP Annals - Manufacturing Technology* 67 (2018): 169–172.

Tao, F., Zhang, H., Liu, A., and Nee, A. (2018b). Digital twin in industry: state of the art. *IEEE Transactions on Industrial Informatics* 15 (4) (2018): 2405–2415.

Tao, F., Sui, F., Liu, A. et al. (2019). Digital twin-driven production design framework. *International Journal of Production Research* 57 (12): 3935–3953.

Terkaj, W., Gaboardi, P., Trevisan, C. et al. (2019). A digital factory platform for the design of roll shop plants. *CIRP Journal of Manufacturing Science and Technology* 26 (2019): 88–93.

Tilson, D., Lyytinen, K., and Sorensen, C. (2010). Digital infrastructures: the missing IS research agenda. *Information Systems Research* 21 (4): 748–759.

Uhlemann, T.H.J., Lehmann, C., and Steinhilper, R. (2017). The digital twin: realizing the cyber-physical production system for industry 4.0, the 24th CIRP Conference on Life Cycle Engineering. *Procedia CIRP* 61 (2017): 335–340.

Umeda, Y., Ota, J., Kojima, F. et al. (2019). Development of an education program for digital manufacturing system engineers based on 'digital triplet' concept. *Procedia Manufacturing* 31 (2019): 363–369.

Viriyasitavat, W., Xu, L., Bi, Z.M., and Hoonsopon, D. (2019a). Blockchain technology for applications in Internet of things – mapping from system design perspective. *IEEE Internet of Things Journal* 6 (5): 8155–8168.

Viriyasitavat, W., Xu, L., and Bi, Z.M. (2019b). Blockchain and internet of things for modern business process in digital economy – the state of the art. *IEEE Transactions on Computational Social Systems* 6 (6): 1420–1432.

Viriyasitavat, W., Xu, L., Bi, Z.M. et al. (2019c). Managing QoS of internet-of-thing services using blockchain. *IEEE Transactions on Computational Social Systems* 6 (6): 1357–1368.

Viriyasitavat, W., Xu, L., and Bi, Z.M. (2020a). Specification patterns of service-based applications using blockchain technology. *IEEE Transactions on Computational Social Systems* 7 (4): 886–896.

Viriyasitavat, W., Xu, L.D., Bi, Z.M., and Sapsomboon, A. (2020b). Blockchain-based business process management (BPM) framework for service composition in Industry 4.0. *Journal of Intelligent Manufacturing* 31 (7): 1737–1748.

Wagg, D.J., Worden, K., Barthorpe, R.J., and Gardner, P. (2020). Digital twins: state-of-the art and future directions for modelling and simulation in engineering dynamics applications. *ASCE-ASME Journal of Risk and Uncertainty in Engineering Systems, Part B: Mechanical Engineering* 6: 030901-1–030901-17.

Wagner, R., Schleich, B., Haefner, B. et al. (2019). Challenges and potentials of digital twins and industry 4.0 in product design and production of high performance products. *Procedia CIRP* 84 (2019): 88–93.

Wang, P. and Luo, M. (2021). A digital twin-based big data virtual and real fusion learning reference framework supported by industrial internet towards smart manufacturing. *Journal of Manufacturing Systems* 58 (2021): 16–32.

Wang, G., Zhang, G., Guo, X., and Zhang, Y. (2021a). Digital twin-driven service model and optimal allocation of manufacturing in shared manufacturing. *Journal of Manufacturing Systems* 59 (2021): 165–179.

Wang, X., Wang, Y., Tao, F., and Liu, A. (2021b). New paradigm of data driven smart customization through digital twin. *Journal of Manufacturing Systems* 58 (2021): 270–280.

Wang, Z., Zhang, Y., Zheng, Y., and Bi, Z.M. (2024). Multineighbor ensemble preserving embedding for process monitoring. *IEEE Transactions on Instrumentation and Measurement* 73: 1–10, Art no. 3500310 https://doi.org/10.1109/TIM.2023.3334335.

World Bank Group (2024). *Digital Progress and Trends Report 2023.* Washington, DC: World Bank https://doi.org/10.1596/978-1-4648-2049-6. License: Creative Commons Attribution CC BY 3.0 IGO.

Xiang, F., Zhang, Z., Zuo, Y., and Tao, F. (2019). Digital twin driven green material optimal-selection towards sustainable manufacturing. *Procedia CIRP* 81 (2019): 1290–1294.

Zhang, L., Zhou, L., Ren, L., and Laili, Y. (2019). Modeling and simulation in intelligent manufacturing. *Computers in Industry* 112 (2019): 103123.

Zhou, J., Li, P., Zhou, Y. et al. (2018). Toward new generation intelligent manufacturing. *Engineering* 4 (2018): 11–20.

Zhou, J., Zhou, Y., Wang, B., and Zang, J. (2019). Human-cyber-physical systems (HCPSs) in the context of new-generation intelligent manufacturing. *Engineering* 5 (2019): 624–636.

10

Cyber-Physical Systems

10.1 Introduction

Cyber-physical systems (CPSs) are the extensions of traditional mechatronic systems as shown in Figure 10.1. A CPS differs from a mechatronic system in the sense that interfaces of mechatronic functional modules (MFMs) or subsystems in CPS can be enabled by either hardware or software (Hehenberger and Bradley 2016; Penas et al. 2017). By integrating virtual resources, CPSs combine the capabilities of cyber and physical systems to solve problems that neither cyber and physical systems can solve the problems by themselves alone (Platzer 2018). Despite that the concept of CPS was conceived in the 1980s, the study on CPSs was not an independent discipline until the concept of CPS was introduced by Helen Gill as an integration of computation with physical processes at the National Science Foundation in 2006 (Taha et al. 2021). It was the time to establish the science for the design and application of CPSs.

As shown in Figure 10.2, enhancing interactions and communications with other virtual systems can empower and upgrade mechatronic systems in such a way that a set of physical and virtual MFMs are connected, interacted, and coordinated to fulfill more complex tasks. In the control theory, a shift from a set of stand-alone mechatronic systems to a networked CPS is an attractive research topic in *Cybernetics*. With standardization, reduced cost, and increasing capabilities of off-the-shelf mechatronic components, it becomes more and more prevalent to build various CPSs for complex tasks (Herrmann and Blech 2016). In a CPS, mechatronic components are treated as reactive functional blocks; their behaviors are defined as the activities in *unified modeling language* (UML) together with established semantics and verification.

Traditionally, a CPS was a generalized embedded system since the things that were previously confined in the realm of relatively separate disciplines can be combined as an integrated system. Cyber systems in CPS are associated with computer science, and their physical systems are associated with physical sciences (Taha et al. 2021). From this perspective, CPS is to integrate computation (sensing, understanding, analyzing, and predicting), communication (interfacing, interacting, and intervening), and controls (coordinating, cooperating, and collaborating) as a smart system (Alur 2015; Rawat et al. 2016; Song et al. 2017).

The advance of a CPS from a traditional mechatronic system is its capabilities in integrating both hardware and software MFMs. Figure 10.3 shows that the integration of CPS is implemented by (1) networking sensors to perceive the environment and provide feedback to cyber systems and (2) networking actuators to coordinate and control behaviors of individual actuators to fulfill the tasks at system level (Gunes et al. 2014). Therefore, a CPS can integrate both the onsite and

Mechatronics for Complex Products and Systems: Project-Based Design Approaches for Robots, Cyber-Physical Systems, Digital Twins, and Other Emerging Technologies, First Edition. Zhuming Bi.

Companion website: www.wiley.com/go/bi/mechatronics

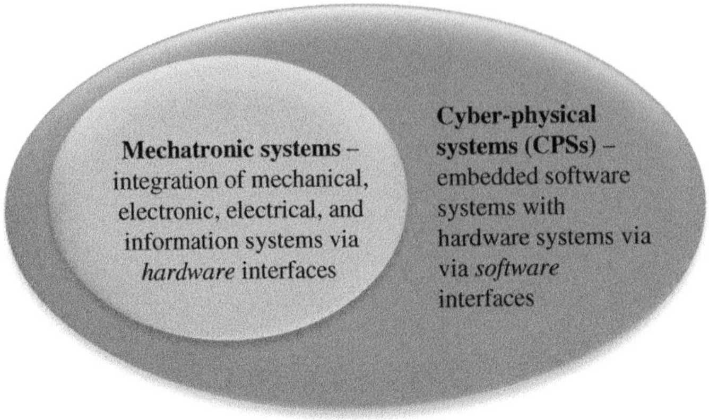

Figure 10.1 Relation of mechatronic system with cyber-physical systems (CPSs)

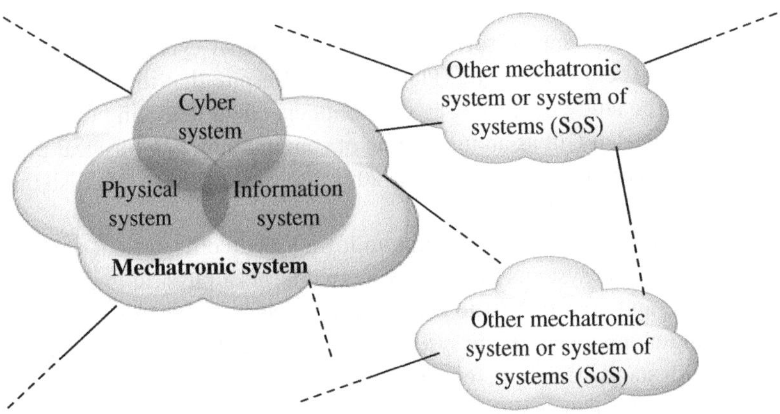

Figure 10.2 From mechatronic systems to *cyber-physical systems* (CPSs)

remote resources; while a conventional mechatronic system only involves the integration of onsite sources. CPSs are engineered physical systems whose operations are controlled, monitored, coordinated, and integrated by embedded cyber systems (Ragulski et al. 2014).

Developing mechatronic systems has led to ubiquitous uses of computers as embedded systems where hardware and software are integrated to fulfill specific requirements. The distinction of a cyber system from the associated physical systems is conceptual and abstracted based on the functions that the system fulfills. In the implementation, they can be separated as respective functional units, or, they can be united as mechatronic units as a whole. Every physical system corresponds to tangible components. However, computation for a cyber system is an abstract notion that can be represented by a mathematical function or relation. A cyber system is traditionally implemented by analog computers or even mechanically. Nowadays, cyber systems are mostly implemented digitally, and microprocessors exhibit both physical and computational characteristics. Plateaux et al. (2016) summarized the difference between mechatronic systems and CPS in Table 10.1.

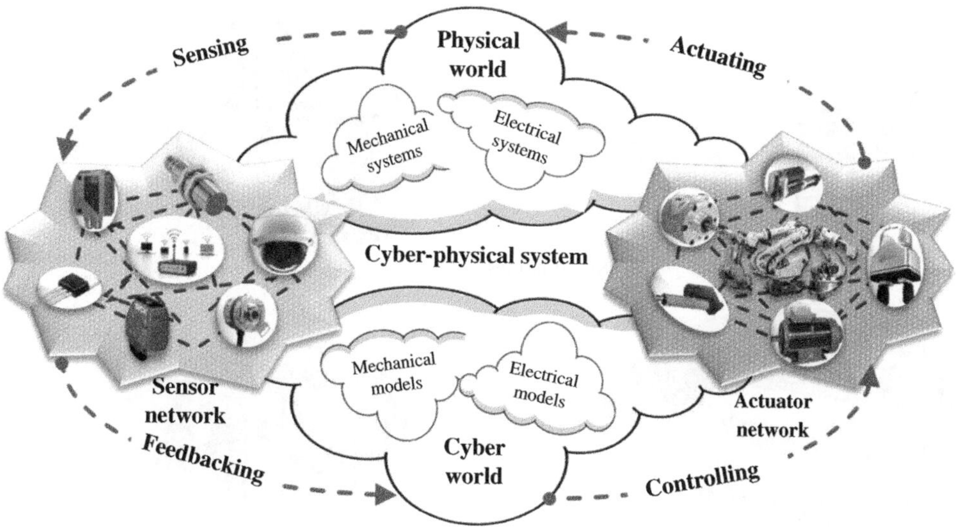

Figure 10.3 Integrating onsite and remote resources by networks in CPS

Table 10.1 Comparison of mechatronic systems and CPSs

Mechatronic systems	Cyber-physical systems
Multi-, cross-, inter-, or trans-disciplinary and heterogeneous systems	
Multi-domain integrated functional systems	
Dynamic physical interactions	
Fixed structure	Adaptive structure
Centralized architecture	Decentralized structure
Physically integrated as a whole and compact unit	Integrated or geographically distributed
	Human in the loop
	Functionally open
	Large communication and real-time processing

Most of the modern engineering products from smartwatches to home appliances are CPSs, and CPSs such as smartphones become ubiquitous. A CPS can be a stand-alone mechatronic component such as *a navigating system* with an onboard GPS for positioning, *a camera* to perceive the environment, *a thermometer* to sense temperature, or *a set of assembled actuators* to replace a human's operations. A CPS can be a full mechatronic system such as a robotic team whose control systems are fully embedded in team members, and the CPS is to coordinate, cooperate, and interact with members for system-level goals in a synergistic manner.

More and more CPSs are introduced to improve the quality of our daily lives. Figure 10.4 shows some typical examples of CPSs including smart homes, segways, electric cars, and robots. In addition, the applications of CPSs have been widely explored in the manufacturing, energy sector, defense, consumer markets, and healthcare industry. Taking an example of the applications in

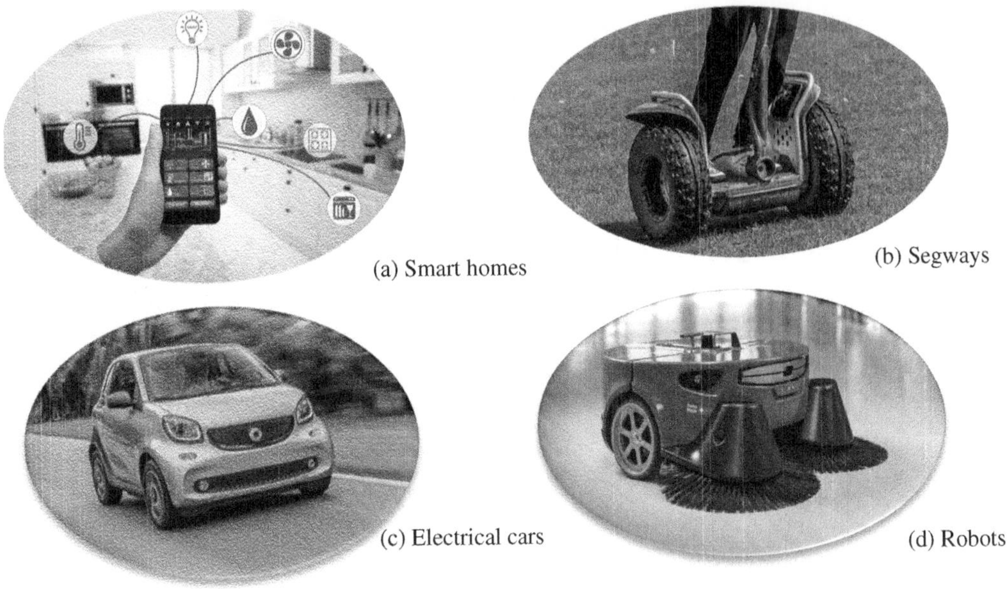

Figure 10.4 Examples of typical CPSs (*Source:* (a) Andrey Popov/Adobe Stock; (b) Adventure Sports Ltd / https://www.adventuresport.co.uk/activities/packages/twin-activity / Last accessed on July 19, 2024)

healthcare, more and more CPSs are developed to replace conventional products such as pacemakers, insulin pumps, personal assistance, diagnosis, fitness, rehabilitation, surgeries, and prosthetics.

Due to its multi-, inter-, and cross-disciplinary nature, the studies on how to design and operate cost-effective CPSs have attracted numerous researchers in multiple disciplines. This chapter discusses the fundamentals and principles for designing, modeling, and using various CPSs. Since the roots of CPSs are mechatronics, designs of CPSs are mostly relevant to *model-based system engineering* (MBSE), *concurrent engineering* (CE), and *distributed systems.*

10.2 Characteristics of CPSs

CPSs are rooted from mechatronic systems. Therefore, CPSs share the commonalities of mechatronic systems in the following three aspects (Penas et al. 2017):

1) **Heterogeneity and Multi- and Cross-domains.** Mechatronic systems integrate solutions from different disciplines in implementing powerful automated systems. Similarly, CPSs are the systems that are integrated from mechanical, electrical and electronic, sensors, actuators, and networks. The solutions to a CPS are viewed as an aggregation of numerous sub-solutions in respective disciplines such as physics, communication, mathematics, system theory, and design theory.
2) **Integration.** A CPS is an integration of cyber systems with physical systems. A CPS consists of system modules, components, subsystems, and multiple systems. A CPS faces a similar challenge with a mechatronic system in developing an integrated solution based on available MFMs.
3) **Dynamic Interactions.** A CPS is designed to interact with computational components by physical inputs and outputs over the Internet; this is different from a stand-alone mechatronic system.

However, both mechatronic systems and CPSs are capable of responding to the detected changes and uncertainties dynamically.

System architecture is a formal way to describe components, interactions, and evolving rules and regulations of the system (Bi et al. 2014; Wang et al. 2014). System architecture can also be used to characterize a system. Many architectures have been developed for CPSs. Here, CPS architecture by Sampayo and Pecas (2022) was introduced to show the basic components and relevant terminologies of CPSs (see Figure 10.5 and Table 10.2).

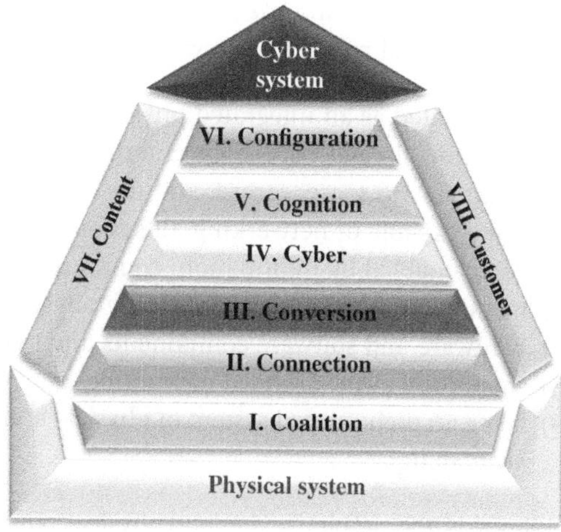

Figure 10.5 Architecture of CPS

Figure 10.5 shows that architecture of CPSs can be featured by 8C, that is, *Connection*, *Conversion*, *Cyber*, *Cognition*, *Configuration*,

Table 10.2 List of basic components and relevant terminologies of CPSs

Basic component		Relevant terminologies
I.	*Coalition*	• Integration of value-chain • stakeholder collaboration • adaptability to changes • dynamic reconfiguration • e-value-chains
II.	*Connection*	• Data collection • sensor network • data transmission • data acquisition • data storage • data distribution • connection • protocols • wearables • interfaces • human intelligence• integration • tele/electronic communications
III.	*Conversion*	• Data-to-information transformation • algorithms • deductions • self-consciousness machines
IV.	*Cyber*	• Data mining • database • network of machines and devices • information center • history records • automatic updates • cloud computing • virtual platforms • collaborative processing • similarity analysis • self-comparison/prevention • Big Data • IoT
V.	*Cognition*	• Supervision • monitoring • intelligence • service framework • information structuring and organization • optimization • intelligence • smart monitoring • visual treatment • dashboards • smart scheduling tools • decision-making support • predictive maintenance • smart scheduling • machine learning • learning techniques for workers
VI.	*Configuration*	• Control actions • application of decisions • smart control • centralization • self-configuration • self-adaptation • smart machine • 3D-printing • autonomous robots
VII.	*Content*	• Enabling access to detailed information for the entire VC (supply, logistics, production, marketing, etc.) • production parameters • product traceability record • smart product • demand forecasts • product life cycle management • business planning • smart design
VIII.	*Customer*	• Customers' role in design • production and after sales services • mass customization • stakeholder participation in processes • flexible specification modification • order tracking • product-centric manufacturing

Source: Sampayo and Pecas (2022)/with permission of Elsevier.

Coalition, *Content*, and *Customers*. According, the characteristics of CPSs that are distinguished from traditional mechatronic systems including the followings (Lozano and Vijayan 2020):

1) be operated as an integrated physical and virtual engineering system;
2) be formed by heterogeneous and distributed MFMs;
3) being able to sense, record, and use data to adapt changes at a higher level;
4) being capable of networking all system resources in a global network;
5) being enabled by information, communications, and technologies (ICT);
6) manipulating objects or performing physical processes by actuations;
7) hardware and software components and system are reconfigurable;
8) supporting event-oriented computing;
9) being an orchestrated system of physical and virtual MFMs to enhance reliability, security, and adaptability.

Since a CPS combines cyber and physical capabilities, it is not enough for engineers to understand cyber and physical systems separately. There is a need to understand (1) how cyber and physical systems interact and (2) what interfaces are used to implement interactions.

CPSs are all about integration and synergies (Platzer 2018). Taha et al. (2021) introduced two mathematic notions to (1) represent the similarities among a wide scope of things in different disciplines, (2) identify the similarities that can be potentially used to understand the causes how things work, find, and evaluate alternative solutions to the same problem effectively.

Note that the origin or nature of a mathematic concept is not the most important thing, while the importance is how a mathematic concept can be used to model and simulate a CPS. In modeling and simulating a CPS, the following three basic laws should be satisfied (Taha et al. 2021).

1) **Condition Laws** apply to the conditions when specified constraints are satisfied. For example, a triggered event when a temperature exceeds a specified degree; a laptop is shut down when the battery level is below a certain threshold.
2) **Discrete Change Laws** apply to the conditions when one or more events occur at a discrete time. For example, loading or unloading a workpiece based on the detected states of a workstation, fixtures, tools, and/or operator.
3) **Rate Change Laws** apply to the conditions of the changing rate of one or more quantities. For example, a cruise control to drive a car at 60 (miles/hour) or set an acceleration (i.e., the rate of a velocity change) to be 10 (m/s^2).

The above three laws can be used to represent all types of phenomena in diversified disciplines. In similar ways, these rules can be used to represent the interactions of multidisciplinary elements in an integrated system.

10.3 Basic Features of Cyber System of CPS

This section looks into the difference of an integrated cyber system of CPS with a conventional information system in Computer Science. Since classic theories of information systems focus on the solutions to deal with computational complexity, the solutions tend to be structured programming. Instead, a cyber system of CPS is embedded in a physical system, and computational complexity must be tackled differently due to some specific *functional requirements* (FRs) at the following aspects (Marwedel 2021; Penas et al. 2017).

10.3.1 Reactive Computation

A classic computation model generates an output when an input is provided, the model can be verified by the corresponding mathematic model, and the computation model can be treated as a function to convert inputs into outputs. The study on complexity and computability answers whether the function is computable and at what level of efficiency. Existing programming techniques can be used to program a solver to a complex problem by composing a set of sub-solvers to some basic problems. The cyber system of CPS interacts with the surrounding environment via inputs and outputs dynamically; the system responds to high-level instructions by generating control outputs for expected outcomes. The behaviors of the cyber system of CPS are *dynamic*, *reactive*, and *data driven*.

10.3.2 Parallel Computing

A traditional computation model consists of a set of instructions that are executed once a time in a sequence. In contrast, parallel computing in a cyber system computes multiple threads. Each thread corresponds to its respective process or component, and multiple threads are executed simultaneously so that they can interact with each other to achieve collective goals. Parallel computing is critical to CPS. Take an example of a robotic team, each robot has its own sensors, processors, controller, and respective tasks to be completed, computing on these components must be performed simultaneously to fulfill the system-level objectives. Distribution and parallel computation are critical to a CPS.

10.3.3 Feedback Controls

The cyber system of CPS is data driven. The physical system is controlled based on the feedback acquired by sensors from physical systems. Take an example of a smart home, users set up a desired temperature in a room, the cyber system reads the actual temperature from the thermometer installed in the location of interest, compares the desired and actual temperatures, and determines if an air conditioner should be turned on or off. In addition, the controller of the physical system is required to model the relations of controlled and controllable variables. The control theory discussed in Chapter 8 can be used to design, model, and analyze a dynamic control system. Note that traditional control theory is mainly for a continuous-time system; while the cyber systems of CPSs are mostly digital. Multiple mechatronic elements operate simultaneously and cooperate and collaborate with one another in different modes of operations. Therefore, a CPS is usually a mix of continuous and discrete dynamics.

10.3.4 Realtime-Ness

Realtime-ness refers to the need to respond to external changes subject to certain timing constraints. Therefore, the real-time requirement varies with the specified timing constraints. Realtime-ness in a machine control means that the updating rate of motion commands must be less than that of motors and actuators; an automatic trading agent operates under strict timing constraints to respond to market changes. Realtime-ness must be assured for the worst-case run-time requirements.

Processor architecture, operating systems, or programming platforms typically do not satisfy real-time requirements while real-time performance is critical for most of CPSs. Take an example of a robotic control, control commands sent to the motors must be real time to smoothen the movements of the robot. This implies all relevant tasks such as computation or communication must be completed within every control cycle. When a time delay occurs in a CPS, it becomes necessary to predict the effect of such a delay on timing-dependent protocols, resource allocations, and system performance.

10.3.5 *Dependability, Reliability, and Safety Assurance*

A CPS should be dependable, it provides its intended service with a high probability, and it must not cause any harm to users. *Dependability* is associated with (1) the security to preserve confidentiality, integrity, and availability of information; (2) the confidentiality where information is confidential and not made available or disclosed to unauthorized individuals, entities, or processes; and (3) the safety as the absence of unacceptable risk of physical injury or of damage to the health of people, either directly or indirectly as a result of damage to property or to the environment (Marwedel 2021).

Reliability refers to a system's ability to continue performing its functions when one or some system components fail. Reliability can be achieved by incorporating redundant capabilities in a system. In addition, probabilistic methods should be used to improve the reliability of CPSs effectively.

Differing from a dedicated control system that has been fully verified and validated before it is applied in the real world, the cyber system of CPS must be capable of dealing with anticipated and unanticipated changes in its application. While it is typically impractical to verify and validate its performance for all possible controlling scenarios, a critical requirement for a CPS is to ensure the safety of both anticipated and unanticipated situations. The safety requirement must be prioritized over other criteria such as versatility, speed, and cost. In particular, all CPSs in transportation, health-care industries, and constructions must prioritize their safety requirements.

CPSs are the systems people could bet their lives on. Platzer (2018) gave an example of CPSs to fly an airplane to avoid collisions. When an airplane is too close to another, it would be critical for the pilot to get some good advice on how to maneuver the airplane to avoid a collision, such advice must be promptly and safely. It is impractical to evaluate all possible trajectories and an immediate response is most critical; the computation must lead to reliable and safe separation of the airplane from others under all relevant scenarios of when and how exactly the pilot will follow the controls, and this is achieved by marking out unsafe zones in Figure 10.6.

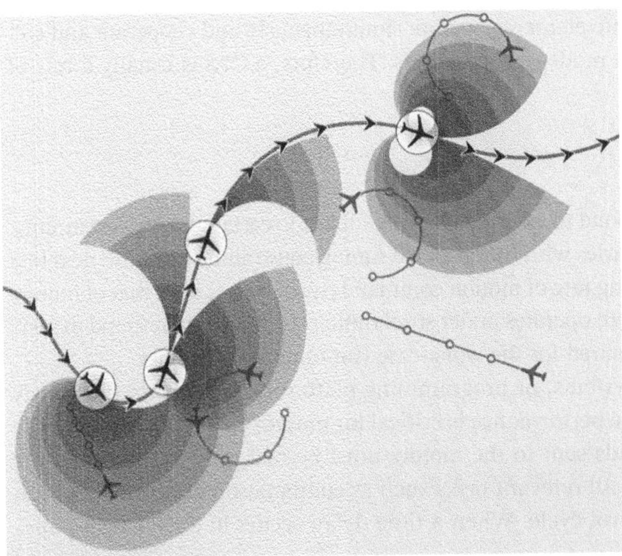

Figure 10.6 CPS example in navigating an airplane for collision avoidance (*Source:* Platzer (2018)/ Springer Nature/CC BY 4.0)

10.3.6 Biological Intelligence

A CPS is similar to a biological system that presents its physical manifestation and exhibits various physical phenomena that can be mechanical, optical, electromagnetic, and chemical. Note that a biological system possesses the intelligence to respond to environments and communicate with others on a regular basis, humans' intelligence can be great inspiration sources in developing cyber systems of CPSs. On the other hand, advanced CPSs provide better tools for human beings to understand biological systems. For example, *artificial intelligence* (AI) was inspired by biological systems to solve complex problems heuristically. However, one limitation of biological intelligence is that it has no clear notation to tell what an acceptable solution should be.

10.3.7 Hybrid Systems

A hybrid system is characterized by both continuous and discrete behaviors. It is a dynamic system that combines discrete dynamics and continuous dynamics. Most CPSs are mathematically modeled as hybrid systems; models of physical systems can be continuous, discrete, or a combination of continuous and discrete systems. For example, a bouncing ball acts in a hybrid system since (1) bouncing events are *discrete* and (2) rising or falling processes are *continuous*. Taking another example, the behaviors of quantum are based on probability, such behaviors cannot be simply classified as discrete or continuous events. In contrast, computers are mostly digitalized but they are implemented by integrating continuous electronic circuits whose reliability can only be assured digitally. In summary, hybrid systems refer to abstracted mathematical models of CPSs rather than their physical implementations.

10.3.8 Embedded Computation

When a computational system is embedded in a physical system, it is called an embedded system. A CPS includes an embedded system in general; while the difference between a CPS and a traditional embedded system is that the later focuses on the implementation of a computational system; the former focuses on the integration of the computation system with physical systems. Design of an embedded system focuses on the optimization of hardware and software for energy efficiency, minimized cost and size, and functional verifications. Design of a CPS focuses on the interfaces of the computational system with physical systems, and the main considerations are how to satisfy the functionalities of physical systems from the control perspective.

10.3.9 Standards of Cyber Systems

A cyber system should follow the IEC61499 to achieve the characteristics of automated systems (Parant et al. 2023). The standards include the requirements for (1) the portability to connect and interact with other software tools, (2) the configurability to ensure software components can be selected, and interconnected and parameterized by various vendors, and (3) the interoperability to ensure work with others in distributed and decentralized applications.

10.4 Design of CPSs

In this section, the theory and methods for designing, analyzing, and modeling CPSs are discussed.

Ever-increasing complexity of CPSs has brought more and more challenges in designing, developing, and operating systems. A CPS usually involves in heterogeneous MFMs that are developed as

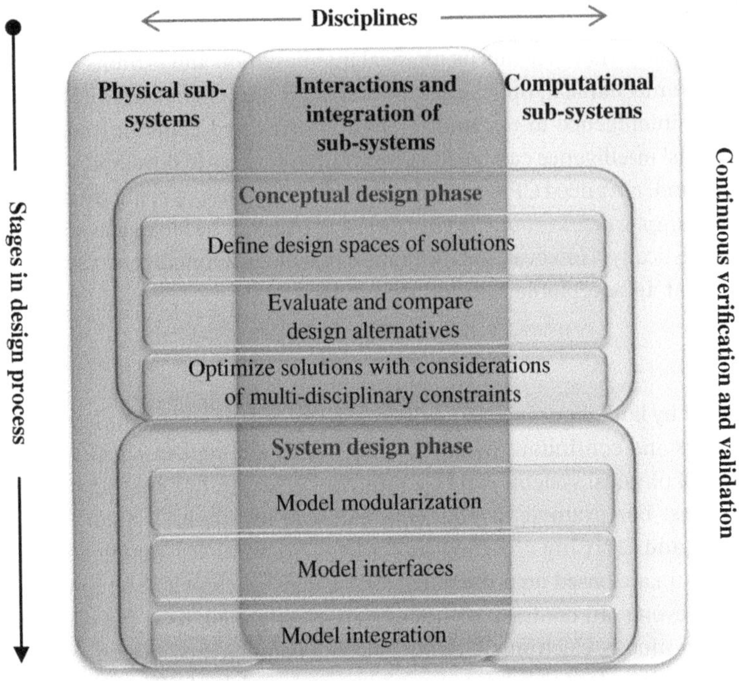

Figure 10.7 Design of a CPS

the solutions to fulfill expected functions in different domains, it is extremely difficult and even infeasible to take into consideration of all *FRs* in detail at the same level. Sophisticated design methodologies are needed to design CPSs (Bouskela et al. 2022).

Due to multidisciplinary nature, applicable methodologies for design of a CPS include *formal models*, MBSE, CE, and distributed intelligence. Note that the system design begins with the establishment of an abstract model to make system complexity manageable. An abstract system model includes block diagrams, code fragments, state machines, and governing differential equations. As shown in Figure 10.7, design of a CPS is multidisciplinary, and it covers the designs of respective physical subsystems, computational systems, and the interactions of subsystems.

Physical systems are directly associated with the functions to be fulfilled, in sense that the states of physical systems must be monitored to implement closed-loop real-time control. Especially, sensing and detecting a system's states play an irreplaceable role in a CPS in comparison with its role in a conventional mechatronic system. Computational subsystems are embedded and strongly coupled with physical systems; software programs become much complex and larger than those of traditional embedded systems. The computational and physical subsystems are integrated by a collaborative design process, and this assures the compatibilities of the interactions of system components in multiple disciplines (Hehenberger et al. 2016).

10.5 Mathematical Modeling

Physics plays an important role in design CPS. Physics covers mechanics, optics, electromagnetics, and thermodynamics that are usually involved in the implementation of CPS. Therefore, understanding the physics behind these phenomena help designers to designing and analyzing real-world problems and establishing mathematic models to control systems for desired outcomes. The working principles of a CPS will be modeled mathematically by differential equations, systems of equations, and vector calculus. Moreover, a CPS is an integration of cyber system with physical systems as a whole. Each physical component is an implementation of a conceptual or logical system such as a computation model and algorithm. The state of a system is dynamic and evolves over time; each physical component has its mathematical properties that can be used to develop the corresponding model; the model can be analyzed and simulated for designers to gain the insights of system (Lee and Seshia 2017).

A model is abstracted representation of a phenomenon, a relationship, a structure, or system which can be represented graphically, mathematically, or physically (Amazon 2024). *Mathematical modeling* is to represent a targeted object or system mathematically. A mathematic model represents most important characteristics of a system. A mathematical model must be aligned with the governing physical laws of a system. When the governing physical laws of a dynamic system cannot be defined completely, developing an explicit mathematical model for the system will be impractical. In addition, when the equations in a mathematic model include some unknown or undetermined parameters or variables, they must be measured or evaluated physically. A process to identify system parameters is called *an identification process*. Figure 10.8 classifies mathematic models into linear/nonlinear, static/dynamic, explicit/implicit, discreate/continuous, deterministic/stochastic, detective/inductive/floating, black/gray/white box, and strategic/non-strategic models. Moreover, different methods should be selected to solve differential equations effectively.

In designing a CPS, one has to determine if certain sub-problem should be solved mechanically, electrically or computationally, and a *system engineering* (SE) approach should be used to plan and schedule design tasks in the whole product lifecycle. MBSE is one of most popular SE approaches, and it can be used to model all system components of a CPS. As shown in Figure 10.9, MBSE serves as a multi-view and multilevel modeling tool in design of CPS; it should be used with others including *system modeling language* (SysML), Digraph, graph-based modeling, ontology web language (OWL), and agent technologies to establish mathematical models of a CPS in seven main design stages (Penas et al. 2017).

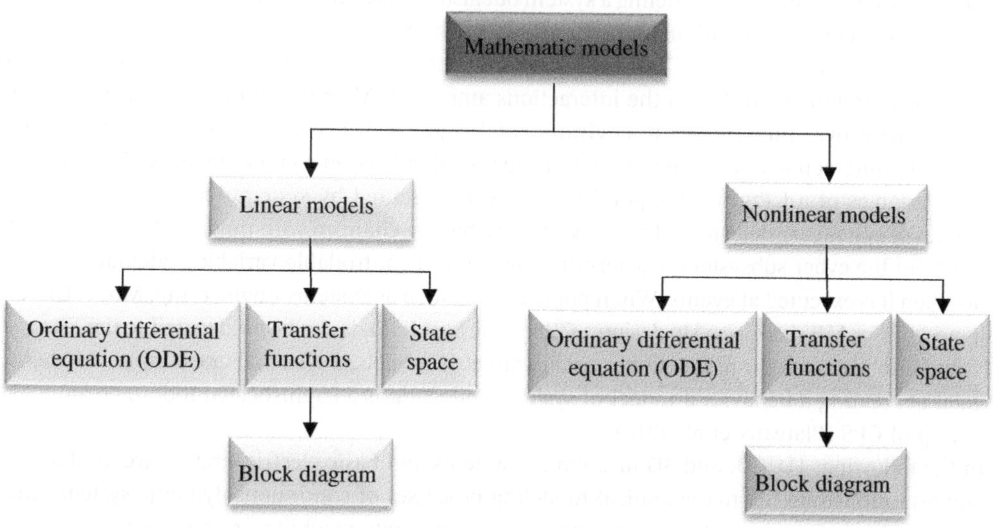

Figure 10.8 Linear and nonlinear differential equations and solutions

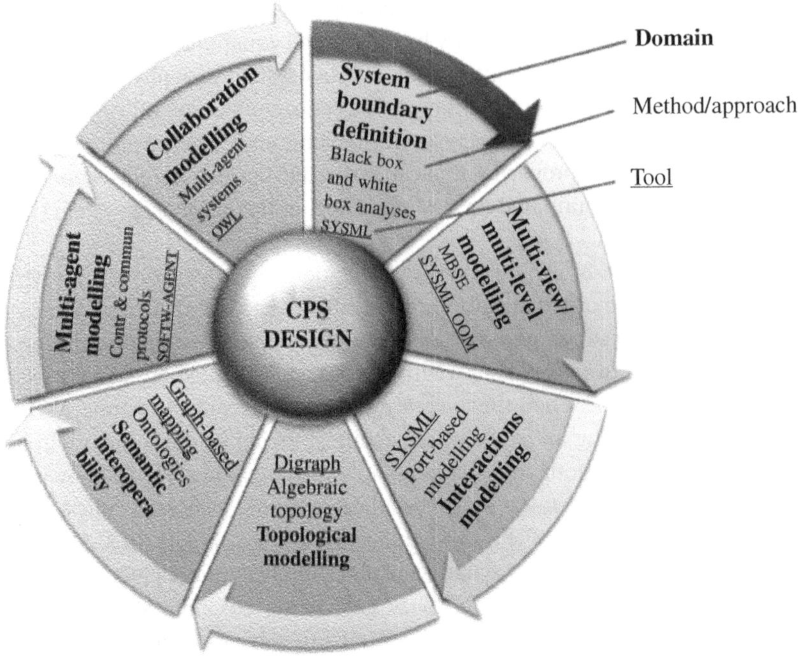

Figure 10.9 MSBE as multi-view and level modeling tools in design of CPS (*Source:* Penas et al. (2017)/ with permission of Elsevier)

SysML was extended from *UML* by the *International Council of Systems Engineering* (INCOSE). It is now an important part of MBSE for mathematical modeling in designing, analyzing, verifying, and validating various dynamic systems (Jensen 2010). Modelica is an advanced option of SysML; it is an *object oriented* (OO), declarative, multi-domain modeling language that is specialized in modeling various CPSs; note that a CPS may contain mechanical, electrical, electronic, hydraulic, thermal, control, electric power, or process-oriented subsystems. In comparison with SysML, Modelica has its advantages of not only modeling a system but also transferring the model into its corresponding set of mathematical equations for the purpose of simulation.

The challenges in designing a CPS lie not only in the heterogeneity of physical and virtual MFMs and changes and uncertainties of the interactions among MFMs but also in the needs of reconfiguring systems over time subject to environmental changes. It becomes necessary to model and analyze the interactions of computational and physical subsystems cross-domains. The changes of the variables of a CPS in a temporal space are represented by system dynamic models, and the states of system are determined by the variables that are changed with time. On the other hand, the state of the cyber subsystem is determined by a set of controllable variables calculated by the code when it is executed at events. When popular tools such as Systems Engineering, Model Driven Engineering, or MBSE are used to design a CPS, some SysML tools can be used to define the FRs of a CPS and establish structural, functional, and parametric models of system. For examples, Modelica is good at creating a behavioral model of system, and Catia is a sophisticated tool to create a 3D mock-up of CPS (Plateaux et al. 2016).

In the following, 1D, 2D, and 3D mechanical systems and basic electric circuits are used as the examples to illustrate the mathematical modeling processes of continuous dynamic systems, *discrete event dynamic systems* (DEDS), hybrid systems, and state machines, respectively.

10.5.1 Modeling Continuous Dynamics

This section aims to model continuous dynamic systems with a focus on mechanical systems. The developed modeling process is easily extendable to continuous dynamic systems such as chemical processes, electronic circuits, and biological processes.

In classical mechanics, the motion of a mechanical part is modeled by differential or integral equations. The motion is smooth when the equation is linear, time invariant, and continuous. When the motion is not smooth, abrupt transition occurs at two or more distinct modes of operations. Taking an example, a collision between two objects or among three or more objects should be modeled as a discrete and instantaneous event appropriately. In addition, a mechanical system with both smooth and unsmooth motions is called a hybrid dynamic system.

Figure 10.10 shows an abstract model of a continuous dynamic system with the sets of inputs (X) and outputs (Y) by a transform function of S. An abstract model is also referred as an *actor model* of a dynamic system in MBSE.

The inputs (X) and (Y) are the physical quantities that are changed over time as,

$$X : R \rightarrow R_x; Y : R \rightarrow R_y \tag{10.1}$$

where R represents the domain spanned by the spatial and temporal variables; R_x and R_y are the domains formed by the functions from input to output variables, respectively. The mapping function (S) from X and Y can be expressed as,

$$S(p_1, p_2, ...) : X \rightarrow Y \tag{10.2}$$

where (p_1, p_2, ...) are the set of system parameters in an MFM that may or may not be changed over time.

Example 10.1 Develop an actor model for the airplane dynamic system as shown in Figure 10.11.

Solution

Newton's second law and Euler's law can be used to describe the motion of the airplane dynamic system as shown in Figure 10.12.

In Figure 10.12, two input variables are elevator deflection angle δ_e and engine thrust force δ_T. u denotes the forward velocity; u and q denote the angular velocities around x and y axes, respectively; θ is the Euler angle, x_e and h are intermediate variables determined by a transfer matrix L_{Eb} and output variable u and w as (Cao et al. 2020),

$$\begin{bmatrix} x_e \\ h \end{bmatrix} = L_{Eb} \begin{bmatrix} u \\ w \end{bmatrix} \tag{10.3}$$

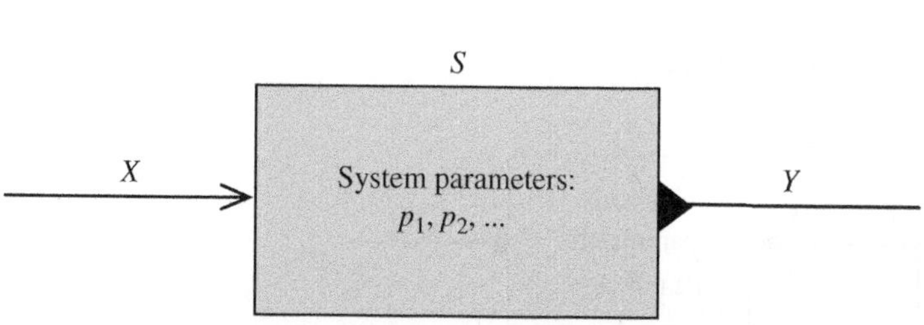

Figure 10.10 Abstract model of a continuous dynamic system

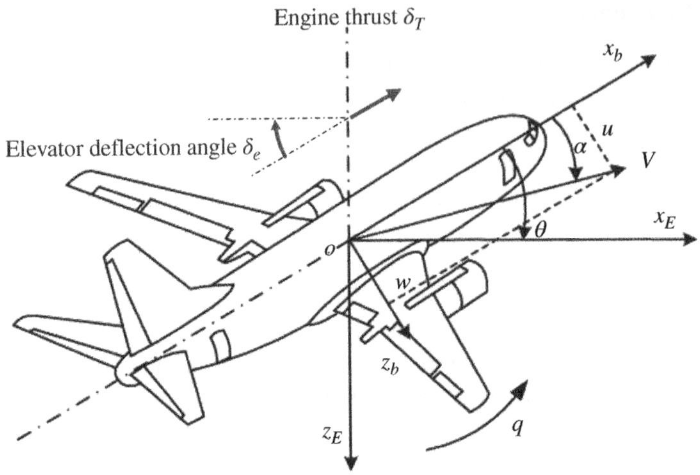

Figure 10.11 Airplane dynamic system

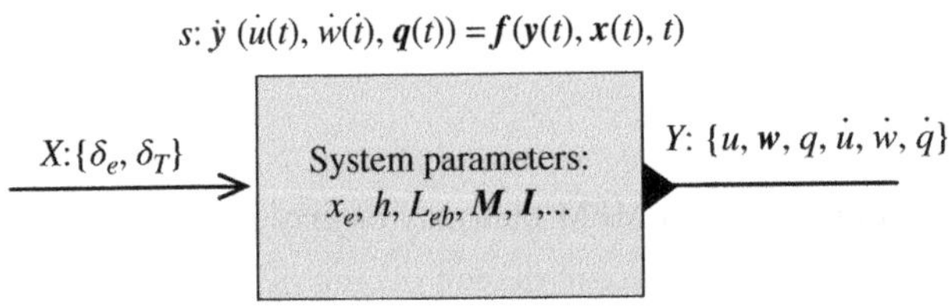

Figure 10.12 Abstract model of airplane dynamic system

Accordingly, the mapping function S from X to Y are determined as,

$$S : \dot{\boldsymbol{y}}\left(\dot{u}(t), \dot{w}(t), q(t)\right) = \boldsymbol{f}(\boldsymbol{y}(t), \boldsymbol{x}(t), t)$$

where

$$\boldsymbol{y}(t) = [u, w, q, \theta, x_e, h]^T \text{ and } \boldsymbol{x}(t) = [\delta_e, \delta_T]^T$$

As shown in Figure 10.13, two actor models (S_1 and S_2) can be connected as one cascade actor model when the inputs of the second actor model are the outputs of the first actor model, that is, $Y_1(t) = X_2(t)$ when $t \in R$

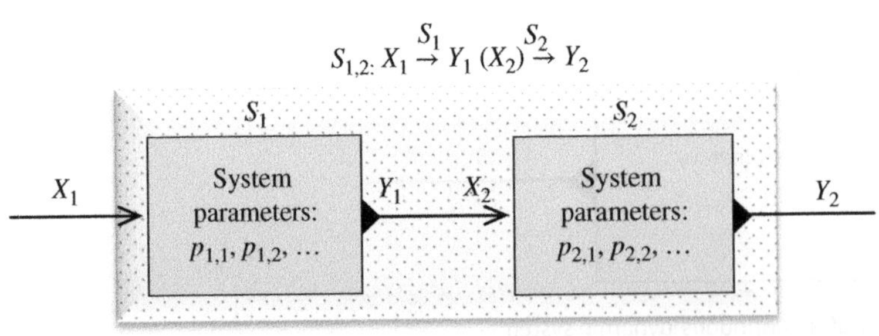

Figure 10.13 A cascade actor model consists of two actor models

Example 10.2 Consider a helicopter is its horizontal posture in Figure 10.14 (Mettle et al. 2000). It has two rotors, one main rotor that provides a lift force and another tail rotor that is used to balance the torque when the main rotor rotates. Note that if a tail rotor does not generate a force to balance the torque around *z*-axis, the helicopter might spin. In addition, the flying direction of the helicopter can be controlled by the net torque from both rotors. Use a cascade actor model to represent the corresponding dynamic system.

Solution

The Euler law is used to describe the relations of new driving torque and the rotation along *z*-axis as,

$$T_{z,main}(t) - T_{z,tail}(t) = I_{zz} \cdot \ddot{\theta}_Z \tag{10.4}$$

Correspondingly, the flying direction of the helicopter can be determined by taking two integrals of Eq. (10.4) as,

$$\theta_z(t) = \theta_z(0) + \int_0^t \dot{\theta}_z(0)dt + \frac{1}{I_{zz}} \int_0^t [T_{z,main}(t) - T_{z,tail}(t)]dt \tag{10.5}$$

Therefore, a cascade actor model can be constructed by representing Eqs. (10.4) and (10.5) into respective actor models as shown in Figure 10.15.

The state of a dynamic system evolves over time, and system behaviors can generally be represented by one or more differential equations. A system is *continuous* when the outputs of system variables can be defined in any specific time. Some examples of continuous dynamic systems are airplanes, satellites, and the responses of op-amp filters to input signals.

Differential equations are used to model the behaviors of continuous dynamic systems.

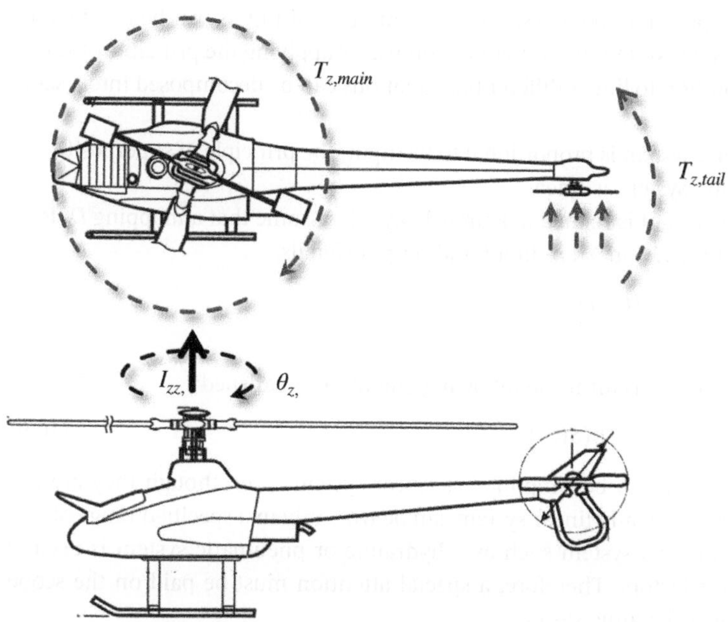

Figure 10.14 Simplified model for directional control of helicopter

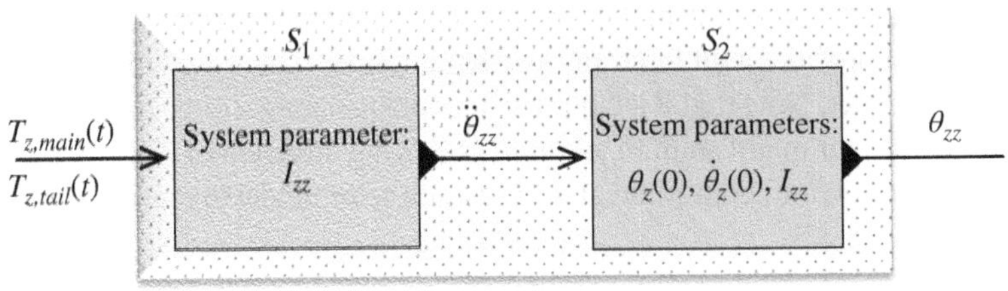

Figure 10.15 A cascade actor model for direction control of helicopter

1) *Linear* and *time-invariant differential equations*

A simple differential equation is *linear* and *time invariant*. The differential equation of an example linear and time-invariant system is,

$$A\frac{d^2x(t)}{dt^2} + B\frac{dx(t)}{dt} + Cx(t) = 0 \tag{10.6}$$

where A, B, and C are the constants and $x(t)$ is the output variable. In a linear and time-varying differential equation (10.6), variable $x(t)$ and its derivatives including $x(t), x'(t), x''(t), \ldots, x^{(n)}(t)$ are combined linearly. However, a coefficient in the combination can be a constant or vary with another independent variable.

The principle of superposition can be applied to a linear system. Assume that a system S has a mapping of $X \to Y$ from input signals (X) to output signals (Y), it is linear if it satisfies the superposition property of,

$$\forall x_1, x_2 \in X \text{ and } \forall a, b \in R, \mathrm{S}(\mathrm{a}x_1 + \mathrm{b}x_2) = \mathrm{aS}(x_1) + \mathrm{aS}(x_2) \tag{10.7}$$

As shown in Eq. (10.7), the principle of superposition states that the respective solutions to two linear systems can be combined as one solution to the combined linear systems.

Consequentially, the output of a linear system with a number of inputs can be obtained by adding all outputs of the linear system for individual inputs. By applying the principle of superposition, the combined solution to linear differential equations can be decomposed into a set of simple solutions.

If the output of a dynamic system is proportional to its input, the principle of superposition is valid and the corresponding system is linear.

The definition of time invariant is related to a time delay (τ). Assume that a mapping D_τ for a time delay from $X \to Y$ where X and Y are input and output signals,

$$\forall x \in X \text{ and } \forall t \in R, D_\tau(x) = x(t - \tau) \tag{10.8}$$

where τ is a delay actor.

A system S: $X \to Y$ is time invariant if the following condition is satisfied:

$$\forall x \in X \text{ and } \forall \tau \in R, S(D_\tau(x)) = D_\tau(S(x)) \tag{10.9}$$

Most of the physical systems can be simplified as linear systems even though they are not strictly linear. As a matter of fact, any linear system can be true only in a specified range of system operations; while a physical system such as a hydraulic or pneumatic system is usually affected by some nonlinear factors. Therefore, a special attention must be paid on the scope of the application of a linear dynamic system.

2) *Linear* and *time-variant differential equations*

   As a comparison, an example of a linear and time-variant differential equation can be,

$$A(t)\frac{d^2x(t)}{dt^2} + B(t)\frac{dx(t)}{dt} + C(t)x(t) = 0 \tag{10.10}$$

   where $A(t)$, $B(t)$, and $C(t)$ are variables that are changed with respect to time t.

   Note that if Eq. (10.10) is treated as a linear equation, it should not include any powers or other functions of dependent variables.

3) *Nonlinear and time-variant differential equations*

   An example of a nonlinear and time-variant differential equation can be,

$$A(t)\frac{d^2x(t)}{dt^2} + B(t)\left(\frac{dx(t)}{dt}\right)^2 + C(t)\cos(x(t)) = 0 \tag{10.11}$$

   where $\left(\frac{dx(t)}{dt}\right)^2$ and $\cos(x(t))$ are two nonlinear terms; $A(t)$, $B(t)$, and $C(t)$ are time-variant terms.

   The principle of superposition cannot be applied to a nonlinear dynamic system. Therefore, the procedure to solve a nonlinear differential system becomes more complicated. The most practical approach to searching for the solution to a nonlinear system is to linearize a nonlinear system as a set of linear dynamic systems in the discretized ranges.

Once the nonlinear system is decomposed into a set of linear models, the techniques including the principle of superposition for linear systems can be applied to find the solution to the original nonlinear system numerically *step by step*.

The most common types of nonlinear differential equations are *ordinary differential equations* (ODEs), *partial differential equations* (PDEs), and *differential algebraic equations* (DAEs).

An ODE is associated with one physical quantity that depends on a single independent variable such as a time or displacement or its derivatives. The system can be affected by both internal and external factors.

For an example, the first-order ODE can be formulated as,

$$\dot{x}(t) = f(x(t), u, \mathbf{p}, t) \tag{10.12}$$

where x is the physical quantity of interest; t denotes the time; u is an external variable; $\mathbf{p}$ is a set of system parameters (constants). Generally, an ODE with the nth order implies that the highest order of the derivatives of $x(t)$ is n. Moreover, the linearity of an ODE depends on whether the function of f is *linear* or *nonlinear*.

A PDE is associated with a physical quantity that depends on both time and space. Typically examples of physical systems with PDEs are heat transfer, fluid flow, elastic solids, wave propagation, electromagnetic field, and some biological phenomena such as population immigration, tumor growth, and epidemic propagation. For an example, a one-dimensional transient heat transfer problem can be modeled as a PDE as,

$$\frac{\partial T}{dt}(x, t) = \alpha\frac{d^2T}{dx^2}d(x, t) + \rho(x, t) \tag{10.13}$$

where x and t are the variables for space and time, respectively; α is the conductivity coefficient; $\rho(x, t)$ relates to the function of heat generation. It is usually impractical to solve PDEs analytically; therefore, PDEs are solved by numerical simulation by default.

A DAE stands for a differential equation whose derivative terms cannot be isolated explicitly from others. The first-order DAE can be expressed as,

$$f(\dot{x}(t), x(t), u, \boldsymbol{p}, t) = 0 \tag{10.14}$$

where x is the physical quantity of interest; t denotes the time; u is an external variable; $\boldsymbol{p}$ is a set of system parameters (constants). Equation (10.14) can be generalized for an nth DAE that includes the highest n-order of the derivative $x(t)$ as $x^{(n)}(t)$ in the inexplicit function f. Due to the strong coupling, Eq. (10.14) or alike is usually solved by numerical simulation in default.

10.5.2 *Discrete Event Dynamic System (DEDS)*

In *a discrete-time dynamic system* or a *DEDS*, the state change of one or more variables occurs only in a discrete instant of time. The discrete instant often refers to a moment when one or more physical quantities are measured and processed. The system is required to respond to the detected changes. Moreover, the inputs of a discrete system can be digital signals or continuous-time signals for discretized differential equations.

A DEDS has two distinguished features. *First*, the state space is discrete, and a change of system states is event-driven rather than time driven. This implies that differential equations are inapplicable to model DEDS. *Second*, events occur asynchronously, in general, and each event causes a jump from one state to another in the discrete state space. The set of events, denoted by E, is a discrete set (Lafortune 2019).

Examples of DEDS are protocol-based communications, material-handling systems in automated production lines, and planning, scheduling, and controls of flexible manufacturing systems. The behaviors of a DEDS can also be abstracted and modeled with respect to space and time. For example, a path or trajectory of a robot in a physical space can be represented by a set of discrete tag points in a space grid, and a robotic motion can be defined as a set of transitions (or events) from one tag point to another.

The behaviors of a DEDS can be modeled as a structure with all possible states and associated transitions from one state to another. The structure of DEDS consists of *circles* and *arrows* where the circles represent system states the arrows represent all possible transitions that can be triggered by one or more events.

Depending on the attributes attached to the states (i.e., circles) and transitions (i.e., arrows), the structures of DEDS can be used to fully model *DEDSs*. Popular modeling tools for DEDSs are *automata*, *labeled transition systems*, and *Petri nets*. Using *Petri nets* is one way to model DEDSs. A Petri net uses discrete structures but system states are not represented explicitly. A Petri net uses *place nodes* and *transition nodes* to model pre- and post-conditions in firing each event. Furthermore, a token mechanism is used to provide more information pre- or post-conditions. Transition nodes may be labeled by events and this leads to a labeled Petri net.

Here, the automaton theory is used to model event-triggered transitions in a DEDS. Automatons are purple automata; an automaton refers to an abstract machine. An automaton consists of the sets of states and transitions that are graphically represented by circles and arrows, respectively.

A *deterministic* automaton is denoted by a six-tuple G as,

$$G = (X, E, f, \Gamma, x_0, X_m) \tag{10.15}$$

where X is the set of states which could be infinite; E is the finite set of the events associated with the transitions; f: $X \times E \to X$ is the transition function; $f(x, e) = y$ is a partial function in its domain. It indicates that there is a transition labeled by event e from state x to y. Γ: $X \to 2^E$ is a feasible event function and $\Gamma(x)$ is the set of all events e where $f(x, e)$ is defined; X_m is the set of the marked states; x_0 is the set of initial states.

A *nondeterministic* automaton is denoted by a six-tuple G as,

$$G = (X, E \cup \{\varepsilon\}, f, \Gamma, x_0, X_m) \tag{10.16}$$

where ε represents an empty string; f refers to the mapping: $f: X \times (E \cup \{\varepsilon\}) \rightarrow 2^X$, i.e., $f(x, e) \subseteq X$; the initial state can be a set of states as $X_0 \subseteq X$.

Automata are used to manipulate formal languages over the set (E) of events.

Let E^* be the set of all finite strings of the elements in E, the transition function f of a deterministic or nondeterministic automaton ($X \times E$) is extended to ($X \times E^*$) recursively as,

$$\left.\begin{aligned} f(x, \varepsilon) &:= x \\ f(x, se) &:= f[f(x, s), e] \text{ for } s \in E^* \text{ and } e \in E \end{aligned}\right\} \tag{10.17}$$

A DEDS can be modeled as an automaton G in the language $\mathcal{L}(G)$ and the marked language $\mathcal{L}_m(G)$ as

$$\mathcal{L}(G) := \{s \in E^* : (\exists x_0 \in X_0) f(x_0, s) \text{ is defined}\} \tag{10.18}$$

$$\mathcal{L}_m(G) := \{s \in \mathcal{L}(G) : (\exists x_0 \in X_0) f(x_0, s) \cap X_m \neq \emptyset\} \tag{10.19}$$

String s is in $\mathcal{L}(G)$ if and only if it represents an admissible path in the transition structure G beginning at an initial state. A prefix of every string in $\mathcal{L}(G)$ must also be in $\mathcal{L}(G)$, that is, $\mathcal{L}(G)$ is prefix – closed. When f is a total function within the domain, $\mathcal{L}(G)$ must satisfy $\mathcal{L}(G) = E^*$.

The set of the states that can be reached by a string from X_0 is called as the set of accessible states of G. $\mathcal{L}_m(G)$ consists of the strings (s) that satisfy $f(x_0, s) \cap X_m \neq 0$, these strings correspond to the paths that terminate at a marked state. Therefore, $\mathcal{L}_m(G)$ is a subset of $\mathcal{L}(G)$, $\mathcal{L}_m(G)$ is not pre-fix – closed unless $X_m = X$.

Marking states help to detect a deadlock or livelock in a transient structure, a deadlock of a deterministic automaton G is a state where $\Gamma(x) = \emptyset$ while $x \notin X_m$; this implies a deadlock state has no possible transition and the corresponding state is unmarked. On the other hand, a livelock state refers to the case when a component contains two or more unmarked states from where unmarked states are reachable in following transitions in G.

It is rare that a complex DEDS is modeled monolithically. A complex system is usually decomposed into a set of subsystems, each subsystem is modeled individually, the over model of DEDS is constructed by assembling the models of subsystems, and the following rules are used to assemble two automata in parallel.

Let two deterministic automata be $G_1 = (X_1, E_1, f_1, \Gamma_1, x_{0,1}, X_{m,1})$ and $G_2 = (X_2, E_2, f_2, \Gamma_2, x_{0,2}, X_{m,2})$ where all of the unreachable states from respective initial states have been deleted. In other words, both G_1 and G_2 are accessible, the composition of G_1 and G_2 in parallel can be found as,

$$G_1 || G_2 := \{X_1 \times X_2, E_1 \cup E_2, f[(x_1, x_2), e], \Gamma_1 || \Gamma_2(x_1, x_2), (x_{0,1}, x_{0,2}), X_{m,1} \times X_{m,2}\} \tag{10.20}$$

where

$$f[(x_1, x_2), e] := \begin{cases} [f_1(x_1, e), f_2(x_2, e)] & \text{if } e \in \Gamma_1(x_1) \cap \Gamma_2(x_2) \\ [f_1(x_1, e), x_2] & \text{if } e \in \Gamma_1(x_1) \backslash E_2 \\ [x_1, f_2(x_2, e)] & \text{if } e \in \Gamma_2(x_2) \backslash E_1 \\ \text{undefined} & \text{otherwise} \end{cases} \tag{10.21}$$

and

$$\Gamma_1||\Gamma_2(x_1,x_2) = [\Gamma_1(x_1) \cap \Gamma_2(x_2)] \cup [\Gamma_1(x_1)\backslash E_2] \cup [\Gamma_2(x_2)\backslash E_1] \tag{10.22}$$

A common event in $E_1 \cap E_2$ in a parallel composition can only be executed simultaneously for both two automata. Otherwise, if an event is private in $(E_2\backslash E_1) \cup (E_1\backslash E_2)$, it can be executed whenever it becomes possible.

When $E_2 = E_1$, the executions of all transitions must be synchronized, and this leads to,

$$\left.\begin{aligned} \mathcal{L}(G_1||G_2) &= \mathcal{L}(G_1) \cap \mathcal{L}(G_2) \\ \mathcal{L}_m(G_1||G_2) &= \mathcal{L}_m(G_1) \cap \mathcal{L}_m(G_2) \end{aligned}\right\} \tag{10.23}$$

When $E_2 \cap E_1 = \emptyset$, no synchronized transition occurs, and $G_1||G_2$ is the concurrent behavior of G_1 and G_2.

In a general case,

$$\mathcal{L}(G_1||G_2) = P_1^{-1}[\mathcal{L}_m(G_1)] \cap P_2^{-1}[\mathcal{L}_m(G_2)] \tag{10.24}$$

The P_i^{-1} $(i = 1, 2)$ in Eq. (10.23) is the inverse of the projection operation P_i and P_i is,

$$P_i : (E_1 \cup E_2)^* \rightarrow E_i^* \ (i = 1, 2) \tag{10.25}$$

where

$$\left.\begin{aligned} &P_i(\varepsilon) := \varepsilon \\ &P_i(e) := \begin{cases} e & \text{if } e \in E_i \\ \varepsilon & \text{if } e \notin E_i \end{cases} \\ &P_i(se) := P_i(s)P_i(e) \text{ for } s \in (E_1 \cup E_2)^*, e \in (E_1 \cup E_2) \end{aligned}\right\} \tag{10.26}$$

The inverse projection becomes $P_i^{-1} : E_i^* \rightarrow 2^{(E_1 \cup E_2)^*}$. For a string $t \in E_i^*$, one has,

$$P_i^{-1}(t) := \{s \in (E_1 \cup E_2)^* : P_i(s) = t\} \tag{10.27}$$

Equation (10.26) shows that $P_i^{-1}(t)$ retunes all the strings in the domain of P_i that are mapped to the projected string t.

10.5.3 Hybrid Modeling

The theory of hybrid systems is used to model mixed continuous and discrete behaviors of dynamic systems. In the theory of hybrid systems, a hybrid system is a combination of differential inclusions for continuous variables and difference inclusions for discrete variables; a hybrid system is subjected to the constraints so-called hybrid inclusions. Hybrid inclusions allow to analyze stability and robustness of CPSs systematically (Rawat et al. 2016). A number of methods, such as hybrid automata and hybrid Petri nets (Ghomri and Alla 2023; Pettersson and Lennartson 2023), have been developed to model hybrid systems.

In this section, hybrid automata by Johansson et al. (2023) are used to show how a hybrid system is modeled mathematically. While differential equations and finite state automata can represent a wide range of physical and computational systems, many CPSs are not purely continuous or discrete systems. It would be helpful to develop a hybrid model to represent both of continuous variables and discrete events at the same time.

A hybrid Automation H is an eight-tuple of,

$$H = (Q, X, f, Init, D, E, G, R) \tag{10.28}$$

where

Q	is a space of discrete states;
$X = R^n$	is a space of continuous states;
$f: Q \times X \rightarrow R^n$	is a vector field;
$Init \subset Q \times X \rightarrow R^n$	is a set of initial states;
$D: Q \rightarrow P(X)$	is a domain or referred to as an invariant set;
$E \subset Q \times Q$	is a set of connections/edges;
$G: E \rightarrow P(X)$	is a guard condition;
$R: E \times X \rightarrow P(X)$	is a reset map.

Let $(q, x) \in Q \times X$ be a state of H and $(q_0, x_0) \in Init$, a hybrid automata H must be able to represent all possible states and their evolutions.

A continuous x evolves based on the vector field $f(q_0, \ldots)$ as long as x is still in $D(q_0)$; while the discrete state q remains constant. When x reaches a guard $G(q_0, q_1)$, the discrete state q_0 changes to q_1 if $(q_0, q_1) \in E$ and the continuous state is reset to a certain value in $R(q_0, q_1, x)$. After this discrete transition, the continuous evolution is resumed and the whole process is repeated for one transition after another.

It can be helpful to visualize a hybrid automaton as a direct graph (Q, E) where Q and E are the sets of vertices and edges, respectively. For a vertex $q \in Q$, a set of the associated initial states is written as $Init_q = \{x \in X: (q, x) \in Init\}$, its vector field is denoted as $f(q, \ldots)$, and its domain is denoted as $D(q)$. For an edge $e \in E$, a set of the associated guard is written as $G(e)$ and a reset map becomes $R(e, \ldots)$.

Example 10.3 Use hybrid automata to model a two-tank system as shown in Figure 10.16. Two tanks are denoted as $i\,\{i = 1, 2\}$, x_i denotes the volume of water, v_i denotes the flow rate out of tank i, and w is the flow rate of water that flows into the system. Note that w can only flow one of two tanks at a time instant.

Assume that at its initial states, $x_1 \geq r_1$ and $x_2 \geq r_2$, and the system is controlled so that the water volume in each tank remains about r_i. The control system can be implemented by switching the incoming flow rate w to tank 1 when $x_1 \leq r_1$ and to tank 2 when $x_2 \leq r_2$.

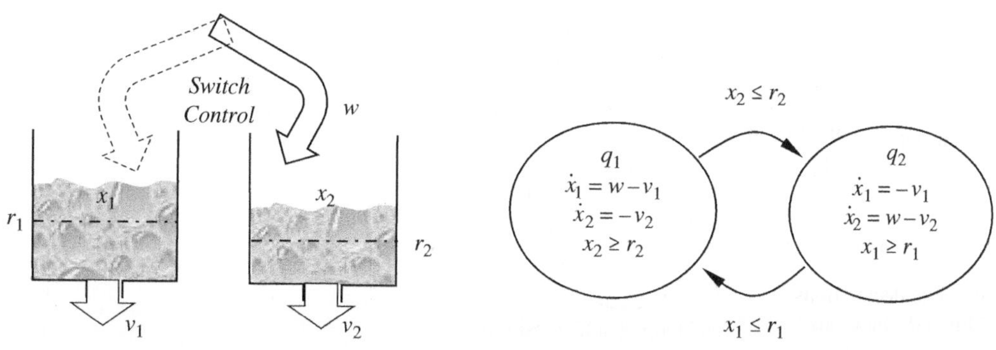

Figure 10.16 Two-tank water system

Solution

Following the definition of hybrid automata in Eq. (10.27), eight-tuple can be defined for the two-tank water system as follows (Figure 10.17).

The set of vertices $Q = \{q\} \rightarrow Q = \{q_1, q_2\}$.
The set of continuous states $X = \{x_1, x_2\}$ and $X \in R^2$.
The set of vector field $f(q_1, x) = (w - v_1, -v_2)$ and $f(q_2, x) = (-v_1, w - v_2)$.
The set of initial states $Init = Q \times \{x \in R^2 : (x_1 \geq r_1) \wedge (x_2 \geq r_2)\}$.
The domain of invariant set $D(q_1) = \{x \in R^2 : (x_2 \geq r_2)\}$ and $D(q_2) = \{x \in R^2 : (x_1 \geq r_1)\}$.
The set of edges $E = \{(q_1, q_2), (q_2, q_1)\}$.
The set of guides $G(q_1, q_2) = \{x \in R^2 : x_2 \leq r_2)\}$ and $G(q_2, q_1) = \{x \in R^2 : x_1 \leq r_1)\}$.
The reset map $R(q_1, q_2, x) = R(q_2, q_1, x) = \{x\}$.

Graphs can be viewed as formal definition of hybrid automata and they provide the same information in the aforementioned definitions (Zheng 2023).

10.5.4 State Machines

A state machine is a model of a system with discrete dynamics, each reaction maps inputs to outputs, and the map depends on the current state.

$$states = \{state\,1, state\,2, \ldots, state\,n\} \tag{10.29}$$

A *finite state machine* (FSM) consists of a finite set of *states, inputs, outputs*, as well as *the transitions* among states. Multiple FSMs may be needed to support the interactions and communications of system elements. In an FSM, the guard determines whether the transition many be taken on a reaction, and the action specifies the outputs are produced on each reaction.

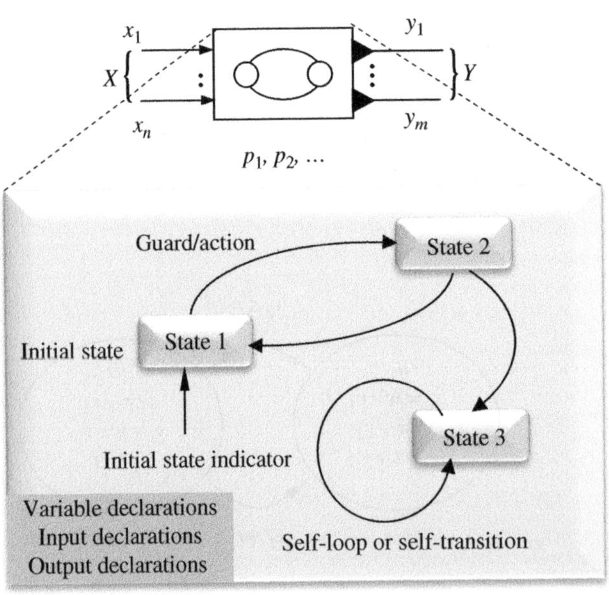

Figure 10.17 Illustration of an FSM and notations

Depending on an application, an action may be triggered when signal states (p_1 and p_2) are changed; an action may also be triggered when signal states (p_1 and p_3) meet some generic conditions related to some constants. Some examples of guards for actions triggered by pure signals are

Guard for action	Explanation
True	Transition is always enabled.
p_1	Transition is enabled when p_1 presents.
$\neg p_1$	Transition is enabled when p_1 is absent.
$p_1 \wedge p_2$	Transition is enabled when both of p_1 and p_2 present.
$p_1 \vee p_2$	Transition is enabled when either of p_1 or p_2 presents.
$p_1 \wedge\neg p_2$	Transition is enabled when p_1 presents and p_2 is absent.
$p_3 = 1$	Transition is enabled when p_3 presents and has a value of 1.
$p_3 = 1 \wedge p_1$	Transition is enabled when p_3 presents and has a value of 1 and p_1 presents.
$p_3 > 10$	Transition is enabled when p_3 presents and has a value larger than 10.

The notation of a FSM is a five-tuple as

$$(States, Inputs, Outputs, Updates, InitialStates) \tag{10.30}$$

where

States	are a finite set of states;
Inputs	are a finite set of inputs;
Outputs	are a finite set of outputs;
Updates	are the set of mapping functions to determine outputs based on inputs, states, and guards for actions, that is, $States \times Inputs \rightarrow States \times Outputs$;
InitialStates	is the set of initial states.

Moreover, a complex state machine can be constructed from a set of simply state machines based on their interactions in applications. For example, let A and B be two FSM with simultaneous actions,

$$\left.\begin{aligned} A &= (States_A, Inputs_A, Outputs_A, Updates_A, InitialStates_A) \\ B &= (States_B, Inputs_B, Outputs_B, Updates_B, InitialStates_B) \end{aligned}\right\}$$

Then, a synchronized side-by-side FSM (C) can be constructed as

$$C = (States_C, Inputs_C, Outputs_C, Updates_C, InitialStates_C) \tag{10.31}$$

where

$$\left.\begin{aligned} States_C &= States_A \times States_B \\ Inputs_C &= Inputs_A \times Inputs_B \\ Outputs_C &= Outputs_A \times Outputs_B \\ InitialStates_C &= (InitialStates_A, InitialStates_B) \end{aligned}\right\}$$

and $Updates_C((s_A, s_B), (x_A, x_B)) = ((s'_A, s'_B), (y_A, y_B))$; all $s_A \in States_A$, $s_B \in States_B$, $x_A \in Inputs_A$, and $y_B \in Inputs_B$ have $(s'_A, y_A) = Updates_A(s_A, x_A)$ *and* $(s'_B, y_B) = Updates_B(s_B, x_B)$.

A dynamic system or component may have some properties that can be characterized by its actor model. In the following, the properties of a state machine including *causality*, *memorylessness*, *linearity*, *time invariance*, and *stability* are discussed.

Causality means the output of one FSM depends only on its current and past inputs. Consider a continuous-time system $S: X \to Y$ where $X = A^R$ and $Y = B^R$ for some sets A and B, the system is *causal* if the following condition is satisfied,

$$x_1|_{t \le \tau} = x_2|_{t \le \tau} \Rightarrow S(x_1)|_{t \le \tau} = S(x_2)|_{t \le \tau} \text{ for all } x1, x2 \in X \text{ and } \tau \in R, \tag{10.32}$$

In other words, the condition for *a causal system* is if two possible inputs x_1 and x_2 are identical up to time τ (including τ), then the corresponding outputs will be identical up to time τ (including τ).

In addition, the dynamic system is *strictly causal* if

$$x_1|_{t < \tau} = x_2|_{t < \tau} \Rightarrow S(x_1)|_{t \le \tau} = S(x_2)|_{t \le \tau} \tag{10.33}$$

Accordingly, the condition for a strictly causal system is if two possible inputs x_1 and x_2 are identical up to time τ (not including τ), the corresponding outputs will be identical up to time τ (including τ). The output at time t of a strictly causal system does not depend on its input at time t but its past values. For example, an integrator actor is strictly *causal*; while an adder actor is *causal* but not *strictly causal*. To construct a feedback system, the actor should be strictly causal (Lee and Seshia 2017).

Memorylessness. A dynamic system has its memory if the output depends on current and historical inputs (or future inputs if the system is not causal). Consider a continuous-time system $S: X \to Y$ where $X = A^R$ and $Y = B^R$ for some sets A and B, the system is *memoryless* if there exists a function $f: A \to B$ to meet the condition of

$$(S(x))(t) = f\,(x(t) \text{ for all } x \in X \text{ and } t \in R \tag{10.34}$$

That is, the output $(S(x))(t)$ at time t depends only on the input $x(t)$ at time t and not on its past values. According to this definition, an integrator is not memoryless; but an adder actor is memoryless.

Superposition. A dynamic system $S: X \to Y$ is *linear* if its satisfies the condition of superposition as,

$$\forall x_1, x_2 \in X \text{ and } \forall\, a, b \in R, S(ax_1 + bx_2) = aS(x_1) + bS(x_2) \tag{10.35}$$

Delay Actor. Let τ be a delay time, and a delay actor $D_\tau: X \to Y$ is defined as,

$$\forall x \in X \text{ and } \forall\, t \in R, (D_\tau(x))(t) = x(t - \tau) \tag{10.36}$$

Time invariance. A dynamic system $S: X \to Y$ is time invariants if,

$$\forall x \in X \text{ and } \forall\, t \in R, S(D_\tau(x)) = D_\tau(S(t)) \tag{10.37}$$

Linear time invariant (LTI). A system is *linear and time invariant* if it meets the conditions of linearity and time invariance in Eqs. (10.34) and (10.36). A LTI system possesses some mathematic properties that can be utilized to optimize a control system. Therefore, it is desirable to design and model a physical system as LTI model; even it becomes infeasible, it is worth to approximate a system as LTI.

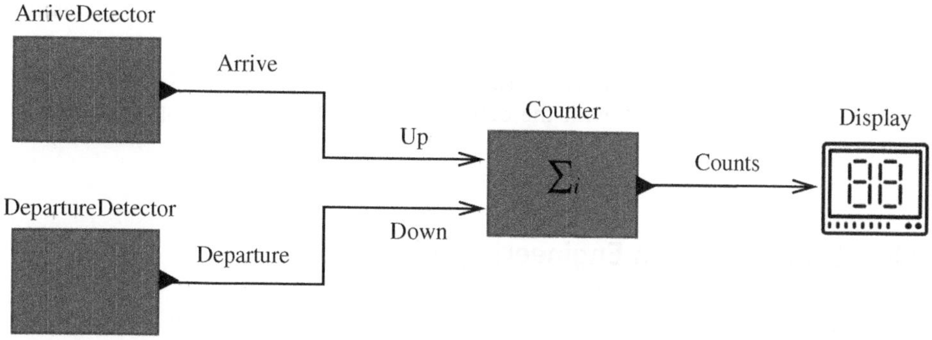

Figure 10.18 Counter in a car garage

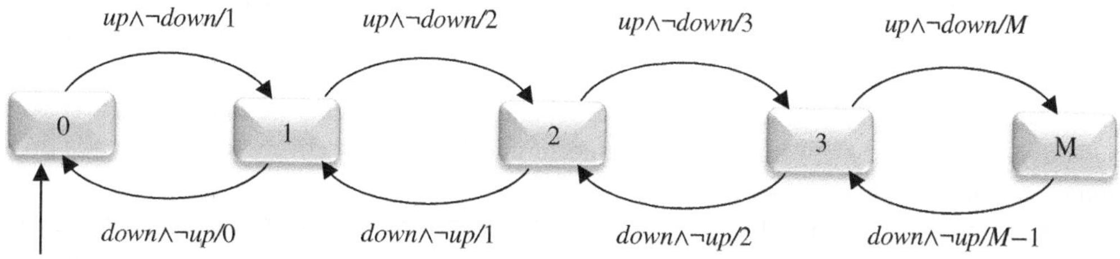

Figure 10.19 FSM of garage car counter

Stability. A system is said to be *bounded-input bounded-output stable* (or simply referring to *stable*) if the outputs are bounded when all inputs are bounded. Consider a dynamic system with an input x and an output y, the input is bounded when there exists a real number $A < \infty$ such as $|x(t)| \leq A$ for any $t \in R$. the output is bounded when there exists a real number $B < \infty$ such as $|y(t)| \leq B$ for any $t \in R$.

Example 10.4 Use FSM to model a counter that counts the number of cars in a parking garage by detecting the events of entering and exiting as shown in Figure 10.18 (Saha 2019).

Solution

In the FSM, assume that: a pure signal *up*: $R \rightarrow \{absent, present\}$; discrete actor: *counter*: $(R \rightarrow \{absernt, present\})^P \rightarrow (R \rightarrow \{absent\} \cup N)$ where $P = \{up, down\}$

The state space of the counter is $states = \{0, 1, 2, ..., M\}$, and Figure 10.19 shows the corresponding FSM that reflects the functions of the car counting system.

10.6 Development Standards

A CPS seems an extension of *software engineering* (SE) from the perspective of cyber systems. Several industry standards are available but they are text and diagram based, and the syntax and semantics are not uniquely designed. IEC 61131-3 was developed for *programming logic control* (PLC) and SysML was used for system modeling. It was clearly that different vendors will adopt the standards differently by the areas of interest, and this affects the collaboration and portability among tools (Buffoni et al. 2021).

Parant et al. (2023) used SysML diagrams and IEC 61499 standard to design CPS and its control systems. They began to design a CPS with the definition of FRs based on the specifications of system.

10.7 Model-Based System Engineering (MBSE)

A CPS is a combination of physical and cyber subsystems. A physical subsystem is an entity for actual implementation in matter, a cyber subsystem such as an algorithm and software is conceptual and logical, and the dynamics of system refers to the change of system state in time. Moreover, it is assumed that the computation and algorithms within an MFM are developed, the cyber system of CPS focuses on the interactions of physical systems rather than cyber implementations of MFMs themselves (Platzer 2018). Design of CPS involves in an iterative process to deal with the couplings of physical and cyber systems (Lee and Seshia 2015). In designing a CPS, the solutions to the following two questions should be found (1) how cyber and physical systems are co-designed to synergize system performance; (2) how to communicate and interact effectively to reduce costs and improve system performance (Song et al. 2017).

Since a comprehensive system like a CPS includes many entities (i.e., system elements, states, and transitions) with complex relations one after another, it requires a hierarchical structure to model system elements, relations, and behaviors. Hierarchy is a key mechanism to manage system complexity in a structured way. MBSE allows to derive system behaviors based on the definitions of inputs, outputs, and attributes of system elements. In other words, if two components are connected, then the events at an interface are predictable. The CPSs in the real world are concurrent, distributed, and decentralized; there is the need to define, specify, and support the concurrency in designing a CPS. Moreover, system components must be able to communicate and synchronize as needed (Marwedel 2006).

10.7.1 Modeling in MBSE

A system model describes the relations of inputs and outputs of system. A high-fidelity model describes behaviors and properties of a system accurately and an abstract model conceptualizes a system but omits details. A CPS should be modeled to meet the requirements of *correct-by-construction* with unambiguity, mathematical rigorousness, and sound semantics of models. This brings challenges to select MFMs, analyze their interactions, and compose the selected MFMs into a CPS to meet the formulated FRs optimally. From this perspective, the methods used to model CPSs should be very flexible to support both of *acausal* and *causal* modeling since a CPS can involve in *continuous, discrete*, on *a combination* of continuous and discrete behaviors (Simko et al. 2013).

MBSE is widely used to model CPSs. MBSE is a model-oriented approach. MBSE provides the abstraction and allows virtual prototyping using mathematic and simulation tool before a physical product is build. To ensure portability, knowledge reuse for a broader applications, open source implementation such as open modelica, bloqqi and Acumen were used to support symloic-numeric modeling, requirement verification, code generation, model debugging, optimization, visualization, and dealing with changes and uncertainties (Buffoni et al. 2021).

MBSE shows the effectiveness in (1) managing the complexity of CPSs, (2) mapping FRs with *design solutions* (DSs) of models at different levels and domains, and (3) providing a complete view of system via the aids of formal semantics and rich visual modeling and tools. MBSE aims to design and analyze the FRs of systems, verify and validate model solutions to sub-FRs, and support all design activities in the whole system lifecycle from the conceptual design phase to the deployment phase, and MBSE makes it feasible to perform a formal verification and validation process in system development (Trehan 2018; Saqui-Sannes et al. 2022). MBSE makes a transition from document-centric into model-centric in SE (Chen et al. 2022).

MBSE was originated from *block diagram modeling* where the implementation and information of a functional block are hidden; only the input and the output of the functional block are clearly defined. Block diagram modeling has been widely used in popular engineering tools such as Labview, Simulink, VisSim, and Scilab; however, block diagram modeling exhibits a number of limitations such as the difficulty in dealing with the interactions in composition and inapplicability for physical modeling. Numerous open-source modeling languages are available and the most popular ones are *Modelica*, *functional mock-up interface* (FMI), *distributed co-simulation protocol* (DCP), and *structure and system parameterization* (SSP). Popular MBSE tools include *object management group* (OMG) visual modeling languages, *UML*, and *SysML*.

The principles of MBSE are illustrated by developing the solutions to representative mechatronic systems such as robotics, distributed systems, and control systems. Note that modeling differs from programming since the former allows the specification of nondeterministic behaviors as well as explicit models about how the environment changes. MBSE focus on (1) structured design where basic mechatronic elements are integrated to perform more complex tasks; in other words, high-level FRs can be decomposed into low-level sub-FRs until the corresponding DS has been identified for each sub-FR; (2) requirement-based design where FRs are specified precisely and used to evaluate, compare, and select design alternatives for implementation. When MBSE is used in designing a CPS, it is expected to (1) model both continuous-time and discrete-time system components; (2) scalable to cope with the scope and complexity; (3) provide multiple views in different physical domains; (4) incorporate with physical properties; (5) supports standardization, model libraries, formal representations, and tools; and (6) model dynamic behaviors of physical systems user-friendly.

10.7.2 Design Stages in MBSE

A V-model in Figure 10.20 can be used to describe all the phases in designing and implementing a CPS from the identification of FRs to conceptual designs, detailed designs, implementations, deployments, and, finally, to use and maintenance of system (Buffoni et al. 2021). Gustafsson (2011) discussed the development of a cyber system of CPS in detail, and its design stages include to (1) define FRs based on given system specifications; (2) determine system architecture by selecting or designing system elements including mechanical systems, sensors, communications, actuators, controllers, and users' interfaces; (3) model, simulate, and evaluate MFMs in design analysis and synthesis; (4) develop cyber systems with the implementations for models, algorithms, and logic; (5) verify and validate the solutions at all stages and domains in hierarchy; (6) perform commissioning and tuning; (7) monitor and assess the system in operation and diagnose deflects when abnormal events occur; and (8) reconfigure and upgrade system when environmental conditions change.

When MSBE is used to implement the V-model in Figure 10.20, the modeling solutions must be available for seven essential tasks: that is, executable specification, modeling and simulation, what-if analysis, model elaboration, virtual prototyping, continuous verification and validation, and knowledge engineering and management (Engin and Engin 2021). *Executable specification* is a

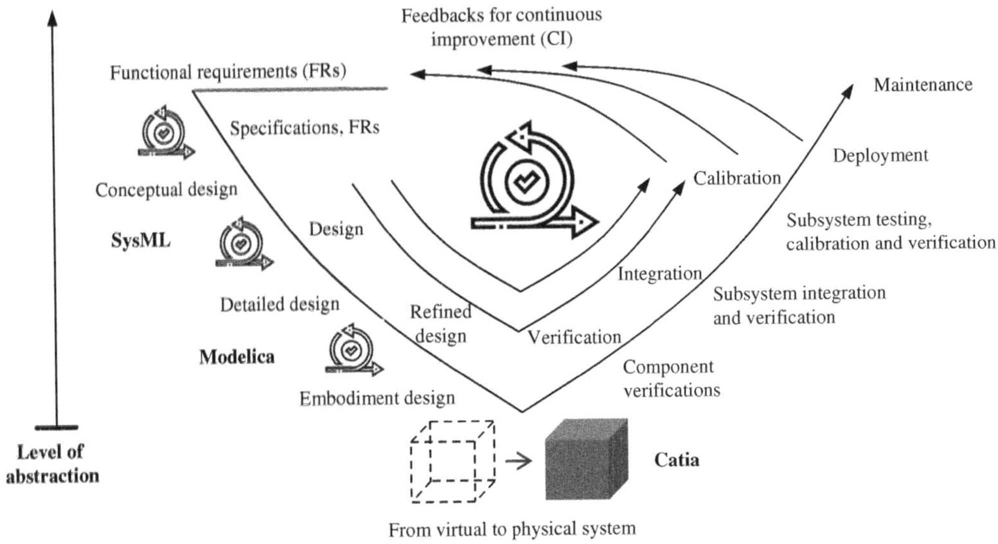

Figure 10.20 All phases of CPS development (Buffoni et al. 2021; Marwedel 2021)

model encapsulated with requirements, intellectual properties, system components, and test scenarios. *Modeling and simulation* is to investigate the interactions of system elements for verification purpose. *What-if analysis* is to test solutions for individual elements, or a part of, or the whole system. *Model elaboration* is based on simulation results, common elaboration approaches include replacing floating points with fixed points, using continuous time instead of discrete time, substituting a detailed actuator model with a behavioral model, or adding signals for diagnostics. *Virtual prototyping* is to reduce development time and cost in evaluating systems. *Continuous V&V* is to identify design defects at the earliest possible time. *Knowledge engineering and management* is to maximize knowledge reuses, avoid reinventing wheels, and reduce V&V efforts to the maximum content.

10.7.3 Acausality Modeling by Modelica

A modeling language is defined by its semantics and syntax. Accordingly, modeling languages can be classified into *nonformal*, *semiformal*, and *formal* languages as shown in Figure 10.21 (Saqui-Sannes et al. 2022). A formal language uses formal and strict semantics to enable formal proofs; however, it takes a long time for users to learn and use semantics, and it limits the flexibility in modeling a CPS. A nonformal language is widely used by engineers to express design concepts with nonformal semantics; however, it takes extra effects to verify and validate models. By comparison, semiformal languages make a good trade-off between efficiency and flexibility of modeling processes; therefore, semiformal languages such as SysML and Modelica are focused in this section.

SysML used to be the de facto standard for MBSE. SysML was extended from *UML* to model engineered systems. Modelica is a modeling language similar to other popular actor modeling languages such as SysML, Simulink, LabView, Scade, StreamIT, and Reactors.

Modelica itself is not modeling tool; but many modeling tools use Modelica as the language to model CPSs. Popular modeling tools for CPSs include *OpenModelica* that is open accessible, *Dymola* by the Dassault System, *Wolfram System Modeler* by Wlfram MathCore, *SimulationX* by ITI,

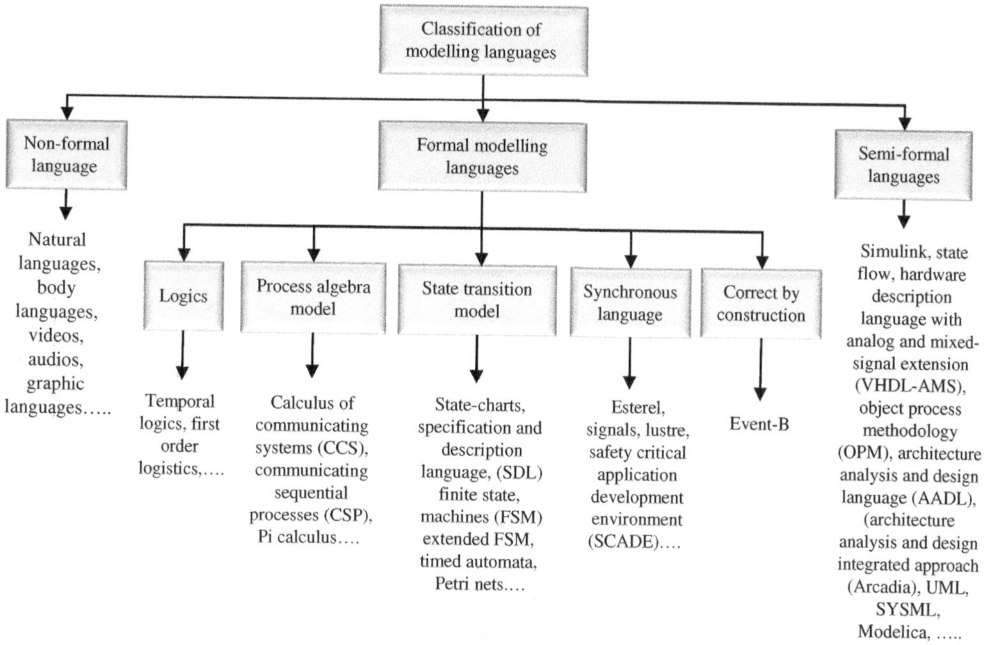

Figure 10.21 Classification of modeling languages

MapleSim by MapleSoft, *AMESIM* by LMP, *MWORKS* by Tongyang, and *IDA Simulation Env* by Equa (Modelica 2024).

In this section, Modelica is used as an example to show how a modeling language is used to model a CPS. In comparison with other modeling languages, Modelica has its unique capabilities to support *acausality modeling*.

Acausality means that an equation is only used to define the relations of input and output variables, the solution does not rely on the causality of inputs and outputs. For example, the Ohm law in an electrical circuit can be equivalent as any one of three equations: $i{:} = v/R$, $v{:} = R \times i$, $R{:} = v/i$. Acausality modeling increases reusability of classes that contain assignment statements to specify inputs and outputs. Figure 10.22 shows the difference between acausality modeling (Modelica) and causality modeling by block-based modeling (Simulink); acausality modeling simplifies decomposing and modeling MFMs of a CPS, defining input–output relations, implementing, and simulating a system model.

By comparing the performance of proprietary coding and block diagram modeling, Figure 10.23 shows the advantages of Modelica in shortening the development time at the stages of "*system definition*," "*system decomposition*," "*modeling subsystems*," "*causality derivation*," "*implementation*," and "*simulation*" (Modelica 2024).

Other than its acausality modeling capabilities, Modelica is also featured as (1) a declarative statically typed language that allows acausal modeling of mathematical equations and functions; it provides the checks on high level specifications and static types; (2) the capability to support modeling in multi-domains such as mechanical, electrical, hydraulic, thermodynamic, biological, and financial systems; (3) the generality that everything can be defined a class; it is objected oriented with the syntax similar to Java and Matlab; (4) the programming capability to construct hierarchical structures; (5) the nonproprietary efficiency to support advanced equation compilation for over 300,000

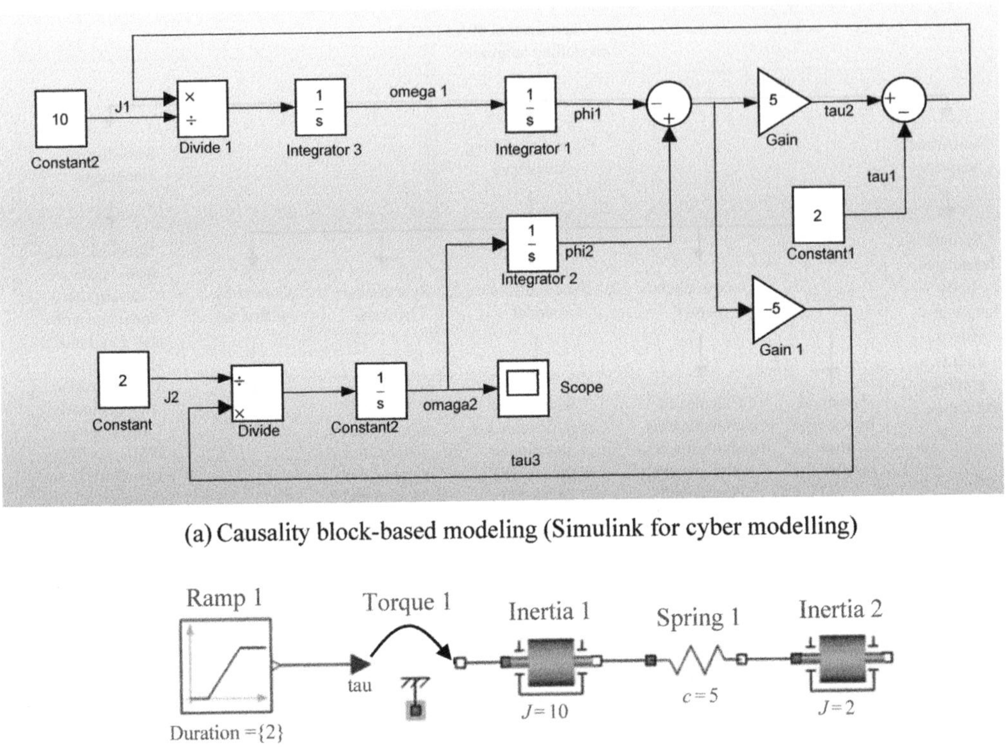

(a) Causality block-based modeling (Simulink for cyber modelling)

(b) Acausality modeling (Modelica for physical modelling)

Figure 10.22 Modelica supports acausality modeling to simplify a cyber system model

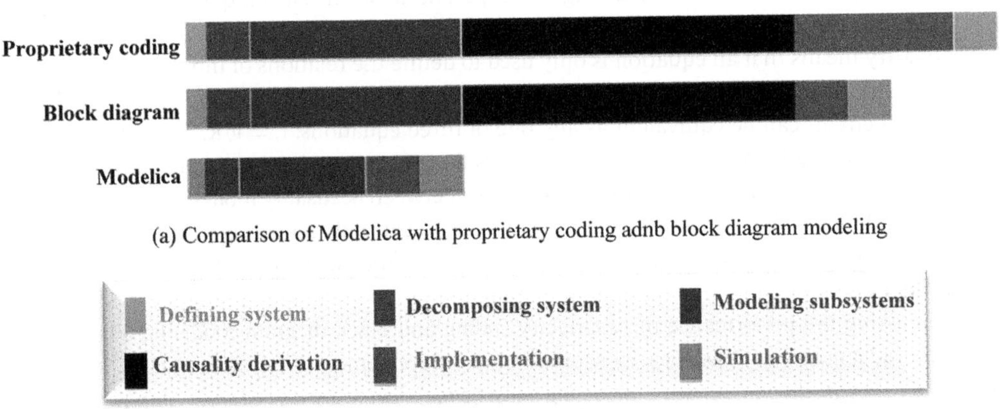

(a) Comparison of Modelica with proprietary coding adnb block diagram modeling

(b) Modeling time at different stages

Figure 10.23 Modelica for shortened development time (*Source:* Adapted from Modelica (2024))

equations; it describes continuous-time behaviors by differential algebraic equations and it represents *DEDS* behaviors by event-triggers.

In a CPS, physical components are integrated with software components. MBE allows to represent and design physical and software components by functional models, and the whole CPS can be simulated before the physical system is implemented and deployed. Most of *Verification and Validation* (V&V) tasks can be carried out virtually or by hardware-in-the-loop simulation. To achieve this objective, modeling languages must be capable of representing both physical and software components. Modelica is widely used in industrial modeling; it can be used to generate embedded control software from the models automatically.

Modelica uses mathematic equations to represent the behaviors of CPS. Modelica is OO and it supports encapsulations of objects and reuses of model components. OO simplifies the design of CPS. The system-level FRs can be decomposed hierarchically, and component-level models can be reused and assembled into large-scale and complex models. Modelica models system behaviors by the equations instead of assignment statements or traditional functional blocks. This allows model components (1) be adaptable to data flow, (2) to increase their reusability, and (3) to improve the efficiency of simulation.

Modelica supports multi-physics simulations; virtual components can be associated with physical ones directly without the needs of signal blocks. Modelica supports hybrid modeling for both continuous-time and discrete-time components; it also supports clocked discrete-time models for improved accuracy and simulation. Modelica has a large set of libraries called *Modelica Standard library* (MSL) that covers more than 1200 functions and 1400 components in multiple domains.

Figure 10.24 shows that Modelica provides both textual and graphic views of the model. The model in the graphic view in Figure 10.24a consists of selected graphical annotations. When the model is moved, both textural and graphic views are preserved and moved.

Modelica can be used to represent physical and cyber components in the same model. Figure 10.25 shows an example mode of a CPS by Modelica. The model example includes (1) a physical electric component; (2) a physical mechanical component; and (3) a cyber control component that is extracted from electrical, mechanical, and block libraries, respectively.

10.7.4 Programming in Modelica

In this section, a few examples are provided to show how Modelica is used to model a CPS based on its governing mathematical models.

```
model RLCircuit
 Modelica.Electrical.Analog.Sources.SineVoltage Source(V = 110)  annotation(
  Placement(visible = true, transformation(origin = {-40, -2}, extent = {{-10, -10}, {10,
10}}, rotation = -90)));
 Modelica.Electrical.Analog.Basic.Resistor R(R = 100)  annotation(
  Placement(visible = true, transformation(origin = {0, 24}, extent = {{-10, -10}, {10,
10}}, rotation = 0)));
 Modelica.Electrical.Analog.Basic.Inductor inductor(L = 0.5)  annotation(
  Placement(visible = true, transformation(origin = {40, -8}, extent = {{-10, -10}, {10,
10}}, rotation = -90)));
 Modelica.Electrical.Analog.Basic.Ground G annotation(
  Placement(visible = true, transformation(origin = {0, -48}, extent = {{-10, -10}, {10,
10}}, rotation = 0)));
equation
 connect(R.p, Source.p) annotation(
  Line(points = {{-10, 24}, {-40, 24}, {-40, 8}}, color = {0, 0, 255}));
 connect(R.n, inductor.p) annotation(
  Line(points = {{10, 24}, {40, 24}, {40, 2}}, color = {0, 0, 255}));
 connect(Source.n, G.p) annotation(
  Line(points = {{-40, -12}, {-40, -38}, {0, -38}}, color = {0, 0, 255}));
 connect(inductor.n, G.p) annotation(
  Line(points = {{40, -18}, {40, -38}, {0, -38}}, color = {0, 0, 255}));
annotation(
  uses(Modelica(version = "4.0.0")));
end RLCircuit;
```

(a) Textural view

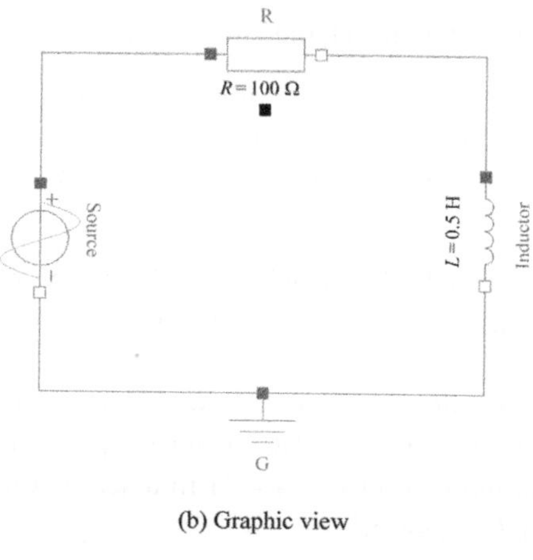

(b) Graphic view

Figure 10.24 Example of textural and graphic views of model in Modelica

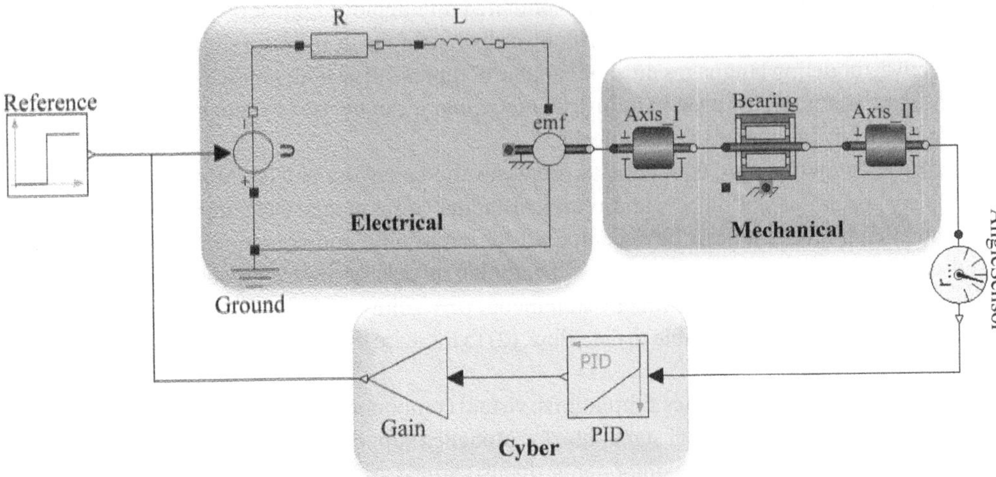

Figure 10.25 Example model of CPS by Modelica

Example 10.5 Create a simple model to plot the result of the following differential equation:

$$\frac{dx}{dt} = -5x, x(0) = 1, t \in [0, 5]$$

Solution

The "TextView" in "OMEdit" of Modelica can be used to create a model as shown in

```
model HelloMedelica          //have a name for the model
Real x(start=1);             //specify initial condition
Parameter Real a = -5;       //specify exponent in differential equation
equation                     // Start to define relations
der(x) = a*x;                //define the differential equation
end HelloMedelica;           //Terminate the definition
```

Save the model and check if there is any error or inconsistency, and this leads to

```
Check of HelloMedelica completed successfully.
Class HelloMedelica has 1 equation(s) and 1 variable(s).
1 of these are trivial equation(s).
```

Simulate the model by specifying the range of $t \in [0, 5]$ and the function of $x(t)$ in this time domain can be plotted as in Figure 10.26.

Example 10.6 Create a Modolica model of a 1D rotational system by a permanent magnet DC. The input of DC power supply is 10 V. The circuit consists of a resistor of $R = 100\,\Omega$, an inductor of $L = 100$ H in a serial connection, and the moment of initial of rotational axis of $J = 2\,\text{kg/m}^2$.

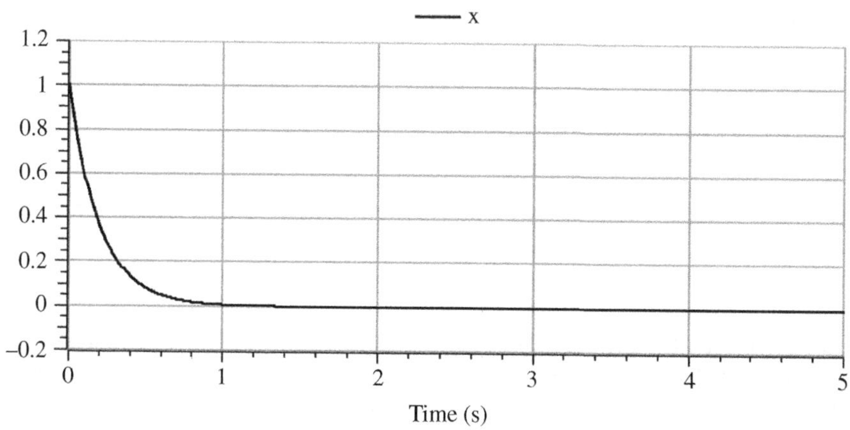

Figure 10.26 The function of $x(t)$ in the time domain of $t \in [0, 5]$ in Example 10.5

Solution

The information in the problem statement can be used to define the corresponding Modelica model as shown in Figure 10.27a graphically; the code version of model can be generated automatically as shown in Figure 10.27b. Once the model is created, it should be checked to see if there is any insistence in connecting a pair of functional modules. If yes, the identified connection must be fixed before the simulation can be performed smoothly.

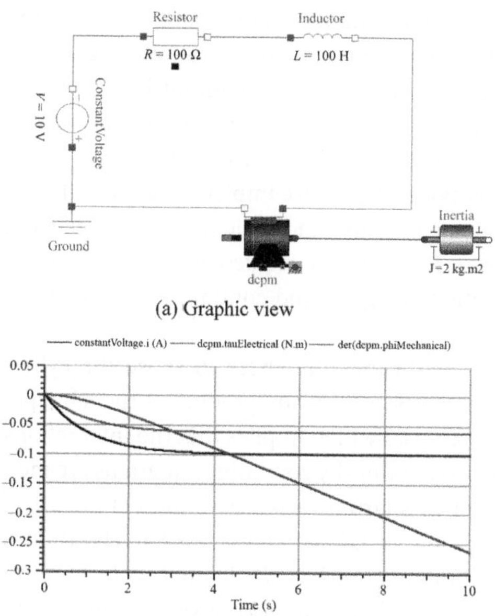

(a) Graphic view

(c) Example of simulation results

```
model DCMOTOR

Modelica.Electrical.Analog.Sources.Constant
Voltage constantVoltage(V = 10);
  Modelica.Electrical.Analog.Basic.Resistor
resistor(R = 100);
  Modelica.Electrical.Analog.Basic.Inductor
inductor(L = 100);
  Modelica.Electrical.Analog.Basic.Ground
ground;

Modelica.Mechanics.Rotational.Components.In
ertia inertia(J = 2);

Modelica.Electrical.Machines.BasicMachines.
DCMachines.DC_PermanentMagnet;
equation
  connect(resistor.n, inductor.p);
  connect(ground.p, constantVoltage.p);
  connect(resistor.p, constantVoltage.n);
  connect(dcpm.flange, inertia.flange_a);
  connect(ground.p, dcpm.pin_an);
  connect(dcpm.pin_ap, inductor.n);
end DCMOTOR;
```

(b) Textual view

Figure 10.27 A one-degree of freedom (DOF) DC rotational system model by Modelica

The check on the code in Figure 10.27b leads to the result of,

```
Check of DCMOTOR completed successfully.
Class DCMOTOR has 209 equation(s) and 209 variable(s).
157 of these are trivial equation(s).
```

The check finds no inconsistence, and the model consists of 209 equations about 209 variables. The simulation can be performed and Figure 10.27c shows the changes of three example variables over time.

Example 10.7 Review a RobotR3 model in Modelica, simulate the forward kinematic model to move the start point ($\theta_1 = -60°, \theta_2 = 20°, \theta_3 = 90°, \theta_4 = 0°, \theta_5 = -110°, \theta_6 = 0°$) to the finish point ($\theta_1 = 60°, \theta_2 = -70°, \theta_3 = -35°, \theta_4 = 45°, \theta_5 = 110°, \theta_6 = 45°$) in the task space, and plot out the joint motions in the joint space. The model is located at Modelica.Mechanics.MultiBody.Examples. Systems.RobotR3.fullRobot.

Solution

The information in the problem statement can be used to define the corresponding Modelica model as shown in Figure 10.28a graphically; the corresponding 3D simulation model can be generated automatically as shown in Figure 10.28c. Once the model is created, the robot model can be simulation to predict joints' motions along with a path in the task space as shown in Figure 10.28c.

10.7.5 Formal Semantics

One critical concept in MBSE is ontology. Ontology is adopted to define the domains with the constitutions and their internal and external relations by formal semantics. This aligns well with the concept of *System of Systems* (SoS) since each mechatronic system is made of a multitude of subsystems. The properties and metadata of a model need to be understood without knowledge of detailed implementation. Therefore, ontology is critical step to achieve standardization and support data exchange for interactions and interoperations among systems (Amazon 2024).

Since a CPS may involve in many heterogeneous components that are implemented in different engineering domains, the concerns raise to trace, manage, verify, and validate their properties, attributes, constraints, and specifications in applications. *Requirement engineering* (RE) is a system engineering approach to discovering, qualifying, tracing, managing, and communicating requirements in an abstract system model.

In a CPS, its physical and cyber subsystems have significant differences in sense the cyber-subsystems are is causal where functional blocks, components, or software produce outputs by specifying inputs. On the other hand, physical systems are acausal which implies that the interactions are nondirectional for inputs and outputs as long as the constraints by governing equations are met. The behaviors of connected components are controlled by sharing variables (Simko et al. 2013). Take a simple example, a resistor has two ports for a voltage and a current, respectively. It does not matter which is input as long as it satisfies the equations of $U_1 - U_2 = R \cdot I_1$ and $I_1 = I_2$. Another challenge of defining semantics for CPS is the difference of realtime-ness of physical systems and discrete time of cyber system; dealing with heterogeneous time domain is nontrivial and involves a number of issues. The third challenge was an algebraic loop in a synchronous system without delays or integrators. There might be no or multiple solution to a set of system equations that lead to ambiguous semantics.

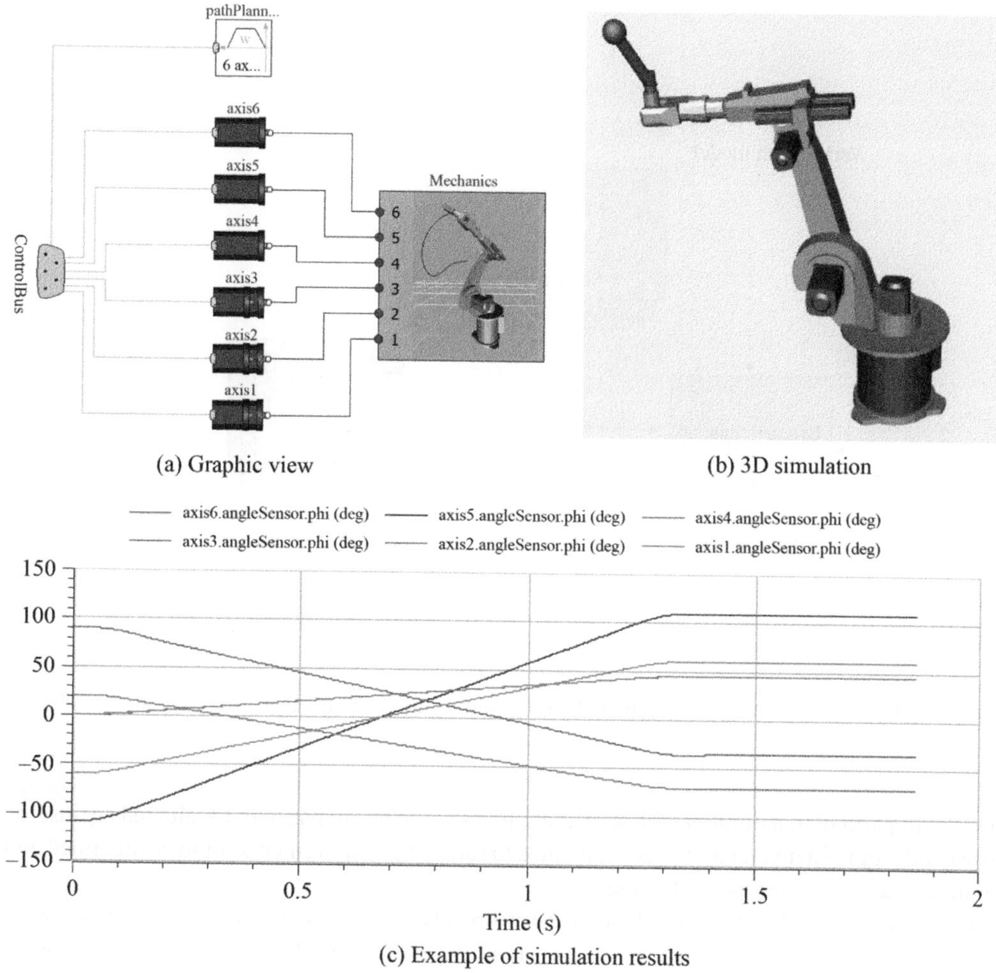

Figure 10.28 A six-DOF robotic system in Modelica

One method is to use formal semantics so that the requirements on system components can be verified when the CPS model is simulated. Developing and incorporating semantics in a modeling language allow to (1) represent conceptual properties of models, (2) bind and automate compositions of models, and (3) trace and verify the outcomes of models. For example, Simko et al. (2013) developed formal semantics called CPS-specific modeling language (CyPhyML), it used constraint rules and corresponding behavioral semantics in a CPS model. A semantic mapping was established based on an executable subset of first-order logic to check the conformance and composition of models.

Bouskela et al. (2022) verified the requirements of a CPS on three model types and their bindings as shown in Figure 10.29. The constraints on a CPS model were associated with a set of *requirements models* that were formal representations of system FRs, an *architectural model* that represented the system structure with different decomposition levels, and a set of the behavioral models which specified the behaviors of system's elements and their interactions. These models were connected by different bindings so that the corresponding interactions could be verified in a system engineering

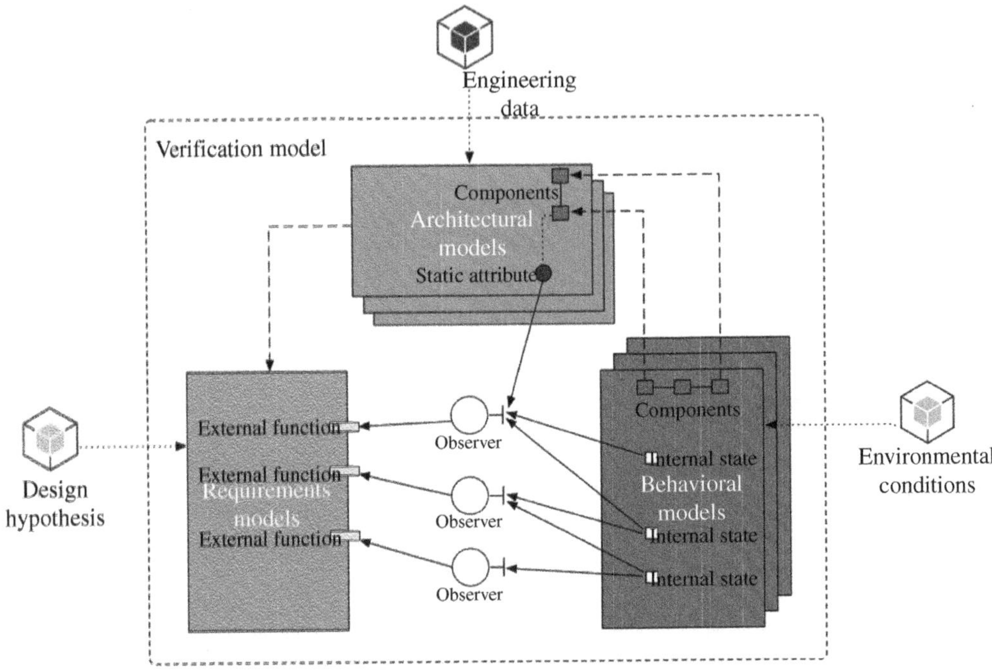

Figure 10.29 Simulation-based verification of requirements (*Source:* Bouskela et al. (2022)/Springer Nature/ CC BY 4.0)

process. In particular, a number of design alternatives could be analyzed and evaluated against the associated requirements, so that those that failed to satisfy the requirements could be identified and eliminated at the earliest possible time.

Mabrok and Ryan (2017) proposed the *category theory* (CT) to tackle with the semantics in a Modelica model. In the CT, any system was viewed as a category, and a category consisted of (1) a set of objects that represented system elements and (2) a set of the arrows that represented the relations of objects. Furthermore, the Ontology log (Olog) was adopted as a formal language to define the requirements of model. CT was rooted from the algebraic topology in which the structural elements in an abstract model were represented by various mathematic fields. Therefore, CT could be incorporated in a system modeling languages such as SysML or Modelica.

Formal semantics can be utilized to verify the relations of system components. The relations of system components can be defined in five different rules: (1) logic rules are based on logical statements such as "if-then" or "in-then-else," (2) mathematic rules that define relations by mathematical equations to comparison symbols, (3) geometry handling rules that are used to govern dimensions, (4) configuration selection rules where logic and math rules are combined to govern the interactions, and (5) communication rules that govern the interactions of system codes with external components (Trehan 2018).

10.7.6 Verification and Validation (V&V)

CPS must be designed reliably and economically; however, using different types of mechatronic components has led to a fragmented space and theories and tools, and the principle of compositionality should be applied to deal with the interactions of system elements (Tripakis 2016). A formal or semiformal language can rely on mathematics and logic for verification and validation. For example, *a reachability analysis* produces a reachability graph to represent all valid execution paths and states a system can reach when initial states are specified. The analysis in a model produces an exhaustive exploration of the state space of model. A stochastic simulation can be used to select the best action or node based on a given reward function in model verification.

The cyber system of CPS has to be verified and validated before it can be deployed. Validation by testing is tedious since the output of one program is turned into the input of the program; therefore, it becomes necessary to include V&V within a program to minimize the need of physical validations. The validations perform four main tasks as,

- *Conformance tests* to verify if the functions meet the corresponding specifications.
- *Interoperability tests* to validate two system elements can be interacted, cooperated, and communicated effectively.
- *Performance tests* to evaluate system's capacities in meeting certain properties un unknown environment.
- *Responsiveness tests* to measure the response time of systems depends on how the system is stimulated.

Finally, a CPS should be defined to assess system security at an early stage of the system lifecycle. Bakirtzis et al. (2018) discussed the need to assure the security of CPSs since safety is prioritized in some critical applications such as military, infrastructure, and transportation. To construct a safety-critical CPS model, a taxonomy of attributes was given to capture essential concepts associated with safety concerns. System modeling language (SysML) and a graphical object-oriented modeling language were used to design and evaluate CPSs.

10.8 Summary

In contrast to traditional mechatronic products or systems where sub-solutions in multiple domains are integrated as integral systems, a CPS is a SoS that emphasizes the dynamic interactions of MFMs with a sophisticated cyber system to deal with the connection, coordination, cooperation, and collaboration of MFMs. Therefore, CPSs are characterized by heterogeneity, dynamic integration, and dynamic interactions of physical and cyber systems. The basic features of CPS are reactive computation, parallel computing, feedback controls, realtime-ness, dependability, biological intelligence, and embedded computation.

The V-model can be used to identify the main tasks and tools in designing a CPS, and the designing process requires to establish mathematical models to analyze system behaviors. There are no universal tools to establish mathematical models of CPSs, different modeling tools should be chosen depending on whether or not a CPS is a continuous, discrete, or hybrid continuous and discrete system.

A CPS is SoS and mostly data driven or event driven, MBSE exhibits its advantages in modeling a CPS. MBSE is a model-oriented approach. MBSE provides the abstraction and allows virtual prototyping using mathematic and simulation tools before a physical product is built. MBSE shows the effectiveness in (1) managing the complexity of CPSs, (2) mapping FRs with DSs of models at different levels and domains, and (3) providing a complete view of the system via the aids of formal semantics and rich visual modeling and tools.

Modelica is a modeling language to support MBSE; it has the capabilities for *acausality modeling*. In comparison with proprietary coding or block diagram modeling, Modelica has shown its great potentials in accelerating the developments of CPSs by reducing the efforts in system definition, system decomposition, modeling and simulation, implement, and mostly importantly, causality derivation.

Problems

10.1 Discuss how mathematical modeling helps design a CPS with examples.

10.2 Discuss the difference of CPS and conventional embedded systems.

10.3 What is a reactive system?

10.4 Develop an actor model for the mass–spring system shown in Figure 10.30.

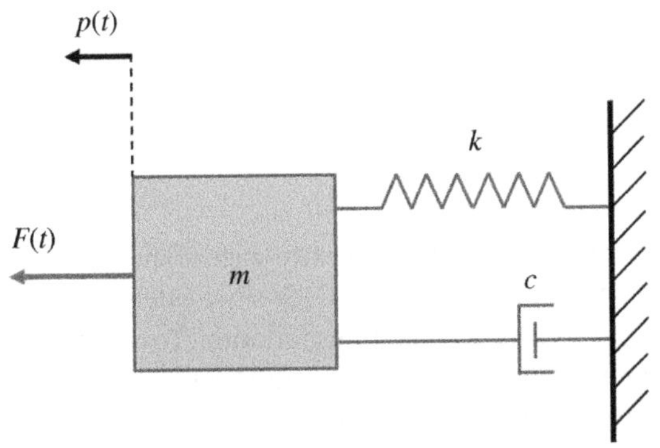

Figure 10.30 One-DOF mass–spring system in problem 10.4

10.5 Develop an actor model for the mass–spring system as shown in Figure 10.31.

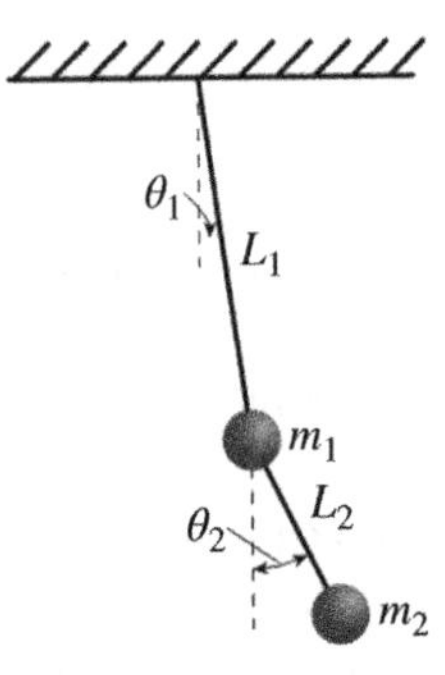

Figure 10.31 Two-DOF pendulum system in Problem 10.5

10.6 Model the mass–spring system in Problem 10.5 as a cascade actor model with three actor models with the output of $\ddot{p}(t)$, $\dot{p}(t)$ and $p(t)$, respectively.

10.7 Use hybrid automata to model a centralized heating system as shown in Figure 10.32. The heater is controlled (i.e., "*on*" and "*off*" when the room temperature is "*below*" or "*above*" the reference temperature of 75°F, respectively).

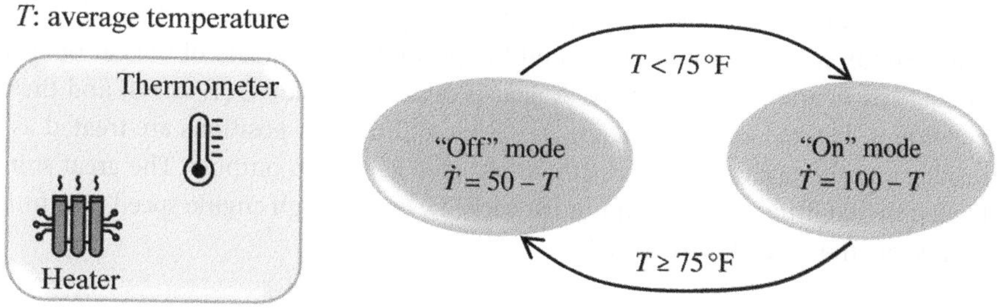

Figure 10.32 A central heating system in Problem 10.7

10.8 Use hybrid automata to model a thermostat and heating system as shown in Figure 10.33a. A thermostat is used to control the temperature in a room. In its implementation, the thermostat turns on the radiator when the temperature is in the range of (16, 18) °C and it turns the radiator off when the temperature is the range of (20, 22) °C. x is used to denoted the temperature and two discrete states are "on" and "off" of the radiator.

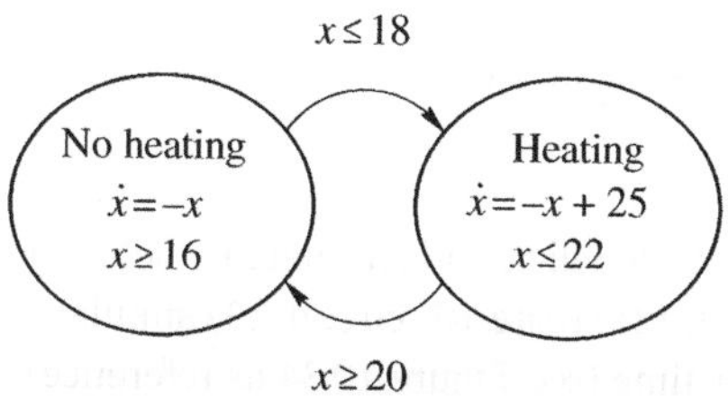

(a) Problem 10.8

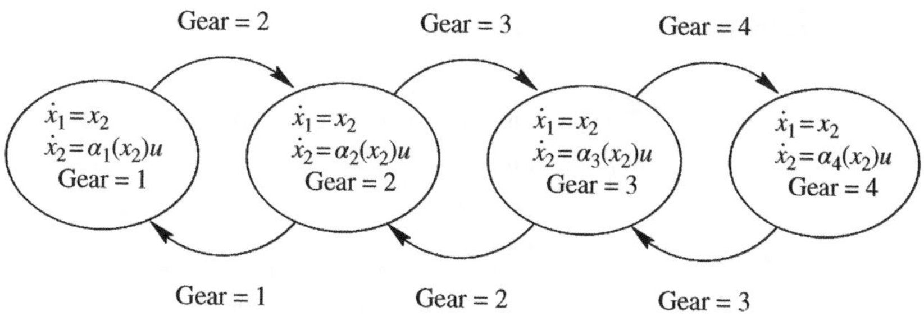

(b) Problem 10.9

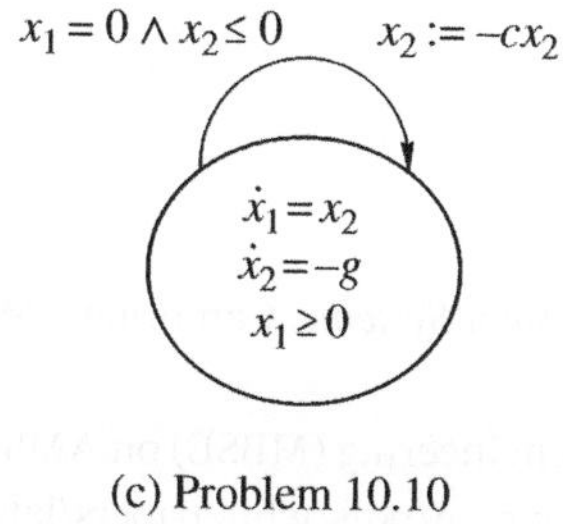

(c) Problem 10.10

Figure 10.33 Illustrations of modeling problems ((a) for 10.8, (b) for 10.9, and (c) for 10.10)

10.9 Use hybrid automata to model a gear shift in the gearbox of car as shown in Figure 10.33b. The gearbox consists of four gears, and the position and velocity of the car on a road are denoted as x_1 and x_2, respectively. A gear is denoted as $gear \in \{1, 2, 3\ 4\}$ and the throttle position is denoted as $u \in \{u_{min}, u_{max}\}$. Gears and throttle positions are treated as system inputs, and the position and velocity are treated as system outputs. The great shit is controlled so that little power is required at very low or very high engine speed. The function α_i represents the efficiency of gear *i*.

10.10 Use hybrid automata to model a bouncing ball as shown in Figure 10.33c. It consists of two continuous states for the vertical position (x_1) and velocity (x_2) of the ball, respectively. The motion of the ball is governed by Newton's second law of motion; it is a differential equation about the position, velocity, and the gravitational acceleration *g*. However, Newton's second law is valid only when the ball is above the ground, that is, $x_1 \geq 0$. The ball bounces when $x_1 = 0.$ and $x_2 \leq 0$. At each bounce, the ball loses a fraction of its energy that is represented by the function of $x_2 := -cx_2$ where $c \in [0, 1]$ in the coefficient of restitution.

10.11 Create a Modelica model to plot the result of the following differential equation,

$$\frac{dx}{dt} = -2x, x(0) = 5, t \in [0, 10]$$

10.12 Create a simple circuit in Modelica and simulate its behavior to verify its correctness (1) install and open OMEdit, (2) create RL circuit, (3) simulate the circuit, and (4) plot the voltages of *R* and *L* over time (see Figure 10.34 as reference).

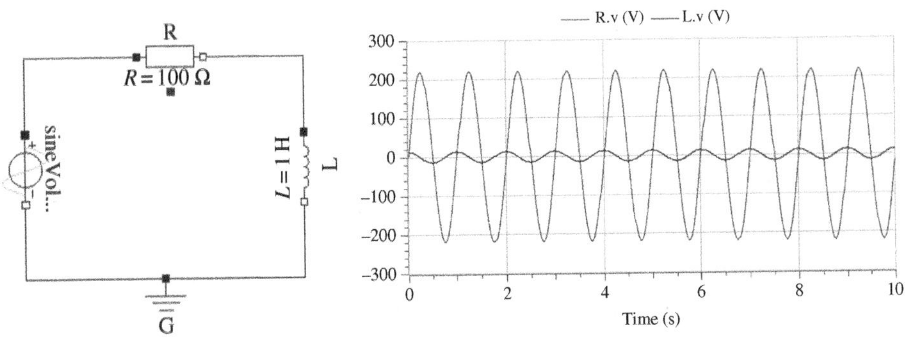

(a) RL circuit

(b) Voltages on R and L over time

Figure 10.34 Reference circuit model and simulation result

References

Alur, R. (2015). *Principles of Cyber-Physical Systems*. Cambridge, Massachusetts, London, England: The MIT Press.

Amazon (2024). Model based systems engineering (MBSE) on AMS: from migration to innovation: AWS whitepaper. https://docs.aws.amazon.com/pdfs/whitepapers/latest/model-based-systems-engineering/model-based-systems-engineering.pdf?did=wp_card&trk=wp_card (accessed 12 September 2024).

Bakirtzis, G., Carter, B.T., Elks, C.R., and Fleming, C.H. (2018). A model-based approach to security analysis for cyber-physical systems. *2018 Annual IEEE International Systems Conference (SysCon)*, Vancouver, BC, Canada, 2018, pp. 1–8. https://doi.org/10.1109/SYSCON.2018.8369518.

Bi, Z.M., Xu, L.D., and Wang, C. (2014). Internet of things for Enterprise Systems of Modern Manufacturing. *IEEE Transactions on Industrial Informatics* 10 (2): 1537–1546.

Bouskela, D., Falcone, A., Garro, A. et al. (2022). Formal requirements modelling for cyber-physical systems engineering: an integrated solution based on FORM-L and Modelica. *Requirements Engineering* 27: 1–20.

Buffoni, L., Ochel, L., Pop, A. et al. (2021). Open source languages and methods for cyber-physical system development: overview and case studies. *Electronics* 10: 902. https://doi.org/10.3390/electronics10080902.

Cao, Y., Tan, W., Su, Y. et al. (2020). The effects of icing on aircraft longitudinal aerodynamic characteristics. *Mathematics* 8 (7): 1171. https://doi.org/10.3390/math8071171.

Chen, J., Chen, Y., Hu, Z. et al. (2022). A semantic ontology based approach to support model-based systems engineering design for an aircraft prognostic health management system. *Frontiers in Manufacturing Technology* 2: 886518.

Engin, D. and Engin, M. (2021). Model based design of cyber physical systems. *Journal of Multidisciplinary Engineering Science Studies* 7 (7): 3936–3938.

Ghomri, L. and Alla, H. (2023). Modelling and analysis using hybrid Petri nets. https://arxiv.org/ftp/arxiv/papers/0706/0706.1716.pdf (accessed 12 September 2024).

Gunes, V., Peter, S., Givargis, T., and Vahid, F. (2014). A survey on concepts, applications, and challenges in cyber physical systems. *KSII Transactions on Internet and Information Systems* 8 (2): 4242–4268.

Gustafsson, T. (2011). Introduction to Modelica. http://www.sm.luth.se/seminar/slides/Thomas.pdf (accessed 12 September 2024).

Hehenberger, P. and Bradley, D. (2016). *Mechatronic Futures: Challenges and Solutions for Mechatronic Systems and their Designers*. Springer International Publishing Switzerland.

Hehenberger, P., Vogel-Heuser, B., Bradley, D. et al. (2016). Design, modelling, simulation and integration of cyber physical systems: methods and applications. *Computers in Industry* 82: 273–289.

Herrmann, P. and Blech, J.O. (2016). Formal model-based development in industrial automation with reactive blocks. In: *Software Technologies: Applications and Foundations. STAF 2016*, Lecture Notes in Computer Science, vol. 9946 (ed. P. Milazzo, D. Varró, and M. Wimmer). Cham: Springer https://doi.org/10.1007/978-3-319-50230-4_19.

Jensen, J.C. (2010). Elements of model-based design. https://www2.eecs.berkeley.edu/Pubs/TechRpts/2010/EECS-2010-110.pdf (accessed 12 September 2024).

Johansson, K.H., Lygeros, J., and Sastry, S. (2023). Modeling of hybrid systems. https://people.kth.se/~kallej/papers/eolss_6.43.28.1.pdf (accessed 12 September 2024).

Lafortune, S. (2019). Autonomous systems discrete eavent systems: modeling, observation and control. *Annual Review of Control, Robotics, and Autonomous Systems* 2: 141–1510.

Lee, E.A. and Seshia, S.A. (2017). *Introduction to Embedded Systems, a Cyber-Physical Systems Approach*, 2e. MIT Press.

Lozano, C.V. and Vijayan, K.K. (2020). Literature review on cyber physical systems design. *Procedia Manufacturing* 45: 295–300. https://doi.org/10.1016/j. promfg.2020.04.020.

Mabrok, M.A. and Ryan, M.J. (2017). Category theory as a formal mathematical foundation for model-based systems engineering. *Applied Mathematics & Information Sciences* 11 (1): 43–51.

Marwedel, P. (2021). *Embedded System Design Embedded Systems Foundations of Cyber-Physical Systems, and the Internet of Things*, 4e, https://doi.org/10.1007/978-3-030-60910-8. Spring International Publishing.

Mettle, B., Kanade, T., and Tischler, M.B. (2000). System identification modelling of a model-scale helicopter. https://www.ri.cmu.edu/pub_files/pub2/mettler_bernard_2000_1/mettler_bernard_2000_1.pdf (accessed 12 September 2024).

Modelica (2024). Modelica tools. https://modelica.org/tools (accessed 12 September 2024).

Parant, A., Gellot, F., Zander, D. et al. (2023). Model-based engineering for designing cyber-physical systems for product specifications. *Computers in Industry* 145: 103808.

Penas, O., Plateaux, R., Patalano, S., and Hammadi, M. (2017). Multi-scale approach from mechatronic to cyber-physical systems for the design of manufacturing systems. *Computers in Industry* 86: 52–610.

Pettersson, S. and Lennartson, B. (2023). Hybrid modelling focused on hybrid Petri nets. https://citeseerx.ist.psu.edu/document?repid=rep1&type=pdf&doi=6bccc994d96ac878d24bf377c4b9d251ff25f92a (accessed 12 September 2024).

Plateaux R, Penas O, Choley JY, Mhenni F, Hammadi M, Louni F (2016) Evolution from mechatronics to cyber physical systems: an educational point of view. *Mechatronics - REM 2016* (15–17 June 2016).

Platzer, A. (2018). *Logical Foundations of Cyber Physical Systems*. Springer.

Ragulski, M., Jiang, H., Quan, Q. et al. (2014). Mechatronics and mechanical engineering in cyber-physical systems. *Advances in Mechanical Engineering* 2014: Article ID 5916210.

Rawat, D.B., Rodrigues, J.J.P.C., and Stojmenovic, I. (2016). *Cyber-Physical Systems from Theory to Practice*. CRC Press, Taylor & Francis Group.

Saha, D. (2019). Cyber-physical systems. https://www.albany.edu/faculty/dsaha/teach/2019Fall_ECE553/slides/08_discreteDynamics.pdf, last accessed on Sept. 12, 2024.

Sampayo, M. and Pecas, P. (2022). CPSD2: a new approach for cyber-physical systems design and development. *Journal of Industrial Information Integration* 28 (2022): 100348.

de Saqui-Sannes, P., Vingerhoeds, R.A., Garion, C., and Thirioux, X. (2022). A taxonomy of MBSE approaches by languages, tools and methods. *IEEE Access* 10: 120936.

Simko, G., Lindecker, D., Levendovszky, T. et al. (2013). Specification of cyber-physical components with formal semantics – integration and composition. In: *Model-Driven Engineering Languages and Systems*. MODELS 2013, Lecture Notes in Computer Science, vol. 8107 (ed. A. Moreira, B. Schätz, J. Gray, et al.). Berlin, Heidelberg: Springer https://doi.org/10.1007/978-3-642-41533-3_29.

Song, H., Rawat, D.B., Jeschke, S. et al. (2017). *Cyber Physical Systems Foundations, Principles, and Applications*. Elsevier Inc.

Taha WM, Taha AEM and Tundberg J (2021) *Cyber-Physical Systems: A Model-Based Approach*, Springer, Gewerbestrasse 11, 6330 Cham, Switzerland.

Trehan, V. (2018). A process model in platform independent and neutral formal representation for design engineering automation. Doctoral thesis, Birmingham City University. https://www.open-access.bcu.ac.uk/8873/7/PhD%20Thesis%20-%20Vibhor%20Trehan.pdf (accessed 12 September 2024).

Tripakis, S. (2016). Compositionality in the science of system design. *Proceedings of the IEEE* 104 (5): 960–972.

Wang, C., Bi, Z.M., and Xu, L.D. (2014). IoT and cloud computing in automation of assembly modeling systems. *IEEE Transactions on Industrial Informatics* 10 (2): 1426–1434.

Zheng, H. (2023). CIS 4930/6930: principles of cyber-physical systems. https://cse.usf.edu/~haozheng/teach/PrinciplesCPS/slides/3_chap5-composition.pdf (accessed 12 September 2024).

11

Internet of Things

11.1 Introduction

The concepts of *Internet of Things* (IoT), mechatronic systems, and *cyber-physical systems* (CPSs) are complementary. IoT provides a symbiotic environment to integrate both onsite and remote *mechatronic functional modules* (MFMs). IoT extends mechatronic systems or CPSs in sense that

1) A traditional mechatronic system integrates mechanical, electrical and electronic, and control systems, while IoT expands the scope of integration by networking physical systems with any accessible virtual resources over the Internet. This allows the system to (a) take advantages of virtual resources to expand system's functionalities and capacities and (b) improve data-driven controls by collecting, analyzing, and using real-time data. The connectivity over the Internet provides numerous opportunities to access, interact, coordinate, cooperate, and collaborate with networked things.
2) A mechatronic system depends on sensors to perceive the environment and learn changes. IoT expands a mechatronic system by accessing numerous sensors for a wide range of data from environmental factors such as temperature, humidity, pressure, motion, and internal states. The enriched data can be processed and utilized to optimize decision-making activities.
3) IoT enables controlling and monitoring MFMs remotely over the Internet. Engineers can monitor a mechatronic system or MFMs in real-time mode and take action promptly when abnormal events occur. Engineers can adjust control variables and troubleshot some problems with no physical presence.
4) The data from IoT can be abundant. IoT-enabled *machine learning* (ML) and *artificial intelligence* (AI) can enhance the capabilities of a mechatronic system to respond to environmental changes promptly, monitor system states, perform preventive maintenance, and implement data-driven controls to optimize system performance.
5) IoT improves *human–machine interactions* (HMIs) when humans are the components of advanced mechatronic systems. For example, more and more *personal assistive devices* (PADs) such as laptops, tablets, and smartphones can be utilized to access, monitor, operate, and control mechatronic systems over the Internet.

Traditional mechatronic systems or CPSs are enclosed systems. Therefore, the number and types of physical elements in a system limit the scale, functions, and capabilities of the system. A natural extension for enhanced capabilities is to network other mechatronic systems, virtual resources, and smart things dynamically. In particular, the Internet can be used as information infrastructure to

Mechatronics for Complex Products and Systems: Project-Based Design Approaches for Robots, Cyber-Physical Systems, Digital Twins, and Other Emerging Technologies, First Edition. Zhuming Bi.

Companion website: www.wiley.com/go/bi/mechatronics

connect anything, anywhere at any time. On the Internet, things can be connected to one another through standardized communication protocols such as *hypertext transfer protocol* (HTTP) and *simple mail transfer protocol* (SMTP). With the rapid growth of the Internet, more and more mechatronic devices are connected, interacted, cooperated, and collaborated to form a *system of systems* (SoS) for more complex and advanced missions.

The scale of the IoT has been growing rapidly due to a number of critical factors such as the trend of ubiquitous connectivity, the prevalent uses of *Internet Protocol* (IP), AI, computing economics, the miniaturization of physical things, and the advancement of *Big Data Analytics* (BDA), the emergence of *blockchain technology* (BCT), *cloud computing* (CC), and *edge computing* (EC).

As shown in Figure 11.1, conventional IoT refers to the middleware to connect smart things by *machine-to-machine* (M2M), *supervisory control and data acquisition* (SCADA), *radio-frequency identification* (RFID), and *wireless sensor networks* (WSN). IoT supports networking small and inexpensive things with unique identities; the connected things are capable of locating themselves, and the computing algorithms behind IoT can process, analyze, and mine big data to support decision-making activities at any domain or level of businesses in an IoT-based system as shown in Figure 11.2 (Vermesan and Fries 2014; Zhou 2013).

11.1.1 IoT Concepts

The concept of IoT was introduced in 1995 by Bill Gates in his book *The Road Ahead*; it was then cited by International Communication to create an era of pervasive and ubiquitous smart things over the Internet. IoT was originally used to track parcels by RFID in supply chains. It became rapidly popular due to influential work by Kevin Ashton at the auto-ID center. Nowadays, IoT is changing the ways for people to manage and operate smart devices, production, distribution, transportation, service and maintenance, and recycling of products (Wang et al. 2014). IoT is the foundation to connect smart things such as sensors, actuators, software tools, controllers, and other smart technologies as a network. An IoT-based application can be understood as a

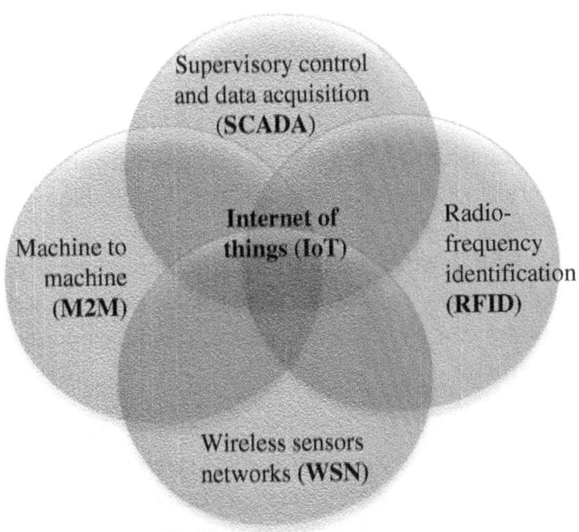

Figure 11.1 Conventional Internet of Things (IoT) as the middleware

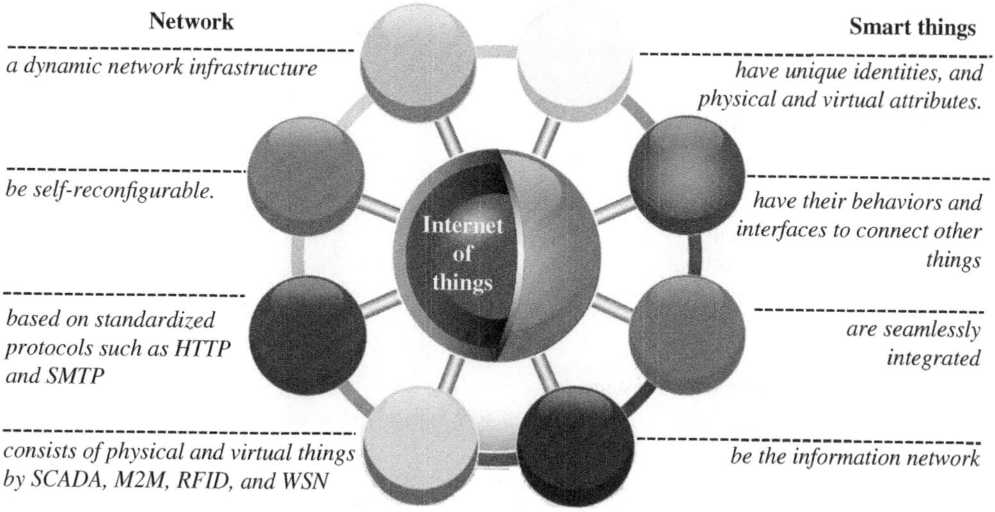

Figure 11.2 Description of Internet of Things (IoT)

distributed measurement and control system, it connects smart objects and other devices to enable a broad range of functions and applications.

IoT aims to network smart things with interacting capabilities globally, so that any entity in IoT can produce, acquire, access, process, and use data at the right place and the right time. Due to the importance of abundant, heterogeneous, and real-time data on prompt and reliable decision-making supports for various IoT applications, the number of networked things on the Internet has been increased exponentially as shown in Figure 11.3 (Castellanos et al. 2022).

The concept of IoT has been defined in a number of ways; this chapter used the description by Vermesan and Fries (2014, 2016) in Figure 11.2. More specifically, an IoT-enabled mechatronic system is a SoS that consists of several networked mechatronic systems to fulfill some aggregated functions of the system. *On the one hand*, an IoT-enabled mechatronic system is built upon a dynamic

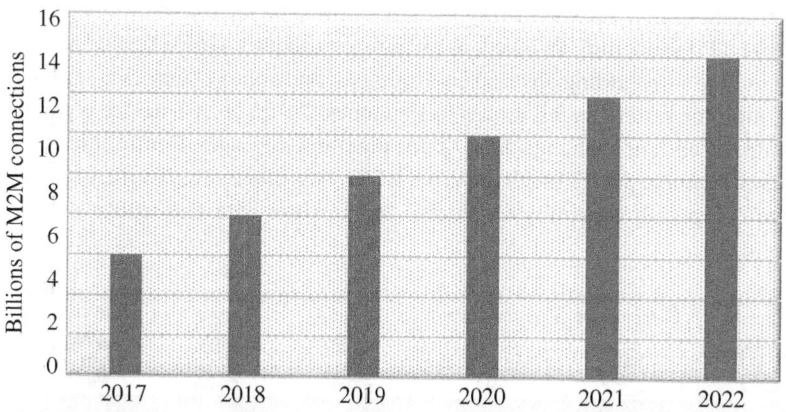

Figure 11.3 Number of networked things over years (*Source:* Castellanos et al. (2022)/Universidad de San Buenaventura)

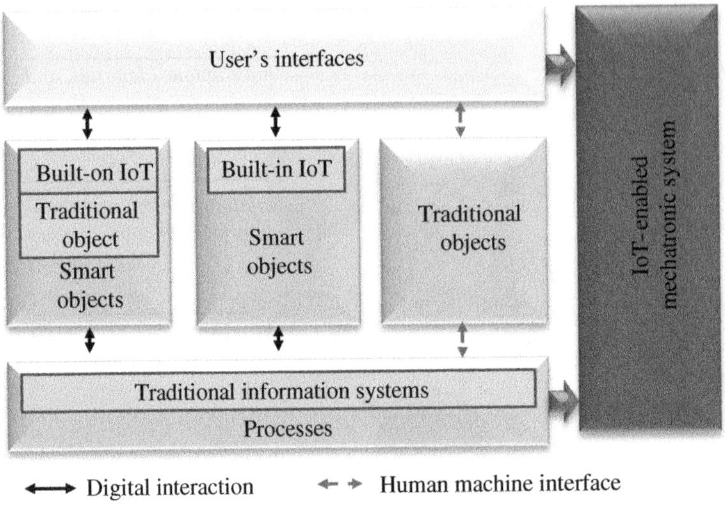

Figure 11.4 Connecting traditional and smart things in an IoT-enabled mechatronic system

network, it is portable and self-reconfigurable by following standardized communication protocols, and each of its system elements is an MFM with integrated physical, cyber, and virtual things. *On the other hand*, each connected thing in the system has its intelligence and is referred to as *a smart thing*. A smart thing has its unique identity together with its physical and virtual attributes, and it has smart interfaces to connect and network other smart things seamlessly (Figure 11.4).

11.1.2 Smart Things

In this chapter, *a smart thing* refers to an MFM that possesses certain intelligence such as perceiving the environment, communicating with others, and making decisions to respond to changes and uncertainties autonomously in its applications.

As shown in Figure 11.5, a networked thing in an IoT-enabled mechatronic system should have some basic functions as a mechatronic module (Grammatikis et al. 2019). *First*, it must be identified itself by its unique identification. In a network, every IoT device must be identified uniquely, and the protocols such as IPv4 and IPv6 identify an object by assigning a unique address. *Second*, an IoT device must be able to sense the surrounding physical environment, one or more sensors are used to sense the environment, monitor system states, and detect changes. *Third*, an IoT device communicates with others to send, receive, or exchange data, assets, files, and messages. Communications are enabled by *Information Technologies* (IT) such as Wi-Fi, RFID, and Bluetooth. *Fourth*, each IoT device possesses its own computing capability to process and analyze data and utilize data to

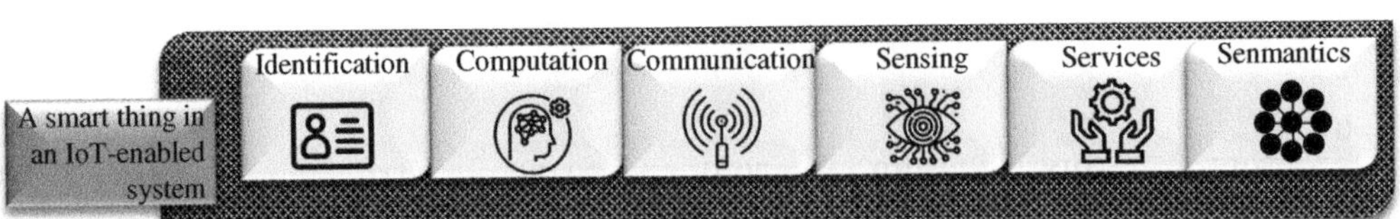

Figure 11.5 Typical smart things in an IoT-enabled system

support local decision-making activities. *Fifth*, an IoT device is capable of responding to requests and providing corresponding services to others. *Finally*, an IoT device has its semantic properties to obtain correct information from the physical environment and provide information as services at the right time, about the right things, and at the right place (Aqeel 2020). From the perspective of an information flow, basic FRs to achieve smartness of a thing over the Internet include to:

1) Communicate with other devices, virtual resources, and even human stakeholders directly or indirectly.
2) Collect data from other devices and process data locally.
3) Transmit real-time data to a server or a cloud-based application to process data for back-end applications.
4) Possess computing capabilities to support decision-making activities and operate devices autonomously.

11.1.3 Communication Protocols

In IoT, smart things connect with each other based on various communication protocols. Using communication protocols ensures that data, assets, and information are exchanged securely and reliably. Communication protocols support seamless connections of sensors, gateways, IoT devices, and backend systems. Figure 11.6 shows that many communication protocols have been developed to facilitate the communications of smart things at four layers, that is, the link layer, the network layer, the transport layer, and the application layer (Bahga and Madisetti 2015).

At the application layer, main protocols include *hypertext transfer protocol* (HTTP), *constrained application protocol* (CoAP), *message queuing telemetry transport* (MQTT), *WebSockets*, *advanced message queuing protocol* (AMQP), *data distributed service* (DDS), and the *universal messaging standard* (XMPP). HTTP is simple to implement, and it is widely used in various IoT applications for data exchange, control, and management. MQTT is lightweight and it is designed for constrained devices with low-bandwidth, high-latency, or unreliable networks. It is used in a client–server architecture where a smart thing publishes data to a broker and others subscribe to specific topics. CoAP is lightweight and designed for constrained devices or networks; it was also

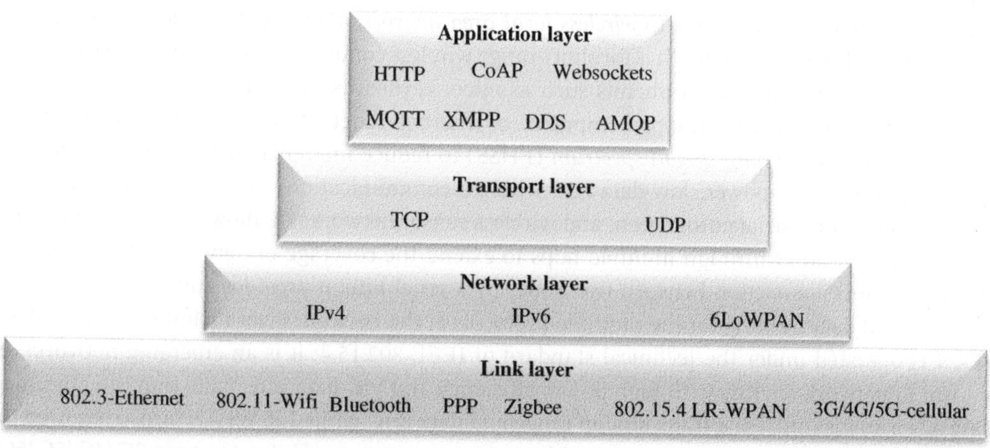

Figure 11.6 Classification of communication protocols

used in a client-server architecture that is operated over *user datagram protocol* (UDP). AMQP is open-standard for message-oriented middle; it supports both node-to-node and publish-subscribe messaging in a distributed system. DDS is a middleware protocol for scalable and high-performance data distributions in a distributed environment. A WebSocket provides two-way and multiple communication channels over a single *transmission control protocol* (TCP) connection; it enables real-time and event-driven communication between a client and a server. XMPP is an open-standard protocol for instant messaging, presence, chatting, video calls, and content syndications.

The transport layer has two main protocols, that is, TCP and UDP. TCP is a set of communication protocols to connect devices to the Internet, and TCP can be used together with Internet Protocol (TCP/IP) to network computers in a private network that is known as an *Intranet* or *Extranet*. TCP/IP serves as an abstraction layer between the application layer and the network layer. It provides the rules about how a message is broken into packets at an emitter, and then be transmitted, addressed, routed, and received at receiver. UDP is a suite of Internet prototypes to send messages as datagrams in packets to others over the Internet. UDP is not required to set up communication channels prior to data transmission or exchanges. UDP uses checksums to assure data integrity, uses port numbers to address different functions, and does not require handshaking to confirm communications. It requires a minimum of prototype mechanisms.

At the network layer, main protocols include IP, *Internet Control Message Protocol* (IMP), *Internet Group Management Protocol* (IGMP), *Address Resolution Protocol* (ARP), *Neighbor Discovery Protocol* (NDP), *Internet Protocol Security* (IPSec), *Border Gateway Protocol* (BGP), and LowPan Neighbor Discovery Extensions (6LoWPAN). IP is fundamental and responsible for addressing and routing packets over the Internet. It provides best-effort delivery services by which each package is routed individually based on IP addresses. IPv4 and IPv6 are two main versions that use 32-bit and 128-bit addresses, respectively. On the other hand, 6LoWPAN supports low-power wireless mesh network where every node is identified with its IPv6 address; it allows a node to be connected to the Internet directly using open standards.

At the link layer, main protocols include *Ethernet*, *Wi-Fi*, *Bluetooth*, *Zigbee*, *Point-to-Point Protocol* (PPP), *Low-Rate Wireless Personal Area Network* (LR-WPAN), and 3G/4G/5G-cellular. Ethernet defines the format of data as *Media Access Control* (MAC) addresses of source and destination, payload, and error-checking information. It uses *Carrier Sense Multiple Access with Collision Detection* (CSMA/CD); it is the most widely used in *local area networks* (LANs). Wi-Fi supports wireless communication based on the standards that are defined in IEEE 802.1. It uses MAC to format data frames, and it connects a device to *wireless local area networks* (WLAN) by radio frequencies in 2.4 GHz or 5 GHz bands. Bluetooth is for short-range wireless communication between two devices. It is widely used to connect peripherals such as mice, keyboards, headphones, personal assisted devices, and smartphones to laptops, computers, or Internet. Bluetooth operates in a 2.4-GHz band and uses *frequency-hopping spread spectrum* (FHSS) to reduce inference and enhance reliability. Zigbee supports a low-power, low-data-rate wireless communication. Zigbee is commonly used in smart homes, industrial automation, and wireless sensor networks. It allows devices to communicate with each other through multiple hops to extend the coverage of network. PPP is used to establish a direct connection between two nodes by a serial link; it provides authentication, error detection, and data framing for the multiple protocols at the network layer including TCP/IP. LR-WPAN is operated under the technical standard of IEEE 802.15.4; it is an emerging networking protocol to network devices with high power efficiency and low data rate. Note that the objective of WLAN is to maximize the transmission range and data rate; while LR-WPAN is also required to save energy in a personal operating space without any construction of infrastructure. 3G/4G/5G/6G are the 3rd, 4th, 5th, and 6th generation of cellular technologies to connect smart devices to the

Internet with different levels of speeds, bandwidth, and accessibility. With 3G, a connected device can access the Internet or make video calls without a delay; with 4G, a mobile user can enjoy online games, watch, high-definition (HD) movies, and participate in video conferences with high quality; with 5G, a mobile user can fully drive a self-driving vehicle, watch ultra-high-definition (HD) mobile streaming, or stream ultra HD security footage directly on smartphones. With 6G, large-scale Metaverse use cases can be fully supported.

11.2 Characteristics of IoT-Enabled Systems

By incorporating various IoT devices in traditional mechatronic systems, the functionalities and capabilities of mechatronic systems can be expanded or enhanced in many aspects as shown in Figure 11.7 and as follows (Bi et al. 2014; Borgia 2014; Ramasamy and Karry 2021; Tran and Djabir 2023).

Interconnectivity. An IoT-enabled system is composed of networked sensors, devices, modules, agents, and systems that can communicate with one another seamlessly. The Internet provides the infrastructure for these smart things to transmit and exchange data in a real-time mode. This enables the coordination and collaboration of mechatronic components to achieve system-level goals. Any object can be connected to access global information.

Ambient Intelligence. It refers to decision-making capabilities to sense and respond to changes in a dynamic environment. On the one hand, mechatronic components are mostly intelligent, autonomous, and interoperable. On the other hand, IoT devices are equipped with sensors and actuators that enable them to sense and perceive environmental factors such as temperature, humidity, motion, and light. This allows an IoT system to monitor and react to changes in a real-time mode.

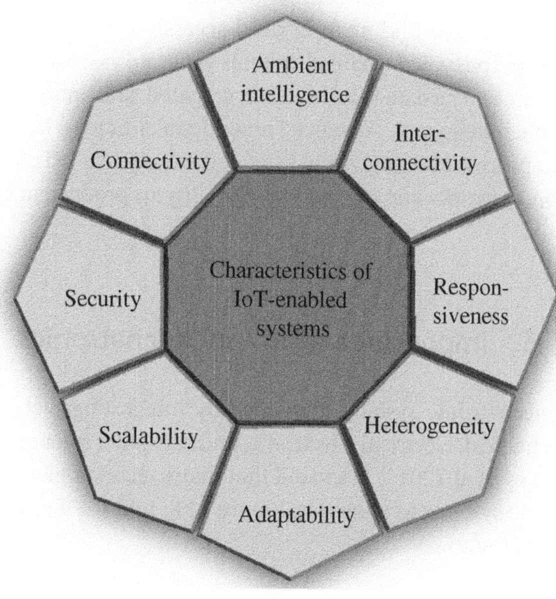

Figure 11.7 Characteristics of IoT-enabled systems

Real-Time Responsiveness. A system with real-time responsiveness enables quick and timely actions based on incoming data and events, it is data-driven. Such capabilities are extremely important to some critical applications such as industrial automation, smart transportation, healthcare monitoring, and emergency response systems. Smart things in IoT can respond to the requests of others subjected to specified constraints such as security assurance and semantic consistency of physical and virtual things.

Heterogeneity. The Internet allows the connection of heterogeneous things. These things have been developed on highly diversified hardware platforms and networks; they can interact with each other by requesting and providing services via the Internet.

Adaptability. An IoT-enabled system is able to (1) sense and perceive environmental changes that can affect behaviors and performance of the system; (2) analyze and utilize data, support decision-making activities, and (3) reconfigure the system to accommodate changes in an ever-changing environment. IoT systems often integrate with existing IT systems, applications, and infrastructure to enable seamless data exchange and interoperability. Standardized protocols and interfaces facilitate integration between different IoT devices, platforms, and ecosystems, allowing for greater interoperability and compatibility.

Scalability. With the rapid growth of new technologies and available smart things, an IoT-enabled system is scalable; it means that the system is flexible to reduce or add system elements to accommodate changes. IoT solutions are designed to be scalable and flexible, allowing them to accommodate a wide range of devices, applications, and use cases. IoT platforms and architectures can be scaled up to support thousands or even millions of devices while remaining flexible enough to adapt to changing requirements and environments.

Safety and Security. The Internet provides numerous opportunities for users to work with untrusted partners. Therefore, safety is one of the primary requirements of both creators and users of IoT-enabled systems. Safety requirements include but are not limited to protect confidential data, virtual and physical assets, and safe access to endpoints and networks. Security and privacy are paramount in IoT-enabled systems due to the interconnections of devices and the sensitivity of collected and transmitted data. IoT solutions incorporate robust security measures, including encryption, authentication, access control, and device management, to protect against cyber threats and unauthorized access.

Connectivity. IoT-enabled systems generate vast amounts of data from sensors, devices, and other sources. The data is collected, aggregated, and analyzed to derive meaningful insights and actionable intelligence. Advanced analytics techniques, including ML, AI, and BDA, can be integrated to process and analyze IoT data efficiently. An IoT-enabled system assures (1) data accessibility on networks and (2) the compatibility to produce and use relevant data.

11.3 Importance of IoT in Mechatronics

A mechatronic system is enabled by integrating technologies in multiple disciplines including mechanical, electrical, electronic, and computer science as a unified solution to a complex problem. Bradley et al. (2015) discussed that many technologies have contributed to advance various mechatronic systems as shown in Figure 11.8, and IoT happened to be one of the most recent and significant ones.

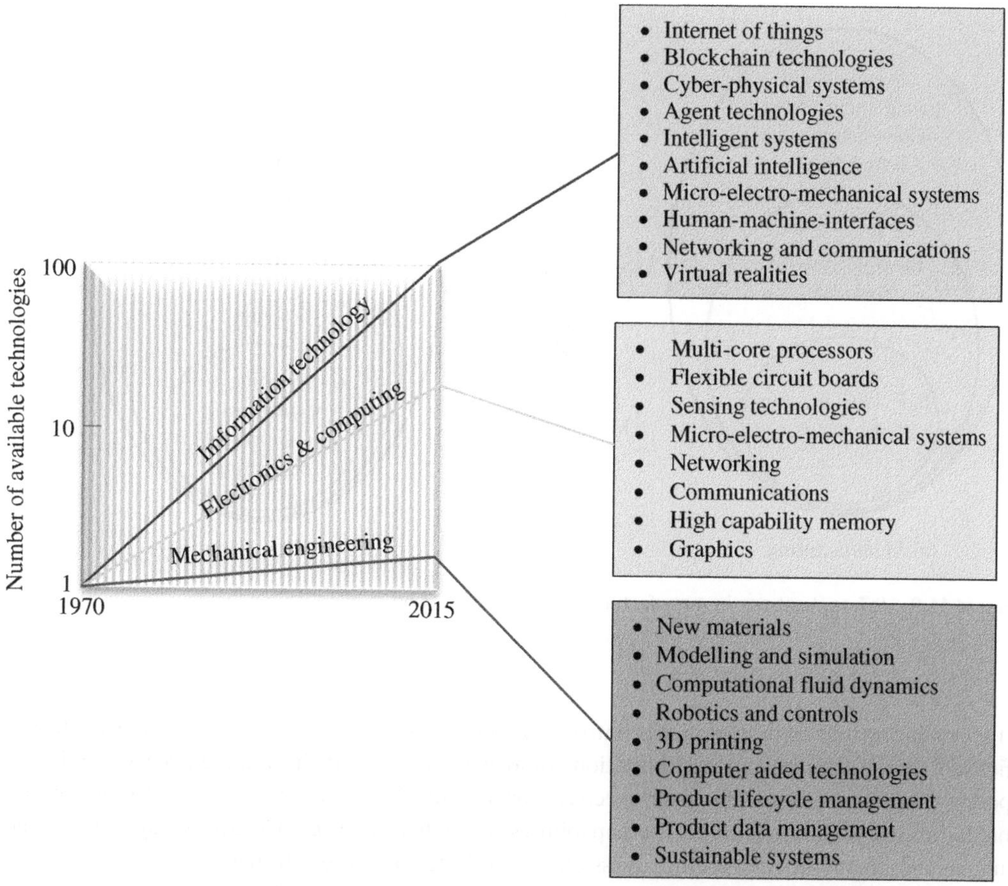

Figure 11.8 Technologies that advanced mechatronic systems (*Source:* Bradley et al. (2015)/with permission of Elsevier)

The capabilities of a mechatronic system will be expanded greatly when open, virtual, and dynamic mechatronic components are integrated over the Internet (Bradley et al. 2020). Both CPS and IoT were spanned from conventional mechatronic systems by connecting physical things with cyber and virtual things. IoT and CPS share the commonality that smart things are equipped with sensors to collect data, and computing units are able to analyze data, and support decision-making to respond to the changes in system states. Moreover, CC can be integrated with CPS to enhance computation for decision-making supports to solve multidisciplinary problems (Alam and Shaddik 2017). However, communications in a CPS are fixed and those in IoT can be either *fixed* or *dynamic* connections.

The vision of IoT-enabled mechatronic systems is to network smart things and synergize the capabilities of virtual and physical resources in the sense that (1) IoT provides the infrastructure to access, share, transmit, and exchange data among smart things; (2) IoT connects smart things to bridge the physical and digital world; computation is embedded in mechatronic systems. This makes it feasible to support (a) local decisions autonomously and (b) data-driven global decisions for the integration, cooperation, coordination, and collaboration of MFMs. Figure 11.9 illustrates

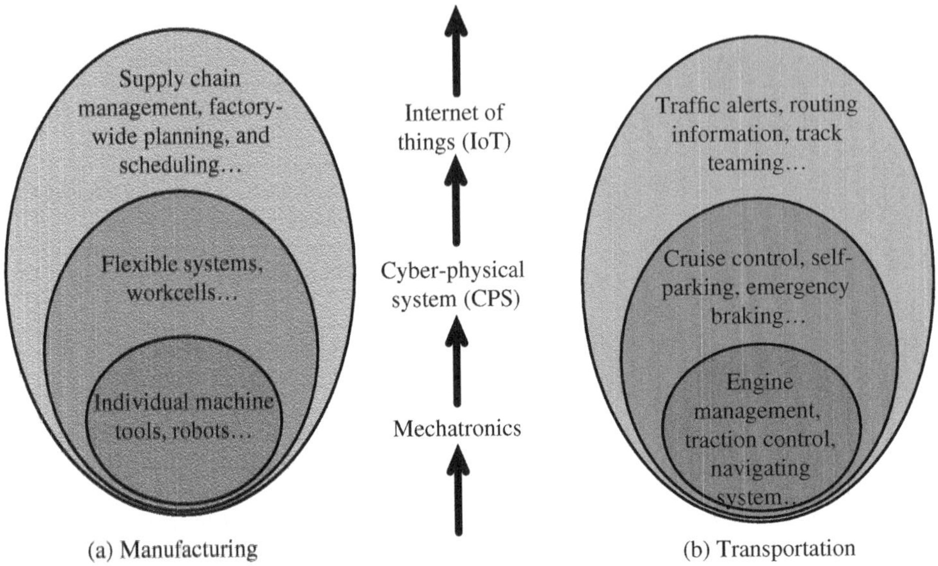

Figure 11.9 IoT applications in manufacturing and transportation

the evolution from *conventional* mechatronic systems to *IoT-enabled mechatronic systems* in the fields of manufacturing and transportation (Bradley et al. 2020). It showed the necessity of incorporating IoT in the next-generation mechatronic systems. While individual mechatronic components limited their capabilities, the capabilities of an IoT-enabled mechatronic system can be augmented greatly by networking accessible smart things over the Internet.

IoT is not a totally new technology. It has utilized the technological developments in numerous fields such as nano-electronics, embedded systems, digital twins (DTs), cloud and edge computing, discovery services, security and privacy, BCT, system integration, mechatronics, software, knowledge-based engineering, communications, and standards. From an engineering perspective, IoT is to connect more and more smart things to empower mechatronic systems. The nature of IoT determines the proliferation of IoT-enabled systems in terms of heterogeneity, scalability, and reconfigurability. In general, the advent of ML and AI has leveraged an IoT-enabled mechatronic system by acquiring and fusing data from heterogeneous sensors and processing and analyzing data to support decision-making activities (Elgazzar et al. 2022).

As shown in Figure 11.10, Hehenberger and Bradley (2016) and Hehenberger et al. (2016) discussed relevant technological developments when conventional mechatronic systems evolved into IoT-enabled mechatronic systems in some typical applications. IoT connects ubiquitous smart things such as actuators, sensors, robots, and mobile phones via unique addressing schemes. This allows more and more MFMs to interact and cooperate with each other for system-level goals (Marwedel 2021, Kamal 2017). Mechatronics used to be the main driver for Industry 2.0 and 3.0 in the 1970s, and CPSs and IoT became the main drivers for Industry 4.0 and 5.0. Moreover, emerging technologies such as next-generation AI, CC, BDA, BCT, and Metaverse are providing more and more opportunities to continuously advance the smartness, scalability, adaptability, and sustainability of next-generation mechatronic systems.

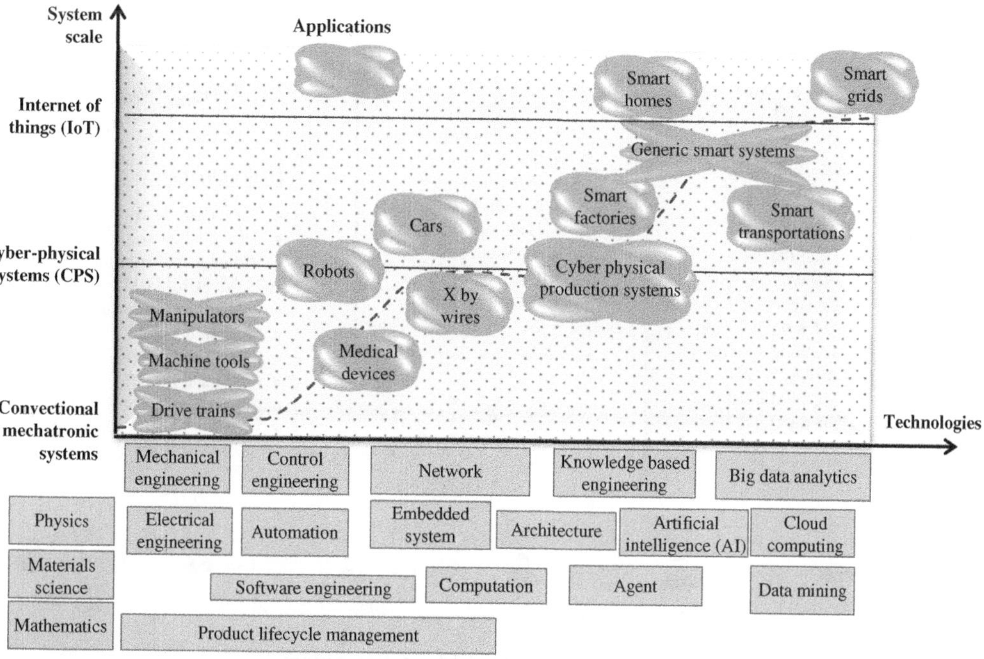

Figure 11.10 Transition from mechatronics to CPS and Internet of Things (IoT) (*Source:* Hehenberger et al. (2016)/Springer Nature)

11.4 Data Flows in IoT-Enabled Systems

System variables represent for the changes of inputs and outputs of the system with respect to time. When the information of a system is insufficient to predict system behaviors, outcomes, and performance, acquiring, analyzing, processing, and utilizing real-time data about the states of system variables become essential to support advanced closed-loop decision-making activities.

Here, the information system of a mechatronic device, component, MFM, sub-system, or SoS is discussed, and Figure 11.11 gives a generalized model of an information system for decision-making supports at specific domains and levels.

An information system is to transfer "*raw data*" into information and knowledge to plan, schedule, and control the operations of physical systems. Accordingly, the tasks in an information system can be classified into three types, that is, *data acquisition*, *data management*, and *data utilization*. In the digital era, some *cutting-edge information technologies* (IT), such as IoT, *cloud manufacturing* (CM) and BDA, and *digital manufacturing* (DM), have been gradually adopted to process and use big data effectively. These technologies are competitive for being one or a few of six enablers (i.e., digitization, open manufacturing, service orientation, shared manufacturing, sustainability, and resilience) in sustainable systems.

1) Outputs of an information system, that is, decisions on the actions of system elements, rely on sufficient, abundant, prompt, and reliable data. IoT and relevant technologies will be used to obtain whatever data the information system needs from networked things such as machines, parts, people, sensors, databases, suppliers, users, and markets. With the rapid growth of smart things in the network, *Volume, Variety,* and *Velocity* (3V) of data have been increased exponentially.

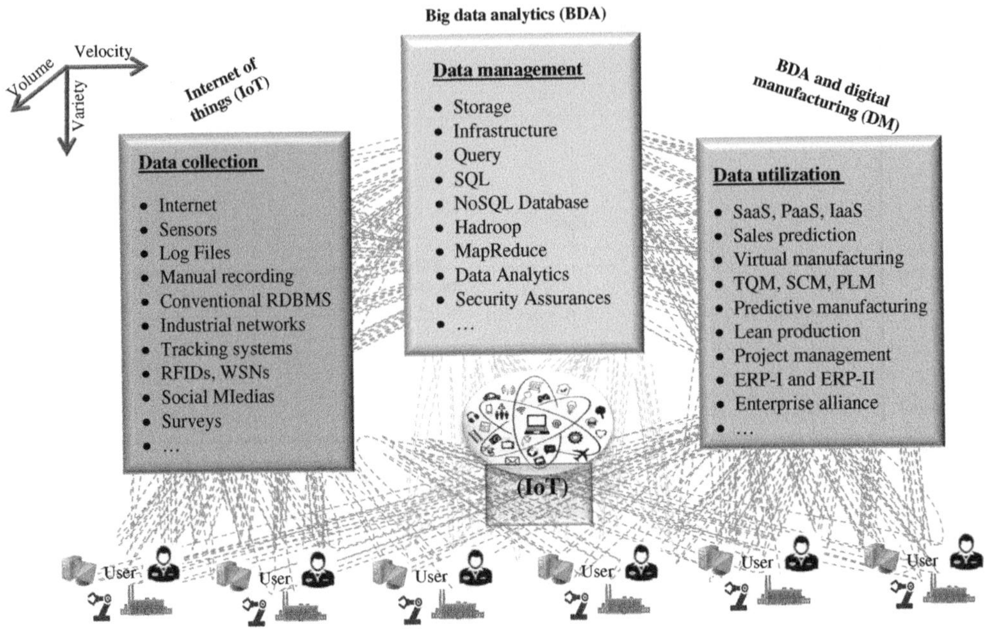

Figure 11.11 Enabling technologies in an IoT-enabled information system

2) When data is local, isolated and with a small size, conventional functional units in *computer-integrated manufacturing* (CIM) are capable of processing data for decision-making supports. Incorporating IoT brings the challenge of an information system to deal with big data promptly. BDA tools could be used to track, sort, transmit, distribute, store, process, and mine data at all levels and domains of system. In particular, the platform technologies in BDA, such as *non-relational database* (NoSQL) and MapReduce, will be used to locate the most relevant data from vast IoT effectively.
3) Every manufacturing system has its specialties in running manufacturing businesses; therefore, the interactions of digital and physical entities must be customized to meet the needs of an individual enterprise. DM is used to bridge the digital and physical worlds seamlessly. DM-enabling technologies allow enterprises to utilize both hardware and software resources over IoT via cloud services.
4) Note that no quantitative tool is available to evaluate the adverse impact of static information architecture on system adaptability and resilience, no artificial constraint is desired to hurdle the direct interactions among smart things. Therefore, structureless architecture can maximize system adaptability, flexibility, agility, and resilience in a dynamic environment.

11.5 IoT-Enabled Capabilities

IoT itself has been supported by a number of technologies including those for sensing, networking, standards, augmented intelligence, and augmented behaviors. On the other hand, IoT serves as the infrastructure for direct interactions of system assets at any domain and level of the system.

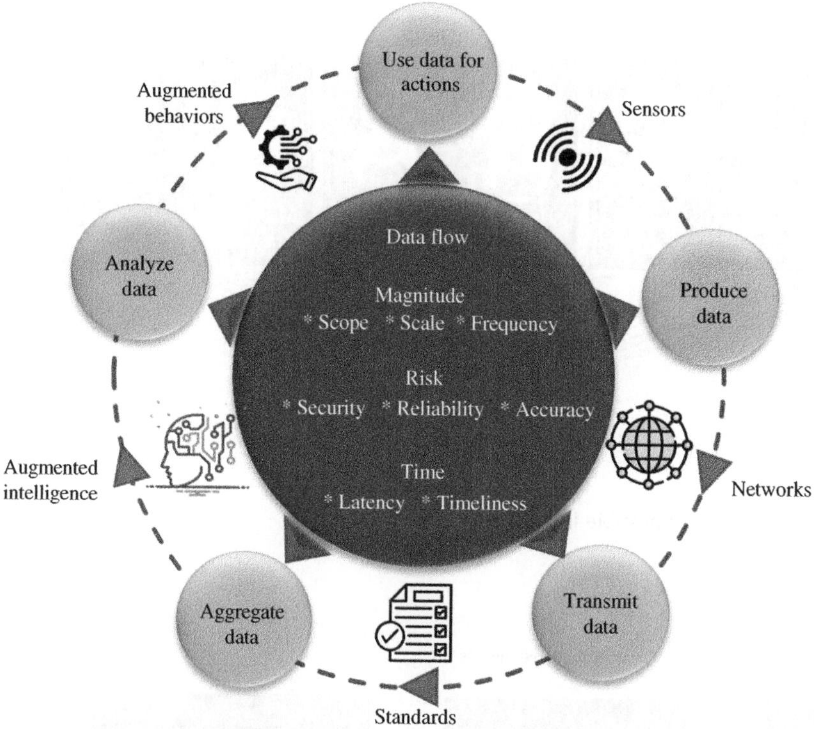

Figure 11.12 IoT as a value driver to run a smart system (*Source:* Adapted from Deloitte (2024))

Through the Internet, IoT networks everything including machines, peoples, materials, parts, and decision-making units at any stage of product lifecycles. As shown in Figure 11.12, IoT becomes a value driver to run a data-driven smart system by (1) generating information based on the data collected from physical objects or systems, (2) transforming data or information from one place to another, (3) fusing information generated by different users, at different sources and different times, (4) discerning patterns and relations to represent, predict, or prescribe actions, and (5) initiating, sustaining, or changing physical events or states (Deloitte 2024).

11.5.1 Interactions

Figure 11.13 shows typical interactions among system elements in a smart system. Typical interactions include the interactions of cyber and physical systems, interactions of machines with humans, infrastructures, environments, other smart things, and humans' interactions. IoT empowers an enterprise system for its sensing and actuating capabilities. In addition, IoT-based information systems can enhance system performance in terms of extendibility, usability, security, vendor-agnostics, non-intrusiveness, and plug-and-play capabilities.

The information technologies have been the most critical drivers to digitize complex systems including next-generation mechatronic systems. The impact of recent IT technologies on the trend of digitization of manufacturing systems can be examined in three aspects as shown in Figure 11.14: (1) *networking* – a direct interaction, collaboration, or interoperation of two elements becomes feasible as long as they are networked; (2) *digitization* – the operations of manufacturing businesses are

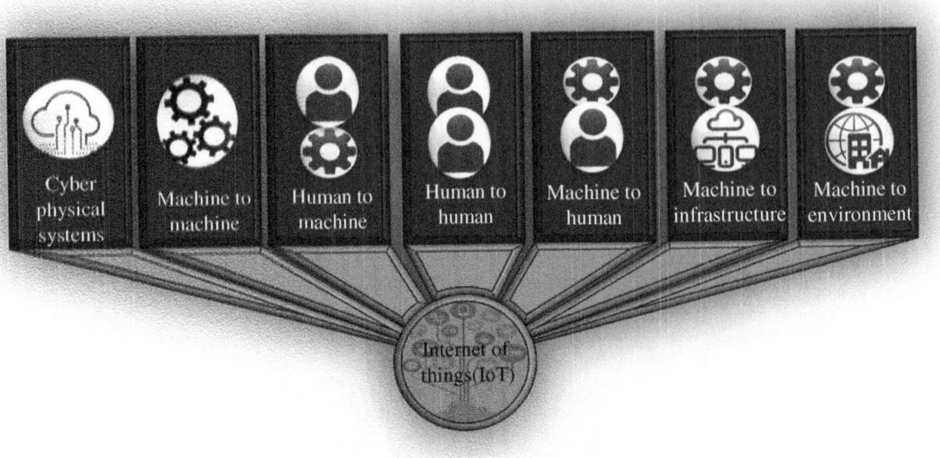

Figure 11.13 IoT makes all types of possible interactions

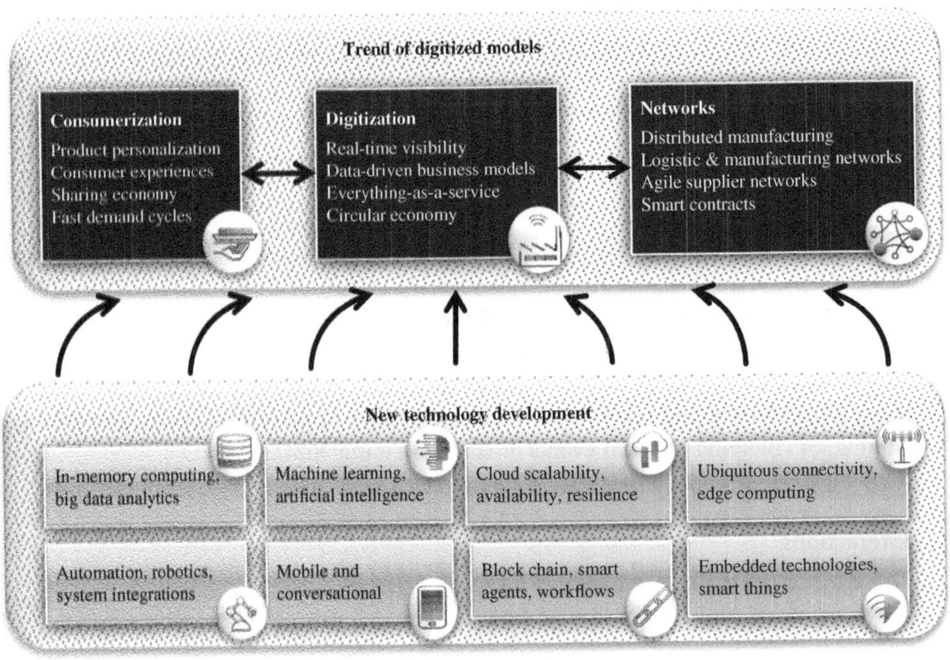

Figure 11.14 Information technologies for digitization

fully data-driven; any physical thing and process are visible in the virtual world; (3) *consumerization* – enhanced intelligence and smartness make it possible for manufacturing resources to fulfill different tasks without extra cost and time; products or services by enterprises can be fully customized.

It is also worth noting that IoT is the foundation for most newly developed information technologies such as *workflow composition*, *EC*, BCT, *in-memory computing*, *cloud scalability*, and *ubiquitous connectivity* (Viriyasitavat et al. 2019a, 2019b, 2020a, 2020b). For example, cloud manufacturing was developed based on IoT. Cloud manufacturing offers convenient, ubiquitous, and on-demand access to shared manufacturing resources over networks (Aghamohammadzadeh et al. 2020; Zhang et al. 2019; Ben-Daya et al. 2019).

11.5.2 Big Data Analytics (BDA)

An information system deals with data and information, and the characteristics of data and information have changed greatly in terms of sources, structures, time sensitiveness, variety, sizes, and distribution natures. While conventional information systems were limited to their capabilities only for structured, static, and small-volume data, BDA tools have enhanced capabilities in classifying, searching, processing, and mining big data into the right information for decision-making supports. BDA is able to process big data at an unprecedented scale (Cloud Security Alliance 2024). In a digital enterprise, BDA aims to make the right decisions promptly for the system to respond to changes and uncertainties. BDA is not a standalone technology. It has been integrated with many other technologies to manage manufacturing resources, peoples, business strategies, and marketing. Sahoo (2022) surveyed most recent progresses of using BDA in manufacturing business management. They suggested expanding BDA into three main clusters. BDA plays a critical role in optimizing *supply chain management* (SCM), logistics management, and inventory management (Maheshwari et al. 2020; Talwara et al. 2021). BDA was used to improve the agility of a manufacturing system through customization, automation, total employee empowerment, and system integration (Gunnasekaran et al. 2018). BDA should include typical data processing tools for data filtering, cleaning, clustering, registering, and fusing and data mining tools for regression analysis, textural analysis, Bayesian analysis, and analysis of decision trees. Cheng et al. (2022) discussed the contribution of BDA to improve the flexibility of supply chains for a high-level sustainability of manufacturing enterprises in the circular economy. BDA brought potential values according to the empirical and quantitative studies by Meriton et al. (2021).

The capabilities of a data management tool determine the performance of an information system. KPMG (2018) argued that the performance of a manufacturing information system could be classified into six levels from *computerization* to *adaptability* shown in Figure 11.15. Adopting more and more advanced data management tools allows the system to (1) automatically collect real-time data from the environment; (2) network all objects and models as smart things; (3) monitor and take preventive actions over products, machines, and processes; (4) predict changes and uncertainties reliably to optimize business plans; (5) reconfigure systems to respond unanticipated changes autonomously.

11.5.3 Digital Manufacturing (DM)

DM is similar to smart manufacturing in the sense that smart manufacturing resources are sensed, networked, and integrated for predictive, adaptive, and sustainable system operations. DM refers to a set of available enabling technologies that are deployed to support decision-making activities at all levels and domains of manufacturing businesses. DM technologies are presented in different forms at different levels, that is, CPSs at the device level (Park et al. 2021), *DTs* at the component or system level, and SoS at the strategic level. In particular, DTs highlight the replica of virtual and physical systems as shown in Figure 11.16 (Siemens 2015, 2019a, 2019b). DTs directly interact seamlessly.

With the integration of IoT, BDA, and DM enabling technologies, no static EA will be necessary in the design, implementation, and operation of manufacturing enterprises. System structure and organization should be flattened for direct interactions of system elements when they are needed, and system operations were data or event driven. Qiao et al. (2020) developed a data-driven CPS to schedule productions with the purpose to improve the adaptability of a smart manufacturing

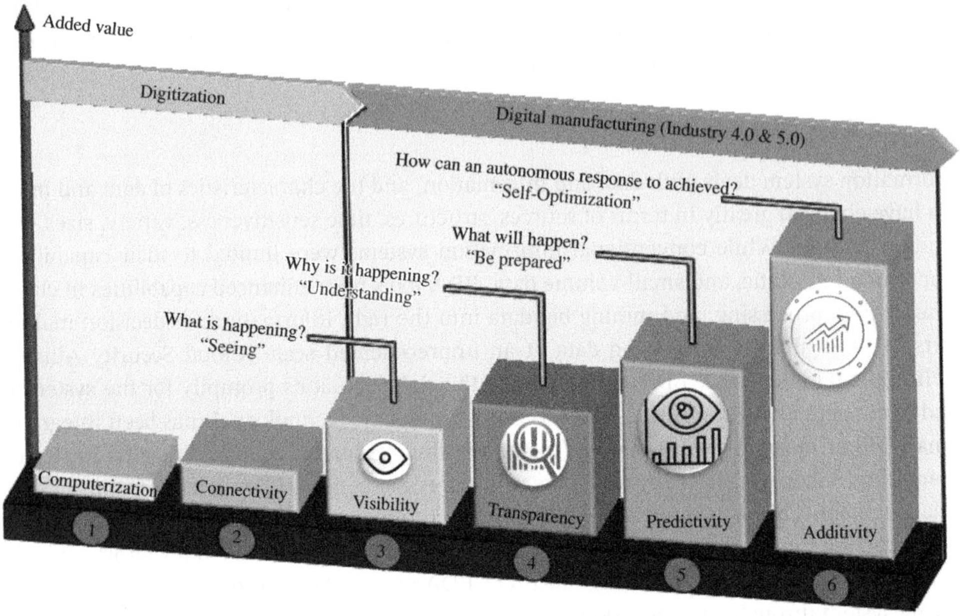

Figure 11.15 Objectives of data analysis in information system

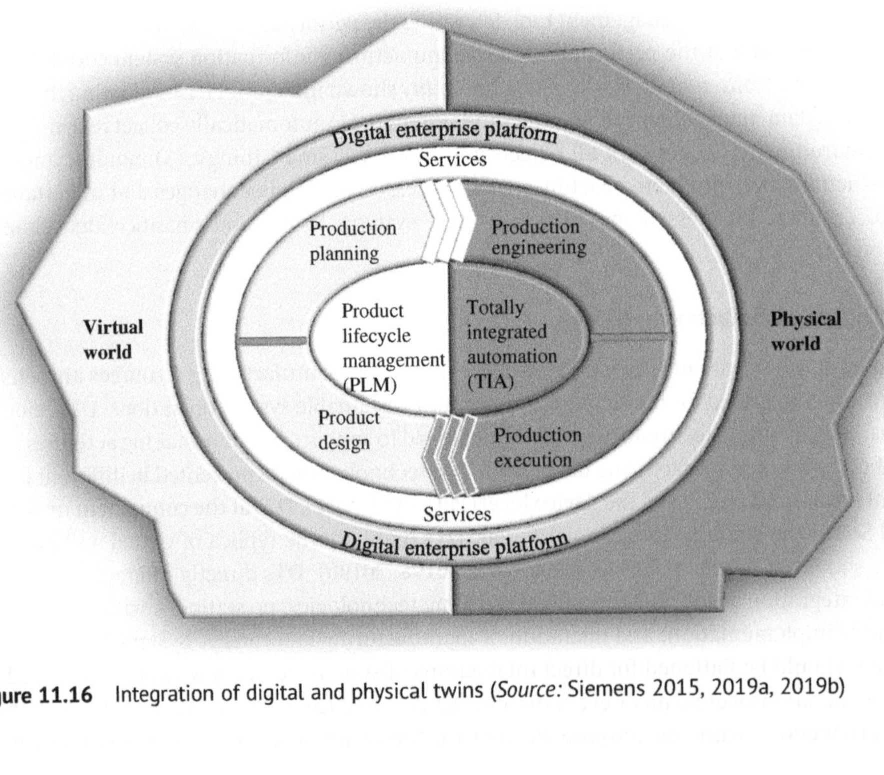

Figure 11.16 Integration of digital and physical twins (*Source:* Siemens 2015, 2019a, 2019b)

system by implementing a four-phase adaptive solution from data acquisition, disturbance identification, strategical adjustment, and generation of schemes. Industry 4.0 is featured by digitization, smart manufacturing, collaborative business, and viable supply chains. Ding et al. (2019) integrated DT and CPS as a DT-based cyber-physical production system to support interactions and interoperations of manufacturing assets. Chae and Olson (2021) investigated the landscape of Industry 4.0 and found that it must be interacted with other technologies, in particular, AI and IoT in applications including healthcare, manufacturing, and financial systems. Ivanov et al. (2020) focused on the applications of Industry 4.0 in operations management. They proposed a conceptual framework that helped understand the state of the art, challenges, and research opportunities of Industry 4.0-based technologies in operations management.

SMEs have limited resources in adopting advanced technologies. For example, some specific challenges related to the digitization of SMEs were discussed by Mittal et al. (2020). By all means, companies should benefit from partial digitization by following the steps of (1) collecting available data from various sources, (2) assessing the readiness of data acquisition, processing, and utilization, (3) raising the awareness of leadership and employees (4) tailoring the visions and missions of smart manufacturing to SMEs, and (5) selecting technological tools for implementation of smart manufacturing paradigm in SMEs. SMEs should begin to digitize their manufacturing systems gradually from monitoring, controls, and optimization to the autonomy of manufacturing systems. Alternatively, high-level goals can include a fast response to self-diagnosing and self-adapting (Moeuf et al. 2018).

Since most IoT-enabled systems are distributed and decentralized systems, CC is needed to address dynamic, exponentially growing demands for real-time, reliable data processing of IoT as shown in Figure 11.17. More specifically, enhanced capabilities include (1) cutting-edge sensing

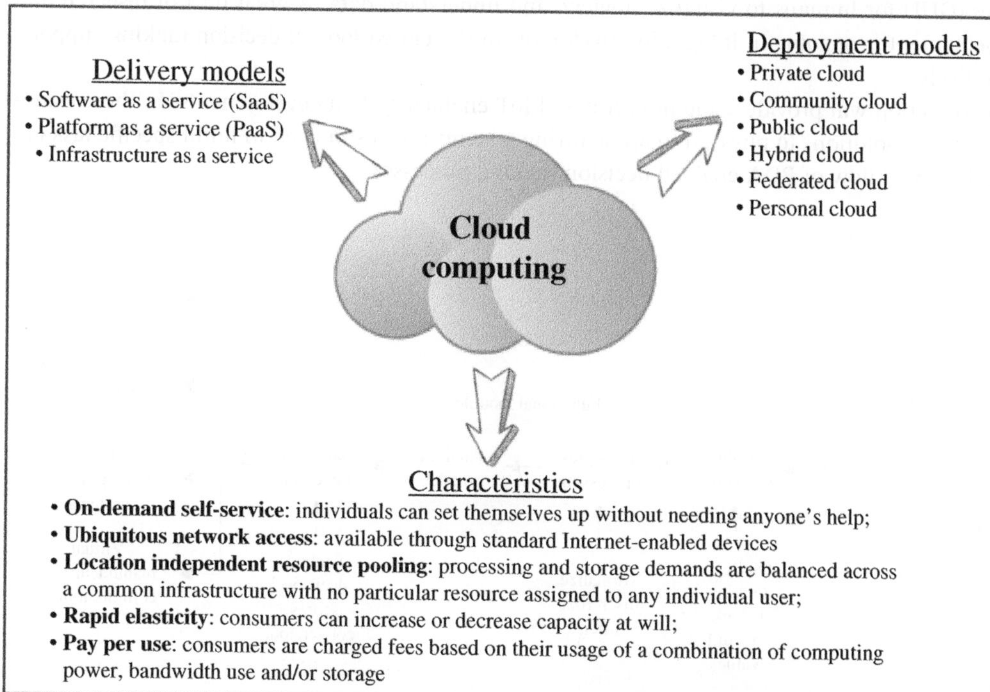

Figure 11.17 Characteristics, delivery, and deployment models of cloud computing (CC)

and actuating technologies that allow "objects" to participate in IoT freely; (2) interoperable service-oriented middleware to share real-world data among heterogeneous devices; (3) networking technologies to interconnect "things"; (4) decision-making applications to store, integrate, and process dynamic data streams from devices in real-time mode. Cloud infrastructure provides the storage and computing capabilities to process big data over the IoT.

11.6 Project-Based IoT-Enabled System Development

Advanced systems rely more and more on newly developed technologies including IoT, BDA, and DM to enhance system performance such as adaptability, robustness, and resilience. There is a high research demand to utilize existing technologies to increase productivity, profitability, and sustainability of systems in the aspects of *vertical integration, decentralization, smart sensing and actuating, autonomy and self-organization*, and the development of *semantic models* (Dotoli et al. 2019).

BDA supports decision-making activities at all domains and levels of the system. Figure 11.18 shows the generic data analytics workflow in a decision-making unit (Rachuri 2019). *First*, raw data is the input of BDA, which has to be collected and delivered to functional modules where data is used. *Second*, the acquired data is fused and processed to eliminate invalid, un-interpretable, or faulty data. *Third*, the data is interpreted and mined, so that it can be understood in computer models or algorithms. *Fourth*, mathematical models and algorithms are applied to mine and use data for decision-making supports. *Fifth*, the obtained information and decisions are sent to physical systems for execution in applications. When a functional unit is not fully automated and humans are involved in a decision-making procedure, the system is usually equipped with *graphic user interfaces* (GUI) for humans to visualize, review, and understand data. A great deal of research and development is ongoing in integrating BDA tools in the closed-loop of decision-making supports effectively.

This section will provide some project-based IoT-enabled system developments. In these examples, digital solutions to physical manufacturing systems are customized to fulfill specific tasks in the data workflow of BDA-enabled decision-making processes.

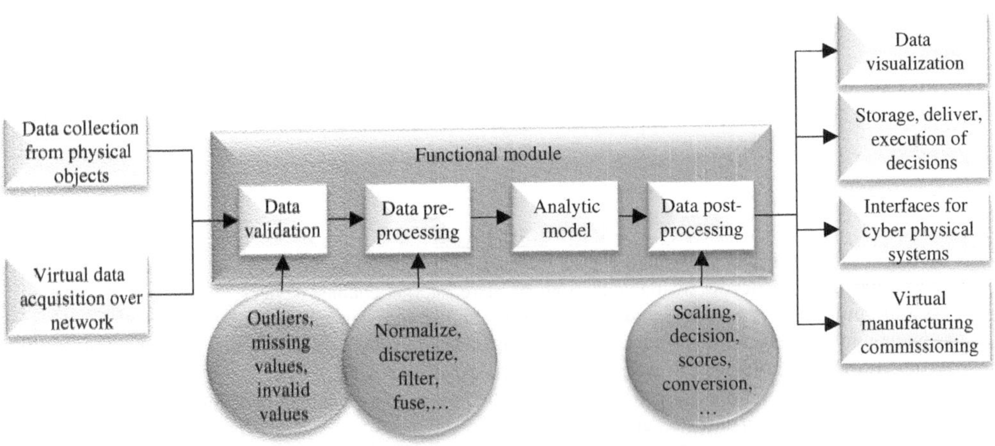

Figure 11.18 BDA workflow in information systems

11.6.1 Ubiquitous Sensing

Due to the incompleteness and dynamics of available information from physical systems, running a digital manufacturing system must be data driven. The intelligence level of the manufacturing system depends on the availability and utilization of the collected data from the physical system. The better manufacturing resources are instrumented, the more data with higher reliability can be collected and utilized, and the better decisions can be made. A great deal of research endeavors have been reported to develop cost-effective sensing methods for smart things.

Figure 11.19 shows an example where a standard gripper is instrumented to obtain force feedback in a gripping process. To develop a cost-effective solution, strain gauges are used to sense the response of the materials to the gripping forces in a real-time manner. Note that some high-ending commercial grippers are fully instrumented to obtain dynamic loads when operating but at a high cost; they are not good options for the majority of the grippers' applications. Most industrial robots use kinematic controls where the dynamic behaviors of robots are not considered.

11.6.2 Fusing and Integrating Data from Heterogeneous Sources

Manufacturing processes and operations are generally complicated and involve many design variables. The optimization of a manufacturing process over the time domain relies on the abundant and reliable data in closed-loop controls. The data is collected from the things involved in the manufacturing process, and then processed, fused, and utilized to support the decision-making activities. Therefore, to develop a smart manufacturing system, it is necessary to have a high level of intelligence of the decision-making supports in the system, and this requires to fuse and integrate data from multiple sources to control manufacturing processes or operations in a real-time mode.

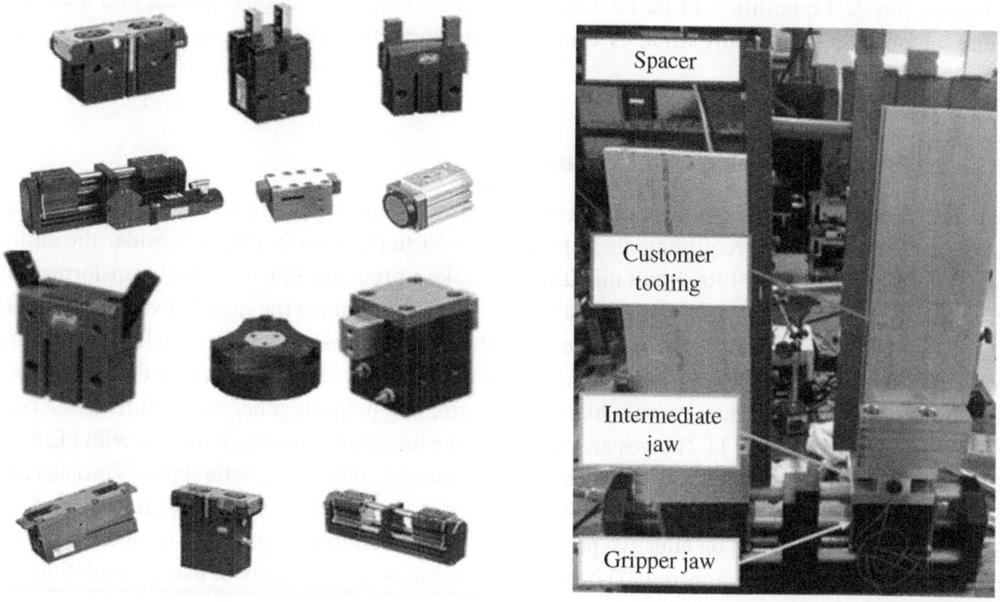

(a) Various grippers

(b) Cost-effective force sensing

Figure 11.19 Developing of cost-effective solution for ubiquitous sensing (*Source:* Bi et al. (2018)/Elsevier)

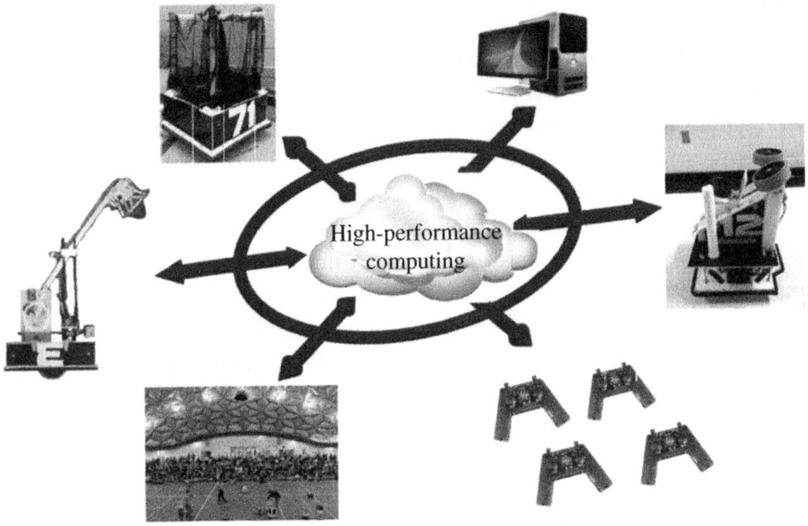

Figure 11.20 Instrumenting things for autonomous decision-making supports

Figure 11.20 shows an example of a prototyped robotic team for collaborative missions (Bi et al. 2017). In this team, robots are equipped with various sensors (e.g., range sensors, vision sensors, and force sensors) to collect sufficient information related to the requirements of tasks to collaborate in the team. By acquiring, processing, fusing, and utilizing real-time data from these sensors, decisions are made to coordinate, cooperate, collaborate, and control individual robots. Depending on the complexity of collaborative tasks, adequate sensors should be selected to detect events, or measure physical quantities, in particular, relative distances, orientations, speeds, and accelerations of objects; such information is critical to support effective collaborations among team members.

11.6.3 Methods of Coping with Big Data

Other than functionalities, quality and cost of products, customizing products, and shortening lead-time become two critical requirements in production. Virtual design is able to consider the entire set of the aforementioned FRs simultaneously and make a product right at the first transformation from the digital to the physical world. Meanwhile, it helps minimize the need for verification and validation by physical tests. When big data is involved, virtual design will face challenges to copy with increasing *Volume, Variety, and Velocity* (3V) of datasets. Innovative methods can narrow down big data into the dataset with a manageable size, this helps to generate a solution in a reasonable timeframe. Figure 11.21 gives an example of the innovative approach to cope with big data (Särestöniemi et al. 2019). It is a simulation model to illustrate how a magnetic signal is propagated in a human body. The results can be used to evaluate the impact of the body constitutions such as skin, fat, bones, and sterns on effective propagating paths. A simplified parametric model is developed to replace reverse-engineered 3D body, and this brings the feasibility of using simulations to evaluate antenna designs.

In IoT-enabled manufacturing, a large amount of data is collected from smart things in the material flow in the physical world. For a mechatronic functional unit, the data may be "*relevant*" or

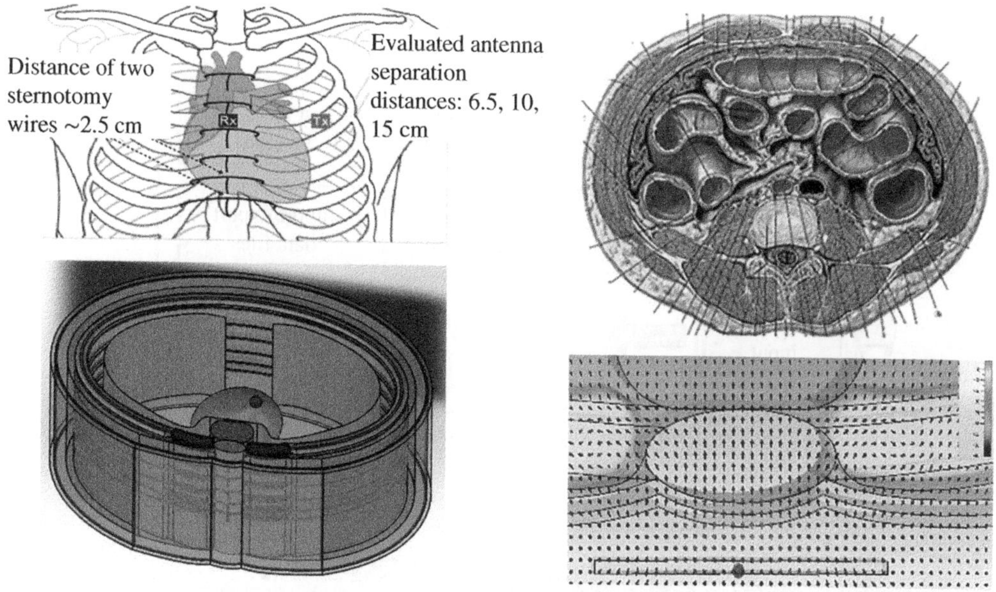

(a) Simplified body model

(b) Visualization of signal propagation

Figure 11.21 Coping with big data (*Source:* Adapted from Säreströniemi et al. (2019))

"*irrelevant*" and "*interpretable*" or "*uninterpretable*"; moreover, the data can be "*structured*" or "*unstructured.*" Effective analytical methods can classify and process data and make the size of the dataset manageable. Cochran et al. (2016, 2017) proposed to use the axiomatic design theory (ADT) to classify and sort data collected from the material flow and utilize the data in high-level decision-making units.

11.6.4 Surveillance and Data Visualization

Even in the digital era, human beings play an irreplaceable role in overseeing manufacturing systems. To incorporate human beings in digital manufacturing, user-friendly GUIs should be available to retrieve and visualize data, so that users can understand the data, assess the scenarios, and make their decisions based on the real-time data from the environment. As shown in Figure 11.22, Bi et al. (2016) prototyped a surveillance system for users to sort, visualize, and utilize the collected data in machine shops for preventive maintenance and smart energy management. Figure 11.23 shows an emergency response system where raw data was collected from IoT, and big and heterogeneous data was processed to support group decision-making in responding large-scale, urgent, and severe disasters such as earthquakes, flooding, and wars (Li et al. 2014).

11.6.5 Workflow Composition

IoT and BDA are reshaping the business models of digital manufacturing. Traditional value-chain business models have been shifted to digital platform models; especially, enterprises run their business by collaborating with strategic partners over the Internet. The enterprise boundary becomes very vague, and the capabilities of performing certain manufacturing operations are treated as services, which become available over the Internet. For an emerging business opportunity, the

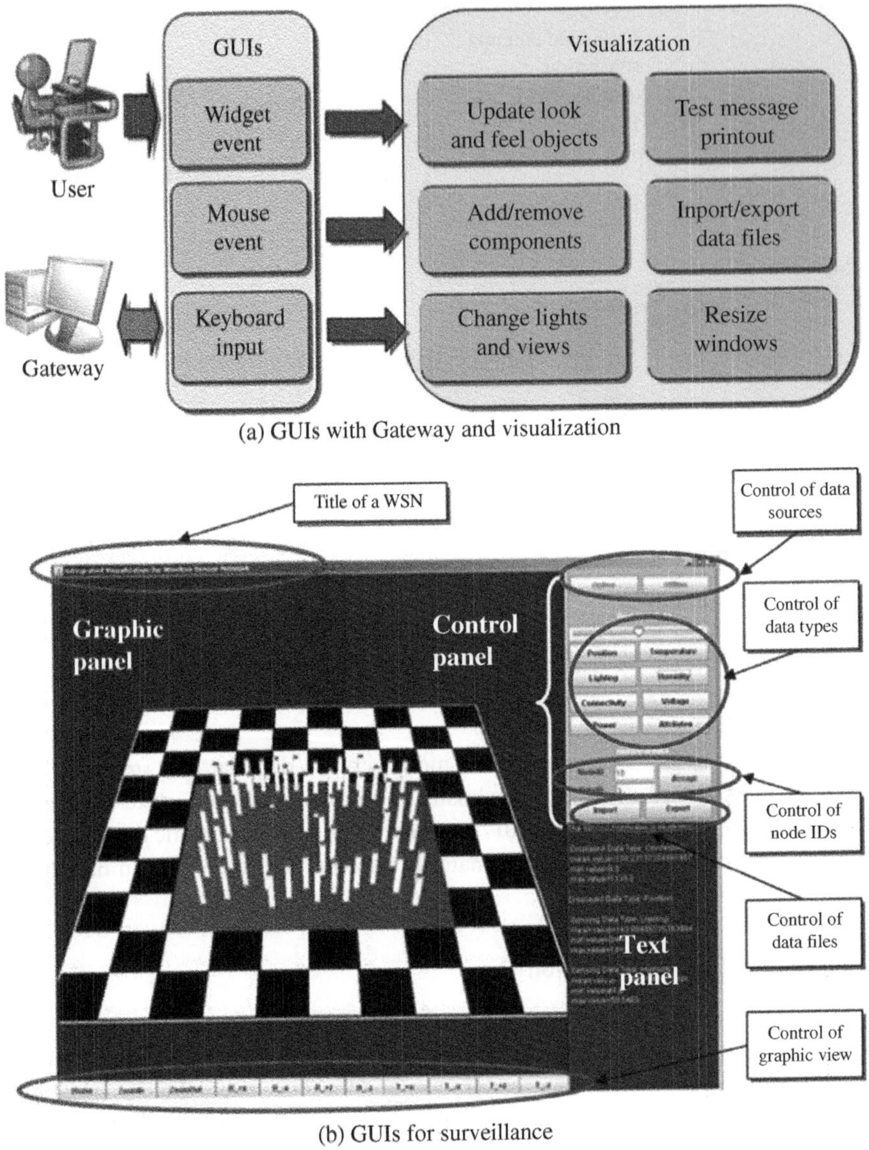

Figure 11.22 IoT-enabled surveillance and data visualization (*Source:* Bi et al. (2016)/Emerald Group Publishing Limited)

enterprise has to find service suppliers, and composite the service as the workflows for specific engineering projects. A great deal of research effort is needed for developing innovative methods and algorithms in defining service specifications, comparing and selecting services, and composing the selected services as the workflows for specific engineering projects optimally.

It poses a big challenge to assure privacy, safety, and security of data in an open environment. BCT was widely explored to embed trustworthiness and visibility in digital manufacturing. Figure 11.24 gives a few exemplifying works on the algorithms to select partners and compose them

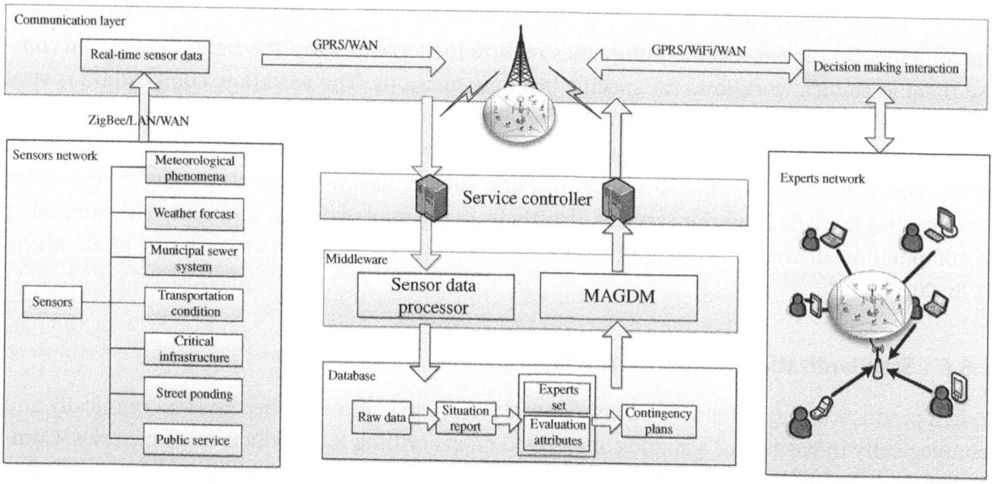

Figure 11.23 IoT-enabled emergence response system (*Source:* Li et al. (2014)/Springer Nature)

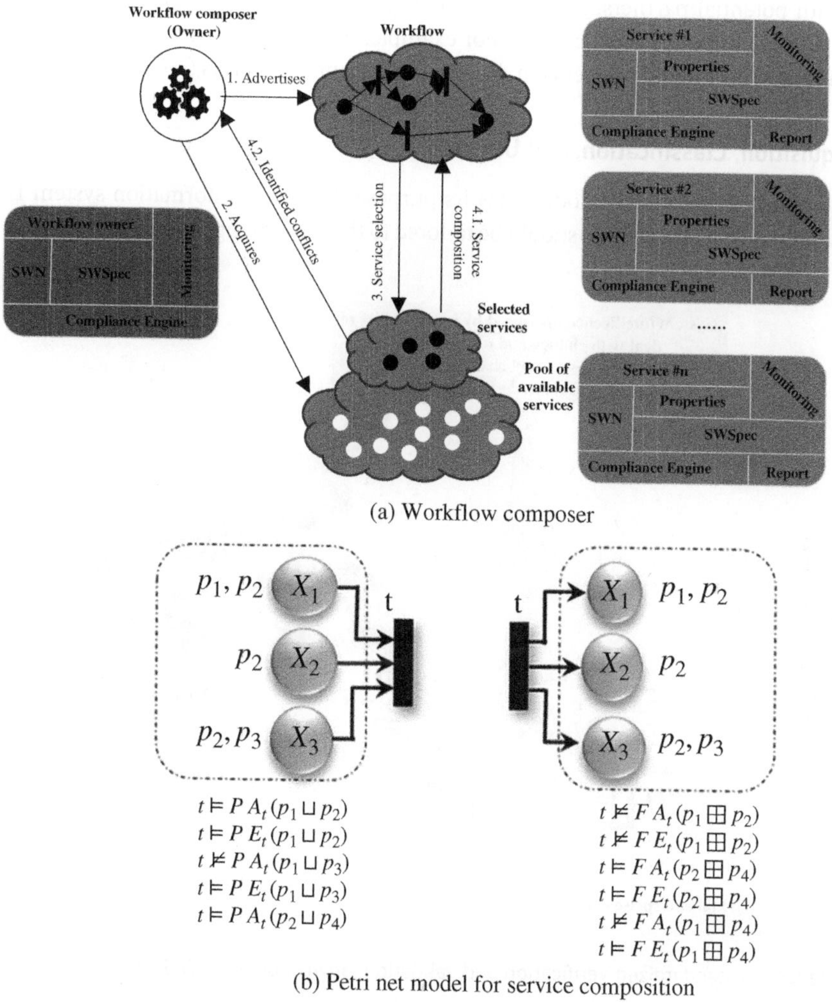

Figure 11.24 Selection and composition of services in virtual enterprise

as virtual enterprises over the Internet. When more and more manufacturing assets with similar functionalities become accessible, companies require the tools to select the right services and compose them as service workflows for specific business missions. The workflow composition is very complex due to the ever-increasing available resources and their dynamic nature over the network. To manage the complexity of service selection and composition, the formalized semantics were defined based on the workflow specifications, and the composition algorithm was developed to eliminate unqualified or inferior services directly from the pool of design solutions. This simplified the computation in workflow computation and compliance checking (Viriyasitavat et al. 2019a, 2019b, 2019c).

11.6.6 Standardization of Specifications

Modern products and systems are becoming more and more complex. It becomes economically and technologically infeasible for a single company to do everything for products in its factories. Companies are pressured to collaborate others in the decentralized environment to catch emerging market needs; while companies are facing with the challenges to select the right business partners due to a number of issues such as (1) an increasing number of eligible business partners, (2) the lacking of mutual trusts with potential partners, and (3) the lacking of standards in specifying the functional requirements (FRs) and evaluation criteria. For example, Bi et al. (2021) discussed the challenges in specifying the FRs of robotic systems in dealing with changeable tasks (see Figure 11.25).

11.6.7 Data Acquisition, Classification, and Utilization

Each information system or functional module has its purpose. When an information system is driven by big data, data analytical methods should be tailored to the underlying mathematic models

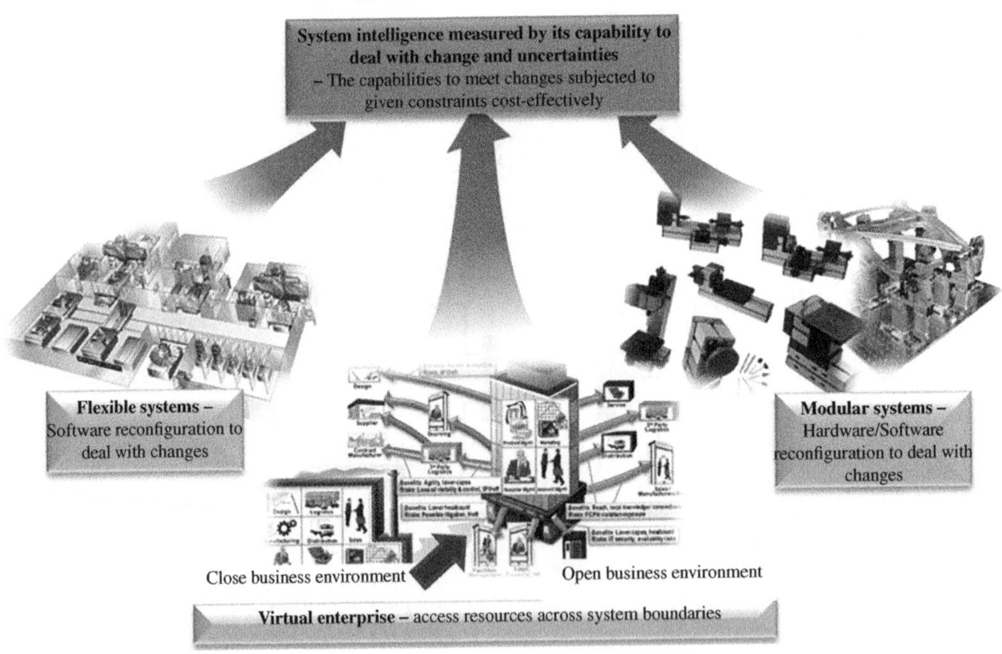

Figure 11.25 Challenges in standardized verification and validation (*Source:* Bi et al. (2021)/IEEE)

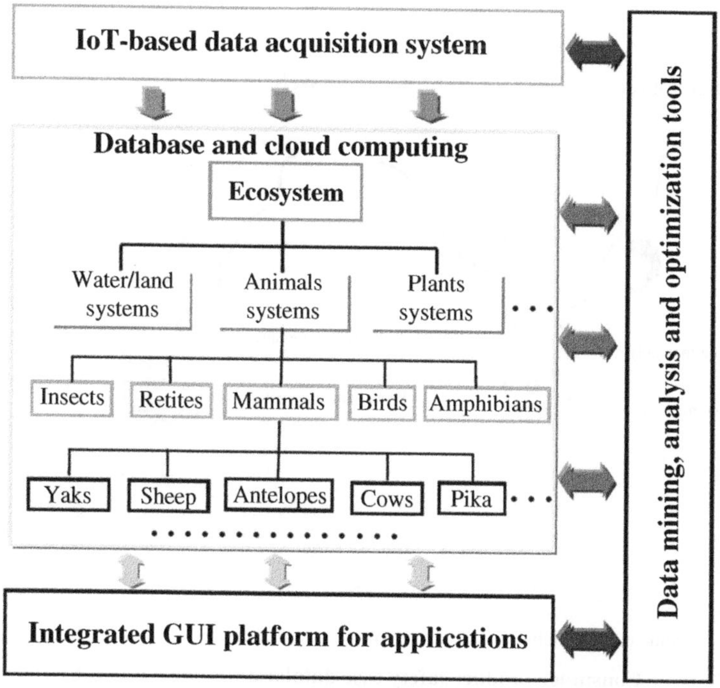

Figure 11.26 IoT-enabled ecosystem information management (*Source:* Adapted from Liu et al. (2015))

that govern system behaviors. In other words, analytical methods should be customized to specific applications, so that the big data from IoT can be acquired, classified, and utilized optimally. For example, Liu et al. (2015) developed an information system architecture to reason the decision-making supports for the protection of antelopes in the Tibet plateau (see Figure 11.26). Pan et al. (2014) integrated a suite of development tools for decision-making supports of complex security management systems (see Figure 11.27).

11.7 Summary and Conclusion

This chapter introduces the concept of IoT and its roles to enhance the capabilities of conventional mechatronic systems. IoT allows networking of smart things on the Internet, so that all of the networked virtual and physical resources can be accessed by others for coordination, collaboration, cooperation, and controls.

IoT provides rich and heterogeneous data to the cyber system of a complex mechatronic system, and this enhances system capabilities in dealing with changes and uncertainties. IoT-enabled information systems are characterized by connectivity, interconnectivity, ambient intelligence, real-time responsiveness, heterogeneity, adaptability, scalability, and safety and security. IoT is not a standalone technology; it requires integrating with other cutting-edge technologies, in particular, AI, DT, BDA, BCT, CC, and edge computing, to implement big data-driven decision-making supports in applications of sustainable mechatronic systems.

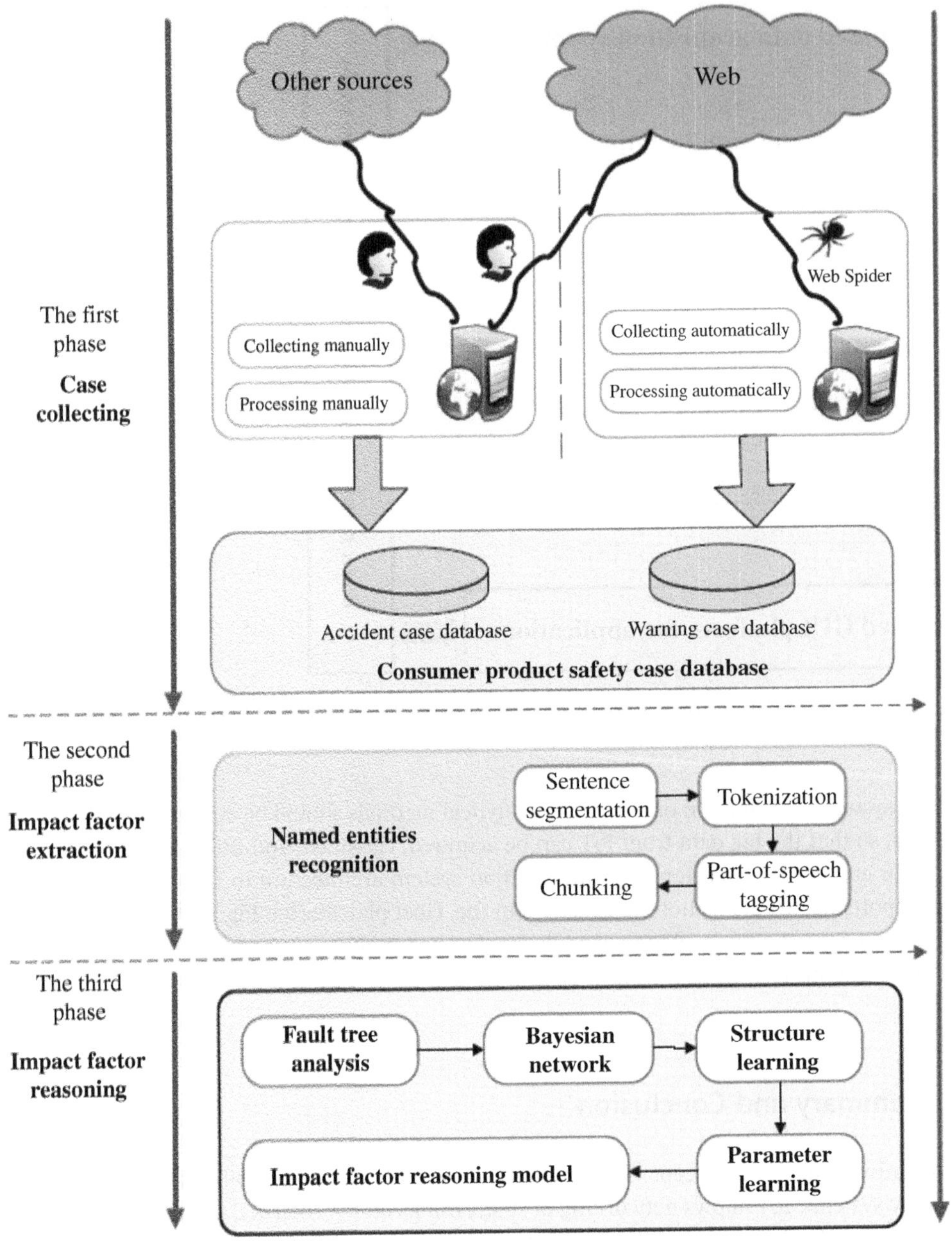

Figure 11.27 Framework for IoT-enabled safety management system (*Source:* Pan et al. (2014)/John Wiley & Sons)

Any IoT-enabled system should be customized to specific needs for its application. To incorporate IoT in a mechatronic system, the solutions to acquire, fuse, process, and mine data should be customized, and project-based methods can be adopted to decompose system-level FRs into module-level FRs, so that the solutions to data acquisition, data analytics, and data-driven decision-making can be developed with minimized couplings.

Problems

11.1 Explain the concept of the Internet of Things.

11.2 Discuss three main components and their roles in an IoT application with examples.

11.3 Discuss the applications of IoT in manufacturing with examples.

11.4 Discuss the relations between IoT and big data and explore the roles of BDA in digital manufacturing.

11.5 List some units in measuring data sizes and explain their relations.

11.6 What are the challenges of using IoT devices? Give some examples of security threats and vulnerabilities due to Internet access.

11.7 Discuss and compare commonly used networking solutions.

11.8 Design an IoT-enabled autonomous watering system for plants at offices.

11.9 Design an IoT-enabled navigation system for an autonomous mower at home.

11.10 Design an IoT-enabled evacuation guide system for residences in high-rise building.

References

Aghamohammadzadeh, E., Malek, M., and Valilai, O.F. (2020). A novel model for optimisation of logistics and manufacturing operation service composition in cloud manufacturing system focusing on cloud-entropy. *International Journal of Production Research* 58 (7): 1987–2015.

Alam, K.M. and Shaddik, A.E. (2017). C2PS: a digital twin architecture reference model for the cloud cyber-physical systems. *IEEE Access* https://doi.org/10.1109/ACCESS.2017.2657006.

Aqeel, M. (2020). Internet of things systematic literature review of security and future research. https://www.diva-portal.org/smash/get/diva2:1469432/FULLTEXT01.pdf (accessed 13 September 2024).

Bahga, A. and Madisetti, V. (2015). A hands-on approach. https://profile.iiita.ac.in/bibhas.ghoshal/IoT_2019/Lecture_Slides/Chapter-1_Introduction.pdf, (accessed 13 September 2024).

Ben-Daya, M., Hassini, E., and Bahroun, Z. (2019). Internet of things and supply chain management: a literature review. *International Journal of Production Research* 57 (15–16): 4719–4742.

Bi, Z.M., Xu, L.D., and Wang, C. (2014). Internet of things for enterprise systems of modern manufacturing. *IEEE Transactions on Industrial Informatics* 10 (2): 1537–1546.

Bi, Z.M., Wang, G., and Xu, L. (2016). A visualization platform for internet of things in manufacturing applications. *Internet Research* 26 (2): 377–401.

Bi, Z.M., Wang, G., Xu, L.D. et al. (2017). IoT-based system for communication and coordination of football robot team. *Internet Research* 27 (3): 162–181.

Bi, Z.M., Liu, Y.F., Krider, J. et al. (2018). Real-time force monitoring of smart grippers for Internet of things (IoT) applications. *Journal of Industrial Information Integration*, https://doi.org/10.1016/j.jii.2018.02.004 2018.

Bi, Z.M., Miao, Z.H., Zhang, B., and Zhang, W.J. (2021). Framework for performance assessment of heterogeneous robotic systems. *IEEE Systems Journal* 15 (1): 1191–1200.

Borgia, E. (2014). The Internet of things vision: key features, applications and open issues. *Computer Communications* 5491: 1–31.

Bradley, D., Russel, D., Ferguson, I. et al. (2015). The Internet of things – the future or the end of mechatronics. *Mechatronics* 27: 57–74.

Bradley, D., Russel, D., Hehenberger, P. et al. (2020). From mechatronics to the cloud. In: *Reinventing Mechatronics: Developing Future Directions for Mechatronics* (ed. X.T. Yan, D. Bradley, D. Russell, and P. Moore), 17–33. Cham: Springer.

Castellanos, W., Macias, J., Pinilla, H., and Alvarado, J.D. (2022). Internet of things: a multiprotocol gateway as solution of the interoperability problem. https://arxiv.org/pdf/2108.00098.pdf (accessed 13 September 2024).

Chae, B. and Olson, D. (2021). Technologies and applications of Industry 4.0: insights from network analytics. *International Journal of Production Research* https://doi.org/10.1080/00207543.2021.1931524.

Cheng, T.C., Kamble, S.S., Belhadi, A. et al. (2022). Linkages between big data analytics, circular economy, sustainable supply chain flexibility, and sustainable performance in manufacturing firms. *International Journal of Production Research* 60 (22): 6908–6922.

Cloud Security Alliance (2024). Big data analytics for security intelligence. https://downloads.cloudsecurityalliance.org/initiatives/bdwg/Big_Data_Analytics_for_Security_Intelligence.pdf (accessed 13 September 2024).

Cochran, D., Kinard, D., and Bi, Z.M. (2016). Manufacturing system design meets big data analytics for continuous improvement. *Procedia CIRP* 50: 647–652.

Cochran, D., Yong-Suk, K., Joseph, F., and Bi, Z.M. (2017). Use of the manufacturing system design decomposition for comparative analysis and effective design of production systems. *International Journal of Production Research* 55 (3): 870–890.

Deloitte University Press (2024). Inside the Internet of things (IoT). https://www2.deloitte.com/content/dam/insights/us/articles/iot-primer-iot-technologies-applications/DUP_1102_InsideTheInternetOfThings.pdf (accessed 13 September 2024).

Ding, K., Chan, F.T.S., Zhang, X. et al. (2019). Defining a digital twin-based cyber-physical production system for autonomous manufacturing in smart shop floors. *International Journal of Production Research* 57 (20): 6315–6334.

Dotoli, M., Fay, A., Miskowicz, M., and Seatzu, C. (2019). An overview of current technologies and emerging trends in factory automation. *International Journal of Production Research* 57 (15–16): 5047–5067.

Elgazzar, K., Khalil, H., Alghamdi, T. et al. (2022). Revisiting the internet of things: new trends, opportunities and grand challenges. *Frontiers in the Internet of Things* 1: 1073780. https://doi.org/10.3389/friot.2022.1073780.

Grammatikis, R., Sarigiannidis, P., and Moscholios, I. (2019). Securing the Internet of things: challenges, threats and solutions. *Internet of Things* 5: 41–70.

Gunnasekaran, A., Yusuf, Y.Y., Afeleye, E.O., and Papadopoulos, T. (2018). Agile manufacturing practice: the role of big data and business analytics with multiple case studies. *International Journal of Production Research* 56 (1–2): 385–397.

Hehenberger, P. and Bradley, D. (2016). *Mechatronic Futures: Challenges and Solutions for Mechatronic Systems and their Designers*. Switzerland: Springer International Publishing.

Hehenberger, P., Vogel-Heuser, B., Bardlety, B. et al. (2016). Design, modelling, simulation and integration of cyber physical systems: methods and applications. *Computers in Industry* 82 (2016): 273–289.

Ivanov, D., Tang, C.S., Dolgui, A. et al. (2020). Researchers' perspectives on Industry 4.0: multidisciplinary analysis and opportunities for operations management. *International Journal of Production Research* 59 (7): 2055–2078.

Kamal, R. (2017). *Internet of Things: Architecture and Design Principles*. McGraw Hill Education Private Limited.

KPMG (2018). Supply chain big data series part 1. https://assets.kpmg.com/content/dam/kpmg/au/pdf/2017/big-data-shaping-supply-chains-of-tomorrow.pdf (accessed 13 September 2024).

Li, N., Sun, M., Bi, Z.M. et al. (2014). A new methodology to support group decision-making for IoT-based emergency response systems. *Information Systems Frontiers* 16: 953–977.

Liu, F., Bi, Z.M., Xu, E.L. et al. (2015). An integrated systems approach to plateau ecosystem management – a scientific application in Qinghai and Tibet plateau. *Information Systems Frontiers* 17: 337–350.

Maheshwari, S., Gautam, P., and Jaggi, C.K. (2020). Role of big data analytics in supply chain management: current trends and future perspectives. *International Journal of Production Research* 59 (6): 1875–1900.

Marwedel, P. (2021). *Embedded System Design Embedded Systems Foundations of Cyber-Physical Systems, and the Internet of Things*, 4e, https://doi.org/10.1007/978-3-030-60910-8. Spring International Publishing.

Meriton, R., Bhandal, R., Graham, G., and Brown, A. (2021). An examination of the generative mechanisms of value in big data-enabled supply chain management research. *International Journal of Production Research* 59 (23): 7283–7310.

Mittal, S., Khan, M.A., Purohit, J.K. et al. (2020). A smart manufacturing adoption frameworks for SMEs. *International Journal of Production Research* 58 (5): 1555–1573.

Moeuf, A., Pellerin, R., Lamouri, S. et al. (2018). The industrial management of SMEs in the era of Industry 4.0. *International Journal of Production Research* 56 (3): 1118–1136.

Pan, S., Wang, L., Wang, K. et al. (2014). A knowledge engineering framework for identifying key impact factors from safety-related accident cases. *Systems Research and Behavioral Science* 31: 383–397.

Park, K.T., Son, Y.H., and Noh, S.D. (2021). The architectural framework of a cyber-physical logistics system for digital-twin-based supply chain control. *International Journal of Production Research* 59 (19): 5721–5742.

Qiao, F., Liu, J., and Ma, Y. (2020). Industrial big-data-driven and CPS-based adaptive production scheduling for smart manufacturing. *International Journal of Production Research* 59 (3): 1–21.

Rachuri, S. (2019). Focus group: big data analytics for smart manufacturing systems. https://assets.kpmg.com/content/dam/kpmg/au/pdf/2017/big-data-shaping-supply-chains-of-tomorrow.pdf (accessed 13 September 2024).

Ramasamy, L.K. and Karry, S. (2021). Chapter 1: blockchain in the industrial Internet of things. https://store.ioppublishing.org/page/detail/Blockchain-in-the-Industrial-Internet-of-Things//?k=9780750336611 (accessed 13 September 2024).

Sahoo, S. (2022). Big data analytics in manufacturing: a bibliometric analysis of research in the field of business management. *International Journal of Production Research* 60 (22): 6793–6821.

Särestöniemi, M., Pomalaza-Ráez, C., Bi, Z.M. et al. (2019). Comprehensive study on the impact of sternotomy wires on UWB WBAN channel characteristics on the human chest area. *IEEE Access* 7 (1): 74670–74682.

Siemens (2015). Industrie 4.0 – digitalization strategy. https://www.iotone.com/files/pdf/vendor/Digital_Factory_(Siemens)_Industrie_4.0_Digitalization_Strategy_2015.pdf (accessed 13 September 2024).

Siemens (2019a). Digital manufacturing – a holistic approach to the complete product lifecycle. https://www.industry.usa.siemens.com/automation/us/en/formsdocs/Documents/AMAR_UTO1_0715_Digital_Manufacturing.pdf (accessed 13 September 2024).

Siemens (2019b). Digital manufacturing. https://www.plm.automation.siemens.com/global/en/our-story/glossary/digital-manufacturing/13157 (accessed 13 September 2024).

Talwara, S., Kaurbc, P., Wambad, S.F., and Dhirefc, A. (2021). Big data in operations and supply chain management: a systematic literature review and future research agenda. *International Journal of Production Research* 59 (11): 3509–3534.

Tran, T.A. and Djabir, D.A. (2023). The fundamental characteristics of IoT for controlling daily pollution. *Advances in Pollution Research* https://doi.org/10.1016/B978-0-323-95876-9.00011-2.

Vermesan, O. and Fries, P. (2014). *Internet of Things – From Research and Innovation to Market Deployment*. Aalborg Ø Denmark: River Publishers.

Vermesan, O. and Friess, P. (2016). *Digitising the Industry Internet of Things Connecting the Physical, Digital and Virtual Worlds*. River Publishers.

Viriyasitavat, W., Xu, L., Bi, Z.M., and Hoonsopon, D. (2019a). Blockchain technology for applications in internet of things - mapping from system design perspective. *IEEE Internet of Things Journal* 6 (5): 8155–8168.

Viriyasitavat, W., Xu, L., and Bi, Z.M. (2019b). Blockchain and Internet of things for modern business process in digital economy – the state of the art. *IEEE Transactions on Computational Social Systems* 6 (6): 1420–1432.

Viriyasitavat, W., Xu, L., Bi, Z.M. et al. (2019c). Managing QoS of Internet-of-thing services using blockchain. *IEEE Transactions on Computational Social Systems* 6 (6): 1357–1368.

Viriyasitavat, W., Xu, L., and Bi, Z.M. (2020a). Specification patterns of service-based applications using blockchain technology. *IEEE Transactions on Computational Social Systems* 7 (4): 886–896.

Viriyasitavat, W., Xu, L.D., Bi, Z.M., and Sapsomboon, A. (2020b). Blockchain-based business process management (BPM) framework for service composition in Industry 4.0. *Journal of Intelligent Manufacturing* 31 (7): 1737–1748.

Wang, C., Bi, Z., and Xu, L.D. (2014). IoT and cloud computing in automation of assembly Modeling systems. *IEEE Transactions on Industrial Informatics* 10 (2): 1426–1434.

Zhang, Q., Lu, X., Peng, Z., and Ren, M. (2019). Perspective: a review of lifecycle management research on complex products in smart-connected environments. *International Journal of Production Research* 57 (21): 6758–6779.

Zhou, H. (2013). *The Internet of Things in Cloud – A Middleware Perspective*. Taylor & Francis Group, LLC CRC Press is an imprint of Taylor & Francis Group, an Informa business.

12

Robotics

12.1 Introduction

With the rapid technological development, robotic systems become ubiquitous around our daily lives, and the proliferation of robotic systems has enhanced humans' competencies and skills to fulfill challenging tasks dramatically and thus improved the quality of our lives.

There is a trend to develop and use more and more advanced robots in all levels and domains of future workforces (Ahmed and La 2019). Figure 12.1 shows some popular robots that have been widely used in different applications. A Roomba robot cleans floors at home; a Starship robot delivers food to doors; a DJ drone takes photos in sky; a medical robot by the Medtronic Plc assists doctors to do spine surgeries; an Nao robot interacts with a kid or disabled who needs assistance; an underwater robot explores the deep sea that is dangerous to divers.

For the purpose of education, robotics is highly multidisciplinary and most relevant to mechatronics. It has been widely taught in different engineering programs including *mechanical engineering* (ME), *electrical engineering* (EE), *industrial engineering* (IE), *system engineering* (SE), *manufacturing engineering* (ME), and *computer science* (CS). In this chapter, a robot is treated as a mechatronic system since it consists of various mechatronic functional modules (MFMs) for sensing, actuating, controlling, and coordinating, cooperating, and collaborating at high levels.

Most robots are also viewed as tightly coupled *cyber-physical systems* (CPSs) where sensors and computing resources are embedded and integrated seamlessly to coordinate physical and computational systems to achieve specific tasks as shown in Figure 12.2 (Zhou 2013).

The concept of the robot was coined by the Czech novelist Capek (2004) to represent an artificially created slave worker. It was so widely used now; but different people understand the concept of robot differently. For children, robots are versatile, educational, and funny accompanies (see examples in Figure 12.3). For adults, robots are mostly smart tools that are designed to replace humans in entertainments, personal assistance, and home services (see examples in Figure 12.4).

For engineers at the factories, robots are intelligent machines to replace operators in performing various manufacturing or assembly processes and transporting materials (see examples in Figure 12.5). For software engineers, robots are computing algorithms, packages, or systems that support decision-making activities for specified missions (see examples in Figure 12.6). For public, robotic technologies mostly stand for the cutting-edge developments of science and technology on the front edge of human civilizations (see examples in Figure 12.7).

Mechatronics for Complex Products and Systems: Project-Based Design Approaches for Robots, Cyber-Physical Systems, Digital Twins, and Other Emerging Technologies, First Edition. Zhuming Bi.

Companion website: www.wiley.com/go/bi/mechatronics

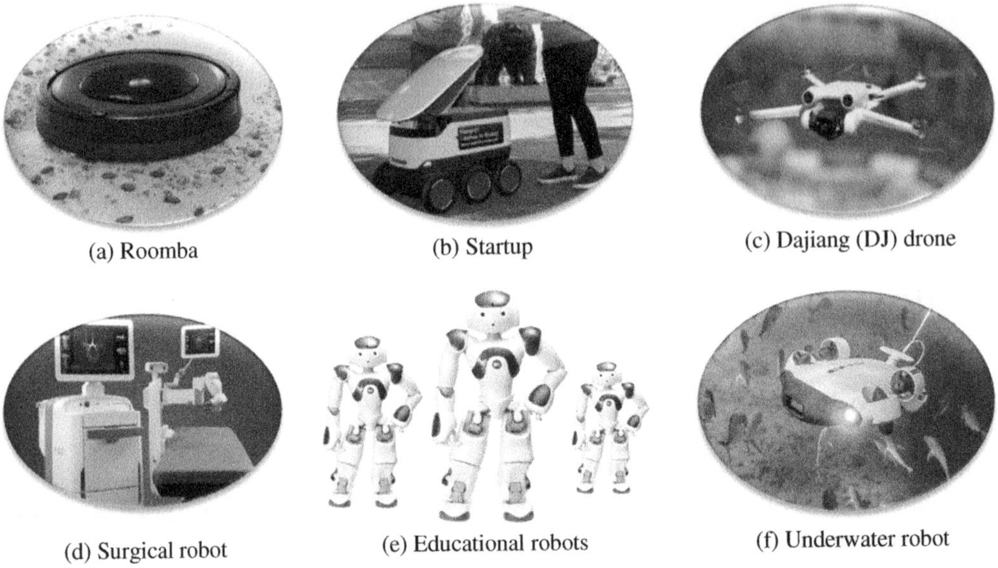

Figure 12.1 Examples of popular robots (*Sources:* (b) Singapore Management University / https://www.facebook.com/smudallas/videos/starship-robots-have-landed-at-smu/1079601849491425/ / Last accessed on July 19, 2024; (c) Darrell Etherington / https://techcrunch.com/2022/05/11/dji-mini-3-pro-drone-review/ / Last accessed on July 19, 2024; (d) Giovanni Dapri (2022)/with permission of Elsevier)

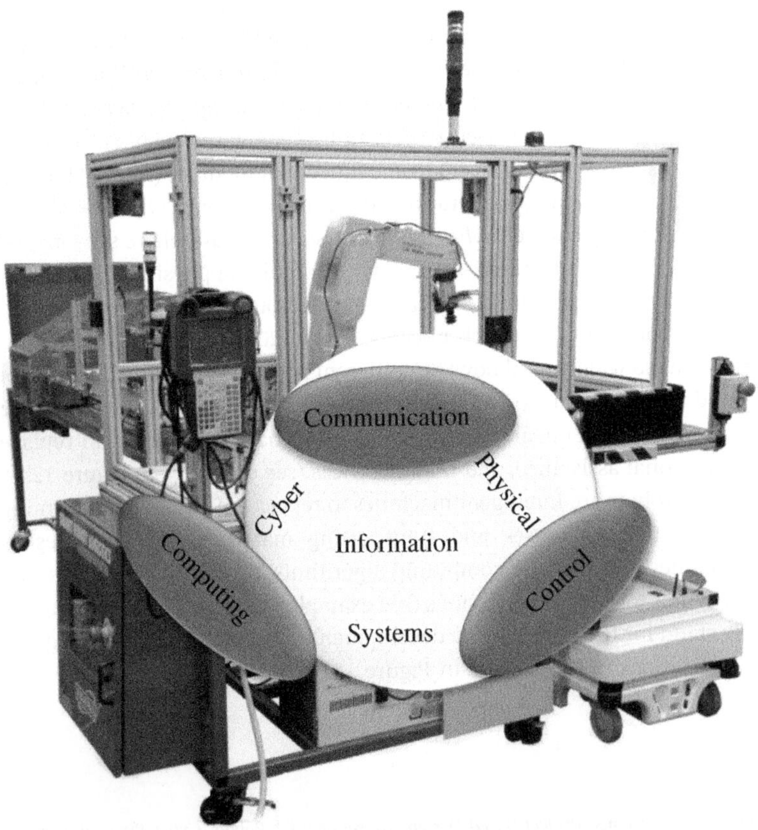

Figure 12.2 A robotic system featured as a CPS (*Source:* Technical Laboratory Systems, Inc / https://tech-labs.com/videos/amatrol/amatrols-smart-robot-logistics-palletizing-workcell / Last accessed on July 19, 2024)

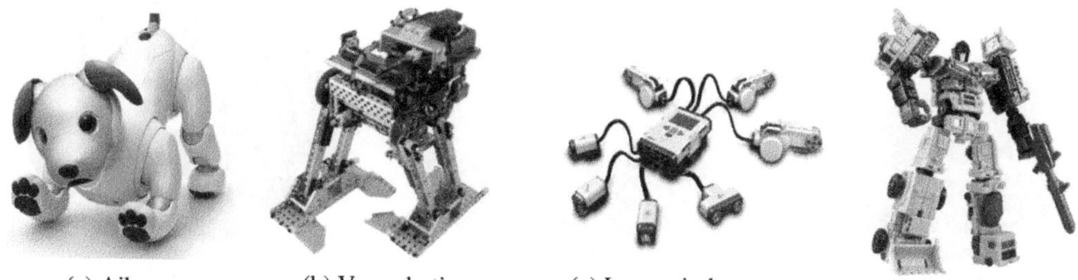

(a) Aibo (b) Vex robotics (c) Lego mindstorm (d) Transform robot

Figure 12.3 Examples of robots to children (*Source:* (d) ABS Plastic / Last accessed on July 19, 2024)

(a) Avatar (b) Social robot (c) Robotic mower (d) Robotic wheelchair

Figure 12.4 Examples of robots to adults (*Sources:* (a) hamara/Adobe Stock; (b) Mary Mark Ockerbloom / Wikimedia Commons / CC BY SA 4.0; (c) Robot Lawn Mower / https://robotlawnmower.nz/blogs/batterypowered/what-to-consider-before-buying-a-robot-lawn-mower / Last accessed on July 19, 2024; (d) domi002/Adobe Stock Photos)

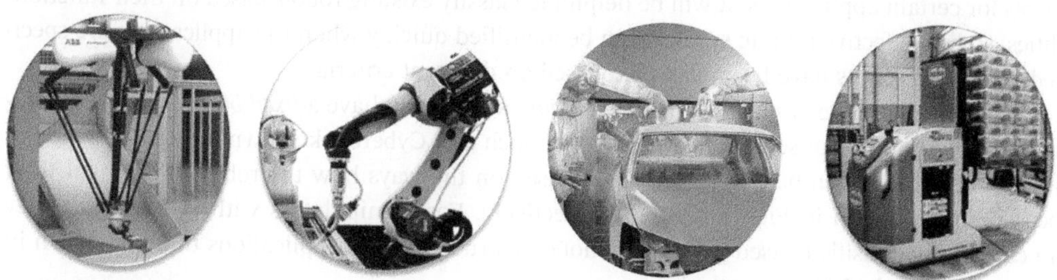

(a) Delta robot (b) Welding robot (c) Painting robot (d) AGV

Figure 12.5 Examples of robots at factories (*Sources:* (c) jeson/Adobe Stock; (d) Cassioli / https://www.cassioli.com/intralogistics-division/automated-vehicles/ / Last accessed on July 19, 2024)

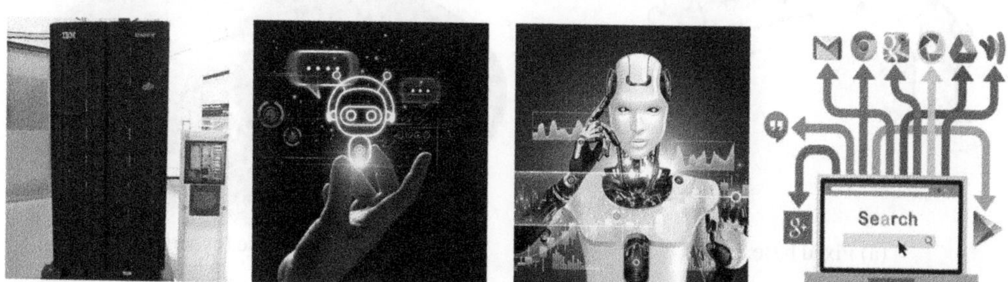

(a) Delta blue (b) Chatbot (c) Trading bot (d) Search bot

Figure 12.6 Examples of robots in information technologies (IT) (*Sources:* (a) Jim Gardner / Flickr / CC BY 2.0; (b) ipopba/Adobe Stock; (c) AndSus/Adobe Stock)

(a) Canadian arm (b) Spirit (c) Lander

Figure 12.7 Examples of robots in News (*Sources:* (a) NASA / https://spaceflightnow.com/2014/10/12/winds-sensor-opens-door-for-earth-science-from-iss/ Last accessed on July 19, 2024; (b) Dima Zel/Shutterstock; (c) ASTROBOTIC TECHNOLOGY / https://www.floridatoday.com/picture-gallery/tech/science/space/2024/01/18/falling-to-earth-major-pieces-of-space-debris-come-tumbling-down/72268583007/ / Last accessed on July 19, 2024)

By all means, whether or not a robot is *simple* or extremely *complex*, a robot must have essential features described by the *International Standards Organizations* (ISO): a robot is a reprogrammable machine that is designed so that the programmed motions or auxiliary functions can be changed without physical alteration (ISO 2021).

12.2 Classifications

Nowadays, robotic systems are becoming ubiquitous in our daily lives. To design and select the right robots for certain applications, it will be helpful to classify existing robots based on their functionalities, so that effective robotic systems can be identified quickly when the applications are specified. However, robots have been classified based on different criteria.

From the perspective of *mobility* and *locomotion*, a robot can have a *fixed base* such as a manipulator in Figure 12.8a or serve as *a mobile device* such as a Cyberbuck shown in Figure 12.8b. For mobile robots, they can be further classified based on the ways how the robots implement their locomotion as shown in Figure 12.9a–e, respectively. For manipulators with a fixed base, they are commonly classified based on the functions in structures and applications that are shown in Figures 12.10 and 12.11, respectively.

(a) Fixed base (b) Mobile base (cybertruck)

Figure 12.8 Classification I – fixed or mobile base

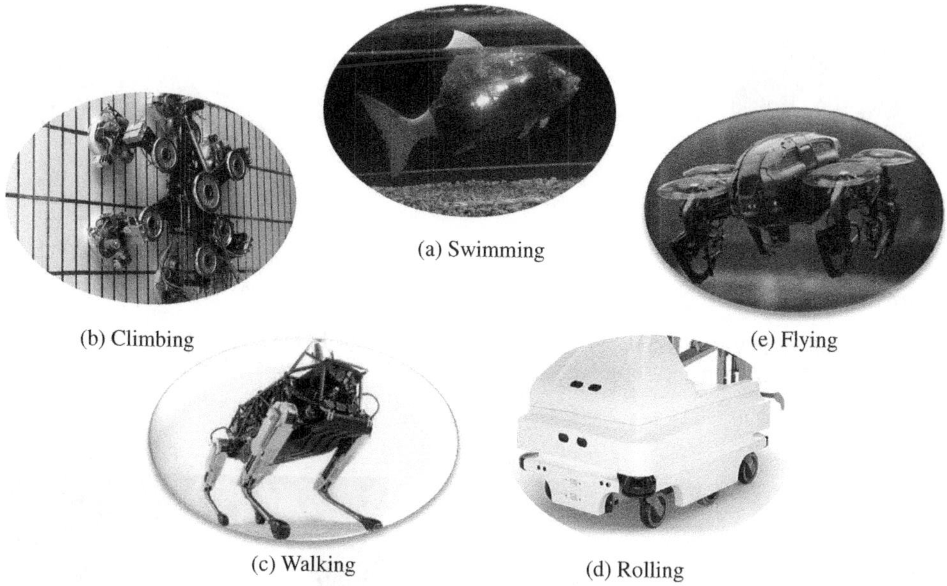

Figure 12.9 Classification II – ways of locomotion (*Sources:* (a) Media Whale Stock/Adobe Stock; (b) Xin Li / Kaige Shi / https://www.eurekalert.org/multimedia/693824 / Last accessed on July 19, 2024; (c) Boston Dynamics / https://www.wtxl.com/syndication/walking-robot-maker-prepares-to-unleash-its-dog-like-machine/article_3273e8d7-8218-5720-b1d9-faade180986b.html / Last accessed on July 19, 2024; (e) italoparnaiba5 / Freepik)

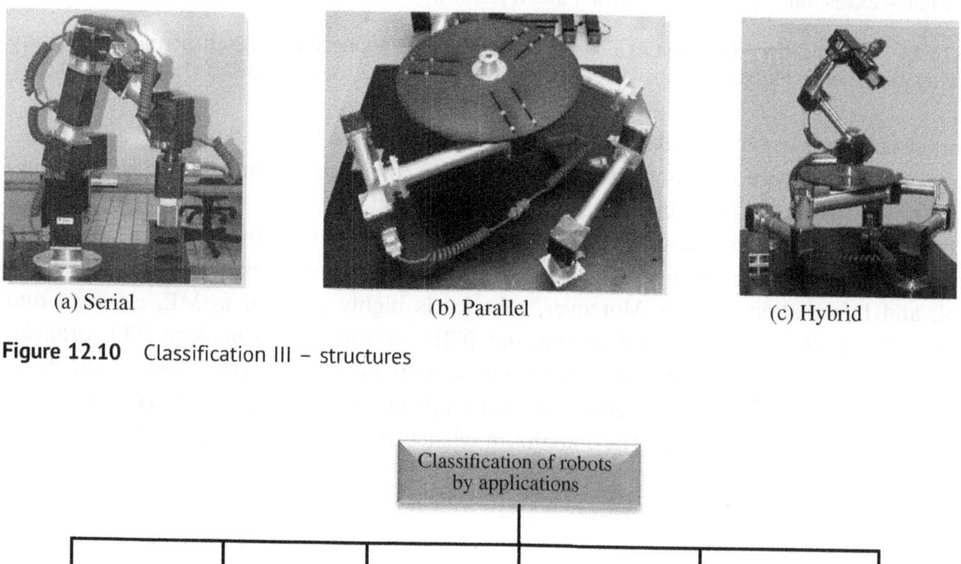

Figure 12.10 Classification III – structures

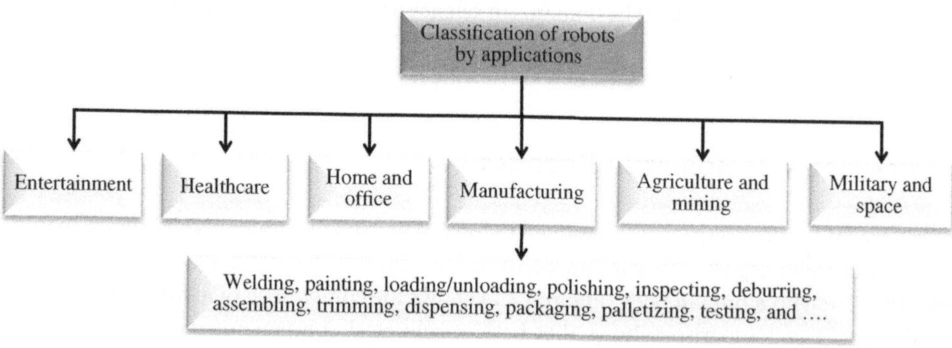

Figure 12.11 Classification IV – applications

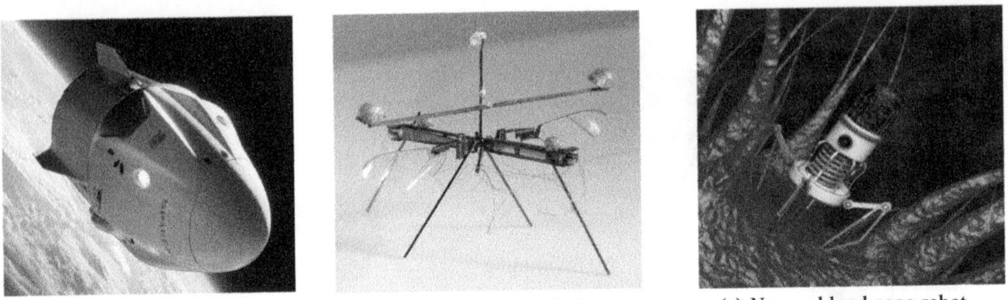

(a) *Marco* - SpaceX (b) *Micro* - robofly (c) *Nano* – blood nano robot

Figure 12.12 Classification V– sizes (*Sources:* (a) NASA / Public domain; (b) Xreality / https://www.xreality1.com/smart-industry/robotic/emergence-of-insect-sized-flying-robots/ / Last accessed on July 19, 2024; (c) doomu/Adobe Stock)

(a) Rigid – exkeleton (b) Compliant – micro platform (c) Soft – soft robot

Figure 12.13 Classification VI – rigidities (*Sources:* (a) Gorodenkoff/Adobe Stock; (b) With permissions of Larry Howell; (c) The Science Museum / https://ocean.si.edu/octopus-robot / Last accessed on July 19, 2024)

In other occasions, robots are also classified based on sizes into "*macro*," "*micro*," and "*nano*" robots as shown in Figure 12.12 and whether or not those constitutive elements are "*rigid*," "*compliant*" or "*soft*" as shown in Figure 12.13.

To design and use robots, the theory of robotics is multidisciplinary that goes far beyond the fundamentals in traditional individual disciplines including physics, mathematics, automated control, and information science. Moreover, robotics is highly relevant to ME, *electrical and electronic engineering* (EEE), *control engineering* (CE), *information technology* (IT), *artificial intelligence* (AI), physics, mathematics, and human–machine interfaces (HMI). Due to the trend in promoting *Science, Technology, Engineering, and Mathematics* (STEM), the courses relevant to robotics are widely spread in most of the domains in science and technologies and at different levels of education.

12.3 Basic Terminologies in Robotics

A robot is a mechatronic system that is designed to replace or expand a human's abilities in manipulating an object with purposes. To learn the constitutive elements of a robot, it is better to compare a robotic structure with a human body. Figure 12.14 shows the comparison of a human body and a humanoid robot. The comparison shows that basic mechanical components of a robot are *rigid*

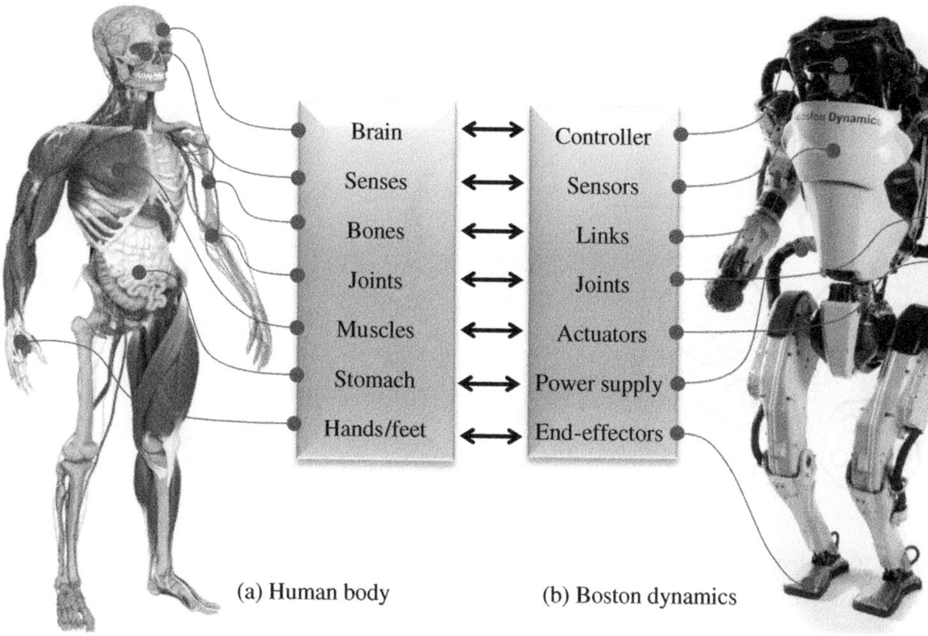

Figure 12.14 Basic components of a robot when compared with human body (*Source:* Boston Dynamics / https://news.mit.edu/2021/cheetah-noids-humanoids-benjamin-katz-1214/ / Last accessed on July 19, 2024)

bodies, joints, and *end-effectors*; basic electrical components are *actuators* and *power supplies*; basic information components are *sensors, networks,* and *controllers*.

The number and types of basic components may vary from one robot to another. For an example, a robot with a high degrees of freedom (DOF) has more rigid bodies and joints than a robot with a low DOF. By all means, a robot must have all types of mechanical, electrical and electronic, and information components, and a robot is basically a mechatronic system.

12.3.1 Mechanical Structure

Figure 12.15 shows the mechanical structure of a robot. It consists of three basic types, that is, (1) *rigid bodies* or *links*, (2) *joints*, and (3) *end-effector* (s). *A rigid body* or *link* possesses a set of the physical properties such as geometries, density, mass, and moments of inertias. One special link is *a ground link* whose location is fixed statically in an application and only a manipulator has a ground link. *A joint* is an intersection of two or more links; a joint brings one or more constraints to connected links. Common types of joints are *Revolute* (*R*), *Prismatic* (*P*), *Universal* (*U*), *Cylindrical* (C), and *Spherical* (*S*).

A manipulator is a robot with a fixed base or platform; it is designed to fulfill certain functions on fixed sites. Figure 12.16a shows that most industrial robots are mounted on a fixed site; most industrial robots are manipulators. An *end-effector* is a tool, gripper, or another integral tool, it is mounted on a robotic body, and it makes a direct contact with the object to be manipulated in applications; Figure 12.16b shows that the DOF of an end-effector varies depending on the complexity of an object to be manipulated.

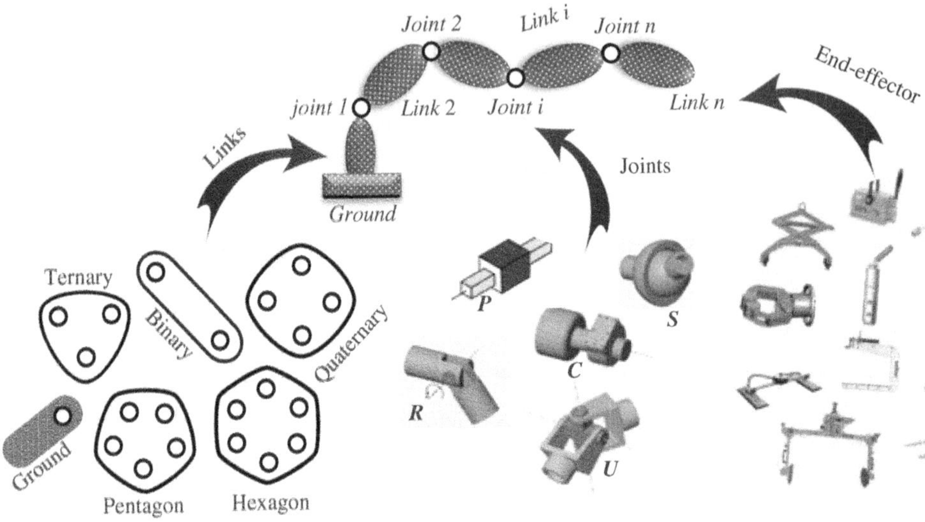

Figure 12.15 Basic elements in mechanical structures of robots

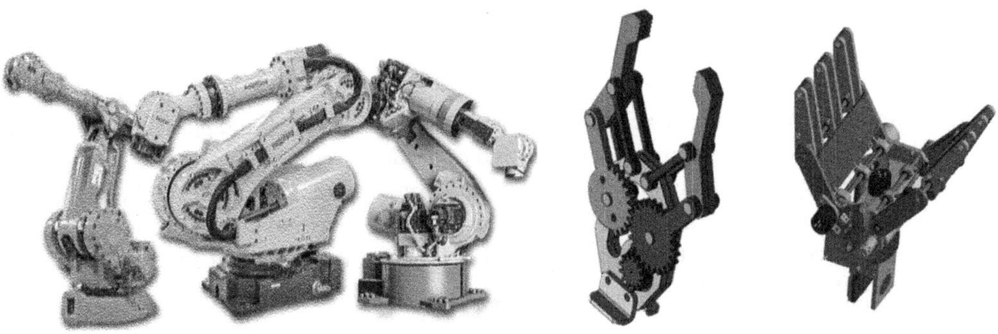

Figure 12.16 Examples of manipulator and end-effector (*Source:* (a) robots.com / https://www.evsint.com/industrial-robotic-arm-cost/ / Last accessed on July 19, 2024)

12.3.2 *Degrees of Freedom* (DOF)

DOF is also called *mobility*. DOF of an object, a robot, or a system refers to the number of independent motions that are used to describe the corresponding motion of object. Take an example of the robot in Figure 12.17, it has six joints, each joint has 1-DOF rotational motion along its corresponding rotational axis; therefore, the robot has DOF = 6 as a whole. From the perspective of robotic structure, the DOF of a robot depends on the number of rigid bodies, and the types and number of joints that are used to connect these rigid bodies. The DOF of a rigid body is the number of independent variables to specify its position and orientation uniquely. Figure 12.18 shows five example objects with their DOF of the objects subjected to given constraints in applications.

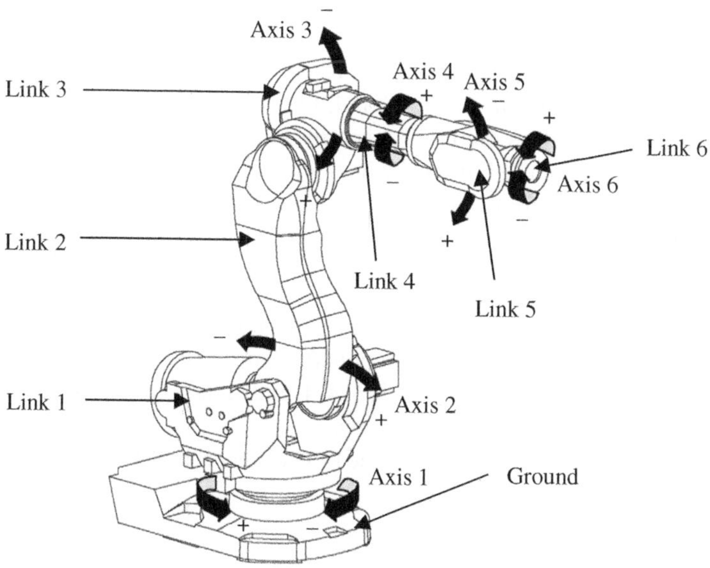

Figure 12.17 IRB 6600 robot with six 1-DOF revolute joints

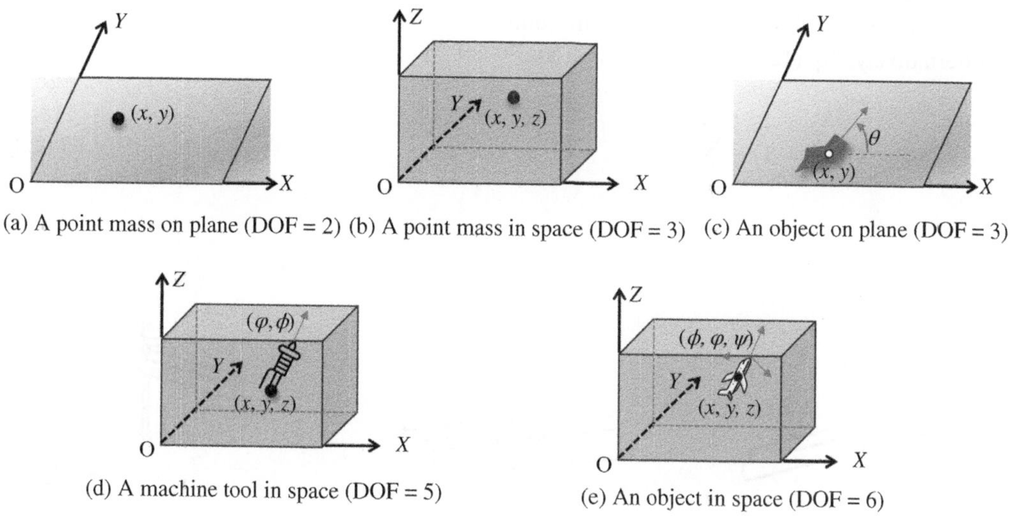

Figure 12.18 DOF of objects to be manipulated

1) To locate a point mass on a two-dimensional plane in Figure 12.18a, the DOF of the point mass is 2 (i.e., DOF = 2), since two independent variables are used to specify x- and y-coordinates at a point in the two-dimensional plane, respectively.
2) To locate a point mass in a three-dimensional space in Figure 12.18b, the DOF of the point mass is 3 (i.e., DOF = 3), since three independent variables are used to specify the x-, y-, and z-coordinates at a point in the three-dimensional space, respectively.

3) To locate a two-dimensional object on a plane in Figure 12.18c, the DOF of the 2-D object is 3 (i.e., DOF = 3), since two independent variables are used to specify *x*- and *y*-coordinates at a point, and one variable is used to specify its orientation with respect to the *x*-axis.
4) To locate an axially symmetric tool in space in Figure 12.18d, the DOF of the tool is 5 (i.e., DOF = 5), since three independent variables are used to specify *x*-, *y*-, and z-coordinates of its origin, and two additional variables are used to specify its orientation of the tool.
5) To locate an arbitrary object in space in Figure 12.18e, the DOF of the object is 6 (i.e., DOF = 6), since three independent variables are used to specify *x*-, *y*-, and z-coordinates of its origin, and three additional variables are used to specify a local coordinate system (CS) completely.

A joint brings motion constraints to the connected bodies. When two bodies are joined, one or more constraints are introduced to constrain their relative motion(s). Figure 12.19 shows that the number of constrained motions by *R, P, H, C, U*, and *S* are 5, 5, 5, 4, 4, and 3, respectively. Accordingly, the DOF of these joints are 1, 1, 1, 2, 2, and 3, respectively.

The *Chebychev–Grübler–Kutzbach criterion* is used to determine DOF of a mechanism including a robotic structure, and it is described as,

$$DoF = 6(n-1) - \sum_{i=1}^{j}(6-f_i) \tag{12.1}$$

where DoF is the degrees of freedom of the mechanism; *n* and *j* are the numbers of rigid bodies and joints, respectively; f_i is the DOF for the *i*th joint.

Alternatively, Eq. (12.1) can be rewritten as,

$$DoF = 6(n-j-1) - \sum_{i=1}^{j} f_i \tag{12.2}$$

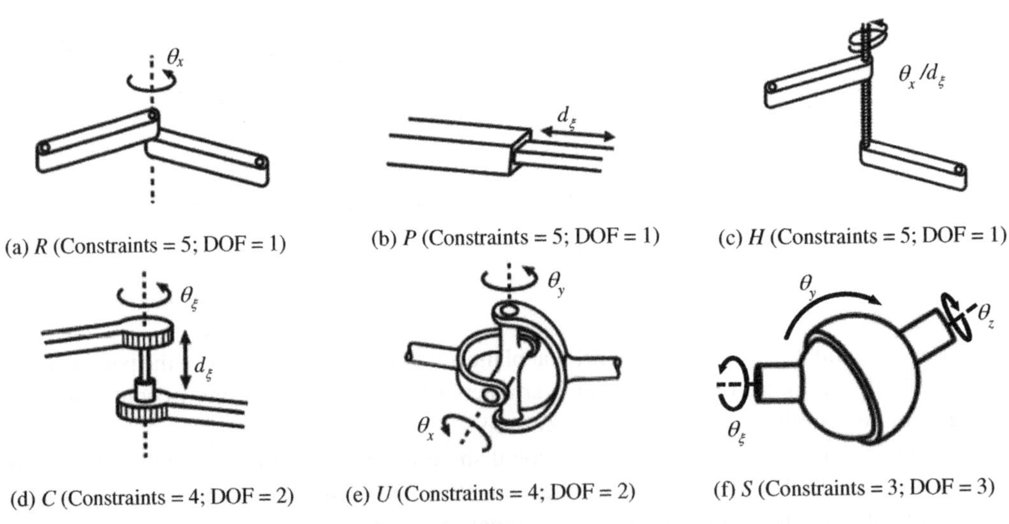

(a) *R* (Constraints = 5; DOF = 1) (b) *P* (Constraints = 5; DOF = 1) (c) *H* (Constraints = 5; DOF = 1)

(d) *C* (Constraints = 4; DOF = 2) (e) *U* (Constraints = 4; DOF = 2) (f) *S* (Constraints = 3; DOF = 3)

Figure 12.19 Constraints and DOF of commonly used joints

Example 12.1 Confirm DOF for the robot as shown in Figure 12.17.

Solution

Checking the robotic structure in Figure 12.17 find that the number of links including the ground is 7, that is, $n = 7$; the number of joints is 6, that is, $j = 6$; all six joints are revolute (R), that is, $f_1 = f_2 = f_3 = f_4 = f_5 = f_6 = 1$. Therefore, applying Eq. (12.1) gets,

$$DoF = 6(7-1) - \sum_{i=1}^{6}(6-f_i) = 36 - 6 \times (6-1) = 6$$

Example 12.2 Determine the DOF of a hexapod robot as shown in Figure 12.20 (You et al. 2022).

Solution

Checking the structure of the robot in Figure 12.20 find that the number of links including the end-effector platform is $n = 3 \times 6 + 1 = 19$; the number of joints is $j = 3 \times 6 = 18$; all joints are revolute (R), that is, $f_1 = f_2 = \ldots = f_{18} = 1$. Therefore, applying Eq. (12.1) gets,

$$DoF = 6(19-1) - \sum_{i=1}^{18}(6-f_i) = 108 - 18 \times (6-1) = 18$$

A robot has its design purpose to transfer a set of the input motions into output motions. By comparing the DOF of input motions *in the joint space* and the DOF of the output motions *in the task space*, the required DOF of output motions can be *under-actuated, exactly actuated*, or *over-actuated*. Accordingly, robots are classified into *under-actuated robots, non-redundant* robots, and *redundant* robots. Figure 12.21a shows that an under-actuated robot has less DOF of the input motions than that of the output motions required by tasks. Figure 12.21b shows that a non-redundant robot has the same DOF of the input and output motions. Figure 12.21c shows that a redundant robot has more DOF of that input motions than that of the output motions by tasks.

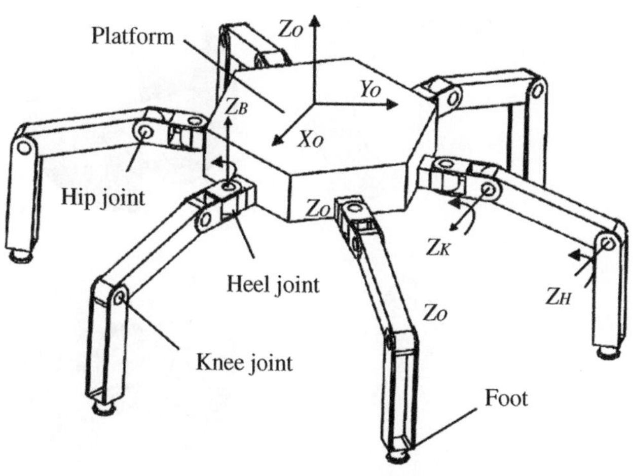

Figure 12.20 Calculate DOF of a hexapod robot

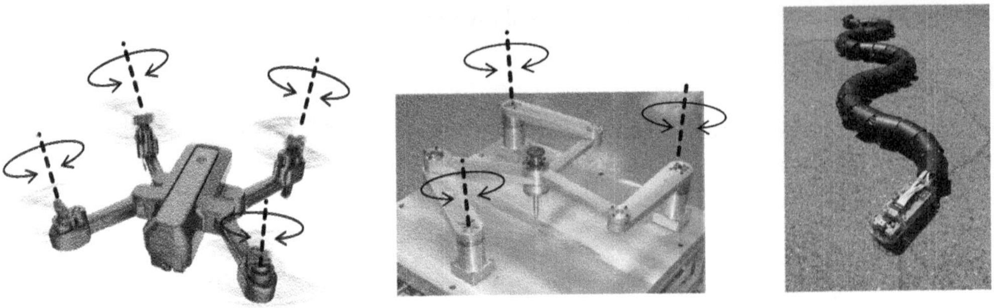

(a) Under-actuated robot (4 < 6) (b) Non-redundant robot (3 = 3) (c) Redundant robot ($n > 6$)

Figure 12.21 Examples of under-actuated, non-redundant, and redundant robots (*Source:* (c) Dr. Gavin Miller / https://blog.adobe.com/en/publish/2017/02/10/gavin-miller-brings-robots-to-squiggling-squirming-life / Last accessed on July 19, 2024)

12.3.3 Workspaces

Workspace is the volume in a task space that the reference point or CS on a robot's end-effector can reach. Table 12.1 shows the workspaces by the positioning mechanisms of *RRR*, *RRP*, *RPR*, *PRR*, *RPR*, *PRP*, *RPP*, and *PPP*, where *R* and *P* in the abbreviations represent a revolute joint and a prismatic joint, respectively.

Table 12.1 Workspaces of eight common positioning mechanisms

Positioning mechanism (R – Revolute; P – Prismatic)		Workspace	
Abbreviation	**Illustration**	**2D cross section**	**3D volume**
RRR	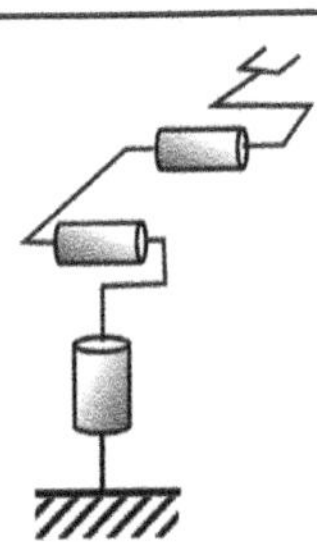	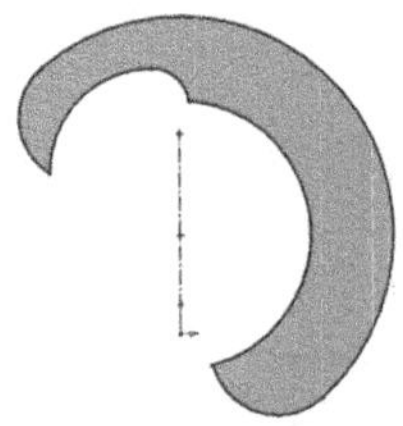	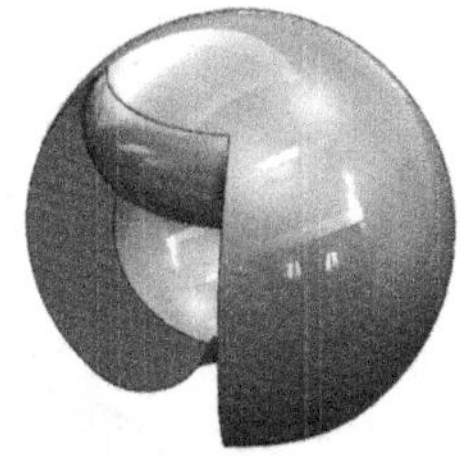
RRP	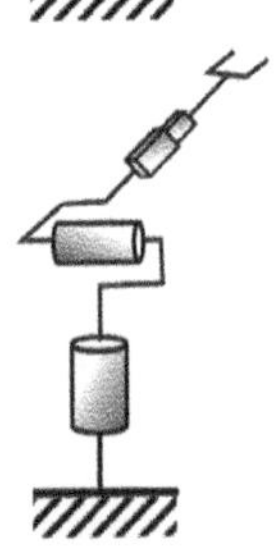	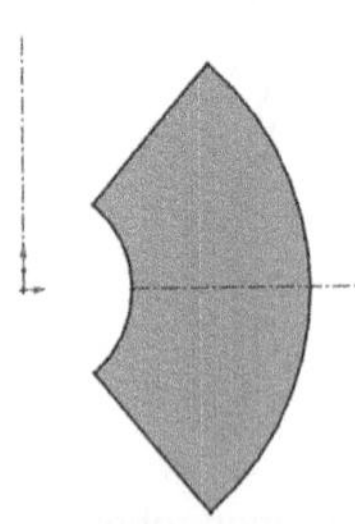	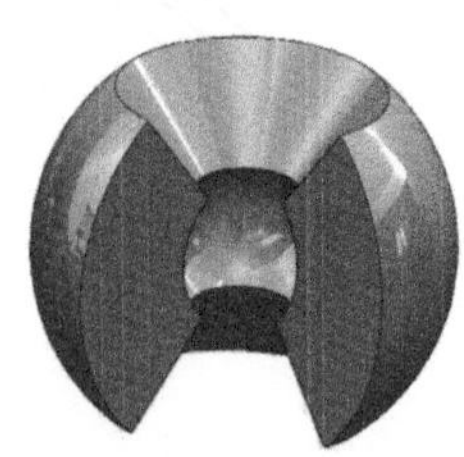

Table 12.1 (Continued)

Positioning mechanism (R – Revolute; P – Prismatic)		Workspace	
Abbreviation	**Illustration**	**2D cross section**	**3D volume**
RPP	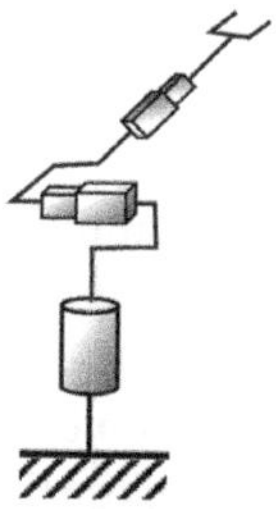	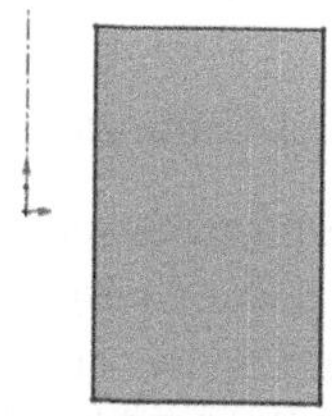	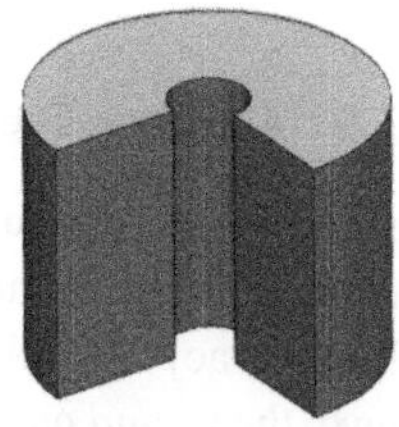
PPP	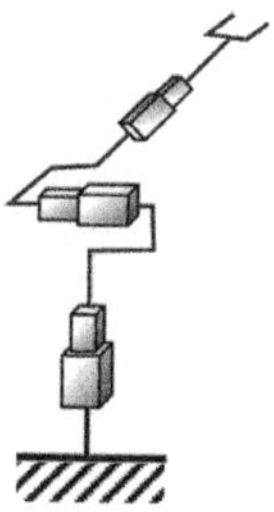	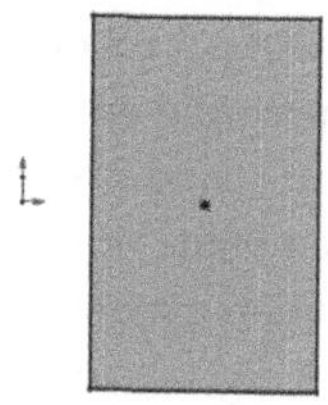	
RPR	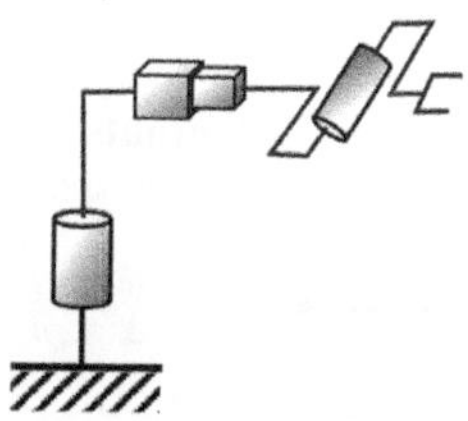	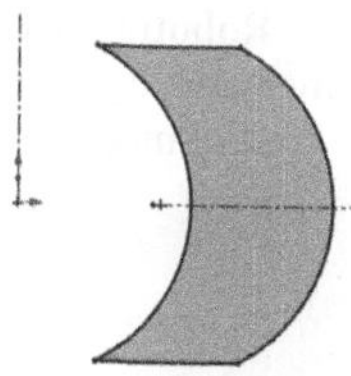	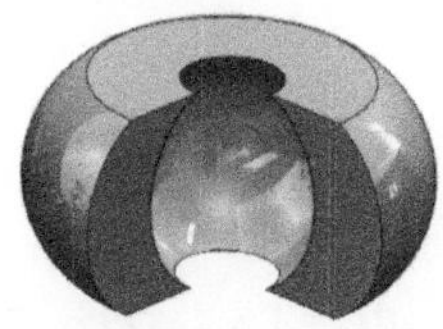
PPR	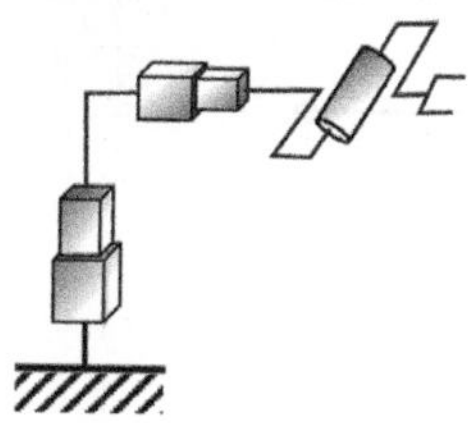	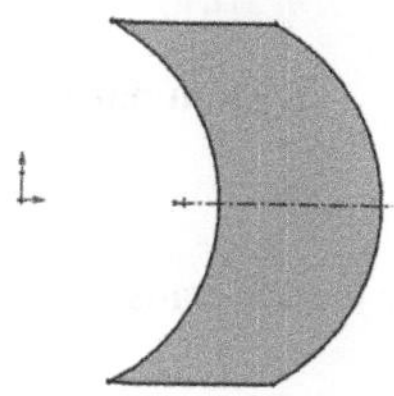	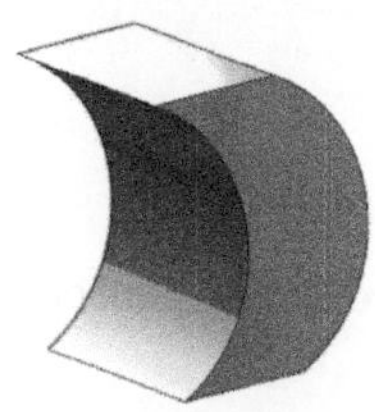
PRR	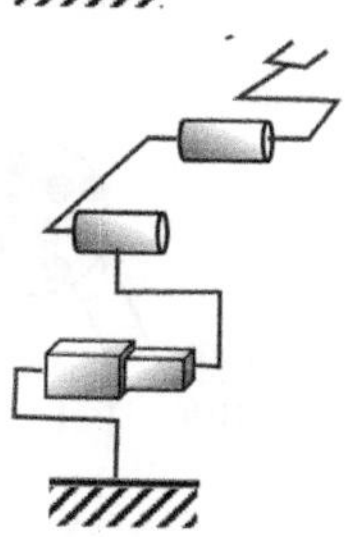	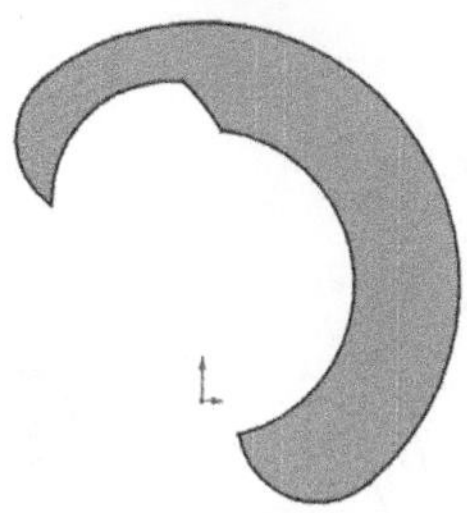	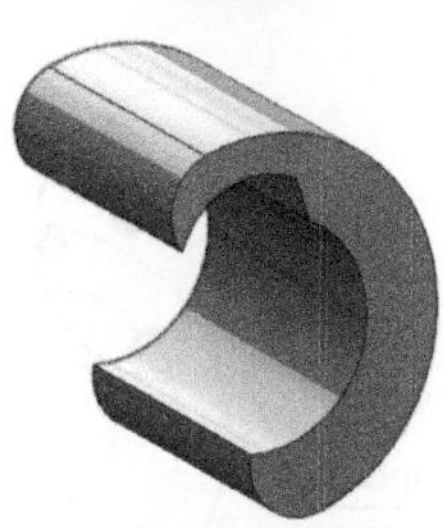
PRP	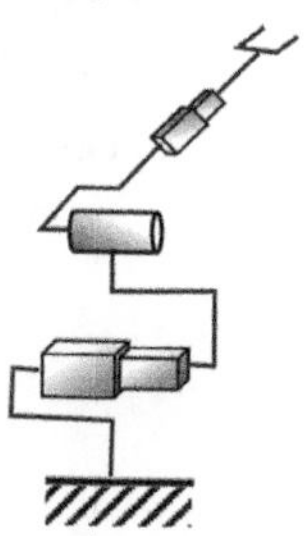	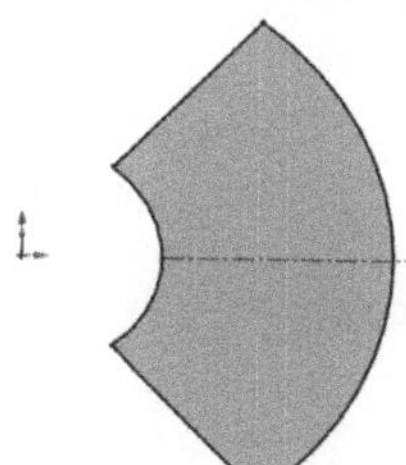	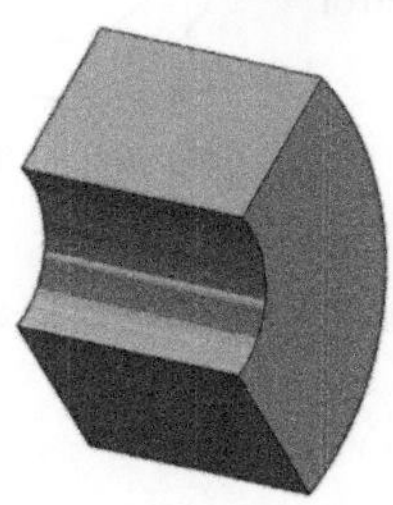

12.3.4 Modeling and Simulation

A robot is a mechatronic system that is designed to transfer the joint motions in the *joint space* into the motion of the end-effector in *a task space*. Figure 12.22 describes a robotic system with its inputs, outputs, system parameters; moreover, kinematic, dynamic, and control models are developed to represent the transformations of input to output motions.

12.3.5 Accuracy, Precision, and Calibration

One criterion to measure the performance of a robot is the accuracy that the robotic end-effector can achieve when it moves to a desired position or when it follows an expected path.

The accuracy of a robot depends on two critical factors. *The first one* is the precision a robot can achieve; *the second one* is how well the calibration process is performed on the robot. Figure 12.23 shows the relations of accuracy and precision. *Precision* represents how close the resulted positions can be from the desired position when a robot performs the same task repeatedly. In contrast, the precision is the variation of the measurements that are taken under the same operating conditions. *Accuracy* represents how close the actual result and desired result is. The accuracy is how much difference is between the averaged value of multiple measured data and the reference value.

Figure 12.23b is the ideal case when a robot is used since the capability (i.e., precision) of robot is fully utilized to achieve the desired high accuracy. The case in Figure 12.23a is undesirable since the

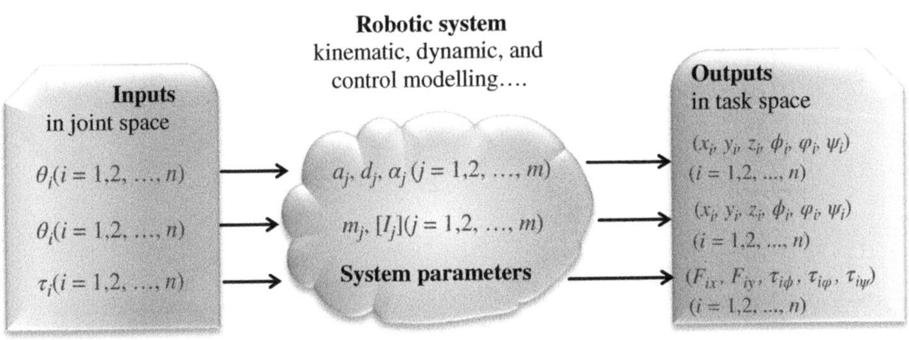

Figure 12.22 A robot model for motion transformation

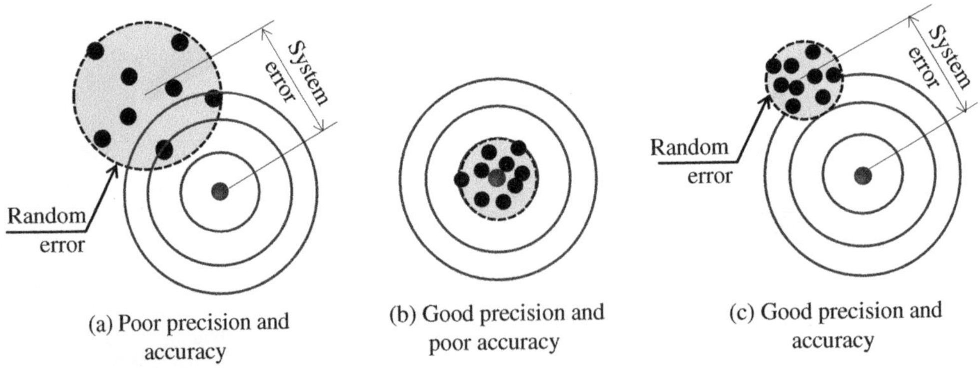

Figure 12.23 Precision and accuracy

robot is not competent to perform the task with expected accuracy. The case in Figure 12.23c shows the need of calibrating a robot to eliminate the system error before the robot can achieve the required accuracy.

Calibration is the process of (1) identifying all main sources of system error such as the derivations of actual geometrical and physical parameters from their normal values and (2) compensating errors by eliminating the effects of these identified derivations. The robot in Figure 12.23c must be calibrated to reduce the system error and achieve the desired accuracy as the case in Figure 12.23b. Calibrating a robot is critical to achieve its accuracy that is equivalent to the characterized precision.

12.3.6 Other Specifications

When a robotic task is given, a competent robot is selected based on its given specifications of robots. The main specifications of a robot include DOF, the workspaces in both the joint space and task space, respectively, maximum speeds and loads, and the precision of end-effector movement as exemplified in Figure 12.24 (Yaskawa 2024).

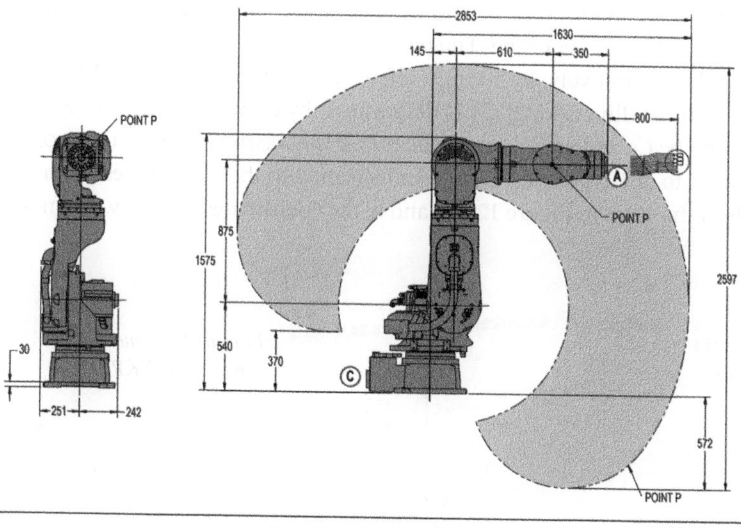

SIA50D specifications			
Structure	Articulated	*Allowable output moment*	Axis 5: 377 N·m Axis 6: 377 N·m Axis 7: 147 N·m
Mounting	Ground		
Motion axes	7	*Allowable output Moment of inertia*	Axis 5: 29.6 kg·m^2 Axis 6: 29.6 kg·m^2 Axis 7: 12.5 kg·m^2
Payload	50 kg		
Repeatability	±0.0001 m		
Vertical reach	2.597 m	*Power rating*	8.7 kVA
Horizontal reach	1.630 m	*Total mass*	~ 640 kg
Maximum joint ranges	Axis 1: (–180°, +180°) Axis 2: (+125°, – 60°) Axis 3: (–170°, +170°) Axis 4: (+215°, –35°) Axis 5: (–170°, +170°) Axis 6: (–125°, +125°) Axis 7: (–180°, +180°)	*Maximum joint speed*	Axis 1: 170°/second Axis 2: 130°/second Axis 3: 130°/second Axis 4: 130°/second Axis 5: 130°/second Axis 6: 130°/second Axis 7: 200°/second

Figure 12.24 Main specifications of SIA50D (*Source:* Adapted from Yaskawa (2024))

12.4 Kinematic Modeling

Kinematics is the study of geometric and timing aspects of a robot motion without considering how the motion is actually produced. In kinematic modeling, a robot is treated as a kinematic chain or structure. A kinematic chain consists of a set of the rigid bodies that are interconnected by the joints such as revolute joints, universal joints, spherical joints, and prismatic joints. Kinematic modeling is essential to determine the workspace of a robot, calibrate a robot to eliminate system errors, plan a robotic path or trajectory, and program a robot for a specified task.

Figure 12.25 shows that kinematic modeling is to find the solutions to two fundamental problems, that is, *forward kinematic problem* (FKP) and *inverse kinematic problem* (IKP). An FKP is to find the motion of the end-effector in the task space when the motions of active joints in the joint space are given. In contrast, an IKP is to find the motions of the active joints in the joint space when the motion of the end-effector in the task space in given.

12.4.1 Positions of Points, Links, and Bodies in 2D and 3D Space

The motions of a robot are modeled by positions and orientations of reference points or a CS associated with the links and joints over time. The positions of a point in a 2D or 3D space are described in Figure 12.26a and b, respectively. By taking a reference *CS* of $\{O\text{–}XY\}$ or $\{O\text{–}XYZ\}$, a position in a two-dimensional or three-dimensional CS is determined by (x, y) and (x, y, z), respectively.

A reference CS can be *Cartesian*, *polar*, *cylindrical*, or *spherical*. Moreover, the coordinates of the position with respect to one type of CS can be transformed to the coordinates in another type CS. Take the example of the cases in Figure 12.26a and b, the coordinates of *P* in two-dimensional polar

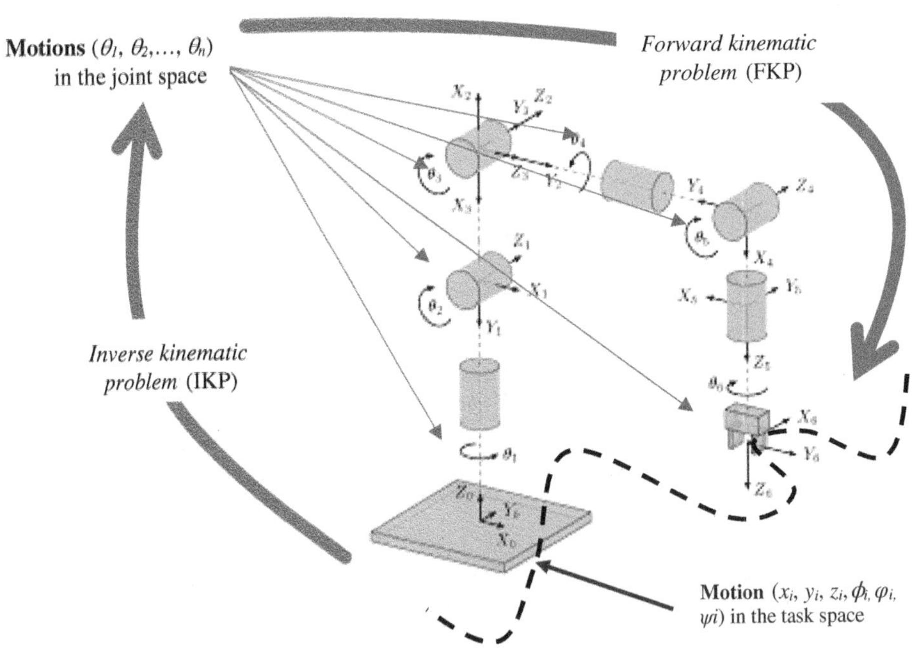

Figure 12.25 Forward kinematic problem (FKP) and inverse kinematic problem (IKP)

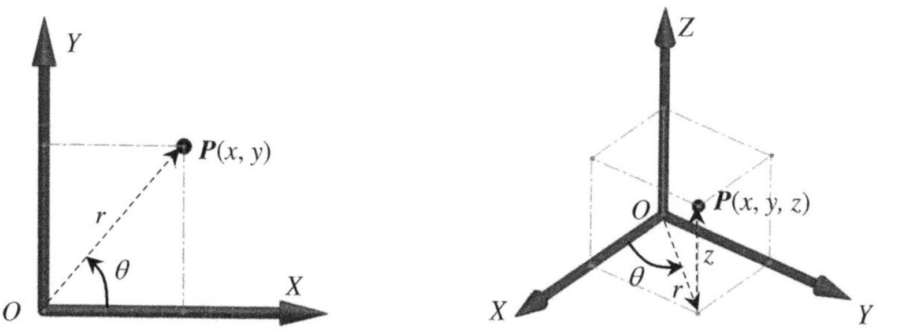

(a) Point P in O-XY plane (b) Point P in the O-XYZ coordinate system

Figure 12.26 Representation of a point in two-dimensional or three-dimensional space

CS (r, θ) and three-dimensional cylindrical CS (r, θ, z) are transferred from (x, y) and (x, y, z) in the Cartesian CS as,

$$\left.\begin{array}{l} \begin{Bmatrix} x \\ y \end{Bmatrix} \text{ in Cartesian } CS \rightarrow \begin{Bmatrix} r \\ \theta \end{Bmatrix} = \begin{Bmatrix} \sqrt{x^2+y^2} \\ \theta = \tan^{-1}\left(\frac{y}{x}\right) \end{Bmatrix} \text{ in polar } CS \\ \begin{Bmatrix} x \\ y \\ z \end{Bmatrix} \text{ in Cartesian } CS \rightarrow \begin{Bmatrix} r \\ \theta \\ z \end{Bmatrix} = \begin{Bmatrix} \sqrt{x^2+y^2} \\ \theta = \tan^{-1}\left(\frac{y}{x}\right) \\ z \end{Bmatrix} \text{ in cylindrical } CS \end{array}\right\} \tag{12.3}$$

The positions of link AB in a 2D or 3D space are modeled in Figure 12.27a and b, respectively. In the reference CS $\{O–XY\}$, the position of link AB is represented by the coordinates of two nodes (x_A, y_A) and (x_B, y_B). Similarly, the position of link AB in the reference CS $\{O–XYZ\}$ is represented by the coordinates of two nodes (x_A, y_A, z_A) and (x_B, y_B, z_B). If a reference CS other than the Cartesian CS is used, the coordinates of two nodes are transformed respectively.

Figure 12.28a shows the representation of a body ABC in the three-dimensional Cartesian CS. The coordinates of three nodes (i.e., A, B, and C) on the body are sufficient to determine the position

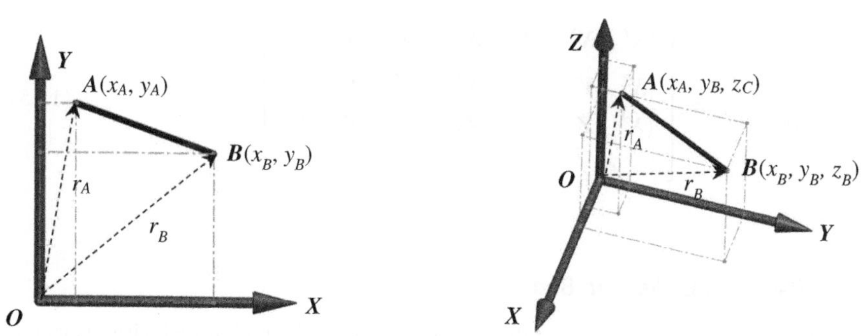

(a) Line A-B in O-XY plane (b) Line A-B in O-XYZ coordinate system

Figure 12.27 Representation of link ab in two-dimensional or three-dimensional space

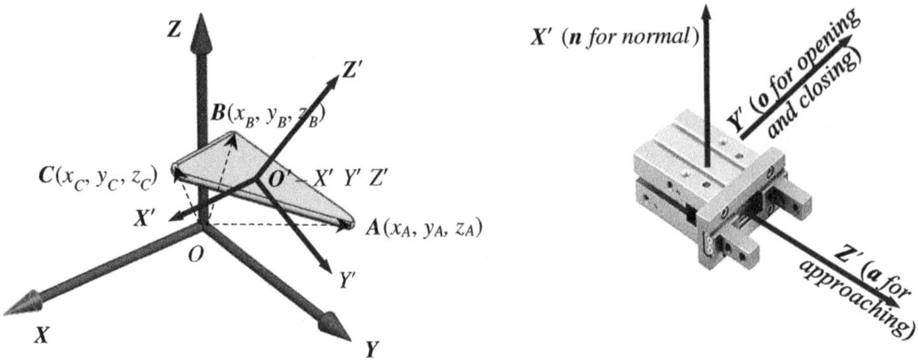

(a) Object in 3D space

(b) Correspondence of x, y, z and n, o, a in tool frame

Figure 12.28 Representation of body ABC in three-dimensional space (*Source:* (b) FTVogue / https://www.amazon.com/Pneumatic-Cylinder-Precision-industrial-cylinders/dp/B0C6SV8D8S / Last accessed on July 19, 2024)

and orientation of the body in space completely. To facilitate the modeling process, a *local CS* (LCS) $\{O' - X'Y'Z'\}$ is introduced and it is attached to the body locally, and its position relative to the *global CS* (GCS) becomes,

$$[\boldsymbol{T}] = \begin{bmatrix} \boldsymbol{R} & \boldsymbol{O'} \\ 0\ 0\ 0 & 1 \end{bmatrix} = \begin{bmatrix} (\boldsymbol{X'} \cdot \boldsymbol{X}) & (\boldsymbol{Y'} \cdot \boldsymbol{X}) & (\boldsymbol{Z'} \cdot \boldsymbol{X}) & O'_x \\ (\boldsymbol{X'} \cdot \boldsymbol{Y}) & (\boldsymbol{Y'} \cdot \boldsymbol{Y}) & (\boldsymbol{Z'} \cdot \boldsymbol{Y}) & O'_y \\ (\boldsymbol{X'} \cdot \boldsymbol{Z}) & (\boldsymbol{Y'} \cdot \boldsymbol{Z}) & (\boldsymbol{Z'} \cdot \boldsymbol{Z}) & O'_z \\ 0 & 0 & 0 & 1 \end{bmatrix} \tag{12.4}$$

where $[\boldsymbol{T}]$ is the homogenous representation of LCS in the GCS $\{O\text{–}XYZ\}$; $[\boldsymbol{R}]$ is a 3×3 matrix to represent the orientation of LCS; $\boldsymbol{O'}$ is a 1×3 vector to represent the origin of LCS. Note that the last row $\{0, 0, 0, 1\}$ is added to make $[\boldsymbol{T}]$ be in a square form so that it can be manipulated more easily.

As shown in Figure 12.28b, if the body is an end-effector, the axes of ($\boldsymbol{X'}$, $\boldsymbol{Y'}$ and $\boldsymbol{Z'}$) of its LCS are commonly denoted as $\boldsymbol{n}$ for a normal formed by $\boldsymbol{o}$ and $\boldsymbol{a}$ following a right-hand rule, $\boldsymbol{o}$ for the direction to open or close the tool, and $\boldsymbol{a}$ for the direction to approach an object to be manipulated, respectively. The vectors of $\boldsymbol{n}$, $\boldsymbol{o}$, $\boldsymbol{a}$ are found as Eq. (12.4) as,

$$[\boldsymbol{T}] = \begin{bmatrix} \boldsymbol{n} & \boldsymbol{o} & \boldsymbol{a} & \boldsymbol{O'} \\ 0 & 0 & 0 & 1 \end{bmatrix} = \begin{bmatrix} (\boldsymbol{X'} \cdot \boldsymbol{X}) & (\boldsymbol{Y'} \cdot \boldsymbol{X}) & (\boldsymbol{Z'} \cdot \boldsymbol{X}) & O'_x \\ (\boldsymbol{X'} \cdot \boldsymbol{Y}) & (\boldsymbol{Y'} \cdot \boldsymbol{Y}) & (\boldsymbol{Z'} \cdot \boldsymbol{Y}) & O'_y \\ (\boldsymbol{X'} \cdot \boldsymbol{Z}) & (\boldsymbol{Y'} \cdot \boldsymbol{Z}) & (\boldsymbol{Z'} \cdot \boldsymbol{Z}) & O'_z \\ 0 & 0 & 0 & 1 \end{bmatrix} \tag{12.5}$$

12.4.2 Motions of Particles, Links, and Bodies

If a point, a link, or a body moves, the position of the corresponding entity is changed with respect to time, the movement can be basically classified into *translational* or *rotational*; the motion of a point is always *translational*.

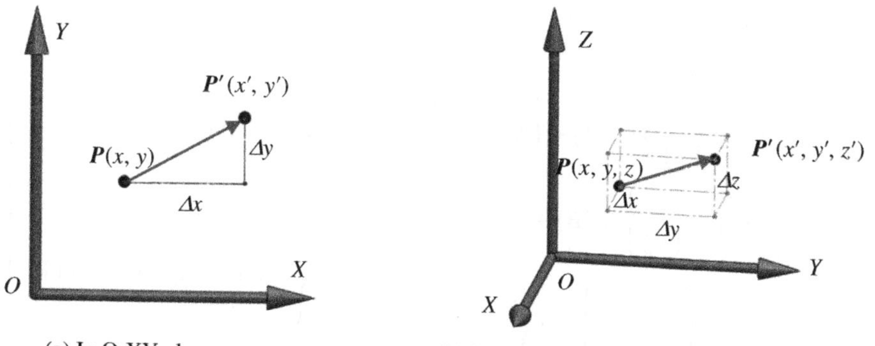

(a) In O-XY plane (b) In O-XYZ coordinate system

Figure 12.29 Motion of a point in two-dimensional and three-dimensional space

Figure 12.29 represents the movement of a particle in a 2D or 3D space with 2- or 3-DOF, respectively. New position $\boldsymbol{P}'$ is updated by a displacement of $\Delta\boldsymbol{P}$ from $\boldsymbol{P}$ as,

$$\left.\begin{array}{l}\{\boldsymbol{P}\} = \begin{Bmatrix} x \\ y \end{Bmatrix} \rightarrow \{\boldsymbol{P}'\} = \begin{Bmatrix} x' \\ y' \end{Bmatrix} = \{\boldsymbol{P}\} + \{\Delta\boldsymbol{P}\} = \begin{Bmatrix} x + \Delta x \\ y + \Delta y \end{Bmatrix} \text{ in 2D space} \\ \{\boldsymbol{P}\} = \begin{Bmatrix} x \\ y \\ z \end{Bmatrix} \rightarrow \{\boldsymbol{P}'\} = \begin{Bmatrix} x' \\ y' \\ z' \end{Bmatrix} = \{\boldsymbol{P}\} + \{\Delta\boldsymbol{P}\} = \begin{Bmatrix} x + \Delta x \\ y + \Delta y \\ z + \Delta z \end{Bmatrix} \text{ in 3D space}\end{array}\right\} \tag{12.6}$$

where $\{\Delta\boldsymbol{P}\}$ is the displacement of movement; $\{\boldsymbol{P}'\}$ is the new position after the movement.

Figure 12.30 describes the movement of link AB in a 2D space. Additional one independent parameter is required to represent its orientation on plane; therefore, the link in the 2D space has 3-DOF. Moreover, it is convenient to describe the motion of link with respect to a LCS attached on the link. Without losing the generality, node A is selected as the origin, and the axis is aligned with the vector from nodes A to B. The movement of link AB can be represented by (Δx, Δy, $\Delta\theta$), where (Δx, Δy) are the translational displacement along axes of X and Y, respectively, and $\Delta\theta$ is the

Figure 12.30 Modeling of link motion in 2-D space

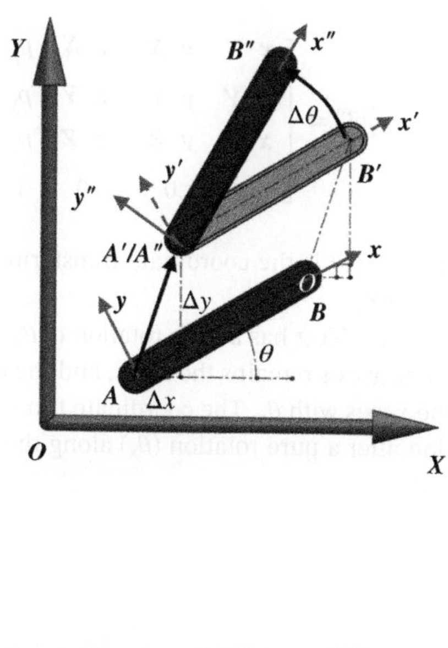

rotational displacement along the axis Z. The new position of the particle $\{P'\}$ with a translation of displacement $\{\Delta P\}$ from $\{P\}$ can be found as,

$$\left.\begin{array}{l} \{A\} = \begin{Bmatrix} x_A \\ y_A \end{Bmatrix} \rightarrow \{A''\} = \begin{Bmatrix} x''_A \\ y''_A \end{Bmatrix} = \{A\} + \{\Delta P_T\} = \begin{Bmatrix} x_A + \Delta x \\ y_A + \Delta y \end{Bmatrix} \quad \text{for A (origin)} \\ \{B\} = \begin{Bmatrix} x_B \\ y_B \end{Bmatrix} \rightarrow \left\{\begin{array}{l} \{B''\} = \begin{Bmatrix} x''_B \\ y''_B \end{Bmatrix} = \{B\} + \{\Delta P_T\} + \{\Delta P_R\} \\ = \begin{Bmatrix} x_B + \Delta x \\ y_B + \Delta y \end{Bmatrix} + L \begin{Bmatrix} \cos(\theta + \Delta\theta) - \cos\theta \\ \sin(\theta + \Delta\theta) - \sin\theta \end{Bmatrix} \end{array}\right. \quad \text{for B (any point on link)} \end{array}\right\} \tag{12.7}$$

where $\{\Delta P_T\}$ and $\{\Delta P_R\}$ are the displacements by translation and rotation, respectively. L is the length of link AB, and θ and $(\theta + \Delta\theta)$ are the directions of link AB before and after the movement, and

$$L = \sqrt{(x_B - x_A)^2 + (y_B - y_A)^2}, \theta = \tan^{-1}\frac{y_B - y_A}{x_B - x_A}$$

Figure 12.31 describes the generic movement of an object in a 3D space. The motion is described by the coordinate transformation of an LCS from its original position $\{\mathbf{o}\text{-}xyz\}$ to the new position $\{\mathbf{o}'\text{-}x'y'z'\}$. The homogenous matrix is introduced to represent the object at its new position as,

$$[\boldsymbol{T}'] = \begin{bmatrix} [\Delta \boldsymbol{R}] & \{\Delta \boldsymbol{P}\} \\ 0 \;\; 0 \;\; 0 & 1 \end{bmatrix} = \begin{bmatrix} \boldsymbol{x}'\cdot\boldsymbol{x} & \boldsymbol{y}'\cdot\boldsymbol{x} & \boldsymbol{z}'\cdot\boldsymbol{x} & \Delta P_x \\ \boldsymbol{x}'\cdot\boldsymbol{y} & \boldsymbol{y}''\cdot\boldsymbol{y} & \boldsymbol{z}'\cdot\boldsymbol{y} & \Delta P_y \\ \boldsymbol{x}'\cdot\boldsymbol{z} & \boldsymbol{y}'\cdot\boldsymbol{z} & \boldsymbol{z}'\cdot\boldsymbol{z} & \Delta P_z \\ 0 & 0 & 0 & 1 \end{bmatrix} \tag{12.8}$$

The rotation of an object along an axis of the reference CS is called a *pure rotation*. Assume that the axes $(\boldsymbol{x}, \boldsymbol{y}, \boldsymbol{z})$ of original LCS are aligned with the axes $(\boldsymbol{X}, \boldsymbol{Y}, \boldsymbol{Z})$ of the reference CS except for its origin at (p_x, p_y, p_z), that is, the position of object before a pure rotation can be represented by,

$${}^{1}_{2}[\boldsymbol{T}] = \begin{bmatrix} \boldsymbol{x}\cdot\boldsymbol{X} & \boldsymbol{y}\cdot\boldsymbol{X} & \boldsymbol{z}\cdot\boldsymbol{X} & p_x \\ \boldsymbol{x}\cdot\boldsymbol{Y} & \boldsymbol{y}\cdot\boldsymbol{Y} & \boldsymbol{z}\cdot\boldsymbol{Y} & p_y \\ \boldsymbol{x}\cdot\boldsymbol{Z} & \boldsymbol{y}\cdot\boldsymbol{Z} & \boldsymbol{z}\cdot\boldsymbol{Z} & p_z \\ 0 & 0 & 0 & 1 \end{bmatrix} = \begin{bmatrix} 1 & 0 & 0 & p_x \\ 0 & 1 & 0 & p_y \\ 0 & 0 & 1 & p_z \\ 0 & 0 & 0 & 1 \end{bmatrix} \tag{12.9}$$

where ${}^{0}_{1}[\boldsymbol{T}]$ is the coordinate transformation for the object from LCS $\{\mathbf{o}\text{-}\boldsymbol{xyz}\}$ to the reference CS $\{\mathbf{O}\text{-}\boldsymbol{XYZ}\}$.

If the object has a pure rotation of θ_x along the x-axis as shown in Figure 12.31a, the unit vector along $\boldsymbol{x}$-axis remains the same, and the unit vectors of along the $\boldsymbol{y}$- and the $\boldsymbol{z}$-axis are rotated along the x-axis with θ_x. The coordinate transformation of object from original position to the new position after a pure rotation (θ_x) along the x-axis is given as,

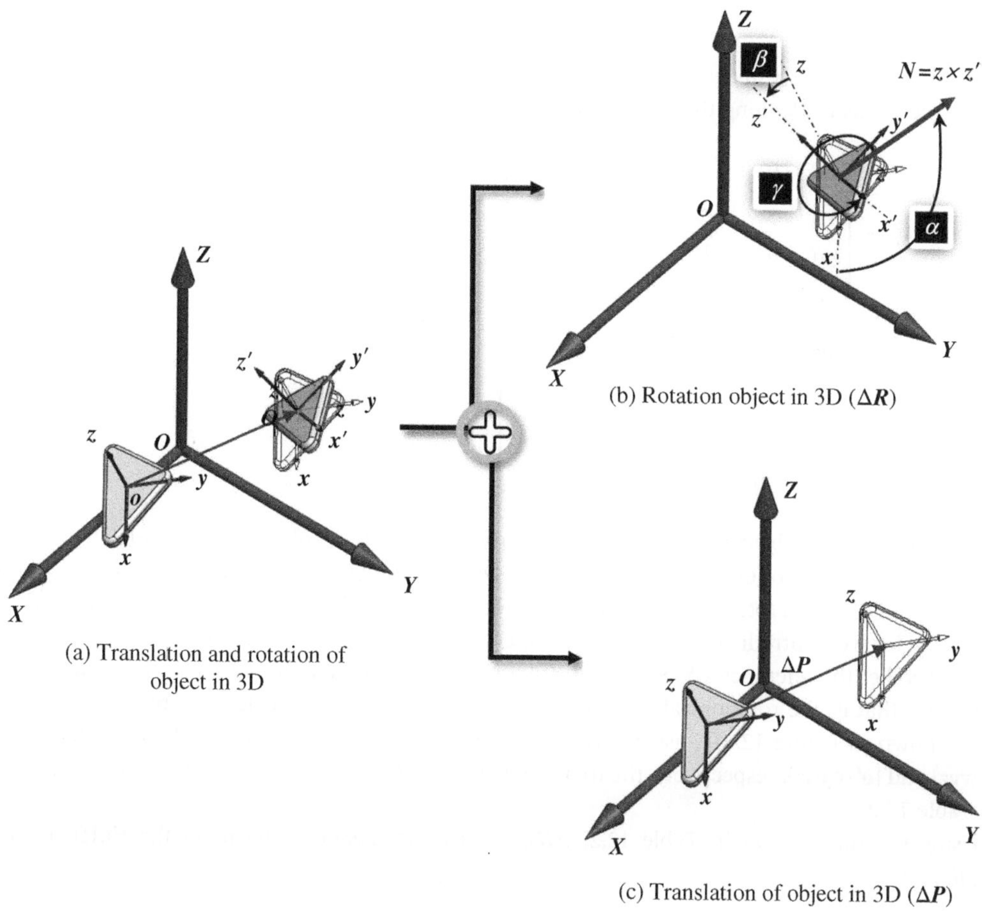

Figure 12.31 General motion of a body in 3D space

$$ {}_{1}^{0}[\boldsymbol{T}] = \begin{bmatrix} \boldsymbol{x}'\cdot\boldsymbol{x} & \boldsymbol{y}'\cdot\boldsymbol{x} & \boldsymbol{z}'\cdot\boldsymbol{x} & 0 \\ \boldsymbol{x}'\cdot\boldsymbol{y} & \boldsymbol{y}'\cdot\boldsymbol{y} & \boldsymbol{z}'\cdot\boldsymbol{y} & 0 \\ \boldsymbol{x}'\cdot\boldsymbol{z} & \boldsymbol{y}'\cdot\boldsymbol{z} & \boldsymbol{z}'\cdot\boldsymbol{z} & 0 \\ 0 & 0 & 0 & 1 \end{bmatrix} = \begin{bmatrix} & & & 0 \\ & [\boldsymbol{R}_x]_{3\times 3} & & 0 \\ & & & 0 \\ 0 & 0 & 0 & 1 \end{bmatrix} = \begin{bmatrix} 1 & 0 & 0 & 0 \\ 0 & c\theta_x & -s\theta_x & 0 \\ 0 & s\theta_x & c\theta_x & 0 \\ 0 & 0 & 0 & 1 \end{bmatrix} \quad (12.10) $$

where "*c*" and "*s*" stand for the *cosine* and *sine* functions, respectively; ${}_{2}^{1}[\boldsymbol{T}]$ is the coordinate transformation for an object from the transformed LCS $\{\boldsymbol{o}'\text{-}\boldsymbol{x}'\boldsymbol{y}'\boldsymbol{z}'\}$to the original LCS $\{\boldsymbol{o}\text{-}\boldsymbol{xyz}\}$. After a pure rotation of the object along $\boldsymbol{x}$-axis, the position of the object in the GCS becomes,

$$ [\boldsymbol{T}]_{\theta_x} = {}_{1}^{0}[\boldsymbol{T}]\cdot{}_{2}^{1}[\boldsymbol{T}] = \begin{bmatrix} [\boldsymbol{R}_x]_{3\times 3} & [\boldsymbol{R}_x]_{3\times 3}\{\boldsymbol{P}\}_{1\times 3} \\ 0\ \ 0\ \ 0 & 1 \end{bmatrix} = \begin{bmatrix} 1 & 0 & 0 & p_x \\ 0 & c\theta_x & -s\theta_x & p_y c\theta_x - p_z s\theta_x \\ 0 & s\theta_x & c\theta_x & p_y s\theta_x + p_z c\theta_x \\ 0 & 0 & 0 & 1 \end{bmatrix} \quad (12.11) $$

where $[\boldsymbol{T}]_{\theta_x}$ is the representation of the object with a pure rotation θ_x along the $\boldsymbol{x}$-axis in the GCS {***O-XYZ***}.

Due to the symmetry in determining the *x*-axis, the representation of an object with a rotation θ_y along the $\boldsymbol{y}$-axis or θ_z along the $\boldsymbol{z}$-axis (shown in Figure 12.31b and c) is readily written as,

$$[\boldsymbol{T}]_{\theta_y} = \begin{bmatrix} c\theta_y & 0 & -s\theta_y & p_x c\theta_y - p_z s\theta_y \\ 0 & 1 & 0 & p_y \\ s\theta_y & 0 & c\theta_y & p_x s\theta_y + p_z c\theta_y \\ 0 & 0 & 0 & 1 \end{bmatrix} \tag{12.12}$$

$$[\boldsymbol{T}]_{\theta_z} = \begin{bmatrix} c\theta_z & -s\theta_z & 0 & p_x c\theta_z - p_y s\theta_z \\ s\theta_z & c\theta_z & 0 & p_x s\theta_z + p_y c\theta_z \\ 0 & 0 & 1 & p_z \\ 0 & 0 & 0 & 1 \end{bmatrix} \tag{12.13}$$

The rotation of an object may be with an arbitrary axis in GCS. An object requires 6-DOF to determine its position and orientation in 3D, that is, 3-DOF are for a translation $\Delta\boldsymbol{P}$ (ΔP_x, ΔP_y, ΔP_z) and the other 3-DOF for a rotation represented by $\Delta\boldsymbol{R}_{3\times3}$ where the Euler angles (α, β, γ) are used. The Euler angles were introduced by Leonhard Euler to represent the orientation of a body in a Cartesian CS. Let the reference CS be {**o-*xyz***}, any orientation can be represented by three elemental rotations, that is, the rotations about three axes of the LCS that is attached on object.

As shown in Figure 12.31c, assume that the starting and finishing CSs attached on object are {**o-*xyz***} and {**o′-*x′y′z′***}, respectively, the rotational matrix $\Delta\boldsymbol{R}_{3\times3}$ is determined by the steps specified in Table 12.2.

Using the information in Table 12.2, $\Delta\boldsymbol{R}_{3\times3}$ can be expressed in terms of the Euler angles (α, β, γ) by,

$$\Delta\boldsymbol{R}_{3\times3} = [\boldsymbol{R}_{z,\alpha}] \cdot [\boldsymbol{R}_{x,\beta}] \cdot [\boldsymbol{R}_{z,\gamma}] = \begin{bmatrix} c\alpha c\gamma - s\alpha c\beta s\gamma & -c\alpha s\gamma - s\alpha c\beta c\gamma & s\alpha s\beta \\ s\alpha c\gamma + c\alpha c\beta s\gamma & -s\alpha s\gamma + c\alpha c\beta c\gamma & -c\alpha s\beta \\ s\beta s\gamma & s\beta c\gamma & c\beta \end{bmatrix} \tag{12.14}$$

where $[\boldsymbol{R}_{z,\alpha}]$, $[\boldsymbol{R}_{x,\beta}]$, $[\boldsymbol{R}_{z,\gamma}]$ are the rotation matrices of (α, β, γ) along ***z***, ***x***, and ***z***, respectively (Figure 12.32).

Table 12.2 Steps of coordination transformation by Euler angles

Step	Actions
1.	define a tentative rotational axis ***N*** as $\boldsymbol{N} = \boldsymbol{z} \times \boldsymbol{z}'$,
2.	take ***z*** as the rotational axis for the rotation of α to align ***x*** with ***N***,
3.	take ***N*** (or new ***x*** (***x′***)) as the rotational axis for the rotation of β to align ***z*** with ***z′***, and
4.	take ***z″*** as the rotational axis for the rotation of γ to align ***N*** with ***x′***,
5.	use the right-hand rule to determine ***y***" from ***x′*** and ***z′***.

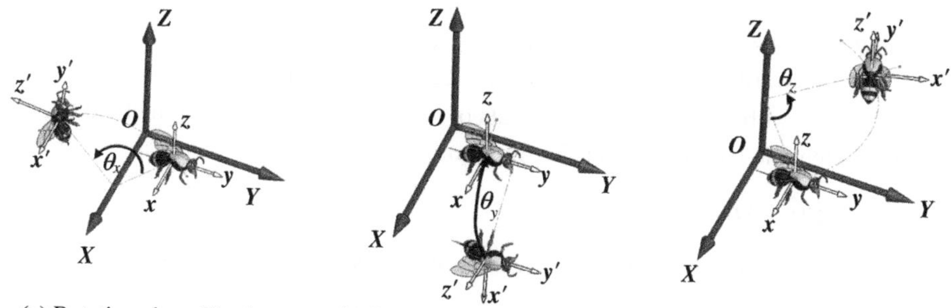

(a) Rotation along X-axis (b) Rotation along Y-axis (c) Rotation along Z-axis

Figure 12.32 Rotation along the axes of CS

12.4.3 Vector-Loop Method for Motion Analysis of Plane Mechanism

The vector-loop method is another approach to model the motion of a planar mechanism. In using a vector-loop method, a link is represented by two or more nodes. Each node is represented by its coordinates, and the constraints on the kinematic chain(s) are expressed as one or more loops of the vectors. Figure 12.33 shows how the vector-loop method is used to model and solve the kinematic problem of a four-bar mechanism.

A mechanism has one or more driving motions as the inputs and the motion of its end-effector as the outputs. A kinematic model is to describe the relations of inputs to outputs mathematically. *A forward kinematic solver* is to obtain the output motions when the input motions are given. *An inverse kinematic solver* is to calculate the input motions when the required output motions are given. Here, the vector-loop method is applied to solve both forward kinematic and inverse kinematic problems of a four-bar mechanism.

Figure 12.33 shows that a four-bar mechanism consists of four links and four revolute joints. The joints are used to connect four links in a loop. Therefore, the mobility of the four-bar mechanism is determined as,

$$M = \lambda(l - j - 1) + \sum_{i=1}^{j} f_i = 3(4 - 4 - 1) + 4 \times 1 = 1 \tag{12.15}$$

Figure 12.33 Vector-loop method in modeling a four-bar mechanism

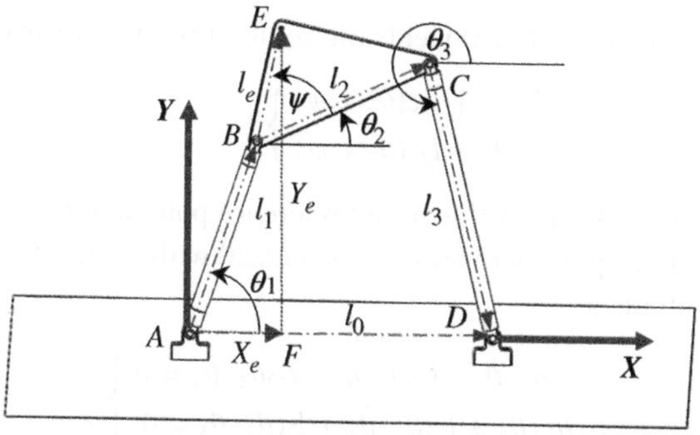

The mechanism has the mobility of 1, that is, DOF = 1. Assume link AB be the driving link, the input motion is θ_1, and the output reference is point $E(X_e, Y_e)$ on the coupler BCE.

To solve the forward kinematic problem, that is, calculating the outputs of $(X_e, Y_e), (\dot{X}_e, \dot{Y}_e), (\ddot{X}_e, \ddot{Y}_e)$ based on the given inputs of $\theta_1, \dot{\theta}_1, \ddot{\theta}_1$. The vector-loop method is used and there is one kinematic loop in the four-bar mechanism, that is,

$$\overrightarrow{\boldsymbol{AB}} + \overrightarrow{\boldsymbol{BC}} + \overrightarrow{\boldsymbol{CD}} - \overrightarrow{\boldsymbol{AD}} = \mathbf{0} \tag{12.16}$$

Letting $\theta_4 = 0$ and using the exponential expression of the vectors in Eq. (12.16) gets,

$$l_1 e^{j\theta_1} + l_2 e^{j\theta_2} + l_3 e^{j\theta_3} - l_0 e^{j0} = \mathbf{0} \tag{12.17}$$

Decomposing the exponential expression of Eq. (12.17) as the projected constraints of kinematic chain on the x-axis and y-axis yields,

$$\left.\begin{aligned} l_1 c\theta_1 + l_2 c\theta_2 + l_3 c\theta_3 - l_0 = 0 \\ l_1 s\theta_1 + l_2 s\theta_2 + l_3 s\theta_3 = 0 \end{aligned}\right\} \tag{12.18}$$

Equation (12.18) leads to the solutions of intermediate variables θ_2 and θ_3 in terms of θ_1 as,

$$\left.\begin{aligned} \theta_3 &= \tan^{-1}\left(\frac{-K_2 \pm \sqrt{K_2^2 - K_3^2 + K_1^2}}{K_3 - K_1}\right) \\ \theta_2 &= \tan^{-1}\left(\frac{l_1 s\theta_1 + l_3 s\theta_3}{l_1 c\theta_1 + l_3 c\theta_3 - l_0}\right) \end{aligned}\right\} \tag{12.19}$$

where K_1, K_2, and K_3 are intermediate parameters as,

$$\left.\begin{aligned} K_1 &= 2l_3(l_0 - l_1 c\theta_1) \\ K_2 &= 2l_3 l_1 s\theta_1 \\ K_3 &= l_2^2 - (l_0^2 + l_1^2 + l_3^2 - 2l_0 l_1 c\theta_1) \end{aligned}\right\} \tag{12.20}$$

Note from Eq. (12.19) that multiple solutions exist for a closed-loop mechanism.

Once θ_2 and θ_3 are solved, another vector loop can be developed to find the coordinates of $\boldsymbol{E}$ from,

$$\overrightarrow{\boldsymbol{AB}} + \overrightarrow{\boldsymbol{BE}} - \overrightarrow{\boldsymbol{AF}} - \overrightarrow{\boldsymbol{FE}} = \mathbf{0} \tag{12.21}$$

Using the exponential expression of the vectors in Eq. (12.21) gets,

$$l_1 e^{j\theta_1} + l_2 e^{j(\theta_2 + \psi)} - X_e e^{j0^\circ} - Y_e e^{j90^\circ} = \mathbf{0} \tag{12.22}$$

Equation (12.22) yields the solution of the forward kinematic problem as,

$$\left.\begin{aligned} X_e = l_1 c\theta_1 + l_2 c(\theta_2 + \psi) \\ Y_e = l_1 s\theta_1 + l_2 s(\theta_2 + \psi) \end{aligned}\right\} \tag{12.23}$$

where (X_e, Y_e) is the output reference point and θ_2 is calculated from Eq. (12.19) when θ_1 is given.

Taking the derivation of Eq. (12.23) with respect to time gives the velocity relations of inputs and outputs as,

$$\left.\begin{aligned} -l_1 s\theta_1 \cdot \dot{\theta}_1 - l_2 s\theta_2 \cdot \dot{\theta}_2 - l_3 s\theta_3 \cdot \dot{\theta}_3 = 0 \\ l_1 c\theta_1 \cdot \dot{\theta}_1 + l_2 c\theta_2 \cdot \dot{\theta}_2 + l_3 c\theta_3 \cdot \dot{\theta}_3 = 0 \end{aligned}\right\} \tag{12.24}$$

Equation (12.24) gives the dependence of $\dot{\theta}_2$ and $\dot{\theta}_3$ on $\dot{\theta}_1$ as

$$\left.\begin{aligned}\dot{\theta}_2 &= \frac{l_1 s(\theta_3-\theta_1)}{l_2 s(\theta_2-\theta_3)}\dot{\theta}_1\\ \dot{\theta}_3 &= \frac{l_1 s(\theta_2-\theta_1)}{l_3 s(\theta_3-\theta_2)}\dot{\theta}_1\end{aligned}\right\}\tag{12.25}$$

Taking the first derivation of Eq. (12.23) with respect to time and using Eq. (12.25) find the velocity relations of inputs and outputs as,

$$\left.\begin{aligned}\dot{X}_e &= -l_1\left(s\theta_1 + \frac{s(\theta_3-\theta_1)s(\theta_2+\psi)}{s(\theta_2-\theta_3)}\right)\dot{\theta}_1\\ \dot{Y}_e &= l_1\left(c\theta_1 + \frac{s(\theta_3-\theta_1)c(\theta_2+\psi)}{s(\theta_2-\theta_3)}\right)\dot{\theta}_1\end{aligned}\right\}\tag{12.26}$$

Taking the derivation of Eq. (12.26) with respect to time and using Eq. (12.25) give the acceleration relations of inputs and outputs as,

$$\left.\begin{aligned}\ddot{X}_e =& -l_1\left(s\theta_1+\frac{s(\theta_3-\theta_1)s(\theta_2+\psi)}{s(\theta_2-\theta_3)}\right)\ddot{\theta}_1\\ &-l_1\left(c\theta_1\dot{\theta}_1+\frac{c(\theta_2-\theta_3)s(\theta_3-\theta_1)s(\theta_2+\psi)(\dot{\theta}_2-\dot{\theta}_3)}{s^2(\theta_2-\theta_3)}+\frac{c(\theta_3-\theta_1)s(\theta_2+\psi)(\dot{\theta}_3-\dot{\theta}_1)}{s(\theta_2-\theta_3)}+\frac{s(\theta_3-\theta_1)c(\theta_2+\psi)\dot{\theta}_2}{s(\theta_2-\theta_3)}\right)\dot{\theta}_1\\ \ddot{Y}_e =& \, l_1\left(c\theta_1+\frac{s(\theta_3-\theta_1)c(\theta_2+\psi)}{s(\theta_2-\theta_3)}\right)\ddot{\theta}_1\\ &+l_1\left(-s\theta_1\dot{\theta}_1+\frac{c(\theta_2-\theta_3)s(\theta_3-\theta_1)c(\theta_2+\psi)(\dot{\theta}_2-\dot{\theta}_3)}{s^2(\theta_2-\theta_3)}+\frac{c(\theta_3-\theta_1)c(\theta_2+\psi)(\dot{\theta}_3-\dot{\theta}_1)}{s(\theta_2-\theta_3)}-\frac{s(\theta_3-\theta_1)s(\theta_2+\psi)\dot{\theta}_2}{s(\theta_2-\theta_3)}\right)\dot{\theta}_1\end{aligned}\right\}\tag{12.27}$$

To solve the inverse kinematic problem, that is, calculating the inputs of $\theta_1, \dot{\theta}_1, \ddot{\theta}_1$ based on the give outputs of (X_e, Y_e), $(\dot{X}_e, \dot{Y}_e)$, $(\ddot{X}_e, \ddot{Y}_e)$, Eq. (12.17) can be converted into,

$$K_1'\,c\theta_1 + K_2'\,s\theta_1 + K_3' = 0\tag{12.28}$$

where K_1', K_2', K_3' are determined by (X_e, Y_e) as,

$$\left.\begin{aligned}K_1' &= 2l_1X_e\\ K_2' &= 2l_1Y_e\\ K_3' &= -\left(l_1^2 + X_e^2 + Y_e^2\right)\end{aligned}\right\}\tag{12.29}$$

Equation (12.28) can be used to find the solution to the inverse kinematic problem as,

$$\theta_1 = \tan^{-1}\left(\frac{-K_2' \pm \sqrt{(K_2')^2 - (K_3')^2 + (K_1')^2}}{K_3' - K_1'}\right)\tag{12.30}$$

Equation (12.30) is further used to determine input velocity $\dot{\theta}_1$ based on the given output velocity $(\dot{X}_e, \dot{Y}_e)$ as

$$\dot{\theta}_1 = -\frac{\dot{X}_e}{\left[l_1\left(s\theta_1 + \dfrac{s(\theta_3-\theta_1)s(\theta_2+\psi)}{s(\theta_2-\theta_3)}\right)\right]} \quad\text{or}\quad \dot{\theta}_1 = \frac{\dot{Y}_e}{\left[l_1\left(c\theta_1 + \dfrac{s(\theta_3-\theta_1)c(\theta_2+\psi)}{s(\theta_2-\theta_3)}\right)\right]}\tag{12.31}$$

Note that $\dot{X}_e$ and $\dot{Y}_e$ are dependent and either of them can be used to calculate $\dot{\theta}_1$.

Similarly, the input angular acceleration $\ddot{\theta}_1$ subjected to given θ_1 and $\dot{\theta}_1$ can be obtained from that of the outputs $(\ddot{X}_e, \ddot{Y}_e)$ from the forward kinematic solution in Eq. (12.30).

Note that $\ddot{X}_e, \ddot{Y}_e$ are dependent and either of them can be used to calculate $\ddot{\theta}_1$.

The vector-loop method can also be extended to mode the geometric constraints in a 3D space. Here, a tripod-based kinematic machine is used to illustrate how the kinematic chains are constructed and modeled by the vector-loop method.

A tripod machine in Figure 12.34 consists of a base platform, an end-effector platform, and three kinematic chains (Bi et al. 2007; Bi and Wang 2008). Two platforms are connected by three symmetrical legs, and each leg is driven by a linear actuator. Moreover, a passive leg is also included in the middle to confine the translations of the end-effector platform. Figure 12.34a and b shows the typical structure and corresponding structural parameters, respectively. The base and end-effector platforms are denoted as $B_1B_2B_3$ and $E_1E_2E_3$. The base platform is fixed on the ground. The end-effector platform is used to mount a tool. A passive link is installed between the middle platform $M_1M_2M_3$ and the end-effector platform. Active leg D_iE_i is connected to the end-effector platform by a spherical joint at E_i, and to the slide of the active prismatic joint by a universal joint at D_i. The passive leg is fixed on the middle platform at one end and connected to the end-effector platform by a universal joint at the other end.

The tripod machine tool is symmetrical and described with the following parameters: angle α_i $(i = 1, 2, 3)$ between O_bB_i and $\boldsymbol{x}_b$, angle β_i $(i = 1, 2, 3)$ between O_eE_i and $\boldsymbol{x}_e$, the radius of the base platform l_b, the radius of the end-effector platform l_e, the direction of a guide-way γ, and the length of an active leg l_i.

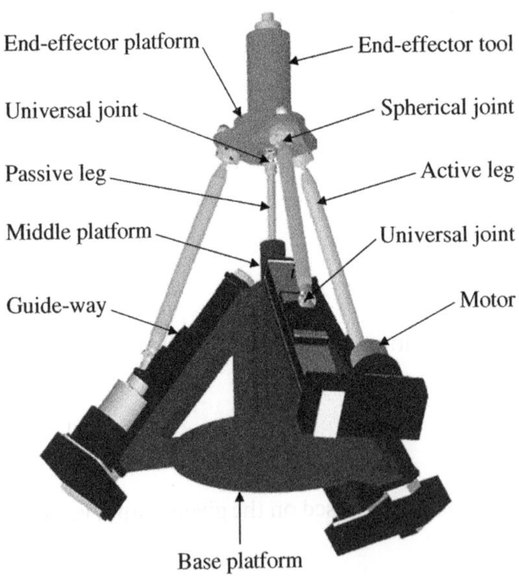

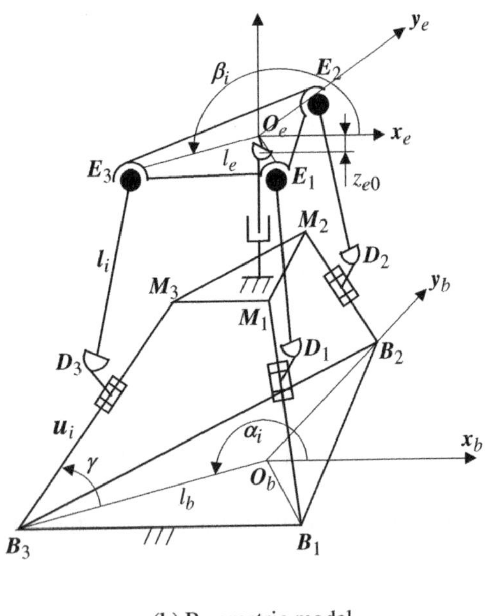

Figure 12.34 A parametric model of a tripod machine tool

12.4.3.1 Kinematic Parameters and Variables

The complete description of the motion of a rigid body requires three rotational parameters (θ_x, θ_y, θ_z) and three translational parameters (x_0, y_0, z_0). However, a tripod machine tool has a motion with 3-DOF. For the sample system, the end-effector motion can be denoted by (θ_x, θ_y, θ_z), where θ_x and θ_y are the rotations along $\boldsymbol{x}_e$ and $\boldsymbol{y}_e$, and z_e is the translation along $\boldsymbol{z}_e$, respectively. The x- and y-translations and z-rotation are eliminated ($x_e = y_e = 0$, $\theta_z = 0$) due to the adoption of a passive leg. The posture of the end-effector is represented by

$$T_e^b = \begin{bmatrix} \boldsymbol{R}_e & \boldsymbol{P}_e \\ 0 & 1 \end{bmatrix} = \begin{bmatrix} c\theta_y & 0 & s\theta_y & 0 \\ s\theta_x s\theta_y & c\theta_x & -s\theta_x c\theta_y & 0 \\ -c\theta_x s\theta_y & s\theta_x & c\theta_x c\theta_y & z_e \\ 0 & 0 & 0 & 1 \end{bmatrix} \tag{12.32}$$

where

c, s denote the cosine and sine functions, respectively,
T_e^b is the posture of the end-effector with respect to $\{\boldsymbol{O}_b - \boldsymbol{x}_b\boldsymbol{y}_b\boldsymbol{z}_b\}$,
$\boldsymbol{R}_e$ is 3×3 orientational matrix of the end-effector, and
$\boldsymbol{P}_e$ is the central position $\boldsymbol{O}_e$ of the end-effector.

12.4.3.2 Inverse Kinematics

Inverse kinematics is to find the joint motions when the end-effector motion (θ_x, θ_y, z_e) is known. The motion of active prismatic joint i is denoted by u_i. The tripod machine includes three independent kinematic loops. The kinematic equation in each loop can be used to find one joint parameter as shown in Figure 12.35. The solution to the inverse kinematic problem is derived based on the condition that the length of support bar i is fixed.

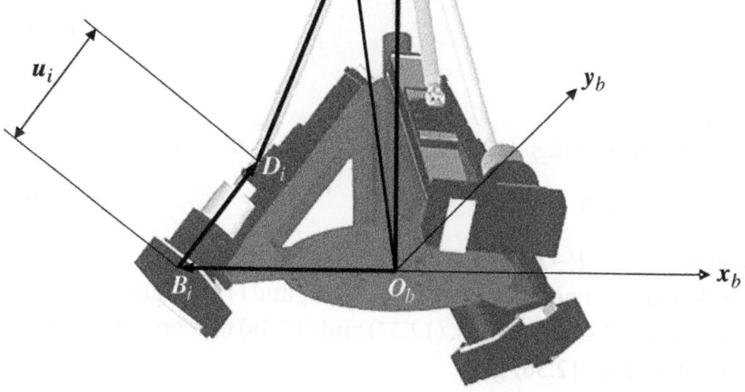

Figure 12.35 Kinematic constraints in tripod machine tool

First, the coordinates of $\boldsymbol{E}_i$ on the end-effector platform are determined by

$$\boldsymbol{P}^b_{e_i} = \begin{bmatrix} x^b_{e_i} \\ y^b_{e_i} \\ z^b_{e_i} \end{bmatrix} = \begin{bmatrix} l_e c\beta_i c\theta_y \\ l_e c\beta_i s\theta_x s\theta_y + l_e s\beta_i c\theta_x \\ -l_e c\beta_i c\theta_x s\theta_y + l_e s\beta_i s\theta_x + z_e \end{bmatrix} \tag{12.33}$$

where

$\boldsymbol{P}^b_{e_i}$ are the coordinates of $\boldsymbol{E}_i$ with respect to $\{\boldsymbol{O}_b - \boldsymbol{x}_b\boldsymbol{y}_b\boldsymbol{z}_b\}$.

Second, the condition that support bar i has a fixed length is applied, that is,

$$|\boldsymbol{O}_b\boldsymbol{E}_i - \boldsymbol{O}_b\boldsymbol{B}_i - \boldsymbol{B}_i\boldsymbol{D}_i| = |\boldsymbol{D}_i\boldsymbol{E}_i| \quad (i = 1, 2, 3) \tag{12.34}$$

Finally, the motion of prismatic joint i from Eq. (12.34) as

$$u_i = \frac{-k_b \pm \sqrt{{k_b}^2 - 4k_c}}{2} \tag{12.35}$$

where

$$k_b = 2\left(c\gamma\left(x^b_{e_i} c\alpha_i + y^b_{e_i} s\alpha_i - l_b\right) - z^b_{e_i} s\gamma\right)$$

$$k_c = \left(x^b_{e_i}\right)^2 + \left(x^b_{e_i}\right)^2 + \left(x^b_{e_i}\right)^2 + l_b^2 - l_i^2 - 2l_b\left(x^b_{e_i} c\alpha_i + y^b_{e_i} s\alpha_i\right)$$

Note that a solution of u_i from Eq. (12.35) should be verified to ensure that none of active or passive joints exceeds its motion limit.

12.4.3.3 Forward Kinematics

Forword kinematics is to find the end-effector motion $(\theta_x, \theta_y, \theta_z)$ when active joint motions u_i $(i = 1, 2, 3)$ are known. The solution to forward kinematic problem is also derived from Eq. (12.34). Equation (12.34) can be rewritten in a way to represent z_e and θ_y by θ_x as,

$$z_e^2 + (A_i s\theta_y + B_i)z_e + (C_i c\theta_y + D_i s\theta_y + E_i) = 0 \quad (i = 1,\ 2,\ 3) \tag{12.36}$$

where coefficients $A_i \sim E_i$ relate to θ_x. Eq. (12.36) yields,

$$c\theta_y = -\frac{Fz_e^2 + Gz_e + H}{Kz_e + L} \tag{12.37}$$

$$s\theta_y = -\frac{Iz_e + J}{Kz_e + L} \tag{12.38}$$

where coefficients $F \sim L$ relate to θ_x.

Besides, there is $c^2\theta_y + s^2\theta_y = 1$; substituting Eqs. (12.37) and (12.38) into this expression gives,

$$M_4 z_e^4 + M_3 z_e^3 + M_2 z_e^2 + M_1 z_e + M_0 = 0 \tag{12.39}$$

where, $M_0 \sim M_4$ relates to coefficients $A_i \sim E_i$, $F \sim L$, and θ_x.

Note that Eq. (12.36) has three independent equations and Eqs. (12.37) and (12.38) have used two of them. The remainder can be derived by substituting Eqs. (12.37) and (12.38) into one of the equations in Eq. (12.36). Substituting $i = 1$ in Eq. (12.36) gets

$$N_3 z_e^3 + N_2 z_e^2 + N_1 z_e + N_0 = 0 \tag{12.40}$$

where $N_0 \sim N_3$ relates to coefficients $A_i \sim E_i$, $F \sim L$, and θ_x.

To assure a common solution of z_e for both Eqs. (12.39) and (12.40), one must have,

$$\begin{vmatrix} M_4 & M_3 & M_2 & M_1 & M_0 & 0 & 0 \\ 0 & M_4 & M_3 & M_2 & M_1 & M_0 & 0 \\ 0 & 0 & M_4 & M_3 & M_2 & M_1 & M_0 \\ N_3 & N_2 & N_1 & N_0 & 0 & 0 & 0 \\ 0 & N_3 & N_2 & N_1 & N_0 & 0 & 0 \\ 0 & 0 & N_3 & N_2 & N_1 & N_0 & 0 \\ 0 & 0 & 0 & N_3 & N_2 & N_1 & N_0 \end{vmatrix} = 0 \tag{12.41}$$

Therefore, solving a forward kinematic problem involves in two steps: (1) calculate θ_x from Eq. (12.41), (2) calculate z_e from Eq. (12.40), and (3) calculate θ_y from Eqs. (12.37) and (12.38).

12.4.4 Denavit–Hartenberg (D–H) Notation

The vector-loop method is effective to derive kinematic models for planar mechanisms or spatial mechanisms with multiple kinematic loops. However, it is not efficient for a spatial mechanism with a serial or hybrid configuration. Since that the vector-loop method represents all the kinematic constraints in terms of the relations of vector magnitudes projected on axes.

The *D–H* notation is popular to model spatial mechanisms. The D–H notation was proposed by Denavit and Hartenberg (1955) to represent the spatial relations of motion axes in space. As shown in Figure 12.36, the relations of one motion axis – *i* with other axes are described by four

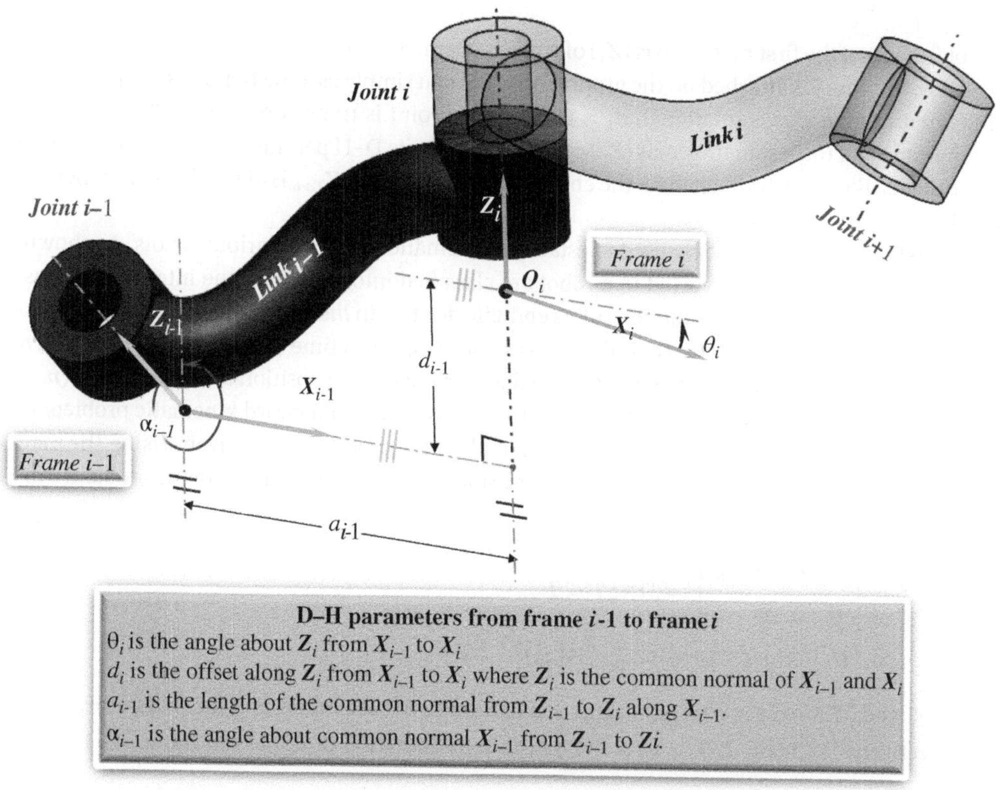

Figure 12.36 D–H convention from frame $i - 1$ to frame i

D–H parameters ($\theta_i, d_i, \alpha_{i-1}, a_{i-1}$). Each motion axis i is associated with a frame i which is attached on link i-1.

In the D–H notion, the frames are established and attached to the joints where the links are connected. Taking an example of link i in Figure 12.36, it is connected to link $i-1$ by joint $i-1$ and to link $i+1$ by joint i, and link i is associated with frame $i-1$ and frame i. To define frame i, the motion axes with link $i-1$ and link $i+1$ are defined as $\boldsymbol{Z}_{i-1}$ and $\boldsymbol{Z}_i$, respectively, and $\boldsymbol{X}_i$ is defined as the common normal of $\boldsymbol{Z}_{i-1}$ and $\boldsymbol{Z}_i$, $\boldsymbol{O}_i$ is at the intersection of $\boldsymbol{X}_i$ and $\boldsymbol{Z}_i$.

The D–H parameters are defined to describe the spatial relation of frame $i-1$ and frame i, and the physical meanings of these parameters. Once the set of D–H parameters ($\theta_i, d_i, \alpha_{i-1}, a_{i-1}$) of motion axis i is found, frame i can be represented with respect to frame $i-1$ as,

$$
\begin{aligned}
{}^{i-1}_{i}\boldsymbol{A} &= \boldsymbol{T}_x(\alpha_{i-1}) \cdot \boldsymbol{T}_x(a_{i-1}) \cdot \boldsymbol{T}_z(d_i) \cdot \boldsymbol{R}_z(\theta_i) \\
&= \begin{bmatrix} 1 & 0 & 0 & 0 \\ 0 & c\alpha_{i-1} & -s\alpha_{i-1} & 0 \\ 0 & s\alpha_{i-1} & c\alpha_{i-1} & 0 \\ 0 & 0 & 0 & 1 \end{bmatrix} \cdot \begin{bmatrix} 1 & 0 & 0 & a_{i-1} \\ 0 & 1 & 0 & 0 \\ 0 & 0 & 1 & 0 \\ 0 & 0 & 0 & 1 \end{bmatrix} \cdot \begin{bmatrix} 1 & 0 & 0 & 0 \\ 0 & 1 & 0 & 0 \\ 0 & 0 & 1 & d_i \\ 0 & 0 & 0 & 1 \end{bmatrix} \cdot \begin{bmatrix} c\theta_i & -s\theta_i & 0 & 0 \\ s\theta_i & c\theta_i & 0 & 0 \\ 0 & 0 & 1 & 0 \\ 0 & 0 & 0 & 1 \end{bmatrix} \\
&= \begin{bmatrix} c\theta_i & -s\theta_i & 0 & a_i \\ s\theta_i c\alpha_{i-1} & c\theta_i c\alpha_{i-1} & -s\alpha_i & -d_i s\alpha_{i-1} \\ s\theta_i s\alpha_{i-1} & c\theta_i s\alpha_{i-1} & c\alpha_i & d_i c\alpha_{i-1} \\ 0 & 0 & 0 & 1 \end{bmatrix}
\end{aligned} \tag{12.42}
$$

For frame 1 of the first motion axis ($\boldsymbol{Z}_1$) of a mechanism, its D–H parameters are affected by frame 0, that is, the base CS attached on the ground link. We can simply assume that $\boldsymbol{Z}_0$ is aligned with $\boldsymbol{Z}_1$; if the first motion axis is rotational, $d_1 = 0$; if the first joint is translational, $\theta_1 = 0$.

For frame n of the last motion axis ($\boldsymbol{Z}_n$) of a mechanism, its D–H parameters are affected by frame $n+1$,that is, the tool CS attached on the end-effector link. We set $\boldsymbol{Z}_{Tool}$ is aligned with $\boldsymbol{Z}_n$; therefore, one has $a_n = 0$ and $\alpha_n = 0$.

The D–H convention is widely used to establish kinematic models of various robots. As shown in Figure 12.37, the motions involved in a robot are classified into two types: the internal motions in *the joint space* and the external motions of its end-effector tool in *the task space*. Internal motions are described by joint variables and their derivatives with respect to time ($q_i, \dot{q}_i, \ddot{q}_i$ and ($i = 1, 2, \ldots, n$)), and external motions are described by the variables for the tooling position and orientation ($\boldsymbol{n}, \boldsymbol{o}, \boldsymbol{a}$ and $\boldsymbol{p}$) as well as their derivatives with respect to time. Note that a forward kinematic problem is to find the position and orientation of tooling when all of joint variables are given. Using the matrix expression Eq. (12.42) of frame i in frame $i-1$, the solution to the forward kinematic problem can be readily defined as,

$$
\begin{aligned}
{}^{B}_{T}\boldsymbol{T} &= \begin{bmatrix} \boldsymbol{n} & \boldsymbol{o} & \boldsymbol{a} & \boldsymbol{p} \\ 0 & 0 & 0 & 1 \end{bmatrix} = {}^{B}_{1}\boldsymbol{T} \cdot {}^{1}_{n}\boldsymbol{T} \cdot {}^{n}_{T}\boldsymbol{T} \\
&= {}^{B}_{0}\boldsymbol{T} \cdot \left[{}^{0}_{1}\boldsymbol{A}(q_1) \cdot {}^{1}_{2}\boldsymbol{A}(q_2) \ldots {}^{n-1}_{n}\boldsymbol{A}(q_n)\right] \cdot {}^{n}_{T}\boldsymbol{T}
\end{aligned} \tag{12.43}
$$

where ${}^{B}_{0}\boldsymbol{T}$ and ${}^{n}_{T}\boldsymbol{T}$ are the matrix representation of frame 0 in base frame, and the tool frame in frame n, respectively. ${}^{i-1}_{i}\boldsymbol{A}$ ($i = 2, 3, \ldots, n$) is the matrix representation of frame i in frame $i-1$.

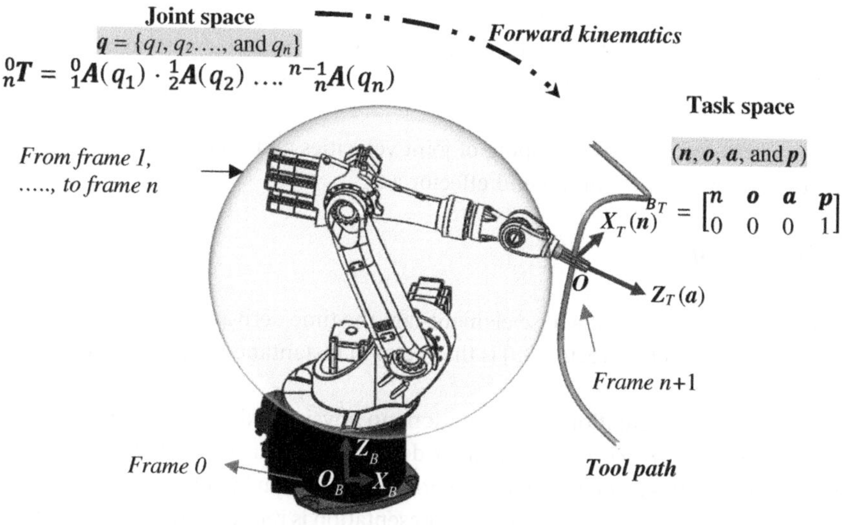

Figure 12.37 D–H notation in modeling of forward kinematic problem.

12.4.5 Jacobian Matrix for Velocity Relations

Let the position and orientation of the end-effector tool be represented by a vector of design variable $\{\boldsymbol{x}\}$, and the set of design variables in the joint space as $\{\boldsymbol{q}\}$, the solution to the forward kinematic problem in Eq. (12.43) can be transformed as a mapping function of,

$$\{x\} = \begin{Bmatrix} x_1 \\ x_2 \\ \vdots \\ x_m \end{Bmatrix} = \begin{Bmatrix} f_1(\boldsymbol{q}) \\ f_2(\boldsymbol{q}) \\ \vdots \\ f_n(\boldsymbol{q}) \end{Bmatrix} = \{\boldsymbol{f}(\boldsymbol{q})\} \tag{12.44}$$

where m and n are the number of variables to describe the motions in task space and joint space, respectively.

Taking the derivation of Eq. (12.44) with respect to time yields,

$$\{\delta \boldsymbol{x}\} = [\boldsymbol{J}]\{\delta \boldsymbol{q}\} = \begin{bmatrix} \dfrac{\partial f_1(\boldsymbol{q})}{\partial q_1} & \cdots & \dfrac{\partial f_1(\boldsymbol{q})}{\partial q_n} \\ \vdots & \ddots & \vdots \\ \dfrac{\partial f_m(\boldsymbol{q})}{\partial q_1} & \cdots & \dfrac{\partial f_m(\boldsymbol{q})}{\partial q_n} \end{bmatrix} \{\delta \boldsymbol{q}\} \quad \text{or} \quad \{\dot{\boldsymbol{x}}\} = [\boldsymbol{J}]\{\dot{\boldsymbol{q}}\} \tag{12.45}$$

where $[\boldsymbol{J}]$ is the Jacobian matrix and is used to map the velocities in a task space to the velocities in a joint space.

Jacobian matrix is not unique since it depends on the representation of the position and orientation of the end-effector tool in the task space. For an example, if cosine functions are used to represent the orientations and the coordinates of tooling in the Cartesian space, the Jacobian matrix has the size of 12×6 for a robot with 6-DOF. If the orientations are represented by the Euler angles, the Jacobian matrix has the size of 6×6 for the same robot. Note that the positional and orientational representation does not affect the kinematic behaviors of a mechanism. It is desirable to have a kinematic model, which is independent of the representation; correspondingly, the Jacobian matrix based on such a kinematic model is unique. It is also called as the basic Jacobian matrix.

The basic Jacobian matrix describes the mapping of joint velocities and uniquely defined linear and angular velocities at a given point on the end-effector as,

$$\begin{Bmatrix} v \\ \omega \end{Bmatrix}_{6\times 1} = [J_0(\boldsymbol{q})]_{6\times n}\dot{\boldsymbol{q}}_{n\times 1} \tag{12.46}$$

where $\{\boldsymbol{v}\}$ is the vector of linear velocities, whose elements are the time derivatives of the Cartesian coordinates of the end-effector position vector. $\{\omega\}$ is the vector of instantaneous quantity related to angular velocities.

It is worth to note that for the orientation representation, no physical axis is associated with the variables such as Euler angles (α, β, γ). Therefore, their derivations against time are not angular velocities along axes. The quantities in $\{\omega\}$ do not have a primitive function but instantaneous quantities. However, the time derivation of any orientation representation is related to the angular velocity. One has to map $\{\boldsymbol{v}\}$ and $\{\omega\}$ into the position and orientation velocities as,

$$\left.\begin{aligned} \dot{\boldsymbol{x}}_p &= \boldsymbol{E}_p(\boldsymbol{x}_p)\boldsymbol{v} \\ \dot{\boldsymbol{x}}_r &= \boldsymbol{E}_r(\boldsymbol{x}_r)\omega \end{aligned}\right\} \tag{12.47}$$

where $\dot{\boldsymbol{x}}_p$ and $\dot{\boldsymbol{x}}_r$ the position and orientation derivations of the end-effector, respectively. The matrics $\boldsymbol{E}_p$ and $\boldsymbol{E}_r$ are dependent on the position or orientation representation. Using $\boldsymbol{E}_p$ and $\boldsymbol{E}_r$ allows converting the basic Jacobian matrix $[\boldsymbol{J}_0]$ into a Jacobian matrix $[\boldsymbol{J}]$ that is associated with specific position and orientation representation.

Figure 12.38 shows a serial kinematic chain. The velocity of a motion axis with respect to the local frame of the proceeding link depends on the joint type at the connection as,

$$\left.\begin{aligned} \boldsymbol{v}_i &= \boldsymbol{z}_i\dot{q}_i \quad \text{joint } j \text{ is prismatic} \\ \omega_i &= \boldsymbol{z}_i\dot{q}_i \quad \text{joint } i \text{ is revolute} \end{aligned}\right\} \tag{12.48}$$

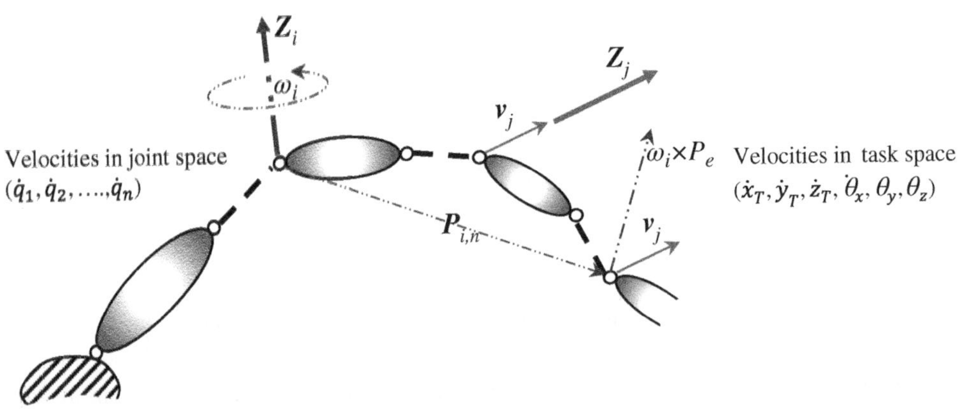

Figure 12.38 Jacobian matrix to modeling relations of velocities

Each motion axis in Figure 12.38 contributes to the motion of the end-effector; while a prismatic joint i only generates a linear velocity $\boldsymbol{v}_i$ that is transferred down to the end-effector. A revolute joint adds both linear and angular velocities to the end-effector. Moreover, the linear velocity depends on the spatial relation of the end-effector and the motion axis, which is calculated by a cross product of $\boldsymbol{\omega}_i$ with the vector toward the reference point of the end-effector. The angular velocity $\boldsymbol{\omega}_i$ is directly transferred down the chain to the end-effector. Therefore, the linear and angular velocities of the end-effector tool are summed as,

$$\left.\begin{aligned}\{\boldsymbol{v}_e\} &= \sum_{i=1}^{n}\left[\epsilon_i \boldsymbol{v}_i + \bar{\epsilon}_i\left(\boldsymbol{\omega}_i \times \boldsymbol{p}_{i,n}\right)\right] \\ \{\boldsymbol{\omega}_e\} &= \sum_{i=1}^{n}\bar{\epsilon}_i \boldsymbol{\omega}_i\end{aligned}\right\} \tag{12.49}$$

Substituting Eq. (12.48) into (12.49) gets,

$$\begin{Bmatrix}\boldsymbol{v}_e \\ \boldsymbol{\omega}_e\end{Bmatrix} = \begin{Bmatrix}\boldsymbol{J}_v \\ \boldsymbol{J}_\omega\end{Bmatrix}\{\dot{q}\} = \begin{bmatrix}\epsilon_1 \boldsymbol{v}_1 + \bar{\epsilon}_1\left(\boldsymbol{\omega}_1 \times \boldsymbol{p}_{1,n}\right) & \cdots & \epsilon_n \boldsymbol{v}_n + \bar{\epsilon}_n\left(\boldsymbol{\omega}_n \times \boldsymbol{p}_{n,n}\right) \\ \bar{\epsilon}_1 \boldsymbol{Z}_1 & \cdots & \bar{\epsilon}_n \boldsymbol{Z}_n\end{bmatrix}\begin{Bmatrix}q_1 \\ q_2 \\ \vdots \\ q_n\end{Bmatrix} \tag{12.50}$$

where $\{\boldsymbol{J}_v\}$ and $\{\boldsymbol{J}_\omega\}$ are the components of the Jacobian matrix for linear and angular velocities.

Note that the way to calculate $\{\boldsymbol{J}_v\}$ is not straightforward since all of relative vectors $\boldsymbol{p}_{i,n}$ ($i = 0,1,2, \ldots, n$) must be determined, respectively. An alternative to calculate $\{\boldsymbol{J}_v\}$ is to take a direct differentiation of the Cartesian coordinates of the reference point of the end-effector as,

$$\boldsymbol{v}_e = \begin{Bmatrix}\dot{x}_e \\ \dot{y}_e \\ \dot{z}_e\end{Bmatrix} = \dot{\boldsymbol{x}}_e = \frac{\partial \boldsymbol{p}_{0,n}}{\partial q_1}\dot{q}_1 + \frac{\partial \boldsymbol{p}_{0,n}}{\partial q_2}\dot{q}_2 + \cdots + \frac{\partial \boldsymbol{p}_{0,n}}{\partial q_n}\dot{q}_n \tag{12.51}$$

where $\boldsymbol{p}_{0,n}$ is the reference point on the end-effector in frame 0.

Therefore, we have,

$$[J] = \begin{bmatrix}\boldsymbol{J}_v \\ \boldsymbol{J}_\omega\end{bmatrix} = \begin{bmatrix}\dfrac{\partial \boldsymbol{p}_{0,n}}{\partial q_1} & \dfrac{\partial \boldsymbol{p}_{0,n}}{\partial q_2} & \cdots & \dfrac{\partial \boldsymbol{p}_{0,n}}{\partial q_n} \\ \bar{\epsilon}_1 {}_1^0\boldsymbol{Z}_1 & \bar{\epsilon}_2 {}_2^0\boldsymbol{Z}_2 & \cdots & \bar{\epsilon}_n {}_n^0\boldsymbol{Z}_n\end{bmatrix} \tag{12.52}$$

where $\bar{\epsilon}_i$ = "0" or "1" (i = 1, 2,..., n) where the motion joint i is prismatic or revolute, respectively. ${}_i^0\boldsymbol{Z}_i$ (i = 1, 2, ..., n) is the direction of the motion axis i that is expressed in frame 0. $[J]$ is the basic Jacobian matrix.

Example 12.3 Find the forward kinematic solution for the 2-DOF robot in Figure 12.39. The robot consists of the ground link, two other links in the body, and two joints to connect these links sequentially. In addition, the tool tip is attached to the last link directly.

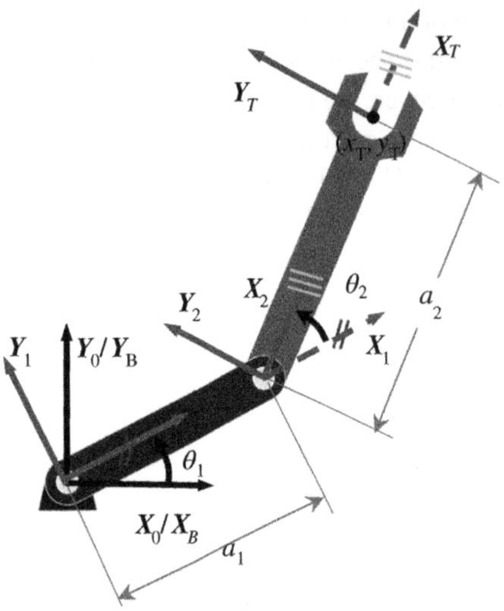

Figure 12.39 D–H convention for forward kinematic problem

Solution

The D–H parameters for internal two links are determined as follows:

Link	θ_i	d_i	α_{i-1}	a_{i-1}
1	θ_1	0	0°	0
2	θ_2	0	0°	a_1
Tool	0	0	0°	a_2

Accordingly, the D–H matrices for these two links and the tool are

$$ {}_1^0\boldsymbol{A}(\theta_1) = \begin{bmatrix} c\theta_1 & -s\theta_1 & 0 & 0 \\ s\theta_1 & c\theta_1 & 0 & 0 \\ 0 & 0 & 1 & 0 \\ 0 & 0 & 0 & 1 \end{bmatrix} \tag{12.53} $$

$$ {}_2^1\boldsymbol{A}(\theta_1) = \begin{bmatrix} c\theta_2 & -s\theta_2 & 0 & a_1 \\ s\theta_2 & c\theta_2 & 0 & 0 \\ 0 & 0 & 1 & 0 \\ 0 & 0 & 0 & 1 \end{bmatrix} \tag{12.54} $$

$$ {}_T^2\boldsymbol{A}(\theta_1) = \begin{bmatrix} 1 & 0 & 0 & a_2 \\ 0 & 1 & 0 & 0 \\ 0 & 0 & 1 & 0 \\ 0 & 0 & 0 & 1 \end{bmatrix} \tag{12.55} $$

In addition, we assume that the base frame is aligned with frame 0, this gives,

$$ {}_0^B\boldsymbol{T} = I_{4\times 4} $$

where $I_{4\times 4}$ is a the unit matrix with the size of 4 by 4.

Further, using Eq. (12.43) yields the solution to the forward kinematic problem as,

$$
{}^{B}_{T}\boldsymbol{T} = \begin{bmatrix} \boldsymbol{n} & \boldsymbol{o} & \boldsymbol{a} & \boldsymbol{p} \\ 0 & 0 & 0 & 1 \end{bmatrix} = {}^{B}_{0}\boldsymbol{T} \cdot \left[{}^{0}_{1}\boldsymbol{A}(q_1) \cdot {}^{1}_{2}\boldsymbol{A}(q_2) \cdot {}^{2}_{T}\boldsymbol{A}(0)\right]
$$

$$
= \begin{bmatrix} c\theta_1 & -s\theta_1 & 0 & 0 \\ s\theta_1 & c\theta_1 & 0 & 0 \\ 0 & 0 & 1 & 0 \\ 0 & 0 & 0 & 1 \end{bmatrix} \cdot \begin{bmatrix} c\theta_2 & -s\theta_2 & 0 & a_1 \\ s\theta_2 & c\theta_2 & 0 & 0 \\ 0 & 0 & 1 & 0 \\ 0 & 0 & 0 & 1 \end{bmatrix} \cdot \begin{bmatrix} 1 & 0 & 0 & a_2 \\ 0 & 1 & 0 & 0 \\ 0 & 0 & 1 & 0 \\ 0 & 0 & 0 & 1 \end{bmatrix}
$$

$$
= \begin{bmatrix} c(\theta_1+\theta_2) & -s(\theta_1+\theta_2) & 0 & a_1c\theta_1 + a_2c(\theta_1+\theta_2) \\ s(\theta_1+\theta_2) & c(\theta_1+\theta_2) & 0 & a_1s\theta_1 + a_2s(\theta_1+\theta_2) \\ 0 & 0 & 1 & 0 \\ 0 & 0 & 0 & 1 \end{bmatrix} \tag{12.56}
$$

Equation (12.56) can be written specifically as the position and orientation of the tool as,

$$
\boldsymbol{n} = \begin{Bmatrix} c(\theta_1+\theta_2) \\ s(\theta_1+\theta_2) \\ 0 \end{Bmatrix}, \quad \boldsymbol{o} = \begin{Bmatrix} -s(\theta_1+\theta_2) \\ c(\theta_1+\theta_2) \\ 0 \end{Bmatrix}, \quad \boldsymbol{a} = \begin{Bmatrix} 0 \\ 0 \\ 0 \end{Bmatrix}, \quad \boldsymbol{p} = \begin{Bmatrix} a_1c\theta_1 + a_2c(\theta_1+\theta_2) \\ a_1s\theta_1 + a_2s(\theta_1+\theta_2) \\ 0 \end{Bmatrix} \tag{12.57}
$$

After the displacement of $\boldsymbol{p}$ is calculated from the angular displacements of two joints in Eq. (12.57), the velocity and acceleration of $\boldsymbol{p}$ can be obtained by taking the derivatives of Eq. (12.57) with respect to time as,

$$
\begin{aligned}
\dot{\boldsymbol{p}} &= \begin{Bmatrix} \dot{x}_T \\ \dot{y}_T \end{Bmatrix} = [\boldsymbol{J}_1]\begin{Bmatrix} \dot{\theta}_1 \\ \dot{\theta}_2 \end{Bmatrix} \\
\ddot{\boldsymbol{p}} &= \begin{Bmatrix} \ddot{x}_T \\ \ddot{y}_T \end{Bmatrix} = [\boldsymbol{J}_1]\begin{Bmatrix} \ddot{\theta}_1 \\ \ddot{\theta}_2 \end{Bmatrix} + [\boldsymbol{J}_2]\begin{Bmatrix} \dot{\theta}_1 \\ \dot{\theta}_2 \end{Bmatrix}
\end{aligned} \tag{12.58}
$$

where

$$
[\boldsymbol{J}_1] = \begin{bmatrix} -a_1s\theta_1 - a_2s(\theta_1+\theta_2) & -a_2s(\theta_1+\theta_2) \\ a_1c\theta_1 + a_2c(\theta_1+\theta_2) & a_2c(\theta_1+\theta_2) \end{bmatrix}
$$

$$
[\boldsymbol{J}_2] = \begin{bmatrix} -a_1\dot{\theta}_1c\theta_1 - a_2(\dot{\theta}_1+\dot{\theta}_2)c(\theta_1+\theta_2) & -a_2(\dot{\theta}_1+\dot{\theta}_2)c(\theta_1+\theta_2) \\ -a_1\dot{\theta}_1s\theta_1 - a_2(\dot{\theta}_1+\dot{\theta}_2)s(\theta_1+\theta_2) & -a_2(\dot{\theta}_1+\dot{\theta}_2)s(\theta_1+\theta_2) \end{bmatrix}
$$

The solution of inverse kinematic problem to get θ_1 and θ_2 from x_T and y_T in Eq. (12.57) is as follows:

$$\left.\begin{aligned} \theta_1 &= \tan^{-1}\left(\frac{-K_2 \pm \sqrt{K_2^2 - K_3^2 + K_1^2}}{K_3 - K_1}\right) \\ \theta_2 &= \tan^{-1}\left(\frac{y_T - a_1 s\theta_1}{x_T - a_1 c\theta_1}\right) - \theta_1 \end{aligned}\right\} \tag{12.59}$$

where K_1, K_2, and K_3 are dependent intermediate parameters as the following:

$$\left.\begin{aligned} K_1 &= 2a_1 x_T \\ K_2 &= 2a_1 y_T \\ K_3 &= a_2^2 - \left(a_1^2 + x_T^2 + y_T^2\right) \end{aligned}\right\} \tag{12.60}$$

The solution of inverse kinematic problem for velocity and acceleration relation is found from Eq. (12.57) as,

$$\begin{aligned} \dot{\boldsymbol{\theta}} &= \begin{Bmatrix} \dot{\theta}_1 \\ \dot{\theta}_2 \end{Bmatrix} = [\boldsymbol{J}_1]^{-1} \begin{Bmatrix} \dot{x}_T \\ \dot{y}_T \end{Bmatrix} \\ \ddot{\boldsymbol{\theta}} &= \begin{Bmatrix} \ddot{\theta}_1 \\ \ddot{\theta}_2 \end{Bmatrix} = [\boldsymbol{J}_1]^{-1} \begin{Bmatrix} \ddot{x}_T \\ \ddot{y}_T \end{Bmatrix} - [\boldsymbol{J}_1]^{-1}[\boldsymbol{J}_2] \begin{Bmatrix} \dot{\theta}_1 \\ \dot{\theta}_2 \end{Bmatrix} \end{aligned} \tag{12.61}$$

where $[\bullet]^{-1}$ in the option for an inverse of matrix $[\bullet]$.

Example 12.4 Figure 12.40 shows the structure of a PUMA 560, find (i) the solution of the forward kinematic problem and (ii) the solution of the inverse kinematic problem.

Solution

The D–H parameters of the Puma 560 can be determined as follows.

Link	θ_i	d_i	α_{i-1}	a_{i-1}
1	θ_1	0	0°	0
2	θ_2	0	−90°	0
3	θ_3	d_3	0°	a_2
4	θ_4	d_4	−90°	a_3
5	θ_5	0	90°	0
6	θ_6	0	−90°	0

Using Eq. (12.43) gets the forward kinematic solution as,

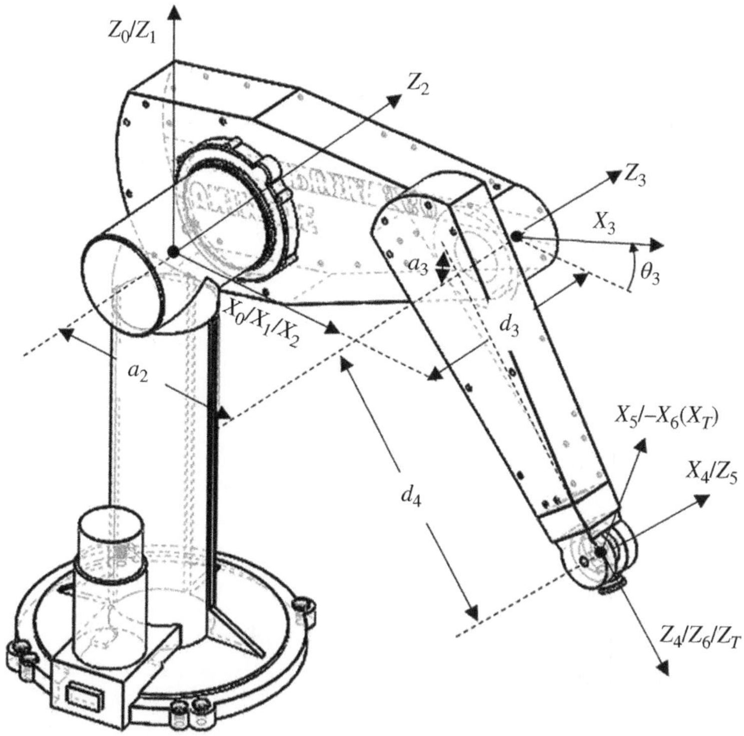

Figure 12.40 Structure of Puma 560

$$
{}^{0}_{1}\boldsymbol{A} = \begin{bmatrix} c\theta_1 & -s\theta_1 & 0 & 0 \\ s\theta_1 & c\theta_1 & 0 & 0 \\ 0 & 0 & 1 & 0 \\ 0 & 0 & 0 & 1 \end{bmatrix} \qquad {}^{1}_{2}\boldsymbol{A} = \begin{bmatrix} c\theta_2 & -s\theta_2 & 0 & 0 \\ 0 & 0 & 1 & 0 \\ -s\theta_2 & -c\theta_2 & 0 & 0 \\ 0 & 0 & 0 & 1 \end{bmatrix}
$$

$$
{}^{2}_{3}\boldsymbol{A} = \begin{bmatrix} c\theta_3 & -s\theta_3 & 0 & a_2 \\ s\theta_3 & c\theta_3 & 0 & 0 \\ 0 & 0 & 1 & d_3 \\ 0 & 0 & 0 & 1 \end{bmatrix} \qquad {}^{3}_{4}\boldsymbol{A} = \begin{bmatrix} c\theta_4 & -s\theta_4 & 0 & a_3 \\ 0 & 0 & 1 & d_4 \\ -s\theta_4 & -c\theta_4 & 0 & 0 \\ 0 & 0 & 0 & 1 \end{bmatrix}
$$

$$
{}^{4}_{5}\boldsymbol{A} = \begin{bmatrix} c\theta_5 & -s\theta_5 & 0 & 0 \\ 0 & 0 & -1 & 0 \\ s\theta_5 & c\theta_5 & 0 & 0 \\ 0 & 0 & 0 & 1 \end{bmatrix} \qquad {}^{5}_{6}\boldsymbol{A} = \begin{bmatrix} c\theta_6 & -s\theta_6 & 0 & 0 \\ 0 & 0 & 1 & 0 \\ -s\theta_6 & -c\theta_6 & 0 & 0 \\ 0 & 0 & 0 & 1 \end{bmatrix}
$$

$$
{}^{0}_{1}\boldsymbol{T} = \begin{bmatrix} {}^{0}_{1}\boldsymbol{n} & {}^{0}_{1}\boldsymbol{o} & {}^{0}_{1}\boldsymbol{a} & {}^{0}_{1}\boldsymbol{p} \\ 0 & 0 & 0 & 1 \end{bmatrix} = \begin{bmatrix} {}^{0}_{1}n_x & {}^{0}_{1}o_x & {}^{0}_{1}a_x & {}^{0}_{1}p_x \\ {}^{0}_{1}n_y & {}^{0}_{1}o_y & {}^{0}_{1}a_y & {}^{0}_{1}p_y \\ {}^{0}_{1}n_z & {}^{0}_{1}o_z & {}^{0}_{1}a_z & {}^{0}_{1}p_z \\ 0 & 0 & 0 & 1 \end{bmatrix} = {}^{0}_{1}\boldsymbol{A}
$$

$$
= \begin{bmatrix} c\theta_1 & -s\theta_1 & 0 & 0 \\ s\theta_1 & c\theta_1 & 0 & 0 \\ 0 & 0 & 1 & 0 \\ 0 & 0 & 0 & 1 \end{bmatrix} \tag{12.62}
$$

$$
{}^{0}_{2}\boldsymbol{T} = \begin{bmatrix} {}^{0}_{2}\boldsymbol{n} & {}^{0}_{2}\boldsymbol{o} & {}^{0}_{2}\boldsymbol{a} & {}^{0}_{2}\boldsymbol{p} \\ 0 & 0 & 0 & 1 \end{bmatrix} = \begin{bmatrix} {}^{0}_{2}n_x & {}^{0}_{2}o_x & {}^{0}_{2}a_x & {}^{0}_{2}p_x \\ {}^{0}_{2}n_y & {}^{0}_{2}o_y & {}^{0}_{2}a_y & {}^{0}_{2}p_y \\ {}^{0}_{2}n_z & {}^{0}_{2}o_z & {}^{0}_{2}a_z & {}^{0}_{2}p_z \\ 0 & 0 & 0 & 1 \end{bmatrix} = {}^{0}_{1}\boldsymbol{T} \cdot {}^{1}_{2}\boldsymbol{A} \tag{12.63}
$$

$$
= \begin{bmatrix} c\theta_1 c\theta_2 & -c\theta_1 s\theta_2 & -s\theta_1 & 0 \\ s\theta_1 c\theta_2 & -s\theta_1 s\theta_2 & c\theta_1 & 0 \\ -s\theta_2 & -c\theta_2 & 0 & 0 \\ 0 & 0 & 0 & 1 \end{bmatrix}
$$

$$
{}^{0}_{3}\boldsymbol{T} = \begin{bmatrix} {}^{0}_{3}\boldsymbol{n} & {}^{0}_{3}\boldsymbol{o} & {}^{0}_{3}\boldsymbol{a} & {}^{0}_{3}\boldsymbol{p} \\ 0 & 0 & 0 & 1 \end{bmatrix} = \begin{bmatrix} {}^{0}_{3}n_x & {}^{0}_{3}o_x & {}^{0}_{3}a_x & {}^{0}_{3}p_x \\ {}^{0}_{3}n_y & {}^{0}_{3}o_y & {}^{0}_{3}a_y & {}^{0}_{3}p_y \\ {}^{0}_{3}n_z & {}^{0}_{3}o_z & {}^{0}_{3}a_z & {}^{0}_{3}p_z \\ 0 & 0 & 0 & 1 \end{bmatrix} = {}^{0}_{2}\boldsymbol{T} \cdot {}^{2}_{3}\boldsymbol{A}
$$

$$
= \begin{bmatrix} c\theta_1 c\theta_{23} & -c\theta_1 s\theta_{23} & -s\theta_1 & a_2 c\theta_1 c\theta_2 - d_3 s\theta_1 \\ s\theta_1 c\theta_{23} & -s\theta_1 s\theta_{23} & c\theta_1 & a_2 s\theta_1 c\theta_2 + d_3 c\theta_1 \\ -s\theta_{23} & -c\theta_{23} & 0 & -a_2 s\theta_2 \\ 0 & 0 & 0 & 1 \end{bmatrix} \tag{12.64}
$$

$$ {}^{0}_{4}\boldsymbol{T} = \begin{bmatrix} {}^{0}_{4}\boldsymbol{n} & {}^{0}_{4}\boldsymbol{o} & {}^{0}_{4}\boldsymbol{a} & {}^{0}_{4}\boldsymbol{p} \\ 0 & 0 & 0 & 1 \end{bmatrix} = \begin{bmatrix} {}^{0}_{4}n_x & {}^{0}_{4}o_x & {}^{0}_{4}a_x & {}^{0}_{4}p_x \\ {}^{0}_{4}n_y & {}^{0}_{4}o_y & {}^{0}_{4}a_y & {}^{0}_{4}p_y \\ {}^{0}_{4}n_z & {}^{0}_{4}o_z & {}^{0}_{4}a_z & {}^{0}_{4}p_z \\ 0 & 0 & 0 & 1 \end{bmatrix} = {}^{0}_{3}\boldsymbol{T} \cdot {}^{3}_{4}\boldsymbol{A} \tag{12.65} $$

where

$$
\begin{aligned}
&\theta_{23} = \theta_2 + \theta_3 \\
&{}^{0}_{4}n_x = c\theta_1 c\theta_{23} c\theta_4 + s\theta_1 s\theta_4 \\
&{}^{0}_{4}n_y = s\theta_1 c\theta_{23} c\theta_4 - c\theta_1 s\theta_4 \\
&{}^{0}_{4}n_z = -s\theta_{23} c\theta_4 \\
&{}^{0}_{4}o_x = -c\theta_1 c\theta_{23} s\theta_4 + s\theta_1 c\theta_4 \\
&{}^{0}_{4}o_y = -s\theta_1 c\theta_{23} s\theta_4 - c\theta_1 c\theta_4 \\
&{}^{0}_{4}o_z = s\theta_{23} s\theta_4 \\
&{}^{0}_{4}a_x = -c\theta_1 s\theta_{23} \\
&{}^{0}_{4}a_y = -s\theta_1 s\theta_{23} \\
&{}^{0}_{4}a_z = -c\theta_{23} \\
&{}^{0}_{4}p_x = a_3 c\theta_1 c\theta_{23} - d_4 c\theta_1 s\theta_{23} + a_2 c\theta_1 c\theta_2 - d_3 s\theta_1 \\
&{}^{0}_{4}p_y = a_3 s\theta_1 c\theta_{23} - d_4 s\theta_1 s\theta_{23} + a_2 s\theta_1 c\theta_2 + d_3 c\theta_1 \\
&{}^{0}_{4}p_z = -a_3 s\theta_{23} - d_4 c\theta_{23} - a_2 s\theta_2
\end{aligned}
$$

$$ {}^{0}_{5}\boldsymbol{T} = \begin{bmatrix} {}^{0}_{5}\boldsymbol{n} & {}^{0}_{5}\boldsymbol{o} & {}^{0}_{5}\boldsymbol{a} & {}^{0}_{5}\boldsymbol{p} \\ 0 & 0 & 0 & 1 \end{bmatrix} = \begin{bmatrix} {}^{0}_{5}n_x & {}^{0}_{5}o_x & {}^{0}_{5}a_x & {}^{0}_{5}p_x \\ {}^{0}_{5}n_y & {}^{0}_{5}o_y & {}^{0}_{5}a_y & {}^{0}_{5}p_y \\ {}^{0}_{5}n_z & {}^{0}_{5}o_z & {}^{0}_{5}a_z & {}^{0}_{5}p_z \\ 0 & 0 & 0 & 1 \end{bmatrix} = {}^{0}_{4}\boldsymbol{T} \cdot {}^{4}_{5}\boldsymbol{A} \tag{12.66} $$

where

$$
\begin{aligned}
&{}^{0}_{5}n_x = (c\theta_1 c\theta_{23} c\theta_4 + s\theta_1 s\theta_4) c\theta_5 - c\theta_1 s\theta_{23} s\theta_5 \\
&{}^{0}_{5}n_y = (s\theta_1 c\theta_{23} c\theta_4 - c\theta_1 s\theta_4) c\theta_5 - s\theta_1 s\theta_{23} s\theta_5 \\
&{}^{0}_{5}n_z = -s\theta_{23} c\theta_4 c\theta_5 - c\theta_{23} s\theta_5 \\
&{}^{0}_{5}o_x = -(c\theta_1 c\theta_{23} c\theta_4 + s\theta_1 s\theta_4) s\theta_5 - c\theta_1 s\theta_{23} c\theta_5 \\
&{}^{0}_{5}o_y = -(s\theta_1 c\theta_{23} c\theta_4 - c\theta_1 s\theta_4) s\theta_5 - s\theta_1 s\theta_{23} c\theta_5 \\
&{}^{0}_{5}o_z = s\theta_{23} c\theta_4 s\theta_5 - c\theta_{23} c\theta_5 \\
&{}^{0}_{5}a_x = c\theta_1 c\theta_{23} s\theta_4 - s\theta_1 c\theta_4 \\
&{}^{0}_{5}a_y = s\theta_1 c\theta_{23} s\theta_4 + c\theta_1 c\theta_4 \\
&{}^{0}_{5}a_z = -s\theta_{23} s\theta_4 \\
&{}^{0}_{5}p_x = -d_4 c\theta_1 s\theta_{23} + a_2 c\theta_1 c\theta_2 - d_3 s\theta_1 \\
&{}^{0}_{5}p_y = -d_4 s\theta_1 s\theta_{23} + a_2 s\theta_1 c\theta_2 + d_3 c\theta_1 \\
&{}^{0}_{5}p_z = -a_3 s\theta_{23} - d_4 c\theta_{23} - a_2 s\theta_2
\end{aligned}
$$

$$ {}^{0}_{6}\boldsymbol{T} = \begin{bmatrix} {}^{0}_{6}\boldsymbol{n} & {}^{0}_{6}\boldsymbol{o} & {}^{0}_{6}\boldsymbol{a} & {}^{0}_{6}\boldsymbol{p} \\ 0 & 0 & 0 & 1 \end{bmatrix} = \begin{bmatrix} {}^{0}_{6}n_x & {}^{0}_{6}o_x & {}^{0}_{6}a_x & {}^{0}_{6}p_x \\ {}^{0}_{6}n_y & {}^{0}_{6}o_y & {}^{0}_{6}a_y & {}^{0}_{6}p_y \\ {}^{0}_{6}n_z & {}^{0}_{6}o_z & {}^{0}_{6}a_z & {}^{0}_{6}p_z \\ 0 & 0 & 0 & 1 \end{bmatrix} = {}^{0}_{5}\boldsymbol{T} \cdot {}^{5}_{6}\boldsymbol{A} \tag{12.67} $$

where

$$
\begin{aligned}
{}^0_6n_x &= c\theta_1(c\theta_{23}(c\theta_4c\theta_5c\theta_6 - s\theta_4s\theta_6) - s\theta_{23}s\theta_5c\theta_6) - s\theta_1(s\theta_4c\theta_5c\theta_6 + c\theta_4s\theta_6)\\
{}^0_6n_y &= s\theta_1(c\theta_{23}(c\theta_4c\theta_5c\theta_6 - s\theta_4s\theta_6) - s\theta_{23}s\theta_5c\theta_6) + c\theta_1(s\theta_4c\theta_5c\theta_6 + c\theta_4s\theta_6)\\
{}^0_6n_z &= -s\theta_{23}(c\theta_4c\theta_5c\theta_6 - s\theta_4s\theta_6) - c\theta_{23}s\theta_5c\theta_6\\
{}^0_6o_x &= c\theta_1(-c\theta_{23}(c\theta_4c\theta_5s\theta_6 + s\theta_4c\theta_6) + s\theta_{23}s\theta_5s\theta_6) - s\theta_1(-s\theta_4c\theta_5c\theta_6 + c\theta_4c\theta_6)\\
{}^0_6o_y &= s\theta_1(-c\theta_{23}(c\theta_4c\theta_5s\theta_6 + s\theta_4c\theta_6) + s\theta_{23}s\theta_5s\theta_6) + c\theta_1(-s\theta_4c\theta_5s\theta_6 + c\theta_4c\theta_6)\\
{}^0_6o_z &= s\theta_{23}(c\theta_4c\theta_5s\theta_6 + s\theta_4c\theta_6) - c\theta_{23}s\theta_5s\theta_6\\
{}^0_6a_x &= -c\theta_1(c\theta_{23}c\theta_4s\theta_5 + s\theta_{23}c\theta_5) - s\theta_1s\theta_4s\theta_5\\
{}^0_6a_y &= -s\theta_1(c\theta_{23}c\theta_4s\theta_5 + s\theta_{23}c\theta_5) + c\theta_1s\theta_4s\theta_5\\
{}^0_6a_z &= s\theta_{23}c\theta_4s\theta_5 - c\theta_{23}c\theta_5\\
{}^0_6p_x &= c\theta_1(-d_4s\theta_{23} + a_3c\theta_{23} + a_2c\theta_2) - d_3s\theta_1\\
{}^0_6p_y &= s\theta_1(-d_4s\theta_{23} + a_3c\theta_{23} + a_2c\theta_2) + d_3c\theta_1\\
{}^0_6p_z &= -d_4c\theta_{23} - a_3s\theta_{23} - a_2s\theta_2
\end{aligned}
$$

Using the results of Eq. (12.62) to Eq. (12.67) in Eq. (12.52) of the Jacobian matrix gets,

$$
[J_v] = \begin{bmatrix} \dfrac{\partial \mathbf{p}_{0,6}}{\partial\theta_1} & \dfrac{\partial \mathbf{p}_{0,6}}{\partial\theta_2} & \dfrac{\partial \mathbf{p}_{0,6}}{\partial\theta_3} & \dfrac{\partial \mathbf{p}_{0,6}}{\partial\theta_4} & \dfrac{\partial \mathbf{p}_{0,6}}{\partial\theta_5} & \dfrac{\partial \mathbf{p}_{0,6}}{\partial\theta_6} \end{bmatrix} \tag{12.68}
$$

where

$$
\frac{\partial \mathbf{p}_{0,6}}{\partial\theta_1} = \begin{Bmatrix} -s\theta_1(-d_4s\theta_{23} + a_3c\theta_{23} + a_2c\theta_2) - d_3c\theta_1 \\ -c\theta_1(-d_4s\theta_{23} + a_3c\theta_{23} + a_2c\theta_2) - d_3s\theta_1 \\ 0 \end{Bmatrix} \qquad
\frac{\partial \mathbf{p}_{0,6}}{\partial\theta_2} = \begin{Bmatrix} c\theta_1(-d_4c\theta_{23} - a_3s\theta_{23} - a_2s\theta_2) \\ s\theta_1(-d_4c\theta_{23} - a_3s\theta_{23} - a_2s\theta_2) \\ d_4s\theta_{23} - a_3c\theta_{23} - a_2c\theta_2 \end{Bmatrix}
$$

$$
\frac{\partial \mathbf{p}_{0,6}}{\partial\theta_3} = \begin{Bmatrix} c\theta_1(-d_4c\theta_{23} - a_3s\theta_{23}) \\ s\theta_1(-d_4c\theta_{23} - a_3s\theta_{23}) \\ d_4s\theta_{23} - a_3c\theta_{23} \end{Bmatrix} \qquad
\frac{\partial \mathbf{p}_{0,6}}{\partial\theta_4} = \frac{\partial \mathbf{p}_{0,6}}{\partial\theta_5} = \frac{\partial \mathbf{p}_{0,6}}{\partial\theta_6} = \begin{Bmatrix} 0 \\ 0 \\ 0 \end{Bmatrix}
$$

$$
[J_\omega] = [\, {}^0_1\mathbf{Z}_1 \;\; {}^0_2\mathbf{Z}_2 \;\; {}^0_3\mathbf{Z}_3 \;\; {}^0_4\mathbf{Z}_4 \;\; {}^0_5\mathbf{Z}_5 \;\; {}^0_6\mathbf{Z}_6 \,] = [\, {}^0_1\mathbf{a}_1 \;\; {}^0_2\mathbf{a}_2 \;\; {}^0_3\mathbf{a}_3 \;\; {}^0_4\mathbf{a}_4 \;\; {}^0_5\mathbf{n}_5 \;\; {}^0_6\mathbf{n}_6 \,] \tag{12.69}
$$

Using the Jacobian matrix to map the velocities of joint motion into that of the end-effector tool is as follows:

$$
\begin{Bmatrix} {}^0_6\dot{p}_x \\ {}^0_6\dot{p}_y \\ {}^0_6\dot{p}_z \\ {}^0_6\omega_x \\ {}^0_6\omega_y \\ {}^0_6\omega_z \end{Bmatrix} = \begin{bmatrix} J_v \\ J_\omega \end{bmatrix} \begin{Bmatrix} \dot{\theta}_1 \\ \dot{\theta}_2 \\ \dot{\theta}_3 \\ \dot{\theta}_4 \\ \dot{\theta}_5 \\ \dot{\theta}_6 \end{Bmatrix} \tag{12.70}
$$

12.5 Dynamic Modeling

Dynamics is to study the *cause-and-effect* relationship between the forces and motions of a machine, and a dynamic model describes the relations of applied forces/torques and the resulting motions. Dynamic models are used in design, simulation, and control of a machine. Similar to kinematic modeling, *an inverse dynamic problem* is to obtain the driving forces at active joints when the motion and external forces of the end-effector are given. *A forward dynamic problem* is to find the motion and loading of the end-effector when the driving forces at joints are given.

In a forward dynamic problem, the terms of $\ddot{\boldsymbol{q}}(t)$ (and then of $\boldsymbol{q}$(t) and $\dot{\boldsymbol{q}}(t)$) are computed when the vector of generalized forces (torques and/or forces) $\boldsymbol{\tau}(t)$ applied to the joints and end-effectors are given. Assume the initial conditions be $\boldsymbol{q}(t = t_0)$, $\dot{\boldsymbol{q}}(t = t_0)$, the forward dynamic problem is described as,

$$\ddot{\boldsymbol{q}} = f(\boldsymbol{q}, \dot{\boldsymbol{q}}, \boldsymbol{\tau}) \tag{12.71}$$

where $\ddot{\boldsymbol{q}}$ is the vector of motion accelerations. $\boldsymbol{\tau}$ is the vector of torques or forces at actuators and end-effector. $\boldsymbol{q}$ and $\dot{\boldsymbol{q}}$ are the vectors of the motion displacements and velocities in current moment of time t. The displacements and velocities are then updated as,

$$\dot{\boldsymbol{q}} = \int \ddot{\boldsymbol{q}} dt, \quad q = \int \dot{\boldsymbol{q}} dt \tag{12.72}$$

In an inverse dynamic problem, the generalized forces (torques and/or forces) $\boldsymbol{\tau}(t)$ are computed when the machine motion $(\boldsymbol{q}, \dot{\boldsymbol{q}}, \ddot{\boldsymbol{q}})$ are given. The inverse forward dynamic model is described as,

$$\boldsymbol{\tau} = f^{-1}(\boldsymbol{q}, \dot{\boldsymbol{q}}, \ddot{\boldsymbol{q}}) = g(\boldsymbol{q}, \dot{\boldsymbol{q}}, \ddot{\boldsymbol{q}}) \tag{12.73}$$

12.5.1 Inertia and Moments of Inertia

When the relation of force and motion is considered, the material properties of solids must be known. *Inertia* is a quantity of the materials that affect rotations. It is a material property by which it continues in its existing state of rest or uniform motion in a straight line, unless that state is changed by an external force.

As shown in Figure 12.41a, when a force (F) acts of a mass, the mass will be accelerated proportionally, and the proportion is the *inertia* of mass (m). As shown in Figure 12.41b, when a force acts on a mass in a distance, it turns into a torque (T) on the mass. The mass will be angularly

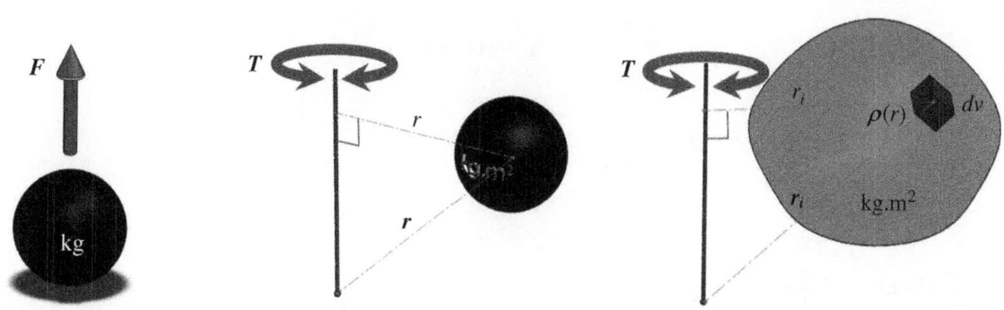

(a) Inertia for linear motion

(b) Moment of inertia of particle for angular motion

(c) Moment of inertia of body for angular motion

Figure 12.41 Inertia and moment of inertia of mass and body accelerated proportionally. The proportion is the *moment of inertia* (I), which is determined by both the mass and the spatial position of the mass and acting force as,

$$I = m \cdot r^2 \tag{12.74}$$

where I is the moment of inertia, m is the mass, and r is the distance of the mass to the axis of acting torque.

Equation (12.74) shows that the moment of inertia relates to the distances of a mass to the axis of the acting torque. When the mass of body (i.e., $\rho(r)$) is distributed in Figure 12.41c, every mass particle in the body has its own distance to the acting torque, and the moment of inertia should be modified as,

$$I = \int \rho(r) \cdot r^2 dv \tag{12.75}$$

where I is the moment of inertia of body with respect to the acting axis of torque, $\rho(r)$ is the density function, r is the distance of the mass particle to the acting axis of torque, and dv is the volume of mass particle.

Equation (12.74) or (12.75) indicates that the moment of inertia varies with the axis of the acting torque. When a torque is applied on an arbitrary axis, it is convenient to determine the vector of angular acceleration using a tensor of the moment of inertia ($\boldsymbol{I}$). In a Cartesian CS {O-XYZ}, a tensor of the moment of inertia ($\boldsymbol{I}$) is defined as,

$$[\boldsymbol{I}] = \begin{bmatrix} I_{xx} & -I_{xy} & -I_{xz} \\ -I_{xy} & I_{yy} & -I_{yz} \\ -I_{xz} & -I_{yz} & I_{zz} \end{bmatrix} = \int_V \rho(x,y,z) \begin{bmatrix} y^2+z^2 & -xy & -xz \\ -xy & x^2+z^2 & -yz \\ -xz & -yz & x^2+y^2 \end{bmatrix} dxdydz \tag{12.76}$$

where I_{xx}, I_{yy}, and I_{zz} are *the mass products of inertia* along X, Y, and Z axes, respectively. I_{xy}, I_{xz}, and I_{yz} are *the cross products of inertia* over XY, XZ, and YZ planes, respectively.

The tensor of the moment of inertia ($\boldsymbol{I}$) in Eq. (12.76) includes six independent parameters, that is, I_{xx}, I_{yy}, I_{zz}, I_{xy}, I_{xz}, I_{yz}; these parameters depend on the selected CS {O-XYZ}. There exists a special CS called *a principal coordinate system* where all of the cross products of inertia are zeros, that is, $I_{xy} = I_{xz} = I_{yz} = 0$. The axes in such a principal CS are *the principal axes* of body.

Figure 12.42 shows the impact of the coordinate transformation on the tensor of the moment of inertia ($\boldsymbol{I}$). For the block, when the origin of the CS is moved to the geometric center of block, all of

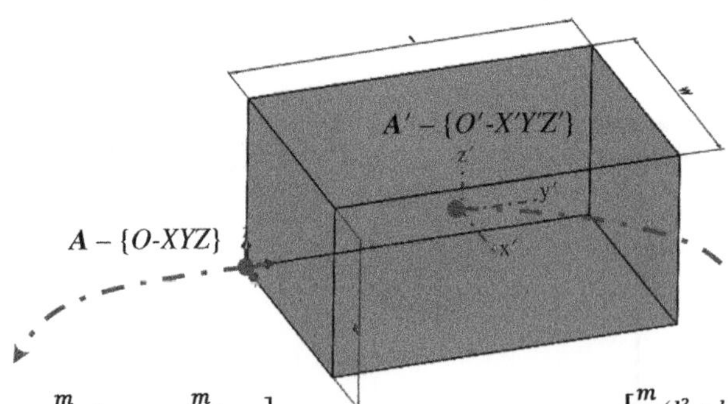

$$[\boldsymbol{I}] = \begin{bmatrix} \frac{m}{3}(l^2+h^2) & -\frac{m}{4}wl & -\frac{m}{4}hw \\ -\frac{m}{4}wl & \frac{m}{3}(w^2+h^2) & -\frac{m}{4}hl \\ -\frac{m}{4}hw & -\frac{m}{4}hl & \frac{m}{3}(l^2+w^2) \end{bmatrix} \qquad [\boldsymbol{I}]' = \begin{bmatrix} \frac{m}{12}(l^2+h^2) & 0 & 0 \\ 0 & \frac{m}{12}(w^2+h^2) & 0 \\ 0 & 0 & \frac{m}{12}(l^2+w^2) \end{bmatrix}$$

Figure 12.42 Example of the principal coordinate system

the terms I_{xy}, I_{xz}, I_{yz} in the tensor become zero. For the block, the principal CS has its origin at its geometric center, and its principal axes are aligned with the directions of length, width, and height of block.

12.5.2 Newton–Euler Formulation

A robot is designed to fulfill a task defined as a motion of the end-effector. The end-effector is driven and moved by the actuators at the joints of robot. It is necessary to know what forces or torques are required to produce expected motion of the end-effector subject to given loads. A dynamic model is usually represented by a set of differential equations called *equations of motion*. Two common dynamic modeling methods are (1) the Newton–Euler formulation and (2) the Lagrangian formulation. In the *Newton–Euler formulation*, a robot is decomposed into rigid bodies, and *Newton's second law* and *the Euler law* are applied for each rigid body to establish the relationship between the driven forces and the accelerations of motion.

Figure 12.43 shows Newton's and Euler laws for the relations of applied force/torque and linear/angular accelerations, respectively. Note that the force ($\boldsymbol{F}$) and toque ($\boldsymbol{T}$) are the summed quantities when all of the internal and external forces/torques are combined. Newton's and Euler laws can be applied to any physical object in a system. Before these laws are applied to a body, its *free-body-diagram* (FBD) is drawn to identify (1) external forces exerted on the body and (2) reactional forces at contacts from other bodies in the system.

Figure 12.44a shows the FBD of link i in a manipulator. The forces on link i are external force/torque ($\boldsymbol{F}_i$ and $\boldsymbol{T}_i$), gravity force ($m_i \cdot \boldsymbol{g}$), torque ($\boldsymbol{T}_i$), reactional force ($\boldsymbol{f}_{i-1,\,i}$) and moment ($\boldsymbol{n}_{i-1,i}$) from link $i-1$, and reactional force ($-\boldsymbol{f}_{i,\,i+1}$) and moment ($-\boldsymbol{n}_{i,\,i+1}$) from link $i+1$. All of the above variables are given in the base CS. By applying Newton's and Euler laws in FBD, the dynamic model of link i can be found as,

$$\boldsymbol{f}_{i-1,i} - m_i\boldsymbol{g} - \boldsymbol{f}_{i,i+1} + \boldsymbol{F}_i = m_i\boldsymbol{a}_i \tag{12.77}$$

$$\begin{aligned} &{}_i^B\boldsymbol{R}^{-1}(\boldsymbol{n}_{i-1,i} - \boldsymbol{n}_{i,i+1}) + \left[{}_i^B\boldsymbol{R}^{-1}\boldsymbol{f}_{i-1,i}\right] \times \boldsymbol{r}_{i-1,Ci} - \left[{}_i^B\boldsymbol{R}^{-1}\boldsymbol{f}_{i,i+1}\right] \times \boldsymbol{r}_{i,Ci} + {}_i^B\boldsymbol{R}^{-1} \cdot \boldsymbol{T}_i \\ &= \boldsymbol{I}_i\left[{}_i^BR^{-1}\dot{\boldsymbol{\omega}}_i\right] - \left[{}_i^BR^{-1}\boldsymbol{\omega}_i\right] \times \boldsymbol{I}_i \cdot \left[{}_i^BR^{-1}\boldsymbol{\omega}_i\right] \end{aligned} \tag{12.78}$$

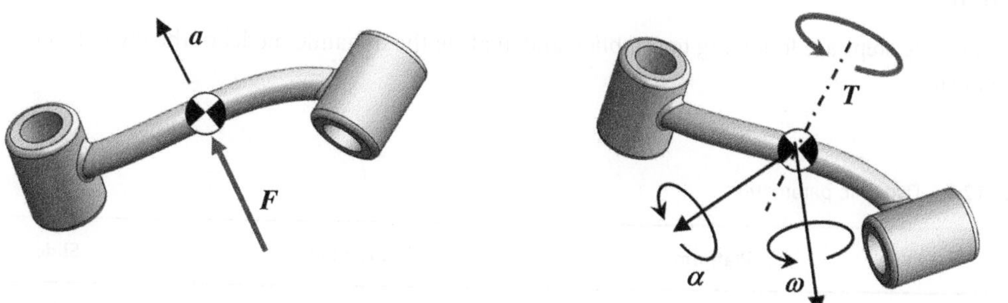

(a) Force causes linear acceleration ($\boldsymbol{F} = m \cdot \boldsymbol{a}$)

(b) Torque causes angular acceleration ($\boldsymbol{T} = I\alpha + \omega \times I\omega$)

Figure 12.43 Newton's and Euler's laws

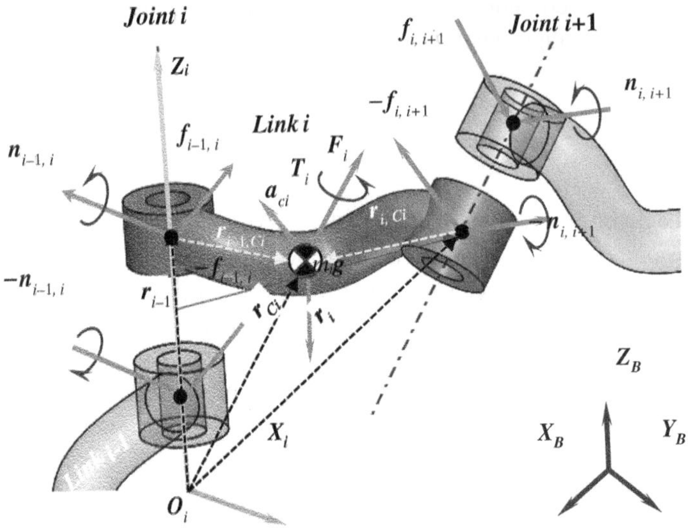

Figure 12.44 *Free-body-diagram* (FBD) of link *i* in frame *i*

In Eq. (12.77), the tensor of the moments of inertia is usually defined in local principal CS built at the center of mass. The variables under the base CS have to be transformed to the local CS by an inverse transformation matrix from frame i to base reference CS as ${}^{B}_{i}R^{-1}$.

The system model is assembled from the dynamic models of all individual bodies. The variables included in the equations of motion from the Newton–Euler method include internal reaction forces and driving forces occurring to joints. Assume that the information of all of displacements, velocities, and accelerations of bodies is known, the dynamic model of a system corresponds to a set of linear equations about the variables of internal reactional forces and external driving forces.

Example 12.5 Table 12.3 gives the values of structural dynamic parameters using the example mechanism in Figure 12.45 (Bi et al. 2007). Establish and analyze the dynamic model of the given robotic mechanism.

Solution

The following steps are following to establish and analyze the dynamic model of the given robotic mechanism.

Table 12.3 Dynamic parameters

	Platform	Support bar	Slide
Mass (kg)	1.348	0.16165	0.2135
Center of mass	O_e	F_i ($l_c = 0.3l_i$)	D_i
Moment of inertia (N.m) (I_{xx}, I_{yy}, I_{zz})	(0.002706, 0.002030, 0.004736)	(0.000602, 0.000602, 0.000032)	N/A

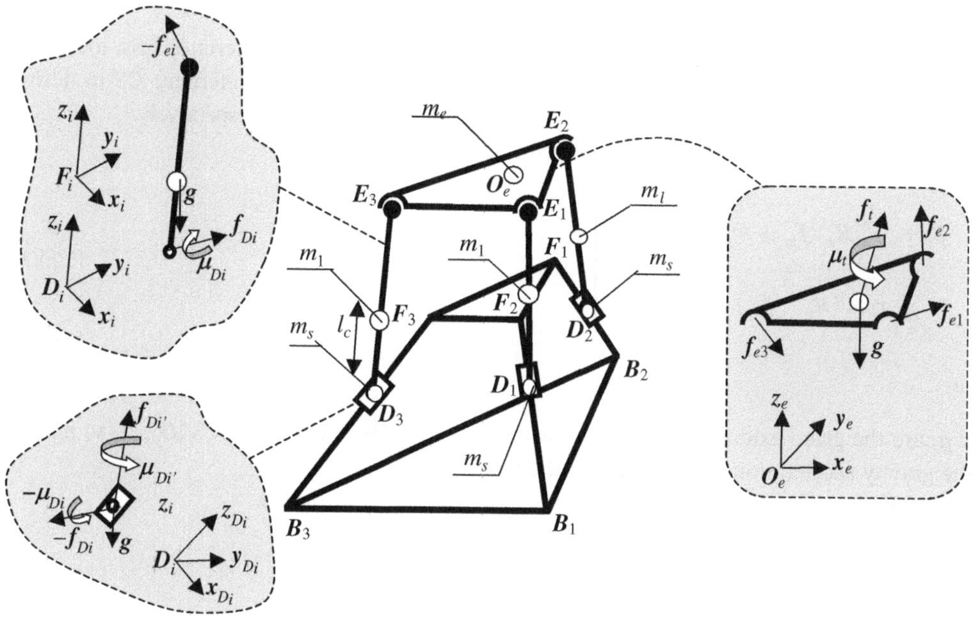

Figure 12.45 Mass centers and force analysis of a tripod mechanism

Step 1: Inertia Force/Moment

As shown in Figure 12.45, local CSs $\{O_{e-}\,x_e y_e z_e\}$, $\{F_{i-}\,x_i y_i z_i\}$, and $\{D_{i-}\,x_{Di} y_{Di} z_{Di}\}$ are assumed to be an inertia CS of the end-effector, support bar i, and slide i, respectively. According to Newton's law and Euler's equation, the inertia force/moments of the elements can be calculated as

$$\left.\begin{aligned} \boldsymbol{h}_e &= m_e \boldsymbol{R}_e^{-1}\boldsymbol{a}_e \\ \boldsymbol{n}_e &= \boldsymbol{I}_e\left(\boldsymbol{R}_e^{-1}\boldsymbol{\varepsilon}_e\right) + \left(\boldsymbol{R}_e^{-1}\boldsymbol{\omega}_e\right) \times \left(\boldsymbol{I}_e\left(\boldsymbol{R}_e^{-1}\boldsymbol{\omega}_e\right)\right) \end{aligned}\right\} \tag{12.79}$$

$$\left.\begin{aligned} \boldsymbol{h}_i &= m_l \boldsymbol{R}_i^{-1}\boldsymbol{a}_{F_i} \\ \boldsymbol{n}_i &= \boldsymbol{I}_l\left(\boldsymbol{R}_i^{-1}\boldsymbol{\varepsilon}_i\right) + \left(\boldsymbol{R}_i^{-1}\boldsymbol{\omega}_i\right) \times \left(\boldsymbol{I}_l\left(\boldsymbol{R}_i^{-1}\boldsymbol{\omega}_i\right)\right) \end{aligned}\right\} \quad (i = 1, 2, 3) \tag{12.80}$$

where

$$\boldsymbol{R}_i = \begin{bmatrix} \boldsymbol{x}_i & \boldsymbol{y}_i & \boldsymbol{z}_i \end{bmatrix} = \begin{bmatrix} -s\alpha_i & -c\varphi_i c\alpha_i & s\varphi_i c\alpha_i \\ c\alpha_i & -c\varphi_i s\alpha_i & s\varphi_i s\alpha_i \\ 0 & s\varphi_i & c\varphi_i \end{bmatrix}$$

$$\left.\begin{aligned} \boldsymbol{h}_{D_i} &= m_s \boldsymbol{R}_{D_i}^{-1}\boldsymbol{a}_{D_i} \\ \boldsymbol{n}_{D_i} &= 0 \end{aligned}\right\} \quad (i = 1, 2, 3) \tag{12.81}$$

where

$$\boldsymbol{R}_{D_i} = \begin{bmatrix} \boldsymbol{x}_{D_i} & \boldsymbol{y}_{D_i} & \boldsymbol{z}_{D_i} \end{bmatrix} = \begin{bmatrix} -s\alpha_i & -s\gamma c\alpha_i & c\gamma c\alpha_i \\ c\alpha_i & -s\gamma s\alpha_i & c\gamma s\alpha_i \\ 0 & c\gamma & s\gamma \end{bmatrix}$$

Step 2: Force Equilibrium Equations

As shown in Figure 12.45, based on the principles of the selection of reference CSs, local CSs $\{O_{e-}\ x_e\ y_e\ z_e\}$, $\{D_{i-}\ x_i\ y_i\ z_i\}$, and $\{D_{i-}\ x_{Di}\ y_{Di}\ z_{Di}\}$ are assumed to be the reference CS to derive the equilibrium equations for the end-effector, support bar i, and slide i, respectively.

The equilibrium equations for the end-effector are derived as

$$\left.\begin{aligned} &\boldsymbol{f}_t + \sum_{i=1}^{3} \boldsymbol{R}_e^{-1}\boldsymbol{f}_{e_i} + m_e\boldsymbol{R}_e^{-1}\mathbf{g} = \boldsymbol{h}_e \\ &\sum_{i=1}^{3} \boldsymbol{r}_i \times \left(\boldsymbol{R}_e^{-1}\cdot\boldsymbol{f}_{e_i}\right) + \boldsymbol{\mu}_t = \boldsymbol{n}_e \end{aligned}\right\} \tag{12.82}$$

where

$\boldsymbol{f}_t$ and $\boldsymbol{\mu}_t$ are the given external force/torque with respect to the end-effector CS $\{O_{e-}\ x_e\ y_e\ z_e\}$.
$\mathbf{g}$ is the gravity acceleration with respect to the world CS.
$\boldsymbol{f}_{ei}$ is the unknown force from support bar i with respect to the world CS.
$\boldsymbol{h}_e$ and $\boldsymbol{n}_e$ are given at Eq. (12.79).

The equilibrium equations for support bar i are derived as

$$\boldsymbol{f}_{D_i} - \boldsymbol{R}_i^{-1}\boldsymbol{f}_{e_i} + m_i\left(\boldsymbol{R}_i^{-1}\mathbf{g}\right) = \boldsymbol{h}_i \quad (i = 1, 2, 3) \tag{12.83}$$

$$\boldsymbol{\mu}_{D_i} - \boldsymbol{r}_{D_iE_i} \times \left(\boldsymbol{R}_i^{-1}\boldsymbol{f}_{e_i}\right) + \boldsymbol{r}_{D_iF_i} \times \left(\boldsymbol{R}_i^{-1}\mathbf{g}\right) - \boldsymbol{r}_{D_iF_i} \times m_i\left(\boldsymbol{R}_i^{-1}\boldsymbol{a}_{F_i}\right) = \boldsymbol{n}_i \quad (i = 1, 2, 3) \tag{12.84}$$

where

$\boldsymbol{r}_{D_iF_i} = l_c(0, 0, 1)^T$
$\boldsymbol{r}_{D_iE_i} = l_i(0, 0, 1)^T$
$\boldsymbol{\mu}_{D_i} = \left(0, \mu_{D_{iy}}, \mu_{D_{iz}}\right)^T$
$\boldsymbol{f}_{Di}$ is the unknown force from slide i with respect to the system $\{D_{i-}\ x_i\ y_i\ z_i\}$.
$\boldsymbol{h}_i$ and $\boldsymbol{n}_i$ are given at Eq. (12.80).

The equilibrium equations for slide i are derived as

$$-\left(\boldsymbol{R}_{D_i}^{-1}\boldsymbol{R}_i\right)\boldsymbol{f}_{D_i} + \boldsymbol{f}_{D_i'} + m_s\boldsymbol{R}_{D_i}^{-1}\mathbf{g} = \boldsymbol{h}_{D_i} \quad (i = 1, 2, 3) \tag{12.85}$$

where

$\boldsymbol{f}_{D'i}$ is the unknown force from the base with respect to the CS $\{D_{i-}\ x_{Di}\ y_{Di}\ z_{Di}\}$.
$\boldsymbol{h}_{Di}$ is given at Eq. (12.81).

Reaction force $\boldsymbol{f}_{D'i}$ includes the contribution of the driving force of joint motor i. Therefore, the driving force of joint i is the projection of $\boldsymbol{f}_{D'i}$ on the translational direction of the prismatic joint

$$\xi_i = \boldsymbol{f}_{D_i'} \cdot \boldsymbol{z}_{D_i} \quad (i = 1, 2, 3) \tag{12.86}$$

where

ξ_i is the force of joint i.

Step 3: Dynamic Model and Solution

Combination of Eq. (12.82) and Eq. (12.84) gives a group of 15 linear equations with 15 unknown parameters, that is, 3 × 3 parameters for $\boldsymbol{f}_{ei}$ (i = 1,2,3) and 2 × 3 parameters for $\boldsymbol{\mu}_{D_i}$ (i = 1,2,3). This group of linear equations can be solved so that $\boldsymbol{f}_{ei}$ and $\boldsymbol{\mu}_{D_i}$ are obtained.

Reaction force $\boldsymbol{f}_{Di}$ and $\boldsymbol{f}_{D'i}$ can derived easily using Eqs. (12.82) and (12.85) sequentially once $\boldsymbol{f}_{ei}$ are known, and Eq. (12.86) is used finally to calculate the driving forces of joints.

Figure 12.46 shows an example of required acceleration for a given path. Using the dynamic model, the required joint forces are calculated as shown in Figure 12.47.

A validating experiment is conducted, and the actual joint forces are measured. The discrepancy between theoretical and measured results is illustrated in Figure 12.48. The maximum discrepancy is 6.1% and the average discrepancy is 1.83%. Note that some simplifications, such as the assumption of a centralized mass of a link body, have been made in dynamic modeling, the discrepancy is very reasonable and acceptable.

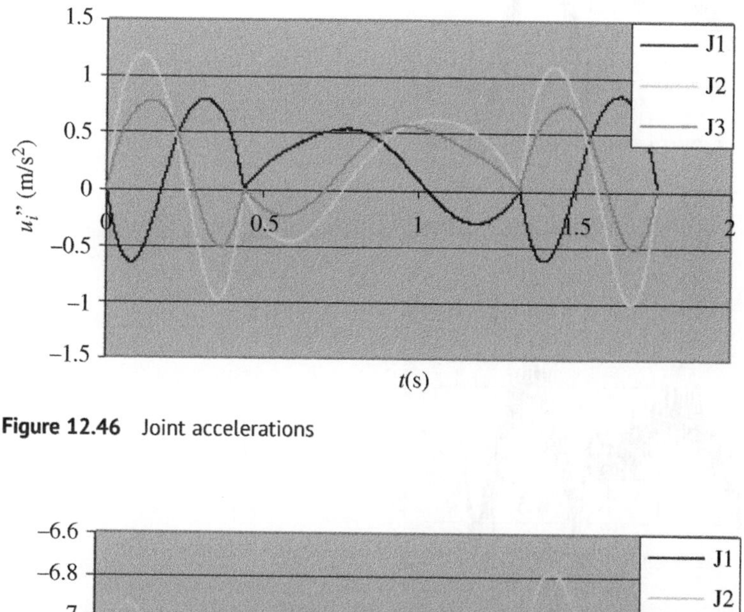

Figure 12.46 Joint accelerations

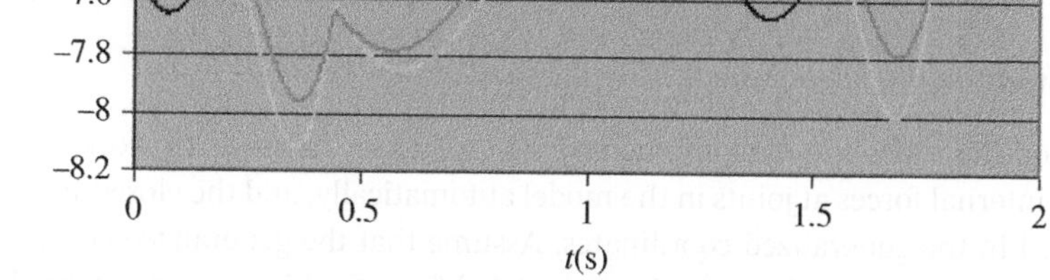

Figure 12.47 Joint forces calculated from the dynamic model

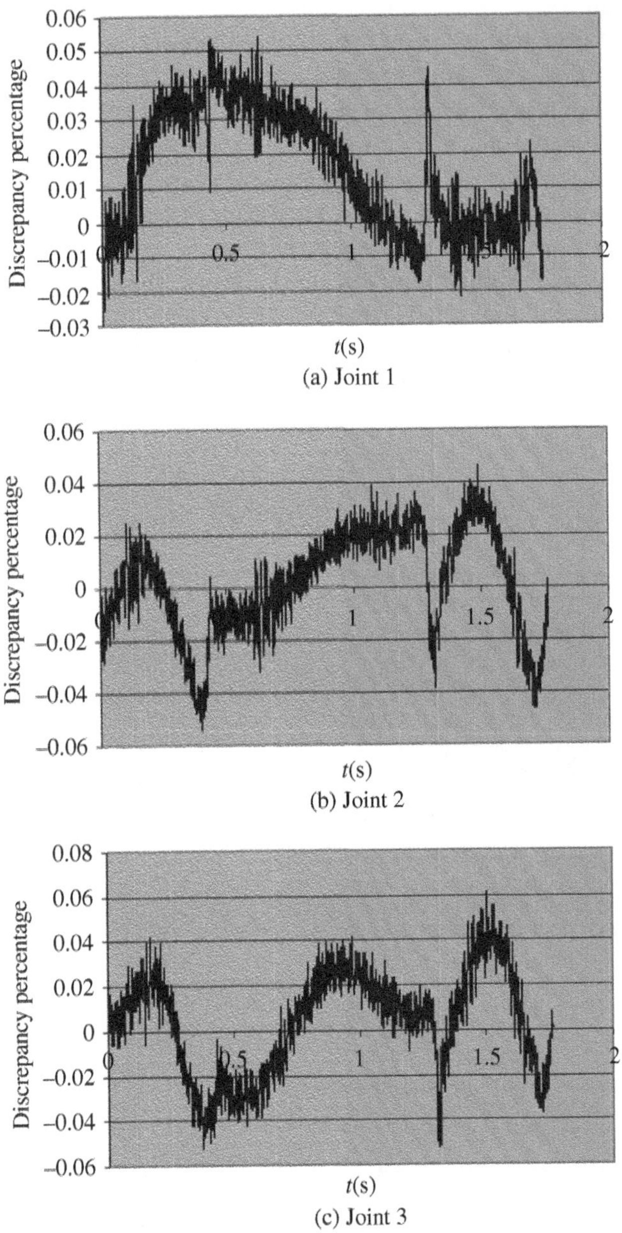

Figure 12.48 Discrepancy of theoretical and measured results of the joint forces

12.5.3 Lagrangian Method

The Lagrangian formulation describes the dynamic behavior of a system in terms of work and energy; this eliminates the internal forces at joints in the model automatically, and the closed-form dynamic models are derived in the generalized coordinates. Assume that the generalized coordinates of system be $q_1, q_2, \dots, q_n$, the Lagrangian term L is calculated from the kinetic and potential energy (K and P) of system as,

$$L(q_i, \dot{q}_i) = K(q_i, \dot{q}_i) - P(q_i) \tag{12.87}$$

where the potential energy P is a function of the generalized coordinates q_i; the kinetic energy is the function of both q_i and $\dot{q}_i$. The equations of dynamic motion of the system can be defined as,

$$\frac{d}{dt}\frac{\partial L}{\partial \dot{q}_i} - \frac{\partial L}{\partial q_i} = Q_i \quad i = 1, 2, \ldots, n \tag{12.88}$$

where Q_i is the generalized force corresponding to the generalized coordinate q_i. Taking into consideration the virtual work by nonconservative forces yields to the generalized forces exerted on the system.

A 2-DOF robotic arm in Figure 12.49 is used to illustrate how the Lagrangian method can be used to establish the dynamic model of a robot.

The angular velocities of the links are

$$\left.\begin{aligned} \omega_1 &= \dot{\theta}_1 \\ \omega_2 &= \dot{\theta}_1 + \dot{\theta}_2 \end{aligned}\right\} \tag{12.89}$$

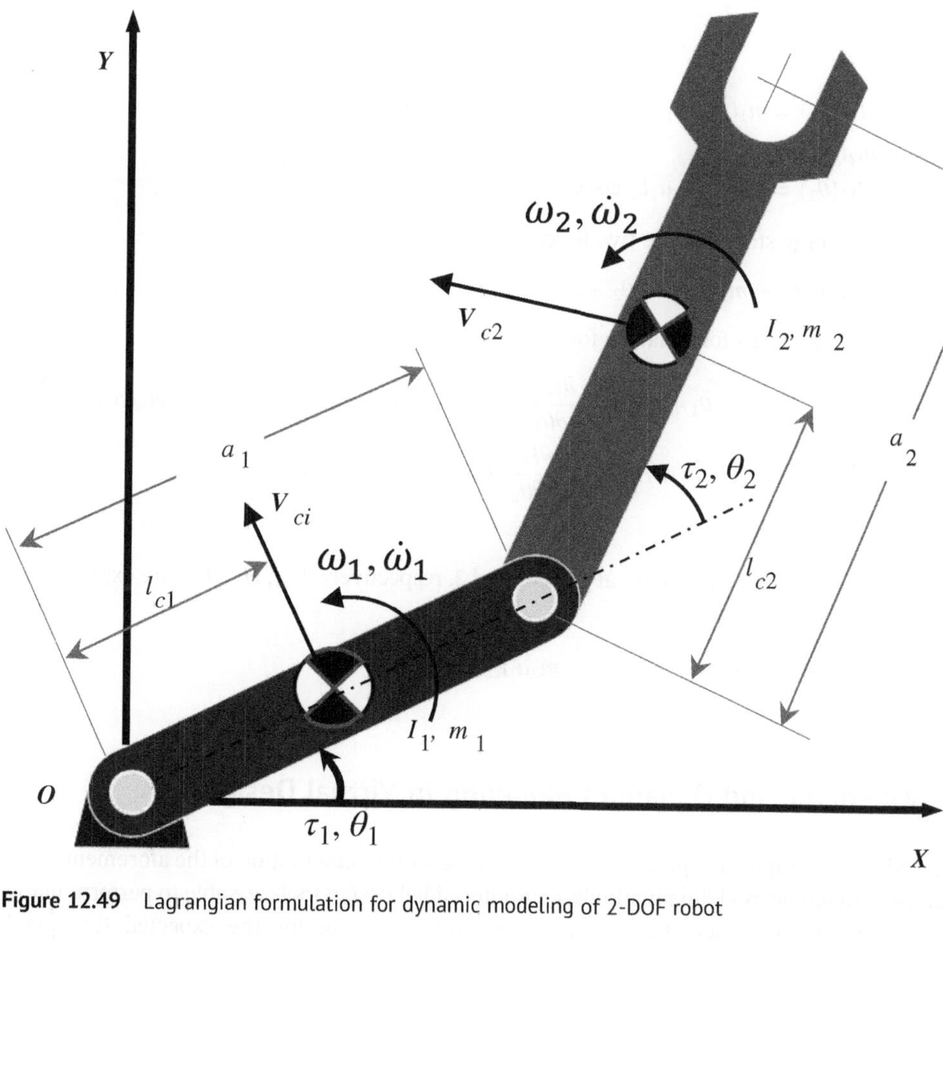

Figure 12.49 Lagrangian formulation for dynamic modeling of 2-DOF robot

The linear velocities of links at their CoGs are

$$\boldsymbol{V}_{c1} = \begin{Bmatrix} -l_{c1}\sin\theta_1 \\ l_{c1}\cos\theta_1 \end{Bmatrix}\dot{\theta}_1 \tag{12.90}$$

$$\boldsymbol{V}_{c2} = [\boldsymbol{J}_{c2}]\{\dot{\boldsymbol{q}}\} = \begin{bmatrix} -a_1\sin\theta_1 - l_{c2}\sin\theta_{12} & -l_{c2}\sin\theta_{12} \\ a_1\cos\theta_1 + l_{c2}\cos\theta_{12} & l_{c2}\cos\theta_{12} \end{bmatrix}\begin{Bmatrix} \dot{\theta}_1 \\ \dot{\theta}_2 \end{Bmatrix} \tag{12.91}$$

where $\theta_{12} = \theta_1 + \theta_2$, $\{\dot{\boldsymbol{q}}\} = \begin{Bmatrix} \dot{\theta}_1 \\ \dot{\theta}_2 \end{Bmatrix}$ is the vector of angular velocities, and $[\boldsymbol{J}_{c2}]$ is 2×2 Jacobian matrix relating to the velocity of CoG of link 2. Note that both linear and angular velocities are the functions of angular displacements and velocities, they are dependent.

The total kinematic energy is calculated by,

$$\boldsymbol{K} = \sum_{i=1}^{2}\left(\frac{1}{2}m_i|\boldsymbol{V}_{ci}|^2 + \frac{1}{2}I_i\omega_i^2\right) = \frac{1}{2}H_{11}\dot{\theta}_1^2 + H_{12}\dot{\theta}_1\dot{\theta}_2 + \frac{1}{2}H_{22}\dot{\theta}_2^2$$

$$= \frac{1}{2}\left(\dot{\theta}_1 \quad \dot{\theta}_2\right)\begin{bmatrix} H_{11} & H_{12} \\ H_{12} & H_{22} \end{bmatrix}\begin{Bmatrix} \dot{\theta}_1 \\ \dot{\theta}_2 \end{Bmatrix} \tag{12.92}$$

where

$$\left.\begin{aligned} H_{11} &= H_{11}(\theta_2) = m_1 l_{c1}^2 + I_1 + m_2\left(a_1^2 + l_{c2}^2 + 2a_1 l_{c2}\cos\theta_2\right) + I_2 \\ H_{22} &= m_2 l_{c2}^2 + I_2 \\ H_{12} &= H_{12}(\theta_2) = m_2\left(l_{c2}^2 + a_1 l_{c2}\cos\theta_2\right) + I_2 \end{aligned}\right\} \tag{12.93}$$

The potential energy stored in the two links is given by

$$P = m_1 g l_{c1}\sin\theta_1 + m_2 g\{a_1\sin\theta_2 + l_{c2}\sin\theta_{12}\} \tag{12.94}$$

Applying the Lagrangian formulation for the first and second joints gets,

$$\left.\begin{aligned} \tau_1 &= \frac{d}{dt}\frac{\partial L}{\partial\dot{\theta}_1} - \frac{\partial L}{\partial\theta_1} = H_{11}\ddot{\theta}_1 + H_{12}\ddot{\theta}_2 + \frac{\partial H_{11}}{\partial\theta_2}\dot{\theta}_1\dot{\theta}_2 + \frac{\partial H_{12}}{\partial\theta_2}\dot{\theta}_2^2 - (m_1 g l_{c1}\cos\theta_1 + m_2 g l_{c2}\cos\theta_{12}) \\ \tau_2 &= \frac{d}{dt}\frac{\partial L}{\partial\dot{\theta}_2} - \frac{\partial L}{\partial\theta_2} = (H_{12}\ddot{\theta}_1 + H_{22}\ddot{\theta}_2) + \frac{\partial H_{12}}{\partial\theta_2}\dot{\theta}_1\dot{\theta}_2 - (m_2 g(a_1\cos\theta_2 + l_{c2}\cos\theta_{12})) \end{aligned}\right\} \tag{12.95}$$

where τ_1 and τ_2 are the driving torques at joints 1 and 2, respectively. H_{11}, H_{12}, H_{22} are expressed in Eq. (12.93), and

$$\frac{\partial H_{11}}{\partial\theta_2} = -2m_2 a_1 l_{c2}\sin\theta_2, \quad \frac{\partial H_{12}}{\partial\theta_2} = -m_2 a_1 l_{c2}\sin\theta_2$$

12.6 Kinematic and Dynamics Modeling in Virtual Design

Robotic design is a complicated process. With the computer implementation of the aforementioned kinematic and dynamic modeling methods, computer-aided design tools are able to perform many critical activities for designers to design and optimize a robot for the expected functional

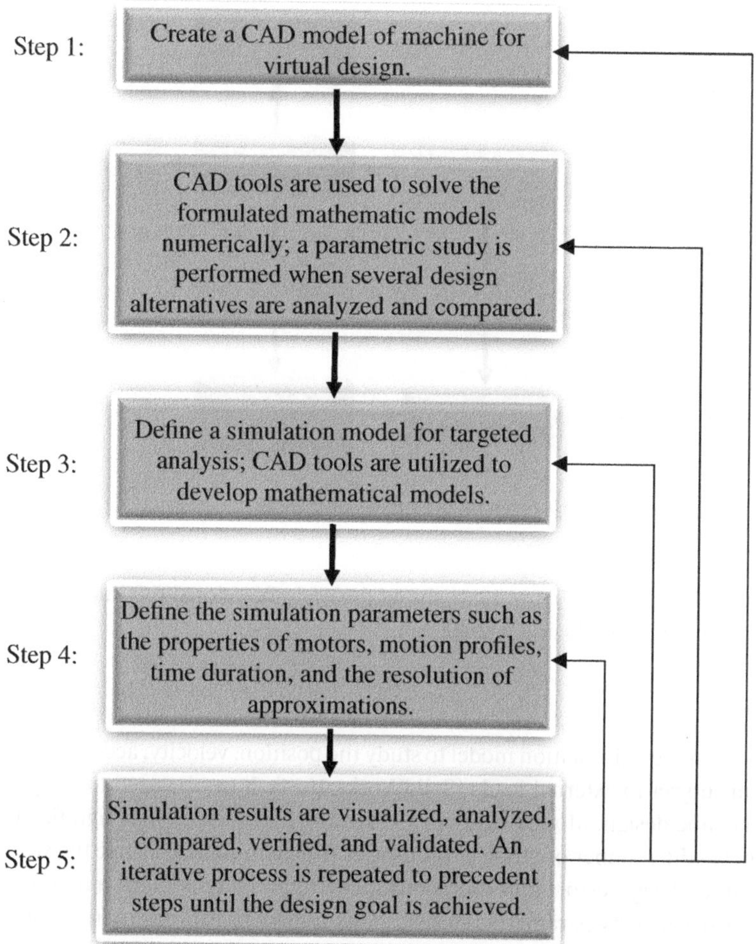

Figure 12.50 Steps of using a computer-aided tool in virtual machine design

requirements. Figure 12.50 shows that a designer could interact and fully utilize the capabilities of commercial CAD tool in virtual design. *First*, the design process begins with the preparation of a CAD model of a machine including the geometry, dimensions, material properties, mating relations, and boundary conditions. *Second*, a simulation model is defined to establish a mathematic model for the targeted simulation. *Third*, designers should specify the model parameters such as the properties of motors, the profiles of expected motions, the duration of simulation, and design variables to be investigated. *Fourth*, the simulation is performed to find the solution to the formulated mathematic model numerically. *Fifth*, the simulation result is analyzed and verified, and an iterative process is repeated to precedent steps until the design and analysis goal is achieved. It should be noted that even CAD tools are available to take over many critical tasks such as developing a mathematical model or solving a system model, sufficient interaction between a designer and computer-aided design tool is essential to a successful machine design.

For a robotic design at different stages, design scopes and goals are different, and different design tools are needed at different stages. The SolidWorks provide a comprehensive toolset to support

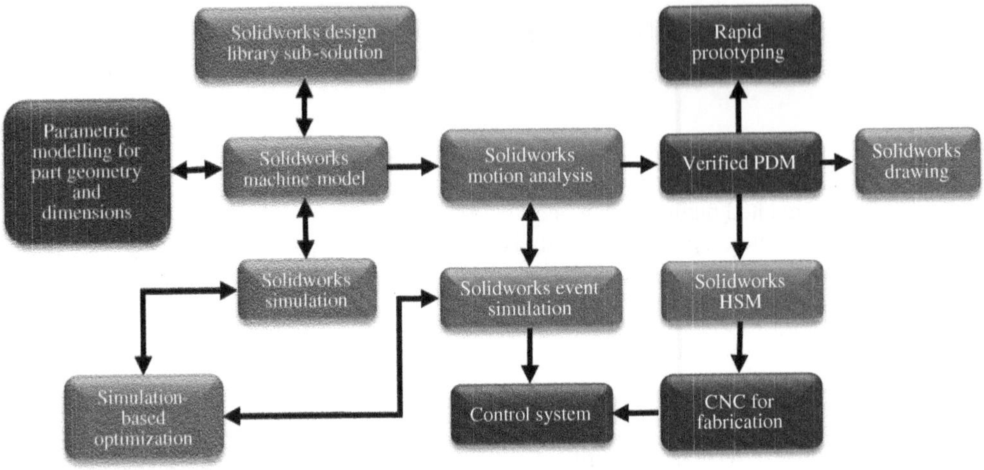

Figure 12.51 Main functional modules for virtual machine design in SolidWorks

virtual machine design (Markkonen 1999). Figure 12.51 shows the main functional modules which can be fully utilized in designing and analyzing a robot.

12.6.1 Motion Simulation

SolidWorks *motion study* is to create a simulation model to study the position, velocity, acceleration, and torque of a mechanism subject to external loads (SolidWorks 2024). It can be used to design, analyze, and optimize a machine design. Motion study in a virtual environment brings significant benefits to machine design: (1) the number of physical prototypes can be minimized since the simulation helps identify potential design errors and omissions; (2) virtual analysis takes much less time than experiments and it supports parametric studies. This allows to explore more design options at very early stage; (3) the simulation-based optimization can be utilized to improve machine design; (4) the simulation is integrated with all of other engineering analyses; it makes possible to gain quantified insights for additional engineering analysis and look into the feasibility of design over the entire product lifecycle.

A motion study model is capable of answering all of the questions related to kinematic and dynamic behaviors such as *what is the cycle time to machine operation? What force is required for given operation? What sizes of motors should be used* for the given task of the machine? *What shapes and dimensions will generate expected motions? What are expected velocities and accelerations* subject to given driving forces and torques? *What are accuracy of motion one can expect? Will parts fail? How long the machine will last* subject to the given operating condition?

12.6.2 Model Preparation

Create a design scheme where the shapes of parts are designed and the nature of their connections are presented in a simplified form; the materials are selected to determine allowable stresses accounting for all the factors that affect the part strength; the forces acting on parts in the process of machine operation are then determined. All of parts or components must assigned with proper material properties.

After modeling each part, the next step is to model the assemblies of parts. If two parts have a static relation in space. These two parts should be grouped since no relative motion is allowed

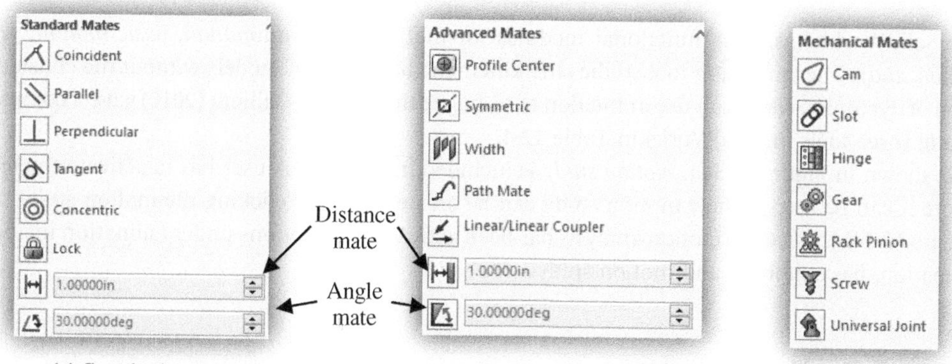

(a) Standard mates (b) Advanced mates (c) Mechanical mates

Figure 12.52 Mates in assembly modeling of SolidWorks

between them. If two parts have a relative motion, a correct joint type must be selected. Assembly relations of a machine are represented by mates in SolidWorks. Figure 12.52 shows the main mate types which are cataloged into (a) standard mates, (b) advanced mates, and (c) mechanical mates. A bonded relation or a joint of two parts is mainly defined as a combination of mates in (a) and (b); other mates occurring to more than two entities or a relative motion with coupling are defined by the mates in (b) and (c).

Figure 12.53 shows an example of assembling modeling of a Yumi collaborative robot. Yumi has two 7-DOF arms. Any one of the 14 joints can be defined as a combination of one "*coincident*" mate of two planes and one "*concentric*" mate of two cylindrical surfaces from two joined parts.

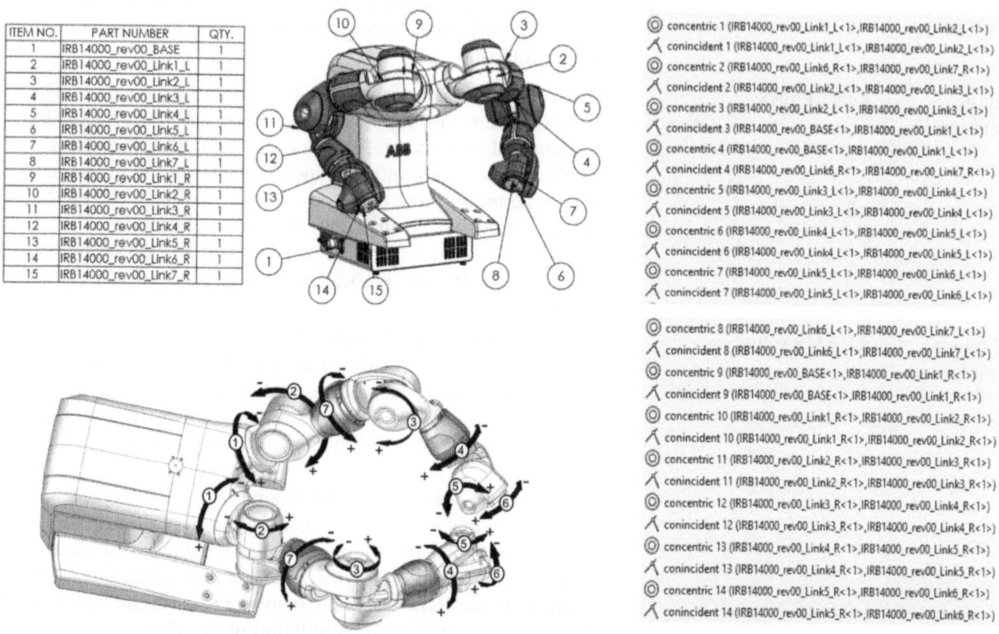

ITEM NO.	PART NUMBER	QTY.
1	IRB14000_rev00_BASE	1
2	IRB14000_rev00_Link1_L	1
3	IRB14000_rev00_Link2_L	1
4	IRB14000_rev00_Link3_L	1
5	IRB14000_rev00_Link4_L	1
6	IRB14000_rev00_Link5_L	1
7	IRB14000_rev00_Link6_L	1
8	IRB14000_rev00_Link7_L	1
9	IRB14000_rev00_Link1_R	1
10	IRB14000_rev00_Link2_R	1
11	IRB14000_rev00_Link3_R	1
12	IRB14000_rev00_Link4_R	1
13	IRB14000_rev00_Link5_R	1
14	IRB14000_rev00_Link6_R	1
15	IRB14000_rev00_Link7_R	1

(a) 14-joints (b) Assembling modeling

Figure 12.53 Assembly modeling of ABB Yumi robot (*Source:* ABB (2018)/ABB Asea Brown Boveri Ltd)

12.6.3 Creation of Simulation Model

The SolidWorks has three functional modules for motion study: *animation*, *basic motion*, and *motion study*. The animation tool studies the kinematic behaviors of models without the consideration of dynamics. Users can use animation to visualize the models. Galliera (2010) gave a comparison of three tools in SolidWorks in Table 12.4.

As shown in Figure 12.54a, *motion study* is included in Add-Ins. A user has to activate the tool before it can be used. A new motion study can be created by right clicking the motion study tab (Figure 12.54b). Since the motion analysis has been activated, the options under animation include animation, basic motion, and motion study.

12.6.4 Define Motion Variables

As shown in Figure 12.55, a motion study includes the motion variables for *motor*, *spring*, *damper*, *force*, *contact*, and *gravity*. For example, motors have to be defined for all of the active joints in a machine. The motion study allows users to set up many parameters for a motor.

Figure 12.56 shows the interface for the definition of motor. A motor motion can be *translational* or *rotational*, and the motion occurs to a moving body relative to a reference body along a specified

Table 12.4 Comparison of SolidWorks animation, basic motion, and motion analysis

Types	Solvers	Description
Animation	3D constraint manager (3DDCM) by D-cube	The 3DDCM solver is capable of positioning parts in an assembly or mechanism. Users can use it to build, modify, and animate the model to visualize the changes comprehensive geometry, dimensions, and design constraints. The animation time can be used to create simple animations where the interpolations are used to specify point-to-point motion of parts in assemblies
Basic motion	Ageia PhysX	The Ageia PhysX is a physics solver primarily for games. The tool simulates how objects behave, move, and react for life-like motion and interaction. Adopting the Ageia PhysX in basic motion makes the simulation look real but not the precise one. The basic motion tool can be used to approximate the effects of motors, springs, collisions, and gravity on assemblies. The physics-based simulation can be generated quickly with less computation. It suits the best for the presentation-worthy animation.
Motion study	ADAMS solver	The ADAMS solver is a sophisticated tool to analyze the complex behavior of mechanical assemblies. The motion analysis in SolidWorks aims at analyzing the forces, torques, contact forces, and power consumption accurately. Users can export any kinematic and dynamic quantities over time from analysis. Motion analysis can be used to simulate and analyze the motion of a machine with the consideration of driving forces, springs, dampers, and frictions. The kinematic solver takes into account of motion constraints, material properties, mass, and component contacts. The result of dynamic loads can be exported directly for discrete event simulation of machine.

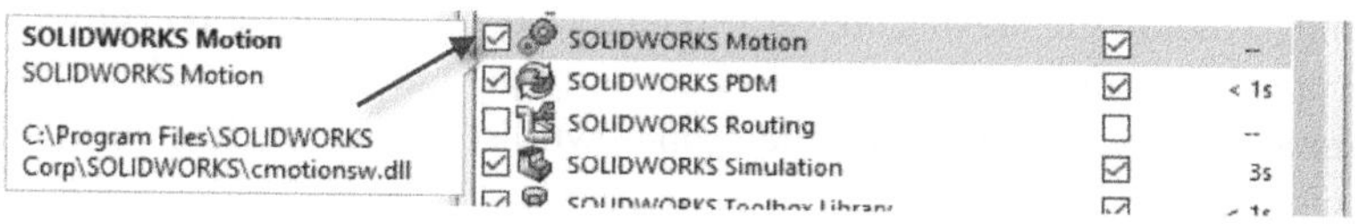

(a) Activate *Motion* in *Adds-Ins*

(b) Create *New Motion Study* by right-clicking the blank area above the *status bar*

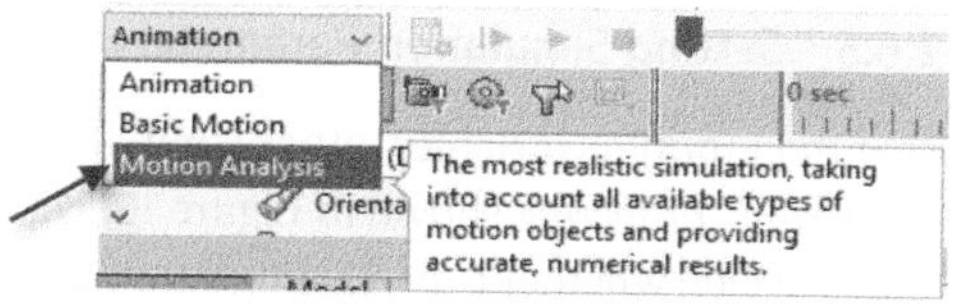

(c) Select *Motion Analysis* in the drop-down menu of *Animation*

Figure 12.54 Create a motion study for a machine model

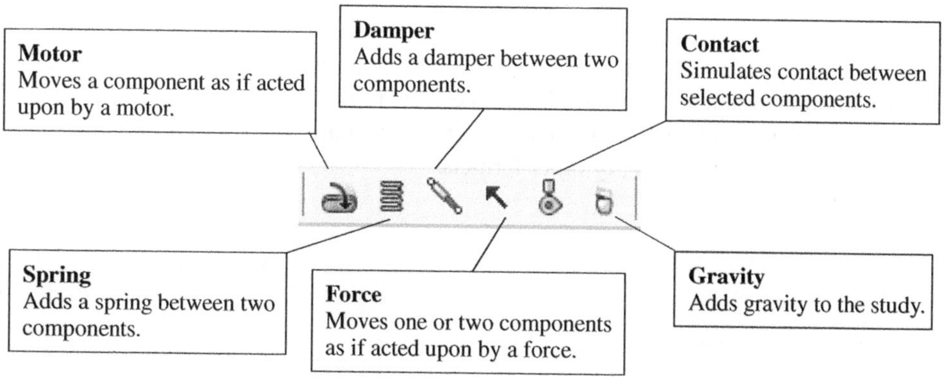

Figure 12.55 Types of motion variables in a motion study

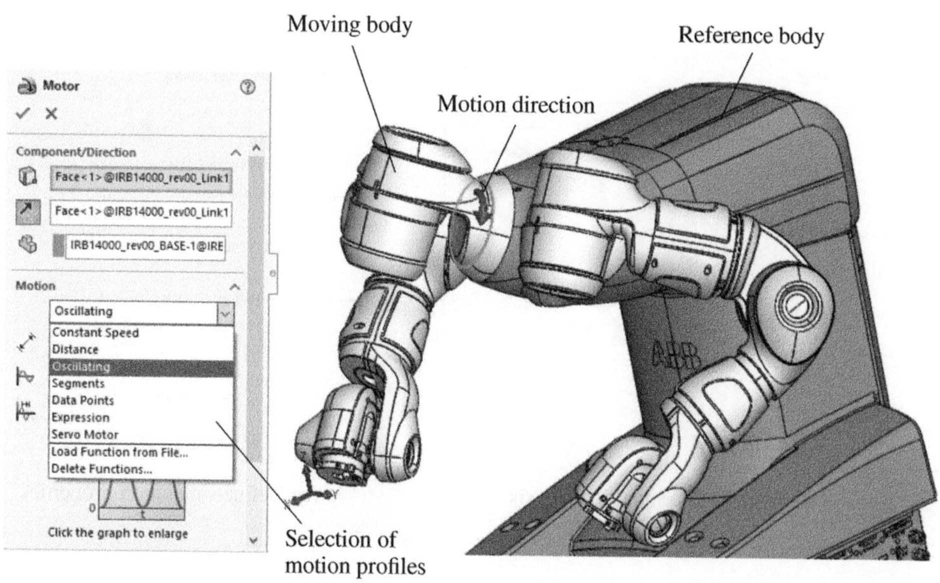

Figure 12.56 Defining a motor in a SolidWorks motion study

direction. In defining a motor, the moving body, the reference body, and the motion direction must be specified. The profile of a motion can be one of *constant speed, distance, oscillating, segments, data points, expression,* or *servo motor* from the drop-down list of *motion*. The direction of the motion must be specified.

12.6.5 Setting Simulation Parameters

Other than motion variables, other simulation parameters of the model can be customized via the *motion study properties* shown in Figure 12.57a. The user can set the number of calculation steps per second, decide whether run the simulation during the calculation, refine the accuracy of 3D contact or of the representation of solid geometry, specify the cycle settings, and select the analysis algorithms and tolerance of terminations as shown in Figure 12.57b.

12.6.6 Run Simulation and Visualize Motion

The motion analysis does the calculation automatically for the simulation model. Note that the calculation will not be updated unless the user click the calculation icon as shown in Figure 12.58. After the calculation is completed, the SolidWorks keeps the results, and the user can use the animation tools to review the animation of motions over time. In addition, the result from the simulation can be saved externally in .AVI or other formats.

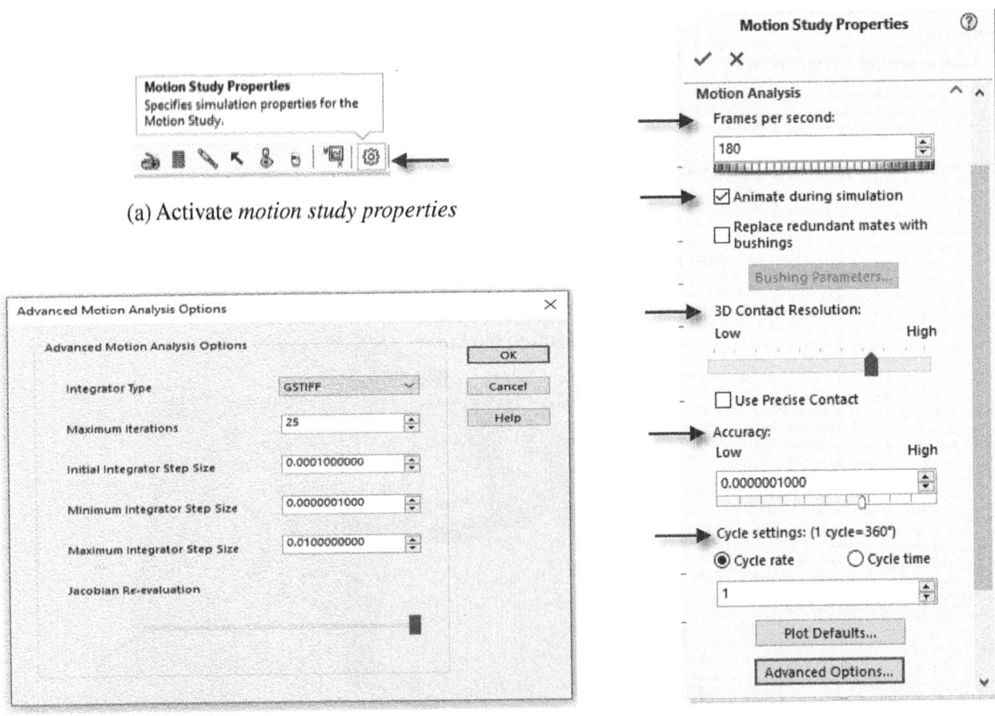

(a) Activate *motion study properties*

(c) Advanced setting for motion analysis

(b) Define simulation properties

Figure 12.57 Interfaces for settings of simulation model

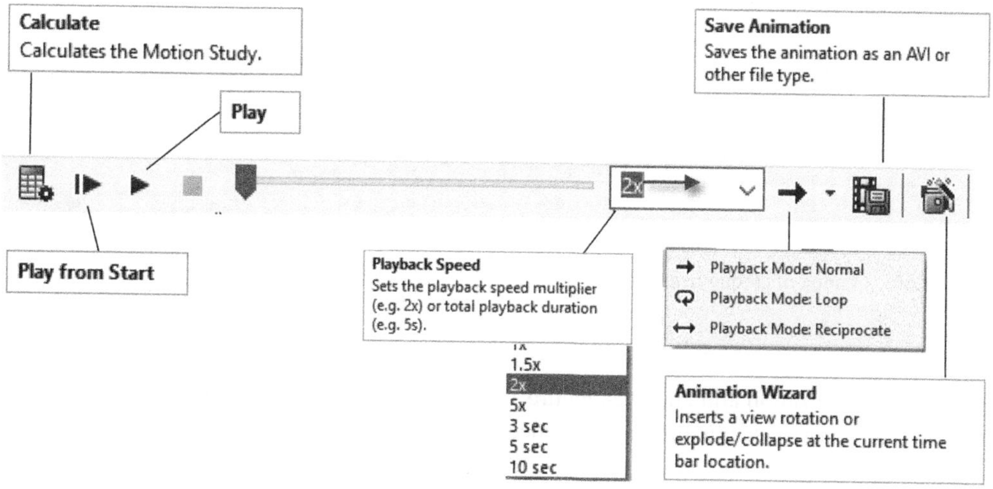

Figure 12.58 Run simulation and visualize motion

A special attention should be paid when the user makes some changes on motion variables or simulation settings after the previous simulation is completed. The user has to run calculate again to update the simulation results.

12.6.7 Analyze Simulation Data

A motion study model involves a large number of motion variables and simulation parameters. It should be very rare case that a user defines all of simulation parameters appropriately at his or her first iteration. The user must possess the knowledge of kinematics and dynamics, be able to understand the simulation results, and make the engineering judgments to determine if the simulation model is defined correctly. Figure 12.59 shows a wide selection of kinematic and dynamic variables in kinematic or dynamic models whose values in simulation with respect to time can be readily

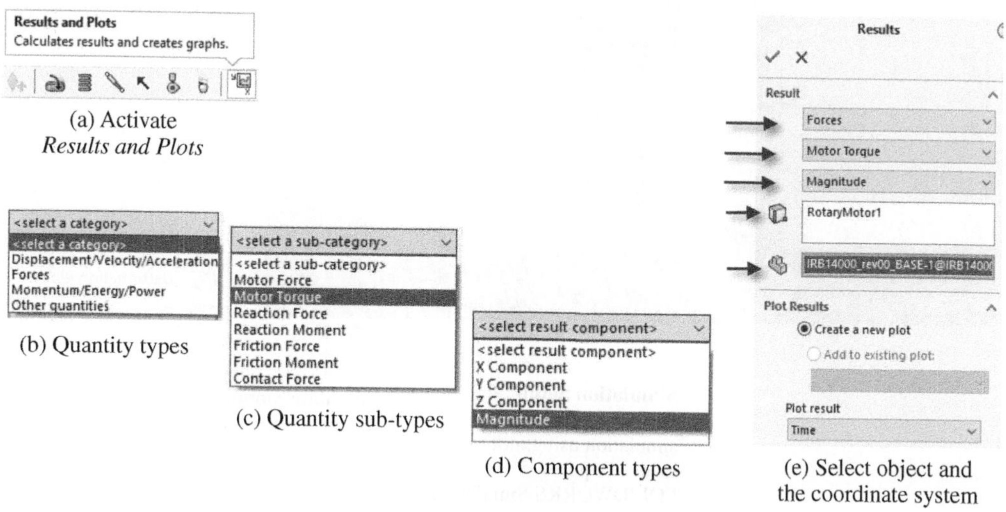

Figure 12.59 Review results of kinematic and dynamic variables

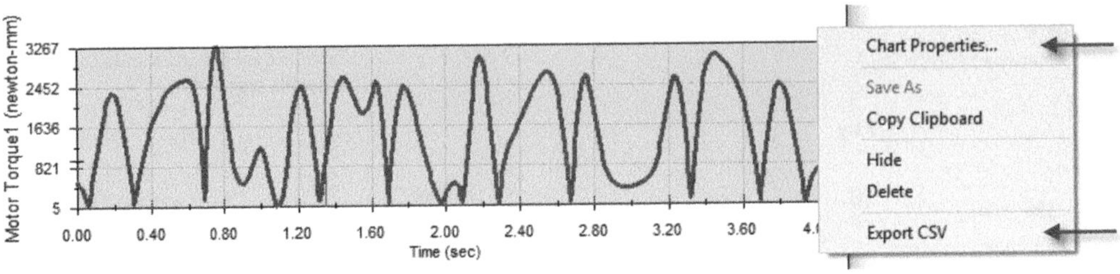

Figure 12.60 Change of torque over time

retrieved for visualization and analyses. Figure 12.60 shows an example plot for the change of the torque over time for a specific motor in the mechanism.

12.6.8 Structural Simulation Using Motion Loads

For a robot with motion, the exerted forces on its constitutive bodies are varied over time, and stress distributions over the bodies are changed accordingly. For a safe design, it is helpful to determine the maximum stress on a body during the motion. The historical loads from a motion analysis simulation can be utilized for structural analysis of the bodies in a machine.

In a structural analysis using motion loads, stresses, factor of safety, or deformation of components can be analyzed without setting up loads and boundary conditions. The loads are imported automatically from the results of a motion study. The user is able to investigate the effects of motion loads on the stress distributions and deformations for one or more components. Figure 12.61 shows

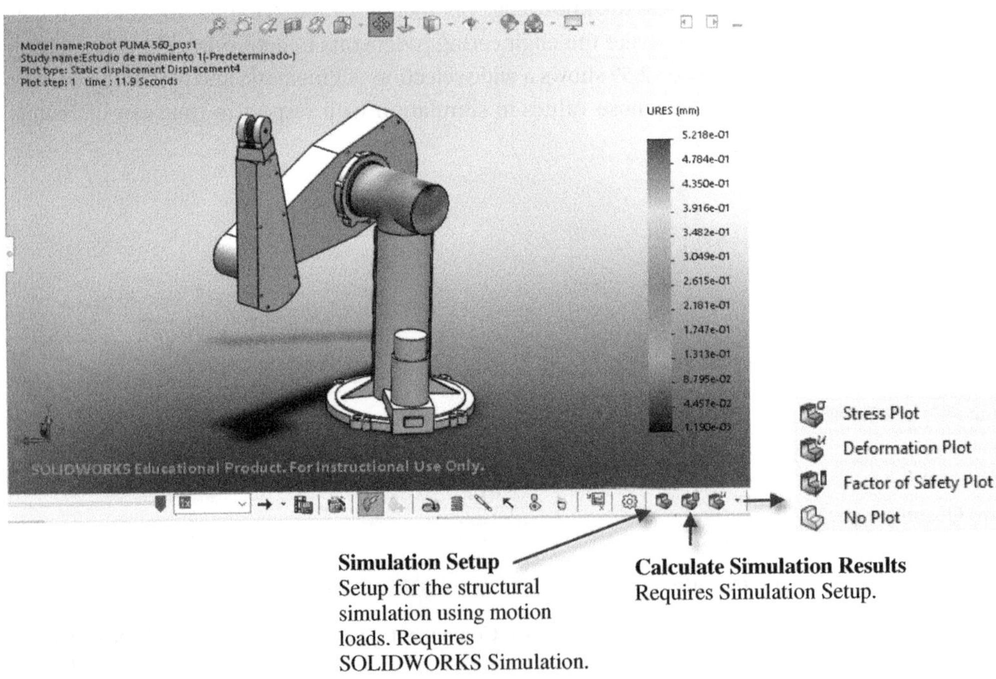

Figure 12.61 Mechanical event simulation in a motion analysis model

an example where the stress and deflection of a component in a robot have been analyzed at the specified time frame where the loads and boundary conditions are automatically defined using a motion study.

The analysis can be performed at specified times and time ranges. The flow of data from a motion study to structural simulation is one directional. The results of the stress analysis do not change the results of motion study. More detailed stress analysis should be performed in SolidWorks simulation as a combined rigid-flexible body analysis where the loads are exported from the motion study at a time (Bi and Zhang 2021).

Example 12.6 Figure 12.62 shows a modularized 3-DOF parallel kinematic machine (PKM) (Bi and Wang 2020). The diameters of base and end-effectors are set as 600 mm and 200 mm, respectively. Each branch of the 3-DOF parallel kinematic machine consists of one 70-mm active rotary joint, two 70-70-110 type-A links, and two 70-mm passive rotary joints. Taking the reference home position of the assembly model, the motions of three active joints are given in Table 12.5.

Assume that the materials are set as 1060 alloy for all components in the assembly model, and the direction of gravity force is along negative Z_b. (1) Create the assembly model for motion analysis. (2) Define and analyze the motion of the end-effector based on the given joint motion. (3) Visualize the trace of the reference (O_e) of the end-effector platform. (4) Export the results of displacements

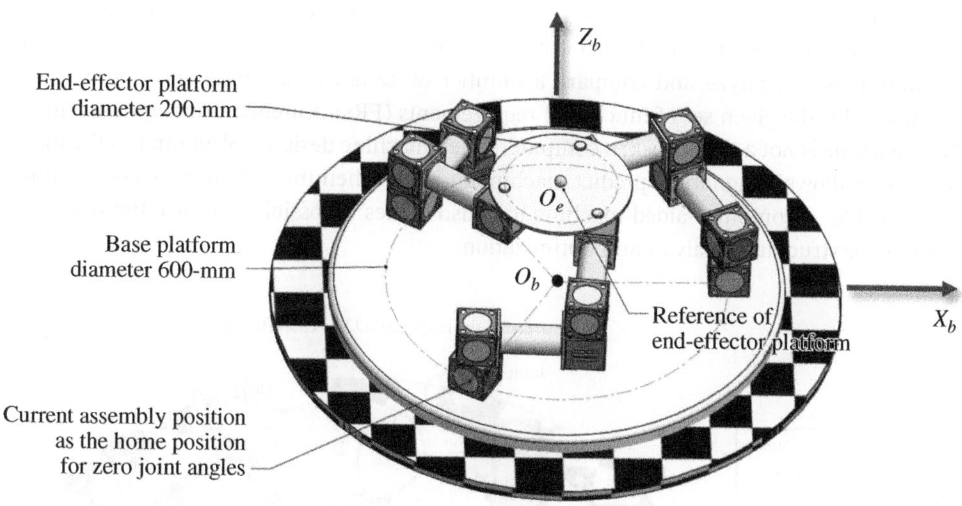

Figure 12.62 Assembly model of 3-DOF parallel kinematic machine

Table 12.5 Specified joint motions

Joint no.	Motion profile	Motion range (°)	Frequency (hertz)
1	*Oscillation*	40	1
2		35	0.5
3		20	1

and driving torques of active joints. (5) Export the results of the displacements of O_e (x_e, y_e, θ_e) with respective time.

Solution

Figure 12.62 shows the assembly model of 3-DOF parallel kinematic machine. It consists of a ground platform, three 3-DOF branches, and an end-effector platform. To create a motion analysis model, all active joints are defined as "motors." In the given 3-DOF parallel kinematic mechanism, each branch is built from two passive rotary joints and one active rotary joint. The first rotary joint in each branch is selected to define the motor for this branch as shown in Figure 12.63. The motion properties of three motors are defined based on Table 12.5 accordingly. The completed model includes three motors; in addition, the direction of gravity acceleration is specified to take into consideration the effect of the weights in motion.

The simulation is calculated, and the simulation results can be visualized, analyzed, and exported. Figures 12.64, 12.65, and 12.66 show the changes of joint displacements, driving torques, and end-effector motions over time, respectively.

12.6.9 Summary on Kinematic and Dynamic Modeling

Kinematics and dynamics lay the foundation to machine design since the motions of a machine are driven by force and power. Kinematics is to study the relation of the motions of rigid bodies in a machine, and dynamics is to study the causes of motions. Robotic design involves in design analysis and synthesis. Design analysis is to predict outcomes for a given configuration of a machine, and design synthesis is to analyze and compare a number of design alternatives and optimize the machine design for the given set of functional requirements (FRs). Kinematic and dynamic modeling for a machine is not a trivial task; a computer-aided machine design tool automates the modeling process. It allows designers to predict machine motions when the virtual model of machine becomes available. A computer-aided machine tool also makes it possible to utilize the results of motion study for structural analysis and optimization.

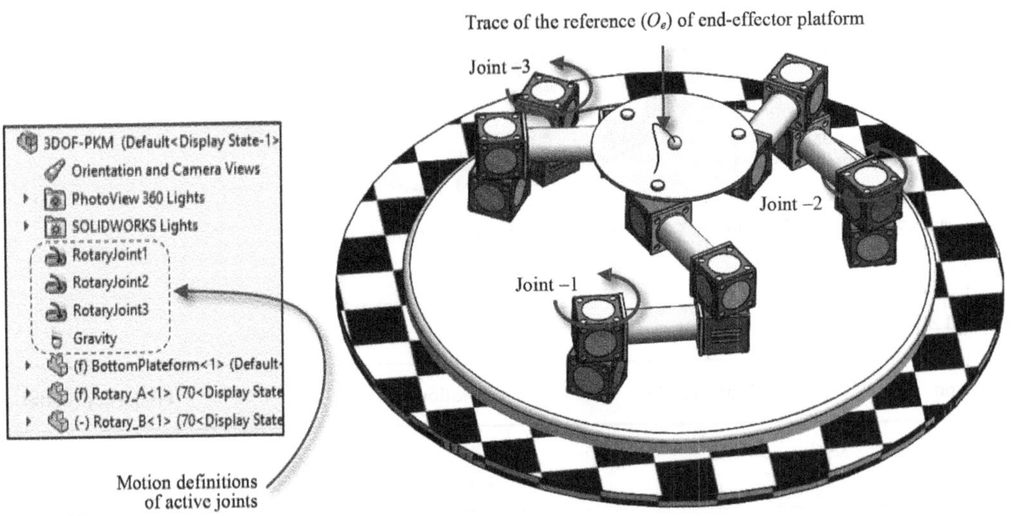

Figure 12.63 The trace of the reference O_e subjected to the given motions of three active joints

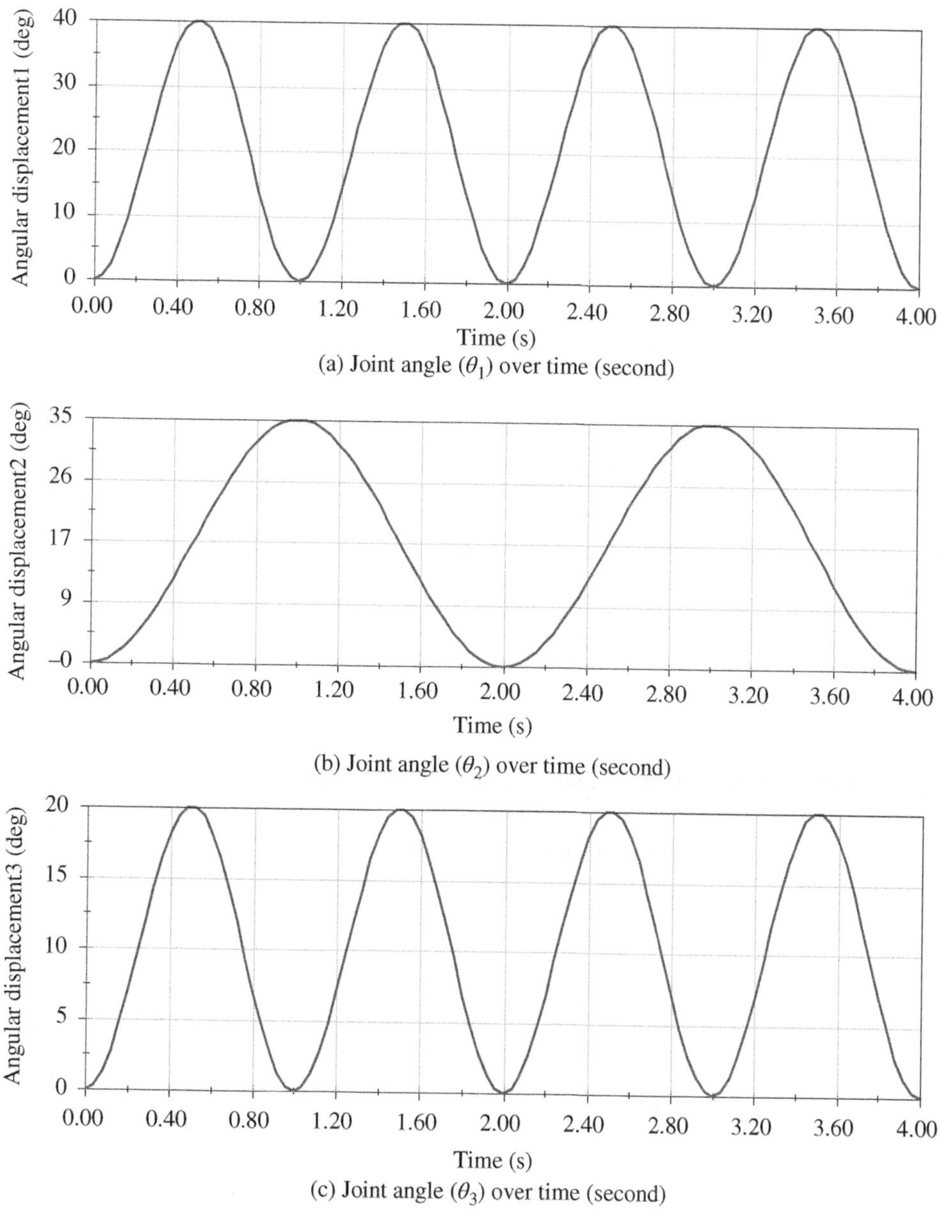

Figure 12.64 Joint displacements over time

12.7 Mobile Robots

Due to their superior abilities in exploring complex environments such as space, transportation, agriculture, rescue operations, mobile robots have received much attention in past decades. They have been widely used to accomplish various tasks autonomously with minimized humans intervenes. Depending on how a locomotion is implemented, mobile robots are classified into *wheeled*, *legged*, and *flying robots*. A wheeled robot traverses on ground by actuated wheels. In comparison

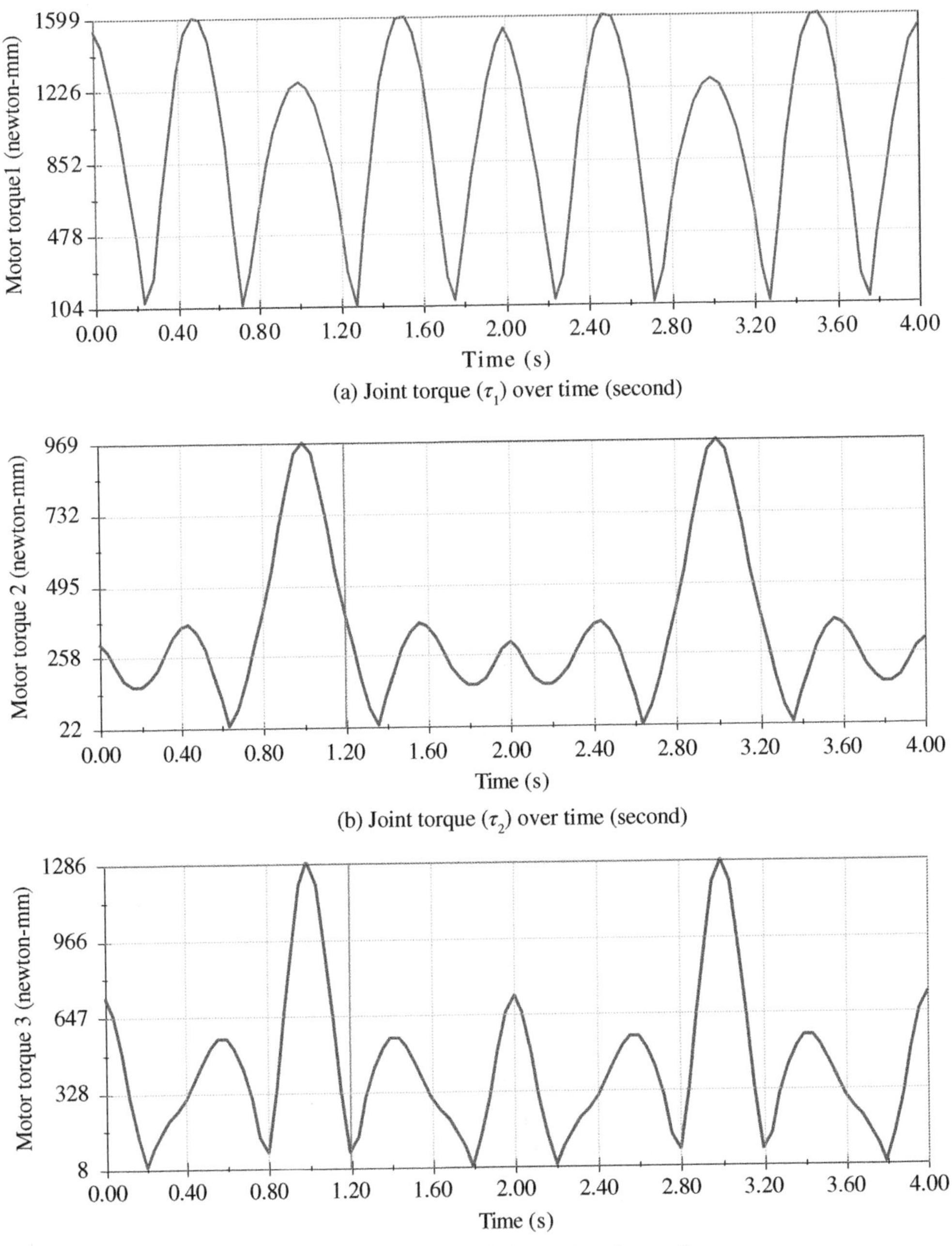

Figure 12.65 Joint torques over time

with a legged or flying robot, a wheeled robot is relatively easier to be designed, built, and operated. Moreover, it is more effective to move on flat and rocky terrain, and it requires few DOF. On the other hand, a wheeled robot faces the challenges to move on a smooth surface, cliff, or a rugged terrain since it needs a friction force to push the robot and the friction force relates to the *coefficient*

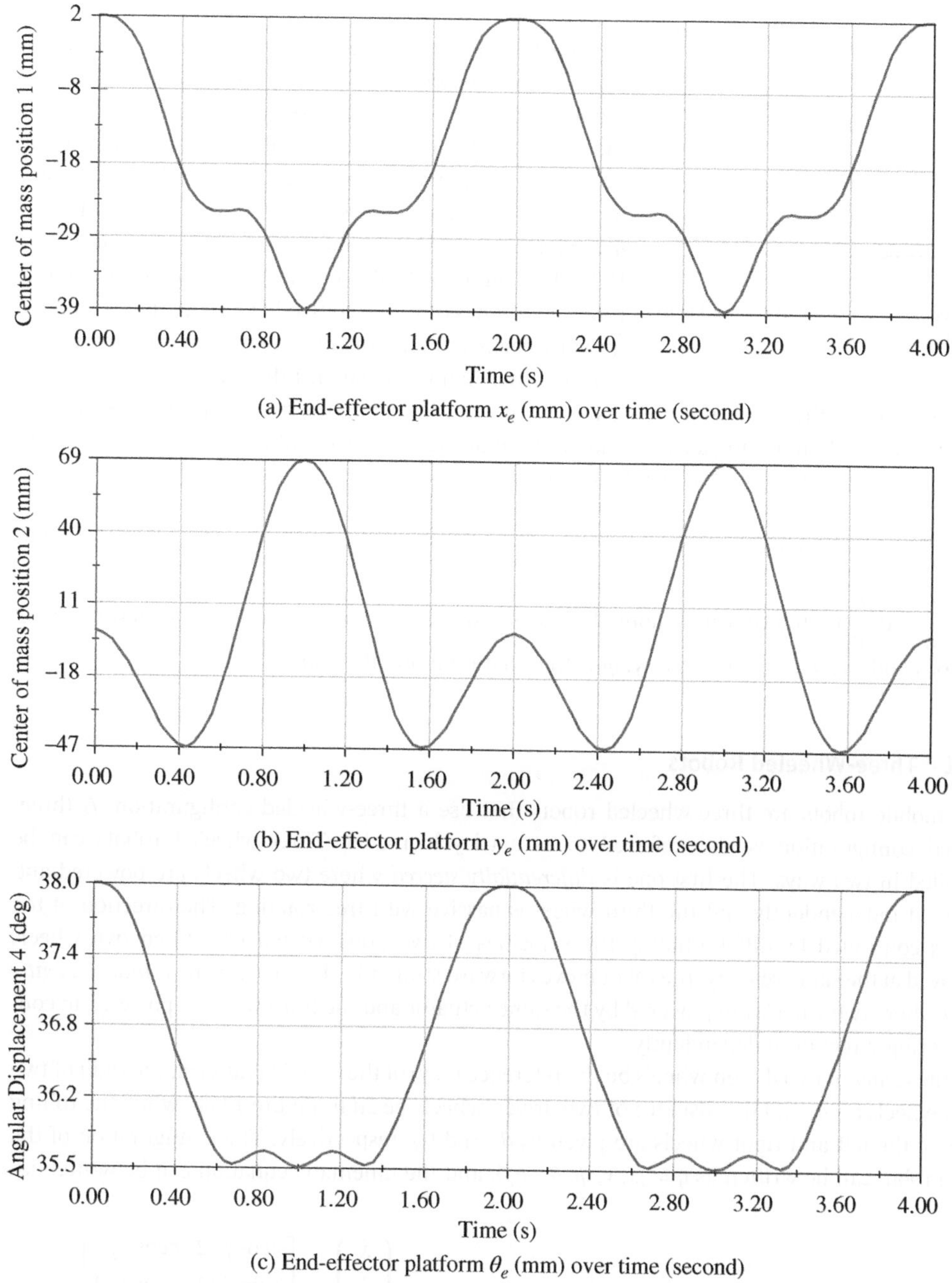

Figure 12.66 The displacements of end-effector over time

of friction (CoF) and the normal force. Nevertheless, due to the simplicity of implementation and low cost, wheeled robots are the most popular applications.

A wheeled robot can have any number of wheels, and there should be a mechanism to achieve a dynamic balance based on the number of wheels. Most of the wheeled robots use three or four

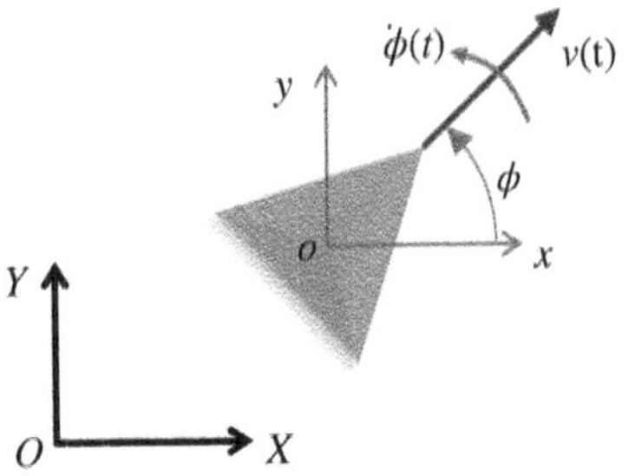

Figure 12.67 Coupling of three DOF in non-holonomic robot

wheels to achieve a static or dynamic balance. Moreover, it is possible that (1) not all wheels are actuated and (2) the number of actuated wheels is more, equal, or less than the required DOF in an application. *A holonomic robot* has the same number of controllable DOF with the total DOF; it can move in any direction freely. A robot built on castor wheels or omni-wheels is a holonomic robot. *A non-holonomic robot* has a smaller number of controllable DOF than the total DOF. A smart car is a non-holonomic robot since it has only two controllable DOF, that is, moving forward/backward and turning angle of the steering wheel. On the other hand, the car has three motions, that is, the translations along the x-axis and the y-axis and the rotation along the z-axis. This causes the coupling of three DOF in driving a car as shown in Figure 12.67. The coupling constraint of the non-holonomic robot in Figure 12.67 can be modeled as,

$$\frac{dx}{dt}\sin\phi - \frac{dy}{dt}\cos\phi = 0 \tag{12.96}$$

where $\frac{dx}{dt}$ and $\frac{dy}{dt}$ are the velocities along the x-axis and the y-axis; x and y are the translations along the x-axis and the y-axis, respectively; ϕ is the rotation along the z-axis.

12.7.1 Three-Wheeled Robots

Many mobile robots are three-wheeled robots that use a three-wheeled configuration. A three-wheeled configuration is also referred to as a tricycle model. Three-wheeled robots can be controlled in two ways. The first one is *differentially steered* where two wheels are powered and controlled independently and the third wheel is passive with free rotating. The direction of the robot is controlled by differentiating the velocities of two powered wheels. When two wheels are moved at the same velocity, the robot moves forward straightly. The second one is *independently steered* where two wheels are powered by the same actuator and the third wheel is powered to control steering direction independently.

Let the *center* of two driven wheels be the reference (x, y) of the mobile platform, the radii of two driven wheels are r, and the distance of two driven wheels be $2d$ in Figure 12.68, when the rolling angles of the left and right wheels are given as θ_L and θ_R, respectively. The configuration of the mobile robot can be written as $q = \{x, y, \phi, \theta_L, \theta_R\}$, and the kinematic equation can be written as,

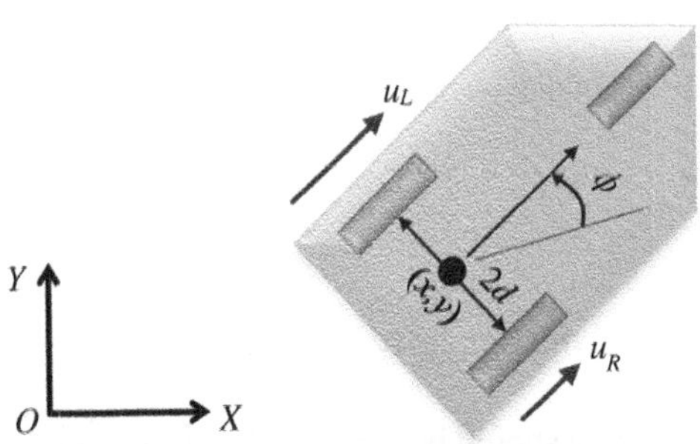

Figure 12.68 Differentially steered robot

$$\dot{q} = \begin{Bmatrix} \dot{x} \\ \dot{y} \\ \dot{\phi} \\ \dot{\theta}_L \\ \dot{\theta}_R \end{Bmatrix} = \begin{bmatrix} r\cos\phi/2 & r\cos\phi/2 \\ r\sin\phi/2 & r\sin\phi/2 \\ -r/(2d) & r/(2d) \\ 1 & 0 \\ 0 & 1 \end{bmatrix} \begin{Bmatrix} u_L \\ u_R \end{Bmatrix} \tag{12.97}$$

where u_L and u_R are the angular velocities of the left and right wheels, respectively.

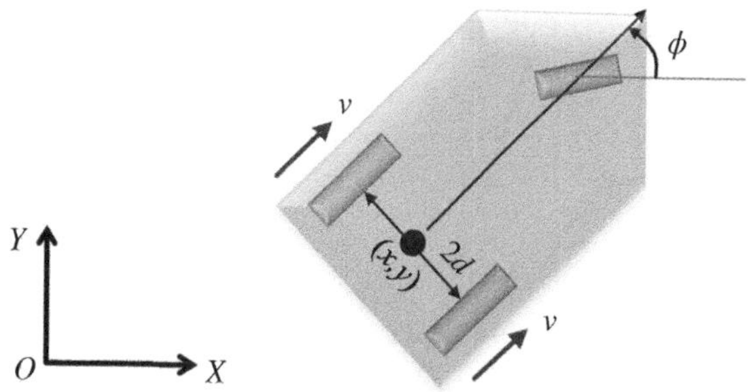

Figure 12.69 Independently steered robot

Let the configuration of an independently steered robot be $q = \{x, y, \phi\}$ in Figure 12.69, the kinematic equation can be written as,

$$\dot{q} = \begin{Bmatrix} \dot{x} \\ \dot{y} \\ \dot{\phi} \end{Bmatrix} = \begin{Bmatrix} v\cos\phi \\ v\sin\phi \\ \omega \end{Bmatrix} \tag{12.98}$$

where v is the linear velocity of two driven wheels and ω is the angular velocity of the steering wheel.

12.7.2 Four-Wheeled Robots

A four-wheeled robot can be one of the three commonly used configurations. *The first configuration* is with two powered wheels and other two wheels with passive but free rotating. This is similar to a differentially steered robot with three wheels; however, a robot with four wheels can be easily balanced since *the center of gravity* (CoG) remains in a rectangle formed by the supports to four wheels instead of a triangle by three wheels. It is worth noting that CoG of a four-wheeled robot should be set as close to the geometric center of the rectangle as possible to maximize the stability when the robot takes a sharp turn or move on rough surface. *The second configuration* is with a pair of two-powered wheels, and two wheels in each pair run at the same speed and turn in the same direction. A speed difference of two pairs allows the robot to change the direction of movement. *The third configuration* is the Ackerman drive that is similar to an independently steered robot with three wheels. In comparison with the previous two configurations, the Ackerman drive has one motor to drive two rear wheels and one servo to steer the direction of movement as shown in Figure 12.70. Both previous methods require two motors for independently controlled rotations at two sides.

Let the configuration of a four-wheeled robot with the Ackerman drive be $q = \{x, y, \phi, \psi\}$, where (x, y) is the location of the midpoint between the rear wheels, ϕ is the heading direction, and ψ is the steering angle defined at a virtual wheel at the midpoint between two front wheels. The kinematic equation can be written as,

$$\dot{q} = \begin{Bmatrix} \dot{x} \\ \dot{y} \\ \dot{\phi} \\ \dot{\psi} \end{Bmatrix} = \begin{bmatrix} \cos\phi & 0 \\ \sin\phi & 0 \\ (\tan\psi)/l & 0 \\ 0 & 1 \end{bmatrix} \begin{Bmatrix} v \\ \omega \end{Bmatrix} \tag{12.99}$$

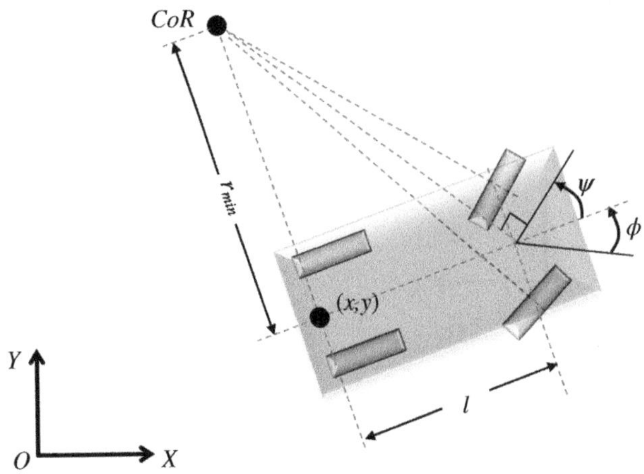

Figure 12.70 Four-wheeled robot with Ackerman drive

where v is the linear velocity of two real wheels and ω is the angular velocity of the virtual steering wheel. l is the distance of the front and rear wheels.

In Eq. (12.99), when the robot with the Ackerman drive controls the steering angle ψ instead of $\dot{\psi}$ or ω, the robotic control can be simplified as,

$$v = v, \quad \psi = \tan^{-1}\left(\frac{l\omega}{v}\right) \tag{12.100}$$

12.7.3 Unmanned Aerial Vehicles (UAVs)

Flying robots are commonly referred to as *unmanned aerial vehicles* (UAVs) or drones. Unlike legged and wheeled robots, flying robots are free to utilize the full six degrees of freedom, allowing for different types of flight for a drone. These are known as *Yaw*, *Pitch*, and *Roll* as shown in Figure 12.71.

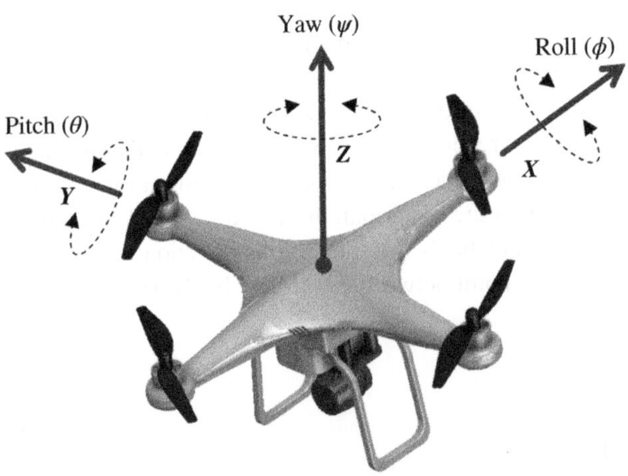

Figure 12.71 Yaw, pitch, and roll of a flying robot

- *Yaw* (ψ) is a rotation of a flying robot's front to either right or left. It spins along the z-axis of the robot. In a remotely piloted robot, this is usually achieved by moving the left throttle stick to left/right.
- *Pitch* (θ) is a rotation of the body of the flying robot to either forward or backward. The forward and backward pitch is generally achieved by pushing a throttle stick forward and backward.
- *Roll* (ϕ) is a rotation of a flying robot on sideways to either right or left. In a remotely piloted robot, this is usually achieved by moving the right throttle stick to left/right.

In this section, a quadrotor in Figure 12.72 is used as an example to illustrate the procedure of kinematic and dynamic modeling (Luukkonen 2011). Note that any object in a 3D space has 6-DOF while a quadrotor only has four rotors, it is under-actuated and it implies that the motion controls in different directions are coupled.

The position of the quadcopter is represented by the coordinates (x, y, z) in the global CS $\{\boldsymbol{O}_G - \boldsymbol{X}_G\boldsymbol{Y}_G\boldsymbol{Z}_G\}$; the orientation of the quadcopter is represented by the Euler angle (ψ, θ, ϕ); the vector $(\boldsymbol{q})$ of linear displacements $(\boldsymbol{\xi})$ and angular displacements $(\boldsymbol{\eta})$ can be represented as,

$$\xi = \begin{bmatrix} x \\ y \\ z \end{bmatrix}, \boldsymbol{\eta} = \begin{bmatrix} \phi \\ \theta \\ \psi \end{bmatrix}, \boldsymbol{q} = \begin{bmatrix} \xi \\ \boldsymbol{\eta} \end{bmatrix} \tag{12.101}$$

Assume that the reference position of the quadrotor is situated at its geometric center C, and the linear velocities $(\boldsymbol{v}_C)$ and angular velocities $(\dot{\boldsymbol{\theta}}_c)$ are denoted as,

$$\boldsymbol{v}_C = \begin{bmatrix} v_x \\ v_y \\ v_z \end{bmatrix}, \dot{\boldsymbol{\theta}}_c = \begin{bmatrix} \dot{\theta}_x \\ \dot{\theta}_y \\ \dot{\theta}_z \end{bmatrix} \tag{12.102}$$

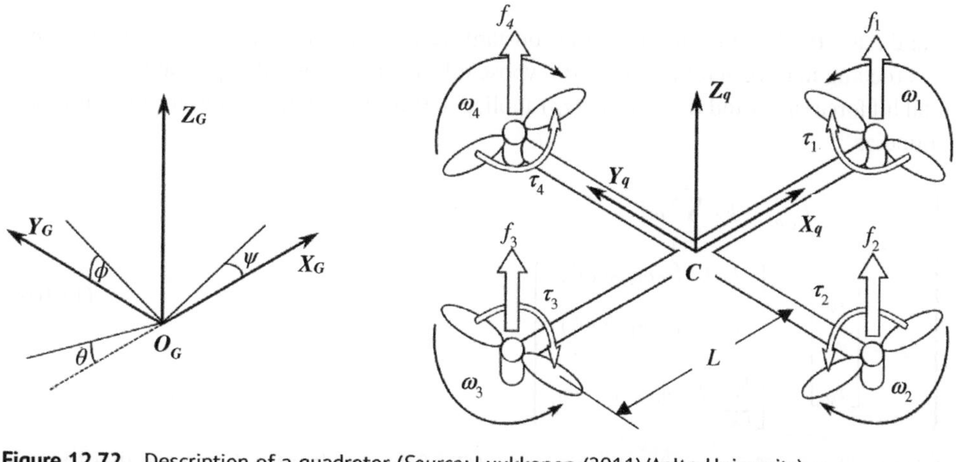

Figure 12.72 Description of a quadrotor (*Source:* Luukkonen (2011)/Aalto University)

The transformation from the local CS $\{\boldsymbol{C}-\boldsymbol{X}_q\boldsymbol{Y}_q\boldsymbol{Z}_q\}$ to the global CS $\{\boldsymbol{O}_G-\boldsymbol{X}_G\boldsymbol{Y}_G\boldsymbol{Z}_G\}$ can be rewritten as,

$$\boldsymbol{T}_q=\begin{bmatrix}\boldsymbol{R} & \xi\\ 0 & 1\end{bmatrix}=\begin{bmatrix}C\varphi\cdot C\theta & C\varphi\cdot S\theta\cdot S\phi-S\varphi\cdot C\theta & C\varphi\cdot S\theta\cdot C\phi+S\varphi\cdot S\theta & x\\ S\varphi\cdot C\theta & S\varphi\cdot S\theta\cdot S\phi+C\varphi\cdot C\theta & S\varphi\cdot S\theta\cdot C\phi-C\varphi\cdot S\theta & y\\ -S\theta & C\phi\cdot C\theta\cdot & C\varphi\cdot C\theta\cdot & z\\ 0 & 0 & 0 & 1\end{bmatrix} \tag{12.103}$$

where S and C denote a sine and cosine function, respectively.

Equation (12.103) can be used to find the transformation of the angular velocities between $\{\boldsymbol{C}-\boldsymbol{X}_q\boldsymbol{Y}_q\boldsymbol{Z}_q\}$ and $\{\boldsymbol{O}_G-\boldsymbol{X}_G\boldsymbol{Y}_G\boldsymbol{Z}_G\}$ as,

$$\dot{\eta}=\begin{bmatrix}\dot{\phi}\\ \dot{\theta}\\ \dot{\psi}\end{bmatrix}=W_\eta^{-1}\dot{\boldsymbol{\theta}}_c=\begin{bmatrix}1 & S\phi\cdot T\theta & C\phi\cdot T\theta\\ 0 & C\phi & -S\phi\\ 0 & S\phi/C\theta & C\phi/C\theta\end{bmatrix}\begin{bmatrix}\dot{\theta}_x\\ \dot{\theta}_y\\ \dot{\theta}_z\end{bmatrix} \tag{12.104}$$

where $T(\cdot)$ is the function of $T(\cdot)$; alternatively, Eq. (12.104) can be written as,

$$\dot{\boldsymbol{\theta}}_c=\begin{bmatrix}\dot{\theta}_x\\ \dot{\theta}_y\\ \dot{\theta}_z\end{bmatrix}=W_\eta\dot{\eta}=\begin{bmatrix}1 & 0 & -S\theta\\ 0 & C\phi & C\theta\cdot S\phi\\ 0 & -S\phi & C\phi/C\theta\end{bmatrix}\begin{bmatrix}\dot{\phi}\\ \dot{\theta}\\ \dot{\psi}\end{bmatrix} \tag{12.105}$$

Assume that the quadcopter has a symmetric structure where four arms are aligned with local x- and y-axes, respectively. The inertia matrix $\boldsymbol{I}$ is diagonal and $I_{xx}=I_{yy}$ as,

$$I=\begin{bmatrix}I_{xx} & 0 & 0\\ 0 & I_{yy} & 0\\ 0 & 0 & I_{zz}\end{bmatrix} \tag{12.106}$$

Denote the angular velocity and resulted lift force of rotor i as ω_i and f_i where $(i = 1, 2, 3, 4)$, respectively, one can determine the lift force and torque based on given angular velocity as,

$$\begin{cases} f_i=k\cdot\omega_i^2\\ \tau_i=b\cdot\omega_i^2+I_m\cdot\acute{\omega}_i\end{cases}\quad(i=1,2,3\text{ or }4) \tag{12.107}$$

where k and b are the lift constant and drag constant, respectively; I_m is the inertia moment of a rotor; $\acute{\omega}_i$ is the angular acceleration of rotor i whose effect can be normally ignorable.

The total lift force (f_{total}) and the torques (τ_C) applied to three axes of the quadrotor by four rotors can be found as,

$$\begin{cases} f_{total}=k\cdot\sum_{i=1}^{4}\omega_i^2\\ \tau_C=\begin{bmatrix}\tau_\phi\\ \tau_\theta\\ \tau_\psi\end{bmatrix}=\begin{bmatrix}L\cdot k\cdot\left(-\omega_2^2+\omega_4^2\right)\\ L\cdot k\cdot\left(-\omega_1^2+\omega_3^2\right)\\ \sum_{i=1}^{4}\left(b\cdot\omega_i^2+I_m\cdot\acute{\omega}_i\right)\end{bmatrix}\end{cases} \tag{12.108}$$

where L is the distance from a robot to the center ($\boldsymbol{C}$) of the quadcopter. It can see that the roll movement is determined by the second and fourth rotors, and the pitch movement is determined by the first and third rotors. A yaw movement is obtained by increasing the angular velocities of two opposite rotors and decreasing the velocities of the others.

The Newton–Euler equations can be used to build the dynamic model of the quadrotor; a dynamic model describes the relations of the motion accelerations with the lift forces applied on four rotors. Applying the Newton law gets,

$$m\dot{\boldsymbol{v}}_C + \dot{\boldsymbol{\theta}}_c \times (m\boldsymbol{v}_C) = R^T\boldsymbol{G} + f_{total} \tag{12.109}$$

where m is the total mass of the quadrotor; $\boldsymbol{G}$ is the vector of the gravity acceleration.

Equation (12.109) can be used to find the linear acceleration of the quadrotor as,

$$\begin{bmatrix} \ddot{x} \\ \ddot{y} \\ \ddot{z} \end{bmatrix} = \frac{f_{total}}{m}\begin{bmatrix} C\psi \cdot S\theta \cdot C\phi + S\psi \cdot S\phi \\ S\psi \cdot S\theta \cdot C\phi - C\psi \cdot S\phi \\ C\theta \cdot C\phi \end{bmatrix} - g\begin{bmatrix} 0 \\ 0 \\ 1 \end{bmatrix} \tag{12.110}$$

where g is the gravity acceleration.

Applying the Euler equation gets,

$$I \cdot \ddot{\boldsymbol{\theta}}_c + \dot{\boldsymbol{\theta}}_c \times \left(I \cdot \dot{\boldsymbol{\theta}}_c\right) + \Gamma = \boldsymbol{\tau}_C \tag{12.111}$$

where Γ is the gyroscopic force.

Equation (12.111) can be used to find the angular acceleration of the quadrotor as,

$$\begin{bmatrix} \ddot{\theta}_x \\ \ddot{\theta}_y \\ \ddot{\theta}_z \end{bmatrix} = \begin{bmatrix} \left(I_{yy} - I_{zz}\right) \cdot \dot{\theta}_y \cdot \dot{\theta}_z / I_{xx} \\ \left(I_{zz} - I_{xx}\right) \cdot \dot{\theta}_x \cdot \dot{\theta}_z / I_{yy} \\ \left(I_{xx} - I_{yy}\right) \cdot \dot{\theta}_x \cdot \dot{\theta}_y / I_{zz} \end{bmatrix} - I_r \begin{bmatrix} \dfrac{\dot{\theta}_y}{I_{xx}} \\ -\dfrac{\dot{\theta}_x}{I_{yy}} \\ 0 \end{bmatrix} \omega_\Gamma + \begin{bmatrix} \tau_\phi / I_{xx} \\ \tau_\theta / I_{yy} \\ \tau_\psi / I_{zz} \end{bmatrix} \tag{12.112}$$

where $\omega_\Gamma = \omega_1 - \omega_2 + \omega_3 - \omega_4$.

Equation (12.104) can be used to transform angular acceleration from the global CS in Eq. (12.112) to local coordination system of the quadrotor as (Luukkonen 2011),

$$\begin{bmatrix} \ddot{\theta}_x \\ \ddot{\theta}_y \\ \ddot{\theta}_z \end{bmatrix}_{local} = \begin{bmatrix} 0 & \dot{\phi} \cdot C\phi \cdot T\theta + \dot{\theta} \cdot S\phi / C^2\theta & -\dot{\phi} \cdot S\phi \cdot C\theta + \dot{\theta} \cdot C\phi / C^2\theta \\ 0 & -\dot{\phi} \cdot S\phi & -\dot{\phi} \cdot C\phi \\ 0 & \dot{\phi} \cdot C\phi / C\theta + \dot{\theta} \cdot S\phi \cdot T\theta / C\theta & -\dot{\phi} \cdot S\phi / C\theta + \dot{\theta} \cdot C\phi T\theta / C\theta \end{bmatrix} \begin{bmatrix} \dot{\theta}_x \\ \dot{\theta}_y \\ \dot{\theta}_z \end{bmatrix}_{global} + W_\eta^{-1} \begin{bmatrix} \ddot{\theta}_x \\ \ddot{\theta}_y \\ \ddot{\theta}_z \end{bmatrix}_{global} \tag{12.113}$$

12.8 Robotic Programming

A robot is programmable to fulfill different tasks and a program is essential to operate a robot. Whether it is a lego system, a drone, or a manipulator, the robot is programmed or coded to perform specific tasks. Many languages have been used to program robots, and most commonly used languages are C+, C++, Python, Java, MATLAB, LabVIEW, or *programmable logic controller* (PLC). In programming, it is necessary to understand a robot by identifying inputs and outputs and developing the mathematic models to transfer inputs into outputs of system. Once the flowchart of the program is available, appropriate language can be selected to program a robot. For example, C or C++ can be used to reduce the execution time at the hardware level. Python is very user-friendly and preferable to be used at the prototyping stage. *MATLAB, LabVIEW*, and *PLC* are graphic-based to support user-friendly programming.

Flowcharts help engineers to design and visualize the logic and execution of program. Figure 12.73 shows five basic elements that are used to construct a flowchart. A flowchart generally flows from left to right and from top to bottom. As shown in Figure 12.74, *pseudocode* is the other common tool to plan a program in which the geometric shapes in a flowchart are replaced by the instructions in plain statements in nature languages.

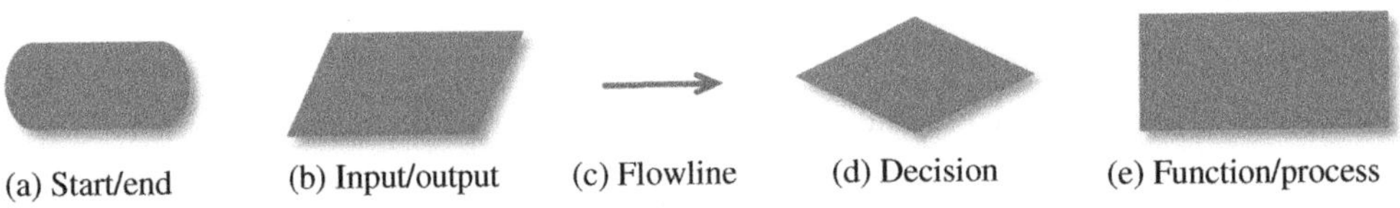

Figure 12.73 Basic elements of flowchart

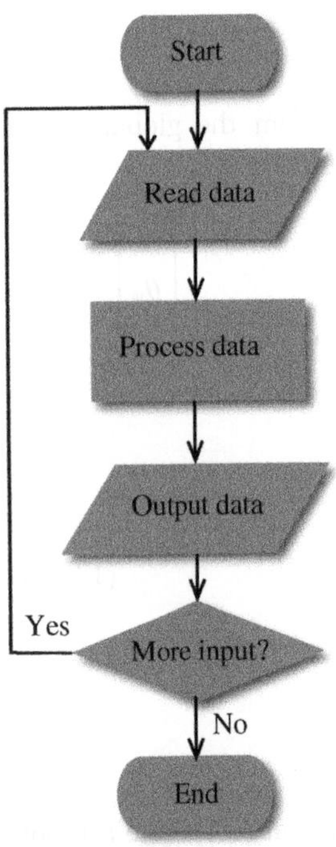

(a) Flowchart

```
Program simple_data_flow
Repeat
    read data;
    process data;
    output the processed data;
    Is there another input?
Until no more input
```

(b) Pseudocode

Figure 12.74 Data-driven robotic control

12.9 Summary

Mechatronics encompasses most intelligent systems where disciplinary solutions in mechanical, electrical, electronic, computer, and control engineering are integrated as mechatronic solutions. Robotics is tied to mechatronics while robots are sophisticated mechatronic systems for autonomation and adaptability to respond to environmental changes.

Nowadays robots are ubiquitous and most mechatronic designs are for robotic systems. Robots are highly diversified, and they have been classified from various aspects including structures, mobilities, rigidities of components, scales, and applications. However, robots are the machines to transfer motions and powers from joints to end-effectors; therefore, robots have some basic specifications such as DOF, workspaces, repeatability and accuracy, and working loads. Robots must be modeled and simulated to determine these specifications.

Mathematic modeling is the fundamental to robotic design. Basic robotic models include kinematic models, dynamic models, and control models. Kinematic models include forward and inverse kinematic models, and commonly used kinematic modeling methods are vector-loop methods, D–H modeling methods, and the screw-theory. Dynamic models include forward and inverse dynamic models, and commonly used dynamic modeling methods are Newton–Euler methods and Lagrangian methods. Control models can be of either kinematic controls or dynamic controls.

Designing and using robots require rigorous *Verification and Validation* (V&V) to eliminate all defects before a virtual design is actually implemented in the physical world and a physical robot is actually used in an expected application. Engineers can benefit greatly to use available tools such as *Automated Dynamic Analysis of Mechanical Systems* (Adams) by the MSC Software Corporation, *Motion Study* in *SolidWorks*; in particular, open-source *Robot Operating System* (RoS) (Moulard 2012; Joseph 2018; Marin-Hernandez 2020; Herath and St-Onge 2022; ROS 2024), to support MBSE designs of robotic systems.

Problems

12.1 What is your definition of a robot?

12.2 What are the key components of a robot?

12.3 Where are the majority of robots applied?

12.4 What are robotic kinematics and dynamics?

12.5 What are the advantages of robots?

12.6 What is "DOF" and how is it calculated for a robot?

12.7 What is robotic control and how does it work?

12.8 What is robot programming?

12.9 What are the key characteristics of an industrial robot?

12.10 What are the limitations of robots?

12.11 Calculate DOF of the mechanisms as shown in Figure 12.75

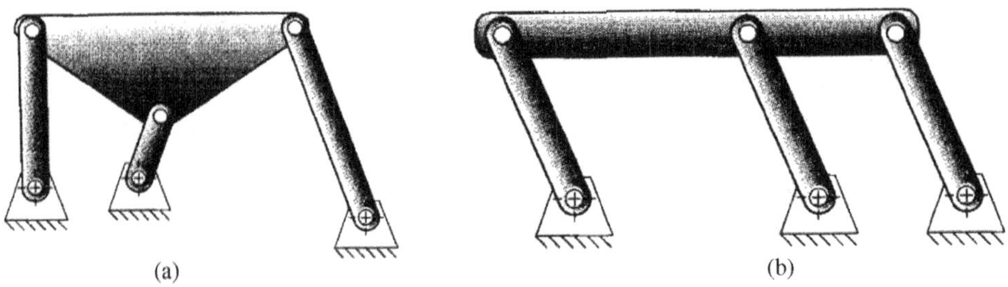

Figure 12.75 Calculate DOF of mechanisms (a) and (b)

12.12 Calculate DOF for a Delta robot as shown in Figure 12.76.

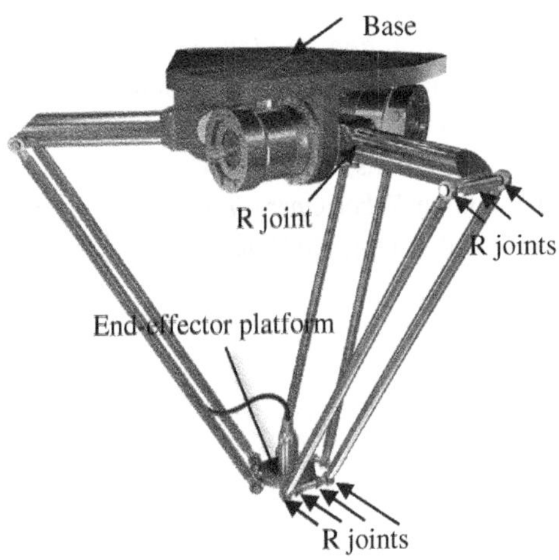

Figure 12.76 Calculate DOF of Delta robot

12.13 Calculate DOF for a hexapod robot showed in Figure 12.77 (Gradcad 2023).

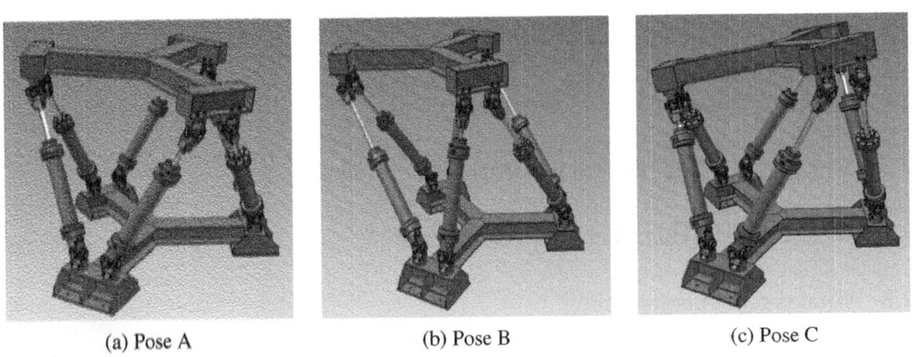

Figure 12.77 Hexapod robot in Problem 12.13

12.14 Develop the forward kinematic model of the 2-DOF robot in Figure 12.78. The robot consists of the ground link, two other links in the body with ($a_1 = a_2 = 0.45$ (m)), and two joints to connect these links sequentially with ($\theta_1 \in [-120°, 120°]$; $\theta_2 \in [-60°, 60°]$). The tool tip is attached on the last link directly: (1) find out the workspace of the robot and (2) find the reference point of the end-effector when $\theta_1 = 30°$ and $\theta_2 = 60°$.

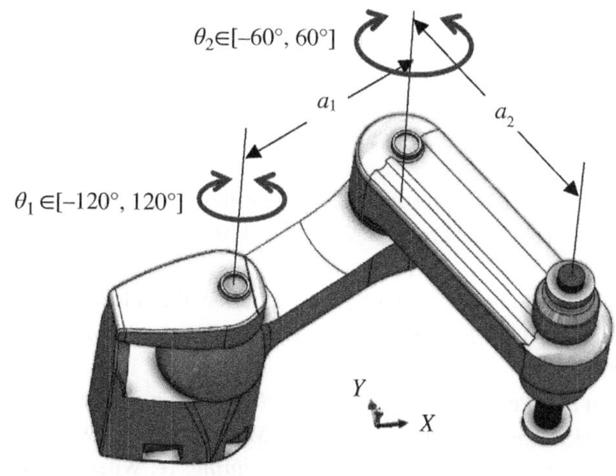

Figure 12.78 2-D robot in Problem 12.14

12.15 Apply SolidWorks motion analysis and discrete event simulation for computer-aided design and analysis of mechanisms (Figure 12.79).

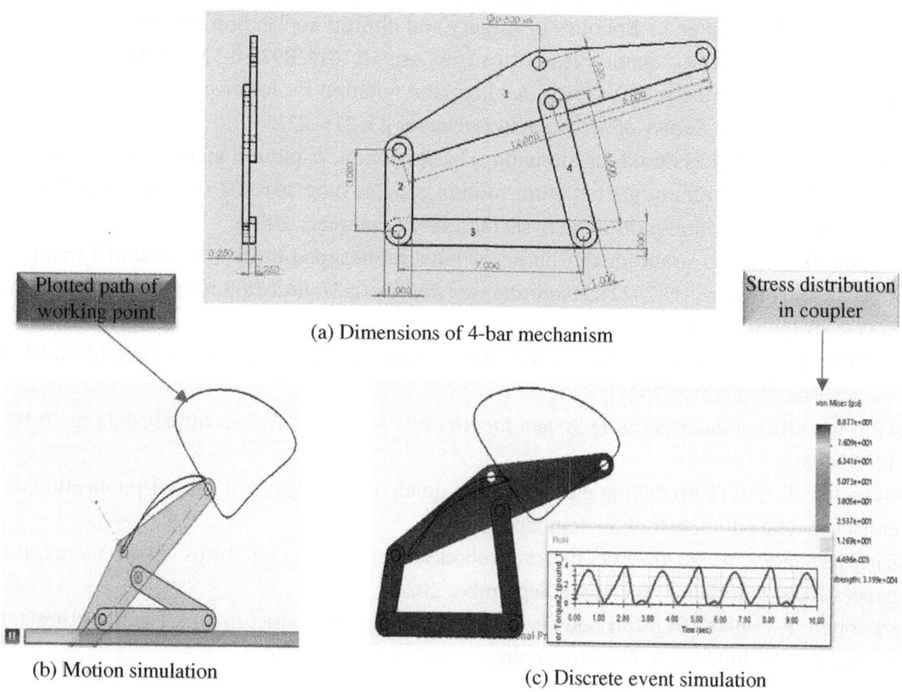

Figure 12.79 Example of *motion study* in SolidWorks

1) Select a device/machine/machine element with motion
2) Create the parts and assembly models for the product
3) Use the motion analysis to determine and plot (i) the trace path of the working point (i.e., tool-tip) and (ii) the driving force/torque of motor
4) Conduct discrete event simulation to evaluate and visualize the deformation, stress, and safety factor of the critical link/body with external load in the product
5) Document your work in your project report

References

ABB (2018). Robotics product specification IRB 14000, https://library.e.abb.com/public/452cf75cf5ce46d9b24c67b20ae9e87a/3HAC052982%20PS%20IRB%2014000-en.pdf (accessed September 2024).

Ahmed, H. and La, H.M. (2019). Education-robotics symbiosis: an evaluation of challenges and proposed recommendations. In *IEEE Integrated STEM Education Conference (ISEC)*, pp. 222–229. https://doi.org/10.1109/ISECon.2019.8881995.

Bi, Z.M. and Wang, L. (2008). *Design of a Reconfigurable Tripod Machine System and its Application in Web-Based Machining in Smart Devices and Machines for Advanced Manufacturing*, 189–218. Springer.

Bi, Z.M. and Wang, X.Q. (2020). *Computer Aided Design and Manufacturing*, 1e. Wiley-ASME Press Series.

Bi, Z.M. and Zhang, W.J. (2021). *Practical Guide to Digital Manufacturing: First-Time-Right for Design of Products, Machines, Processes and System Integration*. Springer.

Bi, Z.M., Lang, S.Y.T., Orban, P., and Verner, M. (2007). An integrated design toolbox for tripod-based parallel kinematic machines. *ASME Journal of Mechanical Design* 129 (8): 799–807.

Capek, K. (2004). *R.U.R. (Rossum's Universal Robots)*. Penguin Classics.

Dapri, G. (2022). Chapter 1 - Robotics in surgery and clinical application. In *Endorobotics* (ed. Luigi Manfredi), pp. 3–21. Academic Press. https://doi.org/10.1016/B978-0-12-821750-4.00001-3.

Denavit, J. and Hartenberg, R.S. (1955). A kinematic notation for lower-pair mechanisms based on matrices. *American Society of Mechanical Engineers* 23: 215–221.

Galliera, J. (2010). Solvers used for animation, basic motion, & motion analysis, https://forum.solidworks.com/community/simulation/motion_studies/blog/2010/02/08/solvers-used-for-animation-basic-motion-motion-analysis (accessed September 2024).

Gradcad (2023). https://grabcad.com/library/octahedral-hexapod-mechanism-parallel-robot.

Herath, D. and St-Onge, D. (2022). *Foundations of Robotics a Multidisciplinary Approach with Python and ROS*. Singapore: Springer Nature.

ISO (2021). ISO 8373:2021 robotics – vocabulary. https://www.iso.org/obp/ui/#iso:std:iso:8373:ed-3:v1:en (accessed September 2024).

Joseph, L. (2018). *Robot Operating System for Absolute Beginners*. Apress https://doi.org/10.1007/978-1-4842-3405-1.

Luukkonen, T. (2011). Modelling and control of quadcopter. https://sal.aalto.fi/publications/pdf-files/eluu11_public.pdf (accessed September 2024).

Marin-Hernandez, A. (2020). ROS tutorial: robotics operation system. https://www.uv.mx/anmarin/papers/ROS_Tutorial.pdf (accessed September 2024).

Markkonen, P. (1999). On multi body systems simulation in product design. Doctoral thesis. Royal Institute of Technology, KTH, Stockholm.

Moulard, T. (2012). ROS tutorials. https://lsi.vc.ehu.eus/pablogn/investig/ROS/ros-tutorial.pdf (accessed September 2024).

ROS (2024). ROSInstallation. https://wiki.ros.org/ROS/Installation (accessed September 2024).

SolidWorks (2024). Introduction to motion studies. https://help.solidworks.com/2021/english/SolidWorks/motionstudies/c_introduction_to_motion_studies.htm (accessed September 2024).

Yaskawa (2024). Solution in motion SIA50D. https://www.motoman.com/getmedia/7c570d49-d235-42e7-ae9b-14cd9a86c454/sia50d.pdf.aspx (accessed September 2024).

You, B., Fan, Y., and Liu, D. (2022). Fault-tolerant motion planning for a hexapod robot with single-leg failure using a foot force control method. *International Journal of Advanced Robotic Systems* https://doi.org/10.1177/17298806221121070.

Zhou, H. (2013). *The Internet of Things in Cloud – A Middleware Perspective*. Taylor & Francis Group.

13

End-Effectors

13.1 Introduction

Robots become ubiquitous to assist or replace humans' operations in various fields to fulfill complex and nonconventional tasks such as performing machining processes, assisting surgeries, searching objects, repairing fiber optic cables in deep sea, or working in extremely harsh environments with high temperature, pressure, or radiation. Most robots accomplish their tasks by grasping, manipulating, or processing objects.

An end-effector is a *mechatronic functional module* (MFM) that is usually mounted at the end of a robot to make direct contact with an object to be manipulated in applications. An end-effector is essential to a robotic system as it performs tasks such as handling, manipulating, probing, or sensing an object. Typical examples of manipulations are loading and unloading a part at a workstation, positioning or rotating an object at hand, performing a process such as painting or welding on a workpiece, and triggering a power tool to do work on an object (Mishkin and Jau 1988).

An end-effector should be customized to fit specific applications; it should usually be portable so that the robotic system can perform a wide range of tasks by changing a sophisticated end-effector for gripping, cutting, welding, painting, or anything else. An end-effector makes direct contact with an object in the environment when the robot executes a task; therefore, the end-effector is a critical MFM for a robot to fulfill its missions in an adaptive, safe, reliable, and effective way. The study on how to design more maneuverable end-effectors has attracted a great deal of attention from robotic engineers (Zhao 2021).

An end-effector, as an *End of Arm Tooling* (EOAT), is a generic concept that covers all mechatronic devices that are mounted at the ends of robots. An end-effector is usually structured as an open or closed kinetic chain to encompass and hold an object; the object is sustained in its position by geometric constraints or a force form (Marandi 2013). An end-effector serves as an interface of a robot with the object to be manipulated in applications, and grasping is the primary functionality of an end-effector (Zhang et al. 2020). It is critical for an end-effector to stabilize grasping operations successfully (Dong et al. 2019). *Grasping* is defined as a tight hold, a firm grasp, or a pressure on an object, in a manner of grasping and holding to secure the object for the robot to fulfil its task.

In the rest of Chapter 13, the theoretical foundations of grasping are introduced in Section 13.2; mechatronic design of an end-effector is discussed in Section 13.3; grasping requirements of configurations are discussed in Sections 13.4 and 13.5, respectively; types of end-effectors are classified in Section 13.6; the main factors in designing an end-effector are introduced in

Mechatronics for Complex Products and Systems: Project-Based Design Approaches for Robots, Cyber-Physical Systems, Digital Twins, and Other Emerging Technologies, First Edition. Zhuming Bi.

Companion website: www.wiley.com/go/bi/mechatronics

Section 13.7; available CAD tools for the designs of end-effectors are briefed in Section 13.8. Section 13.9 summarizes the work of Chapter 13.

13.2 Grasping Theory

Grasping is one of the most fundamental abilities for humans to distinguish themselves from other animals. Most motor cortexes in the human body are developed to grasp and manipulate objects. Designing and analyzing various end-effectors were inspired greatly by studying the grasping behaviors of our hands (Carpin et al. 2016).

An end-effector makes direct contact with an object to be manipulated. It is an active link between the robot and the workpiece to be grasped. The functions of an end-effector depend on the task types. For example, grasping tasks can be to (1) hold a part in the desired position and orientation, (2) apply the right force or moment for a certain process on a part, (3) detect or alter the position and orientation of part with respect to the end-effector or a robot, and (4) perform a certain process on an object (Raval and Patel 2016).

Whatever technique is used to hold and manipulate an object, the end-effector is required to control the forces at contact explicitly or implicitly (Suomalainen et al. 2022; Rao 2023). Grasping an object is achieved by making the contact of the end-effector with an object to restrain the anticipated position, the orientation, and/or the motion of the object.

To analyze the grasping behavior of an end-effector, it is important to investigate the contacts by the end-effector rather than the behaviors of the whole robotic system. The variations of contacts enable diversified manipulations of an object such as tapping, catching, throwing, rolling, pushing, packing, twisting, and grasping (Lynch and Park 2017). Therefore, the grasping theory is established upon the kinematics and dynamics of contacts. This chapter limits its scope to contact kinematics and dynamics of rigid bodies.

In the grasping theory, *kinematics of contacts* studies the relative motions of two bodies without penetration. Possible patterns of relative motion at contacts are approaching, sliding, rolling, and separating. *Contact forces* are normal, tangential, or combined forces occurring in physical contact. Contact forces are transmitted between the bodies on an end-effector and the object by deforming, rolling, and sliding. *Dynamics of contacts* studies the dynamic constraints related to kinematics and contact forces at contacts. *Grasping* is to restrain an object by making contact with the object and applying the right force to ensure certain properties such as the ability to resist external disturbance. Cutkosky (1984, 1989) listed some basic criteria that are used to model and evaluate the process of robotic grasping in Table 13.1.

13.2.1 Contacts on Object

Denote an object as $\mathcal{B}$ that will be grasped and an end-effector as $\mathcal{E}$ that will be used for grasping. The pose of object $\mathcal{B}$ is described in the *global coordinate system* {O-$\boldsymbol{XYZ}$} originated at O, and a *local coordinate system* {B-$\boldsymbol{xyz}$} is attached to $\mathcal{B}$ rigidly, as shown in Figure 13.1.

Object $\mathcal{B}$ in a three-dimensional space has 6 *degrees of freedom* (DOF), i.e., 3 DOF for the translations and the rest of 3 DOF for the rotations along $\boldsymbol{X}$, $\boldsymbol{Y}$, and $\boldsymbol{Z}$ axes, respectively. The pose of object $\mathcal{B}$ can be modeled by a vector $\boldsymbol{u}$ with the size of 6×1 as,

$$\boldsymbol{u} = \begin{Bmatrix} \boldsymbol{p} \\ \boldsymbol{\phi} \end{Bmatrix} \tag{13.1}$$

Table 13.1 Basic criteria for modeling and evaluating grasping processes

Concept	Descriptions
Manipulability	The ability of an end-effector to impart an arbitrary motion on an object. A manipulable grasp must achieve a force closure and have the connectivity of the motion with 6 degrees of freedom (DOF). The rank of the velocity matrix of the end-effector must span the space of the velocities transmitted at contacts.
Stability	The ability of the end-effector to restrain the object to the desired position is subject to the disturbance of an external load. At a low speed, the condition for stable grasping is that the overall stiffness matrix is positive definite; at a high speed, dynamic stability must be taken into consideration.
Grasping forces	The forces at the contacts of the object to be grasped. The grasping forces must achieve the equilibrium conditions of forces or moments on object. There are three and six equations for the equilibrium of 2D and 3D objects, respectively.
Resistance to rolling or slipping	The ability of an end-effector to resist a relative rotation or translation at a contact. The resistance to a relative motion depends on several factors including the configuration of grasping and the number, types, and friction behaviors of contacts.
Form closure	The condition for an end-effector to hold an object by geometric constraints at contacts. A form closure corresponds to a complete kinematic restraint. Generally, at least seven frictionless point contacts.
Force closure	The condition for an end-effector to hold an object by balancing all forces on the object. A grasp meets the conditions of force closure when the rank of the matrix of the wrenches at contacts is 6.
Connectivity	The DOF of the contacts between the end-effector and the object to be manipulated. It corresponds to the number of independent parameters to restrain the position and orientation of the object completely.
Compliance	The response of the object subjected to the forces by the end-effector. A compliance matrix of grasping is a function of the grasping configurations, joints, contacts, and the compliances of contacted bodies.
Isotropy	The permeability of the end-effector to apply a force or moment in an arbitrary direction accurately. When a contact is at its singular configuration, it becomes infeasible to control or motion in a specific direction accurately. The isotropy is a function of the condition number of the Jacobian matrix of grasping.

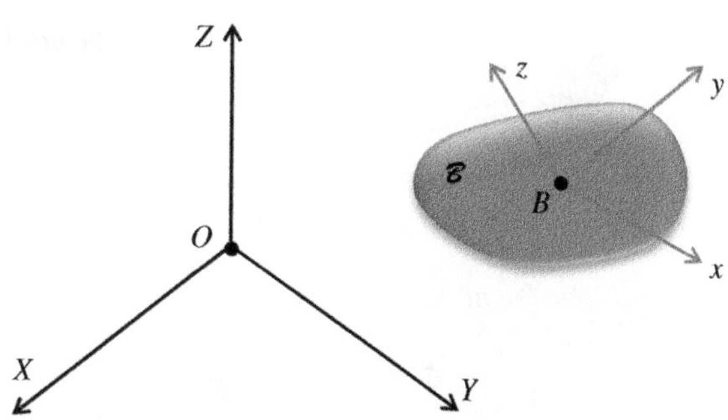

Figure 13.1 Description of an object to be grasped

where $\mathbf{p} = [p_x \ p_y \ p_z]^T$ is the vector to represent the position of $\mathcal{B}$. The Euler angle $\phi = [\varphi \quad \theta \quad \phi]^T$ is the vector to express the orientation of $\mathcal{B}$whereφ, θ,and ϕ are the signed angles of vector $\boldsymbol{u}$ with $\boldsymbol{X}$, $\boldsymbol{Y}$, and $\boldsymbol{Z}$ axes, respectively.

Moreover, the pose of $\boldsymbol{u}$ can be represented as a homogeneous transform matrix as,

$$ {}^{O}_{B}T = \begin{bmatrix} {}^{O}_{B}R|_{3\times 3} & \mathbf{p} \\ \mathbf{0} & 1 \end{bmatrix} = \begin{bmatrix} & & & p_x \\ & {}^{O}_{B}R|_{3\times 3} & & p_y \\ & & & p_z \\ 0 & 0 & 0 & 1 \end{bmatrix} \tag{13.2} $$

where ${}^{O}_{B}R_{3\times 3}$ is the matrix for the orientation of the object $\mathcal{B}$ determined by the Euler angle $\phi = [\varphi \quad \theta \quad \phi]^T$ and $\mathbf{p} = \begin{bmatrix} p_x & p_y & p_z \end{bmatrix}^T$is the the origin B of {B-xyz} in {O-$\boldsymbol{XYZ}$}.

Assume that the end-effector $\mathcal{E}$ has a controllable motion with m-DOF and its configuration is represented by a m-dimensional vector as $\boldsymbol{q} = [q_1, q_2, \dots , q_m]^T$in the joint space of end-effector. Grasping occurs when end-effector $\mathcal{E}$ makes contact with body $\mathcal{B}$. To simplify the modeling process, it is assumed that only point contact is made. There is a total of n_c contact points, and the ith contact point is denoted as $\mathbf{c}_i$ with respect to {O-$\boldsymbol{XYZ}$} or {B-$\boldsymbol{xyz}$}. Each contact point produces a normal reactional force and a tangential friction force. Furthermore, it is helpful to assign a local coordinate system {c_i-$\boldsymbol{x}_i\boldsymbol{y}_i\boldsymbol{n}_i$} at the ith contact point, as shown in Figure 13.2, where $\boldsymbol{n_i}$ is used to represent the normal direction and x_i and y_i are the two principal directions in the tangential plane.

13.2.2 Motions and Forces

The screw theory is introduced to describe the motions and forces of an object. Three basic concepts in the screw theory are *screw*, *twist*, and *wrench*.

A *screw* is introduced to describe the motion of an object. Any displacement occurring to an object can be described as a rotation about an axis followed by a translation along the same axis. Accordingly, the displacement can be modeled by three parameters, that is, an axis $\boldsymbol{a}$, an angle θ, and a scalar displacement d along $\boldsymbol{a}$. It is common that a pitch h is specified instead of d and h is the ratio of d and θ, that is, $d = h\theta$. Therefore, a screw or a screw motion is defined by three parameters ($\boldsymbol{a}$, h, and θ). Note that when h is infinite (i.e., $h = \infty$), the screw motion is pure translation without a rotation. When h becomes zero (i.e., $h = 0$), the screw motion is a pure rotation without a translation.

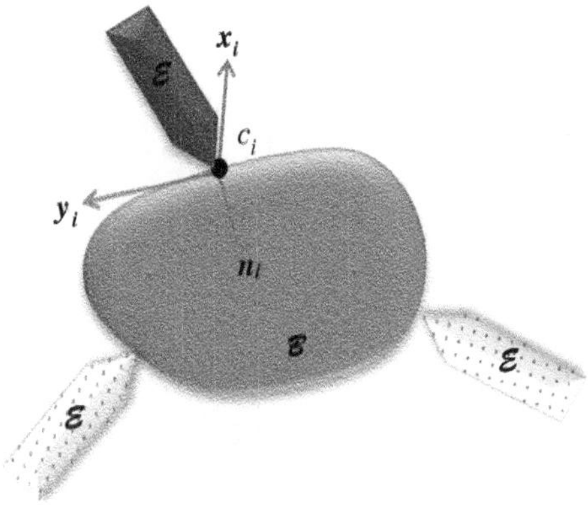

Figure 13.2 Contacts to object to be grasped

A *twist* represents the velocity of an object by a six-dimensional vector as,

$$t = \begin{Bmatrix} v \\ \omega \end{Bmatrix} \tag{13.3}$$

where $\boldsymbol{v}$ and $\boldsymbol{\omega}$ are the vectors for translational and rotational velocities of the object.

A *wrench* is used to model a force $\boldsymbol{f}$ applied to a certain position $\boldsymbol{p}$ on the surface of object $\mathcal{B}$. Position $\boldsymbol{p}$ can be a point of contact or a different point when the force is applied externally. A wrench is also represented by a six-dimensional vector as,

$$\boldsymbol{w} = \begin{Bmatrix} \boldsymbol{f} \\ \boldsymbol{m} \end{Bmatrix} \tag{13.4}$$

where $\boldsymbol{f}$ is the force and $\boldsymbol{m}$ is the moment produced by $\boldsymbol{f}$.

Different coordinate systems can be used to represent a twist or wrench. Especially, a wrench can be modeled with respect to a local coordinate system with its origin at the contact point. In addition, the moment $\boldsymbol{m}$ in a wrench $\boldsymbol{w}$ depends on which reference point is used to calculate the moment.

The concept of a wrench is critical to model a force-closed grasping where the forces between end-effector $\mathcal{E}$ and object $\mathcal{B}$ are exchanged at contact. The forces and wrenches correspond with each other linearly, and this implies that the wrenches at multiple contacts can be superimposed by addition as along as the same reference frame is used in defining these wrenches.

13.2.3 Frictions

Besides a normal reaction force between end-effector $\mathcal{E}$ and object $\mathcal{B}$ at a contact, there is also the friction force to prevent a relative movement of $\mathcal{E}$ and $\mathcal{B}$ at contact. Friction at contact can be characterized into one of three types as below.

Frictionless Contact. Frictionless contact applies to the situation when the contact surfaces are very smooth, the coefficient of friction is very small, and the friction force is ignorable. In this case, the force at the contact is along the normal of two contacted surfaces, and the direction is toward the inside of the object. In other words, the end-effector can only push the object at contact but not pull. Therefore, a frictionless contact introduces a constraint to the contact force, that is, the normal force f_n at the contact must satisfy the condition of $f_n \geq 0$.

Frictional Contact. A friction contact applies to the situation when the coefficient of friction of two contact surfaces cannot be neglectable. In this case, the force at the contact has its components along the normal and tangential directions of two contacted surfaces. Many friction models have been proposed to determine a friction force based on the given normal force, and the most used model is *Coulomb's law*. According to the Coulomb's law, no relative motion occurs as long as the tangential component f_t satisfies the following condition:

$$f_t \leq \mu f_n \tag{13.5}$$

where f_n is the normal component and μ is the static coefficient of friction that characterizes the interaction of two contacted surfaces. Equation (13.5) specifies the maximized friction force when the normal force is given. It is worth noting that the direction of the tangent friction force is undetermined until the direction of possible relative motion is found. Therefore, *a friction cone* is introduced to represent the arbitrary orientation of the friction force in Figure 13.3. The aperture of the cone in the figure is determined by the coefficient of friction, that is, $\theta = \tan^{-1}\mu$.

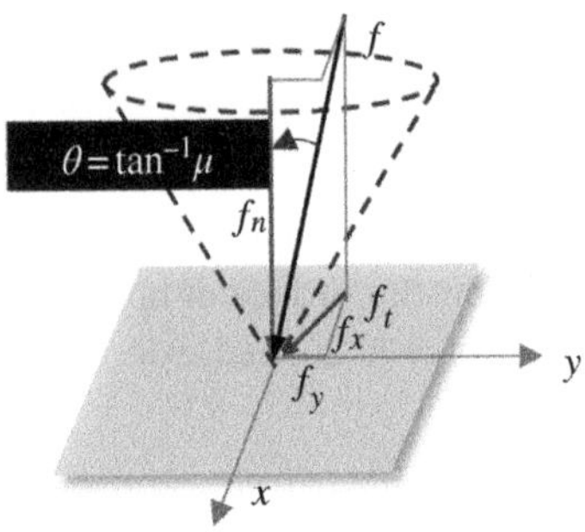

Figure 13.3 Friction cone (FC) in Coulomb's law

Figure 13.3 shows that f_t can also be further decomposed along two axes into f_x and f_y over the tangent plane and $f_t = \sqrt{f_x^2 + f_y^2} \le \mu f_n$. Moreover, the normal force f_n also satisfies the condition of $f_n \ge 0$.

A friction cone (FC) at the contact point $\mathbf{p}_i$ can be represented by:

$$FC(\mathbf{p}_i) = \left\{ \left[f_x f_y f_n \right]^T \in \Re \mid \sqrt{f_x^2 + f_y^2} \le \mu f_n \wedge f_n \ge 0 \right\} \tag{13.6}$$

Equation (13.6) shows that there will not be a relative motion unless the contact force lies outside the friction cone.

Soft Contact: A soft contact applies to the situation when a contact involves not only normal and tangent forces but also a torsional moment about the normal of two contacted surfaces. In this case, the contact is an area instead of a point. A soft contact has an additional condition for the intensity of the torsional moment (τ) as $\tau \le \gamma f_n$.

A contact type determines the number of independent variables in defining a wrench at the contact. A frictionless contact has one independent variable for the contact force normal to the contacted surface. The exerted wrench is modeled as,

$$\boldsymbol{W}_1 = \left\{ \begin{matrix} \boldsymbol{f}_n \\ (\mathbf{c}_1 - \boldsymbol{O}) \times \boldsymbol{f}_n \end{matrix} \right\} \tag{13.7}$$

where $\boldsymbol{O}$ is the reference point to evaluate the moment; it is typically the origin B of the object to be grasped. $\mathbf{c}_i$ is the contact point and $\boldsymbol{f}_n$ is the normal force.

Equation (13.7) can be rewritten into:

$$\boldsymbol{W}_1 = \left\{ \begin{matrix} 0 \\ 0 \\ 1 \\ -p_y \\ p_x \\ 0 \end{matrix} \right\} f_n \tag{13.8}$$

where p_x and p_y are the coordinates of $\mathbf{c}_1$ in the local coordinate system $\{B\text{-}\boldsymbol{xyz}\}$.

A frictional contact ($\mathbf{c}_2$) has three independent variables that include (1) the contact force (f_n) normal to the contacted surface and (2) two force components (f_x and f_y) on the tangential plane. The exerted wrench can be modeled as:

$$\boldsymbol{W}_2 = \begin{bmatrix} \boldsymbol{f}_x & \boldsymbol{f}_y & \boldsymbol{f}_n \\ (\mathbf{c}_2 - \boldsymbol{O}) \times \boldsymbol{f}_x & (\mathbf{c}_2 - \boldsymbol{O}) \times \boldsymbol{f}_y & (\mathbf{c}_2 - \boldsymbol{O}) \times \boldsymbol{f}_n \end{bmatrix} = \begin{bmatrix} 1 & 0 & 0 \\ 0 & 1 & 0 \\ 0 & 0 & 1 \\ 0 & p_z & -p_y \\ -p_z & 0 & p_x \\ p_y & -p_x & 0 \end{bmatrix} \begin{bmatrix} f_x \\ f_y \\ f_n \end{bmatrix} \tag{13.9}$$

where $\boldsymbol{O}$ is the reference point to evaluate the moment; p_x, p_y, and p_z are the coordinates of $\mathbf{c}_2$ in the local coordinate system $\{B\text{-}\boldsymbol{xyz}\}$.

A soft contact ($\mathbf{c}_3$) has four independent variables that include (1) the contact force (f_n) normal to the contacted surfaces, (2) two force components (f_x and f_y) on the tangent plane, and (3) a twisting moment along the normal of the contact surfaces. The exerted wrench can be modeled as:

$$\boldsymbol{W}_3 = \begin{bmatrix} \boldsymbol{f}_x & \boldsymbol{f}_y & \boldsymbol{f}_z & 0 \\ (\mathbf{c}_3 - \boldsymbol{O}) \times \boldsymbol{f}_x & (\mathbf{c}_3 - \boldsymbol{O}) \times \boldsymbol{f}_y & (\mathbf{c}_3 - \boldsymbol{O}) \times \boldsymbol{f}_z & m \end{bmatrix} = \begin{bmatrix} 1 & 0 & 0 & 0 \\ 0 & 1 & 0 & 0 \\ 0 & 0 & 1 & 0 \\ 0 & p_z & -p_y & 0 \\ -p_z & 0 & p_x & 0 \\ p_y & -p_x & 0 & 1 \end{bmatrix} \begin{bmatrix} f_x \\ f_y \\ f_z \\ \tau \end{bmatrix} \tag{13.10}$$

Assuming that there is a total of n_c contact points on object $\mathcal{B}$ and each point corresponds to its own wrench, a grasp matrix $\boldsymbol{G}$ can be constructed by adding all wrenches together. Furthermore, let $n_c = n_l + n_f + n_s$ where n_l, n_f, and n_s are the number of frictionless, frictional, and soft contact points, respectively; according to Eqs. (13.8) to (13.10), the grasp matrix $\boldsymbol{G}$ has six rows and $n = n_l + 3n_f + 4n_s$ number of columns. It can be modeled as a linear mapping of $\boldsymbol{G}: R^n \rightarrow R^6$. Accordingly, all n number of independent force and moment components at n_c contact points can be added as the vector $\boldsymbol{F}_e$ of the exerted forces and moments on object $\mathcal{B}$, and the resulting wrench on the object becomes $\boldsymbol{G} \cdot \boldsymbol{F}_e$. Note that each wrench $\boldsymbol{F}_e$ at contact satisfies the constraints imposed by the generalized friction cone FC shown in Figure 13.4.

13.2.4 Grasping Model

A mechanical model establishes the foundation for an end-effector to grasp, handle, and manipulate an object. A mechanical model is used to analyze the stability, compliance, mobility, and strengths of grasping (Cutkosky 1984).

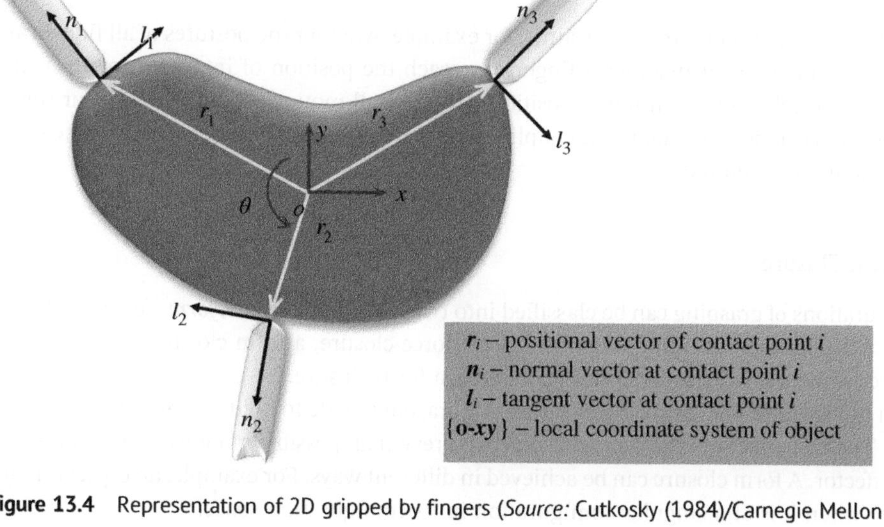

Figure 13.4 Representation of 2D gripped by fingers (*Source:* Cutkosky (1984)/Carnegie Mellon University)

Figure 13.4 shows an object held by an end-effector with m fingers. Assume that each finger has its contact point with the object and the Coulomb friction model is used to represent the friction force at the contact. The total force ($\boldsymbol{f_e}$) and moment ($\boldsymbol{m_e}$) can be found as:

$$\boldsymbol{f_e} = \sum_{i=1}^{m} [f_i(-\boldsymbol{n_i}) + \alpha_i\mu_i f_i(\boldsymbol{l_i})] = \sum_{i=1}^{m} f_i(-\boldsymbol{n_i} + \alpha_i\mu_i\boldsymbol{l_i}) \tag{13.11}$$

$$\boldsymbol{m_e} = \sum_{i=1}^{m} [\boldsymbol{r}_i \times f_i(-\boldsymbol{n_i}) + \boldsymbol{r}_i \times \alpha_i\mu_i f_i(\boldsymbol{l_i})] = \sum_{i=1}^{m} f_i(-\boldsymbol{n_i} \times \boldsymbol{r}_i + \alpha_i\mu_i(\boldsymbol{l_i} \times \boldsymbol{r}_i)) \tag{13.12}$$

When the impact of the self-weight of the objective is neglectable, the contacting conditions in Figure 13.4 are statically indeterminate. Therefore, f_i and $\alpha_i\mu_i\boldsymbol{l_i}$ in Eqs. (13.11) and (13.12) are unknown. μ_i is the coefficient of friction determined by surface properties; α_i is a varying coefficient in a range of [0, 1] and $0 \le \alpha_i\mu_i \le \mu_i$. $\boldsymbol{n}_i$ and $\boldsymbol{l}_i$ are the normal and tangential unit vector of contact point i. However, the direction of $\boldsymbol{l}_i$ cannot be unspecified until the direction of the slip is known. A valid solution to hold an object by an end-effector will be the condition that all α_i satisfy $\alpha_i \le 1$.

The condition of a static indetermination implies that there would be infinite feasible solutions that can achieve a static equilibrium in holding an object; however the solution can be optimized to fulfill a given task.

It is worth noting that the static equilibrium is not the only criterion in manipulating an object, more factors should be taken into consideration in fulfilling a certain task. For example, if the task is to balance the object, the contact points should be close to the geometric center of the object as much as possible. If the task is for light machining, it is preferable to hold the tool on the other end of the operating area. Therefore, the mechanical model should be analyzed further to calculate the deflection of the object subject to gripping forces (Cutkosky 1984),

- Modeling all fingers as elastic structures and treating the object as a rigid body. This is due to the fact that fingers usually consist of several linkages and joints with less stiffness than the object to be manipulated. However, if the object to be manipulated is sold with a low rigidity such as rubber, plastic, or organic materials, the object should also be treated as an elastic one.
- Simplifying the model by ignoring a slipping motion or any dynamic term. A static model can be used to predict when a slip occurs at a certain contact, and it can be further used to determine the maximized force or torque without slippage at contacts. The objective of the simplified model is to analyze stability, stiffness, and resistance to slipping rather than optimizing the griping for the given tasks.
- Ignoring the impact of geometric constraints; for example, whether the postures of all fingers are within their workspaces or whether a finger can reach the position of interest. Moreover, the analysis is applicable only to an initial position with a small motion that allows a linear transformation of force or displacement. The couplings of the motions of the end-effector and the main robotic structure are ignored.

13.2.5 Form Closure

Valid configurations of grasping can be classified into two categories, that is, *form closure* and *force closure*. Since a successful grasping must be at least force-closure, a form closure must also be a force-closure; while a force closure is not necessarily a form-closure.

As shown in Figure 13.5, a form closure refers to a grasping mode to hold an object by geometric constraints. The end-effector's contacts with object $\mathcal{B}$ prevent all possible motions of $\mathcal{B}$ with respect to the end-effector. A form closure can be achieved in different ways. For example, an object is in its form closure when it is at a singular configuration in its configuration space.

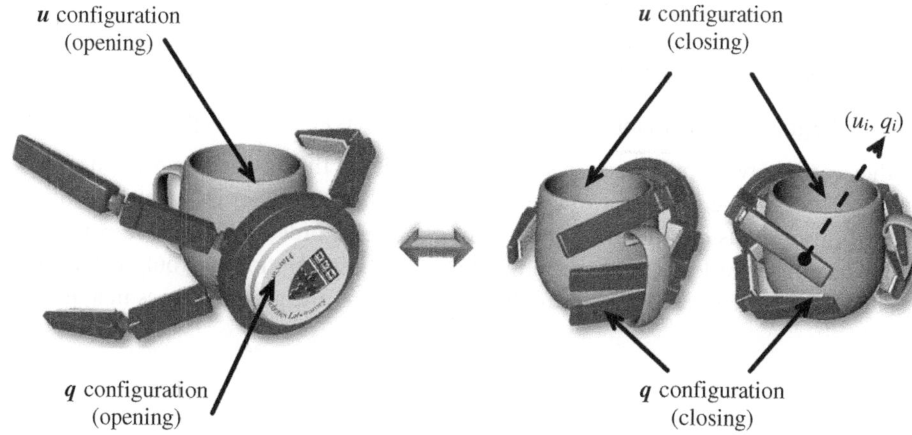

(a) Opening state with $\psi_i(\boldsymbol{u}, \boldsymbol{q}) \geq 0$ (b) Closing state with $\psi_i(\boldsymbol{u}, \boldsymbol{q}) = 0$

Figure 13.5 Illustration of opening and closing in form closure mode

Let the configurations of the object and the end-effector be $\boldsymbol{u}$ and $\boldsymbol{q}$, respectively; assume a grasping process involves in n_c number of contacts and a gap function $\psi_i(\boldsymbol{u}, \boldsymbol{q})$, $1 \leq i \leq n_c$ is defined for each contact, respectively, the gap function measures the distance of the object and the end-effector at a contact point, and the condition for form closure is given as $\psi_i(\boldsymbol{u}, \boldsymbol{q}) = 0$ when the end-effector makes its contacts.

When $\psi_i(\boldsymbol{u}, \boldsymbol{q}) \geq 0$, one or more contacts break; when $\psi_i(\boldsymbol{u}, \boldsymbol{q}) \leq 0$, the end-effector penetrates the object. Therefore, a form closure can be defined as,

$$\psi_i(\boldsymbol{u} + d\boldsymbol{u}, \boldsymbol{q}) \geq 0 \Rightarrow d\boldsymbol{u} = 0 \tag{13.13}$$

Equation (13.13) shows clearly that the condition for form closure is equivalent to the immobility of the object (i.e., $d\boldsymbol{u} = 0$). This occurs at an isolated point in its configuration space.

Since forces are not involved in Eq. (13.13), form closure can be achieved regardless of friction. Therefore, form closure is also a type of frictionless grasp.

In practice, it is important to know (1) how many contacts must be made to hold an object and (2) whether the number of contacts is an essential condition for a form closure?

According to the 3-2-1 principle in the fixturing theory, at least six contacts must be made to restrain all six relative motions between the end-effector and the object; moreover, there must be at least one additional contact to enable the movement of an object in operation; therefore, the minimal number (n_c) of the contacts for form closure must be $n_c \geq 7$. Moreover, having at least seven contacts between the end-effector and the object is sufficient for form closure. There is the need to find other additional constraints to achieve form closure, and this can be addressed by taking the first-order Taylor expansion of Eq. (13.13) as:

$$\psi(\boldsymbol{u} + d\boldsymbol{u}, \boldsymbol{q}) = \psi(\boldsymbol{u}, \boldsymbol{q}) + \left.\frac{\partial \psi(\boldsymbol{u}, \boldsymbol{q})}{\partial \boldsymbol{u}}\right|_{(\boldsymbol{u},\boldsymbol{q})} d\boldsymbol{u} = \left.\frac{\partial \psi(\boldsymbol{u}, \boldsymbol{q})}{\partial \boldsymbol{u}}\right|_{(\boldsymbol{u},\boldsymbol{q})} d\boldsymbol{u} \tag{13.14}$$

where $\psi(\boldsymbol{u}, \boldsymbol{q}) = 0$ in a contact configuration. If $d\boldsymbol{u}$ exists and satisfies the condition that at least one of $\frac{\partial \psi(\boldsymbol{u}, \boldsymbol{q})}{\partial \boldsymbol{u}} d\boldsymbol{u}$ is a positive component, there is a possible movement of the object along the corresponding direction, and the grasping does not achieve form closure. In this direction, the gap function becomes positive, and one of the contacts is broken.

The above condition has its negated equivalent of the implication in Eq. (13.14): for whatever choice of $\boldsymbol{du} \neq 0$, the grasp achieves a form closure if $\frac{\partial\psi(\boldsymbol{u},\boldsymbol{q})}{\partial\boldsymbol{u}}\boldsymbol{du}$ has at least on negative component.

When neither of these two conditions are satisfied, the order of the Taylor expansion has to be increased gradually until one of these two conditions is satisfied. In the grasping theory, this is referred to form closure tests of order n where n is the considered order of the Taylor expansion. The first-order form closure leads to linear programming whose solution not only determines whether a set of contacts achieves form closure but also measures the grasping types in a quantitative way.

13.2.6 Force Closure

Force closure refers to the case that the object is held by external forces, and the condition for force closure is formulated differently.

Assume an external wrench $\boldsymbol{W}_e$ is applied on object $\mathcal{B}$ other than the wrenches ($\boldsymbol{F}_e$) from the contacts by the end-effector. Note that the wrenches by the end-effector at contacts are $\boldsymbol{F}_e$ and the grasp matrix of the end-effector is denoted as $\boldsymbol{G}$, balancing all forces applied on object $\mathcal{B}$ leads to:

$$\boldsymbol{G}\cdot\boldsymbol{F}_e = -\boldsymbol{W}_e \tag{13.15}$$

To achieve the force enclosure, there must be a vector $\boldsymbol{W}_e \in FC$ to meet the condition of Eq. (13.15) for every possible external wrench $\boldsymbol{W}_e$ on object $\mathcal{B}$.

Equation (13.15) can be used to determine if a given grasping matrix achieves a force closure. *First*, it is necessary that the rank of the grasping matrix $\boldsymbol{G}$ must be 6; *second*, any component in $\boldsymbol{F}_e$ must be restricted in its friction cone FC, which is a subset of R^n.

Therefore, the necessary and sufficient condition of a force enclosure is that at any contact point, $\boldsymbol{G}(FC) = R^6$; where $\boldsymbol{G}$(FC) is the set that is obtained by applying the grasping map to every point in the friction cone FC.

In practice, Eq. (13.15) is usually verified numerically by approximating an FC as a regular pyramid with k vertices, as shown in Figure 13.6. The force $\boldsymbol{f}_i$ at the ith contact i lying in the FC can be written as a non-negative combination of force of the discretized vectors $\boldsymbol{d}_i^j$ ($j = 1, 2, \ldots, k-1$) as $\boldsymbol{f}_i = \sum_{j=1}^{k}\alpha_{i,j}\boldsymbol{f}_{i,j}$ where $1 \geq \alpha_{i,j} \geq 0$.

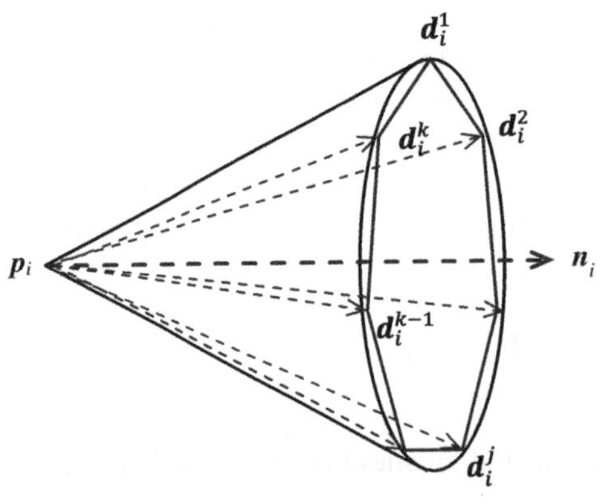

Figure 13.6 Approximation of a friction cone (*Source:* Adapted from Howe (2024))

Accordingly, the wrench at the ith contact can be written as:

$$\begin{aligned}\boldsymbol{W}_i &= \begin{bmatrix}\boldsymbol{f}_i \\ (\boldsymbol{c}_i - \boldsymbol{O})\times\boldsymbol{f}_i\end{bmatrix} \\ &= \begin{bmatrix}\sum_{j=1}^{k}\alpha_{i,j}\boldsymbol{f}_{i,j} \\ (\boldsymbol{c}_i - \boldsymbol{O})\times\sum_{j=1}^{k}\alpha_{i,j}\boldsymbol{f}_{i,j}\end{bmatrix} \\ &= \sum_{j=1}^{k}\alpha_{i,j}\begin{bmatrix}\boldsymbol{f}_{i,j} \\ (\boldsymbol{c}_i - \boldsymbol{O})\times\boldsymbol{f}_{i,j}\end{bmatrix} \\ &= \sum_{j=1}^{k}\alpha_{i,j}\boldsymbol{W}_{i,j}\end{aligned} \tag{13.16}$$

where each $W_{i,j}$ is called an elementary wrench, and all $k \cdot n_c$ number of element wrenches can be assembled into a matrix $\tilde{G}$ to approximate the grasping matrix. The approximating matrix $\tilde{G}$ must span R^6 positively. Force closure can be achieved when each disturbance wrench, $\boldsymbol{W}_e$, can be balanced by the forces that are all within their friction cones.

13.2.7 Quality of Grasping

There is generally an infinite number of solutions to satisfy Eq. (13.13) for a form closure or Eq. (13.15) for a force closure. In controlling an end-effector, it is desirable to evaluate and compare different grasping solutions to optimize a grasping process. Some measures for grasping quality can be defined to evaluate different solutions for best grasping. The most used quality measure is *the Ferrari-Canny metric*. It aims to minimize the grasping forces required to resist external loads applied on objects.

When the external load is not known prior, the force in the worst scenario is taken into consideration, i.e., the grasp should be optimized to resist any external load with the minimized effort. This implies that any wrench on $\mathcal{B}$ must be resisted by the summed wrenches at the n_c contacts as:

$$\boldsymbol{W}_{L_\infty} = CH\left(\prod_{i=1}^{n_c} \{\boldsymbol{w}_{i,1}, \boldsymbol{w}_{i,2}, \dots, \boldsymbol{w}_{i,k}\}\right) \tag{13.17}$$

where $\prod$ is the Minkowski sum and CH stands for the convex hull spanned by the distribution of wrenches. The inner Minkowski sum in Eq. (13.17) produces a set of finite elements.

Therefore, $\boldsymbol{W}_{L_\infty}$ is the convex hull of a finite set of elements in R^6, and the grasp metric Q_∞ can be defined as the distance of the closed facets of $\boldsymbol{W}_{L_\infty}$ from the origin as:

$$Q_\infty = \inf_{x \in \partial \boldsymbol{W}_{L_\infty}} \|x\|_2 \tag{13.18}$$

where $\partial \boldsymbol{W}_{L_\infty}$ is the boundary of $\boldsymbol{W}_{L_\infty}$ or the union of the facets of $\boldsymbol{W}_{L_\infty}$. Q_∞ is defined to minimize the maximum force at any of the contacts to resist an arbitrary external wrench on an object.

13.3 Mechatronic Design of End-Effectors

An end-effector is a typical MFM that consists of a mechanical system, a sensing system, and a control system. As shown in Figure 13.7, an end-effector is usually mounted at the end of a robot by mechanical, electrical, and electronic connections to manipulate objects (Michaud et al. 2004; Rekabi-Bana et al. 2023). From the perspective of a mechatronic device, the implementation of such an MFM relies on (1) mechanical and actuating components to produce required motions; (2) sensor technologies to acquire data about joints, environment, and objects; and (3) controlling components to operate the end-effector in use (Ben-Tzvi and Moubarak 2014).

Hack (2018) gave an example of an end-effector in Figure 13.8. The mechanical and actuating components included a bending mechanism to produce a torque of up to 24.8 Nm, a realigning mechanism to produce a holding torque of 59.4 Nm, a 4-bar mechanism to clamp wires with a stroke of 100 mm, the cylinders that were operated under a pressure of 64 bar, and two stepper motors to drive the wire-feeding mechanism. The sensors include relative encoders to provide precise feedback on the displacements of motors. The controller implemented the closed-loop controls of the end-effector.

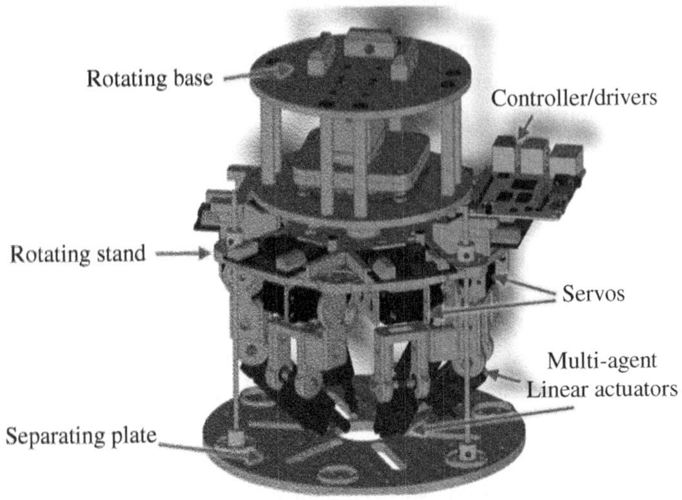

Figure 13.7 Main components of an end-effector (*Source:* Rekabi-Bana et al. (2023)/IEEE)

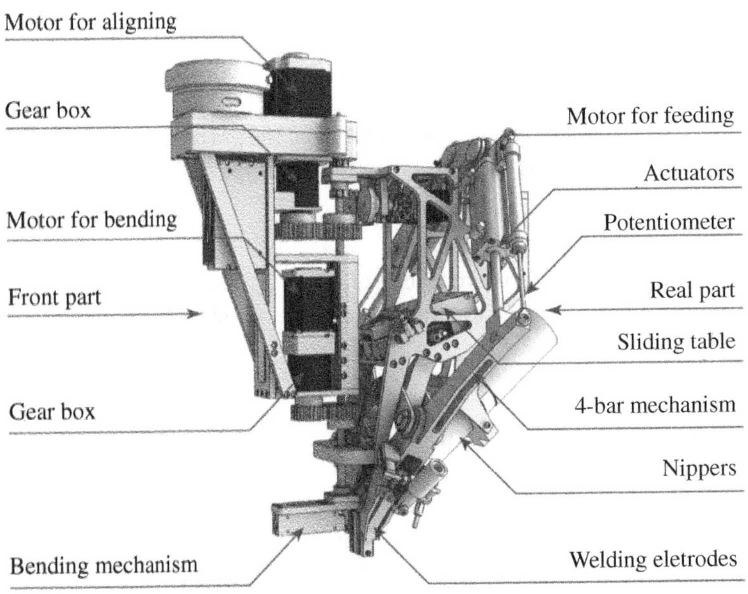

Figure 13.8 Example of an end-effector as a mechatronic functional module (*Source:* Hack (2018)/ETH Zurich)

13.3.1 Mechanical and Actuating Components

Actuating components take up a large percentage of the space in a robotic structure, and the actuating principles and mechanisms of an end-effector are often different from those of joints in the robotic structure. There is a need to investigate the actuations at an end-effector separately from those of the main robotic structure, as actuating mechanisms for end-effectors are differentiated from one another in terms of working principles, structures, complexities, and payloads.

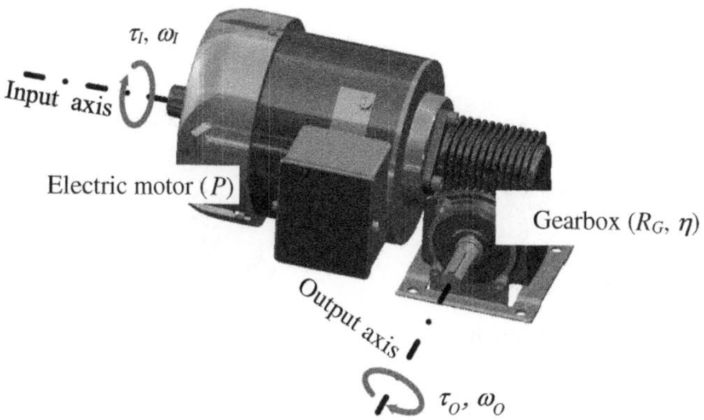

Figure 13.9 Gearbox to speed reduction

Electrical motors are mostly used as the drivers to actuate end-effectors. As shown in Figure 13.9, due to the high speed of an electrical motor (ω_I) and a relatively low operational speed (ω_O) at the output subject to a given output power (P), the input torque (τ_I) by an electrical motor should usually be amplified by a gear box for an increased operating torque of (τ_O) as:

$$\tau_O = \eta\tau_I \frac{\omega_I}{\omega_O} = \eta\tau_I R_G \tag{13.19}$$

where ω_I and ω_O are the input and output speeds, τ_I and τ_O are the input and output torque, respectively. η and $R_G = \frac{\omega_I}{\omega_O}$ are the efficiency and the gear ratio of the gearbox, respectively.

Example 13.1 Assume a DC motor provides a 12 (kg · cm) torque when its rotational speed is 12,000 *revolutions per minute* (RPM), and the gear ratio of the attached gearbox is 200:1. Determine the output speed and torque when the efficiency of the gear trains is 95%.

Solution

The output speed can be determined by the gear ratio and input speed as:

$$\omega_O = \frac{\omega_I}{R_G} = 60\ (\text{RPM})$$

Equation (13.19) can be used to determine the output torque as:

$$\tau_O = \eta\tau_I \frac{\omega_I}{\omega_O} == (0.95)(12)(200) = 2280\ (\text{kg} \cdot \text{cm})$$

Commonly used actuating mechanisms for end-effectors are mechanical grippers, vacuum grippers, magnetic grippers, and servo grippers.

Mechanical Grippers. Gripping can be performed by using mechanical jaws or fingers. These mechanical components are reliable, simple, and robust, and they are used to perform stable and firm grasps such as pick-and-place in loading/unloading or in assembling.

Depending on the types and numbers of jaws or fingers in an end-effector, mechanical grippers are further classified into parallel grippers, angular grippers, and multi-finger grippers, as shown in Figure 13.10.

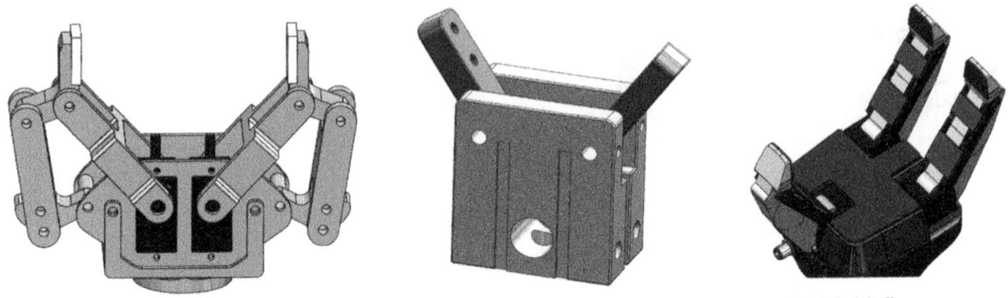

(a) Parallel (b) Angular (c) Multi-finger

Figure 13.10 Types of mechanical grippers

Vacuum Grippers: Objects can be grabbed by suction cups through vacuuming. Vacuuming becomes very effective when an object has a large, flat, and smooth surface(s). Examples of objects with such features are metal sheets, glass panels, and circuit boards. Vacuum grippers are characterized by their fast cycle times and reliable operations; however, they are unsuitable for grasping objects with irregular shapers or heavy loads.

Let an end-effector has n_c number of suction cups and the difference of vacuuming and atmosphere pressure be p_v, the grasping force of the end-effector is:

$$F = n \cdot p_v A = n \cdot p_v \frac{\pi d^2}{4} \tag{13.20}$$

where F is the total grasping force and d and A is the diameter and the effective suction area of a cup.

When the payload (m) of the end-effector is given, the diameter (d) can be determined by:

$$d = \begin{cases} n_f \sqrt{\dfrac{4m}{\pi \cdot n \cdot p_v}} & \text{for horizential grasping} \\ n_f \sqrt{\dfrac{4m}{\pi \cdot n \cdot p_v \cdot \mu}} & \text{for vertical grasping} \end{cases} \tag{13.21}$$

where m is the payload in kg, μ is the coefficient of friction (CoF), and n_f is the factor of safety.

Example 13.2 Assume a vacuum gripper has four suction cups, and the vacuuming pressure is $p_v = 0.4$ bar, the payload of the end-effector is 50 kg, and the coefficient of the friction (CoF) is $\mu = 0.25$, what should be the minimized diameter of the suction cups if the factor of safety (n_f) must be larger than 2? Considering two cases where the payload is placed horizontally and vertically, respectively.

Solution

Equation (13.5) can be used to calculate the minimized diameter of a suction cup. For the case of a horizontally placed payload, the minimized diameter of a suction cup is:

$$d = n_f \sqrt{\frac{4m}{\pi \cdot n \cdot p_v}} = (2)\sqrt{\frac{4(50)}{\pi \cdot (4) \cdot (0.4)}} = 12.6\ (\text{mm})$$

For the case of a horizontally placed payload, the minimized diameter of a suction cup is:

$$d = n_f\sqrt{\frac{4m}{\pi \cdot n \cdot \mu \cdot p_v}} = (2)\sqrt{\frac{4(50)}{\pi \cdot (4) \cdot (0.25) \cdot (0.4)}} = 25.2\,(\text{mm})$$

Commonly used actuating mechanisms for end-effectors are mechanical grippers, vacuum grippers, magnetic grippers, and servo grippers.

Magnetic Grippers: Ferromagnetic objects, such as steel plates, can also be grasped by magnetic forces in a magnetic field. Magnetic grippers become very effective in grasping heavy metal objects with irregular boundary surfaces that cannot be easily handled by mechanical or vacuum grippers. Magnetic grippers have the advantages of high reliability, fast cycle time, and the minimum need for maintenance; however magnetic grippers are inapplicable to non-ferromagnetic materials.

The grasping force of a magnetic gripper can be determined by (Datta et al. 2015):

$$F = k \cdot B \cdot L \cdot I \cdot N = \frac{k \cdot B \cdot L \cdot V \cdot N}{R} \tag{13.22}$$

where k is a constant related to materials and coil design, B is the strength of magnetic field, L is the length of the coil, N is the number of coils, I is the current flow through a coil, and V is the voltage applying at the ends of coils. Equation (13.22) shows that when the magnetic gripper is made, the actual grasping force (F) is proportional to the applied voltage (V).

Servo Grippers: Servo grippers are advanced mechanical grippers that use motors and gearboxes to actuate grasping, moreover, the forces and positions of grasping are controlled precisely. Servo grippers are applied in some applications that require high precision and high flexibility such as inspection or quality control. Servo grippers are versatile in the sense that the grasping force or position of a servo gripper can be adjusted by re-programming to fit the shape and size of the object to be handled.

Every gripper type exists in applications for its own reasons, and Table 13.2 summarizes the advantages and disadvantages of grippers.

13.3.2 Sensing Components

An end-actuator makes physical contact with an object. The smartness of an end-effector or even a robot is determined by the number of sensors that are equipped to monitor operating conditions and perceive the environment.

Table 13.2 Comparison of different grippers

Type	Advantages	Disadvantages
Mechanical	Reliable, cost-effective, and simple	Inapplicable to handle irregular shapes.
Vacuum	Reliable, ease to use, and fast cycle-time	In applicable to grasp a heavy object or object without a smooth or porous surface.
Magnetic	Reliable and minimal maintenance	Inapplicable to handle non-ferromagnetic materials.
Servo	High reliability and high flexibility	Expensive and customized control system.

Table 13.3 Common sensors on end-effectors

Type	Functions
Proximity Sensors	Detect the presence or absence of an object in the proximity of the end-effector. Sensors are selected based on the type of targeted object such as metal, plastic, or fluid flow.
Force Sensors	Measure the amount of force or torque occurring to the contacts of the end-effector with an object; the sensed data is used to determine the required force or torque to grasp, hold, or move the object.
Cameras	Acquire the information of the position, orientation, and movement of an object with respect to the end-effectors. Such information is essential in precise positioning and manipulation, and it is used to recognize, track, and navigate an object. It can be also used as an end-effector to avoid obstacles.
Light Sensors	Detect the presence or absence of light; a light sensor can be used to detect objects in a low-light environment.
Magnetic Sensors	Detect the presence or absence of magnetic fields to locate or orient a magnetic object.
Range Sensors	Measure the distance between the end-effector and the object to be grasped.

Sensors on an end-effector collect the data from the surroundings to gather data about the surroundings, position, orientation of objects, and robotic state. The control system of the end-effector processes and mines information to provide the decision-making support for the end-effector to interact with its environment. Table 13.3 shows some commonly used sensors in an end-effector.

While introducing more sensors in an end-effector increases costs, equipping sensors in an end-effector makes it possible to implement closed-loop control in grasping an object for increased robustness, flexibility, accuracy, and efficiency. The end-effector becomes capable of perceiving the environment, optimizing the grasping process, or accommodating dynamic changes based on the sensed data.

Figure 13.11 shows an example where a number of sensors are mounted at the end-effector to locate objects and move and control the end-effector for integrated structural assembly. Integrating sensing systems in the control helped to reduce the time for development, verification, and regulatory acceptance of new composite materials and design methods.

13.3.3 Control Components

An end-effector is a part of the robotic system; therefore, the control of the end-effector must be coordinated with that of the base robot to achieve a high level of autonomy in system operation.

From this perspective, the control of an end-effector can be treated as an integral part of the whole robot control system. In general, developing a control scheme begins with the establishment of mathematical models for kinematic, dynamic, or combined kinematic and dynamic behaviors of the robotic system including its end-effector. Accordingly, robotic systems can be controlled kinematically, dynamically, or supervisory.

The objective of a kinematic control is to follow an expected trajectory that moves the articulated joints from an initial point to a final point in an optimal configuration, and then execute an operation at a tag point while minimizing a cost function subject to constraints.

In the kinematic analysis and control of robotic manipulation, the first objective is to derive a global transformation matrix that maps the local kinematics (such as position or speed) of a point

Figure 13.11 Integrated structural assembly at NASA's Langley Research Center (*Source:* Photo by Motion Control Tips / WTWH Media LLC)

on a specific link into a global inertial frame 0. Every joint *i* of the manipulator and end-effector is assigned a local frame. For each branch to a finger of the end-effector, the transformation matrix is given by:

$$ {}^{0}_{n}T = \prod_{i=0}^{n-1} {}^{i}_{i+1}T = {}^{0}_{1}T\,{}^{1}_{2}T\,{}^{2}_{3}T \ldots {}^{n-1}_{n}T \tag{13.23}$$

where ${}^{i}_{i+1}T \in R^{4\times 4}$ is the 4 by 4 homogeneous matrix for a spatial relation of joint $i+1$ and joint i.

The objective of dynamic control is to determine a control history that moves the robot along a trajectory from an initial point to a final point. The dynamic model of a branch in an end-effector with n-link can be established by the energy conservation in terms of the generalized coordinates $q(t)$ as:

$$M(q,t)\ddot{q} + F(\dot{q},t)\dot{q} + V(q,\dot{q},t)\dot{q} + G(q,t) = \tau(t) \tag{13.24}$$

where $q(t), \dot{q}(t), \ddot{q}(t) \in R^{n\times 1}$ represent the displacements, velocities, and accelerations of joints. $M(q, t) \in R^{n\times n}$ is the mass matrix; $F(\dot{q}(t), t) \in R^{n\times n}$ is the dissipative term for damping effects; $V(q(t), \dot{q}(t), t) \in R^{n\times n}$ is the Coriolis matrix; $G(q, t) \in R^{n\times 1}$ is the non-dissipative components such as gravity; $\tau(t) \in R^{n\times 1}$ is the vector of input torque.

In kinematic control or dynamic control, the trajectories to move and operate an end-effector should be known and set in prior, and a robotic task must be specified so that the robot can perform it automatically. When a robot is used in an open environment, robotic tasks are too complicated to be programmed without real-time information in the deployed environment.

One alternative to address this concern is to use supervisory control. *Supervisory control* refers to a closed-loop master–slave scheme where control commands are assigned by the robot. The robot user is the master who initiates commands, and the robot is the slave that executes control commands. In supervisory control, the human operator is included in the control loop to supervise the task execution based on his or her perception of the environment. A supervisory control scheme is implemented by *Human–Machine Interaction* (HMI). A data glove catches the operator's motion intents and translates these intents into proportional electrical signals for expected robotic motions (Fahn and Sun 2005).

13.4 Evaluation of Grasping Performance

Like the main robotic structure, an end-effector usually consists of multiple bodies that are connected as open or closed kinematic chains; it faces challenges in dealing with singularities, nonholonomic constraints, and under- or over-constrained motions. Moreover, the contact of the end-effector with an object can be nonlinear and very complex if the object is soft and viscoelastic.

To define grasping requirements in a concise way, Cutkosky (1989) made a list of reasonable assumptions below:

1) Kinematic models are linearized instantaneously.
2) No rolling or sliding occurred to fingertips.
3) Fingertips are ideal, for example, for point-contacts or only linear elastic deformation for soft fingers.
4) Friction models are ideal that ignore the effects of sliding, dirt, or moisture.
5) The deformations of the object to be grasped and fingertips are ignorable.
6) No over-constraints or redundant motions are taken into consideration.

Accordingly, the grasping requirements should be quantified in the aspects of geometry, kinematics, dynamics, and constitutive relations, as shown in Figure 13.12. The grasping theory introduced in Section 13.2 can be used to model an end-effector and evaluate if it can satisfy the grasping requirements.

It is clear that the grasping requirements must be tailored to the tasks of a robotic system. For example, Mishkin and Jau (1988) defined the grasping requirements of the end-effector of a space robot by taking into consideration (1) weights and sizes of parts to be lifted; (2) materials of parts; (3) clearance surrounding a part that could possibly cause an interfere in grasping; (4) robotic paths where the end-effector are attached and moved together; (5) available power supplies. Moreover, the size, shape, and required grasping force of parts play the most critical roles in designing an end-effector.

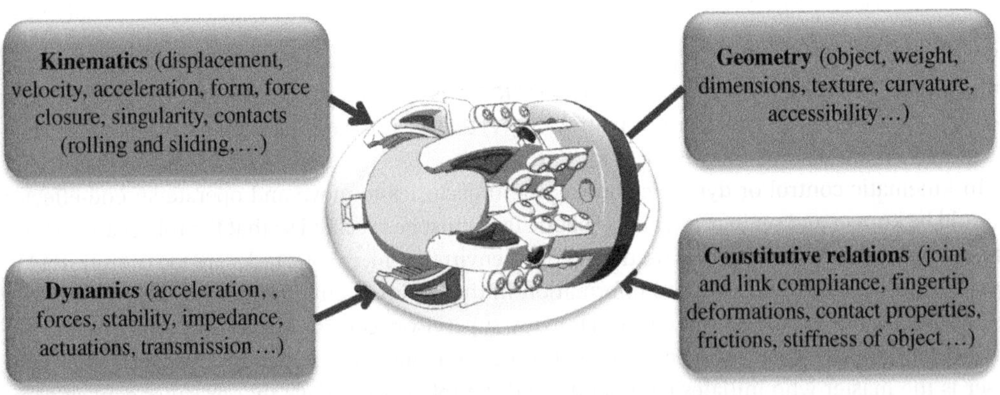

Figure 13.12 Evaluating grasping performance (*Source:* IEEE / https://www.ieee-ras.org/robotic-hands-grasping-and-manipulation / Last accessed on July 19, 2024)

13.5 Grasping Configurations

Numerous solutions exist to hold and manipulate an object. Taking an example, a spherical object can be grasped in one of six configurations, as shown in Figure 13.13 (Raval and Patel 2016). It is unrealistic to select a grasping configuration that is optimized for all criteria discussed in Table 13.1. In applications, finding a feasible solution made from a finite set of grasping configurations makes more than finding an optimal solution by extended searching.

Taking the example of a grasping in manufacturing, a grasping configuration can be feasible if both precision and power requirements can be satisfied. Cutkosky (1989) provided the mapping of grasping configurations to the precision and power requirements of Figure 13.14. This mapping can be used to accelerate a search for a feasible grasping solution for a specific object.

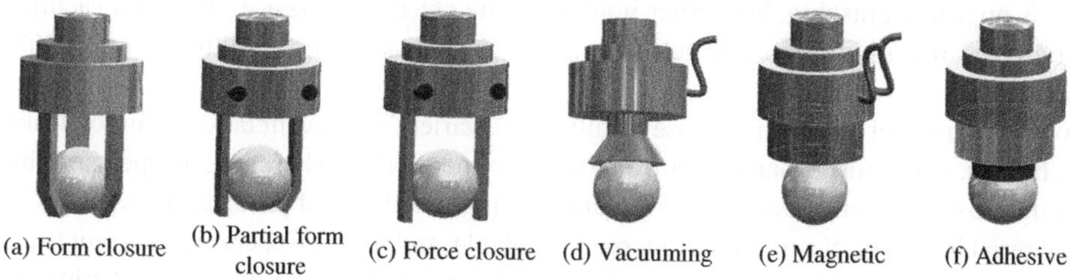

Figure 13.13 Six example configurations to grasp a spherical object

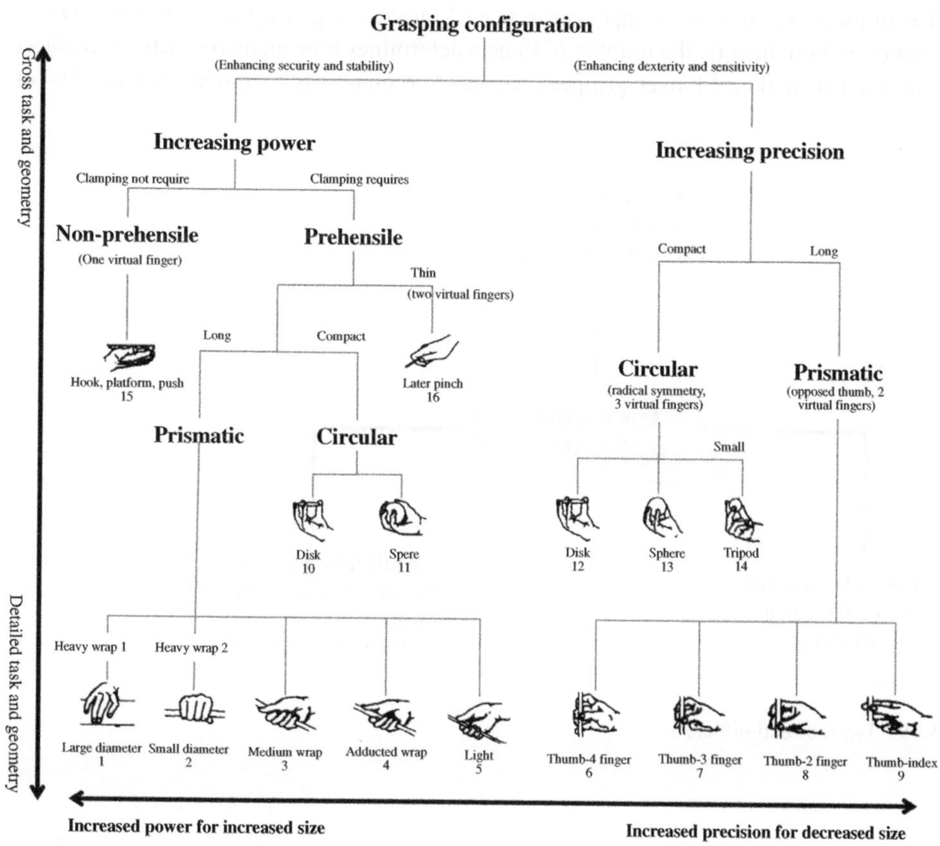

Figure 13.14 Grasping configurations versus power and precision requirements

13.6 Types of End-Effectors

Figure 13.15 shows that end-effectors can be generally classified into (1) *grippers* to pick and place objects; (2) *processing tools* to hold a specific tool for certain operations such as painting, welding, and deburring; and (3) *multifunctional tools* to perform a series of operations for an object at the same setup (Motion Control Tips 2024).

13.6.1 Types of Grippers

Grippers are sophisticated MFMs to provide temporary contact with the object. They ensure the position and orientation of the object when the object is handled or processed. Prehension is achieved by force-producing and form-matching elements. The term "gripper" is also used in cases where there is no actual grasping, but rather holding of the object by a device, such as a vacuum suction, where the retention force can act on a point, line, or surface (Monkman et al. 2007; unidex 2024).

Any robot is designed for a certain purpose by fulfilling a variety of relevant tasks, common tasks are to pick, handle, and transport an object from one place to another. Therefore, grippers are the most commonly used end-effectors for picking-and-placing, loading-and-unloading, assembling, and other material handling. Numerous grippers are available, and these grippers are different from one another in terms of the *numbers of fingers*, *types*, *modes*, *mechanisms*, and *methods of grippers,* as shown in Figure 13.16 (Zhang et al. 2020).

Grippers are highly diversified since they are customized to specific workpieces in various applications. The number of fingers in a gripper varies from *2-finger*, *3-finger*, *4-finger*, *anthropomorphic 5-finger*, and even more fingers; the number of fingers determines how many contacts to an object the end-effector can make. 2-Finger grippers are most common; as a comparison, 60–70% of

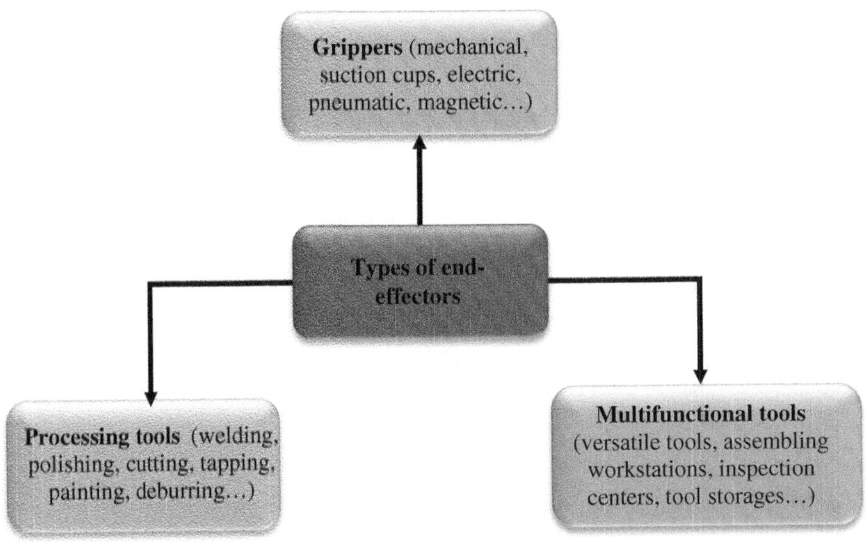

Figure 13.15 Types of end-effectors

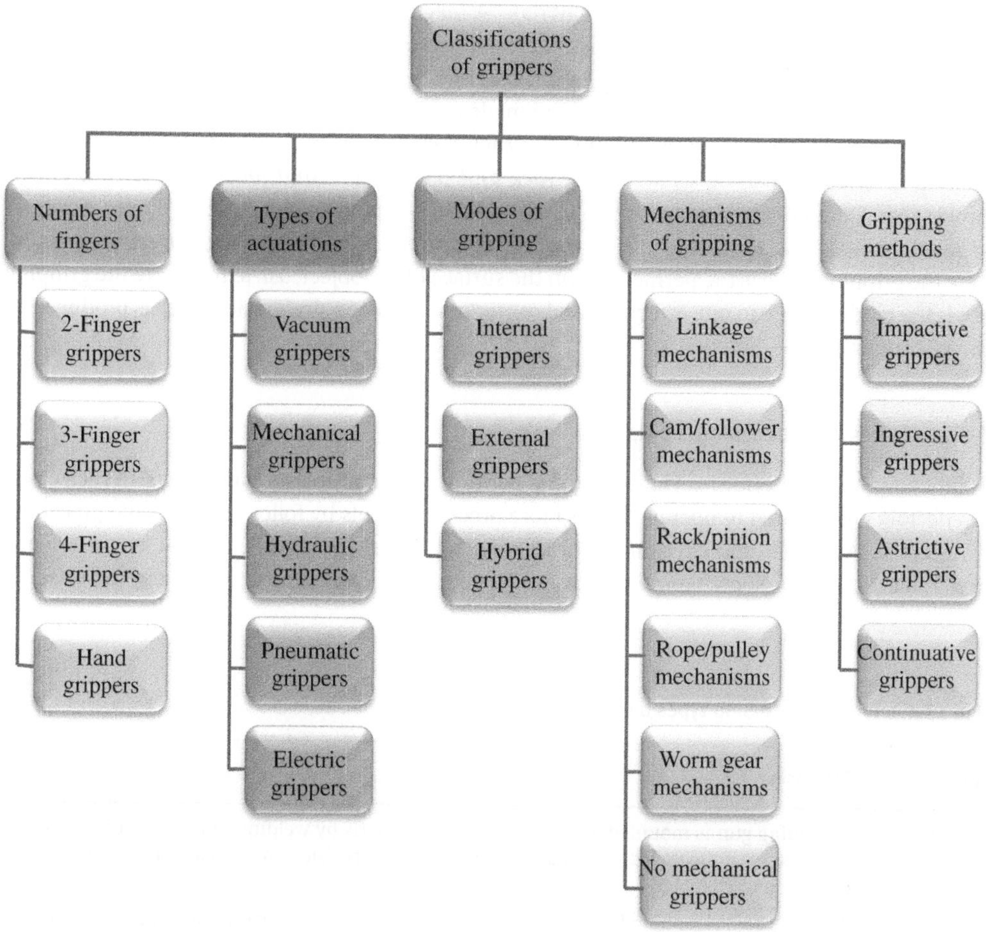

Figure 13.16 Classification of grippers

grasping tasks on cylindrical, parallelepiped, or pyramidal objects by humans are performed by two fingers. On the other hand, multi-finger grippers can grasp an object more securely, but the control system would be more complicated.

Gripping forces of an end-effector can be *vacuum*, *pneumatic*, *hydraulic*, *magnetic*, or *electric*. A vacuum gripper often uses suction cuppers to create a vacuum; the design is very simple and offers great flexibility in handling materials cost-effectively. Some special vacuum grippers use the Bernoulli principle to produce the gripping force by a high-speed air flow that goes through the contacts between the gripper and the object. Vacuum grippers are inapplicable to perforated materials.

Pneumatic grippers and electric grippers are very popular. A pneumatic gripper produces a gripping force by forcing compressed air through a piston, allowing it to open or close a finger parallelly or angularly. A pneumatic gripper has the advantage of a relatively high gripping force. An electric gripper is motor-driven to facilitate positional and speed controls. Electric grippers are adaptable to various applications such as bin picking, machine tending, and material handling in manufacturing. The design of a magnetic gripper is like that of a vacuum gripper, while the gripping force is magnetic and applicable only to ferrous materials. A magnetic gripper is simple and requires minimal maintenance, and it eliminates the risk of dropping materials that occur to an electrical gripper in power outage or a pneumatic gripper in air loss.

From the perspective where the contacts are made on an objective, a gripper can be operated in one of the *internal*, *external*, or *hybrid* modes. When a gripper is implemented mechanically, the actuating mechanism can be *screw driven*, *rack and pinion*, *cam and follower*, *rope and pulley*, or *worm gear*. Different physical principles can be used in grasping an object. Accordingly, a gripper can be *impactive* where the forces in two or more directions apply on the object directly, *ingressive* where the gripping force is the permeation on the surface of the object, *astrictive* where the gripping force is the binding force in one direction, and *continuative* where a prehension force is produced by a direct contact in one direction.

13.6.2 Types of Processing Tools

The functions of end-effectors can be extended to perform certain processes over objects. In such cases, end-effectors are required to operate one or more processing tools by following a specific procedure or trajectory. Available end-effectors as processing tools can be classified based on processing types. Table 13.4 shows some typical processes, where the end-effectors are competent to perform.

Table 13.4 Typical processing types by robotic end-effectors

Type	Processing
Welding	A welding gun is mounted on end-effector to join parts by welding; robotic welding has the advantages of the welds with high consistency and precision in comparison with manual operations.
Painting	A painter tool is mounted on an end-effector to paint on objects; painting on the surface must be even and smooth which poses a challenge to human operators, and the painting environment is unpleasant to humans. Both of these factors make robotic painting attractive.
Picking and placing	An end-effector is to pick up and place objects from one location to another; regardless of the materials, size, and fragility of object, it requires safe handling and precision when the object is picked and placed.
Machining tending	An end-effector is equipped with the tools for pre- or post-processing such as loading and unloading, blowing off debris, inspecting, sorting, and cleaning around a workstation.
Packing and palletizing	An end-effector is used to pack and palletize objects in a short cycle time; these capabilities are highly desirable when the materials to be handled are fragile and the processes must also be of high efficiency with cautious protection.
Assembling	An end-effector is used to assemble parts into components mechanically by adding individual objects one by one following a specific sequence. The end-effector must be able to handle objects of different shapes and sizes.
Quality controls	An end-effector replaces human inspectors by applying consistent loads on products for rigorous testing after they have been machined. Human inspectors may struggle to maintain consistent tension, pressure, and overall precision; while it is easy to program a robot for such tasks when the robot is equipped with the right end-effector.
Surface treatment	An end-effector offers the flexibility for automated surface finishing; surface finishing is a particularly dirty task in the production process, it is also ideal to replace human operators with programmable robots.

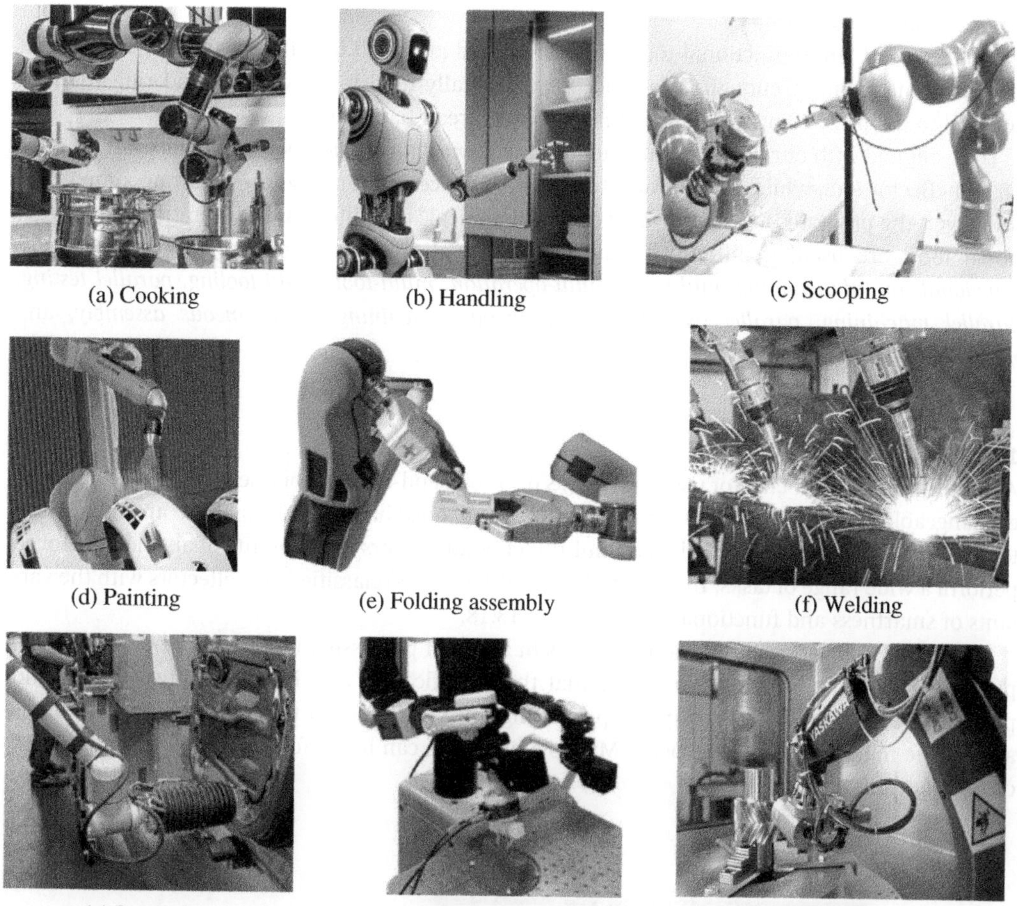

(a) Cooking (b) Handling (c) Scooping

(d) Painting (e) Folding assembly (f) Welding

(g) Inspecting (h) Multi-peg-in-hole (i) Deburring

Figure 13.17 Examples of different processing tools on robots (*Source:* Moley Robotics / https://thegadgetflow.com/blog/moley-robotics-kitchen-ces2021/ / Last accessed on July 19, 2024)

Figure 13.17 shows some examples where end-effectors on robots are used as processing tools to fulfill different tasks. The acting force by a processing tool must be controlled explicitly or implicitly to fulfill the corresponding task appropriately (Suomalainen et al. 2022).

Moreover, a fixtureless tool uses the same end-effector to (1) handle an object and (2) perform a machining process simultaneously (Bi and Zhang 2001). For example, Reddy (2016) and Reddy et al. (2021) proposed a cable-driven gripping system with a spindle to perform lightweight milling on objects with non-ferrous metals.

13.6.3 Multifunctional Tools

A multifunctional tool is a special type of end-effectors that can perform a variety of tasks without tool changes. Since a multifunctional tool can switch a task to be performed without another setup of hardware tools, it reduces operating cycles, improves productivity, and reduces the unit cost of products. This section aims to (1) understand key business drivers for using multifunctional tools on robots; (2) learn available multifunctional tools in various applications, including automotive, pharmaceutical, and electronics; and 3) identify trends in the development of multifunctional tools.

13.6.3.1 Concepts

The concepts of "multifunctional tools" were proposed and used in different fields with different purposes. Interestingly enough, this concept was initially used in software engineering where it referred to a type of computing algorithm or multi-thread hardware.

To be aligned with our interest in manufacturing, *a multifunctional tool* is defined as an MFM or an end-effector tool, which can perform two or more processes at one setup of part or component. Moreover, the processes to be performed can be "identical" or "different." In literature, multifunctional tools are used in different applications with alternative terminologies such as *multi-functional, multifunctioning, multitask, multi-operation, multi-tool, multi-tooling, parallel testing, parallel machining, parallel assembly, simultaneous machining, simultaneous assembly*, and *multi-spindle tools.*

13.6.3.2 Classification

A multifunctional end-effector usually consists of (1) one end-effector (or a set of end-effectors in an exchangeable system) and (2) one or a set of processing tools that are mountable on an end-effector to do works on an object. In particular, multifunctionality refers to the ability of an end-effector to perform a wide range of tasks. In fact, Mishkin and Jau (1988) classified end-effectors with the variants of smartness and functionalities in Figure 13.18.

Figure 13.19 classifies multifunctional tools in terms of processing types, part types, number of processing tools, and applications. Note that the classification also includes software tools and processes such as quad-core computer processing units (CPU) even though these software tools are not the main interest of the book. Multiple processes can be performed in either a *sequential* or *simultaneous* way.

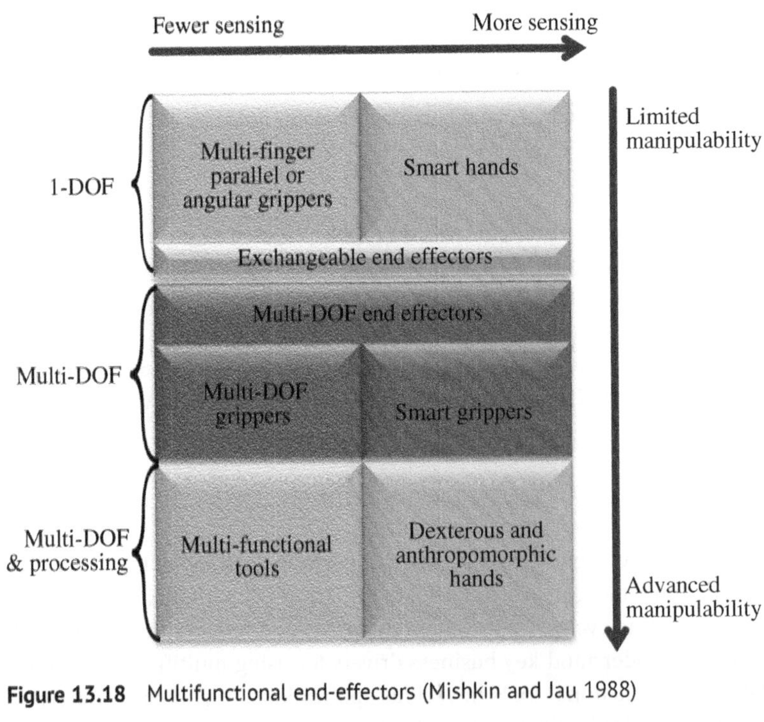

Figure 13.18 Multifunctional end-effectors (Mishkin and Jau 1988)

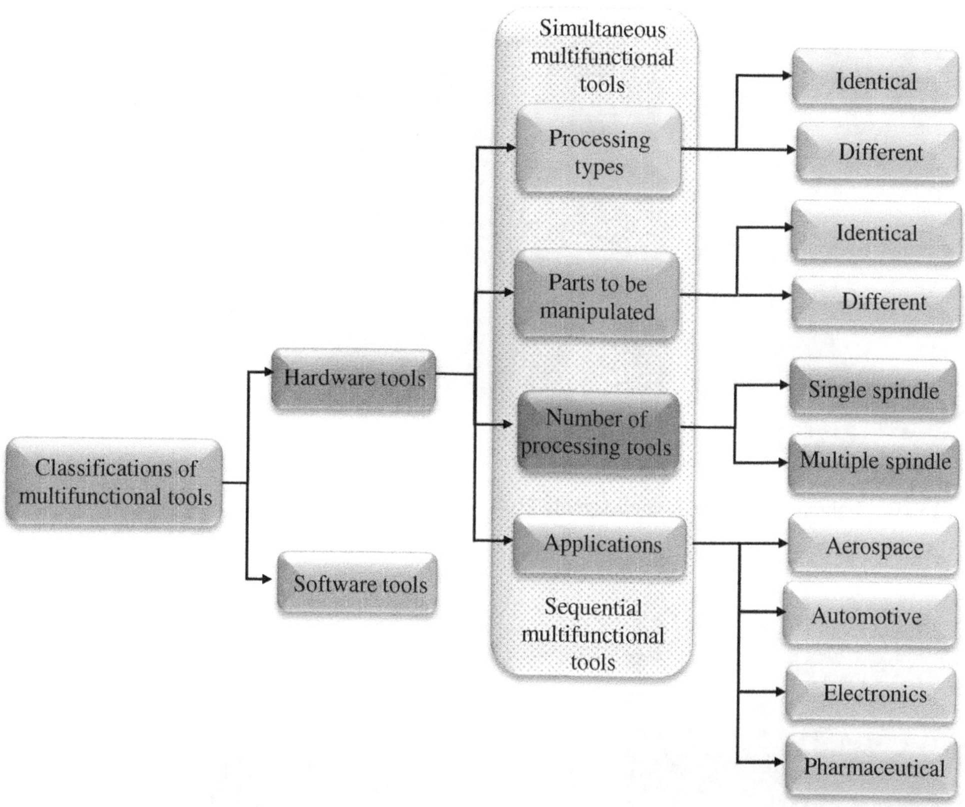

Figure 13.19 Classification of multifunctional tools

Many multifunctional tools perform multiple processes in a sequential way (Bonomi and Oldani 1990; Hatch 1992; Clark and Givler 1993; Braun 1997; Sanchez and Uecker 2005; Elliot et al. 2008; EMAG 2024; Everson and Jones 2008; Rhinoassembly 2024). For example, a multifunctional workstation is capable of performing a turning, drilling, and milling operation but it has to perform these processes one by one in sequence. A multifunctional tool may perform multiple processes simultaneously, and the tools under this category can be classified based on the types of processes, objects, and applications.

13.6.3.2.1 Identical or Different Processes

Multiplied processes can be *identical* or *different*. The majority of the multifunctional machines/tools are targeted at the same processes. The end-effector tool has a set of similar spindles and each of them performs one process. The motions and operations of the group of spindles can be controlled simultaneously. Take an example of the multi-spindle tightening system in Figure 13.20, it uses the same spindles for tightening, the motions of spindles are synchronized, and the same control is issued to all spindles.

A multifunctional tool for different processes is generally complex. Taking an example of the multi-spindle lathe in Figure 13.21, the raw materials are steel rods that are supplied continuously to each axis. A variety of parts shown on the right of the figure can be machined simultaneously. Each position corresponds to a specific process accomplished by one spindle. The same workstation

Figure 13.20 Tightening system (Rhinoassembly 2024)

Figure 13.21 Multi-spindle automatic lathe (*Source:* Photo by Tornos (2024))

can perform a variety of operations such as turning, milling, drilling, and deburring. The part is finished after the axis goes through all of the spindles that surround the main axis of the lathe. This type of lathe is specialized in making small round parts from rod materials.

13.6.3.2.2 Single Part or Multi-parts

Multiplied processes can be on one part or multiple parts. In food, pharmaceutics, and electronic industries, many products are customized with small batch sizes, and they are suitable to be processed in batch by multifunctional tools. For example, an automated filling machine in Figure 13.22

Figure 13.22 Automated filling machine (*Source:* Photo by DirectIndustry / VirtualExpo Group)

Figure 13.23 Multi-drilling on an auto bushing plate (Wright et al. 2014)

can fill water, light oil, and cream for a batch of six containers simultaneously at a speed of up to 100 containers per minute (DirectIndustry 2024). Figure 13.23 shows a mechatronic system where multi-drilling processes are applied on parts simultaneously (Wright et al. 2014).

13.6.3.2.3 Applications

Although different names are used, multifunctional tools are widely applied in various applications, and we can find many examples in our daily lives. For example, Farrelly (1997) invented a two-axis material handling tool to handle parts in a batch. Lee (1991) designed a multi-head electro-discharging machine, each head was able to produce a 3-DOF motion, and all heads worked simultaneously but independently to discharge respective components. Figure 13.21 shows six examples of multifunctional tools used in manufacturing, packaging, and pharmaceutics.

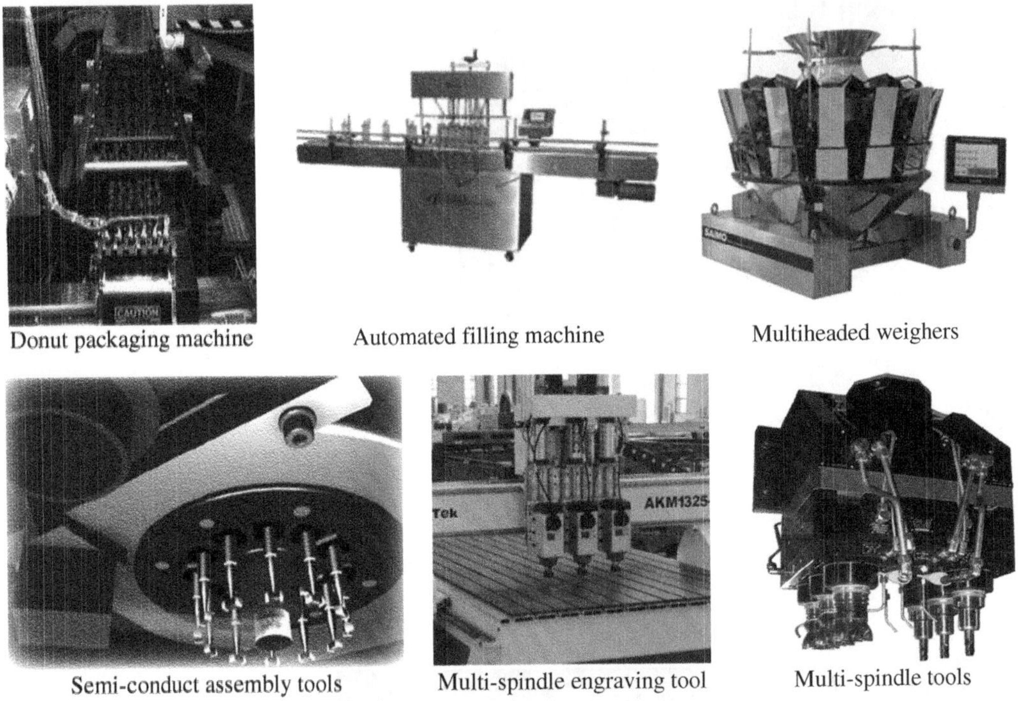

Figure 13.24 Examples of multifunctional tools in manufacturing, packaging, and pharmaceutics (*Source:* Cleco Production Tools (2024)/Apex Tool Group, LLC)

Multifunctional tools are especially popular in automotive manufacturing and assembling due to the strengths of multifunctional tools for high productivity. For example, Rhino (2024) provides a variety of systems for manual and automated fastening. Figure 13.24 shows six application examples of multifunctional tools including multi-spindle tightening machines (Cleco Production Tools 2024). Figure 13.25 shows more examples of using multifunctional tools in the automotive industry. A parallel testing system was developed to test an array of electronic chips simultaneously (NI 2024). Hessbruggen and Walter (2005) patented the method and device for machining parts with two or more end-effectors.

In the aerospace industry, few multifunctional tools are available in the aerospace industry. Some pioneer efforts are ongoing to develop multifunctional tools. For example, Boeing was developing a new track-based positioning system (Buttrick 2003; Thompson et al. 2005) (see Figure 13.26a) and a multi-spindle drilling system (Neugebauer and Hoppe 2006; Thompson et al. 2008) (see Figure 13.26b). A multitask robot was developed at Brotje-Automation GmbH to perform both drilling and fastening operations on aircraft. The end-effector was able to drill, feed, and insert fasteners, seal countersinks, detect holes, and measure the stack on the drill feed unit. A portable *multifunctional end effector* (MFEE) was developed, and it was able to drill and fasten-on longitudinal splices (Reid 2007).

13.6.3.3 Advantages and Disadvantages

While developing a multifunctional tool is technologically feasible, the use of a multifunctional tool should be well justified based on cost-saving, easiness of setup, process time, productivity, and

Intake and exhaust manifold tightening station

Multi-spindle engine assembly station

Cylinder heads and intake manifold assembly station

36 Spindle oil pan assembly fixtures

Figure 13.25 Automotive applications of multi-tool process tools

(a) Flex track with recent effort of carrying multi-spindle drills (Buttrick 2003; Thompson et al. 2005)

(b) Multi-task robot cell for aircraft structure system (Hogen et al. 2003; Thompson et al. 2008; Neugebauer and Hoppe 2006)

Figure 13.26 Examples of development of multifunctional tools in aerospace engineering

Table 13.5 Advantages and disadvantages of multifunctional tools

Advantages	Disadvantages
1) Reduced lead times 2) Reduced manufacturing costs with fewer fixtures, tools, and far less labor 3) Reduced wastes caused by multiple setups, material handling, and queueing 4) Improved product quality by eliminating a tolerance build-up 5) Reduced the need for tooling 6) Increased throughput and profitability.	1) Heavy machine base and large power supply 2) Complexity in coordinating multi-axis motions 3) Limited applications due to the constraints of relative multifunctional operations 4) Complication of tool-wear monitoring and compensation

overall quality. Sandvik (2024) and Mazak (2024) summarize the advantages and disadvantages of using a multifunctional tool in Table 13.5.

13.6.3.4 Selection Principles

The critical features common to different multifunctional tools in Section 13.6.2.4 help engineers identify new areas of multifunctional tools. A multifunctional tool should be explored when an application has the most attributes as below:

1) Multiple identical or similar processes.
2) Features to be machined are patterned.
3) Operations are mostly duplicated.
4) Multiple features are made in a limited footprint.
5) Features are made on a flat or smooth surface with a constant curvature.
6) Machining tools have long lifespans and tool-wears are controllable.
7) The machine base is stable and provides high rigidity.
8) Parts to be machined can be positioned or oriented automatically.
9) Outcomes of machining operations are highly predictable and operation conditions are well maintained with low-level uncertainty and randomness.
10) There is a need to reduce the cycle time for a lengthy process by parallel machining.
11) Operating or machining is limited to those that require no or low executing force such as testing, fastening, placing, inserting, sealing, filling, dispensing, laser cutting, and drilling with small dimensions.

13.6.3.5 Development Trends

Researchers are continuously advancing the technologies to improve the efficiency and productivity of multiple-functional tools and explore their new applications in the following aspects (Rao 2023).

13.6.3.5.1 Micro- and Nano-level Multi-manipulation

A multifunctional tool shows significant advantages when the operational space is highly confined and multiple operations are processed in parallel or sequentially; in addition, operating forces are very light. Note that the required operating force depends on the oversize of the object; the operating force on an object at micro- or nano-level is very small.

Table 13.6 Grasping principles for objects at different scales

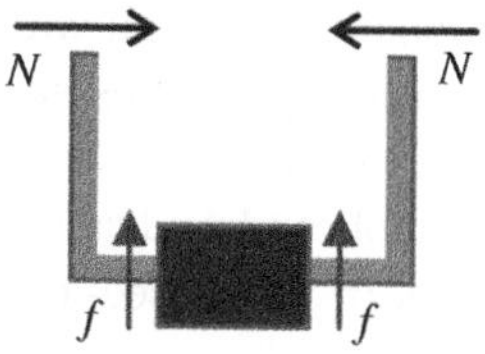

(a). Friction force
Macro/Meso/Micro

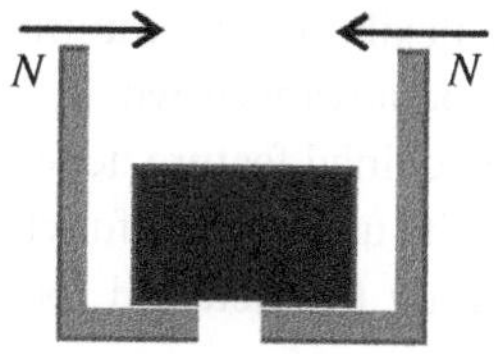

(b). Jawing force
Macro/Meso/Micro

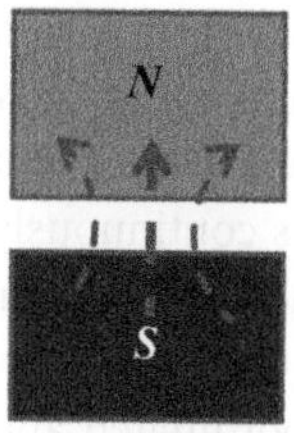

(c). Magnetic forces
Macro/Meso/Micro

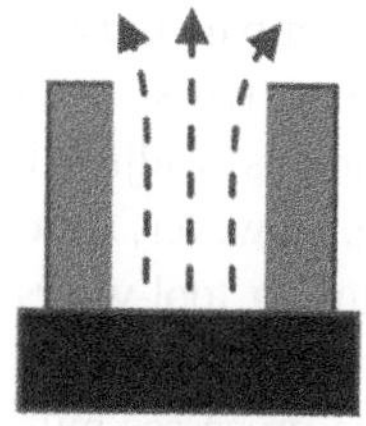

(d). Suction forces
Macro/Meso/Micro

(e). Piecing force by needle
Macro

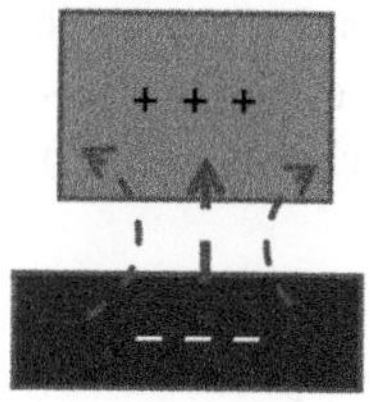

(f). Electrostatic forces
Meso/Micro

(g). Van der Walls forces
Meso/Micro

(h). Attraction by ice
Macro/Meso/Micro

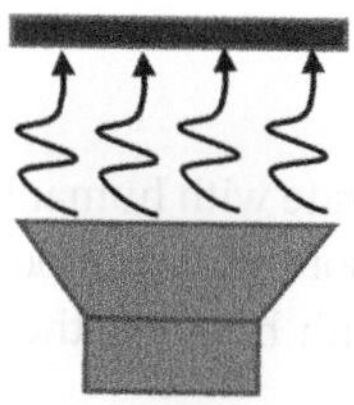

(i). Acoustic force
Macro/Meso/Micro

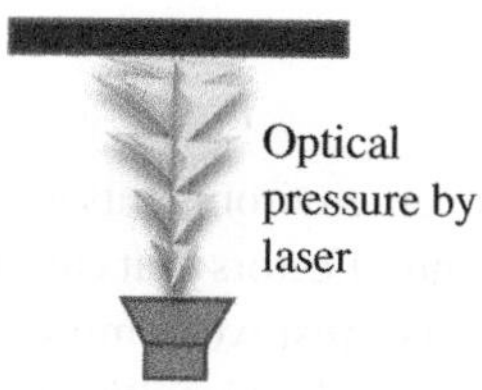

(j). Optical pressure by laser
Micro

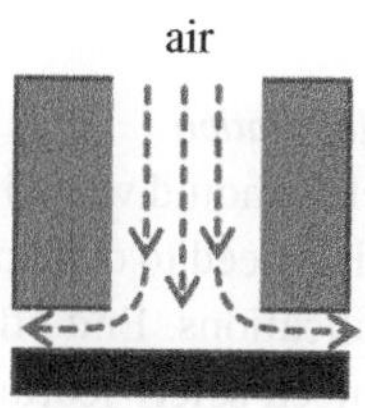

(k). Bernoulli pressure
Meso/Micro

(l). Adhesive forces
Micro

These factors have shown great potential for developing multifunctional tools in electronics and bio-medical fields. For example, Bower et al. (2008) developed a micro-scale device for parallel assembly. Gallagher (1994) patented a nano-level manipulator that was used for parallel fabrication of nanostructure. Kusterer et al. (2008) developed a bioreactor with a number of tubes that were controlled for germ cultivations in parallel.

It is interesting to note that grasping principles at the micro-level are quite different from those for macro-level gripers. Sharma (2015) classified grasping principles that are applied in different scales in Table 13.6.

13.6.3.5.2 *Improved Sensing and Adaptive Control*

One of the key areas of development in end-effectors is improved sensing and control. Advancements in sensing technology will enable end-effectors to better detect and respond to changes in their environment, allowing for more precise and efficient operations. In addition, advancements in control systems will make end-effectors more adaptable to changing tasks and situations.

The requirement of performing multiple processes by one mechatronic device brings challenges in coordinating these processes. For example, the relative position or orientation of two machined features on a part must be fixed; this affects the flexibility of a multifunctional tool to deal with different part types. A multifunctional tool can be advanced when the relative spatial position of two spindles is adjustable. The accuracy of a machined feature depends greatly on the accuracy of tooling. However, a tool wears continuously in its use. A multifunctional tool can be advanced by monitoring tool-wears online and compensating the detected tool-wears in the closed-loop machine controls.

Multiple processes must be coordinated appropriately and such coordination also includes the motions of operating tools. A multifunctional tool can be advanced to coordinate the motion of operation tools in some complex conditions such as (1) the processes to be processes are different, (2) an operational environment involves some obstacles that are randomly shown up. The control system of the multifunctional tool must be advanced to enhance robustness, reliability, and adaptability in complex applications (Rakowski 2002, DP technology 2005).

13.6.3.5.3 Enhanced Dexterity and Mobility

Next-generation end-effectors can be with greater dexterity in handling complex objects and tasks, with better precision and shortened cycle time. It requires advanced materials and manufacturing techniques to create more versatile and lightweight tools. In addition, an end-effector is attached to a robotic system and this limits its range of motion. The mobility can be improved expanding the workspace of an end-effector.

13.6.3.5.4 Human–Robot Collaboration

Next-generation robotic systems should work in open-environments and collaborate with humans directly. There will be a growing need to develop end-effectors that can fully support human–robot collaborations in various applications. End-effectors must work more closely with humans; they must be equipped with enhanced safety features and allow intuitive controls.

13.6.3.5.5 Smartness by Artificial Intelligence

Integrating with AIs brings opportunities for an end-effector to perceive the environment, detect events and changes, and utilize the sensed data by AI in closed-loop controls to improve the flexibility, robustness, and adaptability in dynamic and unstructured environments. End-effectors equipped with AI will be able to learn and adapt to their environment, improving their efficiency and precision over time. They may also be able to work autonomously, freeing up human operators to focus on more complex tasks.

13.7 Main Factors in Designing an End-Effector

Since an end-effector connects the robot body on one side and an object to be manipulated on the other side, the design of an end-effector depends on both robotic structure and the tasks to be performed. For example, the end-effectors of industrial robots in manufacturing should usually be very rigid to carry a high load. It will be equipped with one of the sophisticated tools for light-machining, welding, water-jetting, and painting. On the other hand, when a robot is used in precision machinery, it is critical to avoid damage to the object by the end-effector. In such a case, the end-effector should be compliant, and the grasping force can be attraction or adhesion (Zhao 2021).

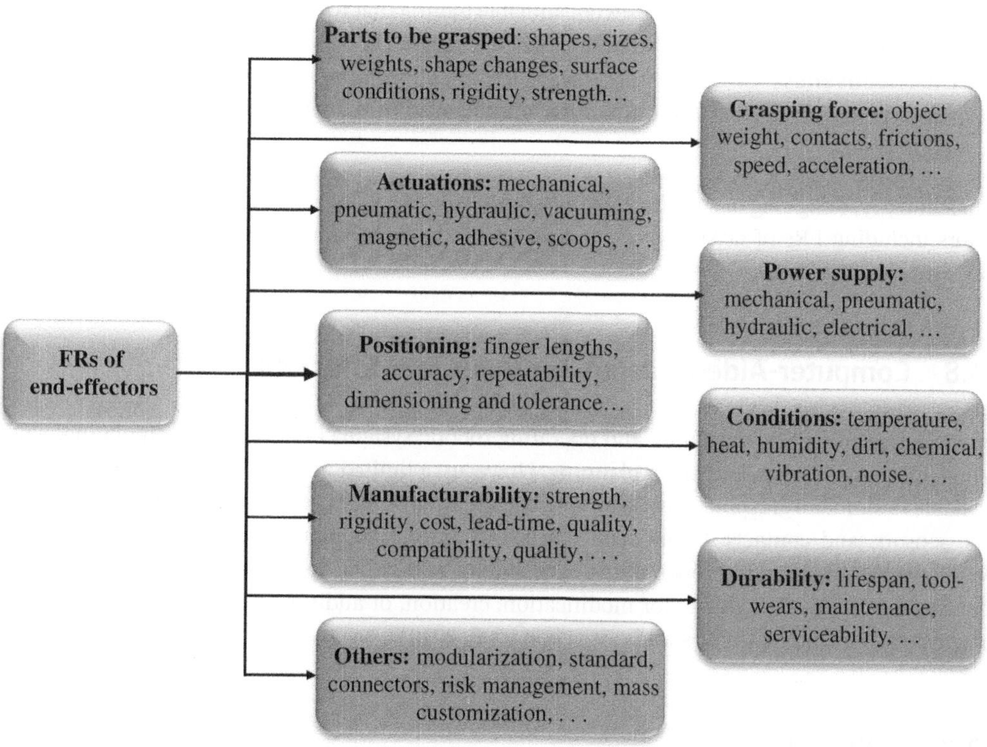

Figure 13.27 FRs of end-effectors

Sharma (2015) discussed the main factors in designing an end-effector, and the functional requirements (FRs) are defined as the aspects of *parts to be grasped*, *actuation*, *energy source*, *grasping force*, *dimensioning and accuracy*, *durability*, *operational conditions*, *manufacturability*, and *miscellaneous factors* such as modularization and standardization (see Figure 13.27).

Like other commonly used mechatronic systems, engineers should use as many off-the-shelf functional modules on markets as many as possible. However, selecting a right end-effector for a particular application can be complex since many criteria must be taken into consideration simultaneously. In addition, the selection of an end-effector can affect the effectiveness and efficiency of the whole robotic system. The following are the key factors in selecting the right end-effector (Rao 2023; UR 2024).

1) **FRs of Task**. The first step in choosing an end-effector is to understand FRs elaborated in Figure 13.27. These include the shape, size, and weight of the objects that the robot will be manipulating, the required precision and speed of the task, and any environmental factors that may affect the end-effector's performance.
2) **Matching Capabilities**. Once the FRs of the task are defined, the next step is to analyze, evaluate, and compare the capabilities of end-effectors and further determine the best one for the given task. For example, a mechanical gripper may be suitable to pick up and hold objects with a smooth surface, but those who understand the capabilities of others know a vacuum gripper would be ideal for some particular tasks.
3) **Compatibility**. The chosen design must be compatible with the specifications of the whole robotic system since the robotic interface is fixed and portable only to certain end-effector types.

4) **Maintenance**. An end-effector uses consumable tooling that will be worn over time. It is important to take into consideration maintenance and repair need in selecting an end-effector.
5) **Cost-effectiveness**. Cost is an important factor in determining affordable capabilities and lifespans. The cost for end-effector and other miscellaneous auxiliaries adds to the overall cost of robotic system.

In summary, designing or selecting an end-effector for a specific application must consider many factors including FRs of tasks, expected capabilities, the compatibility with host robotic system, cost, and maintenance.

13.8 Computer-Aided Design Tools for End-Effectors

Simulation is critical for designing and operating robotic systems involved in manipulating, navigating, motion planning, and grasp planning. However, a limited number of software packages are available to simulate grasping operations in a holistic environment at various aspects such as actuators, sensors, and contacts. OpenGrasp stands out among popular tools including GraspIt and OpenRave for enhanced interoperability, extensibility, and public openness. The software architecture was modularized, allowing for modification, creation, or addition of functional modules and integration of technologies or standards (Leon et al. 2010).

13.9 Summary

End-effectors are an essential component of robotic systems, enabling them to interact with the world around them in a variety of ways. End-effectors are designed to perform specific tasks, such as gripping, sensing, and manipulating objects, and are often customizable to meet the needs of different industries and applications.

The key concepts of this chapter include: (1) end-effectors are highly customizable and differ from one another in terms of gripping principles, sensing, and processing, each type of end-effector has its own set of advantages and limitations; (2) designing and selecting the right end-effector for a given task requires careful consideration of factors such as the size and shape of the objects being handled, the required precision and speed, and the environment in which the robot will be operating; (3) there are the needs to advance end-effectors to improve efficiency, precision, and productivity across many different industries; (4) as robotics technology continues to advance, end-effectors will play an increasingly important role in a wide range of industries, from manufacturing and logistics to healthcare and entertainment.

Problems

13.1 Assume a DC motor provides a 20 (kg · cm) torque when its rotational speed is 10,000 *revolutions per minute* (RPM), and the gear ratio of the attached gearbox is 150:1. Determine the output speed and torque when the efficiency of the gear trains is 90%.

13.2 An end-effector is designed to hold an object with a weight of 80 lbf vertically, and the object is expected to have the maximum acceleration is 386.1 in · s^2, what is the minimum payload that must be specified for the end-effector?

13.3 Assume the friction of coefficient is 0.1, the dimeter of the object is $\phi = 3$ in, what are the minimal clamping force and moment the grippers in Figure 13.28a and b should provide to hold the object?

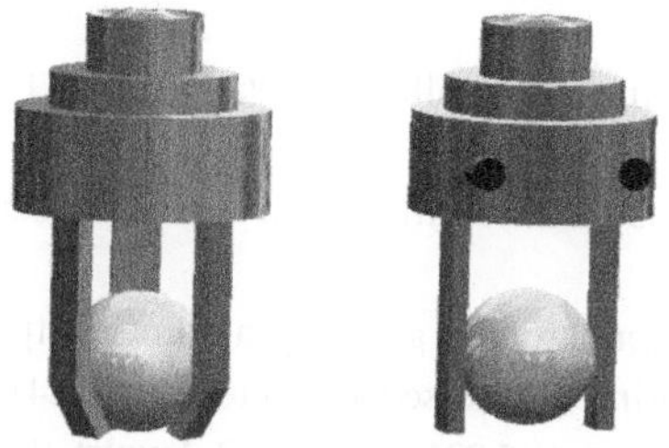

(a) Form closure (b) Force closure

Figure 13.28 Clamping configurations in Problem 13.3

13.4 A gripper includes 10 suction cups as shown in Figure 13.29, the object to be grasped is a metal sheet with a mass of 8 kg, the pressure at a suction cup is 60,000 Pa, and the effective suction area is $A_e = 0.005\ \text{m}^2$, determine the factor of safety of the end-effector in grasping the metal sheet.

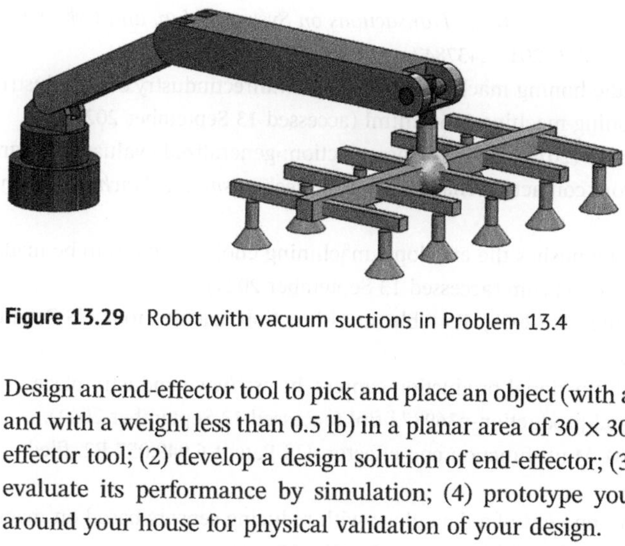

Figure 13.29 Robot with vacuum suctions in Problem 13.4

13.5 Design an end-effector tool to pick and place an object (with a size smaller than $2 \times 2 \times 2\ \text{in}^3$ and with a weight less than 0.5 lb) in a planar area of $30 \times 30\ \text{in}^2$: (1) define FRs of the end-effector tool; (2) develop a design solution of end-effector; (3) create the virtual model and evaluate its performance by simulation; (4) prototype your design using the materials around your house for physical validation of your design.

References

Ben-Tzvi, P. and Moubarak, P. (2014). A mechatronic perspective on robotic arms and end-effectors. https://www.researchgate.net/publication/221910943_A_Mechatronic_Perspective_on_Robotic_Arms_and_End-Effectors (accessed 13 September 2024).

Bi, Z.M. and Zhang, W.J. (2001). Flexible fixture design and automation: review, issues and future directions. *International Journal of Production Research* 39 (13): 2867–2894.

Bonomi, G.B. and Oldani, B. (1990). Multi-task end effector for robotic machining center. US Patent 4,955,119, filed 11 July 1989.

Bower, C.A., Menard, E., and Garrou, P.E. (2008). Transfer printing: an approach for massively parallel assembly of microscale devices. *58th Electronic Components and Technology Conference*, pp. 1105–1109. Lake Buena Vista, FL, USA, 2008.

Braun, H.D. (1997). Machine tool, more particularly for drilling and milling. US Patent, 5,6111,137, filed 29 April 1997.

Buttrick, J. (2003). Flex track drill. SAE International, 2003-01-2955. https://www.jstor.org/stable/44699218 (accessed 13 September 2024).

Carpin, S., Lu, S., Falco, J., and Wyk, K.V. (2016). Multi-fingered robotic grasping: a primer. https://www.nist.gov/publications/multi-fingered-robotic-grasping-primer (accessed 13 September 2024).

Clark, G.L. and Givler, G.C. (1993). Drill/rivet device. US Patent 5,231,747, filed 3 August 1993.

Cleco Production Tools (2024). Automotive manufacturing solutions that maintain line efficiency without compromising safety and quality. https://www.clecotools.com/solutions/automotive (accessed 13 September 2024).

Cutkosky, M.R. (1984). *Mechanical Properties for the Grasp of a Robotic Hand*. Pittsburgh, Pennsylvania: The Robotics Institute, Carnegie-Mellon University https://www.ri.cmu.edu/pub_files/pub3/cutkosky_mark_r_1984_3/cutkosky_mark_r_1984_3.pdf (accessed 13 September 2024).

Cutkosky, M.R. (1989). On grasp choice, grasp models, and the design of hands for manufacturing tasks. *IEEE Transactions on Robotics and Automation* 5 (3): 269–279.

Datta, R., Pradhan, S., and Bhattacharya, B. (2015). Analysis and design optimization of a robotic gripper using mulitobjective generntic algorithm. *IEEE Transactions on Systems, Man, and Cybernetics: Systems* https://doi.org/10.1109/TSMC.2015.2437847.

DirectIndustry (2024). Multi-spindle honing machines. https://www.directindustry.com/industrial-manufacturer/multi-spindle-honing-machine-88652.html (accessed 13 September 2024).

Dong, H., Qiu, C., Prasad, D.K. et al. (2019). Enabling grasp action: generalized evaluation of grasp stability via contact stiffness from contact mechanics insight. *Mechanism and Machine Theory* 134: 625–644.

DP Technology (2005). Multitasking pushes the envelope: machining choices remain to be mad. CAM Software. www.toolingandproduction.com (accessed 13 September 2024).

Elliot, E.R., Messick, C., and Stallings, R. (2008). Portable multi-function machine tool. US Patent 2008/0193239, filed 25 January 2008.

EMAG (2024). VSC vertical multi-functional production centers. https://www.industry-plaza.com/vsc-vertical-multi-functional-production-p36987.html (accessed 13 September 2024).

Everson, C. and Jones, E.E. (2008). Multi-function end effector. US Patent 6,949,057 B2, filed 31 December 2007.

Fahn, C. and Sun, H. (2005). Development of a data glove with reducing sensors based on magnetic induction. *IEEE Transactions on Industrial Electronics* 52 (2): 585–594.

Farrelly, P.J. (1997). Two-axis article loader/unloader. US Patent 5,611,193A, filed 31 January 1995.

Gallagher, A.C. (1994). Microtip-controlled nanostructure fabrication and multi-tipped field emission tool for parallel process nanostructure fabrvication. US Patent 5,294,465A, 1994, filed 14 January 1991.

Hack, N.P. (2018). Mesh mould a robotically fabricated structural stay-in-place formwork system. Doctoral thesis, Diss. ETH No. 25068. https://www.researchgate.net/publication/325362607_Mesh_Mould_A_robotically_fabricated_structural_stay-in-place_formwork_system (accessed 13 September 2024).

Hatch, J.M. (1992). Automatic, programming drilling, gluing and doweling apparatus. US Patent 5,276,953, filed 19 August 1992.

Hessbruggen, N. and Walter, U. (2005). Method and device for machining parts. US Patent 6,845,899 B2, filed 16 January 2001.

Hogan, S., Hartmann, J., Thayer, B. et al. (2003). Automated wing drilling system for the A380-GRAWDE. SAE International, 2003-01-2940, 2003.

Howe, R.D. (2024). Harvad biorobotics lab. https://biorobotics.harvard.edu/robotic_hand_optimization.html (accessed 13 September 2024).

Kusterer, A., Krause, C., Kaufmann, K., and Arnold, M. (2008). Fully automated single-use stirred-tank bioreactors for parallel microbial cultivation. *Bioprocess and Biosystems Engineering* 31: 207–215.

Lee, S.L. (1991). Multi-head type of electro-discharging machine. US 5,075,530A, filed 23 July 1990.

Leon, B., Ulbrich, S., Diankov, R. et al. (2010). OpenGRASP: a toolkit for robot grasping simulations. *SIMPAR 2010*, LNAI 6472 (ed. N. Ando et al.), pp. 109–120.

Lynch, K.M. and Park, F.C. (2017). *Modern Robotics, Mechanics, Planning, and Control*. Cambridge University Press.

Marandi, P. (2013). Development of a hand grip library intended for implementation in an ergonomic simulation tool. https://publications.lib.chalmers.se/records/fulltext/196361/196361.pdf (accessed 13 September 2024).

Mazak (2024). Multi-tasking. https://www.mazakusa.com/machines/process/multi-tasking/ (accessed 13 September 2024).

Michaud, S., Dominguez, M., Nguyen, U. et al. (2004). Eurobot end-effectors. In *Proceedings of the 8th ESA Workshop on Advanced Space Technology for Robotics and Automation, 'ASTRA 2004' ESTEC*, Noordwijk, the Netherlands (2–4 November 2004).

Mishkin AH and Jau BM (1988) *Space-based Multifunctional End-Effector Systems: Functional Requirements and Proposed Designs*, Jet Propulsion Laboratory, California Institute of Technology, Pasadena, CA. https://ntrs.nasa.gov/api/citations/19890001866/downloads/19890001866.pdf (accessed 13 September 2024).

Monkman, G.J., Hesse, S., Steinmann, R., and Schunk, H. (2007). *Robot Gripper*. Wiley-Vch Verlag GmbH & Co. KGaA.

Motion Control Tips (2024). What are grippers and end effectors? Technical Summary. https://www.motioncontroltips.com/electromechanical-grippers (accessed 13 September 2024).

Neugebauer, L. and Hoppe, S. (2006). Full flexible multi task robot cell for complex 360° aircraft structures. SAE International, 2006-01-3124.

NI (2024). Benefits of parallel testing. https://www.ni.com/en/shop/electronic-test-instrumentation/application-software-for-electronic-test-and-instrumentation-category/what-is-teststand/benefits-of-parallel-testing.html (accessed 13 September 2024).

Rakowski, L. (2002). Programming multitasking machines? https://www.productionmachining.com/articles/programming-multitasking-machines (accessed 13 September 2024).

Rao, R. (2023). What are end-effectors in robotics? Types of end-effectors, applications, future. https://www.wevolver.com/article/what-are-end-effectors-in-robotics-types-of-end-effectors-applications-future (accessed 13 September 2024).

Raval, S. and Patel, B. (2016). A review on grasping principle and robotic grippers. *International Journal of Engineering Development and Research* 4 (1): 483–390.

Reddy, G.R. (2016). Design and structural analysis of a robotic arm. ISRN: BTH-AMT-EX--2016/D06—SE, ISRN: BTH-AMT-EX--2016/D06—SE.

Reddy, C.E., Padayachee, J., and Bright, G. (2021). Development and analysis of reconfigurable robotic end-effector for machining and part handling. *South African Journal of Industrial Engineering* 32 (1): 56–70.

Reid, E. (2007). Development of portable and flexible track positing system for aircraft manufacturing processes. SAE International, 2007-01-3781.

Rekabi-Bana, F., Stefanec, M., Ulrich, J. et al. (2023) Mechatronic Design for Multi Robots-Insect Swarms Interactions. https://roboroyale.eu/pdfs/2023_icm_manipulator.pdf (accessed 13 September 2024).

Rhino Tool House (2024). Innovative manufacturing improvements. https://rhinotoolhouse.com/ (accessed 13 September 2024).

Rhinoassembly (2024). Tighening system. http://www.rhinoassembly.com/ (accessed 13 September 2024).

Sanchez, D. and Uecker, D. (2005). Multifunctional handle for a medical robotic system. US 2005/0043719 A1, filed 18 August 2004.

Sandvik (2024). Multi-task machining. https://www.sandvik.coromant.com/en-us/tools/turning-tools/multi-task-machining-tools (accessed 13 September 2024).

Sharma, A. (2015). Design study of end-effectors. *International Journal of Engineering and Advanced Technology* 4 (3): 164–168.

Suomalainen, M., Karayiannidis, Y., and Kyrki, V. (2022). A survey of robot manipulation in contact. *Robotics and Autonomous Systems* 156: 104224.

Thompson, P., Hartmann, J., Feikert, E.D., and Buttrick, J. (2005). Flex track for use in production. SAE International, 2005-01-3318.

Thompson, P., Oberoi, H., and Draper, A. (2008). Development of a multi spindle flexible drilling system for circumferential splice drilling applications on the 777 airplane. SAE International, 2008-01-2298.

Tornos (2024). Multi-spingle automatic lathe. http://www.tornos.com/ (accessed 13 September 2024).

Unidex (2024). Different types of end effectors. https://www.unidex-inc.com/blog/end-effector-what-it-is-how-it-works-types-applications/ (accessed 13 September 2024).

UR (2024). Types of grippers used in manufacturing, universal robots. https://www.universal-robots.com/blog/types-of-grippers-used-in-manufacturing/ (accessed 13 September 2024).

Wright, R.S., Amirehteshami, D.H., and Pringle, J.W. (2014). Method and apparatus for automated multi-drilling and multi-rivet machine. US Patent 9,862,020B2. Filed 16 September 2015.

Zhang, B., Xie, Y., Zhou, J. et al. (2020). State-of-the-art robotic grippers, grasping and control strategies, as well as their applications in agricultural robots: a review. *Computers and Electronics in Agriculture* 177: 105694.

Zhao, W. (2021). Design of robot end-effector for collaborative robot works. *Robotics*. https://hal.science/tel-03747941/document/ (accessed 13 September 2024).

14

Metaverses for Sustainability Mechatronic Systems

14.1 Introduction

In the book *Snow Crash* by Stephenson, the metaverse was referred to as a black planet that was accessible by users through *virtual reality* (VR) equipment and terminals (Stephenson 1992; Bansal et al. 2022). In fact, metaverse is constructed from the words of "*Meta*" and "*Universe*." Later, it was used to represent a hypothetical next-generation Internet; in an application scenario of metaverse, all virtual things are connected to the Internet, and users can interact in the same ways they communicate and interact in our physical worlds (Cheng et al. 2022).

A use case in the metaverse consists of a number of immersive worlds, in which users appear as avatars and can interact with one another without bounds. Therefore, metaverse was also viewed as an evolution of the Internet where VR, *augmented reality* (AR), and *mixed reality* (MR) are used as popular modalities. Nowadays, metaverse is gaining more and more popularity in supporting peoples' interactions and communications virtually and remotely in distributed geographies (Tian 2021; Chen 2022). Various metaverse use cases have been developed in broadened applications such as entertainment, space, education, manufacturing, healthcare, real estate, transportation, agriculture, socialization, and smart communities (Kang et al. 2022; Felice et al. 2023; Li et al. 2023a). From the perspective of sustainability, metaverse is making a growing economic, social, and environmental impact in terms of diversity, inclusion, equity, safety, and accessibility (Zallio and Clarson 2022). The statistics showed that the economic scale of enabling technologies for metaverse, including VR, AR, XR, AI, blockchain technologies (BCTs), and robots, has grown from $106 billion in 2021 to $303 billion in 2028 only in the education sector (Bansal et al. 2022; Barrera and Shah 2023; Kshetri et al. 2022).

Metaverse is not a stand-alone technology or begun completely from scratch. It was advanced from conventional VR, AR, MR, XR, CPS, IoT, AI, and robotics. In fact, people still view a metaverse use case as a *cyber-physical-social system* (CPSS) that is composed of humans and all other smart things in physical, mental, and artificial worlds. A CPSS unifies two essential components of complex science; that is, *the first one* is an emergence in cyberspace and *the second one* is a convergence in physical space for problem-solving or decision-making (Wang 2022a, 2022b).

Metaverse has proven its great potential to expand conventional digital technologies for effective human–machine interactions seamlessly (Dwivedi et al. 2022). The metaverse era was so newly born and most people agreed that it began in 2021 when Roblox was launched as a metaverse use case for users to play games interactively over the Internet. Now, Roblox maintains over 150 million active subscriptions monthly. Most of the Roblox users are young, two-thirds of

Mechatronics for Complex Products and Systems: Project-Based Design Approaches for Robots, Cyber-Physical Systems, Digital Twins, and Other Emerging Technologies, First Edition. Zhuming Bi.

Companion website: www.wiley.com/go/bi/mechatronics

children are aged from 9 to 12, and one-third of children are aged under 16 in the United States (Park and Kim 2022). Other two influential products for metaverse are the Meta platform, which was developed by Facebook, and the Mesh platform, which was created by Microsoft (Dong and Liu 2022). Looking forward, the rapidly advanced computing power and *human and computer interactions* (HCI) have brought opportunities to create various virtual worlds and support the interactions of the avatars with physical worlds in innovative ways (Faraboschi et al. 2022).

This chapter explores the potential applications of metaverse-enabling technologies in sustainable mechatronic systems, in particular, sustainable manufacturing systems. Metaverses use cases have been widely explored in all four industry sectors, such as mining, fishing, hunting, forestry, and agriculture *in primary sector*; manufacturing, construction, chemical engineering, textile, aerospace, shipbuilding, and energy supplies in the *secondary sector*; retail, healthcare, entertainment, banking, tourism, transportation, and legal services in the *tertiary sector*; and research and development, education, information technologies, and financial service and consultancy in the *quaternary sector*. For example, Huang et al. (2023) developed a use case to train pilots in the aerospace engineering and defense industry, it was able to (1) model anticipated or unanticipated events, weather conditions, and terrains and (2) guide, assess, and verify trainees' actions in the virtual world.

In manufacturing, metaverse is a critical component of Industry 5.0 in global industry transformation. Industry 5.0 requires effective human–machine interactions in immersive ways, and metaverse provides most emulation tools to support redesigning, reconfiguring, decommissioning, and optimizing human-centralized systems. Metaverse is a general cyberspace to combine virtual and physical spaces as a unified physical, cyber-enabled, social, and thinking space (Ning et al. 2023); it adds new dimensions of data over the Internet to overcome information blindness in industrial scenarios (Leng et al. 2022).

The rest of the chapter is organized as follows. Section 14.2 discusses the background of sustainable mechatronic systems to identify a set of *functional requirements* (FRs) for the sustainability of the economy, society, and environment. Section 14.3 discusses the recent advancement of metaverses and critical enabling technologies including DT-I, DT-II, IoT, IoDDT, BDA, BCT, AR, HRI, and *data-driven decision-making systems*. Section 14.4 discusses how metaverses can be used to fulfill FRs for system sustainability. Section 14.5 discusses the limitations of existing metaverses and introduces some ongoing research projects to develop metaverse use cases for sustainable systems. Section 14.6 gives a summary of our findings from the presented work.

14.2 FRs of Sustainable Mechatronic Systems

Despite numerous discussions on the importance of system sustainability, the first step to pursue and enhance the sustainability of a mechatronic system is to learn how to measure and evaluate the sustainability of a product or system in a quantifiable way (Bi 2011). Generally, *sustainability* is defined as meeting the present global needs of human society without compromising the well-being of the generations in the future. In the United States, *sustainable development* is a commitment to create and maintain conditions under which humans and nature can exist in productive harmony that permits fulfilling the social, economic, and other requirements of present and future generations (Saari et al. 2022). Many researchers have investigated the models and methods used to evaluate system sustainability. For example, Dogea and Stolt (2021) argued that sustainability can be associated with two most critical objectives, that is, *efficiency* and *productivity*, and more and more digital technologies should be adopted to enhance sustainability by improving efficiency and productivity. Bhat et al. (2023) discussed the requirements of modern manufacturing from an economic perspective and added a set of new requirements for sustainability including the need to (1) reduce latencies; (2) process, analyze, and use multimodal real-world data at an extremely fast-growing rate; (3) assure security, safety, and privacy; and (4) consume less energy. Fielder (2023) corresponded sustainability to system smartness that was in turn related to efficiency, productivity, and competitiveness; the set of identified FRs was (1) digitization to harness data for success since data is the most critical resource in a complex decision-making system; (2) real-time visibility to make data and executions transparent and controllable; (3) agility in responding anticipated and unanticipated, internal or external events; (4) automation and efficiency for minimized operation cost; (5) connectivity for seamless communication, coordination, collaboration, and cooperation; and (6) ease of use to empower the workforce.

Figure 14.1 summarizes the author's findings from the literature survey in regard to basic *functional requirements* (FRs) of sustainable mechatronic systems within the context of manufacturing. The top-level functional requirement (*FR-0*) for the sustainability in economy, society, and environment is decomposed into (1) three relatively independent sub-FRs (i.e., *FR-12*, *FR-13*, and *FR-14*) for the sustainability at these three aspects, respectively, and (2) one sub-FR (FR-11) on the technological sustainability that is strongly coupled with the first set of sub-FRs. Furthermore, the *axiomatic design theory* (ADT) is adopted to define FRs of sustainability and find system solutions to these FRs in a system design (Cochran et al. 2016a, 2016b). Every sub-FR must be decomposed and detailed *level* by *level*, until each of the decomposed sub-FRs finds its corresponding solution. Taking an example of *FR-12* for economic sustainability in traditional manufacturing, it can be further decomposed into a set of the sub-FRs at the third level on products, processes, and manufacturing systems, respectively.

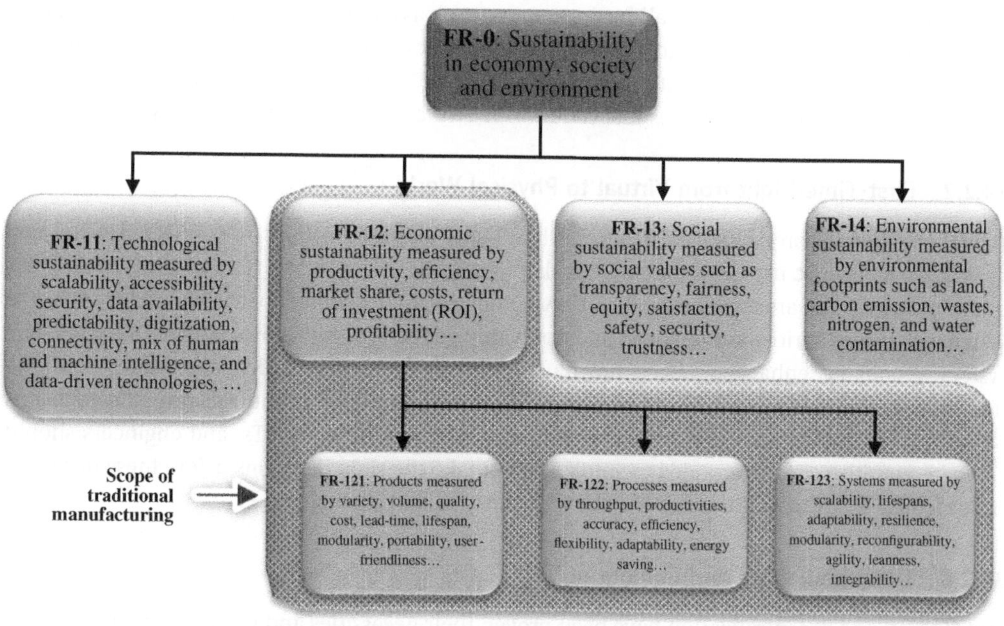

Figure 14.1 Functional requirements (FRs) of system sustainability

Since this chapter discusses the roles of metaverse in enhancing system sustainability, FR-11 is focused on the technological functional requirements of sustainable systems. The specific FRs to which metaverse and relevant enabling technologies are contributing are discussed below.

14.2.1 Scalability, Accessibility, Security, Privacy, and Legal Issues

Scalability refers to the system's capability to add, replace, modify, or remove resource(s) in a cost-effective way to satisfy changes in applications when they are needed. With the growth of the Internet, metaverse has shown its great potential to network with virtual or physical resources since the residual Internet is fully scalable. Cheng et al. (2022) conducted their preliminary investigations on the scalability of metaverse and they showed that social VR platforms such as AltspaceVR and Workrooms were only able to support a limited number of partners. An increase in the number of online partners brought the challenge of meeting an exponentially increased bandwidth demand for uploading and downloading massive data streams. Metaverse use cases must be scalable before they can be applied in scalable sustainable systems.

Accessibility allows a mechatronic system to access any internal or external resource via networking whenever it is needed. In a metaverse use case, everything over the Internet has its virtual model, and virtual models can interact with each other freely. On the other hand, an interaction relies on interface equipment such as a keyboard, mouse, headset, or a haptic device; this might cause some issues such as motion sicknesses or the accessibility to users who need glasses or contact lenses. The interfacing technologies can be advanced for users to interact with others over the Internet with fewer constraints. The accessibility of metaverse affects the effectiveness of the interoperability of associated physical resources in a virtual enterprise.

Security, privacy, and *legality* are basic FRs for an enterprise to protect its assets and *intellectual properties* (IPs) in a decentralized and distributed business environment. When a metaverses use case is developed as the solution to assure security, privacy, and legality, the assets, authentications, and privacies of all businesses must be assured (Viriyasitavat et al. 2022, 2023), and this implies the resolutions to assure security, privacy, and legality in (1) implementing the interfaces of digital twins (DTs) with their physical twins, (2) supporting free movements of avatars, and (3) addressing copyright issues associated with virtual assets of certain entities.

14.2.2 First-Time-Right from Virtual to Physical World

Digital transformation has gone far beyond basic data communications to connected social and healthcare metaverse use cases that are enabled by AI, VR, AR, and MR (Bansal et al. 2022). Metaverse use cases that are enabled by AI, VR, AR, and MR. In the context of a value-added system, a system consists of various value-added and non-value-added operations over system lifecycles. The system should be enhanced effectively by minimizing non-value-added operations; fulfilling non-value-added operations consumes system resources. In digital manufacturing, the decisions on system operations are made by corresponding functional units in DTs, and engineers should compete to utilize computer-aided technologies to make the right decisions before these decisions are executed in the physical world for the first time (Bi and Zhang 2021).

14.2.3 Ubiquitous Data and Computing

Distribution and decentralization have been proven their necessities and values in manufacturing, logistics, governance, and management. Accordingly, enabling technologies should be distributed and decentralized in applications. For example, AI can be decentralized as *edge intelligence* (EI) to

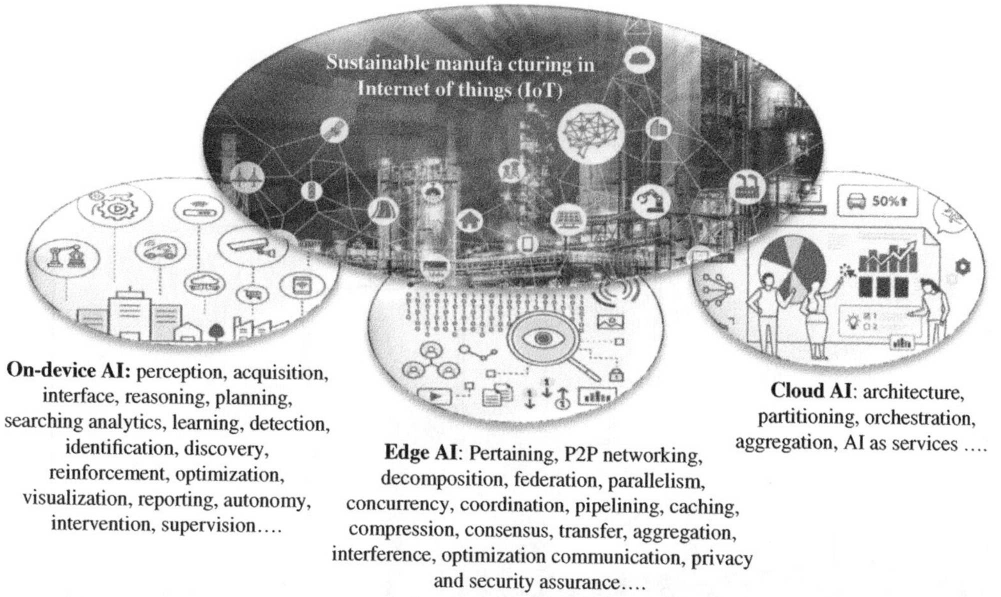

Figure 14.2 Ubiquitous data and computing in sustainable mechatronic systems (Telecom 2019; Cao et al. 2022; Gorilla 2023)

use Web 3.0 and blockchain in metaverse. EC allows for the connection of isolated end devices in IoT and supports decentralized computation, storage, and communication. Two examples of edge intelligence applications are *decentralized autonomous organizations* (DAOs) and *decentralized science* (DeSci). These tools were developed and used in decentralized organizations and they were implemented by cryptocurrency, biomedicine, funding, and professional services. Figure 14.2 illustrates the scenarios where AI, EI, and cloud computing (CC) are used to enable ubiquitous data and computing in sustainable mechatronic systems (Telecom 2019; Cao 2022; Gorilla 2023). As an application of the Internet, every networked resource is connected as a smart thing, and all of the associated data and information are accessible by IoT. Data can be processed and analyzed by EI, and relevant data can be aggregated, processed, and mined to support high-level complex decision-making activities at broadened scopes and longer terms by CC.

14.2.4 Diagonalizability, Predictability, and Adaptability

The performance of a mechatronic system relies largely on the intelligence of its decision-making units at different levels and domains. As shown in Figure 14.3, intelligence can be measured by the capabilities of (1) understanding past and current states of the system; (2) the prediction of future system states and changes in the long term. A sustainable system requires (1) collecting data in a real-time mode, (2) networking relevant objects and models as smart things, (3) predicting the changes to optimize business plans, and (4) reconfiguring systems to respond to anticipated and unanticipated changes automatically (Bi et al. 2021a, 2021b, 2021c).

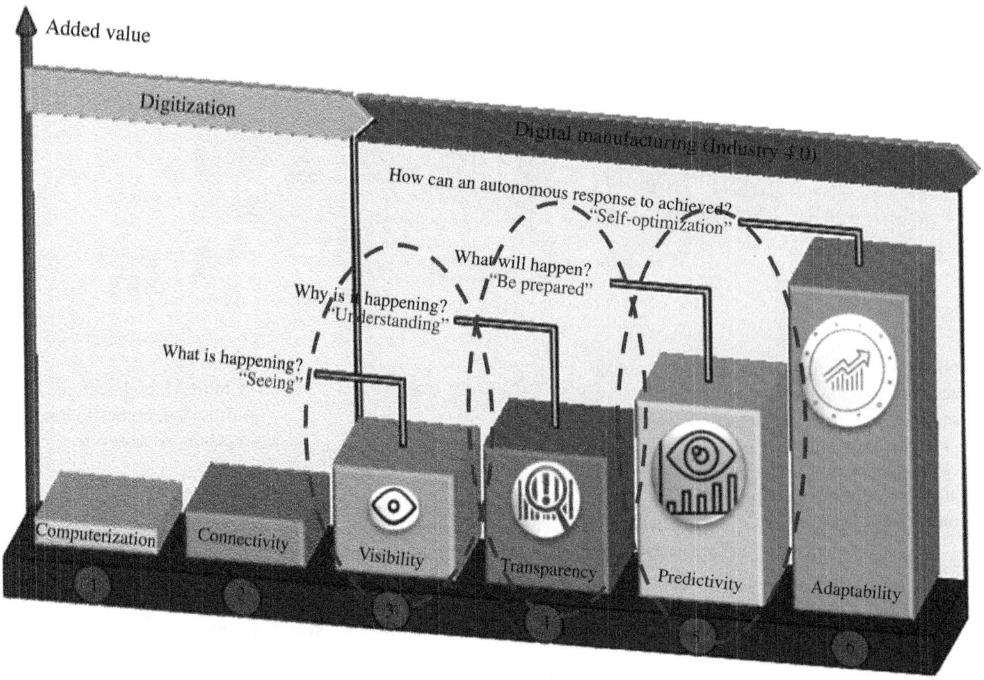

Figure 14.3 Diagonalizability, predictability, and adaptability in sustainable system (*Source:* Bi and Zhang (2021)/Springer Nature)

14.2.5 Human Intelligence for Uncertainty and Changes

Even though full automation becomes feasible, humans still play critical roles due to our adaptability to deal with changes and uncertainties. Traditional machines are generally programmed for sophisticated tasks; they are incapable of responding to unexpected events. The cooperation of a worker and machine helped to improve system adaptability in dealing with the changes. In particular, wearable devices with advanced sensing and haptic technologies fill the gap between human and cyber world for seamless collaboration (Boyce et al. 2019).

As a matter of fact, a central challenge in Industry 5.0 is how to fully utilize human resources to enhance system adaptability and resilience. *Industry 5.0* refers to an integrated application of collective digital technologies and concepts in a value chain organization, and the most critical technologies are CPSs, IoTs, and smart manufacturing (SM). In Industry 5.0, *collaborative robots* (cobots) are able to work with human operators in various tasks that need *human–robot collaborations* (HRC) such as quality controls, inspections, picking and placing, parallelizing, and assembling operations. Alike other conventional automated processes, the benefits of HRC can be justified by reliability analysis, system flexibility, adaptability, and productivity. Bi et al. (2021c) characterized the robots that were utilized to deal with uncertainty and changes in the manufacturing environment in Figure 14.4a, and Bi et al. (2021d) further classified collaborations into human teams, robot teams, HRC, and three types of *Human–Robot Interactions* (HRI) in Figure 14.4b,

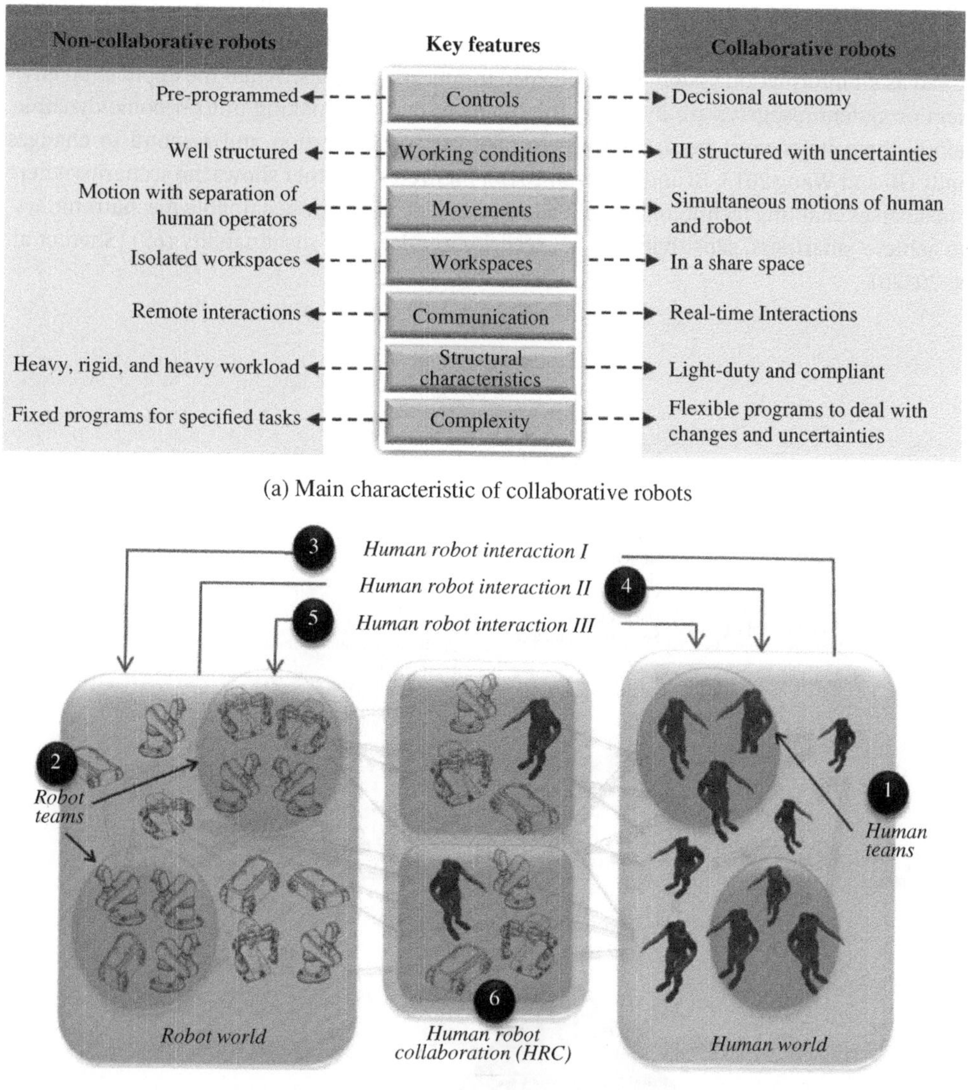

Figure 14.4 Using collaborative robots in sustainable system (*Source:* Bi et al. (2021d, 2021e)/with permission of Taylor & Francis Group)

14.2.6 Data-Driven Decision-Making Supports

Any manufacturing system involves numerous decision-making activities, especially for sustainable manufacturing systems. When a system becomes very dynamic, real-time data collection and analysis are essential to understand the current and future states of the system. The performance of decision-making systems depends greatly on the availability and sufficiency of data, as well as the means to process data and mine the information and knowledge from data. With the rapid adoption of IoT, the structures of information systems in enterprises have shifted gradually from static and centralized structures to dynamic, decentralized, and distributed structures.

As shown in Figure 14.5, the operations of a sustainable system rely on the decision-making units at all domains and levels of system operation. In the information system, each functional unit can be viewed as an information system with specified inputs and outputs. When the application environment or system resources are dynamic, the inputs of decision-making units become dynamic. Therefore, these units must be data-driven to take into consideration and respond to changes promptly (Bi and Wang 2013; Bi and Cochran 2015). Figure 14.6 further shows the scenarios where how metaverses enabling technologies were adopted in a data-driven sustainable mechatronic system to achieve smartness, sensitivity, safety, security, service, and sustainability (6S) (Shen et al. 2022a, 2022b).

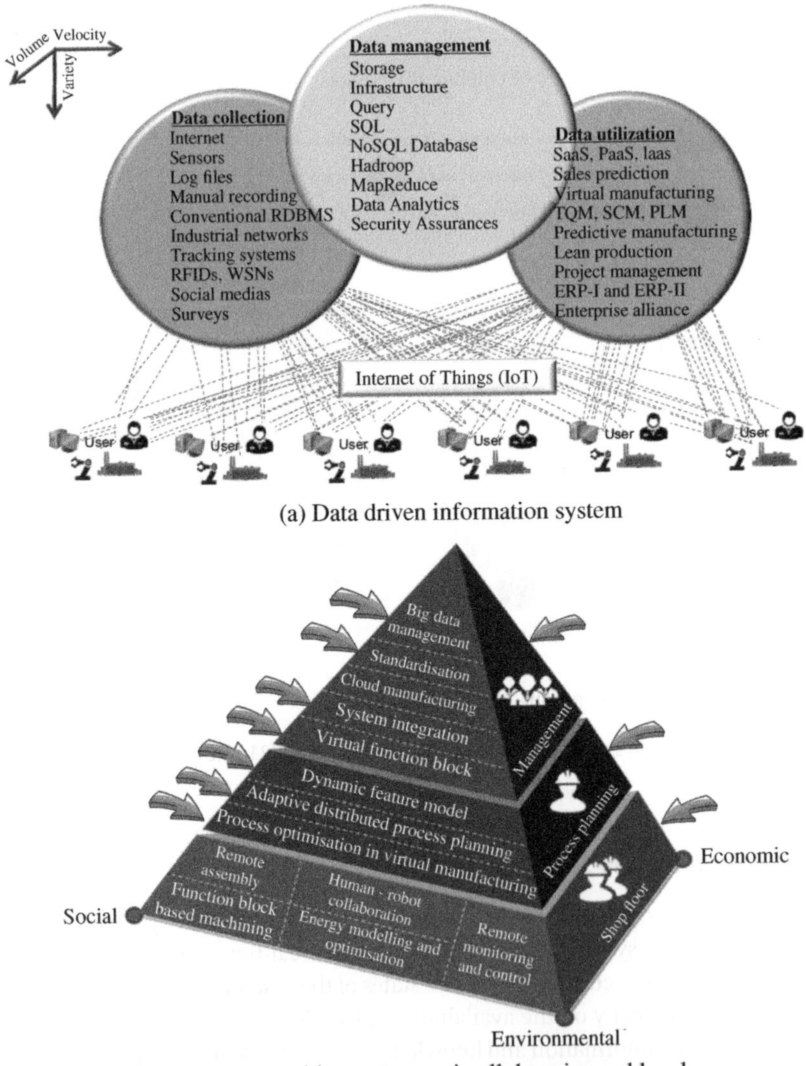

Figure 14.5 Data-driven decision-makings in sustainable systems (Bi and Wang 2013; Bi and Cochran 2015)

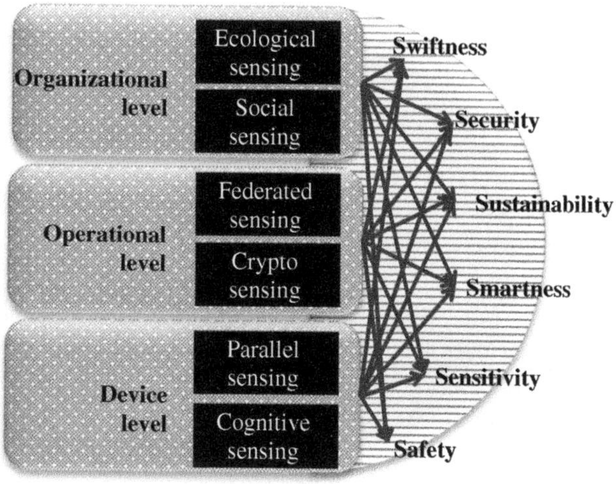

Figure 14.6 Objectives of using metaverses in data-driven information systems (Shen et al. 2022a)

14.3 Metaverse and Relevant Technologies

Among the 13 aspects of enabling technologies identified by Future Today Institute (2022), those that are closely related to metaverse are AI, AR, VR, synthetic media, *home of things*, *logistics*, *robotics* and *transportation*, *decentralization* and *blockchain*, and *telecommunication* and *computing*. The enabling technologies behind existing metaverse use cases have evolved over the years; moreover, creating a countable virtual world in metaverse requires the confluence of the most advanced *information technologies* (ITs) over a trustable, reliable, and efficient network with low latency (Faraboschi et al. 2022). For many developers, a metaverses use case is simply viewed as an enhanced and upgraded version of today's Internet application. Even though people have different expectations from their use cases, and even the definition of metaverses is not universally accepted, some commonly agreed features of metaverse use cases are: (1) VR, AR, MR, or hardware-in-loop is adopted to create illusionary and immersive environments that are used to model, simulate, analyze, and predict behaviors of physical things in a physical world; (2) decentralized functional units interact and security and safety are assured by blockchain and cryptocurrencies; (3) users are allowed to interoperate digital identities and assets in and across multiple platforms; (4) humans present as avatars in metaverse; (5) communications are implemented synchronously in a real-time mode (Kshetri 2022b; Zainab et al. 2022).

Fu et al. (2023) summarized core-enabling technologies behind metaverse as the abbreviation of *BIGANT*, that is, *B* from "***B**lockchain Technology*," *I* from "***I**nteroperation*," *G* from "***G**aming*," *A* from "***A**rtificial Intelligence* (AI)," *N* from "***N**etwork and Cloud Computing*," and *T* from "*Internet of **T**hings*," as shown in Figure 14.7. In the proposed architecture of metaverse use cases, enabling technologies are discussed in the aspects of hardware platforms, operating systems, decision-making supports, software tools, applications, and ecosystems.

Similar to a mechatronic system, implementing a metaverse use case requires integrating the MFMs at various domains and aspects of decision-making supports. From this perspective, metaverse is able to integrate multidisciplinary solutions from smart things over the Internet at the hardware layer, to computing and intelligence and virtual components at the middle layers, and finally to various user services at the upper layers.

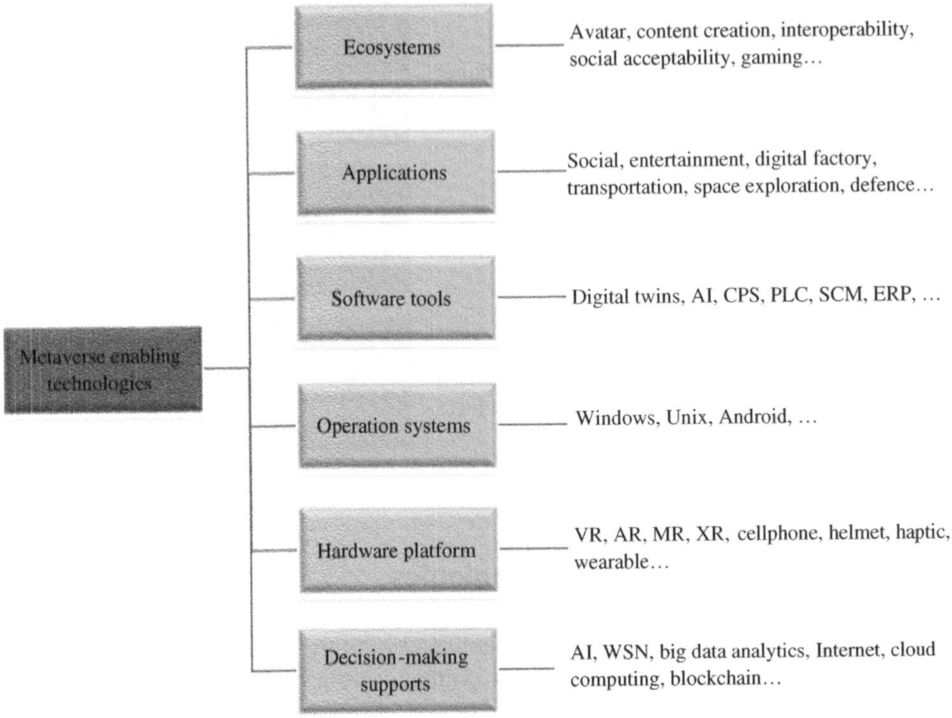

Figure 14.7 Classification of metaverse enabling technologies (Fu et al. 2023)

Metaverse is viewed as an integrated platform that is fused with other cutting-edge information technologies to accomplish their missions in sustainable mechatronic systems. This is aligned with the work by Bhat et al. (2023), as shown in Figure 14.8.

Metaverse is able to eliminate geographical limitations since physical locations become irrelevant and apply no constraints on users' interactions. The economic effect of the metaverse will reach $5 trillion by 2023 (Bansal et al. 2022). Metaverse can be used to gain an immersive experience in a purely virtual world or a mixed virtual and physical environment. This supports human–machine interactions and collaboration in an interactive world, develops vivid applications of using metaverse infrastructure for various applications, and enriches ecosystems for marketing, financial services, and economic and social services (Cao 2022). Metaverse is characterized by its immersiveness, hyper spatiotemporality, sustainability, interoperability, scalability, and heterogeneity, and these characteristics require metaverse to incorporate AI tools in six fields, as shown in Figure 14.9 (Wang et al. 2023a).

14.3.1 Architecture or Framework

According to the International Organization for Standardization (ISO), the architecture of an metaverse use case can be instantiated from the ISO architecture of an information system. It consists of seven layers from the lowest physical-device layer to the highest application layer, as shown in Figure 14.10 (Faraboschi et al. 2022). *The device layer* allows users to collect data and take actions for the metaverse use case via sensors and smart things such as brain–computer interfaces (BCIs).

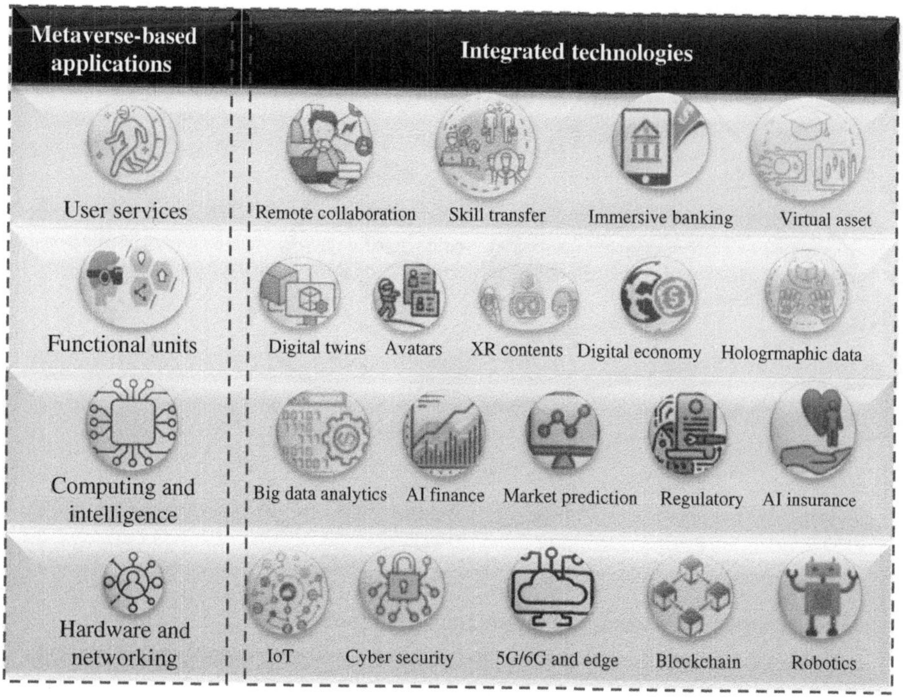

Figure 14.8 Metaverse as platform to integrate advanced technologies

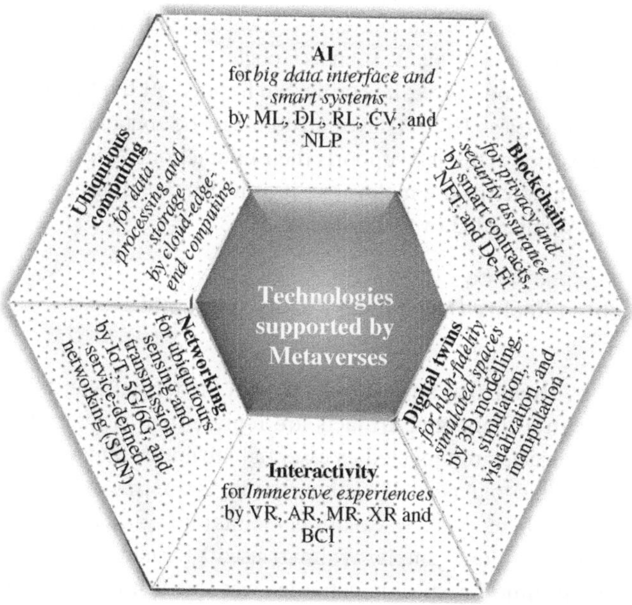

Figure 14.9 AI in metaverse applications (Wang et al. 2023a)

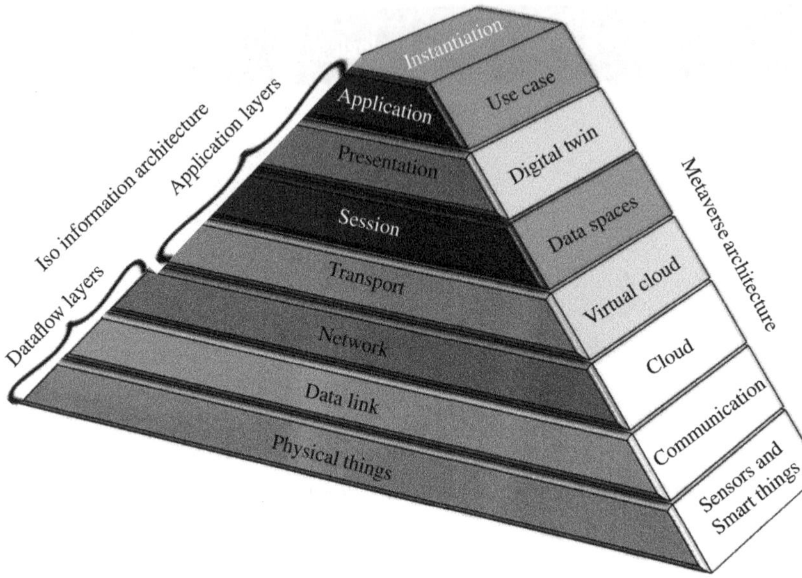

Figure 14.10 Metaverse architecture versus ISO seven-layer architecture

For example, a haptic device can sense touch, an earbud can sense sound, and an accelerometer can measure the acceleration of an object. Physical devices are connected as the sources of data space in *the communication layer*. Current infrastructure based on *the Fourth generation* (4G) of cellular networks is expected to be advanced to *the Fifth generation* (5G) and 6G to support communications in clouds as a mesh architecture. *The cloud layer* hosts data and software; it can be fuzzy and embedded down into device and communication layers and up to a virtual cloud to form *fog* or *edge*. *The virtual cloud layer* is formed by a group of distributed data and software on different clouds; cloud services have been virtualized by a number of IT giants such as Google, Azure, and Amazon. *The data space layer* defines the ontology of the metaverse; it is critical to implement the functionalities of an ecosystem, especially by third parties. The DT layer customizes the metaverse for a specified user's experience. It orchestrates various DTs and supports their interactions. *The use case layer* defines the application where users interact in the metaverse, it provides the landscape for seamless interactions by using the services embedded in the ambient.

14.3.2 Virtual Reality (VR), Augmented Reality (AR), Mixed Reality (MR), and Extended Reality (ER)

Any object in the physical world should be modeled as at least one DT(s) in the virtual world. For example, people are usually presented with avatars, and artifacts are represented as their DTs. Moreover, assets, security, and privacy of objects are modeled as essential attributes that are protected by blockchain technology or nonfungible tokens (NFT); value exchanges are tracked closely by distributed ledges (Far et al. 2022; Faraboschi et al. 2022). Creating, maintaining, and using DTs are usually performed by VR, AR, MR, and ER. For example, Walmart used VR headsets to train their employees; as outcomes, the training time for specific activities was cut from 8 hours to 15 minutes. A study by the UK's National Health Service showed that 92% of VR users gained an adequate understanding of infection controls in contrast to 16% of those without using VR

(Kshetri 2022a, 2022c). The VR-relevant technologies have proven their value in reshaping the physical world into a digital world as metaverse use cases with smart things, avatars, and societies (Tian 2021; Lv et al. 2022a).

A metaverse use case is a fully immersive virtual world in which users can see, feel, move, and act on virtual objects or features interactively. AR incorporates virtual objects in a physical world by various sensory modalities such as auditory, visual, haptic, or olfactory. MR merges virtual and physical worlds with a single display, and it jointly exhibits virtual and physical objects simultaneously in the same display. ER fuses all of the realities from AR, VR, and MR for technology-mediated experiences by sensory interfaces, infrastructures, and applications. extended reality (XR) such as VR, AR, MR, and 360 stereoscopic streaming enable metaverse for interaction and collaboration in virtual space. From a multimedia perspective, the user's Quality of Experience (QoE) depends on the latency, bandwidth, traffic pattern, and immersive media of the network (Dong and Lee 2022). Xiao et al. (2022) developed a hybrid user experience by integrating radio frequency identification (RFID) with VR; a new architecture was proposed to compose RFID and VR promptly for RFID-augmented VR applications.

VR and AR or other media were contributing more and more to digital economics including digital learning (Gugenheimer 2022; Kanemastsu et al. 2022). To support development across different media, Chidambaram et al. (2022) developed a unified platform to capture users' intentions for creations without programming. Berardino and Tucker IV (2022) developed a portable demonstration system to train military space professionals for common operations; immersive experiences were created by MR to facilitate effective decision-making in operations with reduced tactical timelines. Metaverse requires high efficiency of data transmission over the Internet. The video standards of *moving picture experts group* (MPEG) and *low complexity enhancement video code* (LCEVC) were discussed and their applications on *high dynamic range* (HDR), 8K-resolution, XR, immersive video, and metaverse were explored. LCEVC was compliant with other codecs to enhance the compressing efficiency of videos, and it was deployed in a wide spectrum of applications from HDR, to 8K, and to immersive metaverse (Ferrara et al. 2022).

Metaverse is the convergence of VR, MR, AR, ER, XR, DT-I, DT-IIs, BDA, and BCT that enable multisensory interactions with people, objects, and resources in physical and digital worlds (Mourtzis et al. 2022). Metaverse gains experiences from an ultra-realistic, pervasive, and decentralized digital world. Metaverse is the most advanced digital transformation (Bi et al. 2022; Leng et al. 2022). Metaverse is a virtual universe that goes far beyond traditional three-dimensional and immersive environments (Lv et al. 2022b). De Felice et al. (2022) provided an overview of existing technologies, applications, and challenges and focused on the challenges of interoperability and interconnection. A virtual world is a simulated environment where users can inhabit and interact with others through avatars, agents, bots, or digital models. Metaverse distinguished it from other online games in the sense it provides a seamless and persistent virtual world where users can roam around transparently without preset objectives (Kumar et al. 2008). Zhang et al. (2023a) discussed AR services enabled by access edge computing (AEC); it allowed users to locate and communicate simultaneously to request AR services. In developing a virtual world, metaverses can be used to implement both the Ethereum-based Decentraland and the sandbox (Kshetri 2022b). Pena et al. (2023) investigated the impact of customized styles of avatar leaders on subsequent altruistic behaviors and confirmed that business leaders influence individuals' behaviors, while avatar customizations would affect social perceptions and altruism effectively.

Note that incorporating VR and relevant technologies in metaverse needs to address a number of new challenges due to its distributed, decentralized, and dynamic nature. For example, Vondrek et al. (2023) described the security and privacy attacks on VR technologies in the metaverse as man-in-the-room attacks; this new attack type was associated with worming and botnets over social networks.

As far as the hardware is concerned, HoloLens 2 supports untethered MR over Windows Holographic Operation System. It consists of four vision cameras to track the head, two infrared cameras to track the eye, one *time of flight* (ToF) sensor to measure depth, and one inertial measurement unit (IMU) with an accelerator, gyroscope, and magnetometer; it has a Qualcomm Snapdragon 850 processor for fast processing and a short wake-up time. It has a USB-C port as Bluetooth connection (Bansal et al. 2022). Microsoft Dynamics 365 is an MR for users to collaborate in a distributed environment by Microsoft VR equipment running on iOS, Android, HoloLens, and HoloLens 2.

14.3.3 Digital Twins (DTs), Cyber-Physical Systems

Figure 14.11 shows that a metaverse use case is rooted in DTs. Digital models of physical things used to be represented by 2D graphics and AI engines; they would be advanced greatly with sophisticated 3D graphics, kinematic and dynamic models, and AI algorithms to represent their physical behaviors vividly. Key features of virtual models in the metaverse are multimodal input, heterogeneous clients, server scalability, network constraints, object encoding, physics engines, security, privacy, and fairness (Kumar et al. 2008). In the metaverse, a digital entity coexists with its physical counterpart, and it is empowered by AI for effective interactions in virtual and physical worlds (Shi et al. 2023).

DT was proposed in 2003 by Grieves for paired virtual and physical models with their connections. Metaverse consists of a number of virtual sub-worlds operated by corresponding *virtual service providers* (VSPs), and VSPs are enabled by *digital twins* (DTs); each DT establishes the communications and interactions of a virtual twin in the virtual world with its physical twin in the physical world (Han et al. 2023a). DTs made significant contributions to conceptualizing, prototyping, verifying, and vilifying engineering concepts, and DTs have to be empowered by deep learning to deal with an increasing volume, variety, and velocity of data (Huang et al. 2022).

Song and Qin (2022) explored existing metaverse technologies for their applications in healthcare systems; especially, DTs were especially adopted to assess and maintain personal health based on some commonly used factors such as vital signs, behaviors, and other observed symptoms. On the other hand, the quality of DTs was determined by real-time-ness, completeness, sufficiency, and accuracy of monitoring data. Metaverse is a hyperrealistic virtual world where users are able to perform their daily activities such as working, socializing, gaming, entertaining, and shopping. This requires the metaverse to represent physical objects virtually and model their relations, events, processes, states, and activities in a real-time manner. The key facets of metaverse are datasets, simulation engines, artificial intelligence, hardware, and collaboration (Jaimini et al. 2022). Han et al. (2023b) viewed a metaverse as an instance of CPSSs, and they integrated artificial cyberspace,

(a) Phase I: digital twins (b) Phase II: digital native (c) Phase III: Metaverse (surreality)

Figure 14.11 From digital twins to metaverse use case (*Source:* SumitAwinash / Wikimedia Commons / CC BY SA 4.0) parallel execution, and computational experiments as ParaDefnder to defend against cyber threats in the metaverse. Corresponding tasks in the physical and virtual cyberspace were executed parallelly for mutual verification and validation to assure security This parallel execution is event driven, and an event is one of the possible spatial–temporal combinations of security threats in the metaverse. Huynh et al. (2022) adopted DTs in the metaverse to support communication, storage, and computing in edge computing; DTs were used as powerful computing infrastructure to offload tasks for reduced latency. DTs were empowered by edge intelligence to advance the immersive metaverse by improving the allocations and utilization of computing resources. The integration of DTs and CPSs with Industry 5.0 supported humans to immerse in the metaverse and interact with smart things and other avatars. Jagatheesaperumal and Rahouti (2022) proposed a service-oriented architecture for a metaverse platform for effective human-machine interactions. CPSS was incorporated into virtual classrooms to standardize workflow and construct a system from off-the-shelf components. Metaverse was introduced to enable immersion, interactivity, and diversity of education systems (Shen et al. 2022b).

14.3.4 Internet of Things (IoT) and Edge Computing

Metaverse is implemented in Web 3.0. Web technologies have evolved from Web 1.0 to 3.0, as shown in Figure 14.12 (Mourtzis et al. 2022). The metaverse comes with Web 3.0 as one of the next

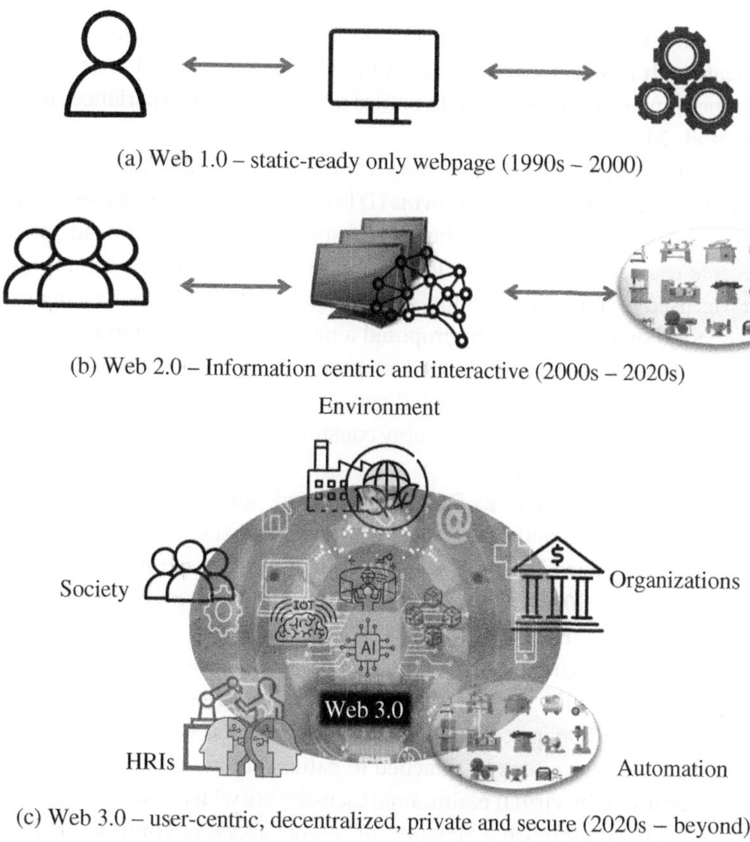

Figure 14.12 From Web 1.0, Web 2.0, and to Web 3.0

big things that are coexisting and evolving rapidly. Web 3.0 has been expected to have over 1 billion users by 2023. From the perspective of Web 3.0, metaverse use cases are the applications in the virtual world that are featured by smart things, avatars, and other functional units. With IoT, these entities can interact with each other freely (Kshetri 2022a, 2022b, 2022c). Moreover, edge-computing becomes essential to support on-device XR to empower metaverse. The convergence of edge intelligence such as deep neural networks has been incorporated into metaverse architecture (Parmar et al. 2023). Especially, newly developed edge-based technologies have been adopted to support metaverse engines (Lam et al. 2022).

IoT helps to collect data from multiple sources in real-time mode. The data becomes big and abundant to support reliable decision-making activities effectively. Dai et al. (2023) discussed the trend of using wearable electromechanical sensors and machine learning (ML) in soft robots and human–robot interactions for metaverse applications. Metaverse enabling technologies should mostly be distributed intelligence; therefore, the interactions among networked things over the Internet rely on intensive communications. The high requirements of metaverse use cases for a fully immersive experience and a large number of distributed and concurrent users have brought some critical challenges to bandwidths, speeds, and reliability of networks. The sixth generation (6G) becomes an ideal option due to its superior characteristics of ultra-low latency, ubiquitous connectivity, superior reliability, reliability, and capability. 6G enables edge intelligence of smart things in IoT. In an IoT application, humans, devices, and smart things interconnect, cooperate, and collaborate anywhere and anytime over the Internet. Metaverse aims to connect billions of people in a shared world (Chang et al. 2022). Multisensory technologies are advanced enough to collect, transfer, and share sensed data over the Internet; digital olfactory stimulation is ready to implement digital multisensory communication. It has been expected to commercialize the Internet of Senses in 2030; at that time, users can be immersed in the intimate nature of multisensory experiences from reality (Panagiotakopoulos et al. 2022).

Tang et al. (2023) proposed a roadmap to advance enabling technologies for communication and networking in metaverse, and they emphasized *digital twins* (DTs), *multi-access edging computing* (MEC), and BCT. Siyaev and Jo (2021) investigated the role of speech communication in space applications, and an metaverse use case was designed to perform aircraft maintenance trainings, and speed communication was used in traditional manuals, 3D models, and simulators, and neuro-symbolic speech executor (NSSE). Shen et al. (2022b) proposed a parallel sensing framework. The framework was composed of concepts, backgrounds, and methods for parallel sensing; sensors were referred to as integrated CPSs to support parallel intelligence. Parallel sensing allows a physical twin to operate in a discrete time domain to reduce energy consumption; it allows a virtual twin to offer compensation data and serve as a guardian.

The synchronization of a physical system with its DT is most critical. Sampling rates and communications in synchronization should be optimized to reduce loads and tracking errors. Meng et al. (2023) used expert systems and deep learning algorithms to optimize sampling rates for enhanced prediction. Abdel-Basset et al. (2022) proposed federated learning with privacy-preserved intelligence to protect data subjected to various cyberattacks. The prevalence of IoT makes it possible to implement various mechatronic systems such as smart actuators, sensors, appliances, and cyber services. Via metaverse, IoT provides users with cyber-virtual experiences in immersive mixed-reality environments; in turn, IoT maps collect data from the physical world into a DT in the virtual world. For example, smart devices can be attached to patients to monitor health conditions, and this elicits users' responses in the virtual realm; smart sensors are widely used to track the exercises of users. IoT data is utilized to establish a network for remote access or control of real or virtual things (Li et al. 2023a).

14.3.5 Big Data Analytics (BDA) and Cloud Computing (CC)

A metaverse use case usually involves in massive amount of real-time data, and such big data is processed and mined in a real-time mode. Bi and Cochran (2015) and Li et al. (2022) discussed existing data analytic methods in solving complex engineering problems from the definitions of *customers' requirements* (CRs) to designs, manufacturing, logistics, and services of products or systems. With an ever-growing scale of IoT, there is an emerging demand to advance BDA to support various decision-making activities.

In regard to the correspondence of BDA and metaverse, metaverse use cases are enhanced DTs, and BDA would be the gateways of digital and physical things (Lv et al. 2022b). To achieve satisfactory immersive experiences, it is critical to render scenes in a real-time mode. This involves intensive computation and poses challenges to support real-time rendering. Jiang et al. (2023) proposed *code distributed computing* (CDC) as a collaborative computing paradigm to offload rendering tasks to other mobile devices. CDC guarantees fault tolerance by aggregating spare computing resources from mobile devices. Moreover, a metaverse use case involves the massive sensed data that is transmitted from the physical world to the virtual world. Wang et al. (2023c) proposed a semantic framework to encode and transmit sensed data based on the semantic information; this helped reduce the amount of data to be transmitted without affecting the quality of data in data mining.

Metaverse use cases distinguished themselves from traditional CPSs by parallel intelligence to enable BCT, smart contracts, Web-3, DAO, and decentralization (Wang et al. 2022b). Ye and Wang (2022) viewed metaverse as an infrastructure for *artificial systems, computational experiments*, and *parallel execution* (ACP). They proposed a parallel population to manage human behaviors in human hybrid systems. People's thoughts were modeled by heterogeneous learning and analyzed by deep evolutionary reasoning. Yang et al. (2022b) further extended ACP in parallel manufacturing where customers' demands were analyzed, extracted, and utilized to plan research, development, and production. Zhang et al. (2023b) presented a future parallel vision of a metaverse use case for an intelligent transportation system (ITS) with enhanced efficiency, effectiveness, and safety.

14.3.6 *Blockchain Technologies (BCTs)*

Metaverse and BCT have been integrated to establish digital spatial models in *building information modeling* (BIM). BIM is a comprehensive process to collect and utilize building information to support decision-making activities such as design, construction, use, and demolition in an enterprise's lifecycle. Adopting blockchains assures security and fairness in transferring resources and assets, protecting data, and executing businesses (Huang et al. 2022). As a decentralized technology, blockchain-alike technologies such as *nonfungible tokens* (NFTs) and cryptocurrencies are critical to sustaining an open and safe business environment (Kshetri 2022a, 2022b, 2022c). When humans are immersed in the metaverse, they are digital entities and interact with each other through *ubiquitous access controls* (UACs). While UAC provides various bridges to exchange data in a virtual world, numerous threats exist. Blockchain-enabled multi-signature locks can alleviate security issues at the access layer in the UAC scheme (Gai et al. 2022).

Metaverse is characterized as immersive interaction, socialization, worlding building, and expandability. Meanwhile, metaverse is exposed to numerous securities or privacy risks such as leakage of personal information, eavesdropping, phishing, broken authentication, and unauthorized accesses (Huang et al. 2023). Tao et al. (2022) modeled a blockchain transaction network in the forms of NFT, *Ethereum* (ETH), and *bitcoin* (BTC) to describe the relations among entities from the perspective of structural identity. A metaverse use case involves multiple flows of synchronized data among smart things; each data flow should be customized to specified quality requirements. Maksymyuk et al. (2022) defined a service quality model in which NFT was used to allocate and select a data flow; this enabled stakeholders' trade in a decentralized mobile environment. BCT was integrated with metaverse to collect users' data and improve services in a decision-making system of sustainable public transportation (Deveci et al. 2022a).

Metaverse differs from conventional communication services such as those by microphones or cameras for cultural or economic businesses. Metaverse use cases are vulnerable to various security threats since services are provided through public channels. Ryu et al. (2022) presented a model to manage and secure users' privacies via mutual authentications by BCT. Chen et al. (2023) considered the adaption of digital fair exchange in metaverse; the trusted third party (TTP) was mostly used to address disputes among stakeholders. Different protocols exist for cryptographic fair exchanges in a decentralized environment. Critical properties of fair exchange protocols include fairness, decentralization, modification-free, privacy, verifiability, efficiency, and interruption. As a fully immersive, hyper-spatiotemporal, and self-sustainable space for users, metaverse is challenged significantly by privacy invasion and security breaches. Since assuring security and privacy is primary in applications, standards should be developed to enhance security and privacy. Two standards dedicated to metaverse applications were (1) ISO/IEC 23005 as a framework for *networked virtual environment* (NVEs) and (2) IEEE 2888 as standardized interfaces to synchronizing cyber and physical worlds, as shown in Figure 14.13 (Wang et al. 2023e).

No generic types of blockchains are applicable to any application scenarios; blockchains with different features and characteristics exist. However, there is a need to communicate, transfer, and exchange values and data among heterogeneous blockchains. Ou et al. (2022) developed

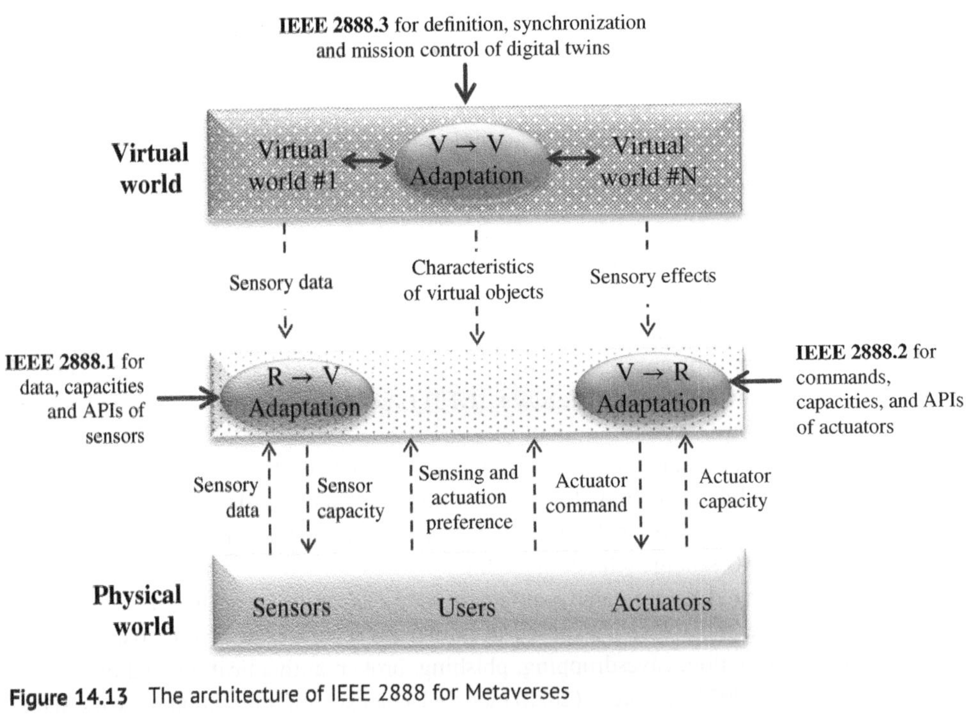

Figure 14.13 The architecture of IEEE 2888 for Metaverses

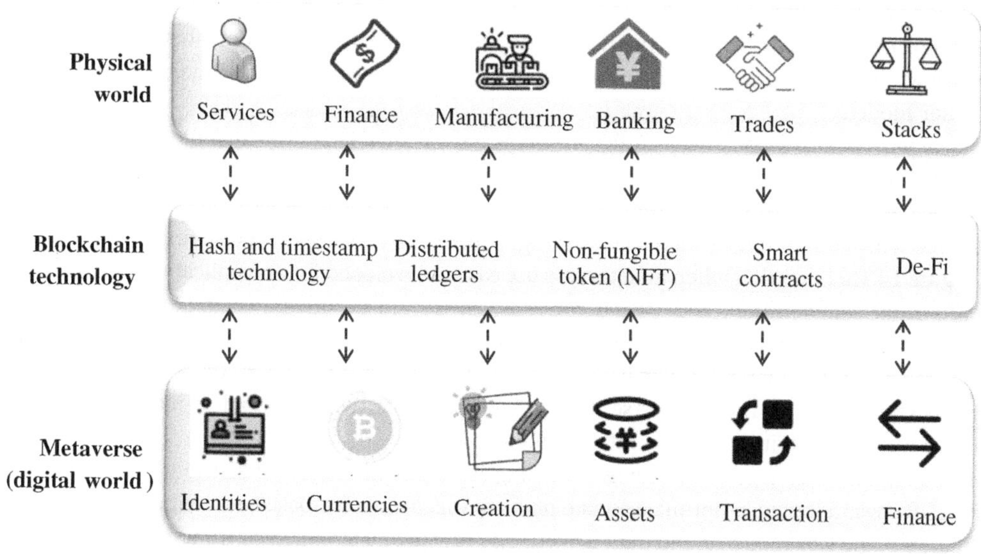

Figure 14.14 BCT as secure and safe connection of metaverse with physical world (*Source:* Adapted from Wang et al. (2023c))

cross-chain technologies to identify the challenges in supporting the interactions in multiple blockchains. Note that the main objective of using BCT is to assure privacy and security of the interactions, as shown in Figure 14.14 (Wang et al. 2023d).

Metaverse improves the efficiency and safety of traffic and reduces pollution and emissions. However, since wireless communication is open, data packets transmitted among vehicles are exposed to risks such as public and concomitant leaks. This might lead to vehicular platoons. Pan et al. (2023) discussed the controls of privacy-preserving platooning of vehicular CPSs. Lv et al. (2022b) integrated BCT with DT where BCT was introduced as an information infrastructure to ensure the security of digital mapping processes over IoT. The nonautogenic Hash Geocoding method was developed to index multidimensional information, avoid information loss in data mining, and facilitate the implementation of metaverse use cases through spatial DT. Users in a metaverse use case interact with each other through digital transformation. Digital transformations among stakeholders were established based on essential conditions to assure the security and privacy of data and assets of users. With a rapid increase in *variety, volume, and velocity* (3V) of data, metaverse faces the challenges of assuring security, privacy, and synchronization of data with low latency. BCT and enabled intelligent networks exhibit the potential to support trusted executions of business processes, data transmission, and data exchanges in distributed, dynamic, and decentralized environments (Fu et al. 2023).

14.3.7 Artificial Intelligence (AI)

A sustainable system is complex due to the requirement for a prolonged lifecycle; it involves numerous decision-making activities at different levels and domains in its lifecycle. Traditionally, computer-aided decision-makings rely greatly on analytical or numerical models. By all means, developing a functional mathematic model becomes impractical when the system involves too many variables, uncertainties, and changes. While the inputs, outputs, and interactions of the system with its environments could not be defined clearly, the system has to be treated as a "*grey*" or "*black*" box rather than a "white" box. In a "white" box model, the information about the system is fully known and all changes can be anticipated, and this allows to establishment of an explicit computer model.

To deal with changes and uncertainties, humans have shown superior capabilities over computers in dealing with incomplete information, complexity, dynamics, and changes. In particular, *artificial intelligence* (AI) artificially mimics intelligence to perform cognitive functions. AI can be associated with human minds such as reasoning, perceiving, learning, coordinating, and interacting. Figure 14.15 shows that when a metaverse use case is developed for an application virtual exhibition, virtual collaboration, and virtual concerts, AI should be integrated to support decision-making activities at different levels; this is due to the fact that the application environment is open and involves in numerous unanticipated events. Yang et al. (2022a, 2022b) discussed the importance of AI and BCT in the metaverse; they suggested that it was essential to integrate AI and BCT in developing a fair, open, and rational metaverse.

Gu et al. (2023) used *deep reinforcement learning* (DRL) to model crowd distribution in an evacuation; DRL was adopted to optimize escape paths in this metaverse-based evacuation system. The evacuation scene was modeled and data-driven. The raw data was collected from multiple sources such as cameras and smoke detectors. The collected data was processed by DRL in the cloud to guide evacuating activities dynamically. Hare and Tang (2023) used a hierarchical multiagent framework where DRL was to augment intelligence in personalizing learning experiences in metaverse. Kang et al. (2022) proposed to adopt DAO and decentralized science (DeSci) in an agricultural metaverse (AgriVerse) in which AI was used to process heterogeneous data, knowledge, and model and support interactions. Zheng et al. (2023) developed a framework with data-knowledge-driven

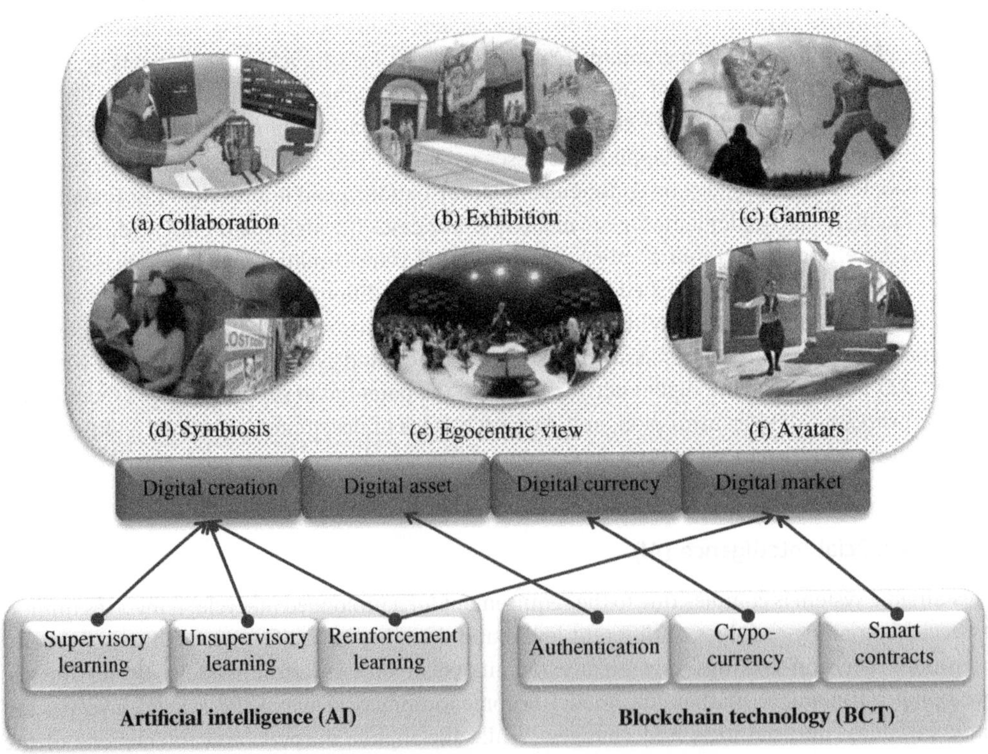

Figure 14.15 AIs in Metaverse use cases (*Source:* Yang et al. (2022a, 2022b)/IEEE/CC BY 4.0)

group intelligence to mine data for wisdom and knowledge. Ubiquitous data is the data source to empower AI tools in performing value-added activities and academic studies; on the other hand, metaverse compromises conventional AI tools in dealing with latency, fragile stability, and security assurance (Chang et al. 2022). Ren et al. (2022) proposed a collective learning and many-to-many matching game-based scheme to cope with insufficient or frequently changed data.

14.3.8 Human–Machine Interactions (HMI)

Even with the rapidly enhanced capabilities of AIs, humans are still playing irreplaceable roles in sustainable mechatronic systems. On the other hand, humans' creativities can be greatly empowered by machine intelligence. Therefore, humans and machines should collaborate in many scenarios to create, disseminate, and trade innovations effectively. Guo et al. (2023) proposed a parallel system in which ML was incorporated to improve humans' creativities, ML was able to make products under humans' guidance, and humans could further perfect automatically generated ideas to improve productivity dramatically.

Mourtzis et al. (2022) presented a human-centralized platform. It was used to personalize value creations in a metaverse use case for sustainable manufacturing in Web 3.0. When metaverse was adopted to support human–machine interactions, conventional manufacturing systems would be elevated as a CPPS. Wang et al. (2023e) proposed *an industrial foundation model* as a data repository to manage, operate, and control computing resources, digital assets, enterprise resources, and I/O interfaces, as shown in Figure 14.16.

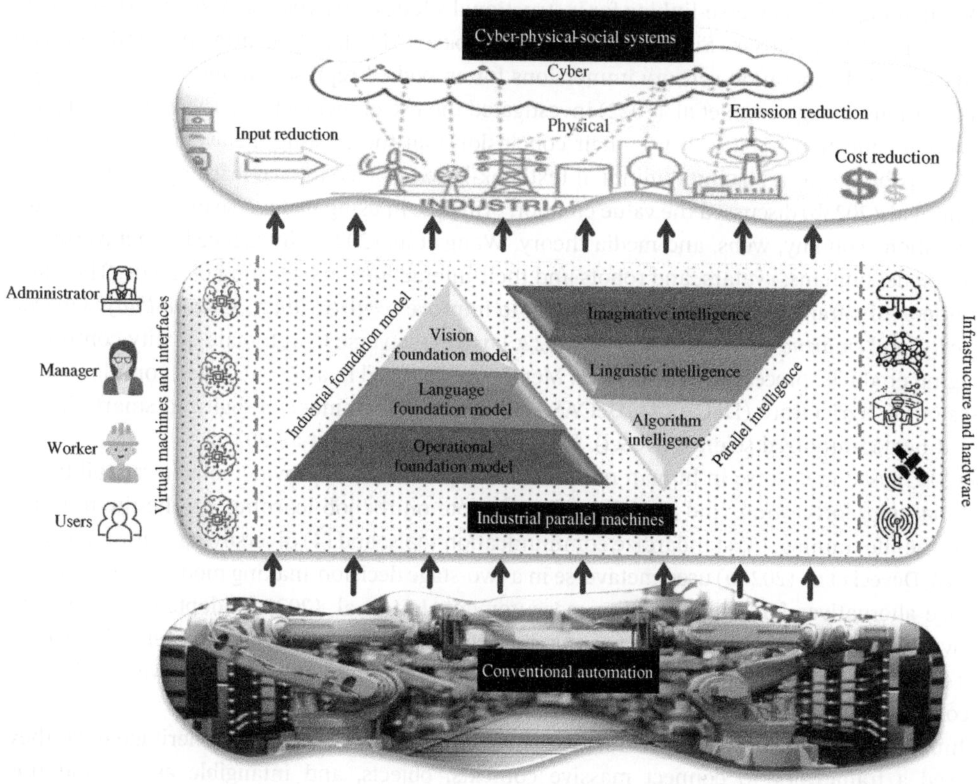

Figure 14.16 Industrial parallel machines for effective HMI (*Source:* Wang et al. 2023e/IEEE/CC BY 4.0)

Fan et al. (2022) adopted the concept of *CPPS* to monitor, simulate, diagnose, and predict road conditions. The metaverse use case was developed in a road maintenance system to assess road surface conditions to provide timely alerts about sudden damages. Liu et al. (2022b, 2023) viewed metaverse use cases as the extensions of CPSSs in planning and managing radar systems. AI tools were integrated with computational experiments and parallel executions for effective decision-making support. Park and Kim (2022) analyzed the approaches for sensing, interactions, and implementation of users' experiences, and they suggested advancing comprehensive recognition and interpretation over multimodal sensors to improve the interactions in metaverse.

14.3.9 Data-Driven Decision-Making Systems

Metaverse was used as an essential infrastructure in implementing some applications such as a hyper-flexible, self-adaptable smart factory, a robotics metalaboratory for neuroergonomics, game-based digitally enhanced advanced services (DEAS), and aircraft maintenance with speech interactions. Furthermore, metaverse will become a shared, persistent, and decentralized immersive environment for e-commerce, networking, education, and the digital economy.

Many researchers have explored the applications of metaverses in data-driven decision-making systems. For example, Wang et al. (2022a) proposed a framework for using metaverse for education. It consisted of an instructional design hub, knowledge hub, research and technology hub, and talent and training hub. The common components of all four hubs were (1) infrastructure and communication, (2) technological access and equity, and (3) data security, privacy policy, and user rights. Exemplifying metaverse platforms for education included Sloodle as an interactive and immersive laboratory to teach programming; immersive journalism; Vortex for collaborative learning; VR-making and metaverse-linking for instructional VR design; as well as VRchat, AltSpaceVR, Virbela, EngageVR, Sansar, Sinespace, Somnium Space, Mozilla, Decentraland, and Horizon World as the 3rd generation sensory immersions for embodied representations, online meetings, and educations. Onggirawan et al. (2022) investigated the use of metaverse in the learning process as a virtual educational space. It was their conclusion that the existing metaverse use cases had limited capabilities in replacing traditional textbook-based or face-to-face learning education.

Wang et al. (2023b) discussed the value creation principle in using metaverse in the crypto industry, attention economy, webs, and media theory. Wang et al. (2023c) introduced a metaverse use case to model humans, avatars, and robots and their interactions in operating oil fields. Their system supported parallel executions and was able to improve *security*, *safety*, *sustainability*, *services*, *sensitivity*, and *smartness* (6S). Wang et al. (2022c) used metaverse to automate quality controls in managing products, processes, and services. Metaverse was used in a decentralized complex adaptive system in agriculture (DeCASA). DeCASA consisted of a number of digitized smart villages enabled by decentralized science (DeSci), DAO, and CPSS (Wang et al. 2022d).

Le (2022) used metaverse in human resource development to (1) provide holistic capabilities as benchmarking, forecasting, and modeling; (2) promote on-the-job learning, on-the-life-training, and on-the-life training; (3) standardize communication tools; and (4) diversifying communication accesses. Deveci et al. (2022b) used metaverse in a two-stage decision-making model to analyze and optimize alternatives for freight fluidity measurement. Liu et al. (2022a) adopted metaverse to improve the efficiency and security of mining; the infrastructure of the proposed metamining consisted of the human world and virtual and physical mining systems. Virtual mining was used to test and correct proposed algorithms and systems.

Rahmani and Tienken (2022) used metaverse to fuse and distribute culture heritage data; they proposed a framework to connect massive contexts, objects, and intangible assets, and this promoted use-driven creation, offset economic models, and rewarded performers and creators. Pamucar et al. (2022) assessed 4 metaverse applications for sustainable transportation based on 13 criteria in terms of operation, efficiency, social and health impact, and legislation and regulation; these metaverse alternatives included auto-driving testing, public transportation operation and safety, traffic operation, and sharing economic applications. Zhang et al. (2023d) integrated metaverse into healthcare systems to optimize the utilization and allocations of healthcare resources, and they used multiple servers to share data and support deduplication securely.

Li et al. (2023b) proposed a six-layer data-driven information system; these six layers were infrastructure, operation, knowledge, intelligence, management, and interactions layers. Li et al. (2023c) proposed an intelligent cockpit system with three levels of perception, cognition, and decision-making supports in interactions. It allowed humans to perceive environments and provide feedback dynamically, and this further enhanced humans' experiences in terms of safety, pleasance, and efficiency. Active empathetic auditory was regulated to detect drivers' affections. Roadside vehicle-to-infrastructure systems were introduced to enable connected and autonomous vehicles (CAVs). Metaverse was used to manage CAVs at signal-free interactions to maximize throughputs (Li et al. 2023d). In using echocardiography to detect congenital heart diseases, Qiao et al. (2022) incorporated metaverse to improve robustness and detection accuracy by suppressing adverse factors such as instinctive fetal movements, artifacts in ultrasound images, and distinctive fetal cardiac structures.

To exchange and share opinions, resources, and interests in distributed environments worldwide, language generation methods can be used to generate languages in a regulated VR environment. Popular language generation methods were transfer learning and neural learning such as RNN, GRU, and GPT-3. Park et al. (2022) proposed a *recurrent voting generator* (RVG); it was used to understand the 3D text and perform emotional analytics in metaverse. Zhang et al. (2023c) discussed the requirements of large-scale metaverse use cases. A metaverse use case may involve billions of hyperparameters; these parameters were trained using thousands of gigabytes of data. A foundation model was used to train a generic system for better adaptation to the changes in large-scale datasets. To manage the risks of data poisoning attacks of the foundation model, Li et al. (2023b) proposed a closed-loop human–computer environment to validate and calibrate the model.

14.4 Metaverses for Sustainability

Metaverse has evolved systematically over the years. Now, it has distinguished greatly from its early virtual products such as Second Life. Three capabilities of metaverse are immersiveness, sociability, and environmental fidelity (Barrera and Shah 2023). The concept of metaverse has grown its popularity rapidly; however, a limited number of metaverse use cases are available. There will be a long way to develop fully immersive, interoperable, and embodied metaverse applications. Developing a full version of the metaverse use case must be edge enabled. There are still challenges associated with computation costs due to the limitations of edge devices. The aspects that need more advancement include multiple accesses for immersive streaming, multisensory and media networking, multimodal semantic communication, edge networks of DTs, edge intelligence and intelligent edges, sustainable resource allocation, smart blockchain, quality of experience, markets and mechanisms of services, and industrial metaverse (Xu et al. 2023).

Metaverse is a scalable and interoperable platform for immersive activities. It supports the interactions between the physical and cyber worlds, across private and public networks, and in open and

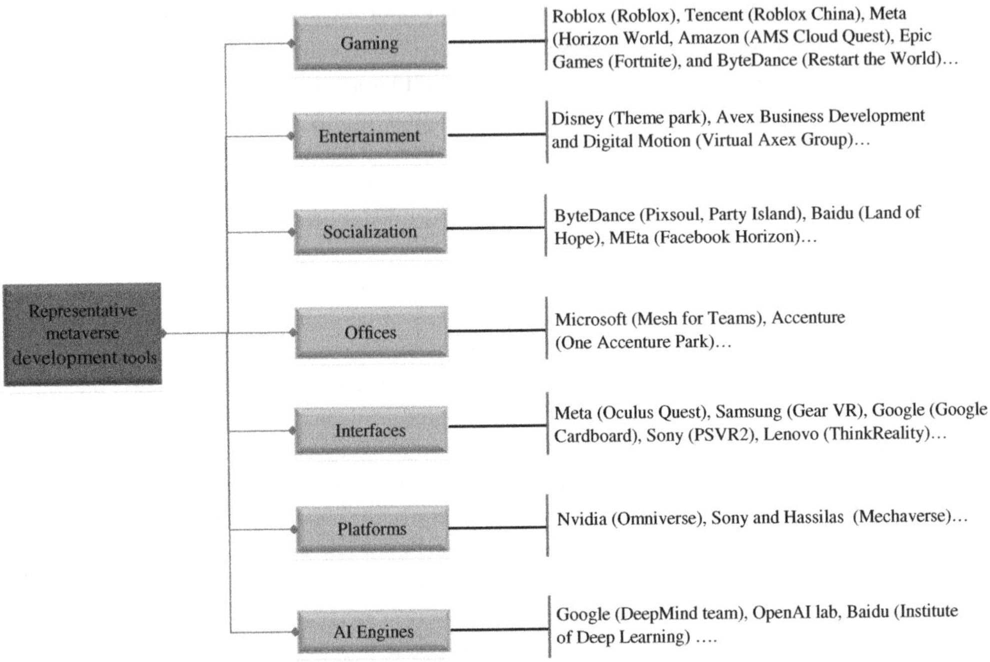

Figure 14.17 Metaverse use cases in representative areas (*Source:* Adapted from Li et al. (2023a))

closed platforms. Many companies have launched their metaverse use cases such as video conferencing, gaming, entertainment, and AI face painting shown in Figure 14.17 (Zhou 2022; Li et al. 2023a). Moreover, metaverses can be developed on different platforms, and it will be beneficial greatly to develop a universal framework, such as NVIDIA Omniverse in Figure 14.18. To share and exchange users' experiences across multiple platforms, NVIDIA Omniverse supports cross-works of sensing and actuating, representation of assets, artificial intelligence, blockchain, and cybersecurity (Dong and Liu 2022).

However, similar to DTs or the Internet, metaverse is just a tool for users to effectively interact with people, machines, society, the environment, and any networked things over the Internet. Metaverse is usually not a standalone technology, it must be incorporated with other technologies to exhibit its full capabilities in various applications. Musamih et al. (2022) explored the use of metaverse in the healthcare sector such as training, marketing, supply chain management, fitness and wellness, and telemedicine. As shown in Figure 14.19, the incorporated technologies to implement a whole metaverse application included distributed computing, AI, IoT, decentralized ledger technology (DLT), and telecommunication. Cao (2022) called such a coexisting of multiple technological environments as a metaverse ecosystem.

14.4.1 Metaverses to Deal with Changes and Uncertainties

Metaverses have been explored to support *HCIs* to enhance the flexibility, survivability, and adaptability of *unmanned aerial vehicles* (UAVs). Bi et al. (2023a, 2023b) proposed a new platform to enable BCIs in machine control. Various BCI devices could be integrated depending on their capabilities to acquire, classify, and utilize users' brain signals. The current version of the platform

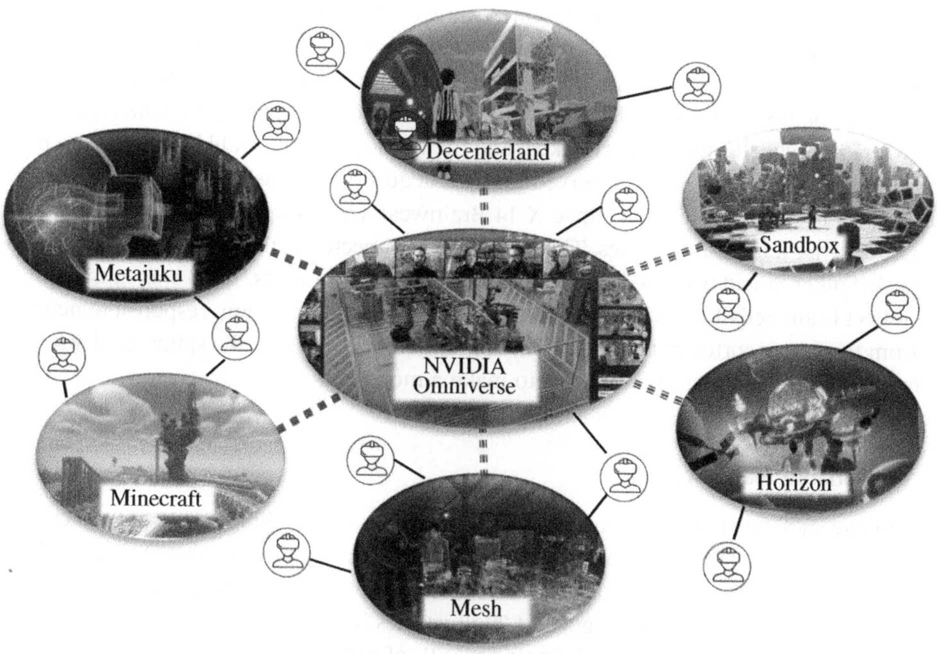

Figure 14.18 Universal framework to share users' experiences across platforms

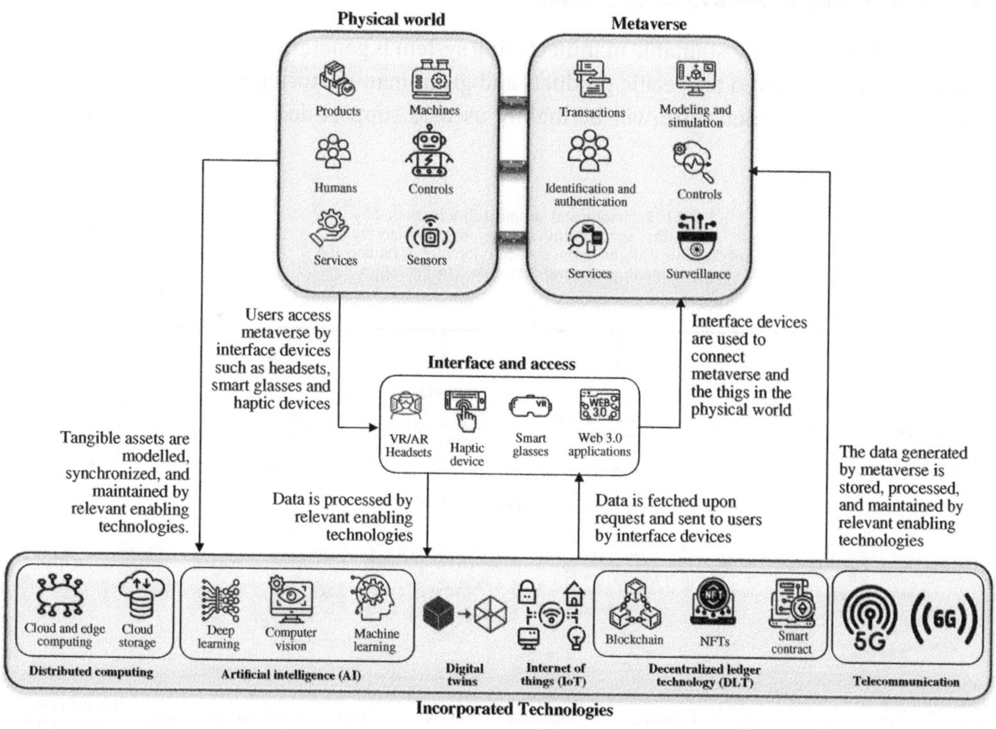

Figure 14.19 Needs to incorporate metaverses with other technologies in applications (Cao 2022; Chen 2022; Musamih et al. 2022)

was equipped with the Epoc X 14 Channel Mobile Brainwear by Emotive and Muscke SpikerShield by Backyard Brains to collect brain signals from users, and customized programs were developed to train users until correct signal patterns could be produced for the expected motions on a robot. Additionally, the platform also included portable drones with cameras to collect data from environments. Moreover, everything was networked, and the whole platform could be viewed as a portable smart thing in the metaverse that served for data acquisition, transmission, processing, and communication. For data acquisition, Epoc X 14 Brainwear was equipped with 14 channels for EEG signals to sense a motion in 9 axes. This device was connected with other smart things wirelessly. Muscle Spiker Shield was used to acquire EMG signals from users. EMG was muscle physiology to reflect brain activities. When the human decides to take an action, responsible neurons traveled from the motor cortex of the brain (i.e., upper motor neurons) to the spinal cord and synapse with lower motor muscles and neurons to make a motor unit.

14.4.2 Sustainable Manufacturing

Similarly, metaverses must be incorporated with other digital technologies for manufacturing enterprises to fully utilize their capabilities in enhancing sustainability. Section 2 discussed the FRs of technological sustainability that were measured in six aspects, Section 3 discussed metaverses and relative technologies, and Figure 14.20 illustrates how these technologies can be used as the *design solutions* (DSs) to fulfill the technological FRs of sustainable manufacturing elaborated in Section 2.

14.4.3 Framework of Metaverse Use Cases

Designing and operating a sustainable manufacturing system is generally very complex since every system has to be customized to specific products and given manufacturing resources and business environment. However, metaverse will be mainly used to support decision-making activities at

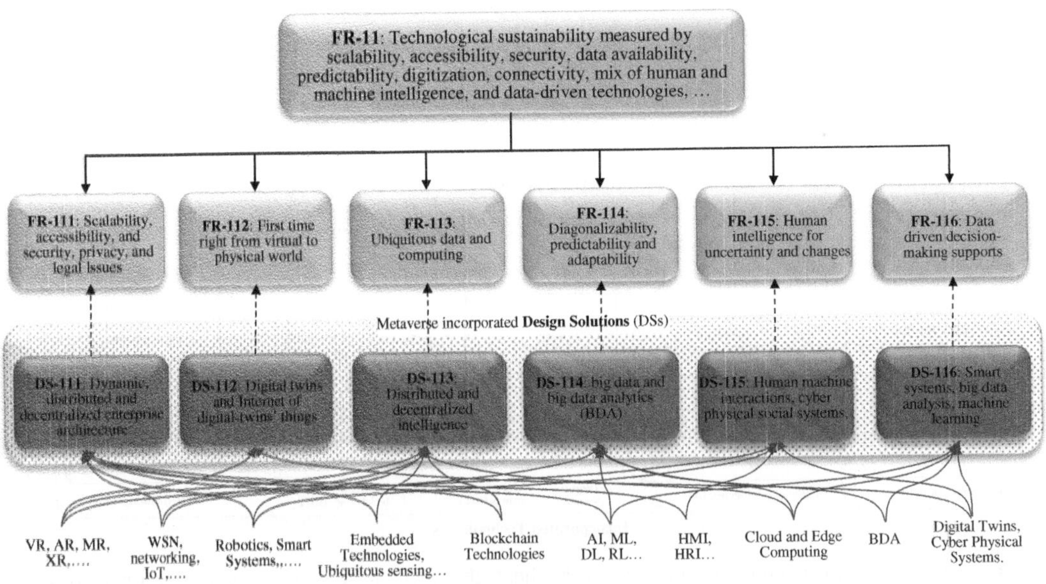

Figure 14.20 Metaverses in satisfying technological FRs of sustainable manufacturing

different levels and domains of manufacturing businesses. A generic framework can be developed to select the right technologies for implementing certain functional units in a system.

Bi et al. (2021a) viewed any decision-making information system as an analogous manufacturing system whose "raw materials" are acquired data from various sources and "products" are the decisions, knowledge, or information required to operate a manufacturing system. Accordingly, the complexity of an information system can be tackled by respective digital technologies as follows:

1) At the phase of *data collection*, metaverse can be adopted to collect users' data, visualize data, and establish the correspondences of relevant data; IoT helps to collect data from networked things such as machines, operators, computers, sensors, and environments. When more and more businesses are involved in a system, the data becomes "big" due to a continuous increase in volume, velocity, and variety (3V).
2) At the phase of *data management*, metaverse can be used to distribute and classify data, and distributed intelligence, edge computing, and CC can be applied to transmit, distribute, share, and process data at all domains and levels of businesses. Some platforms such as MapReduce are adopted to manage distributed and big data and identify the most relevant data efficiently.
3) At the phase of *data utilization*, metaverse can be used to review and visualize data and identify correspondence of data, and big data analytics (BDA) is adopted to analyze and mine knowledge. Edge computing and CC can be applied for users to access and analyze data from anywhere and at any time when it is needed.

In developing a metaverse use case for a functional unit in a sustainable manufacturing system, the generic system design approach, *axiomatic design theory* (ADT), can be applied (Cochran et al. 2016a, 2016b). The system design follows three basic steps as shown in Figure 14.21. *First*, a set of the quantified measures are selected as FRs; *second*, all the accessible technologies including metaverse are identified to determine a space for DSs; *third*, all of the technologies are evaluated against the given FRs until an optimized sub-DS is found for any sub-FR.

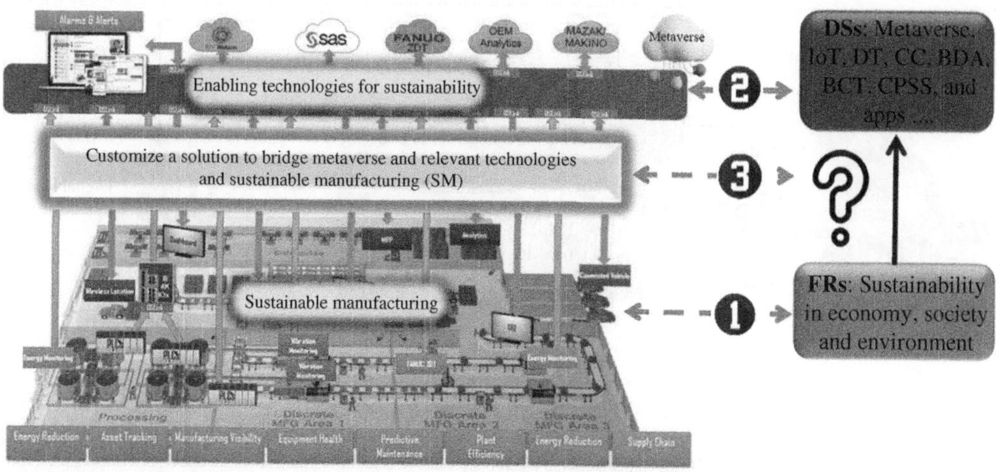

Figure 14.21 Customize technical solution for sustainable manufacturing

14.4.4 Metaverses for Remote Access

Metaverse has exhibited its superior capabilities in accessing resources, collaborating with people, and verifying and validating system solutions remotely and cost-effectively. Sustainable manufacturing can potentially benefit greatly by incorporating metaverses in businesses that require access to remote resources, collaboration with people, or rigorous verification and validation in transforming a virtual model into a physical model.

Figure 14.22 shows a concept project where metaverse is proposed to (1) share a metal printer among geometrically distributed users and (2) verify and validate a design before it is materialized. Similar to other precise machine tools, running a metal printer involves many factors and the printing process can be complex and unreliable. Users must be well trained to set up the machine adequately and operate it safely by following a correct procedure. We proposed to develop a metaverse use case for users to gain hands-on skills in machine operations and be certified for their qualifications that assure the safe operations of machines. As a matter of fact, similar needs have been identified by many scholars when *learning management systems* (LMS) or digital education systems were widely adopted to manage, curate, and deliver courses remotely during the COVID-19 pandemic since 2020; *no LMS* or *digital education platform* has yet been developed to support interactive technology learning for hands-on training, laboratories, experiments, tests, and practices in physical things, virtually.

The available metal printer enables research on the methodologies and tools used to train students and certify them for machine operations. In the proposed metaverse use case, (1) extended reality (XR) will be integrated with metaverse in an immersive training environment; (2) technical solutions will be developed to digitize laboratory devices, instruments, tools, specimens, and other physical things; (3) the publicly accessible robot operating system (ROS) will be adopted to model physical things and their interactions; (4) extended Petri nets will be used to model users' interactions as events; (5) the solutions to items (1)–(4) will be integrated as a metaverse use case to show the feasibility and the effectiveness of using a precision machine tool and teaching engineering labs in an immersive way.

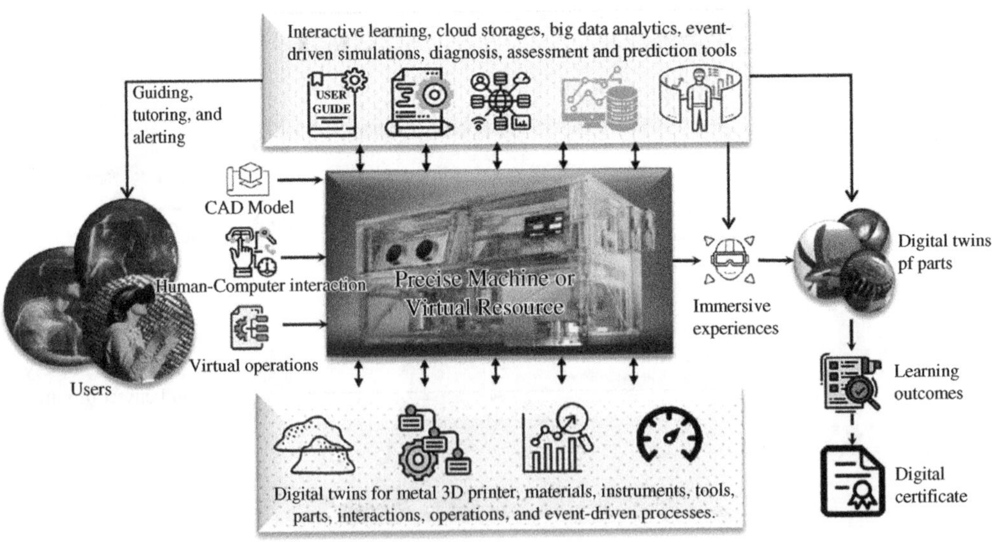

Figure 14.22 Metaverse use case for metal 3D printing training

14.5 Summary and Future Work

This chapter aims to understand the historical development of metaverse and relevant technologies, and the focuses are its state-of-the-art development and its potential to be explored as an enabling technology for enterprises to implement sustainable manufacturing systems.

Metaverse has recently attracted more and more attention which is evidenced by a vast number of recent publications on relevant subjects. Metaverse was identified as one of several disruptive technologies together with IoT, AI, and 6G (Bhat et al. 2023). The research on metaverse was prioritized to advance intelligence, innovation, communication, experiences, consumer behaviors, and policy formulation (Barrera and Shah 2023). Metaverse is widely explored as successors to mobile Internet (Lam et al. 2022).

Developing and implementing seamless and interoperative metaverse use cases are still facing some challenges in sensing, communication, and computing. For example, metaverse brings some security and privacy concerns such as threats to identity, network, or assets. Dunnett et al. (2022) discussed the use of cyber threat intelligence (CTI) to protect privacy and security. CTI is built upon conventional cyber security risk management (CSRM) for timely, relevant, and actional information about possible threats and attacks. There is a need to use reinforcement learning (RL) for awareness of XR contents based on spatial or temporal differences from the perspective of multimedia communications (Dong and Lee 2022). There is a need to use RL for awareness of XR contents based on spatial or temporal differences from the perspective of multimedia communications (Dong and Lee 2022). Moreover, increasing number of users demanding network resources, heavily for graphical extended reality and rendering a plethora of virtual things, an attention-aware network resource allocation scheme is designed to customize metaverse services such as eye-tracking, remote rendering, and QoE analysis (Du et al. 2022). Metaverse must be advanced to support (1) green networking with sustainable energy resources to sense things and exchange data for real-time interactions; (2) heterogeneous management to manage diversified resources and services; (3) technology integration to meet strict requirements of 6G for sensing, communicating, and computing in metaverses; and (4) standard specifications for plug and play of network access, routing, and congestion controls.

Most researchers agree that metaverse has brought numerous new opportunities to enterprises, people, and societies for using cutting-edge digital technologies for immersive interactions, mission-critical services, and omnipresent IoT (Eukdden 2022). However, we have found very limited use cases in which the capabilities of the metaverse are fully explored to benefit manufacturing enterprises. We suggest that metaverse must be advanced and integrated seamlessly with other disruptive technologies, especially DT-II, IoDDT, CC, EC, ML, AI, and BDA. Moreover, further development of metaverse should be needs driven, and we will continue our research to showcase the integration of metaverse with other technologies for enterprises to improve their survivability by developing effective metaverse use cases to support system reconfigurations and extend enterprises' lifespans.

Problems

14.1 What is metaverse?

14.2 What is the relation of metaverse with Internet of Things (IoT)?

14.3 What is the relation of metaverse with VR, MR, ER, and XR?

14.4 What is the relation of metaverse with digital twins (DT-I) and digital triad (DT-II)?

14.5 Discuss some examples of enabling technologies of metaverse.

14.6 Lay out a plan to develop a metaverse use case to teach a lab on the tensile test.

14.7 Lay out a plane to develop a metaverse use case to train the use of a 3D metal printer.

References

Abdel-Basset, M., Hawash, H., Moustaf, N. et al. (2022). Privacy-preserved learning from non-iid data in fog-assisted IoT: a federated learning approach. *Digital Communications and Networks* 2022: https://doi.org/10.1016/j.dcan.2022.12.013.

Bansal, G., Rajgopal, K., Chamola, V. et al. (2022). Healthcare in metaverse: a survey on current metaverse applications in healthcare. *IEEE Access* https://doi.org/10.1109/ACCESS.2022.3219845.

Barrera, K.G. and Shah, D. (2023). Marketing in the metaverse: conceptual understanding, framework, and research agenda. *Journal of Business Research* 155 (A): 113420.

Berardino, R.A. and Tucker, A.O. IV (2022). ARMOUR X: augmented reality mission operations UseR eXperience. *Exploring Immersive Technology at APL* 35 (3): 245–252.

Bhat, J.R., AlQahtani, S.A., and Nekovee, M. (2023). FinTech enablers, use cases, and role of future internet of things. *Journal of King Saud University, Computer and Information Sciences* 35 (2023): 87–101.

Bi, Z. (2011). Revisiting system paradigms for the viewpoint of manufacturing sustainability. *Sustainability* 3 (9): 1323–1340.

Bi, Z.M. and Cochran, D. (2015). Big data analytics with applications. *Journal of Management Analytics* 1 (4): 249–265.

Bi, Z.M. and Wang, L. (2013). Chapter 5: manufacturing paradigm shift towards better sustainability. In: *Cloud Manufacturing*. London: Springer-Verlag.

Bi, Z.M. and Zhang, W.J. (2021). *Practical Guide to Digital Manufacturing – First-Time-Right from Digital Twin to Physical Twin*. Springer International Publishing.

Bi, Z.M., Jin, Y., Maropoulos, P. et al. (2021a). Internet of things (IoT) and big data analytics (BDA) for digital manufacturing (DM). *International Journal of Production Research* https://doi.org/10.1080/00207543.2021.1953181.

Bi, Z.M., Wu, C., Luo, C., and Xu, L. (2021b). Generic design methodology for smart manufacturing systems from a practical perspective, part I – digital triad concept and its application as a system reference model. *Machines* 9 (10): 207. https://doi.org/10.3390/machines9100207.

Bi, Z.M., Wu, C., Luo, C., and Xu, L. (2021c). Generic design methodology for smart manufacturing systems from a practical perspective, part II – systematic designs of smart manufacturing systems. *Machines* 9 (10): 208. https://doi.org/10.3390/machines9100208.

Bi, Z.M., Chen, B., Xu, L.D. et al. (2021d). Security and safety assurance of collaborative manufacturing in industry 4.0. *Enterprise Information Systems* https://doi.org/10.1080/17517575.2021.2008512.

Bi, Z.M., Luo, C., Miao, Z. et al. (2021e). Safety assurance mechanisms of collaborative robotic systems in manufacturing. *Robotics and Computer-Integrated Manufacturing* 67: 102022.

Bi, Z.M., Zhang, W.J., Wu, C., and Lin, L. (2022). New digital triad (DT-II) concept for lifecycle information integration of sustainable manufacturing systems. *Journal of Industrial Information Integration* 26: 100316.

Bi, Z.M., Liu, Y., Quaye, E., and Luo, C. (2023a). Brain computer interfaces for supervisory controls of unmanned aerial vehicles. *2023 IEEE/ASME International Conference on Advanced Intelligent Mechatronics (AIM)*, Seattle, WA, USA, 2023, pp. 1067–1072. https://doi.org/10.1109/AIM46323.2023.10196220.

Bi, Z.M., Mikkola, A., Ip, A.W.H. et al. (2023b). Brain computer interfaces for shared controls of unmanned aerial vehicles. *IEEE Transactions on Aerospace and Electronic Systems* tentatively accepted article.

Boyce MW, Gardony AL, Shorter P, Horner C, Stainrod CR Flynn J, Brunte TT and Amburn CR (2019) Characterizing the cognitive impact of tangible augmented reality. https://doi.org/10.1007/978-3-030-21607-8_32

Cao, L. (2022). Decentralized AI: edge intelligence and smart blockchain, metaverse, web3, and DeSci. *IEEE Intelligent Systems* 2022: https://doi.org/10.1109/MIS.2022.3181504.

Chang, L., Zhang, Z., Li, P. et al. (2022). 6G-enabled edge AI for metaverse: challenges, methods, and future research directons. *Journal of Communicatios and Information Networks* 7 (2): 107–121.

Chen, S.C. (2022). Multimedia research toward the metaverse. *IEEE MultiMedia* 2022: 125–127.

Chen, J., Xiao, H., Hu, M., and Chen, C.M. (2023). A blockchain-based signature exchange protocol for metaverse. *Future Generation Computer Systems, Future Generation Computer Systems* 142: 237–247.

Cheng, R., Wu, N., Chen, S., and Han, B. (2022). Will metaverse be nextG internet? Vision, hype, and reality. *IEEE Network* 2022: 197–204.

Chidambaram S, Reddy, S.S., and Rumple, M. et al. (2022). EditAR: a digital twin authoring environment for creation of AR/VR and video instructions from a single demonstration. *2022 IEEE International Symposium on Mixed and Augmented Reality (ISMAR)*, Singapore, 2022, pp. 326–335. https://doi.org/10.1109/ISMAR55827.2022.00048.

Cochran, D.S., Hendricks, S., Barnes, J., and Bi, Z. (2016a). Extension of manufacturing system design decomposition to implement manufacturing systems that are sustainable. *Journal of Manufacturing Science and Engineering* 138 (10): 101006.

Cochran, D.S., Jafri, M.U., Chu, A.K., and Bi, Z. (2016b). Incorporating design improvement with effective evaluation using the manufacturing system design decomposition (MSDD). *Journal of Industrial Information Integration* 2: 65–74.

Dai, N., Lei, I.M., Li, Z. et al. (2023). Recent advances in wearable electromechanical sensors – moving towards machine learning-assisted wearable sensing systems. *Nano Energy* 105: 108041.

De Felice, F., De Luca, C., Chiara, S.D., and Petrillo, A. (2022). Physical and digital worlds – implications and opportunities of the metaverse. *Procedia Computer Science* 217: 1844–1754.

Deveci, M., Mishra, A.R., Gokasar, I. et al. (2022a). A decision support system for assessing and prioritizing sustainable urban transportation in metaverse. *IEEE Transactions on Fuzzy Systems* https://doi.org/10.1109/TFUZZ.2022.3190613.

Deveci, M., Gokasar, I., Catillo, O., and Daim, T. (2022b). Evaluation of metaverse integration o freight fluidity measurement alternatives using fuzzy dombi EDAS model. *Computers & Industrial Engineering* 174: 118773.

Dogea, R. and Stolt, R. (2021). Identifying challenges related to industry 4.0 in five manufacturing companies. *Procedia Manufacturing* 55: 328–334.

Dong, H. and Lee, J.S.A. (2022). The metaverse from a multimedia communications perspective. *IEEE Multimedia* 2022: 123–127.

Dong, H. and Liu, Y. (2022). Metaverse meets consumer electronics. *IEEE Consumer Electronics Magazine*. https://doi.org/19.1109/MCE.2022.3229180.

Du, H., Wang, J., Niyato, D. et al. (2022). Exploring attention-aware network resource allocation for customized metaverse services. *IEEE Network* 37 (6): 166–175.

Dunnett, K., Pal, S., Jadidi, Z., and Jurdak, R. (2022). The role of cyber thread intelligence sharing in the metaverse. *IEEE Internet of Things Magazine* 6 (1): 154–160.

Dwivedi, Y.K., Hughes, L., Baabdullah, A.M. et al. (2022). Metaverse beyond the hype: multidisciplinary perspectives on emerging challenges, opportunities, and agenda for research, practice and policy. *International Journal of Information Management* 66: 102542.

Eukdden, E. (2022). Future network trends driving universal metaverse mobility. *Ericsson Technology Review* 2–10.

Fan, L., Cao, D., Zeng, C. et al. (2022). Cognitive-based crack detection for road maintenance: an integrated system in cyber-physical-social systems. *IEEE Transactions on Systems, Man, and Cybernetics: Systems*. https://doi.org/10.1109/TSMC.2022.3227209.

Far, S.B., Bamakan, S.M., Qu, Q., and Qingshan, J. (2022). A review of non-fungible tokens applications in the real-world and metaverse. *Procedia Computer Science* 2014: 766–762.

Faraboschi, P., Frachtenberg, E., Laplante, P. et al. (2022). Virtual worlds (metaverse): from Skepticism, to fear, to immersive opportunities. *Computer* 55 (10): 100–106.

Felice, F. de, Luca, C. de, Chiara, S. di, and Petrillo (2023). Physical and digital worlds: implications and opportunities of the metaverse, 4th International Conference on Industry 4.0 and Smart Manufacturing, *Procedia Computer Science*, vol. 217: pp. 1744–1754.

Ferrara, S., Ciccarelli, L., Moreno, A.J. et al. (2022). The next frontier for MPEG-5 LCEVC: from HDR and immersive video to the metaverse. *IEEE Multimedia* 29 (4): 111–122. https://doi.org/10.1109/MMUL.2022.3213879.

Fielder, D. (2023). The functional requirements for smart manufacturing: a guide for leaders. https://www.switchplane.com/the-functional-requirements-for-smart-manufacturing-a-guide-for-leaders (accessed 13 September 2024).

Fu, Y., Li, C., Yu, F.R. et al. (2023). A survey of blockchain and intelligent networking for the metaverse. *IEEE Internet of Things Journal* 10 (4): 3587–3610.

Future Today Institute (2022). Tech trends. https://edisciplinas.usp.br/pluginfile.php/4883861/mod_resource/content/1/Tech%20Trends%20Report%20-%20Future%20Today%20Institute.pdf (accessed 13 September 2024).

Gai, K., Wang, S., Zhao, H. et al. (2022). Blockchain-based multisignature lock for UAC in metaverse. *IEEE Transactions on Computational Social Systems* https://doi.org/10.1109/TCSS.2022.3226717.

Gorilla (2023). Edge AI technology. https://www.gorilla-technology.com (accessed 13 September 2024).

Gu, J., Wang, J., Guo, X. et al. (2023). A metaverse-based teaching building evacuation training system with deep reinforcement learning. *IEEE Transactions on Systems, Man, and Cybernetics: Systems* 53 (4): 2209–2219.

Gugenheimer, J. (2022). Ubiquitous mixed reality: human-computer interaction and our path towards ubiquitous mixed reality. https://www.telecom-paris.fr/wp-content-EvDsK19/uploads/2022/03/JPE-2022-RI-4-RealVirtuelle-J-Gugenheimer.pdf (accessed 13 September 2024).

Guo, C., Dou, Y., Bai, T. et al. (2023). ArtVerse: a paradigm for parallel human–machine collaborative painting creation in Metaverses. *IEEE Transactions on Systems, Man, and Cybernetics: Systems* 53 (4): 2200–2208.

Han, J., Yang, M., Chen, X. et al. (2023a). ParaDefender: a scenario-driven parallel system for defending metaverses. *IEEE Transactions on Systems, Man, and Cybernetics: Systems* 53 (4): 2118–2127.

Han, Y., Niyato, D., Leung, C. et al. (2023b). A dynamic hierarchical framework for IoT-assisted digital twin synchronization in the metaverse. *IEEE Internet of Things Journal* 10 (1): 268–284.

Hare, R. and Tang, Y. (2023). Hierarchical deep reinforcement learning with experience sharing for metaverse in education. *IEEE Transactions on Systems, Man, and Cybernetics: Systems* 53 (4): 2047–2055.

Huang, H., Zeng, X., Zhao, L. et al. (2022). Fusion of building information modelling and blockchain for metaverse: a survey. *IEEE Open Journal of the Computer Society* https://doi.org/10.1109/OJCS.2022.3206494.

Huang, Y., Li, Y., and Cai, Z. (2023). Security and privacy in metaverse: a comprehensive survey. *Big Data Mining and Analytics* 6 (2): https://doi.org/10.26599/BDMA.2022.9020047.

Huynh, D.V., Khosravirad, S.R., Masaracchia, A. et al. (2022). Edge intelligence-based ultra-reliable and low-latency communications for digital twin-enabled metaverse. *IEEE Wireless Communications Letters* 11 (8): 1733–1737.

Jagatheesaperumal, S.K. and Rahouti, M. (2022). Building digital twins of cyber physical systems with metaverse for Industry 5.0 and beyond. *IT Professional* 24 (6): 34–40.

Jaimini, U., Zhang, T., Brikis, G.O., and Sheth, A. (2022). imetaverseKG: industrial metaverse knowledge graph to promoto interoperability in design and engineering applications. *IEEE Internet Computing* 26 (6): 59–67.

Jiang, Y., Kang, J., Niyato, D. et al. (2023). Reliable distributed computing for metaverse: a hierarchical game-theoretic approach. *IEEE Transactions on Vehicular Technology* 72 (1): 1084–1100.

Kanemastsu, H., Barry, D.M., Shirai, T. et al. (2022). Virtual experiences of metaverse using mobile type head-mounted displays and their applicability. *Procedia Computer Science* 207: 4428–4433.

Kang, M., Wang, X., Wang, H. et al. (2022). The development of AgriVerse: past, present, and future. *IEEE Transactions on Systems, Man, and Cybernetics: Systems* https://doi.org/10.1109/TSMC.2022.3230830.

Kshetri, N. (2022a). Web 3.0 and the metaverse shaping organizations' brand and product strategies. *IT Professional* 24 (2): 11–15.

Kshetri, N. (2022b). A typology of Metaverses. *Computer* 55 (12): 150–155.

Kshetri, N. (2022c). Metaverse and developing economies. *IT Professional* 24: 66–69.

Kshetri, N., Rojas-Torres, D., and Grambo, M. (2022). The metaverse and higher education institutions. *IT Professional* 24 (6): 69–73.

Kumar, S., Chhugani, J., Kim, C. et al. (2008). Second life and the new generation of virtual worlds. *Computer* 41 (9): 46–53.

Lam, T., Lah, S., and Pu, Q.L. (2022). Metaverse report – future is here Global XR industry insight, Deloitte China. https://www2.deloitte.com/content/dam/Deloitte/cn/Documents/technology-media-telecommunications/deloitte-cn-tmt-metaverse-report-en-220321.pdf (accessed 13 September 2024).

Le, Z. (2022). Metaverse age: scheduling strategies for digital resource management. *IEEE Consumer Electronics Magazine* https://doi.org/10.1109/MCE.2022.3229840.

Leng, J., Sha, W., Wang, B. et al. (2022). Industry 5.0: prospect and retrospect. *Journal of Manufacturing Systems* 65: 279–295.

Li, K., Cui, Y., Li, W. et al. (2023a). When internet of things meets metaverse: convergence of physical and cyber worlds. *IEEE Internet of Things Journal* 10 (5): 4148–4173.

Li, X., Tian, Y., Ye, P. et al. (2023b). A novel scenarios engineering methodology for foundation models in metaverse. *IEEE Transactions on Systems, Man, and Cybernetics: Systems* 53 (4): 2148–2159.

Li, W. et al. (2023c). Intelligent cockpit for intelligent vehicle in metaverse: a case study of empathetic auditory regulation of human emotion. *IEEE Transactions on Systems, Man, and Cybernetics: Systems* 53 (4): 2173–2187.

Li, B., Cao, D., Tang, S., and Zhang, T. (2023d). Sharing traffic priorities via cyber–physical–social intelligence: a lane-free autonomous intersection management method in metaverse. *IEEE Transactions on Systems, Man, and Cybernetics: Systems* 53 (4): 2025–2036.

Li, X., Liu, D., Sun, J., and Zhu, Z. (2022). Challenges of industrial engineering in big data environment and its new directions on extension intelligence. *9th International Conference on Information Technology and Quantitative Management, Procedia Computer Science* 214: 1561–1567.

Liu, K., Chen, L., Li, L. et al. (2022a). MetaMining: mining in the metaverse. *IEEE Transactions on Systems, Man, and Cybernetics: Systems* https://doi.org/10.1109/TSMC.2022.3233588.

Liu, K., Li, L., Lv, Y. et al. (2022b). Parallel intelligence for smart mobility in Cyberphysical social system-defined Metaverses: a report on the international parallel driving Alliance. *IEEE Intelligent Transportation Systems Magazine* 14 (6): 18–25.

Liu, Y. et al. (2023). RadarVerses in Metaverses: a CPSI-based architecture for 6S radar systems in CPSS. *IEEE Transactions on Systems, Man, and Cybernetics: Systems* 53 (4): 2128–2137.

Lv, Z., Xie, S., Li, Y. et al. (2022a). Building the metaverse using digital twins at all scales, states, and relations. *Virtual Reality & Intelligent Hardware* 4 (6): 459–470.

Lv, Z., Qiao, L., Li, Y. et al. (2022b). BlockNet: beyond reliable spatial digital twins to parallel metaverse. *Patterns* 3 (5): 100468.

Maksymyuk, T., Gazda, J., Bugár, G. et al. (2022). Blockchain-empowered service management for the decentralized metaverse of things. *IEEE Access* 10: 99025–99037.

Meng, Z., She, C., Zhao, G., and De Martini, D. (2023). Sampling, communication, and prediction co-design for synchronizing the real-world device and digital model in metaverse. *IEEE Journal on Selected Areas in Communications* 41 (1): 288–300.

Mourtzis, D., Panopoulos, N., Angelopoulos, J. et al. (2022). Human centric platforms for personalized value creation in metaverse. *Journal of Manufacturing Systems* 65: 653–659.

Musamih, A., Yaqoob, I., Salah, K. et al. (2022). Metaverse in healthcare: applications, challenges, and future directions. *IEEE Consumer Electronics Magazine* https://doi.org/10.1109/MCE.2022.3223522.

Ning, H., Lin, Y., Wang, W. et al. (2023). Cyberology: cyber–physical–social-thinking spaces-based discipline and interdiscipline hierarchy for metaverse (general cyberspace). *IEEE Internet of Things Journal* 10 (5): 4420–4430.

Onggirawan, C.A., Kho, J.M., Kartiwa, A.P., and Gunawan, A.A.S. (2022). Systematic literature review: the adaptation of distance learning process during the covid-19 pandemic using virtual educational spaces in metaverse. *Procedia Computer Science* 216: 274–283.

Ou, W., Huang, S., Zheng, J. et al. (2022). An overview on cross-chain: mechanism, platforms, challenges, and advances. *Computer Networks* 218: 109378.

Pamucar, D., Deveci, M., Gokasar, I. et al. (2022). A metaverse assessment model for sustainable transportation using ordinal priority approach and Aczel-Alsina norm. *Technological Forecasting and Social Change* 182: 121778.

Pan, D., Ding, D., Ge, X. et al. (2023). Privacy-preserving platooning control of vehicular cyber–physical systems with saturated inputs. *IEEE Transactions on Systems, Man, and Cybernetics: Systems* 53 (4): 2083–2097.

Panagiotakopoulos, D., Marentakis, G., Metzitakos, R. et al. (2022). Digital scent technology: toward the internet of senses and the metaverse. *IT Professional* 24 (3): 52–55.

Park, S.M. and Kim, Y.G. (2022). A metaverse: taxonomy, components, applications, and open challenges. *IEEE Access* 10: 4209–4251.

Park, W.H., Qureshi, N.M.F., and Shin, D.R. (2022). An effective 3D text recurrent voting generator for metaverse. *IEEE Transactions on Affective Computing* https://doi.org/10.1109/TAFFC.2022.3216782.

Parmar, V., Sarwar, S.S., Li, Z. et al. (2023). Memory-oriented design-space exploration of edge-AI hardware for XR applications. https://arxiv.org/abs/2206.06780 (accessed 13 September 2024).

Pena, J., Barake, M.A., and Falin, J.M. (2023). Virtual leaders: can customizing authoritarian and democratic business leader avatars influence altruistic behavior and leadership empowerment perceptions? *Computers in Human Behavior* 141: 107616.

Qiao, S., Pang, S., Sun, Y. et al. (2022). SPReCHD: four-chamber semantic parsing network for recognizing Fetal congenital heart disease in medical metaverse. *IEEE Journal of Biomedical and Health Informatics* https://doi.org/10.1109/JBHI.2022.3218577.

Rahmani, R. and Tienken, S. (2022). Distributed meta-learning context based on culture heritage collection data in an immersive reality system. *Procedia Computer Science* 201: 70–77.

Ren, Y., Xie, R., Yu, F.R. et al. (2022). Quantum collective learning and many-to-many matching game in the metaverse for connected and autonomous vehicles. *IEEE Transactions on Vehicular Technology* 71 (11): 12128–12139.

Ryu, J., Son, S., Le, J. et al. (2022). Design of secure mutual authentication scheme for metaverse environments using blockchain. *IEEE Access* 10: 98944–98958.

Saari, L., Heialal, J., Heikkila, T. et al. (2022). Digital product passport promotes sustainable manufacturing. https://cris.vtt.fi/ws/portalfiles/portal/67162320/DPP_white_paper.pdf (accessed 13 September 2024).

Shen, Y., Liu, Y.H., Tian, Y.L., and Na, X.X. (2022a). Parallel sensing in Metaverses: virtual-real interactive smart systems for "6S" sensing. *IEEE/CAA Journal of Automatica Sinica* 9 (12): 2047–2054. https://doi.org/10.1109/JAS.2022.106115.

Shen, T., Huang, S.S., Li, D. et al. (2022b). VirtualClassroom: a lecturer-centered consumer-grade immersive teaching system in cyber–physical–social space. *IEEE Transactions on Systems, Man, and Cybernetics: Systems*. https://doi.org/10.1109/TSMC.2022.3228270.

Shi, H., Liu, G., Zhang, K. et al. (2023). MARL Sim2real transfer: merging physical reality with digital virtuality in metaverse. *IEEE Transactions on Systems, Man, and Cybernetics: Systems* 53 (4): 2107–2117.

Siyaev, A. and Jo, G.S. (2021). Neuro-symbolic speech understanding in aircraft maintenance metaverse. *IEEE Access* 9: 154484–154499.

Song, Y.T. and Qin, J. (2022). Metaverse and personal healthcare. *Procedia Computer Science* 210: 189–197.

Stephenson, N. (1992). *Snow Crash*. New York: Bantam Books.

Tang, F., Chen, X., Zhao, M., and Kato, N. (2023). The roadmap of communication and networking in 6G for the metaverse. *IEEE Wireless Communications* https://doi.org/10.1109/MWC.019.2100721.

Tao, B., Dai, H.N., Xie, H., and Wang, F.L. (2022). Structural identity representation learning for Blockchain-enabled metaverse based on complex network analysis. *IEEE Transactions on Computational Social Systems* https://doi.org/10.1109/TCSS.2022.3233059.

Telecom (2019). What industrial IoT means for the future of the manufacturing sector. https://www.telecomreviewasia.com/news/featured-articles/1773-what-industrial-iot-means-for-the-future-of-the-manufacturing-sector (accessed 13 September 2024).

Tian, X. (2021). Scanning the literature – from AR/VR to metaverse. *IEEE Network* 8–9.

Viriyasitavat, W., Xu, L.D., Bi, Z., and Hoonsopon, D. (2022). User-oriented selections of validators for trust of internet of things services. *IEEE Transactions on Industrial Informatics* 18 (7): 4859–4867.

Viriyasitavat, W., Xu, L.D., Dhiman, G., and Bi, Z. (2023). Blockchain-as-a-service for business process management: survey and challenges. *IEEE Transactions on Services Computing* 16 (3): 2299–2314.

Vondrek, M., Baggili, I., Casey, P., and Mekni, M. (2023). Rise of the metaverse's immersive virtual reality malware and the man-in-the-room attack & defense. *Computers & Security* 127: 102923.

Wang, F.Y. (2022a). Parallel intelligence in Metaverses: welcome to Hanoi. *IEEE Intelligent Systems* 1541–1672. https://doi.org/10.1109/MIS.2022.3154541.

Wang, F.Y. (2022b). The DAO to MetaControl for MetaSystems in Metaverses: the system of parallel control Systems for Knowledge Automation and Control Intelligence in CPSS. *IEEE/CAA Journal of Automatica Sinica* 9 (11): 1899–1908.

Wang, M., Yu, H., Bell, Z., and Chu, X. (2022a). Constructing an edu-metaverse ecosystem: a new and innovative framework. *IEEE Transactions on Learning Technologies* 15 (6): 685–696.

Wang, X., Yang, J., Han, J. et al. (2022b). Metaverses and DeMetaverses: from digital twins in CPS to parallel intelligence in CPSS. *IEEE Intelligent Systems* 37 (4): 97–102.

Wang, Y., Tian, Y., Wang, J. et al. (2022c). Integrated inspection of QoM, QoP, and QoS for AOI Industries in Metaverses. *IEEE/CAA Journal of Automatica Sinica* 9 (12): 2071–2078.

Wang, X., Kang, M., Sun, H. et al. (2022d). DeCASA in AgriVerse: parallel agriculture for smart villages in Metaverses. *IEEE/CAA Journal of Automatica Sinica* 9 (12): 2055–2062.

Wang, Y., Su, Z., Zhang, N. et al. (2023a). A survey on metaverse: fundamentals, security, and privacy. *IEEE Communications Surveys & Tutorials* 25 (1): 319–352.

Wang, C., Yu, C., and Li, Y. (2023b). Toward understanding attention economy in metaverse: a case study of NFT value. *IEEE Transactions on Computational Social Systems* https://doi.org/10.1109/TCSS.2022.3221669.

Wang, J., Du, H., Tian, Z. et al. (2023c). Semantic-aware sensing information transmission for metaverse: a contest theoretic approach. *IEEE Transactions on Wireless Communications* https://doi.org/10.1109/TWC.2022.3232565.

Wang, X., Cheng, X., Lu, J. et al. (2023d). Metaverses-based parallel oil fields in CPSS: a framework and methodology. *IEEE Transactions on Systems, Man, and Cybernetics: Systems* 53 (4): 2138–2147.

Wang, J., Tian, Y., Wang, Y. et al. (2023e). A framework and operational procedures for Metaverses-based industrial foundation models. *IEEE Transactions on Systems, Man, and Cybernetics: Systems* 53 (4): 2037–2046.

Xiao, R., Vianto, A., Shaikh, A. et al. (2022). Exploring the application of RFID for designing augmented virtual reality experience. *IEEE Access* 10: 96840–96851.

Xu, M., Ng, W.C., Lim, W.Y.B. et al. (2023). A full dive into realizing the edge-enabled metaverse: visions, enabling technologies, and challenges. *IEEE Communications Surveys & Tutorials* 25 (1): 656–700.

Yang, Q., Zhao, Y., Huang, H. et al. (2022a). Fusing Blockchain and AI with metaverse: a survey. *IEEE Open Journal of the Computer Society* 3: 122–136.

Yang, J., Wand, W., and Zhao, Y. (2022b). Parallel manufacturing for industrial Metaverses: a new paradigm in smart manufacturing. *IEEE/CAA Journal of Automatica Sinica* 9 (12): 2063–2070.

Ye, P. and Wang, F.Y. (2022). Parallel population and parallel human – a cyber-physical social approach. *IEEE Intelligent Systems* 37 (5): 19–27.

Zainab, H.E., Bawany, N.Z., Imran, J., and Rehman, W. (2022). Virtual dimension – a primer to metaverse. *IT Professional* 24 (6): 27–33.

Zallio, M. and Clarson, P.J. (2022). Design the metaverse: a study on inclusion, diversity, equity, accessibility and safety for digital immersive environment. *Telematics and Informatics* 75: 1019099.

Zhang, H., Mao, S., Niyato, D., and Han, Z. (2023a). Location-dependent augmented reality services in wireless edge-enabled metaverse systems. *IEEE Open Journal of the Communications Society* 4: 171–183.

Zhang, H., Luo, G., Li, Y., and Wang, F.Y. (2023b). Parallel vision for intelligent transportation systems in metaverse: challenges, solutions, and potential applications. *IEEE Transactions on Systems, Man, and Cybernetics: Systems* 53 (6): 3400–3413.

Zhang, S., Lim, W.Y.B., Ng, W.C. et al. (2023c). Towards green metaverse networking: technologies, advancements and future directions. *IEEE Network* https://doi.org/10.1109/MNET.130.2200510.

Zhang, T., Shen, J., Lai, C.F. et al. (2023d). Multi-server assisted data sharing supporting secure deduplication for metaverse healthcare systems. *Future Generation Computer Systems* 140: 299–310.

Zheng, W., Yan, L., Zhang, W. et al. (2023). D→K→I: data-knowledge-driven group intelligence framework for smart service in education metaverse. *IEEE Transactions on Systems, Man, and Cybernetics: Systems* 53 (4): 2056–2061.

Zhou, M. (2022). Editorial: evolution from AI, IoT and big data analytics to metaverse. *IEEE/CAA Journal of Automatica Sinica* 9 (12): 2041–2042.

15

Human Cyber-Physical Systems (HCPS)

15.1 Introduction

With continuous growth of scale and complexity, more and more mechatronic systems require collaborating and cooperating with humans closely to fulfill their missions. In regard to the designs of interdisciplinary systems, there is a need for collaborative research across disciplines. Mechatronic systems, robots, and *cyber-physical systems* (CPSs) have integrated with other emerging technologies such as *artificial intelligence* (AI), *distributed computing*, and the *Internet of Things* (IoT) to create numerous breakthroughs in many sectors such as healthcare, manufacturing, entertainment, transportation, and space exploration. Especially, more and more CPSs are cooperating, coordinating, and collaborating with humans seamlessly, giving birth to a new field known as *human-cyber-physical systems* (HCPSs).

Figure 15.1 illustrates the correspondence between *Industry Revolution* (IR) and mechatronic systems in their evolutions (Santoso et al. 2022). Industry 1.0 and 2.0 made it prevalent to develop mechanical systems and electrical systems, respectively. Industry 3.0 emphasized the integration of physical–mechanical systems, electrical systems, and cyber systems as stand-alone mechatronic systems. The Internet accompanying Industry 4.0 brought the opportunities to integrate mechatronic systems and advanced cyber systems as various *CPSs* over the Internet. In Industry 5.0, some innovative technologies such as AI and metaverse allow CPSs to further integrate with people in human societies as *HCPSs*.

In Figure 15.2, one can see the trend that conventional mechatronic systems have been integrated gradually with more and more cyber systems including HCPSs to deal with the growing systems' complexity and scale in a dynamic environment. Nowadays, mechatronic systems in the real world are closely *interlinked* as a *system of systems* (SoS) to enhance the capabilities of the integrated system in terms of adaptability, efficiency, productivity, and cost-effectiveness. HCPSs can fully utilize emerging information technologies such as *digital twins* (DT-I). *AI*, *edge computing* (EC), *digital triads* (DT-II), *Internet of Things* (IoT), *Internet of Digital Triad Things* (IoDDT), and *big data analytics* (BDA).

What does the integration of a human with a CPS mean to users? *First*, it allows users to deal with unanticipated changes, randomness, and uncertainties in applications promptly, and this implies that cyber systems in an HCPS must be *data driven*; every and each functional module in HCPS must be capable of acquiring and analyzing data and utilizing data effectively for decision-making supports at all domains and levels that are illustrated in Figure 15.3. *Second*, humans still exhibit their advantages over automated systems in terms of agility, adaptability, and responsiveness, and HCPS must ensure that humans are favorable options to fulfill some competent tasks in HCPS.

Mechatronics for Complex Products and Systems: Project-Based Design Approaches for Robots, Cyber-Physical Systems, Digital Twins, and Other Emerging Technologies, First Edition. Zhuming Bi.

Companion website: www.wiley.com/go/bi/mechatronics

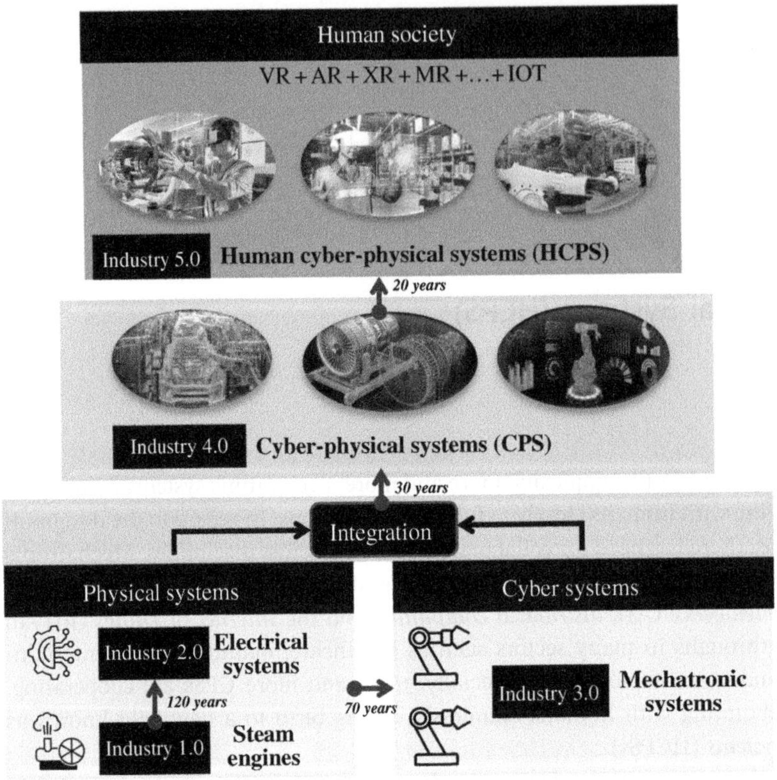

Figure 15.1 Development of human cyber-physical systems (HCPSs) (*Source:* Chen et al. (2020); Santoso et al. (2022)/Gorodenkoff/Adobe Stock)

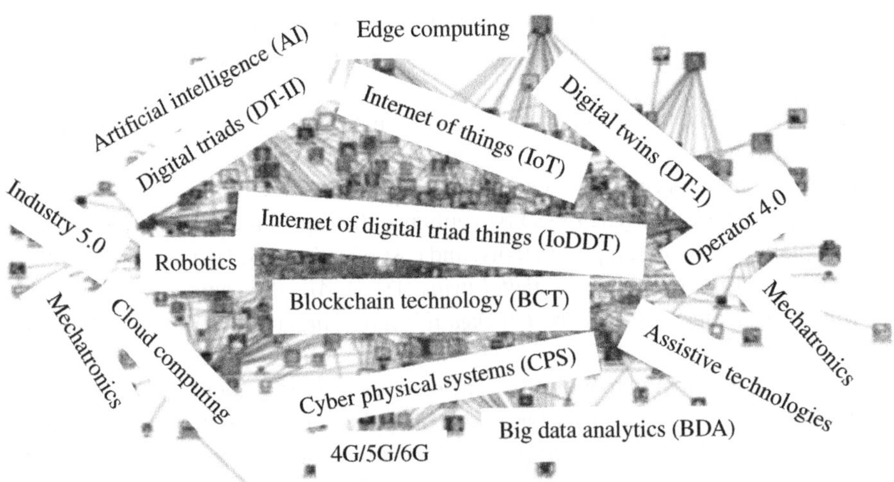

Figure 15.2 Fusion of mechatronics with other advanced technologies over Internet leads to HCPS

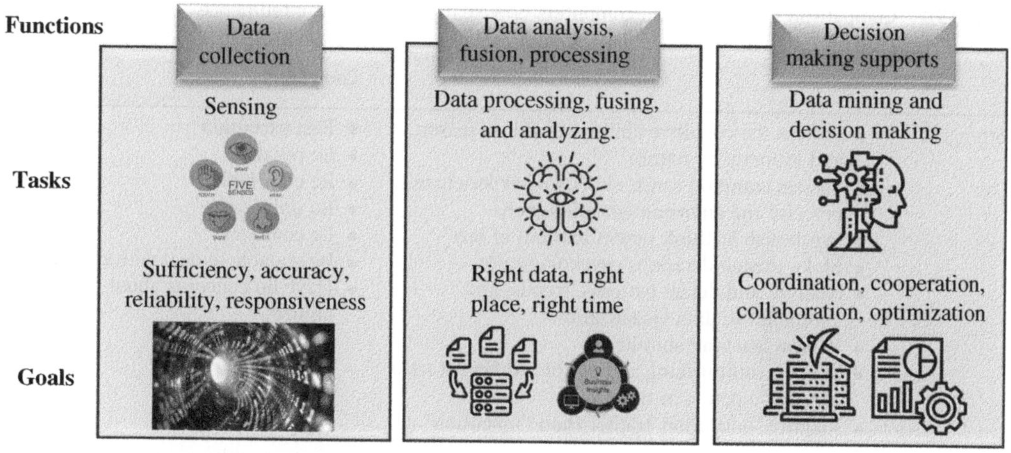

Figure 15.3 Adaptability and flexibility of cyber systems in HCPS (*Source:* R Studio/Adobe Stock Photos)

15.2 Humans' Roles in CPS

Despite the growing maturity of manufacturing automation, computers are still facing the challenges of automating many human tasks cost-effectively. In fact, fewer than 5% of remaining manual tasks and operations are replaceable by machines and only 60% of workers have automated over 30% of their tasks (Santoso et al. 2022).

Even with the adoption of most cutting-edge technologies, robots are just machines that are distinguished greatly from human beings (Fortunati 2021). A robot has no infancy that will be raised by a community of adults; a robot does not feel emotions, amuse itself, pretend to care or be cared; a robot does not share feelings or belonging as humans do; a robot does not have goals or objectives for their livings; a robot does not perceive sociocultural, emotional, or spiritual environments.

Table 15.1 shows a comparison of humans and computers in the aspects of visual perception and intelligence where both humans and computers have their strengths and weaknesses that are not easily replaceable with each other (Sinha and Pathak 2019; Simplilearn 2024). Moreover, Simplilearn (2024) discussed the differences between humans and machine intelligence in the aspects of operations, social networking, flexibility, adaptability, perfection, decision-making, learning capabilities, operational paces, essence, functionalities, and evolution. It becomes clear that CPS systems can be beneficial greatly by collaborating with humans so that both humans and machines can excel at their strengths to the maximum context.

In developing the solution to a complex system, it is beneficial to integrate humans into automated decision-making systems, so that both humans and machines can excel at their strengths to the maximum context. Cummings (2014) characterizes humans and machines in human–machine interactions (HMIs) (see Table 15.2).

Alternatively, it seems more promising to develop HCPSs where humans play a central role while allowing their cooperation and collaboration with machines seamlessly. As shown in Figure 15.4, HCPS is a concept introduced in *Industry 5.0* by integrating humans into CPSs. In an HCPS, real-time data is collected from mechatronic modules and the deployed environment. The raw data is then processed and utilized in the HCPS to keep humans in the loop of system controls. To support effective interactions between humans and machines, immersive technologies such as *virtual reality* (VR), *augmented reality* (AR), *extended reality* (ER), and *mixed reality* (MR) are widely adopted,

Table 15.1 Comparison of humans and computers on visual perception in intelligence

	Humans	Computers
Strengths	• Possess the cognitive abilities to think, reason, and evaluate by nature • Possess common sense, skills, and experiences to perceive the environment initiatively • Synthesize or think new rules "out of box" • Make trusted decisions psychologically • Analyze and detect patterns, trends, and anomalies in data visualization • Possess learning abilities • Support multitasking and perform different tasks at specific points in time • Require "education" rather than "invention"	• Fast execution • Be reliable • Be endurable • Be unbiased • Be consistent • Be scalable to deal with complexity • Have no concerns about fatigue failure
Weaknesses	• Easily bored and tired • Biased and inconsistent • Be error-prone • Decisions are not on a scientific base • Limited capabilities to deal with raw data	• Lack of ability to synthesize new rules • Knowledge is confined by available database • No support for common sense

Table 15.2 Characterizing humans and machines in human–machine interactions (HMI)

Attribute	Human	Machine
Speed	Slow	Fast
Power	Limited	Scalable and stable
Consistency	Unreliable learning and fatigue effects	Consistent and repetitive actions
Information capacity	Primarily single channel	Multichannel
Memory	Better for principles and strategies, access is versatile and innovative	Ideal for literal reproduction, access restricted, and formal
Reasoning computation	Inductive, easier to program, slow, accurate, and good error correction	Detective, tedious to program, fast and accurate, and poor error correction
Sensing	Wide range, multifunction, judgment	Good at quantitative assessment and poor at pattern recognition
Perceiving	Copes with variation better and susceptible to noise	Copes with variation poorly and susceptible to noise.

Source: Adapted from Cummings (2014).

and HCPSs are empowered to deal with internal and external changes such as climate crises, societal changes, and the pandemic by coronavirus disease 2019 (CDC 2024).

HCPSs were widely explored to improve systems' resilience, adaptability, and sustainability in the era of Industry 5.0. Taking an example of their manufacturing applications, as shown in Figure 15.4 (Bi 2018), both humans and computers play important roles in accomplishing corresponding activities. Moreover, humans and computers should collaborate effectively to perform more tasks that are challenging. The strengths of computers and humans can be synergized by

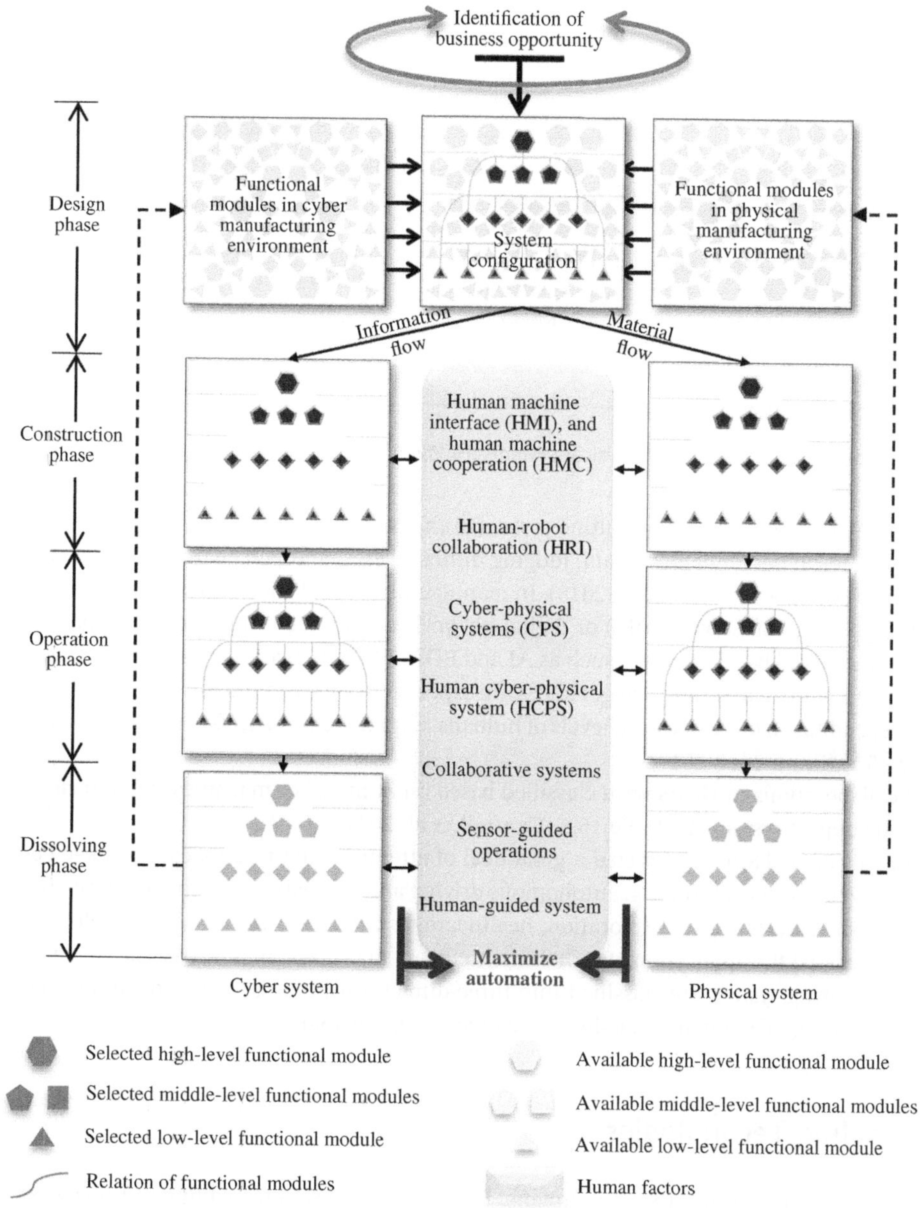

Figure 15.4 Humans' roles in manufacturing automation (*Source:* Bi (2018)/with permission of Elsevier)

developing HCPS. Humans still have irreplaceable roles, and humans' needs and interests must be placed at the center of the production process (Bitsch 2022). Humans are part of decision-making supports in HCPSs where humans are integrated with cyber and physical systems seamlessly (Taylor et al. 2023).

In a typical HCPS, physical and cyber systems work together as the agents to execute tasks and humans are the "master" behind the system. CPS is used to complete and augment humans' capabilities. On the one hand, machines excel at their best in performing repetitive tasks, collecting and

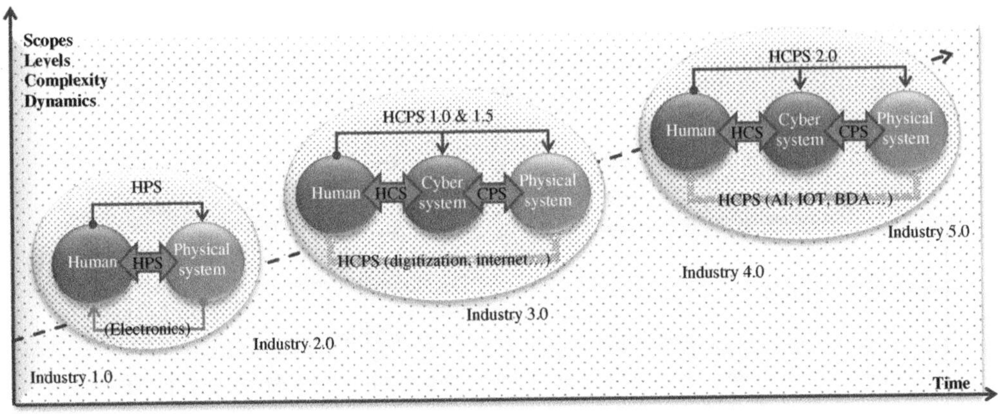

Figure 15.5 Evolution of human cyber-physical systems (HCPSs)

processing big data, and performing routine tasks. On the other hand, humans do what they are good at, that is, resolving ambiguous data, judging difficult cases, and perceiving and responding to social society (Daugherty and Wilson 2018). In manufacturing, HCPS is an advanced system elevated from digital manufacturing (DM) or digital-networked manufacturing since it adopts most emerging technological breakthroughs such as AI and BDA. With an increasing demand for smartness and sustainability of manufacturing, HCPSs are expected to evolve gradually to HCPSs 1.0 and HCPSs 2.0 depending on the scopes and levels of humans' engagements in decision-making systems (see Figure 15.5) (Zhou et al. 2019).

Types of collaborations in HCPS were classified based on humans' or machines' roles in closed-loop decision-support systems in Table 15.3 (Cantrell et al. 2024).

The potentials of HCPS have attracted a great deal of attention, and its applications have been explored widely in transportation (e.g. autonomous driving and navigation), national security (e.g., unmanned aerial vehicles), space exploration, healthcare, and manufacturing. For example, Zhou et al. (2019) discussed the application of HCPSs in intelligent manufacturing whose roles in enhancing a manufacturing system are classified into three dimensions, that is, value-added, technical upgradation, and organizational scalability, as shown in Figure 15.6.

15.3 Enabling Technologies

HCPS is formed by integrating CPS with humans. This section focuses on enabling technologies to model and implement the interactions among system components of an HCPS. According to Wang et al. (2022), the relations of system elements in HCPS can be classified into the following:

$$\begin{cases} \mathfrak{R}^{X}: X \times \mathrm{X} \rightarrow \mathfrak{R} \ \text{ where } X \text{ be } 'H', 'C', \text{or } 'P' \\ \mathfrak{R}^{XY}: X \times \mathrm{Y} \rightarrow \mathfrak{R} \ \text{ where } X \text{ and } Y \text{ be } 'H', 'C', \text{or } 'P' \text{ and } X \neq Y \\ \mathfrak{R}^{HCP}: H \times \mathrm{C} \times \mathrm{P} \rightarrow \mathfrak{R} \end{cases} \tag{15.1}$$

Equation (15.1) shows that seven types of relations in an HCPS are $\mathfrak{R}^{H}$, $\mathfrak{R}^{C}$, $\mathfrak{R}^{P}$, $\mathfrak{R}^{HC}$, $\mathfrak{R}^{HP}$, $\mathfrak{R}^{CP,}$ and $\mathfrak{R}^{HCP}$. Other than the functional requirements (FRs) of HCPS for the implementations of MFMs at the component level, the FRs of HCPS also include the implementations of seven types of interactions among humans, cyber, and physical systems.

Table 15.3 Types of collaborations based on humans' roles in HCPS

	Machines' roles	**Definition**
Collaboration types	Supervisor	Machines allocate tasks to humans and evaluate the performances of humans' work autonomously. For example, a taxi company uses AI to assign requests to drivers who can accept or deny a request without knowing a fare or destination.
	Prioritizer	Machines prioritize a set of tasks based on potential values and importance, and humans perform the tasks in order with guidance from machines.
	Personal coach	Machines identify the strengths of humans and the improvements in tasks, and humans continuously improve their performances with their engagements with computers.
	Muse	Humans provide creative initiatives, and machines produce generative solutions in detail.
	Collaborative decision-maker	Humans and machines collaborate with effective human–machine interfaces to make complex decisions by selecting, enumerating, making trade-offs, or synthesizing existing solutions for the highest possibility of success.
	First pass at a task	Machines perform preliminary works for the solutions to given problems, and humans are responsible for reviewing, approving, improving, and rejecting resolutions.
	Triage nurse	Machines assess problems such as symptoms and failures to decide the need for human consultations; if not, machines run systems automatically.
	Doppelganger	Machines learn from humans by mimicking their actions and making initiative decisions; humans are replicated digitally.
	Subordinate	Machines perform menial and structured tasks; human supervises processes and evaluates outcomes.

Source: Adapted from Cantrell et al. (2024).

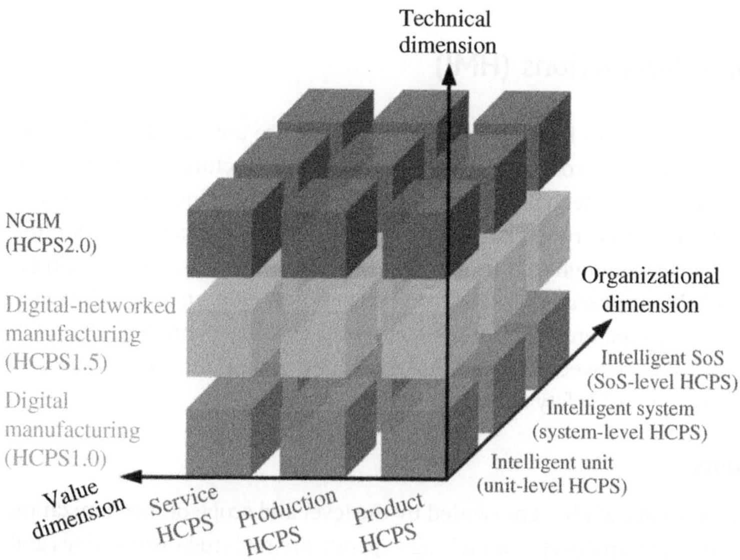

Figure 15.6 Humans' roles in enhancing the intelligence of mechatronic systems (*Source:* Zhou et al. (2019)/ with permission of Elsevier)

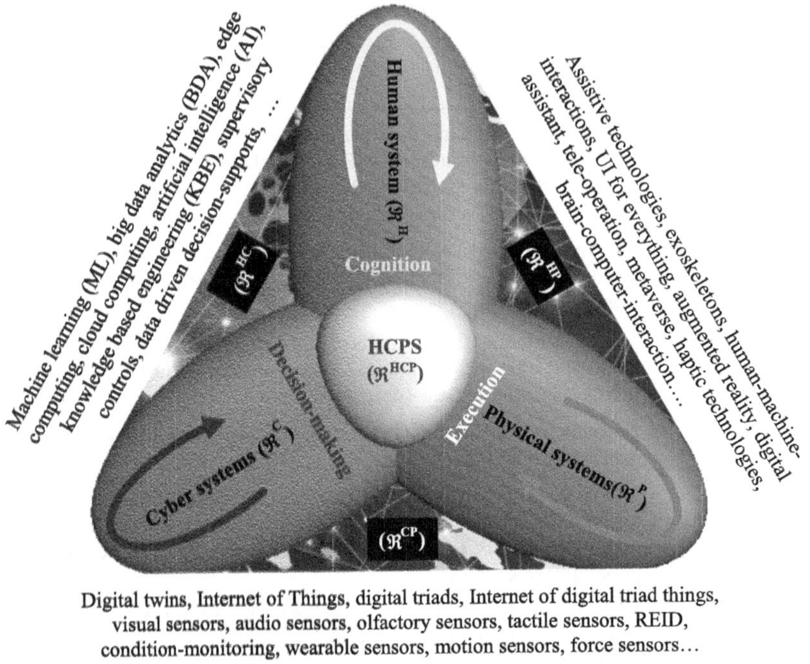

Figure 15.7 Enabling technologies for human cyber-physical systems (HCPSs)

Figure 15.7 shows some prevalent technologies to implement humans' interactions with cyber and physical systems in HCPS. For example, the technologies to support humans' interactions with physical systems (Long et al. 2020; Agote-Garrido et al. 2023) include *assistive technologies* (ATs), wearable systems, exoskeletons, HMIs, user interface (UI) for everything, *AR*, digital assistant, teleoperation, metaverse, haptic technologies, and *brain–computer interaction* (BCI). In the following sections, three case studies will be developed to illustrate the technical developments on HMI, AT, and BCI for HCPSs, respectively.

15.4 Human–Machine Interactions (HMI)

Intelligent machines that collaborate with humans in an open workspace are called *collaborative robots, or cobots.* Differing from industrial robots that are placed in well-structured environments, cobots must be equipped with additional mechanisms to assure humans' safety since the workspace is shared by machines and humans. It is critical in manufacturing since an industrial robot often carries an object or tool that is heavy enough to hurt people when contact occurs. This section (1) covers the concepts of HMI, collaborations, and standardizations for safety assurances; (2) discusses the applications and functional requirements of HMIs; (3) examines existing technologies and the limitations to assure safe interactions between humans and machines; and (4) introduces two case studies to exemplify the development of safety assurance of HMI for HCPS.

15.4.1 Collaborative Robots

The evolution of mechatronic systems can be represented by the level and scope of mechanization and automation where labor and humans' decision-making supports are replaced by machines and computers, respectively.

With the advancement of sensing and information technologies, computers are playing more and more important roles in freeing humans from repetitive, boring, or dangerous tasks. However, it becomes common sense that machines and robots cannot fully replace humans in many applications; it is advantageous to keep humans in the closed-loop controls of systems since humans still show superior capabilities in dealing with uncertainties, dynamics, and randomness. By comparing machines, humans have our advantages for intuitive decisions and enhanced responsiveness, agility, and adaptability. Cobots are specialized to support humans' interactions with machines in a shared workspace. In general, a cobot is designed to utilize humans' agility, flexibility, and adaptability in an automated system. The cobot concept is closely associated with HMIs and has gained popularity recently. For example, Lloyd (2019) conducted a statistical study and concluded that the market share of cobots increased from 3% in 2018 to 34% in 2025, and the total market value of cobots increased from $9 billion in 2018 to $12 billion in 2025 with a rate of annual growth of 57%.

With the oversaturation of global manufacturing, companies are facing the challenges of making products with a small volume but large variety at the lowest possible cost. As shown in Figure 15.8, the *market share* of a company relies on the *varieties* and *volumes* of products, and the company's performance can be evaluated by the *utilization rate* of available *manufacturing capabilities*. A high rate corresponds to the company's high potential to increase its market share. The utilization rates are applicable to all manufacturing resources: (1) conventional capabilities including labor for low production, *rigid automation* for mass production, and *cellular manufacturing* for mass customization and (2) advanced capabilities including *smart/soft automation*, and *HMI* to improve system flexibility for more product varieties by machines and humans collaboratively.

Figure 15.8 shows that companies must select appropriate system paradigms to fully utilize manufacturing capabilities. (1) *Rigid automation* is used in mass production to make products in a high volume but with few variants. (2) *Manual operations* are used in a craft system paradigm to make products in small volumes but large variants. (3) *Batch production* or *reprogrammable systems* are

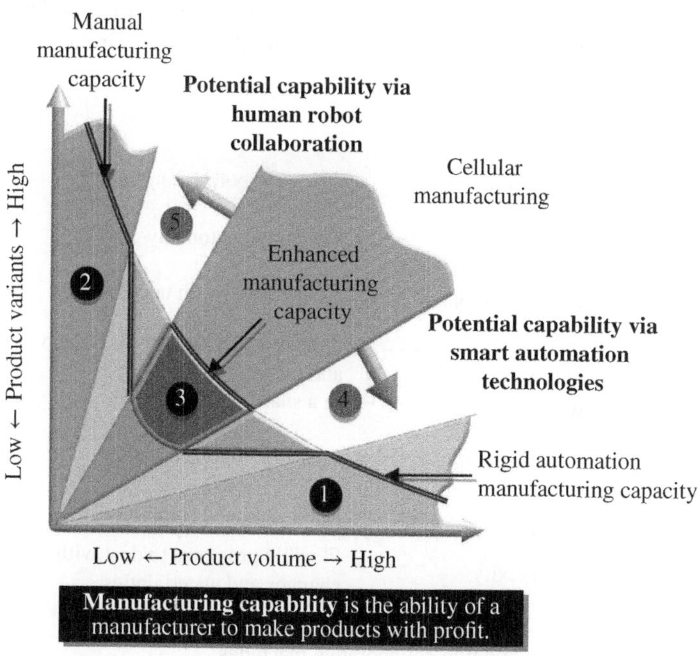

Figure 15.8 HMI as an effective way to gain competitiveness

used to make the products with mixed variants and volumes. (4) *Smart automation* technologies are used in mass customization to make more products with smaller volumes; typical enablers of smart automation are DT-I, AR, BDA, and IoT. (5) HMIs and relevant technologies are used in *flexible manufacturing* to make products with more variants in shortened lead times and at reduced costs. In summary, HMI makes complex manufacturing or assembling processes practical that could never be fully automated. HMI allows cobots to outperform conventional robots for enhanced decision-making support.

15.4.2 Types of HMIs

When a machine shares the workspace with people, it must be capable of sensing and identifying environmental changes; the machine needs advanced control to respond to detected changes adequately; for example, to avoid potential interference when a human approaches the machine. From this perspective, a machine in HMI is more advanced since additional mechanisms are incorporated to detect changes and assure safety.

Figure 15.9 gives a comparison of traditional and collaborative machines in the aspects of structures, operating conditions, movements, controls, communications, workspaces, and complexity (Guiochet et al. 2017). *For controls*, the program in a non-collaborative machine dominates and has the authority to accept an operator's inputs in a real-time mode. *In the operating phase*, a collaborative machine allows a human operator to access the shared workspace. *For the freedom of movement*, a collaborative machine works with a human interactively; in other words, human's interactions are in the closed control loop of the collaborative machine. *For the human's distance from a machine*, a collaborative machine and human share a workspace, while a non-collaborative machine is isolated in its own workspace. *For communication*, HMI is established to support human and machine interactions interactively. *For overall structures,* a non-collaborative machine can be rigid, heavy, and with a high workload; a collaborative machine is usually light-duty and sometimes the structure can be compliant to reduce damage when a collision occurs. *For complexity*, a collaborative machine requires advanced capabilities to sense, detect, and respond to changes in tasks or environmental factors.

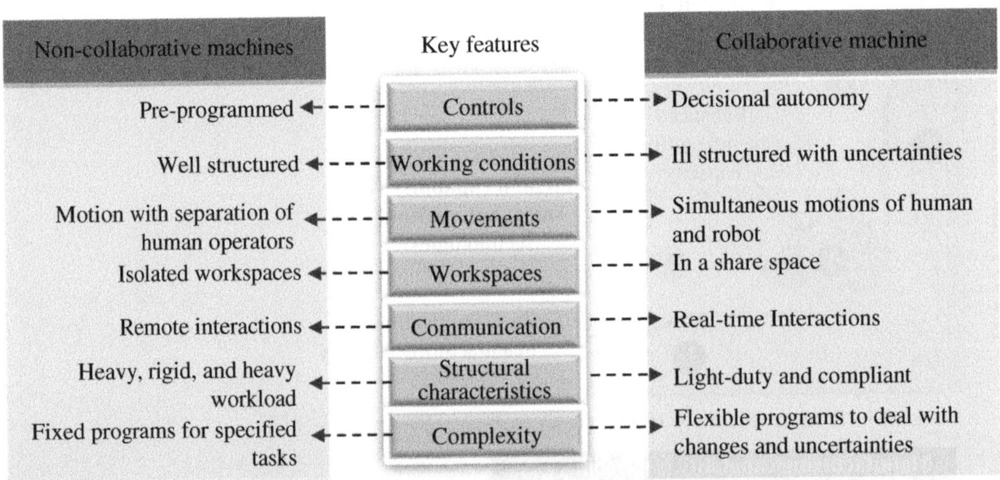

Figure 15.9 Comparison of collaborative machines and non-collaborative machines

Table 15.4 Modes of *human–machine interactions* (HMIs)

Safety by separation:	A machine stops working when a human is at a safe distance from the workspace, or a machine is completely separated from human operators. A human is not allowed to reach the machine's workspace, and no interaction is supported for collaboration.
Safety by supervision:	The human takes an active role to monitor, supervise, or control the machine, and the human can make contact with the machine with precaution.
Harmless collaboration:	The human and machine take actions simultaneously in the shared workspace without the safety risk to human or machine.

Mostly, safety assurance is the key for HMIs. The safety of HMI is assured by incorporating safety devices such as instrumentations to measure distances or forces of objects and protective elements for passive compliance and overload protection. For example, *a force monitor* can measure the force at a contact of interest and stop a cobot when the force exceeds the safety limit. *Passive compliance* is designed to absorb impact energy when the joint force goes beyond the limit. An *overwork detection* releases the overload simply by reversing the motion direction. From the perspective of safety assurance, a machine can work with humans in one of the following three modes in Table 15.4.

To quantify the performance of HMI, different measures such as *safe distance, risk index, state component index, performance level*, and *cost* can be used in the design of HMIs.

15.4.3 Collaborative Machines in Manufacturing

From now on, collaborative machines refer alternatively to collaborative robots or cobots. The study on cobots is relatively new in comparison with those in the fields of conventional robots. The first commercially available cobot was the universal robot (UR), which occurred in 2008 (UR 2024a). However, the applications of cobots have been explored broadly; there are more than 40 cobot series of cobots from lead robotic manufacturers including FANUC, ABB, Rethink Robotics, Kuka, Feston, Espon, MGS, Motoman, Pmron, and MGS Machine. Moulieres-Seban et al. (2020) estimated that the market size of cobots was growing at an annual rate of 45% from 2019 to 2025 from \$649 million in 2019. Cobots were expected to be widely adopted in *small or medium-sized enterprises* (SMEs) to automate manufacturing or assembly processes. Table 15.5 gives some examples of applications where cobots were used to support HMI in open environments.

15.4.4 Critical Requirements of Cobots

In manufacturing, cobots are used to assist humans in improving the flexibility and productivity of manufacturing or assembly processes. Other than the FRs for the processes themselves, the requirements for safety and trust assurance, ergonomics, and the flexibility to deal with the changes are essential to HMIs. By comparing a non-collaborative robot that was pre-programmed, a cobot is required to sense and respond human's interactions by means of vision, sound, or the recognition of gazes and gestures. Additional FRs of cobots for safe, smooth, and effective HMIs were discussed by many scholars (Kragic et al. 2018; Villani et al. 2018; Robotiq 2024), and Table 15.6 summarizes these requirements.

Table 15.6 shows that by comparing non-collaborative robots, cobots are required to meet additional requirements of safety assurances. This is due to the fact that humans and robots interact

Table 15.5 Cobots' applications in HMIs

Application types	Cobots' applications
Manufacturing processes	• Wadelar et al. (2018) introduced a work cell where a cobot was used to apply sealants in fuel tanks. • Antonelli and Astanin (2016) developed a welding cell where humans and robots worked together in operations. • Realyvásquez-Vargas et al. (2019) proposed a generic procedure for designing HMIs for manufacturing applications. • Khalid et al. (2018) explored the feasibility of using a cobot as s payload robot with additional features for safety assurance. • Bilberg and Malik (2019) adopted an object-oriented method to model a work cell with cobots. • Villani et al. (2018) discussed enabling technologies to support cognitive and physical collaborations of humans and machines.
Assembly processes	• Wang et al. (2017) investigated existing cobots that were used in assembly automation and developed a new assembly system to verify the performance of cobots. • Michalos et al. (2015, 2018) discussed the functional requirements of cobots in the aspects of controls, safety assurances, and specifications in three assemblies of three components, that is, rear suspensions, dashboards, and refrigerators. • Bobka et al. (2016) evaluated different safety-assurance strategies where genetic algorithms were used to detect collisions with reduced computation in a virtual environment. • Wang et al. (2019) proposed to adopt voice processing, gesture recognition, and context awareness identifications to support symbiotic HMI in collaborative assemblies.
Materials-handling	• Cobots used at DHL (Bonkenburg 2024) possessed some critical functions similar to human operations, that is, sensing systems like *eyes* for observing things, manipulating systems like *hands* for handling objects and transportation systems like *feet* for locomotion, and controlling systems like *brains* for decision-making supports. It was found that only 15% of warehouse transportations were mechanized; therefore, cobots would have great applications in material handling. • Gil-Vilda et al. (2017) used a cobot to reduce changeover time in a restricted workspace in lean production.
Personal assistance	• Levratti et al. (2019) used a cobot to assist an operator in a tire work cell; it was capable of balancing tires, lifting components, and rotating tires. By collaboration, the operator could concentrate on the tasks where his/her skills and expertise were highly valued. • The cobot by Khalid et al. (2016, 2018) assisted people with cognitive disabilities in assembling complex products; the cobot served as a guide to highlight the connecting points.
Inspection	• Magrini et al. (2024) introduced a cobot in inspection, and the user collaborated with the cobot to evaluate the results. The interaction was enabled by recognizing the user's gestures.
Human-guided operations	• Stefan et al. (2024) surveyed the cobots in medical applications including human-assisted surgeries. • Bi et al. (2008) designed a cobot to move heavy medical examinations in emergency facilities. The driving force was from human operators, and the cobot guided the path passively.
Agricultural processes	• Cobots were used to improve productivity and profitability in agriculture. Cobots were competent to assist farmers in harvesting, cutting, grafting, detecting, and transplanting in greenhouse businesses (Vasconez et al. (2019).
Others	• Cobots were used in construction for material handling, mixing, joining, and shaping of timber, concrete, and metal materials (Afsari et al. 2024). • Cobots were widely explored in other applications such as *entertainment, military environments, angiography, healthcare*, and *research and education* (Beaupre 2024)

Table 15.6 Additional FRs of cobots for safety assurances

Criterion	Non-collaborative robots	Collaborative robots
Safety	Safeguards by separating robots from human's reaches	Safety mechanisms to ensure safe collaborations in a shared workspace
Proximity to humans	Prohibited	Allowed when the procedure of safety assurance is followed
Footprint of robot	Large for structured protection	Small for close collaboration
Skills for programming	Sophisticated	Intuitive
Payload	Medium to high	Low to medium
Utilization of workspace	High	Limited due to obstacles
Productivity	High	limited
Product volume	Used for products with a high volume	Used for products with lower volume and high variants
Adaptability and flexibility	Hard automation by programs	Soft automation by interactions

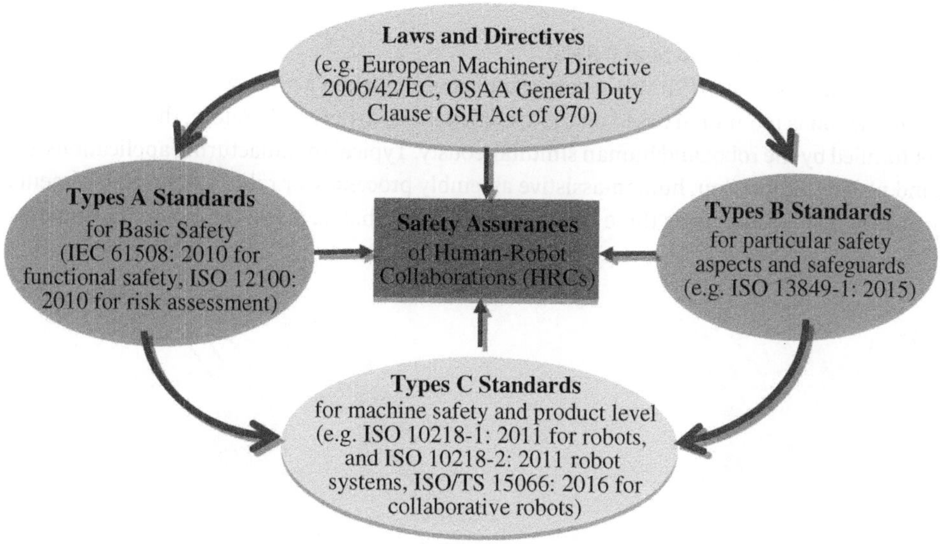

Figure 15.10 Safety assurance standards and regulations for human and machine collaboration (*Source:* Wolf and Schunk (2018, 2024))

with each other in an open environment and potential hazards must be prevented when humans and cobots share their workspaces. Moreover, safety mechanisms must be designed to meet corresponding industrial standards.

Figure 15.10 shows the most relevant standards related to HMIs. *First*, a cobot as a general machinery must meet the safety standards such as the European Machinery Directive (2006/42/EC) and the Occupational Safety and Health Administration (OSHA) General Duty Clause OSH Act of 970. *Second*, mechatronic components in a cobot must meet type A standards of functional modules, for example, functional safety by IEC 62508:2010, risk assessment by ISO 12100:2010, and safety devices and active controls by ISO 12100. *Third*, cobots' applications must meet type B standards; for example, manufacturing applications must meet the standards of ISO 11161:2006. *Finally*, cobots as products must meet type C standards; for example, robotic work cells must meet the standards of ISO 10218-2:2011 and 10 218-1:2011 (Wolf and Schunk 2018, 2024).

Note that the standards for human–robot collaborations (HRCs) are available in ISO 10218-1:2011. It specifies the safety requirements related to safeguard zones, workspaces, collisions, object detections, and velocity and force monitoring (Koch 2019; Beaupre 2024). ISO 10218-1:2011 was supplemented by ISO 15066:2016 to specify the speed limits of cobots in human–robot interactions (HRIs) (Land 2024). Since a cobot directly interacts with humans, ergonomic considerations are also important even though there is no corresponding standard. Maurice et al. (2017, 2019) assessed cobots from the perspective of ergonomics, with a set of performance indicators to represent the biomechanical requirements of human operations.

15.4.5 Safety Assurance Mechanisms for Cobots

A cobot must be designed to prevent any interference of the cobot with humans, obstacles, tools, or any other object in its workspace. The mechanisms for safety assurance are classified into four types in ISO 10218-1/2 that are illustrated in Figure 15.11 (Afsari et al. 2024; Land 2024; Luca 2024; Magrini et al. 2024).

15.4.5.1 Safety-Rated Monitored Stop (SRMS)

In a *safety-rated monitored stop* (SRMS), a cobot stops running when a human interacts with it. The human's interaction is monitored based on the specified safety criteria. Therefore, the collaborative task is not fulfilled by the robot and human simultaneously. Typical manufacturing applications are picking and placing, inspection, human-assistive assembly processes, or robot-assistive placements of heavy objects. The SRMS needs three functional modules, that is, *a sensing system* for humans,

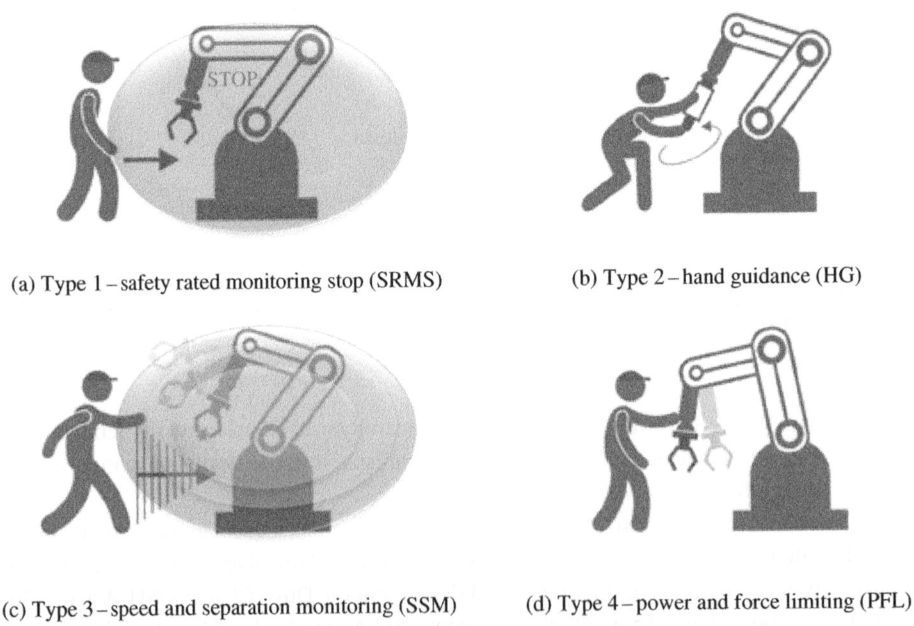

(a) Type 1 – safety rated monitoring stop (SRMS)

(b) Type 2 – hand guidance (HG)

(c) Type 3 – speed and separation monitoring (SSM)

(d) Type 4 – power and force limiting (PFL)

Figure 15.11 Four types of mechanisms for safety assurance

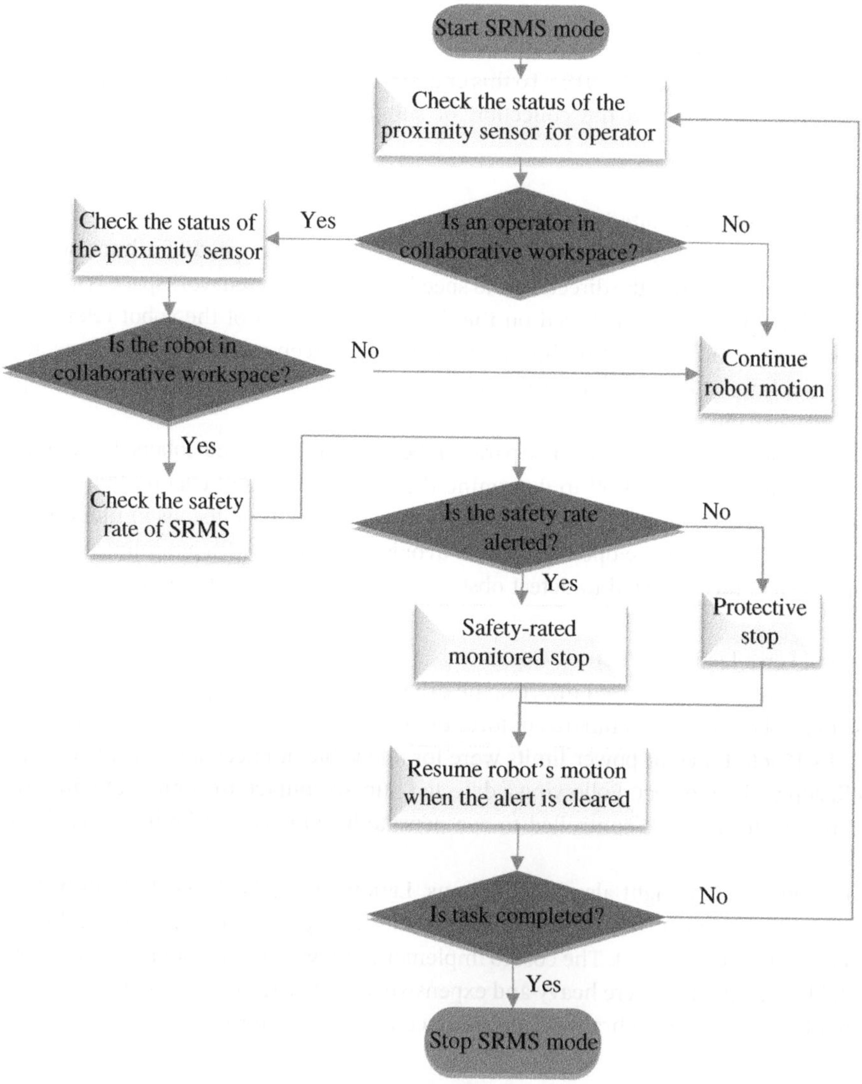

Figure 15.12 Flowchart of SRMS implementation

a sensing system for the robot, and *a monitoring system* to evaluate the safety level of HRIs. Accordingly, the cobot is controlled in one mode *continuing motion*, *resuming motion*, *protective stop*, and *safety-rated stop* (Heydaryan et al. 2018; Heydaryan 2024). Figure 15.12 shows the flowchart of SRMS implementation (Heydaryan 2024).

15.4.5.2 Hand Guiding (HG)

In *hand guiding*, a human guides the cobot by contacting, and the cobot is programmed on the fly by teaching. To facilitate teaching, the control system compensates for driving forces at joints; moreover, the control system compensates for the gravity force of the cobot in the reachable workspace.

HG needs an interactive control to achieve compliance and asymptotic stability, and the power or forces at joints are monitored and controlled to prevent overloads. For example, Raiola et al. (2024)

implemented impedance control to limit the power level of a Kuka cobot. In addition, the control system should be reconfigurable, so that human's skills can be fully utilized to program cobots at the task level directly (Schou et al. 2017, 2018). To this end, the task of a cobot was defined as a set of skills, and the control program was the collection of sub-programs generated by kinesthetic teaching.

15.4.5.3 Speed and Separation Monitoring (SSM)

In *speed and separation monitoring* (SSM), the relative distance and the speed of a human with respect to a cobot are used to control the direction and speed of cobot. A shared workspace is divided into regions with different safety levels based on the distance and speed of the cobot relative to humans. The cobot runs at full speed if no human is within the green zone, the cobot runs at a reduced speed when a human is approaching the robot (in the yellow zone), and the cobot stops when the human reaches the red zone.

SSM safeguards people by a safe distance of separation between humans and cobots; the criteria for productivity and safety are traded off to determine the motion speeds of cobots (Byner et al. 2019). Moreover, the controls over cobots are optimized to avoid possible collisions in motion trajectories. Zanchettin et al. (2019) developed a collision-avoidance method with a minimized travel time; vision systems were incorporated to detect obstacles in the working environment.

15.4.5.4 Power and Force Limiting (PFL)

In *PFL*, contact forces are limited to avoid injuries, the driving forces at active joins were limited at the design phase of a cobot, and the standards of force or power limits for safe collaborations were specified in ISO/TS 15066. Force or power limits were for allowable biomechanical loads when a human has a collision with an object. Following a detected human contact, the brakes at joints are activated, and forces or torques are controlled to compensate for gravity and limit contact force or power.

Other accessory components might also hurt humans. Lammle (2019) discussed the impact of inertial forces from tooling; they further developed a magnetorheological clutch to reduce collision energy by the end-effector on the cobot. The cost to implement PFL was another identified concern since conventional brakes at joints were heavy and expensive. Lee et al. (2018) developed an elastic solenoid latch in a brake to improve braking performance in cost and lifespan.

15.4.6 Cobotic Systems

Many cobots are commercially available as standalone mechatronic products. In real-world applications, cobots must be integrated with other mechatronic functional modules to fulfill system-level tasks collaboratively. Figure 15.13 shows different levels of HRCs when cobots are centralized in HMI (Bi et al. 2018; Omron 2024). Even a cobot can used as a standalone machine like a traditional robot, and its superior features can fully excel only when it collaborates with humans.

In the HCPS described by Khalid et al. (2018), as shown in Figure 15.14, safety assurance is also applied to the cyber system of cobot because data exchanges and communications in an information flow are vulnerable to cyber-attacks. Therefore, an advanced security framework is required to deal with both physical and cyber threats to cobot systems. Yao et al. (2018a, 2018b) indicated that the combined strengths of a robot's productivity and human cognition empowered industrial robots for more complex and dynamic tasks through HMI.

Many researchers have contributed to integrating cobots as HCPSs. Brown and Pierson (2018) developed a system with path planners, simulators, graphic user interfaces, and external sensors

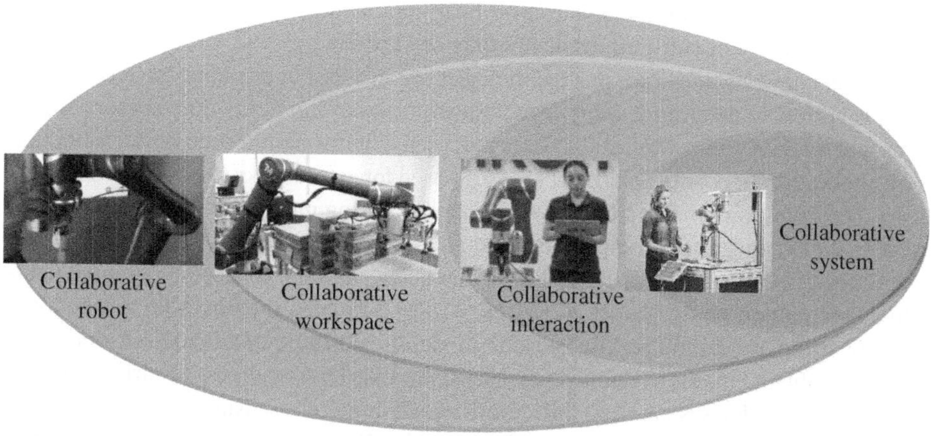

Figure 15.13 Four collaboration levels of cobots are in systems (*Source:* Omron Industrial Automation EMEA / https://m.youtube.com/watch?v=WuJtvbUwxe0 / Last accessed on July 19, 2024)

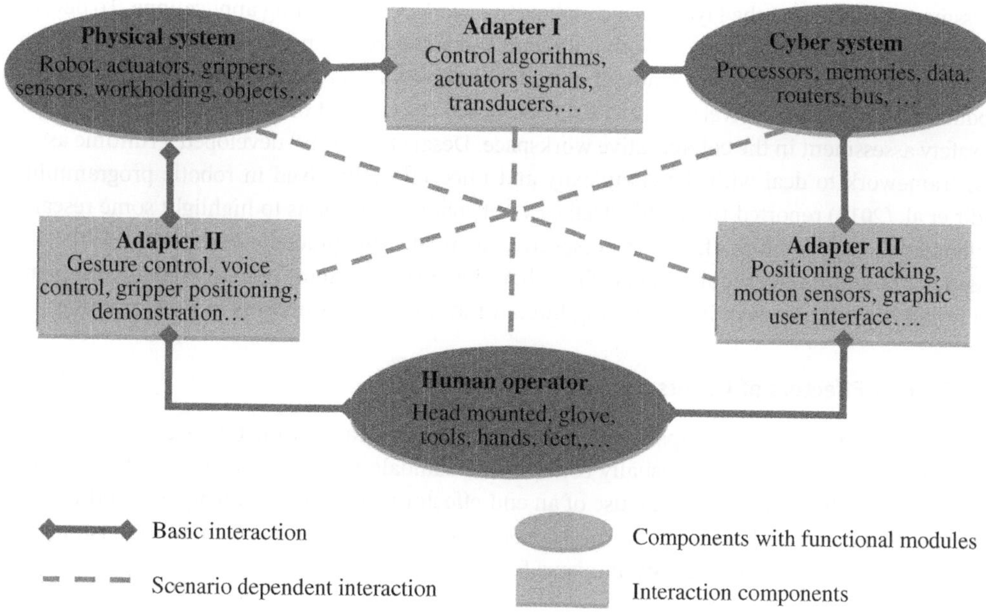

Figure 15.14 Cyber safety on HCPS

for cobots to support HMI in HCPS. To make an HCPS successful, three important factors were (1) sufficiency of real-time data, (2) effective safety assessment, and (3) adaptive system control (Oberc et al. 2019). Gustavsson et al. (2018) classified existing mechanisms for HMIs, and the criteria for classifications included (1) the ways HMI were accessed (e.g., wearable, portable, wireless, and handhold), (2) the types of valid data (e.g., sensing data, commands, warnings, demos, and guides), and (3) the ways to acquire real-time data (e.g., vision or glove sensors for *gesture recognition,* head-mounted or distant microphones for *speech recognition*, articulated devices, probes or end-effectors for *haptic control* head-mounted or stationary devices for *gaze recognition,* and stationary or soft-buttons as *joysticks or buttons*).

Many researchers have advanced sensing technologies for HCPSs. For example, the torque/force data at joints is critical to determine safety level; however, mounting heavy-duty sensors at harmonic drives will weaken joints. Min et al. (2019) proposed a lightweight torque sensor with high rigidity as an alternative for cobots. Solanes et al. (2018) used force feedbacks to prioritize the transporting tasks. The safety was ensured by non-conventional sliding mode. Safeea and Neto (2019) and Safeea et al. (2019) integrated a range finder and an inertial measurement unit on a cobot and human, respectively, to estimate spatial closeness dynamically, and the concept of potential fields was used to adjust motion paths. Yao et al. (2018a, 2018b) estimated the forces on a cobot by a human operator using a dynamic and quasi-static model to substitute additional sensors in the application with ambient vibration. Halme et al. (2018) discussed the readiness of various sensing technologies including vision systems for HMIs; even though numerous sensing solutions are available, a very limited number of them were successfully used in HCPSs. Trimble et al. (2019) acquired reliable real-time data in collaboration by presenting new methods to condition and process positional and torque data at the wrist of a cobot. Vogel et al. (2017) relied on external cameras to define a dynamic safe workspace in cooperative processes; they used speed and separation monitoring methods to separate humans and cobots in the workspace. Wadelar et al. (2018) conducted a risk assessment in designing the layout of a collaborative work cell for sealing applications. To develop an integrated collaborative system, both hardware and software components should be reconfigurable to deal with task changes; Schou et al. (2017) selected assistive components based on given products and processes. Polverini et al. (2017) proposed the concept of the kinetostatic safety field for safety assessment in the collaborative workspace. Desai et al. (2019) developed a runtime assurance framework to deal with the complexity and uncertainty involved in robotic programming. Kadir et al. (2018) reported three industrial cases of cobot applications to highlight some research challenges and opportunities from the perspective of engineering practice. Zaatari et al. (2019) discussed the programming requirements of collaborative robotic systems in industrial environments and found the difficulties of incorporating human intuition and awareness in programming.

15.4.7 End-Effectors of Cobots

End-effectors are essential components for any robotic applications including cobots. Unfortunately, an end-effector does not usually come with a standalone cobot in which safety assurance mechanisms are integrated. The safe use of an end-effector is often assured when the end-effector is designed and mounted in HCPS. In this section, two case studies are presented to illustrate the development of safety assurance mechanisms for end-effectors.

15.4.7.1 Affordable Force Monitoring

Grippers are the most widely used end-effectors. Grippers are selected based on the types, geometries, and dimensions of manipulated objects, and Figure 15.15a classified commercial grippers based on the criteria of *types of motions*, *contacts*, *actuations*, *sizes*, *gripping forces*, and *configurations*. To ensure the safe use of a gripper in a cobot, gripping forces must be monitored to understand the contacting condition between the robot and the human. However, commercial medium- and low-cost grippers are open-loop functional components without force sensing. It is desirable to develop sensing solutions to measure and monitor forces for decision-making support in HCPS. Figure 15.15b shows an example of developing a force sensor for a gripper. To reduce cost and weight, strain gauges are adopted to measure the strains, and the data processing system converts

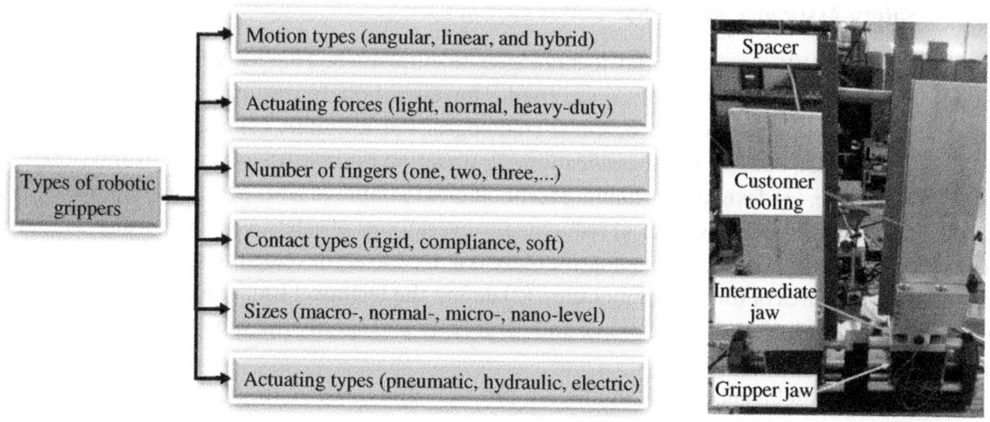

(a) Grippers for cobots (b) Real-time force sensing

Figure 15.15 Gripping with force sensing for HMI (*Source:* Bi et al. (2018)/with permission of Elsevier)

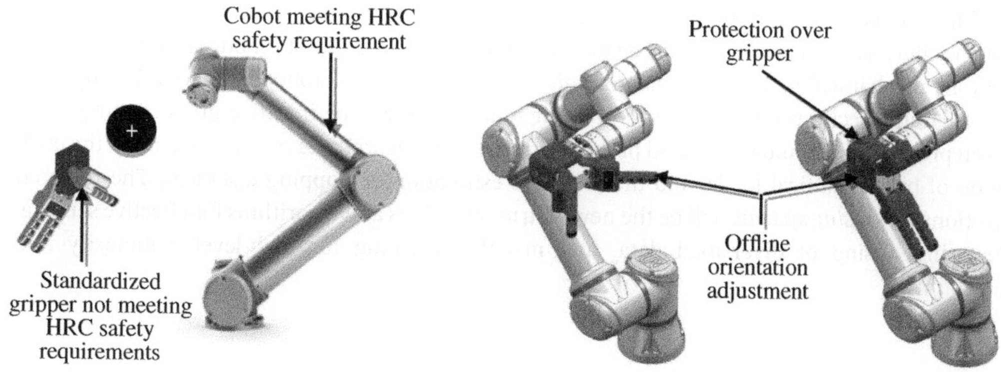

(a) Industrial gripper as an accessory device in HRC (b) Protected gripper with additional posture change

Figure 15.16 Protection of accessory components in HCPS. (a) Industrial gripper as an accessory device in HRC. (b) Protected gripper with additional posture change

strains into gripping forces by external objects. The sensing system is prototyped, verified, and validated with satisfactory force measurement (Bi et al. 2018).

15.4.7.2 Ergonomic Protection of Grippers

A gripper on a cobot must match the safety and protection level of the cobot itself. It is impractical to consider the safety assurance of grippers when a cobot is designed, as grippers are customized to specific applications. One alternative way is to build the protection after a gripper has been chosen. Figure 15.16a illustrates a case where a standard gripper GRB-32 (PhD 2024a, 2024b) is adopted as an end-effector on a universal robot (UR 2024b). Figure 15.16b is an integrated cobotic system where new protection is made to cover all of the undesirable features, which cause potential harm to humans when contact is made. This specific design also incorporated a new feature, which allows adjusting the direction of gripping operations in the mount.

15.4.8 Safety Assurance in HCPSs

HMIs, HRIs, and HRCs are essential technologies to implement HCPSs; while safety must be assured when HMIs, HRIs, and HRCs are used. However, how to ensure the safety of human and cobots still pose a number of technical challenges. For example, (1) acquiring, processing, and fusing diversified data for risk classification; (2) updating the control to avoid any interference in a real-time mode; (3) developing enabling technologies to improve HMI performance, especially, workloads and working speeds; and (4) reducing the overall cost of safety assurance features. From the perspective of safety assurance standards, there is a need to (1) develop the standards in expressing the safety features of functional modules, (2) define the technical implementations to enforce corresponding guidelines and regulations, and (3) classify and specify the ways of the recognitions for hazard scenarios.

To improve the intelligence levels of modern products and systems, more and more automated components such as industrial robots, end-effectors, and smart sensing systems will be used in open and ill-structured environments where humans present and collaborate with automated systems. Despite some industrial standards being available as the guidance for the safety assurance of HRCs, a few commercial systems have been reported for HRCs in the manufacturing sector. Safety assurance must be implemented to cover all of the components in collaborative robotic systems to eliminate hazard factors for humans. In designing and building a cobot or a collaborative robotic system, safety assurance factors must be incorporated into the designs of robotic structures, motion controls, mechanical systems, virtual commissioning, and human–robot interfaces. Examples of the research directions on *system programming and control* will be intuitive programming, task-driven programming, and skill-based programming, which incorporate risk management, the evaluation of biomechanical loads, and the real-time estimation of stopping distances. The research directions on *sensing systems* will be the new instrumentations and algorithms for effective sensing, processing, fusing of diversified data, and machine learning for high-level complexity and uncertainty.

15.5 Example of Assistive Technologies

ATs are another application of collaborative robots (e.g., cobots) where robots are used to assist users in accomplishing specific tasks collectively. This section presents a case study where a new cobot is designed, modeled, and simulated as an example of HCPS as an AT.

When a cobot is used to assist users in an open environment, the cobot prioritizes to assure users' safety and it is equipped with user-friendly interfaces for users to collaborate intuitively. In this section, a cobot is designed to guide and assist a user in moving a heavy object in a given trajectory. The proposed cobot uses omniwheels instead of steering wheels to alleviate the difficulties to (1) follow a trajectory with a curvature larger than that of the base platform, (2) mount encoders on steering wheels due to self-spinning of the wheels, and (3) quarantine dynamic control performance since it is purely kinematic control.

15.5.1 Cobots in Healthcare

The first cobot in healthcare was invented at Northwestern University by Peshkin and Colgate (1996). It was used as a guide to assist doctors in moving heavy medical equipment along a predefined path; the moving power was from doctors. The cobot was passive in the sense that the power to move an object was from operators. The wheels of a cobot were actuated to control the direction of movement along the path. It was typical collaborative system, as the cobot worked with doctors in a shared workspace (Paula 1997). The cobot allowed doctors to move an object safely and effectively at their will (Peshkin et al. 2001; Jabre et al. 2002).

The cobots in healthcare were primarily for two purposes: (1) assisting caregivers in material handling subjected to physical constraints, and (2) assisting patients or disabilities in transportation, tasking, or rehabilitation. Peshkin et al. (2000) developed a variety of cobots for materials handling, with some commercially available products (Stanley 2003). Hodgson and Emrich (2002) and Townsend et al. (2004) patented a cobot for surgeons to move heavy equipment such as an X-ray machine in which the surgeon was in full control of the moving path of the equipment. Ulrich and Borestein (2001) developed a smart guide cane to navigate visually impaired patients, and Nejatbakhsh et al. (2005) introduced a passive walker to assist patients in rehabilitation.

Most mobile cobots used steering wheels; the wheels roll freely, which is different from those on a conventional mobile robot. Therefore, the control of a cobot was more complicated than a conventional robot. The research groups led by Peshkin developed the most comprehensive control models for tricycle cobots where a path was parametrized and the speeds of steering wheels were measured to determine control variables accordingly (Peshkin and Colgate 1996; Wannasuphoprasit et al. 1997; Wannasuphoprasit 1999; Gillespie et al. 2001; Peshkin et al. 2001). The number of wheels in a mobile cobot was 2, 3, or even more. For example, the cobot by Ulrich and Borestein (2001) had two wheels, and it was used to assist visually impaired patients in walking. However, using steering wheels in a cobot might cause some issues including (1) the difficulty in following a trajectory with a curvature larger than that of the base platform, (2) the difficulty of mounting encoders on steering wheels because of self-spinning, and (3) the difficulty to quarantine the dynamic control performance, as it relied purely on kinematic control.

To overcome the aforementioned drawbacks, omniwheels were used to replace steering wheels, and the motion direction of the cobot was controlled by applying braking torques instead of steering a set of conventional wheels. Omniwheels were adopted to improve the flexibility of directional changes in many robotic systems. For example, Nejatbakhsh et al. (2005) designed an assistant walker in which four omniwheels were used to facilitate directional change, while their main interests were the system identification and dynamic characterization for specific patients. In the rest of the section, a cobot with three omniwheels was proposed; kinematic, dynamic, and control models were developed; and the feasibility of the proposed cobot was verified via simulation.

15.5.2 Conceptual Design of Cobot

Figure 15.17 shows the conceptual design of the proposed cobot, which consists of three omniwheels that were placed symmetrically. Each omniwheel was equipped with a motor to apply a braking torque for directional control. The sensing system at the center of the cobot tracked the location and orientation of the cobot. The driving force was applied by the user at the handler. The force sensor embedded in the handler measures the external force along six *degrees of freedom* (DOF).

Figure 15.18 gives the parametrized model of the proposed cobot where $\{O_w - X_wY_w\}$ is the world coordinate system; $\{x_c, y_c, \theta_c\}$ is for the position and orientation of the cobot to be controlled; $\{O_i - X_iY_i\}$ is local coordinate system attached on the cobot where O_i is the center of the ith omniwheel. The center of the ith omniwheel was denoted as $\{x_i, y_i, \theta_i\}$ where $i = 1, 2,$ and 3; in $\{O_c - X_cY_c\}$, the location of the ith omniwheel is determined by L_i and α_i; m_p and m_i are the masses of the platform and the ith omniwheel; I_p and I_i are the inertias of the platform and the ith omniwheel, respectively.

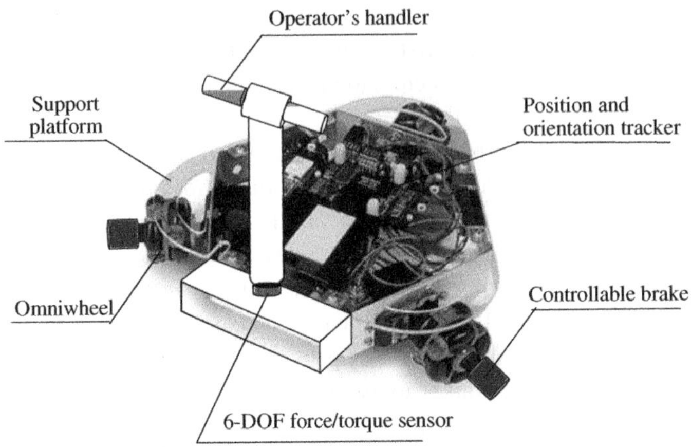

Figure 15.17 Design of cobot with three omniwheels

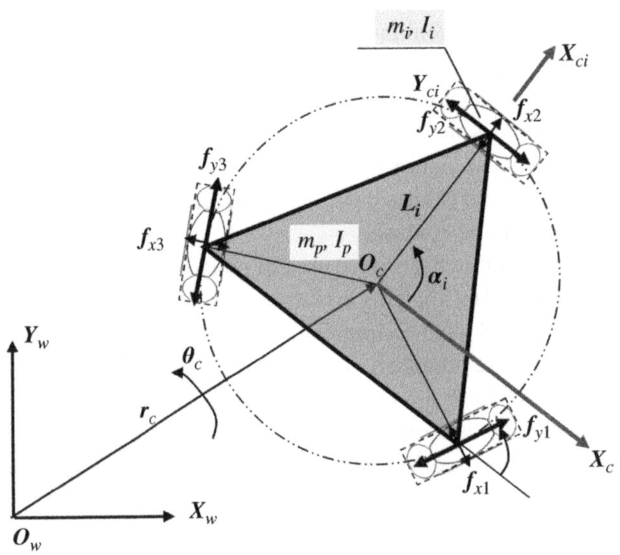

Figure 15.18 Parametrized cobot model

15.5.3 Kinematic Model

A kinematic model describes the motion relations of the omniwheels and the cobot. A kinematic model is either a forward kinematic model or an inverse kinematic model. Here, the forward kinematic model is developed to correspond a prescribed path of cobot to the motions at three omniwheels, as shown in Figure 15.18. The motion of the cobot is denoted as $(\boldsymbol{r}_c, \theta_c)$, where $\boldsymbol{r}_c = [x_c, y_c]^T$ and θ_c is the rotation angle along the z-axis in $\{O_w - X_w Y_w\}$.

A prescribed trajectory is defined as:

$$x_c = x_c(t), y_c = y_c(t), \theta_c = \theta_c(t) \quad \text{or} \quad x_c = x_c(s), y_c = y_c(s), \theta_c = \theta_c(s) \tag{15.2}$$

where t is the time variable and s is the arc length along the prescribed trajectory; $\boldsymbol{r}_c = [x_c, y_c]^T$ is the position of the reference origin of the cobot; θ_c is the orientation of the cobot platform.

The velocities and accelerations of cobot with respect to the time t are defined as:

$$\dot{x}_c = \frac{dx_c(s)}{ds}\dot{s}, \dot{y}_c = \frac{dy_c(s)}{ds}\dot{s}, \dot{\theta}_c = \frac{d\theta_c(s)}{ds}\dot{s} \tag{15.3}$$

$$\left.\begin{aligned} \ddot{x}_c &= k_{x_c,1} + k_{x_c,2}\ddot{s} \\ \ddot{y}_c &= k_{y_c,1} + k_{y_c,2}\ddot{s} \\ \ddot{\theta}_c &= k_{\theta_c,1} + k_{\theta_c,2}\ddot{s} \end{aligned}\right\} \tag{15.4}$$

where

$$k_{x_c,1} = \frac{d^2x_c(s)}{ds^2}, k_{x_c,2} = \frac{dx_c(s)}{ds}, k_{y_c,1} = \frac{d^2y_c(s)}{ds^2}, k_{y_c,2} = \frac{dy_c(s)}{ds}, k_{\theta_c,1} = \frac{d^2\theta_c(s)}{ds^2}, \text{and } k_{\theta_c,2} = \frac{d\theta_c(s)}{ds}$$

Note that $\frac{dx_c(s)}{ds}, \frac{dy_c(s)}{ds}, \frac{d\theta_c(s)}{ds}$ and $\frac{d^2x_c(s)}{ds^2}, \frac{d^2y_c(s)}{ds^2}, \frac{d^2\theta_c(s)}{ds^2}$ depend only on the prescribed trajectory and they are obtained directly from Eq. (15.2) when the trajectory is given. Equations (15.3) and (15.4) also show the dependences on linear velocity $\dot{s}$ and acceleration $\ddot{s}$ along a trajectory.

15.5.4 Motion for Arbitrary Explicit Trajectory

It is common that a trajectory is defined based on a variable other than *time* or *arc length*. Therefore, the conversion between two models for the same trajectory becomes necessary. Let an arbitrary parameter for a trajectory be q, and a trajectory is defined as:

$$x_c = x_c(q),\ \ y_c = y_c(q),\ \ \theta_c = \theta_c(q) \quad \text{and } q \in [q_{start}, q_{end}] \tag{15.5}$$

where $[q_{start}, q_{end}]$ is the varying range of q for the whole trajectory.

Taking the derivations of Eq. (15.5), we get:

$$\left.\begin{aligned} \dot{x}_c = \frac{dx_c(q)}{dq}\frac{dq}{dt}, \dot{y}_c = \frac{dy_c(q)}{dq}\frac{dq}{dt}, \dot{\theta}_c = \frac{d\theta_c(q)}{dq}\frac{dq}{dt} \\ \ddot{x}_c = \frac{d^2x_c(q)}{dq^2}\left(\frac{dq}{dt}\right)^2 + \frac{dx_c(q)}{dq}\frac{d^2q}{dt^2}, \ddot{y}_c = \frac{d^2y_c(q)}{dq^2}\left(\frac{dq}{dt}\right)^2 + \frac{dy_c(q)}{dq}\frac{d^2q}{dt^2} \\ \ddot{\theta}_c = \frac{d^2\theta_c(q)}{dq^2}\left(\frac{dq}{dt}\right)^2 + \frac{d\theta_c(q)}{dq}\frac{d^2q}{dt^2} \end{aligned}\right\} \tag{15.6}$$

Moreover, the velocities and accelerations in Eq. (15.6) met the constraints on the given trajectory:

$$\left.\begin{aligned} \frac{ds}{dt} &= \sqrt{\left(\frac{dx_c}{dt}\right)^2 + \left(\frac{dy_c}{dt}\right)^2} = \sqrt{\dot{x}_c^{\,2} + \dot{y}_c^{\,2}} \\ \frac{d^2s}{dt^2} &= (\dot{x}_c\ddot{x}_c + \dot{y}_c\ddot{y}_c)/\sqrt{\dot{x}_c^{\,2} + \dot{y}_c^{\,2}} \end{aligned}\right\} \tag{15.7}$$

Since $\dot{x}_c$, $\dot{y}$, $\ddot{x}_c$, and $\ddot{y}_c$ are determined in Eq. (15.6), using these variables in Eq. (15.7) gives: Substituting Eq. (15.5) into Eqs. (15.6) and (15.7) results in:

$$\left.\begin{aligned} \frac{dq}{dt} &= \frac{ds/dt}{\sqrt{\left(\frac{dx_c}{dq}\right)^2 + \left(\frac{dy_c}{dq}\right)^2}} \\ \frac{d^2q}{dt^2} &= \left(\sqrt{\left(\frac{dx_c}{dq}\right)^2 + \left(\frac{dy_c}{dq}\right)^2}\frac{d^2s}{dt^2} - \left(\frac{dx_c}{dq}\frac{d^2x_c}{dq^2} + \frac{dy_c}{dq}\frac{d^2y_c}{dq^2}\right)\left(\frac{dq}{dt}\right)^2\right) / \left(\left(\frac{dx_c}{dq}\right)^2 + \left(\frac{dy_c}{dq}\right)^2\right) \end{aligned}\right\} \tag{15.8}$$

where $\frac{dq}{dt}$ and $\frac{d^2q}{dt^2}$ are the derivations of q with respect to time t, respectively.

15.5.5 Motions of Omniwheels

The position $\boldsymbol{r}_i$ of the ith omniwheel is modeled by taking the center O_c of the cobot as:

$$\boldsymbol{r}_i = \begin{Bmatrix} x_i \\ y_i \end{Bmatrix} = \begin{Bmatrix} x_c + L_i\cos(\theta_c + \alpha_i) \\ y_c + L_i\sin(\theta_c + \alpha_i) \end{Bmatrix} \quad (i = 1, 2, 3) \tag{15.9}$$

Taking the derivation of Eq. (15.9) with the time t gives the velocity of the ith omniwheel as:

$$\frac{d\boldsymbol{r}_i}{dt} = \begin{Bmatrix} \dot{x}_i \\ \dot{y}_i \end{Bmatrix} = \begin{Bmatrix} \dot{x}_c - L_i\dot{\theta}_c\sin(\theta_c + \alpha_i) \\ \dot{y}_c + L_i\dot{\theta}_c\cos(\theta_c + \alpha_i) \end{Bmatrix} = \begin{bmatrix} 1 & 0 & -L_i\sin(\theta_c + \alpha_i) \\ 0 & 1 & L_i\cos(\theta_c + \alpha_i) \end{bmatrix} \begin{Bmatrix} \dot{x}_c \\ \dot{y}_c \\ \dot{\theta}_c \end{Bmatrix} \quad (i = 1, 2, 3) \tag{15.10}$$

where $\{\dot{x}_c \quad \dot{y}_c \quad \dot{\theta}_c\}^T$ are calculated from Eq. (15.6).

Similarly, the acceleration of the ith omniwheel is found by taking another derivation of Eq. (15.11) as:

$$\frac{d^2\boldsymbol{r}_i}{dt^2} = \begin{Bmatrix} \ddot{x}_i \\ \ddot{y}_i \end{Bmatrix} = \begin{bmatrix} 1 & 0 & -L_i\sin(\theta_c + \alpha_i) \\ 0 & 1 & L_i\cos(\theta_c + \alpha_i) \end{bmatrix} \begin{Bmatrix} \ddot{x}_c \\ \ddot{y}_c \\ \ddot{\theta}_c \end{Bmatrix} - \begin{Bmatrix} L_i\cos(\theta_c + \alpha_i) \\ L_i\sin(\theta_c + \alpha_i) \end{Bmatrix} \dot{\theta}_c^2 \quad (i = 1, 2, 3) \tag{15.11}$$

The displacement, velocity, and acceleration of the ith omniwheel in Eqs. (15.9)–(15.11) are with respect to the world coordinate system $\{O_w - X_wY_w\}$. It is more convenient to describe the motion in the local coordinate system $\{O_i - X_iY_i\}$ associated with the ith omniwheel as:

$$\left.\begin{aligned} \begin{Bmatrix} x_i^L \\ y_i^L \end{Bmatrix} &= \begin{bmatrix} \cos\alpha_i & \sin\alpha_i \\ -\sin\alpha_i & \cos\alpha_i \end{bmatrix} \begin{Bmatrix} x_i \\ y_i \end{Bmatrix} \\ \begin{Bmatrix} \dot{x}_i^L \\ \dot{y}_i^L \end{Bmatrix} &= \begin{bmatrix} \cos\alpha_i & \sin\alpha_i \\ -\sin\alpha_i & \cos\alpha_i \end{bmatrix} \begin{Bmatrix} \dot{x}_i \\ \dot{y}_i \end{Bmatrix} \\ \begin{Bmatrix} \ddot{x}_i^L \\ \ddot{y}_i^L \end{Bmatrix} &= \begin{bmatrix} \cos\alpha_i & \sin\alpha_i \\ -\sin\alpha_i & \cos\alpha_i \end{bmatrix} \begin{Bmatrix} \ddot{x}_i \\ \ddot{y}_i \end{Bmatrix} \end{aligned}\right\} \quad (i = 1, 2, 3) \tag{15.12}$$

where $\begin{Bmatrix} x_i^L \\ y_i^L \end{Bmatrix}$, $\begin{Bmatrix} \dot{x}_i^L \\ \dot{y}_i^L \end{Bmatrix}$, and $\begin{Bmatrix} \ddot{x}_i^L \\ \ddot{y}_i^L \end{Bmatrix}$ are the displacement, velocity, and acceleration of the ith omniwheel in $\{O_i - X_iY_i\}$, respectively.

15.5.6 Dynamic Control Model

Braking torques on omniwheels were controlled to confine the cobot to move along a prescribed trajectory. The acceleration of the cobot depended on the applied force by the user and the braking torques on omniwheels. To develop a dynamic control model, the forces on omniwheels and cobot were analyzed as follows, respectively.

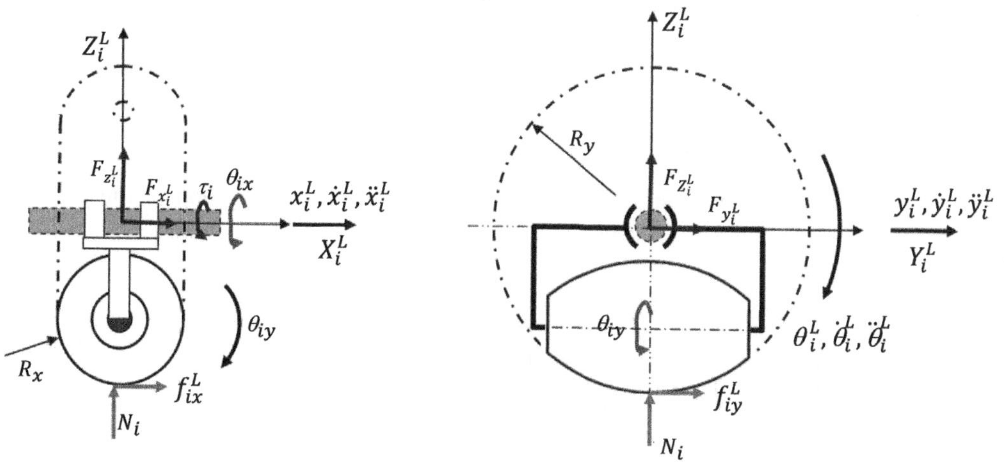

Figure 15.19 Free-body-diagram (FBD) of the ith omniwheel

15.5.6.1 Analyses of Force on Omniwheels

Figure 15.19 gives the *free-body-diagram* (FBD) of an omniwheel. The omniwheel was mounted to the cobot in the direction X^L of its primary rotation. The reacting force and torque from the cobot platform to the omniwheel are $\boldsymbol{F}\left(F_{x_i^L}, F_{y_i^L}, F_{z_i^L}\right)$ and τ_i. Besides, the ith omniwheel had an external normal reaction force N_i and friction force $\left(f_{x_i^L}, f_{y_i^L}\right)$ from ground, respectively.

Taking into consideration the force balance of the ith omniwheel yields:

$$\begin{Bmatrix} F_{x_i^L} \\ F_{y_i^L} \\ F_{z_i^L} \end{Bmatrix} + \begin{Bmatrix} f_{x_i^L} \\ f_{y_i^L} \\ N_i - m_r g \end{Bmatrix} = \begin{Bmatrix} m_r \ddot{x}_i^L \\ m_r \ddot{y}_i^L \\ 0 \end{Bmatrix} \quad (i = 1, 2, 3) \tag{15.13}$$

where $\boldsymbol{F}\left(F_{x_i^L}, F_{y_i^L}, F_{z_i^L}\right)$ is the reacting force from the cobot to the ith omniwheel; $\left(f_{x_i^L}, f_{y_i^L}\right)$ is the friction force from the ground; N_i is the normal reacting force from the ground; $(\ddot{x}_i', \ddot{y}_i')$ $(\ddot{x}_i^L, \ddot{y}_i^L)$ is the translational acceleration of the ith omniwheel defined in Eq. (15.12); and m_r is the mass of the ith omniwheel.

Taking into consideration the torque balance gives

$$\begin{Bmatrix} \tau_i \\ 0 \end{Bmatrix} + \begin{Bmatrix} f_{y_i^L} R_y \\ f_{x_i^L} R_x \end{Bmatrix} = \begin{Bmatrix} I_{x_i^L} \ddot{\theta}_i^L \\ I_{y_i^L} \ddot{\theta}_i^L \end{Bmatrix} \quad (i = 1, 2, 3) \tag{15.14}$$

where τ_i is the torque applied to x^L axis of the ith omniwheel; $I_{x_i^L}$ and $I_{y_i^L}$ are inertia moments of the ith omniwheel with respect to x^L and y^L axes, respectively; $\ddot{\theta}_{x_i^L}$ and $\ddot{\theta}_{y_i^L}$ are the angular accelerations of the ith omniwheel with respect to x^L and y^L, respectively; and R_x and R_y are the radii of the ith omniwheel and its rollers, respectively.

Equations (15.13)–(15.14) give the friction and reaction forces along its rotational axis of the ith omniwheel:

$$\begin{Bmatrix} f_{x_i^L} \\ F_{x_i^L} \end{Bmatrix} = \begin{Bmatrix} I_{y_i^L}\ddot{x}_i^L / R_x^2 \\ \left(m_r - I_{y_i^L}/R_x^2\right)\ddot{x}_i^L \end{Bmatrix} \quad (i = 1, 2, 3) \tag{15.15}$$

Similarly, the friction and reaction forces along the perpendicular direction of the ith axis become:

$$\begin{Bmatrix} f_{y_i^L} \\ F_{x_i^L} \end{Bmatrix} = \begin{Bmatrix} \left(\dfrac{I_{x_i^L}\ddot{y}_i^L}{R_y} - \tau_i\right)/R_y \\ m_r\ddot{y}_i^L + \left(\dfrac{I_{x_i^L}\ddot{y}_i^L}{R_y} + \tau_i\right)/R_y \end{Bmatrix} \quad (i = 1, 2, 3) \tag{15.16}$$

Furthermore, a slip can be prevented by satisfying the condition:

$$\sqrt{f_{x_i^L}^2 + f_{y_i^L}^2} \le \mu_{friction} N_i \quad (i = 1, 2, 3) \tag{15.17}$$

Finally, the total reacting force (F_e) from the ground to the cobot is determined from the following equation:

$$\left.\begin{aligned} \sum_{i=1}^{3} N_i \boldsymbol{r}_i + F_z \boldsymbol{r}_e = 0 \\ \sum_{i=1}^{3} N_i = (m_p + 3m_w)g + F_z \end{aligned}\right\} \tag{15.18}$$

where m_p and m_w are the masses of the platform and an omniwheel, respectively, and g is the gravity acceleration.

15.5.6.2 Analyses of Force on Cobot Platform

The cobot in Figure 15.20 was applied to uncontrollable external force F_e and torque τ_e by the user, while a force sensor was mounted to measure these forces in a real-time mode. When the platform was isolated from three omniwheels, the reacting forces $(-F_{x_i^L}, -F_{y_i^L}, -F_{z_i^L})$ and torque $-\tau_i$ from three omniwheels were "external" forces to the platform. Having the equilibrium condition of the forces over the plane of X_wY_w yields:

$$\boldsymbol{F}_e = m_p \begin{bmatrix} \cos\theta_c & -\sin\theta_c \\ \sin\theta_c & \cos\theta_c \end{bmatrix} \ddot{\boldsymbol{r}}_e + \sum_{i=1}^{3} \begin{bmatrix} \cos\alpha_i & -\sin\alpha_i \\ \sin\alpha_i & \cos\alpha_i \end{bmatrix} \begin{Bmatrix} F_{x_i^L} \\ F_{y_i^L} \end{Bmatrix} \tag{15.19}$$

Having the equilibrium torque of the platform along Z_w gives:

$$\tau_e = I_p\ddot{\theta}_c - \boldsymbol{F}_e \times \boldsymbol{r}_e + \sum_{i=1}^{3} \begin{bmatrix} \cos\alpha_i & -\sin\alpha_i \\ \sin\alpha_i & \cos\alpha_i \end{bmatrix} \begin{Bmatrix} F_{x_i^L} \\ F_{y_i^L} \end{Bmatrix} \times \boldsymbol{r}_e \tag{15.20}$$

where

I_p is the inertial moment of the platform in $\{O_c - X_cY_cZ_c\}$.
$\boldsymbol{r}_e$ is a vector from the center of the cobot to the position of interest where the force is applied.
τ_e is the external torque along by a human operator.

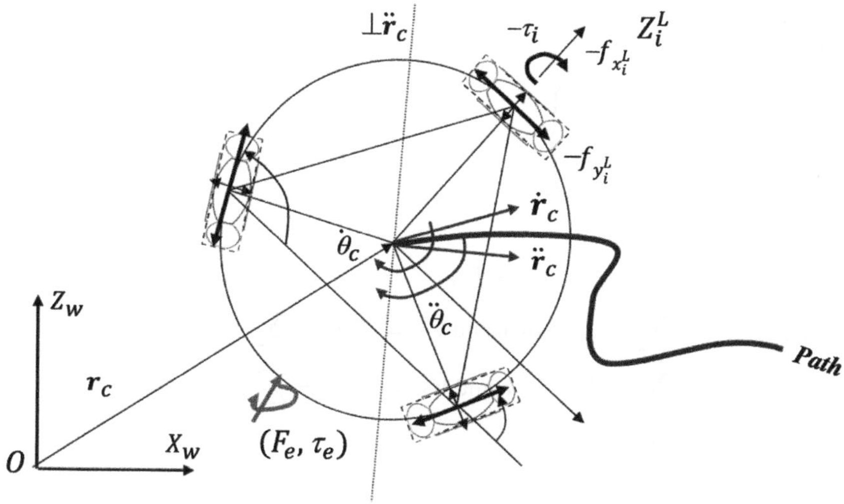

Figure 15.20 Free-body-diagram (FBD) of cobot platform

Since the results in Eqs. (15.16) and (15.17) are in their respective local coordinate systems, they must be transformed into the local coordinate systems of the cobot platform and Eqs. (15.19) and (15.20) can be reformatted as:

$$\left.\begin{aligned} U_{11}\tau_1 + U_{12}\tau_2 + U_{13}\tau_3 + U_{14}\ddot{s} = U_{15} \\ U_{21}\tau_1 + U_{22}\tau_2 + U_{23}\tau_3 + U_{24}\ddot{s} = U_{25} \\ U_{31}\tau_1 + U_{32}\tau_2 + U_{33}\tau_3 + U_{34}\ddot{s} = U_{35} \end{aligned}\right\} \tag{15.21}$$

where U_{ij} ($i = 1, 2, 3; j = 1, 2, 3, 4, 5$) are the efficiencies determined by the calculations from Eqs. (15.3) to (15.12) (Bi et al. 2008).

Solving Eq. (15.22) found the solutions of applied torque τ_i ($i = 1, 2, 3$) with respect to $\ddot{s}$ as,

$$\tau_i = k_{\tau 1,i} + k_{\tau 2,i}\ddot{s} \qquad (i = 1, 2, 3) \tag{15.22}$$

where $k_{\tau 1,1} = \begin{vmatrix} U_{15} & U_{12} & U_{13} \\ U_{25} & U_{22} & U_{23} \\ U_{35} & U_{32} & U_{33} \end{vmatrix} /\Delta$, $k_{\tau 2,1} = \begin{vmatrix} U_{14} & U_{12} & U_{13} \\ U_{24} & U_{22} & U_{23} \\ U_{34} & U_{32} & U_{33} \end{vmatrix} /\Delta$, $k_{\tau 1,2} = \begin{vmatrix} U_{11} & U_{15} & U_{13} \\ U_{21} & U_{25} & U_{23} \\ U_{31} & U_{35} & U_{33} \end{vmatrix} /\Delta$,

$k_{\tau 2,2} = \begin{vmatrix} U_{11} & U_{14} & U_{13} \\ U_{21} & U_{24} & U_{23} \\ U_{31} & U_{34} & U_{33} \end{vmatrix} /\Delta$, $k_{\tau 1,3} = \begin{vmatrix} U_{11} & U_{12} & U_{15} \\ U_{21} & U_{22} & U_{25} \\ U_{31} & U_{32} & U_{35} \end{vmatrix} /\Delta$, $k_{\tau 2,3} = \begin{vmatrix} U_{11} & U_{12} & U_{14} \\ U_{21} & U_{22} & U_{24} \\ U_{31} & U_{32} & U_{34} \end{vmatrix} /\Delta$

and $\Delta = \begin{vmatrix} U_{11} & U_{12} & U_{13} \\ U_{21} & U_{22} & U_{23} \\ U_{31} & U_{32} & U_{33} \end{vmatrix}$

15.5.6.3 Constraints to Maintain Contacts to Ground

The torques applied on the axes of omniwheels affected the distributions of the normal reacting forces from the ground, and the reacting forces must meet the condition:

$$m_p g + F_{e,z} - \sum_{i=1}^{3} F_{z_i^L} = 0 \tag{15.23}$$

where g is the gravity acceleration and is the external force by a human operator.

In the local coordinate system of the cobot platform, no rotation was allowed along x_c or y_c axis; this led to the constraint:

$$\left.\begin{aligned} V_{11}\tau_1 + V_{12}\tau_2 + V_{13}\tau_3 + V_{14}\ddot{s} = V_{15} \\ V_{21}\tau_1 + V_{22}\tau_2 + V_{23}\tau_3 + V_{24}\ddot{s} = V_{25} \\ V_{31}\tau_1 + V_{32}\tau_2 + V_{33}\tau_3 + V_{34}\ddot{s} = V_{35} \end{aligned}\right\} \tag{15.24}$$

where

$$V_{1i} = 1(i = 1, 2, 3), V_{14} = 0 \text{ and } V_{15} = m_p g + F_{e,z}$$

$$V_{2i} = r_{i,x}(i = 1, 2, 3), V_{24} = \sum_{i=1}^{3} k_{\tau 2,i} \cos \alpha_i \text{ and } V_{25} = F_{e,z} r_{e,x} - \sum_{i=1}^{3} k_{\tau 1,i} \cos \alpha_i$$

$$V_{3i} = r_{i,y}(i = 1, 2, 3), V_{34} = \sum_{i=1}^{3} k_{\tau 2,i} \sin \alpha_i \text{ and } V_{25} = F_{e,z} r_{e,y} - \sum_{i=1}^{3} k_{\tau 1,i} \sin \alpha_i$$

Combining Eqs. (15.23) and (15.24) gives:

$$F_{e,z}\begin{Bmatrix} r_{e,x} \\ r_{e,y} \end{Bmatrix} - \sum_{i=1}^{3} F_{z_i^L}\begin{Bmatrix} r_{i,x} \\ r_{i,y} \end{Bmatrix} - \sum_{i=1}^{3} \tau_i \begin{Bmatrix} \cos \alpha_i \\ \sin \alpha_i \end{Bmatrix} = 0 \tag{15.25}$$

The normal reacting force is calculated as:

$$F_{z_i^L} = k_{z_1^L} + k_{z_2^L}\ddot{s} \qquad (i = 1, 2, 3) \tag{15.26}$$

where $k_{z_1^L}$ *and* $k_{z_2^L}$ are two coefficients derived from Eq. (15.25). It must be ensured that $F_{z_i^L}$ be positive to contact the ground. Moreover, Eq. (15.17) for the no-slip condition must be verified once $(f_{x_i^L}, f_{y_i^L})$ are found based on $\ddot{s}$.

15.5.6.4 Strategies of Cobot Controls

Differing from a cobot with steering wheels whose undesired motions were confined by steering wheels, a cobot with omniwheels was free to move along its sideways if an external force was applied in this direction. However, the motion direction of the cobot was controlled by the braking torques applied on the axes of omniwheels.

Figure 15.21 shows the schematic of cobot controls. The output of control was the required acceleration of the motion along the planned trajectory, and the inputs of control included:

1) The prescribed trajectory could be revised based on real-time feedback.
2) The detected velocities and accelerations of the cobot.
3) The measured force and torque applied by a human operator.

The established models showed that motion accelerations of components and motor torques of omniwheels were all associated with$\ddot{s}$; four controllable variables were ($\tau_i(i = 1, 2, 3)$ and $\ddot{s}$). Since Eq. (15.24) only included three equations for these variables, the number of control variables was redundant by comparing the DOF of cobot.

System redundancy was understandable since motor torques on the omniwheels not only impact on motion directions of omniwheels but also the magnitude of the accelerations along the trajectory. Note that motor torques were applied to generate appropriate ground frictions to the omniwheel and their values could be positive or negative. Generally, a minimized torque was just enough to maintain the required motion of the omniwheel, and the maximized torques made a partial contribution to accelerate the motion along the trajectory. In addition, the motion torques should be selected to avoid lifting or sliding.

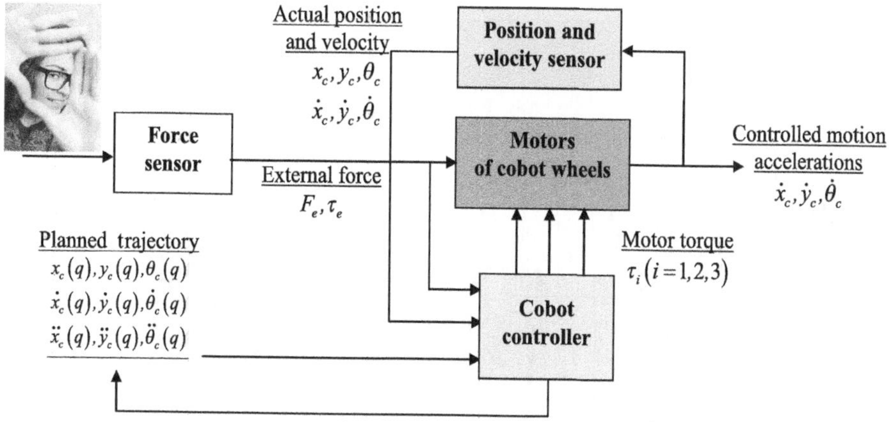

Figure 15.21 Schematic of cobot controls (*Source:* aeroking/Adobe Stock)

15.5.7 Simulation

In this case study, the structural parameters of cobot are defined as follows:

1) The omniwheels are mounted on the cobot symmetrically, and the main dimensions are:

$$L_i = 0.3m, \quad (i = 1, 2, 3) \text{ and } \alpha_1 = -\pi/6, \quad \alpha_2 = \pi/2, \quad \alpha_3 = 7\pi/6$$

2) The mass of the platform is $m_p = 600$ kg and $I_{pz} = m_p L_i^2/4$.
3) The force sensor is measured at $r_e = (0, \quad L_i/2, \quad 0)$
4) The main parameters of an omniwheel are $R_y = 0.02m$, $m_o = 0.5$ kg, $I_{ox} = m_o R_y^2/4$
5) The parameters of a roll on an omniwheel are $R_x = 0.005m$, $m_r = 0.05$ kg, $I_{ox} = m_r R_x^2/4$

The required trajectory is defined in terms of the trajectory parameter q as:

$$\left.\begin{aligned} x_c &= 0.125q + 0.8\cos(q) - 1.0 \\ y_c &= 0.8\sin(q) \\ \theta_c &= 0.1q \end{aligned}\right\} \tag{15.25}$$

The external forces are assumed as:

$$\left.\begin{aligned} \angle(\boldsymbol{F}_e, \boldsymbol{Y}_c) &= \frac{\pi}{4}(1.0 - 2.0^* random()) \\ F_{ex} &= [100 + 50^*(1.0 - 2.0^* random())]\cdot \cos\angle(\boldsymbol{F}_e, \boldsymbol{Y}_c) \\ F_{ey} &= [100 + 50^*(1.0 - 2.0^* random())]\cdot \sin\angle(\boldsymbol{F}_e, \boldsymbol{Y}_c) \\ F_{ez} &= -100 - 50^*(1.0 - 2.0^* random()) \\ \tau_e &= 0.05^*(1.0 - 2.0^* random()) \end{aligned}\right\} \tag{15.26}$$

where

random() represents a randy number between (0.0, 1.0)
$\angle(\boldsymbol{F}_e, \boldsymbol{Y}_c)$ is the angle between the projected $\boldsymbol{F}_e$ on plane $\boldsymbol{X}_c\boldsymbol{Y}_c$ and $\boldsymbol{Y}_c$.

The gravity acceleration and the friction coefficient are set to 9.81 kg/m^2 and 0.1, respectively. Figures 15.22 to 15.26 show the simulated results subjected to the given constraints.

The above results verified the feasibility of applying brake torques on omniwheels for a cobot to follow a specified trajectory. The applied torques were transformed into reacting friction forces at contacts, and reacting friction forces would be combined with the applied external forces and

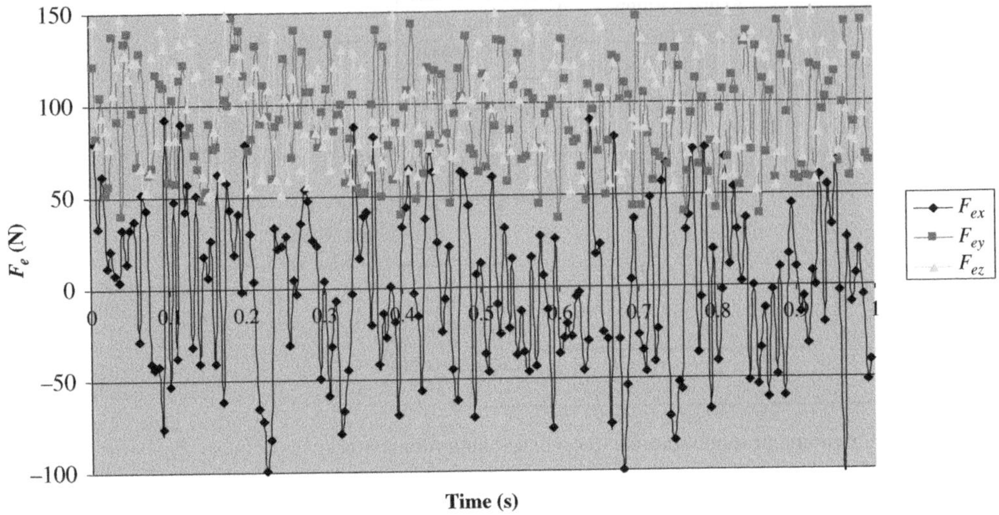

Figure 15.22 External force over time

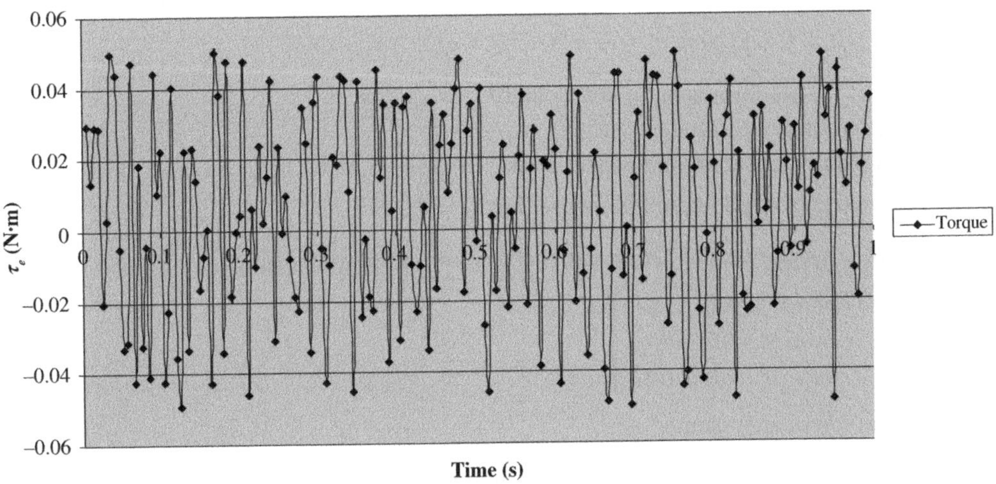

Figure 15.23 External torque over time

torque for the motion on the prescribed trajectory. The applied torque was not always to decelerate the rotation of the wheel and the applied braking torque could be positive or negative depending on the calculated results from the kinematic and dynamic models. Figure 15.27 visualizes the simulation results when the cobot was required to follow a spiral curve.

15.5.8 Summary of HCPS as Assistive Technologies

Section 15.5 introduced a cobot as an example application of HCPS in ATs. It was equipped with omniwheels to improve manipulability. Its kinematic and dynamic models were established, and the simulation verified its feasibility. The cobot was proposed to move a heavy object such as a

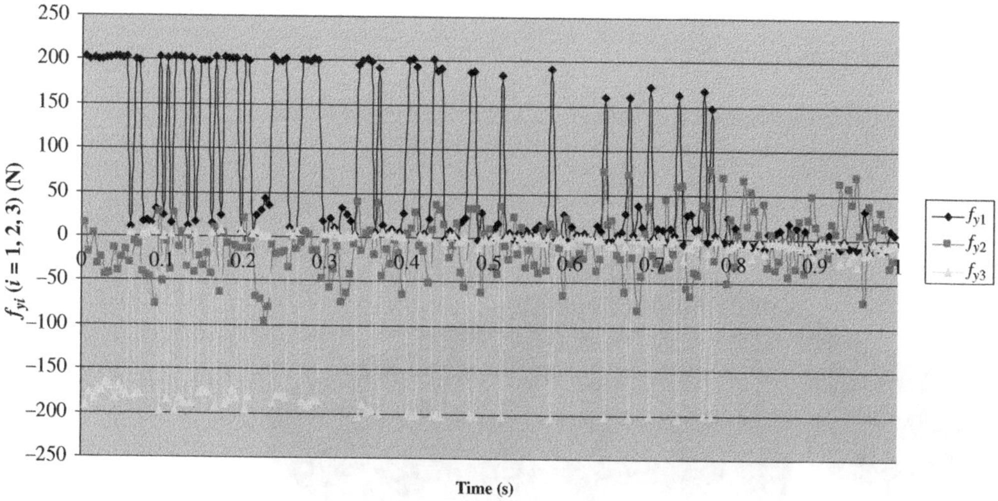

Figure 15.24 Frictions occurring to omniwheels over time

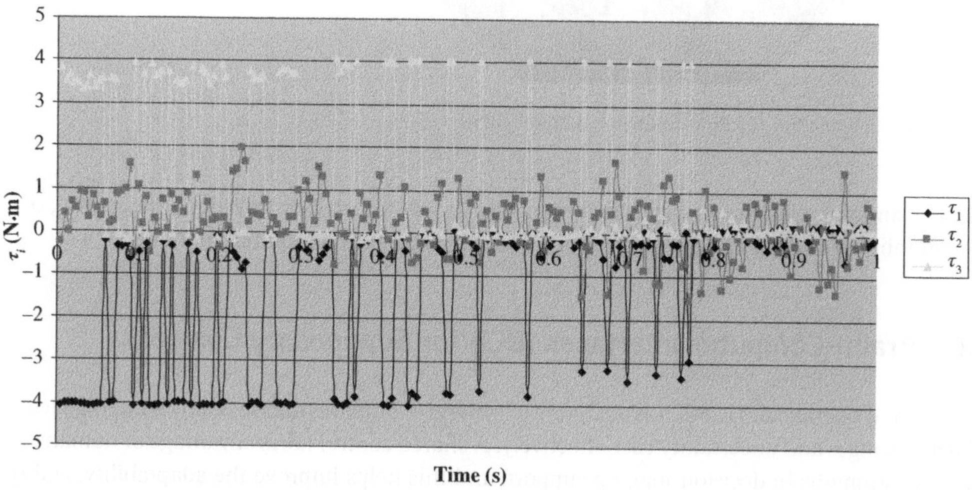

Figure 15.25 Torques applied to omniwheels over time

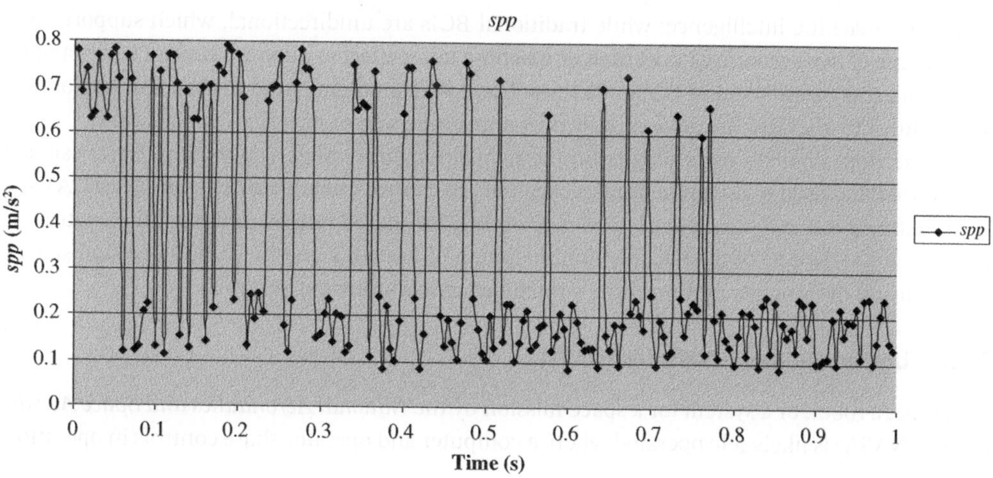

Figure 15.26 Acceleration along a trajectory over time

Figure 15.27 Simulation result in following a spiral curve

medical examining machine in a specified trajectory. It was featured by its redundant control, that is, the number of control variables was more than the DOF of the required motion.

15.6 Brain–Computer Interfaces (BCI) for Supervisory Controls

To design and operate an HCPS, it is desirable to synergize human's and machine's intelligence to deal with changes and uncertainty cost-effectively. A shared control takes advantage of human and computers strengths in decision-making support, and this helps improve the adaptability, agility, reliability, responsiveness, and resilience of the system.

Since the decision spaces for human's thinking and machine intelligence are quite different, challenges occur to fuse human intelligence and machine intelligence effectively. A *BCI* can bridge human and machine intelligence; while traditional BCIs are unidirectional, which support interaction in one of two scenarios: (1) human or machine takes effect at different control layers and (2) either human or machine takes effect at a time. There is an emerging need to close the loop of BCI-based control to alleviate the adverse effects of a machine's error or a human's mistake.

In this section, available technologies for the acquisition, processing, and mining of brain signals are reviewed; the need to integrate human capability to control *unmanned aerial vehicles* (UAVs) is elaborated; and research challenges in advancing BCI for shared human and machine control are discussed at the aspects of data acquisition, mapping of human's and machine's decision spaces, and the fusion of human's and machine's intelligence in automated controls.

15.6.1 Unmanned Aerial Vehicles (UAVs)

A machine, a robot, or a system for a space mission by the *National Aeronautics and Space Administration* (NASA) is likely tele operated where a computer and operator share controls in operation.

Conventionally, computer and human shared control is implemented as a *human-machine interface* (HMI) where an operator gives control commands through an intermediate broker such as a *graphic user interface* (GUI), a haptic device, or even a pedal, a driving wheel, or a speed shift for an *advanced driver assistance systems* (ADASs); using an intermediate broker in an HMI involves in some issues, in particular, *a unidirectional interaction* that might lead to a severe consequence when a human makes a wrong decision. Therefore, BCI should be advanced so that HMIs can be *bi-directional*; electrical signals about brain activities can be acquired, processed, and mined to extract the subject's intents and evaluate the subject's mental state for decision-making activities; both types of information are further integrated with machine intelligence to control a complex machine reliably. It will be desirable to advance BCI. In this paper, a new BCI is proposed to integrate human's intelligence and computer's intelligence seamlessly in controlling a smart machine. UAVs are used as targeted applications to elaborate the proposed platforms, associated tasks, and schematic of the advanced BCI-enabled HMIs.

UAVs are ideal examples of intelligent machines that require HMIs. The first UAV was the lightning bug developed by the US Air Force as a target drone in the 1960s. Due to its unique capabilities in comparison with other systems for similar missions, UAVs have been rapidly advanced in terms of their variety and quantity, and more and more UAVs have been successfully developed and deployed in various applications such as surveillance, reconnaissance, defenses, and attacks. This means that the technologies to control a UAV have exposed their limitations in flexibility, adaptability, and reliability to deal with changes and uncertainty in a highly turbulent environment. Especially, a UAV can be very vulnerable to extreme weather conditions and an attack by no-kinetic and kinetic weapons. This is true for low-altitude, slow-moving, and large UAVs since they are not so reliable or adaptable in hostile environments. The statistics showed that 75% of accidents of UAVs were induced by human factors in HMIs (Williams 2024). UAVs in the next generation must improve survivability and airworthiness to be aligned with the priorities of space missions. It is hypothesized that adverse effects of human factors in HMIs can be alleviated by (1) advancing BCI for human operators to control UAVs with their brain signals directly and (2) developing and implementing an arbitrating mechanism to sense the mental state of a human operator, weight the control authority of the human operator, and fuse human's and machine's intelligence for shared control of UAV.

15.6.2 UAV Controls

The US Army force began to deploy UAVs in the 1960s; for example, around 3500 lightning bugs were in the Vietnam War for some missions of tactical reconnaissance. Earlier UAV control systems such as Compass Arrow programs or D-21 Tagboard/Senior Bowl programs have shown some disadvantages such as low reliability, overrun cost, and unvalidated requirements. In fact, the technologies for UAVs were not developed at a fast pace until the early 1980s when some Israel UAV control systems were used to fight against opponents' air defenses successfully. US Force and Israeli Force made a joint effort to develop the RQ-1 Predator model, it began its use as an *intelligence, surveillance, and reconnaissance* (ISR) in 1996. RQ-1 models were latter upgraded as MQ-1 Predator models and RQ-4 Global Hawks models in 1998 and 2004, respectively.

Nowadays, UAVs have been highly demanded to fulfill various space missions and fight against terrorists, and typical operations include searching, tracking, identifying, locking, and attaching specific locations or objects. The amount of investment in the technologies for UAVs has increased steadily. For example, the US Department of Defense (DoD) has tripled its investment in developing UAVs in the last decade. It was estimated in 2017 that the scale of the global market of UAVs was expanded to a total of 7.3 billion in expenditure; the US Air Force was the greatest contributor to the fast growth of UAV markets. Moreover, it was expected that UAVs would expand their civil applications rapidly in the coming decade (Lake 2007). The US Navy and Special Operations Command (USSOCOM) expanded their funds for continuous development of unmanned aerial systems (US Air Force 2013). More and more new UAVs have been developed. For example, the US Air Force and the German Air Force are partnered to develop a "Euro Hawk" as an *unmanned combat air vehicle* (UCAV).

Even with the great improvement of modern UAVs in comparison with their earlier versions, there still are challenges to achieving high reliability in operations due to human factors and limited capability of machine intelligence in dealing with changes and uncertainty. Humans and machines do not collaborate so effectively for a UAV to deal with extreme weather conditions to weapon attacks, and this limits the survivability of a UAV, in particular, of a low altitude, low speed, and large-size UAV. It was found that while the overall accident rate of UAVs has decreased gradually, the average accident rate of some UAVs such as RQ-1/MQ-1 Predators was still high; there were about 24 mishaps/100,000 hours of flying in conservative years. For the cases of predators and Global Hawks, their accident rates were in an order of magnitude greater than any of a manned aircraft Therefore, the technologies for UAV controls must be improved at a number of aspects.

15.6.3 BCI for Effective HMI

To advance BCI for the improved effectiveness of HMIs in controlling a UAV, the human's role in the shared control of UAV is re-examined. Traditionally, BCIs provide the brain with new output channels that depend on brain activities rather than on peripheral nerves and muscles. A BCI is defined as systems that allow people to use a computer, an electromechanical arm, or various neuroprostheses without the use of motor nervous systems; BCI is a communication pathway used to establish direct communication between the brain and a peripheral interface, and it allows the transfer of information in one or both directions (Saha et al. 2021). BCI has been widely explored for the feasibility of controlling intelligent machines such as robots at users' will. For example, Lafleur et al. (2013) proposed using BCI to control a quadcopter, where EEG was acquired from an operator, the signal was analyzed to identify intents, and it was then fused with the control commands by a program to implement asynchronous controls. However, it was not a complete solution to the formulated problem in a number of aspects: (1) the user and the computer have respective roles in controller, and the entire control system does not have redundant capabilities to improve survivability and airworthiness; (2) the concern on a user's poor decision-making activities at low mental state have not been addressed which is the main cause of UAV accidents; (3) no systematic solution was presented to fill the gap of low-frequency of users' brain activities and high-frequency of real-time machine controls; and (4) users' mental conditions are not monitored and utilized in implementing an arbitration in shared control.

The schematic of the conventional control of a UAV is shown in Figure 15.28 in which a *multiple-layer hierarchical architecture* is used to control the UAV. The controls at *the lowest level* are for the operations of basic motion elements such as motors and actuators. The controls at the top level translate humans' intents such as *Deep Search, Wide Search, Direct Search, Return and Search, and Go Home* (Akagi et al. 2021) into the actions and movements at sub-systems.

The controls at *intermediate levels* translate the mission of the whole UAV into the behaviors of sub-systems, components, and motor-level. Some examples of commands at intermediate levels are "*Accelerating,*" "*Turning Left,*" "*Decelerating,*" "*Turning Left,*" "*Ascending,*" "*Descending,*" "*Taking off,*" "*Cruising,*" and "*Landing.*" A hierarchical architecture allows shared control where the humans' decisions are at the top level of control, and machine intelligence is deployed to translate

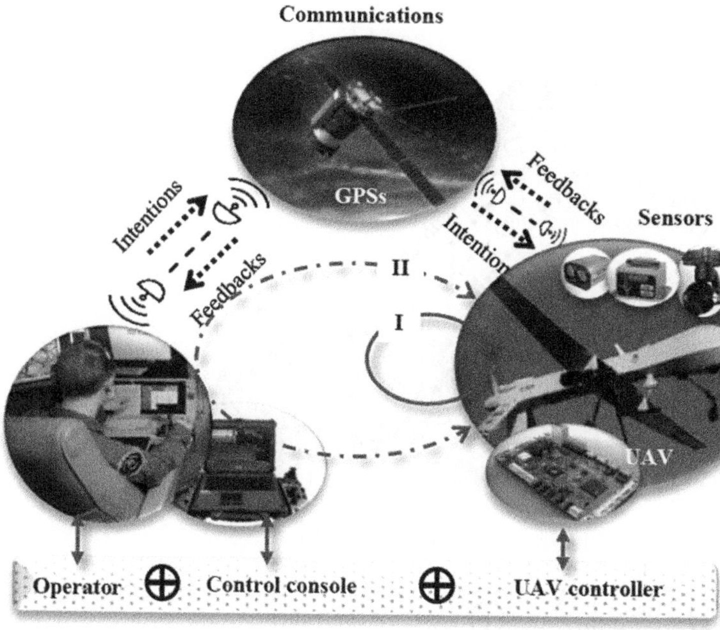

Figure 15.28 Traditional human–machine interface (HMI)

humans' commands into actions at lower levels of control. Note that communications in the control architecture are supported by sophisticated wireless networks and *global positioning systems* (GPS). One can observe clearly from Figure 15.28 that the control loop at the top level (see loop II in a dashed line) is not closed; it implies that the human operator has their own justifications about the difference between expected and actual states of a UAV to issue control commands in next steps. Therefore, there is no effective mechanism in a conventional control architecture to correct human's mistakes when human's control commands are input via a broker such as a control console, GUI, or a haptic device.

Figure 15.28 shows that an unsatisfactory level of survivability or reliability of a UAV is partially attributed to the limited capability of HMI in responding to human's mistakes; there is the need to empower the interfaces between an operator and computer in justifying if (1) the human operator is at the full capability of *situation awareness* (SA) and decision-making, (2) the human's commands are reasonable and no violent to common senses or emerging situations. Under conventional control architecture, the controller of a UAV will execute human commands even if it would be a mistake or if a correct decision is input incorrectly, and this will fail the expected mission.

To address the aforementioned concerns in conventional control architecture, we propose a new schematic of a shared control, as shown in Figure 15.29. In comparison with HMI in a conventional UAV, we propose to introduce three new components to fix human errors and avoid severe sequences due to human factors. *The first component* is the instrument to monitor the human's brain activities and acquire corresponding brain signals; multimodal brain signals will be processed and analyzed to (1) extract the human's intents as control commands directly to avoid possible errors in issuing a command by a broker indirectly and (2) monitor and classify the human's physical or physiological states and determine if the human or computer should dominate the control at top-level. *The second component* is an arbitrating mechanism to make the trade-off of human and computer roles based on the detected mental states of human operations; its core function is for the computer to take over the human's control authority when the acquired brain signals show the

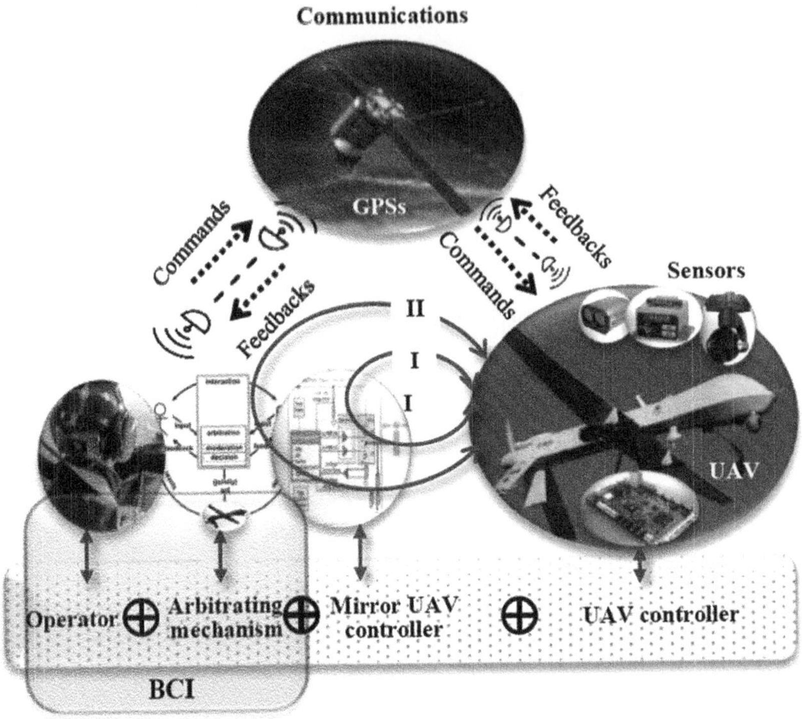

Figure 15.29 Schematic of proposed supervisory controls

human is not in full capability to make correct decisions. *The third component* is the digital twin of the physical controller of the UAV that can substitute the human to generate control commands at the top level when the human is incapable of controlling a UAV in certain circumstances. Integrating these components in the proposed architecture will close the control loop at high level (see the control loop II in solid line in red); the closed control loop empowers HMI to avoid adverse effects of human mistakes.

15.6.4 Development of BCIs

Two critical enhancements in the proposed control architecture are that (1) machine intelligence takes over human's intelligence when the human is detected not in full capability of SA and decision-making activities; (2) an arbitrating mechanism that is capable of (i) extracting human's intents from brain signals and translate into control commends and (ii) making the trade-off of human's and machine's intelligence in the shared control based on quantified human's affection that reflects the state of the human's physical and physiological conditions.

The inputs of the arbitrating mechanism are raw data acquired from human subjects; raw data is acquired, processed, and analyzed by BCI to (1) extract human's intents and (2) quantify human's affection. The basic structure of BCI includes four main components: *acquisition of brain signals*, *feature extraction*, *state classification*, and *human–machine interactions*, respectively. It is clear that the feasibility of the proposed shared control relies on the advancement of existing BCI technologies. In the following section, existing technologies of BCIs are discussed briefly.

15.6.4.1 Brain Signals

In Figure 15.30, BCI is incorporated to (1) extract human intents about UAV controls and (2) classify human's physical and physiological conditions based on brain signals acquired from a human operator. In theory, both types of brain signals are direct outcomes of the human's brain activities, and they should be translated appropriately to obtain the required information. However, the intricate relationships between brain signals and brain activities raise two questions for a BCI application: (1) how to choose and acquire the most relevant brain signals to the outcomes of interest? (2) How to acquire and interpret such signals from the subject in a safe, feasible, and cost-effective way with minimized disturbance to the subject?

An adult brain has nearly 100 billion neurons that are networked by over 500 trillion connections; every neuron is connected with more than 10,000 adjacent neurons. As a whole, the overall brain capability of memory is about 2.5 petabytes or 2.5×10^{15} bytes. The communications among neurons are electrical impulses called *action potentials* that are transmitted over hundreds of billions per second (UC-Olli 2024). From this perspective, it seems technically infeasible to be thoroughly instrumented to gain a full understanding of brain activities. It is also worth mentioning that a human brain consists of a number of sections that fulfill different duties, and Figure 15.31 shows a number of main sections and their corresponding functions. Firstly, a brain can be divided into a *forebrain*, a *midbrain*, and a *hindbrain*, where the hindbrain is further divided into a *brain stem*, an *upper spinal cord*, and a *cerebellum*, which is referred to as a wrinkled tissue ball. A forebrain is usually the most developed and largest section of the brain and it consists of the cerebrum and supporting structures below.

In Figure 15.31, half of the cerebrum is illustrated, and it is divided by a deep fissure into six main sections, including *motor cortex*, *frontal brain*, *parietal lobes*, *occipital lobes*, *somatosensory cortex*, and *temporal lobes*. Their associated activities are also illustrated in Figure 15.5; for example, the motor cortex takes charge of planning, controlling, and executing voluntary or intended motions.

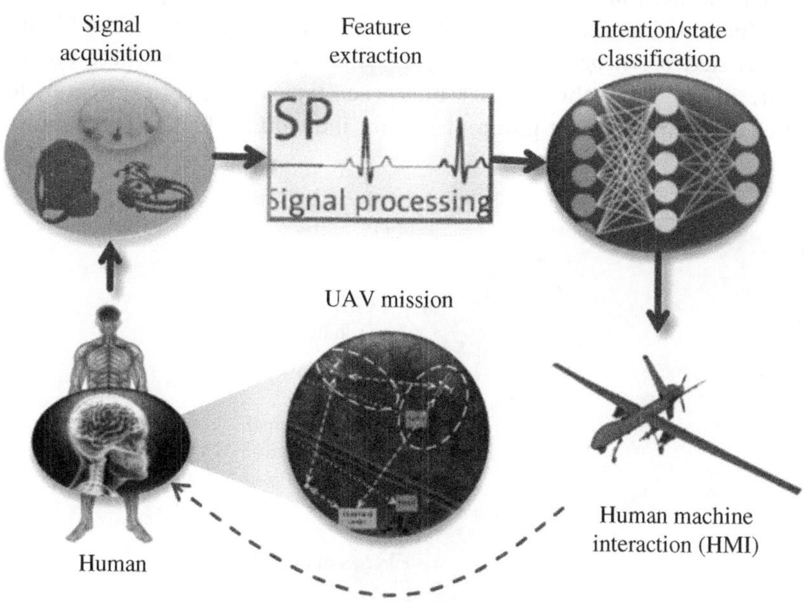

Figure 15.30 Proposed BCI structure

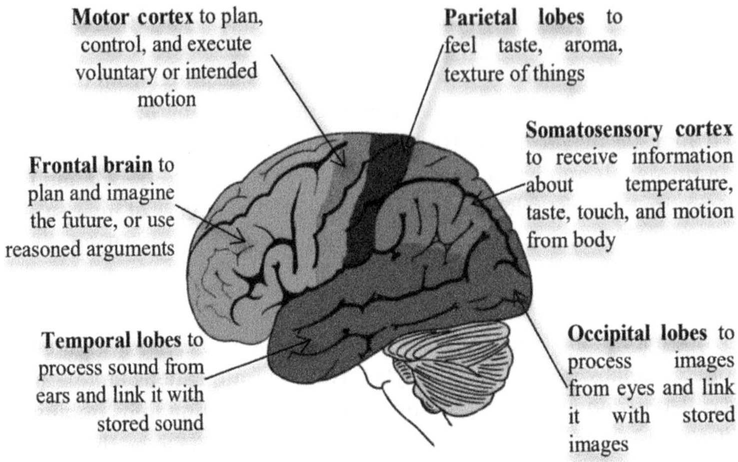

Figure 15.31 Main sections and types of brain signals (Source: Bi et al. 2024/IEEE)

15.6.4.2 Data Acquisition

More than 30 companies provide commercial hardware and software systems that can be used to develop BCI applications, while BCI applications are mainly limited to the entertainment industry and healthcare where real-time performance is not prioritized (UC-Oilli 2024). From the perspective of the invasion need in data acquisition, the techniques to acquire brain signals from human subjects can be generally classified into three categories: (1) noninvasive technique, where the sensors are placed on the subject's scalp and brain signals, are (i) measured electrical potentials that are called as *electroencephalogram* (EEG) or (ii) measured magnetic fields that are called as *electromyogram* (MEG); (2) *semi-invasive* techniques where the electrodes of sensors are mounted on the exposed surface of the subject's brain to obtain brain signals that are called as *electrocorticography* (ECoG); (3) *invasive* techniques where microelectrodes of sensors are inserted into the subject's brain (i.e., typically in cortex section) to measure brain activities occurring to a set group of neuron individuals. Figure 15.32 shows

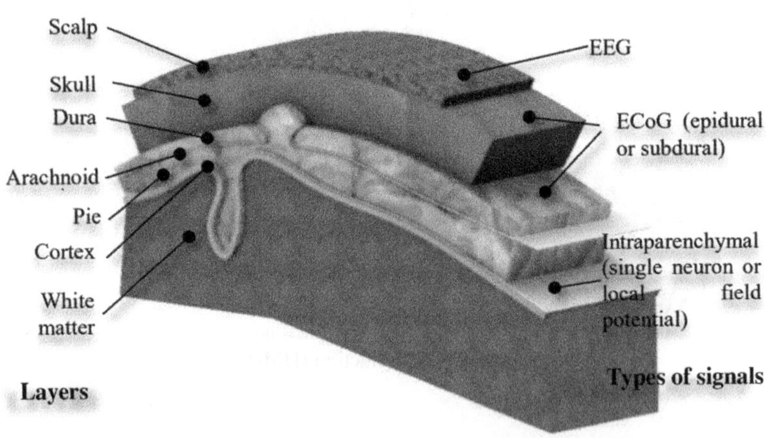

Figure 15.32 Brain layers and types of signals

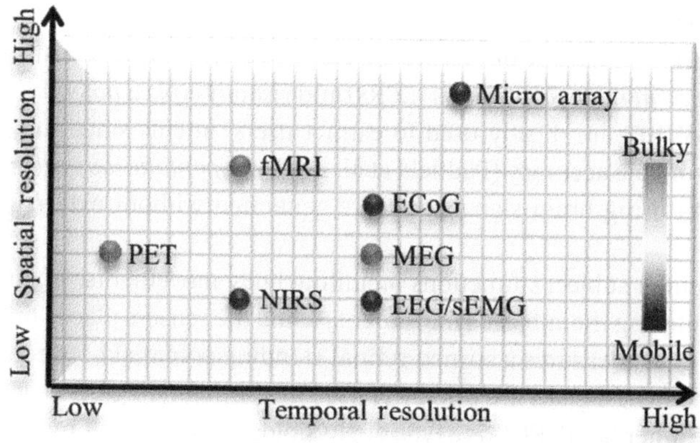

Figure 15.33 Types of brain signals versus temporal and spatial resolutions

the types of brain signals that can be obtained from different regions of a brain, and these signals are associated with different types of brain activities elaborated in Figure 15.31.

Brain signals obtained by different techniques can also be distinguished from each other by spatial and temporal resolutions of brain signals that are compared in Figure 15.33 (Lin and Do 2021). It should also be noted that even the quality of a specific type of brain signal such as electroencephalography (EEG), magnetoencephalography (MEG), electrooculography (EOG), and electrocardiography (ECG) can be affected by many other factors in data acquisition. Commonly used electrophysiological signals that are used to measure brain activities are *positron emission tomography* (PET), *micro-array*, *ECG*, *functional Near-InfraRed spectroscopy* (fNIRs), *functional magnetic resonance imaging* (fMRI), *EOG*, *surface electromyography* (sEMG), EEG, MEG, and *galvanic skin response* (GSR) (Bi et al. 2023). The selection of certain techniques to acquire brain signals also takes into consideration of many aspects such as safety, cost, and reliability. For example, fMRI signals have better spatial resolutions but lower temporal resolutions, while MEG or EEG signals are just the opposite. A software tool to process brain signals is another concern. Generally, a brain signal acquired by a noninvasive technique is affected by many random factors that affect the quality of the brain signal; therefore, a brain signal from a noninvasive instrument must be preprocessed to remove noises before it can be further processed to classify and extract intents. Another factor is *portability*, which is important when BCI is used in a mobile environment. Some instruments, such as the systems to acquire fMRI or MEG signals from subjects, are very expensive and too bulky for mobile applications; on the other hand, an instrument for EEG is very attractive since it is light, portable, and very affordable (Hekmatmanesh et al. 2021).

EMG is mostly used by researchers in HMIs to understand humans' intents for motion controls. The instruments for EMG became available in the 1990s. EMG signals reflect electrical potentials that are generated by muscle cells. Early EMG signals were collected from several muscles on a subject's forearm; the collected EMG signals were then processed by a series of tasks from *sampling* to *high-pass filtering*, *low-pass filtering*, *amplifying*, and then identifying the subject's intents for hand movements such as *an opening*, *a closing*, or just *a relaxation* of hand. EMG signals were considered reliable means to understand a human subject's intent for hand movements (Choudhary et al. 2022). In comparison with MEG and fMRI, EEG has its advantages in the aspects of temporal resolution, portability, and cost-effectiveness for data acquisition, processing, and mining. EEGs can be used in two alternative ways in systems, that is, "evoked" systems or "spontaneous" systems.

An evoked system is exogenous and it uses external stimulations such as sensing, visual, or auditory sensors, and the responses to these external signals by brains are analyzed to extract users' intents. Two commonly used evoked signals are *event related potentials* (EPRs) and *visually evoked potentials* (VEPs), especially, *steady-state VEP* (SSVEP) are mostly used to extract users' intents since they need minimum training to obtain a reasonable accuracy (Padfield et al. 2019). *A spontaneous system* is endogenous, and it controls a system using brain signals that are generated without external simulations.

Since a brain controls most of the human subject's functions including body movements, brain activities could also be acquired by monitoring the subject's movements using the sensors placed on the corresponding parts. For example, EOG measures a corneo-retinal standing potential between the front and back of a human's eye, and EOG signals are primarily used in ophthalmological diagnosis to analyze eye movements. On the other hand, ECG is a type of physiological signal that represents electrical activities by the subject's heart. To acquire ECG, the instrument is usually set to place nine sensors on the arm and legs to acquire ECG signals in real-time mode. Alternatively, *photoplethysmography* (PPG) measures brain activities by reading microvascular blood volume in tissues; the change in blood volume corresponds to electrical signals from the subject's brain. Similarly, a *galvanic skin response* (GSR) can be used to measure the subject's response to the changes in the environment. GSP is also known as *skin conductance* (SC) or *electrodermal activity* (EDA). Note that SC is not under control by the subject's conscious.

In general, a brain signal by a noninvasive technique involves a high-level noise. Therefore, a noninvasive brain signal may include some abnormalities that affect the overall performance of BCI. Sagha et al. (2015) discussed the causes of anomalous data, and some common factors included wrongly placed electrodes, loosened connections, and degraded impedance. To detect an abnormality, the deviation of data in each channel was measured, and it was counted as the weight to fuse the datasets from a problematic channel. Finally, not all brain signals can be monitored continuously.

Multiple sensors can be deployed to collect multimodal brain signals from various sources. To obtain a holistic view of brain activities, multimodal data must be fused to investigate the correspondence of multimodal data; since multiple sources can be in a distributed environment and sharing and fusing of datasets are performed over the Internet, how to process and utilizing multimodal datasets brings some challenges to assure privacy of subjects, security, and reliability (Greenhorn 2020). Kuan et al. (2015) discussed the scenario of collecting and utilizing big data of brain signals over the Internet, and they developed a data telemetry system over a *wireless sensor network* (WSN) to guarantee both high-rate data transmission and the affordability of power consumption over an implantable device. Cui et al. (2017) developed a framework for an information system to fuse multimodal signals including EEG, EMG, and MMG; the system was able to extract the subject's intent to move multiple joints of a lower limb.

15.6.4.3 Feature Classification and Detection

To implement a BCI, one must select and acquire the correct brain signals to control an object in a specific application. The acquired brain signals must be associated with (1) a subject's intent to control the object or (2) a subject's mental, psychological, or physical status that may affect the subject's ability in decision-making. To classify and extract features from brain signals, one must understand raw brain signals thoroughly in terms of their characteristics in temporal and spatial domains, as well as the effects of some random factors. The most critical characteristic of a brain signal is its fluctuating frequency over time in the temporal domain; the time factor is of the most importance since the majority of BCI applications require real-time controls.

Table 15.7 Classification of brain signals by frequency ranges

Type	Frequency range	Characteristics
Delta (δ)	$f < 4$ Hz 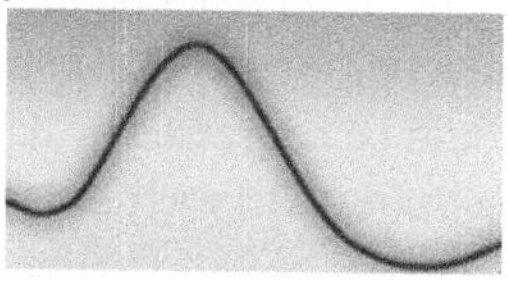	A wave in the slowest frequencies for a strong sense of *empathy and intuition*; the growth hormone of the human subject is released in multiple frequencies.
Theta (θ)	4 Hz $< f < 7$ Hz 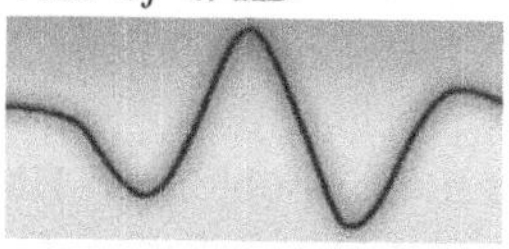	A wave in a frequency range of (4 to 7) Hz for *deep relaxation and meditation* is associated with stress relief and memory recollection.
Alpha (α)	8 Hz $< f < 12$ Hz 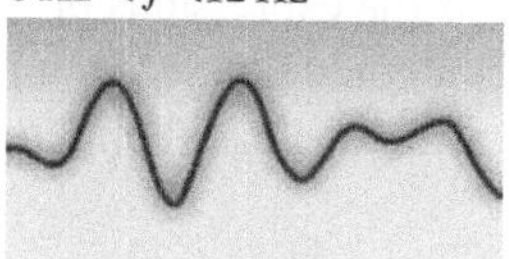	A wave in a frequency range of (8 to 12) Hz for *creativity and relaxation* is present mostly in awake relaxation when the subject's eyes are closed and the subject's brain is at rest.
Beta (β)	12 Hz $< f < 30$ Hz 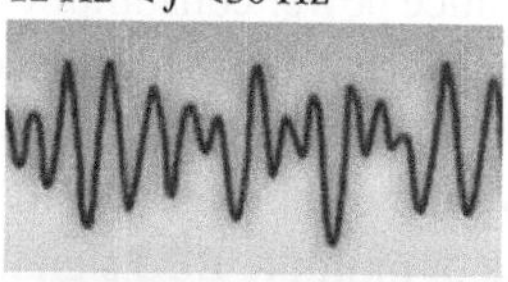	A wave in a frequency range of (12 to 30) Hz for *alertness and concentration* is produced when the human subject exhibits alertness or anxiety; the wave is a dominant rhythm.
Gamma (γ)	$f > 30$ Hz 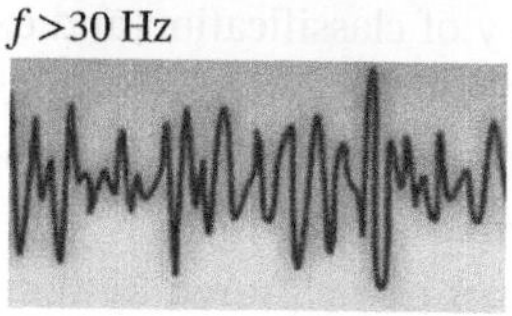	A wave in the fastest frequency over 30 Hz for *regional learning, memory and language processing, and ideation*; it is produced when the human subject responds to an abnormal condition or a mental disorder occurs to the subject.

Table 15.7 classifies brain signals based on their ranges of fluctuated frequencies in a temporal domain. The explanations in the right column clearly show the correspondence of the subject's mental status with the frequencies' ranges of brain signals. When a conventional data processing method, such as *spectral estimation*, *time-frequency analysis*, *system identification*, or *statistical signal processing*, is used to process brain signals, it does not show satisfactory accuracy and reliability in classifying and extracting human intents due to two main aspects, i.e., the complexity of brain signals and the difference of human's and machine's decision spaces.

To improve the adaptability and reliability of conventional data processing methods, *artificial intelligence* (AI) and *machine learning* (ML) have been incorporated and the integrated systems have shown their applicability in classifying and extracting important features from brain signals. Many researchers have applied AI and ML to decode brain signals and use the identified human intents to implement BCIs. Owusu et al. (2021) discussed various AI-enabled techniques to detect facial expressions of a human subject based on the measurements of humans' emotions by an instrument for physiological signals such as ECG, EEG, MEG, EMG, MRI, MRI, PET, fMRI, or NIRS. Hwaidi and Chen (2022) established a *convolutional neural network* (CNN) to analyze brain signals and classify human's intents based on *motor imagery* (MI) in a brain signal. CNN was integrated to classify EEG signals as an autoencoder. Multimodal brain signals are often used to classify brain signals based on the correspondence of the brain signals collected from different sources.

For example, Fu et al. (2017) used both EEG and *near-infrared spectroscopy* (NIRS) in decoding brain signals that were generated in the sensorimotor areas; NIRS signals were acquired when the human subject imagined applying force to clench a hand. The extracted features from EEG and NIRS will be combined as a *state vector machine* (SVM) that is used to classify imagined force and speed at three levels. Note that the accuracy of an AI or ML algorithm relies on the sufficiency, abundance, and completeness of brain signals; it may cause a challenge for practical applications. Li et al. (2022) discussed this concern and argued that the issue could be addressed by data augmentation. This helped to fully utilize existing datasets. *Contrastive learning* (CL) was used to extract the features from insufficient labeled data.

Physiological signals controlled by a subject's brain can also be used to detect the subject's intents or emotions for HMIs, and examples of typical applications are autonomous driving, UAVs, gaming, and healthcare. For example, Hu et al. (2016) collected psychophysiological signals from human subjects and used these to determine the trustiness of human subjects in HMI. Ikenishi et al. (2010) collected both vision and physiological signals to detect a driver's intent of action on a pedal and a steering wheel. Jose and Loper (2015) recorded the expressions of the lower lip of a human subject when a human interacted with a computer. It could be used by a patient with tetraplegia to control an input device, and its performance was evaluated based on *Fitts' Law* in an ISO standard.

The power supply of an electronic device is always a big challenge, and it becomes worse if a large number of sensors are used in a distributed environment. To reduce power consumption in collecting and processing brain signals over an embedded platform, Kragic et al. (2018) designed an instrument called *parallel ultra-low power system-on-chip*; it was equipped with nine RISC-V cores and capable of performing *canonical correlation analysis* (CCA) over the collected SSVEP signals. In analyzing multimodal signals, establishing the correspondence of multiple EEG signals that were collected from different sectors helped to enhance the accuracy of classification and extraction of features. Wang et al. (2022) adopted *partial directed coherence* (PDC) to explore the functional connectivity of multimodal signals and establish a causal network to represent the correspondence; it was then used to extract features based on *common spatial patterns* (CSP).

Brain activities relevant to the subject's limb movements can be measured by motor imagery (MI), and MI signals are widely used in the healthcare industry for patients to control haptic, prosthetic, and neuro-rehabilitative devices. Jeong et al. (2020) developed a control system to detect the human's intents for the rotational movement of a forearm, and it used EEG signals collected from associated muscles on the forearm. Padfield et al. (2019) surveyed available techniques to acquire and process EEG signals. Minati et al. (2016) designed a wearable device capable of collecting EMG, EEG, and EOG signals about jaw and head movements simultaneously. It was expected to control an assistive robot by a BCI. In their preliminary study, a 6-DOF manipulator was selected as the device to be controlled in four control modes, and the control commands were based on the identified human's intents from multisource brain signals. In collecting EEG signals for the classification of human intents at a high level of accuracy, Wang et al. (2015) provided tactile stimuli on the human subject's forearm, and the subject's responses were recorded and decoded in the form of EEG signals.

Even though MI signals are closely relevant to the movements of a human subject's limbs, the machine to be controlled usually has a different kinematic structure, and it is impractical to control a machine based on extracted intents from a human directly. It is common knowledge that one has to decode brain signals for human intents and then translate these into the motion commands of the machine with different kinematic structures. Mao et al. (2019) developed a control system whose inputs were SSVEP and P300 to detect human intents. Fuzzy logic and AI were used to process multisource images. Human and machine intelligence were fused to (1) reduce a human's mental stress and workload and (ii) improve the control performance of the robotic system. Moritz et al. (2016) introduced some constraints in enabling a BCI to restore motor and sensory functions for people with disabilities; especially, three main hurdles were to (1) detect neural activities of interest appropriately, (2) interpret the subject's intents of movements in the presence of plasticity, and (3) take into the consideration social and ethical concerns in market entrances.

15.6.5 BCI Development Platform

To explore the feasibility of fusing human and machine intelligence to enhance the adaptability and survivability of UAVs, we propose a new platform to develop BCIs for machine controls, as shown in Figure 15.34. Various instruments will be introduced into the platform depending on the types of brain signals to be collected. Currently, the platform has two types of instruments that are used to acquire brain signals: *the first one* is Epoc X 14 Channel Mobile Brainwear by Emotive and *the second one* is Muscke SpikerShield by Backyard Brains. From the perspective of available machines, the platform includes a small drone with an HD camera by Thames. Moreover, everything is networked, and the whole platform is viewed as an instance of IoT for data acquisition, transmission, processing, and communication (Levratti et al. 2019; Lloyd 2019; Long et al. 2020; Lin and Do 2021; Li et al. 2022).

For data acquisition, Epoc X 14 Brainwear is equipped with 14 channels for EEG signals, and it can also sense 9-axis motions. This device can be connected with any smart things in IoT by *wireless sensors networks*. Muscke SpikerShield is used to monitor brain activities in the form of EMG signals; EMG is muscle physiology that is closely related to brain activities but it can be from the subject's skin rather than the brain directly. When the human subject decides to move a part of his/her body, responsible neurons travel from the motor cortex of the brain (i.e., upper motor neurons) to the spinal cord and synapse with lower motor muscles and neurons to make a motor unit.

Figure 15.35 shows an example of typical EMG signals by Muscke SpikerShield. When a subject was required to make some movements on hand, different movements corresponded to different

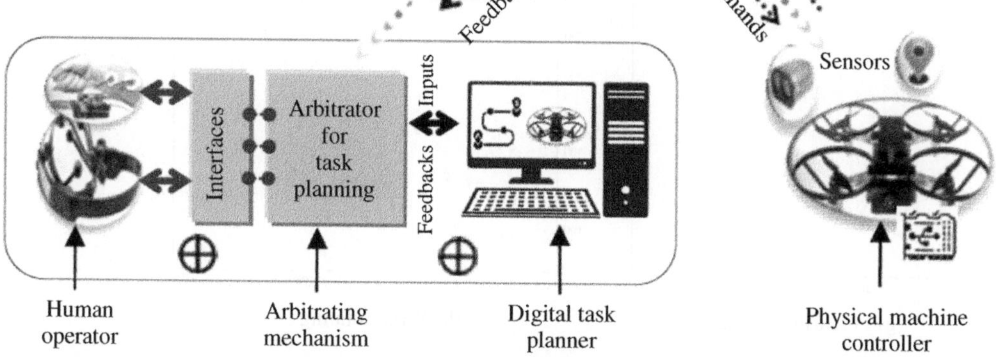

Figure 15.34 BCI development platform for supervisory controls

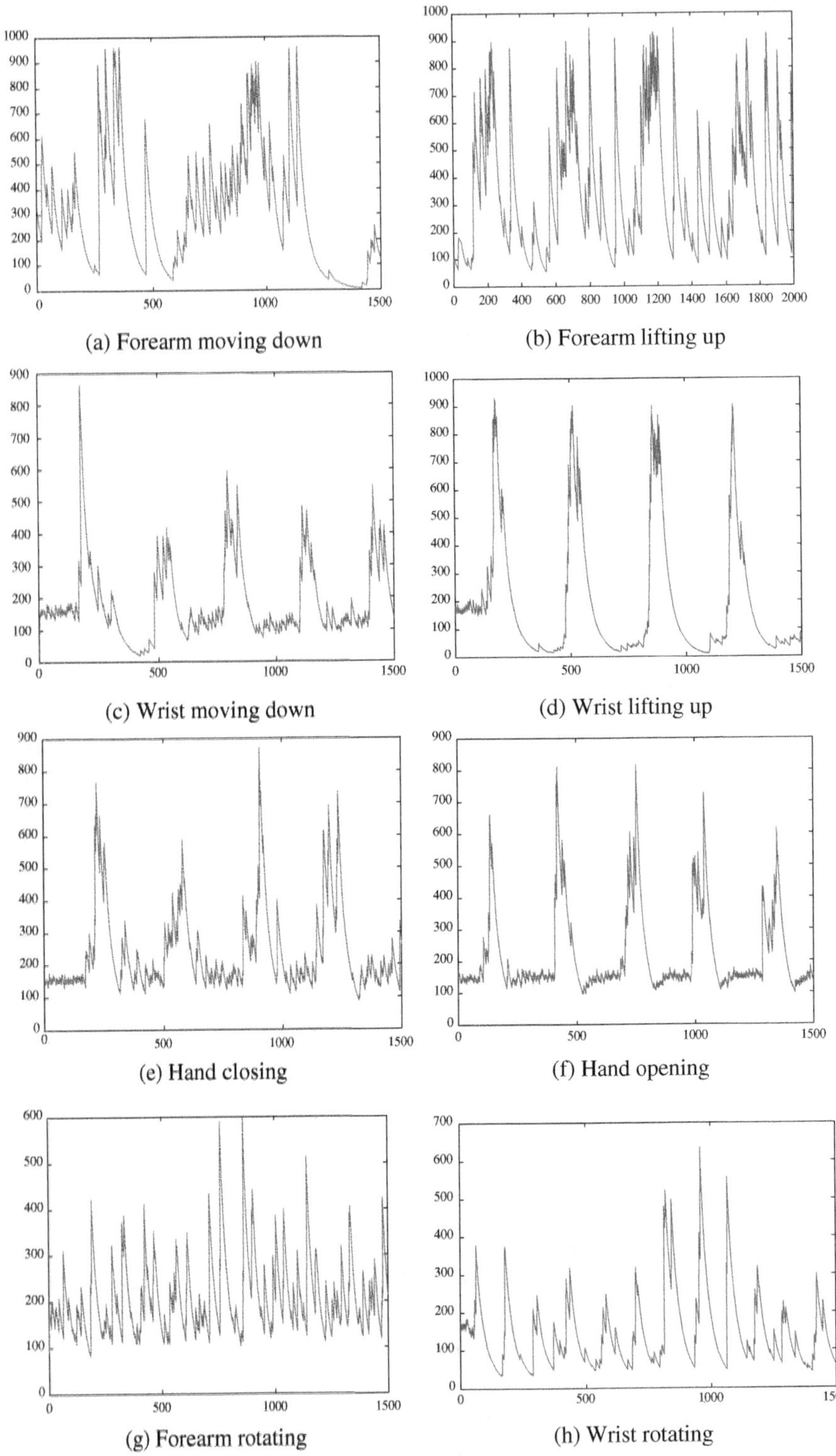

Figure 15.35 Examples of acquired EMG signals for different motions

patterns in original EMG signals. Therefore, ML can be applied to classify and extract human's intents using a sufficient amount of training data. Technically, the number of the intents to be detected depends on the number of actions that the muscles can accomplish. The probes of Spike-Shield were mounted on these muscles. However, it takes *seconds* or even *minutes* for BCI to obtain the next new command for a robotic control, while most machine tools require real-time performance. There is a need to fill the gap between the decision spaces of human intelligence and machine's intelligence.

Figure 15.36 shows the proposed supervisory control to fuse human intelligence and machine intelligence for advanced controls of UAVs in a dynamic environment. In comparison with conventional unidirectional BCI shown in Figure 15.28, the following three new functional components have been introduced:

1) A comprehensive BCI will be developed to acquire brain signals from multiple sources; ML algorithms will be incorporated to classify and extract important features from brain signals; many researchers have developed similar ML algorithms to decode brain signals and identify human intents in BCIs. In the developed platform, the features in brain signals will be extracted to detect human's intents in operating a machine for a specific mission based on the subject's *situation awareness* (SA).
2) Collected brain signals will also be analyzed to quantify the subject's affections that are associated with the trustiness and quality of human decisions. This requires the development of another ML algorithm to cluster the subject's mental states based on physiological signals. Researchers have shown the feasibility of detecting mental states based on various bio-signals.

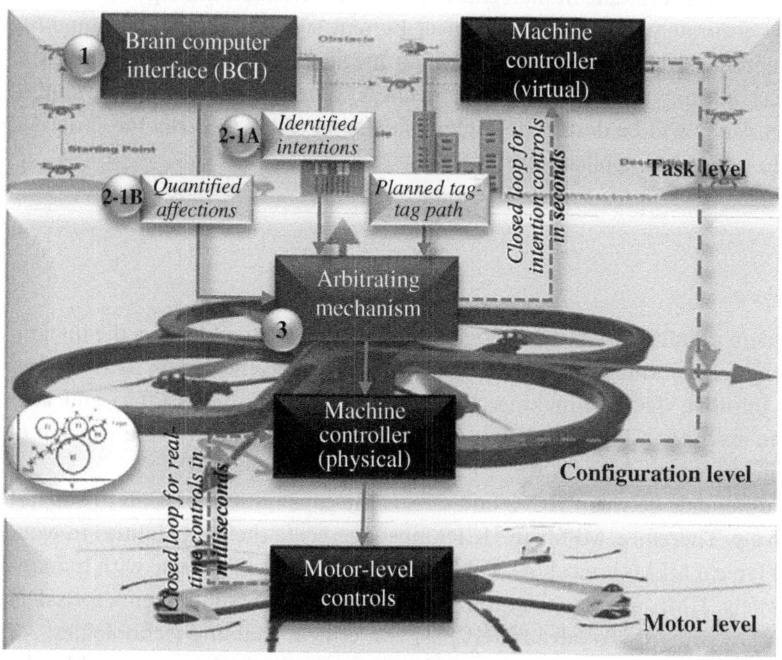

Figure 15.36 BCIs for supervisory machine controls

3) An arbitrating mechanism will be developed to (i) fuse human intelligence and machine intelligence in transferring control commands from the system level to the configuration level and finally to the motor level; (ii) determine whether or not the human dominates the control rather than a computer at the system level. A number of AI algorithms have been developed to make the trade-off between the authorities of humans and computers in a supervisory control.

The recent development in the controls of UAVs has been investigated, and the techniques for BCIs were focused on enhancing the adaptability and survivability of UAVs. It was found that even though significant progress was made in AI, ML, IoT, CPSs, BDA, WSN, CC, and BCT, existing BCI techniques must be advanced to reduce the average accident ratio in operating UAVs; available BCI techniques have exposed their limitations in (1) fusing human's intelligence and machine's intelligence and (2) lacking mechanisms to deal with humans' errors due to unidirectional controls. To overcome these two limitations, we proposed a new platform for the development of BCIs for UAVs. The platform encourages acquiring abundant brain signals from multiple sources; it will be equipped with a number of ML algorithms to detect human subject's intents and quantify human's affection in making decisions on machine controls. Most importantly, an arbitrating mechanism will be developed to fuse human intelligence and machine intelligence collectively based on quantified human affection. The arbitrating mechanism can transfer control commands at a low frequency at a high level to these commands at a high frequency at a low level for real-time performance. The proposed platform is expected to deal with human mistakes or human decisions with poor-quality decisions when the subject is not in good condition for SAs or decision-making.

The platform is still in its preliminary stage of development (Bi et al. 2024). It requires intensive research effort in a number of aspects as follows: *first*, more cutting-edge instruments should be introduced and evaluated to ensure appropriateness and sufficiency of brain signals to (1) classify and extract control intents and (2) evaluate human emotions; *second*, available ML algorithms must be advanced to reliably, accurately, and promptly detect human subject's intents and quantify human's affections comprehensively. *Third*, the arbitrating mechanism will be implemented to reduce the adverse effects of human mistakes or ineffectiveness. *Fourth*, a number of case studies for a number of UAV missions will be designed, implemented, and tested to verify the proposed development platform in real-world applications.

15.7 Summary

With the rapid development of information technologies, the next extension of CPSs is the integration of humans into social society. Integrating humans in CPSs enhances the system's capabilities in cognitive thinking, intuitive decision-making, innovative learning, multitasking, and self-correction of errors. The extended CPSs are HCPSs in which automated systems are capable of working with humans in shared workspaces.

The most significant difference between HCPS and CPS is the capabilities of HCPS to support HMIs safely and effectively. Therefore, MFMs in HCPS must aggregate their capabilities to work with humans, and most typical MFMs are cobots. Cobots are designed to collaborate with humans for HMIs directly and safely. HMIs are classified into four types, and the requirements for safety assurances of HMI are standardized. To advance HCPSs, two critical enabling technologies, AT and BCI, have been especially discussed to illustrate their potential roles in designing and implementing HCPSs.

Problems

15.1 What is a human cyber-physical systems (HCPSs), and what is its difference from a CPS?

15.2 What is human–robot interaction (HRI)? How is it relevant to the mechatronic system? Have an example of HRI to support your answers.

15.3 What is a collaborative robot (cobot)? How is a collaborative robot relevant to HCPS?

15.4 Classify collaboration types based on humans' roles in HCPSs.

15.5 What are the advantages of integrating humans with CPSs?

15.6 Discuss the basic requirements of a cobot.

15.7 What are the basic types of human–machine interfaces?

15.8 What are the basic safety mechanisms for a cobot?

15.9 What is assistive technologies (ATs)? How is it relevant to the mechatronic system? Have an example of AT to support your answers.

15.10 What is brain–computer interfaces (BCIs)? How is it relevant to the mechatronic system? Have an example of BCI to support your answers.

15.11 What is your perception on the next-generation development of mechatronic systems?

References

Afsari, K., Gupta, S., Afkhamiaghda, M., and Lu, Z. (2024). Application of collaborative industrial robots in building construction. *54th ASC Annual International Conference Proceedings*, http://ascpro0.ascweb.org/archives/cd/2018/paper/CPRT178002018.pdf (accessed 13 September 2024).

Agote-Garrido, A., Martin-Gomez, A.M., and Lama-Ruiz, J.R. (2023). Manufacturing system design in industry 5.0: incorporating sociotechnical systems and social metabolism for human-centered, sustainable, and resilient production. *Systems* 11: 537.

Akagi, J., Morris, T.D., Moon, B. et al. (2021). Gesture commands for controlling high-level UAV behavior. *SN Applied Science* 3: 603.

Antonelli, D. and Astanin, S. (2016). Qualification of a collaborative human-robot welling cell. *Procedia CIRP* 41 (2016): 352–357.

Beaupre (2024). Collaborative robot technology and applications. https://www.robotics.org/userAssets/riaUploads/file/4-KUKA_Beaupre.pdf (accessed 16 September 2024).

Bi, Z.M. (2018). *Finite Element Analysis Applications: A Systematic and Practical Approach*, 1ste. Amsterdam: Academic Press.

Bi, Z.M., Liu, Y., Krider, J. et al. (2018). Real-time force monitoring of smart grippers for internet of things (IoT) applications. *Journal of Industrial Information Integration* 11 (2018): 19–28.

Bi, Z.M., Liu, Y., Quaye, E., and Luo, C. (2023). Brain computer interfaces for supervisory controls of unmanned aerial vehicles. *2023 IEEE/ASME International Conference on Advanced Intelligent Mechatronics (AIM)*, Seattle, WA (28–30 June 2023).

Bi, Z.M., Mikkola, A., Ip, A.W.H. et al. (2024). Brain computer interface (BCI) for shared controls of unmanned aerial vehicles (UAVs). *IEEE Transactions on Aerospace and Electronic Systems*, 10.1109/TAES.2024.3368402.

Bi, Z.M., SYT, L., and Wang, L. (2008, 2008). Improved control and simulation models of a tricycle collaborative robot. *Journal of Intelligent Manufacturing* 19 (6): 715–722.

Bilberg, A. and Malik, A.A. (2019). Digital twin driven human-robot collaborative assembly. *CIRP Annals – Manufacturing Technology* 68 (2019): 499–502.

Bitsch, G. (2022). Conceptions of man in human-centric cyber-physical productions systems. *Procedia CIRP* 107: 1439–1443.

Bobka, P., Germann, T., Heyn, J.K. et al. (2016). Simulation platform to investigate safe operation of human-robot collaboration systems. *Procedia CIRP* 44 (2016): 187–192.

Bonkenburg, T. (2024). Robots in logistics – a DPDHL perspective on implications and use cases for the logistics industry. https://www.dhl.com/content/dam/downloads/g0/about_us/logistics_insights/dhl_trendreport_robotics.pdf (accessed 13 September 2024).

Brown, S. and Pierson, H.A. (2018). A collaborative framework for robotic task specification. *Procedia Manufacturing* 17 (2018): 270–277.

Byner, C., Matthias, B., and Ding, H. (2019). Dynamic speed and separation monitoring for collaborative robot applications – concepts and performance. *Robotics and Computer Integrated Manufacturing* 58 (2019): 239–252.

Cantrell, S., Davenport, T.H., and Kreity, B. (2024). Strengthening the bonds of human and machine collaboration. https://www2.deloitte.com/us/en/insights/topics/talent/human-machine-collaboration.html (accessed 13 September 2024).

CDC (2024). Coronavirus disease 2019 (COVID-19) | CDC. https://www.cdc.gov/coronavirus/2019-ncov/index.html (accessed 13 September 2024).

Chen. X., Eder, M.A., and Shihavuddin, A.S.M. (2020). A concept for human-cyber-physical systems of future wind turbines towards industry 5.0. TechRxiv. Preprint. https://doi.org/10.36227/techrxiv.13106108.v1 (accessed 13 September 2024).

Choudhary, M., Lokhande, M., Borse, R., and Bhute, A. (2022). Chapter 6 – a machine learning approach to aid paralysis patients using EMG signals. In: *Hybrid Computational Intelligence for Pattern Analysis, Advanced Data Mining Tools and Methods for Social Computing* (ed. S. De, S. Dey, S. Bhattacharyya, and S. Bhatia), 107–125. Academic Press.

Cui, C., Bian, G.B., Hou, Z.G. et al. (2017). A multimodal framework based on integration of cortical and muscular activities for decoding human intentions about lower limb motions. *IEEE Transactions on Biomedical Circuits and Systems* 11 (4): 889–899.

Cummings, M. (2014). Man verse machine or man + machines? *IEEE Intelligent Systems* 29 (5): 1–9.

Daugherty, P.R. and Wilson, H.J. (2018). *Human + Machine: Reimaging Work in the Age of AI*. Boston, MA: Harvard Business Review Press.

Desai, A., Ghose, S., Seshia, S.A. et al. (2019). SOTER: a runtime assurance framework for programming safe robotics systems. *49th Annual IEEE/IFIP International Conference on Dependable Systems and Networks (DSN)*, Portland, OR, pp. 138–150.

Fortunati, L. (2021). Human machine communication. Volume 2, ISSN 2638-602X. https://stars.library.ucf.edu/cgi/viewcontent.cgi?article=1050&context=hmc (accessed 13 September 2024).

Fu, Y., Xiong, X., Jiang, C. et al. (2017). Imagined hand clenching force and speed modulate brain activity and are classified by NIRS combined with EEG. *IEEE Transactions on Neural Systems and Rehabilitation Engineering* 25 (9): 1641–1652.

Gillespie, R.B., Colgate, J.E., and Peshkin, M.A. (2001). A general framework for cobot control. *IEEE Transactions on Robotics and Automation* 17 (4): 391–401.

Gil-Vilda, F., Sune, A., Yague-Fabra, J.A. et al. (2017). Integration of a collaborative robot in a U-shaped production line: a real case study. *Procedia Manufacturing* 13 (2017): 109–115.

Greenhorn, S.C. (2020). Toward a secure platform for brain-connected devices – issues and current solutions. *IEEE Transactions on Technology and Society* 1 (3): 161–172.

Guiochet, J., Machin, M., and Waeselynck, H. (2017). Safety-critical advanced robots: a survey. *Robotics and Autonomous Systems* 94 (2017): 43–52.

Gustavsson, P., Holm, M., Syberfeldt, A., and Wang, L. (2018). Human-robot collaboration – towards new metrics for selection of communication technologies. *Procedia CIRP* 72 (2018): 193–128.

Halme, R.J., Lanz, M., Kamarainen, J. et al. (2018). Review of vision-based safety systems for human-robot collaboration. *Procedia CIRP* 72 (2018): 111–116.

Hekmatmanesh, A., Nardelli, P.H.J., and Handroos, H. (2021). Review of the state-of the-art of brain-controlled vehicles. *IEEE Access* 9: 110172.

Heydaryan, S. (2024). Human-robot collaboration in automotive industry. https://iris.polito.it/retrieve/handle/11583/2710819/204370/SAHAR%20HEYDARYAN-PHD%20THESIS.pdf (accessed 13 September 2024).

Heydaryan, S., Bedolla, J.S., and Belingardi, G. (2018). Safety design and development of a human-robot collaboration assembly process in automotive industry. *Applied Science* 8 (2018): 344.

Hodgson, A.J. and Emrich, R. (2002). Control of a minimally constrained cobot. *Journal of Robotic Systems, Special Section on Robot and Human Integration* 19 (7): 299–314.

Hu, W.L., Akash, K., Jain, N., and Reid, T. (2016). Real-time sensing of trust in human machine interactions. *IFAC-Papers Online* 49-32: 048–053.

Hwaidi, J.F. and Chen, T.M. (2022). Classification of motor imagery EEG signals based on deep autoencoder and convolutional neural network approach. *IEEE Access* 10: 48071–48081.

Ikenishi, T., Kamada, T., and Nagai, M. (2010, 2010). Parallel factor analysis of driver' EEG in lane change maneuver for cooperative driver assistance systems. *IFAC Proceedings* 43 (13): 105–110.

Jabre, L., McGrew, R., Gillespie, R.B., and Goleski, P. (2002). An assistive cobot for aid in self care activities. *Second IFAC Conference on Mechatronic Systems*, Berkeley, CA (December 2002).

Jeong, J.H., Lee, B.H., Lee, D.H. et al. (2020). EEG classification of forearm movement imagery using a hierarchical flow convolutional neural network. *IEEE Access* 8: 66941–66950.

Jose, M.A. and Loper, R.D. (2015). Human-computer interface controlled by the lip. *IEEE Journal of Biomedical and Health Informatics* 19 (1): 302–308.

Kadir, B.A., Broberg, O., and Souza da Conceicao, C. (2018). Designing human-robot collaborations in industry 4.0: explorative case studies. *DS92: Proceedings of the DESIGN 2018 15th International Design Conference*. https://doi.org/10.21278/idc.2018.0319, pp. 601–610.

Khalid, A., Kirisci, P., Ghrairi, Z., and Thoben, K.D. (2016). A methodology to develop collaborative robotic cyber physical systems for production environments. *Logistics Research* 9 (1): 1–15. http://dx.doi.org/10.1007/s12159-016-0151-x.

Khalid, A., Kirisci, P., Khan, Z.H. et al. (2018). Security framework for industrial collaborative robotic cyber-physical systems. *Computer in Industry* 97 (2018): 132–145.

Koch, T. (2019). Approach for an automated safety configuration for robot configurations. *Procedia CIRP* 84 (2019): 896–901.

Kragic, D., Gustafson, J., Karaoguz, H. et al. (2018). Interactive, collaborative robots: challenges and opportunities. https://www.ijcai.org/proceedings/2018/0003.pdf (accessed 13 September 2024).

Kuan, Y.C., Lo, Y.K., Kim, Y. et al. (2015). Wireless gigabit data telemetry for large-scale neural recording. *IEEE Journal of Biomedical and Health Informatics* 19 (3): 949–957.

LaFleur, K., Cassady, K., Doud, A. et al. (2013). Quadcopter control in three-dimensional space using a noninvasive motor imagery-based brain-computer interface. *Journal of Neural Engineering* 10 (4): 046003.

Lake, J.P. (2007). Continuously available battlefield surveillance. https://apps.dtic.mil/sti/pdfs/ADA497522.pdf (accessed 13 September 2024).

Lammle, A. (2019). Development of a new mechanic safety coupling for human robot collaboration using magnetorheological fluids. *Procedia CIRP* 81 (2019): 908–913.

Land, N. (2024). Towards implementing collaborative robots within the automotive industry. https://pdfs.semanticscholar.org/a0ec/daf226e6f12d969c71f45d617a0176715f35.pdf (accessed 13 September 2024).

Lee, W.B., Park, H.C., Ahn, K.H., and Song, J.B. (2018). Safe robot joint brake based on an elastic latch module. *Mechatronics* 56 (2018): 67–72.

Levratti, A., Riggio, G., Fantuzzi, C. et al. (2019). Tirebot: a collaborative robot for the tire workshop. *Robotics and Computer Integrated Manufacturing* 57 (2019): 129–137.

Li, R., Wang, L., Suganthan, P.N., and Sourina, O. (2022). Sample-based data augmentation based on electroencephalogram intrinsic characteristics. *IEEE Journal of Biomedical and Health Informatics* https://doi.org/10.1109/JBHI.2022.3185587.

Lin, C.T. and Do, T.T.N. (2021). Direct-sense brain-computer interfaces and wearable computers. *IEEE Transactions on Systems, Man, and Cybernetics: Systems* 51 (1): 298–312.

Lloyd's (2019). Emerging risk report 2019 technology – taking control robots and risk. https://assets.lloyds.com/assets/pdf-taking-control-robotics-2019-final/1/pdf-taking-control-robotics-2019-final.PDF (accessed 13 September 2024).

Long, F., Padovano, A., and Umbrello, S. (2020). Value-oriented and ethical technology engineering in industry 5.0: a human-centric perspective for the design of the factory of the future. *Applied Science* 10: 4182.

Luca, A.D. (2024). Experiences with a control architecture enabling safe human-robot collaboration. http://robotsafety-iros18.com/wp-content/uploads/2018/10/IROS18_Workshop_RobotSafety_ADL.pdf (accessed 13 September 2024).

Magrini, E., Ferraguti, F., Ronga, A.J. et al. (2024). Human-robot coexistence and interaction in open industrial cells. *Robotics and Computer Integrated Manufacturing* 61 (2020): 101846.

Mao, X., Li, W., Lei, C. et al. (2019). A brain-robot interaction system by fusing human and machine intelligence. *IEEE Transactions on Neural Systems and Rehabilitation Engineering* 27 (3): 533–542.

Maurice, P., Malaisé, A., Amiot, C. et al. (2019). Human movement and ergonomics: an industry-oriented dataset for collaborative robotics. *The International Journal of Robotics Research* 38 (14): 1529–1537.

Maurice, P., Padois, V., Measson, Y., and Bidaud, P. (2017). Human-oriented design of collaborative robots. *International Journal of Industrial Ergonomics* 57 (2017): 88–102.

Michalos, G., Kousi, N., Karagiannis, P. et al. (2018). Seamless human robot collaborative assembly – an automotive case study. *Mechatronics* 55 (2018): 194–211.

Michalos, G., Makris, S., Tsarouchi, P. et al. (2015). Design considerations for safe human-robot collaborative workplaces. *Procedia CIRP* 37 (2015): 248–253.

Min, J.K., Ahn, K.H., Park, H.C., and Song, J.B. (2019). A novel reactive-type joint torque sensor with high-torsional stiffness for robot applications. *Mechatronics* 63 (2019): 102365.

Minati, L., Yoshimura, N., and Koike, Y. (2016). Hybrid control of a vision-guided robot arm by EOG, EMG, EEG biosignals and heat movement acquired via a consumer-grade wearable device. *IEEE Access* 4: 9528–9541.

Moritz, C.T., Ruther, P., Goering, S. et al. (2016). New perspectives on neuroengineering an neurotechnologies: NSF-DFG workshop report. *IEEE Transactions on Biomedical Engineering* 63 (7): 1354–1367.

Moulieres-Seban T, Solotti JM, Claverie B, and Bitonneau D (2020) Classification of cobotic systems for industrial applications, https://fja.sciencesconf.org/conference/fja/Moulieres.pdf (accessed 13 September 2024).

Nejatbakhsh, N., Hirata, Y., and Kosuge, K. (2005). Passive omnidirectional walker – design and control, *12th International Conference on Advanced Robotics, ICAR'05* (18–20 July 2005), pp. 518–523.

Oberc, H., Prinz, C., Glogowski, P. et al. (2019). Human robot interaction – learning how to integrate collaborative robots into manual assembly lines. *Procedia Manufacturing* 31 (2019): 26–31.

Omron (2024). Omron collaborative robot safety guide – information, products, and services for the complete collaborative robot safety solutions. https://assets.omron.com/m/31fba745c05ce84e/original/Omron-Cobot-Safety-Guide.pdf (accessed September 2024).

Owusu, E., Kumi, J.A., and Appati, J.K. (2021). On facial expression recognition benchmarks. *Applied Computational Intelligence and Soft Computing* 2021: 20.

Padfield, N., Zabalza, J., Zhao, H. et al. (2019). EEG-based braincomputer interfaces using motor imagery: techniques and challenges. *Sensors* 19: 1423. https://doi.org/10.3390/s19061423.

Paula, G. (1997). Cobots for the assembly line. http://www.memagazine.org/backissues/october97/features/cobots/cobots.html (accessed 13 September 2024).

Peshkin, M. and Colgate, J.E. (1996). Cobots' work with people. *IEEE Robotics and Automation Magazine* 3 (4): 8–9.

Peshkin, M.A., Colgate, J.E., Akella, P., and Wannasuphoprasit, W. (2000). Cobots in materials handling. http://www.mech.northwestern.edu/colgate/Website_Articles/Book_Chapters/Peshkin_2000_CobotsInMaterialsHandling.pdf (accessed 13 September 2024).

Peshkin, M.A., Colgate, J.E., Wannasuphoprasit, W. et al. (2001). Cobot architecture. *IEEE Transactions on Robotics and Automation* 17 (4): 377–390.

PhD Inc. (2024a). PhD solutions to industrial automation. https://www.phdinc.com (accessed 13 September 2024).

PhD Inc. (2024b). GRB pneumatic 180° angular gripper. https://www.phdinc.com/pdf/GRB.pdf (accessed 13 September 2024).

Polverini, M.P., Zanchettin, A.M., and Rocco, P. (2017). A computationally efficient safety assessment for collaborative robotics applications. *Robotics and Computer-Integrated Manufacturing* 46 (2017): 25–37.

Raiola, G., Cardenas, C.A., Tadele, T.S. et al. (2024). Development of a safety and energy aware impedance controller for collaborative robots. https://iit-dlslab.github.io/papers/raiola18ral.pdf (accessed 13 September 2024).

Realyvásquez-Vargas, A., Arredondo-Soto, K.C., Garcia-Alcaraz, J.L. et al. (2019). Introduction and configuration of a collaborative robot in an assembly task as a means to decrease occupational risks and increase efficiency in a manufacturing company. *Robotics and Computer Integrated Manufacturing* 57 (2019): 315–328.

Robotiq (2024). Cobots ebook. https://neffautomation.com/wp-content/uploads/COBOT-EBOOK-FINAL6.pdf (accessed 13 September 2024).

Safeea, M. and Neto, P. (2019). Minimum distance calculation using laser scanner and IMIs for safe human-robot interaction. *Robotics and Computer Integrated Manufacturing* 58 (2019): 33–42.

Safeea, M., Neto, P., and Bearee, R. (2019). On-line collision avoidance for collaborative robot manipulators by adjusting off-line generated paths: an industrial use case. *Robotics and Autonomous System* 119 (2019): 278–288.

Sagha, H., Perdikis, S., Millan, J.R., and Chavarriage, R. (2015). Quantifying electrode reliability during brain-computer interface operation. *IEEE Transactions on Biomedical Engineering* 62 (3): 858–864.

Saha, S., Mamun, K.A., Ahmed, K. et al. (2021). Progress in brain computer interface: challenges and opportunities. *Frontiers in Systems Neuroscience* 15: https://doi.org/10.3389/fnsys.2021.578875.

Santoso, H.B., Baroroh, D.K., and Darmawan, A. (2022). Future application of multisensory mixed reality in the human cyber – physical system. https://hdl.handle.net/10520/ejc-indeng_v32_n4_a5 (accessed 13 September 2024).

Schou, C., Andersen, R.S., Chrysostomou, D. et al. (2018). Skill-based instruction of collaborative robots in industry setting. *Robotics and Computer Integrated Manufacturing* 53 (2018): 72–80.

Schou, C., Hansson, M., and Madsen, O. (2017). Assisted hardware selection for industrial collaborative robots. *Procedia Manufacturing* 11 (2017): 174–184.

Simplilearn (2024). Artificial intelligence vs. human intelligence. https://www.simplilearn.com/artificial-intelligence-vs-human-intelligence-article (accessed 13 September 2024).

Sinha and Pathak (2019). Artificial intelligence versus natural intelligence – global challenge for human rights. *International Journal of Applied Engineering Research* 14 (7): 19–21.

Solanes, S.E., Gracia, L., Munoz-Benavent, P. et al. (2018). Human-robot collaboration for safe object transportation using feedback. *Robotics and Autonomous Systems* 107 (2018): 196–208.

Stanley (2003). Cobotics: new solutions in material handing. http://www.cobotics.com/pdf/cobotics_corporate_broch.pdf (accessed 13 September 2024).

Stefan, P., Traub, J., Hennersperger, C. et al. (2024). Chapter 40: Challenges in computer assisted interventions. In: *Handbook of Medical Image Computing and Computer Assisted Intervention* (ed. G. Fichtinger, D. Rueckert, and S.K. Zhou), 979–1012. London: Elsevier https://doi.org/10.1016/b978-0-12-816176-0.00045-4.

Taylor, N.C., Kruger, K., and Bekker, A. (2023). A human cyber-physical system for human-centered computing in seafaring. *Journal of Ambient Intelligence and Humanized Computing* 14: 7871–7884.

Townsend, B., Radford, M., and Groves, A. (2004). Collaborative robotics with medical applications. Project report 0401, ENPH 459, University of British Columbia, Canada.

Trimble, S., Naeem, W., and McLoone, S. (2019). Slip signal on a Baxter robot. *IFAC Paper OnLines* 52-11 (2019): 19–24.

UC-Olli (2024). An introduction to the brain-computer interface (BCI) and companies working on it. https://www.uc.edu/content/dam/refresh/cont-ed62/olli/olli_docs/brain-computer-interfaces.pdf (accessed 13 September 2024).

Ulrich, I. and Borestein, J. (2001). The guidecane – applying mobile robot technologies to assist the visually impaired. *IEEE Transactions on Systems, Man, Cybernetics – Part A: Systems and Humans* 31 (2): 131–136.

UR (2024a). How to boost productivity with automation. https://www.universal-robots.com (accessed 13 September 2024).

UR (2024b). Industrial universal robots. http://www.universal-robots.com (accessed 13 September 2024).

US Air Force (2013). Unmanned aircraft system (UAS) service demean 2015–2035. https://irp.fas.org/program/collect/service.pdf (accessed 13 September 2024).

Vasconez, J.P., Kantor, G.A., and Cheein, F.A. (2019). Human-robot interaction in agriculture: a survey and current challenges. *Biosystem Engineering* 179 (2019): 35–48.

Villani, V., Pini, P., Leali, F., and Secchi, C. (2018). Survey on human-robot collaboration in industrial settings: safety, intuitive interfaces and applications. *Mechatronics* 55 (2018): 248–266.

Vogel, C., Walter, C., and Elkmann, N. (2017). Safeguarding and supporting future human-robot cooperative manufacturing processes by a projection- and camera-based technology. *Procedia Manufacturing* 11 (2017): 39–46.

Wadelar, P., Gopinath, V., and Johansen, K. (2018). Safe layout design and evaluation of a human-robot collaborative application cell through risk assessment – a computer aided approach. *Procedia Manufacturing* 25 (2018): 602–611.

Wang, B., Zheng, P., Yin, Y. et al. (2022). Toward human-centric smart manufacturing: a human-cyber-physical systems (HCPS) perspective. *Journal of Manufacturing Systems* 63: 471–490.

Wang, D., Liu, Y., Hu, D., and Blohm, G. (2015). EEG-based perceived tactile location prediction. *IEEE Transactions on Autonomous Mental Development* 7 (4): 342–348.

Wang, L., Gao, R., Vancza, J. et al. (2019). Symbiotic human-robot collaborative assembly. *CIRP Annals – Manufacturing Technology* 68 (2019): 701–726.

Wang, X., Kemeny, Z., Vancza, J., and Wang, L. (2017). Human-robot collaborative assembly in cyber-physical production: classification framework and implementation. *CIRP Annals – Manufacturing Technology* 66 (2017): 5–8.

Wannasuphoprasit, W. (1999). Cobots: collaborative robots. PhD thesis. Northwestern University.

Wannasuphoprasit, W., Gillespie, W., Colgate, J.E., and Peshikin, M. (1997). Cobot control. https://www.researchgate.net/publication/3701604_Cobot_control (accessed 13 September 2024).

Williams, K.W. (2024). A summary of unmanned aircraft accident/incident data: human factors implications. DOT/FAA/AM-04/24, Office of Aerospace Medicine, Washington, DC, 20591, https://apps.dtic.mil/sti/pdfs/ADA460102.pdf (accessed 13 September 2024).

Wolf, A. and Schunk, H. (2018). Chapter 3: Understanding gripping. In: *Grippers in Motion*. Munich, CN: Varl Hanser Verlag.

Wolf, A. and Schunk, H. (2024). Grippers in motion – the fascination of automated handling tasks. https://www.hanser-elibrary.com/doi/pdf/10.3139/9781569907153.fm (accessed 13 September 2024).

Yao, B., Zhou, Z., Wang, L. et al. (2018a). Sensorless and adaptive admittance control of industrial robot in physical human-robot interaction. *Robotics and Computer-Integrated Manufacturing* 51 (2018): 158–168.

Yao, B., Zhou, Z., Wang, L. et al. (2018b). A function block based cyber-physical production system for physical human-robot interaction. *Journal of Manufacturing Systems* 48 (2018): 12–23.

Zaatari, S.E., Marei, M., Li, A., and Usman, Z. (2019). Cobot programming for collaborative industrial tasks: an overview. *Robotics and Autonomous Systems* 116 (2019): 162–180.

Zanchettin, A.M., Rocco, P., Chiappa, S., and Rossi, R. (2019). Towards an optimal avoidance strategy for collaborative robots. *Robotics and Computer Integrated Manufacturing* 59 (2019): 47–55.

Zhou, J., Zhou, Y., Wang, B., and Zang, J. (2019). Human-cyber-physical systems (HCPSs) in the context of new-generation intelligent manufacturing. *Engineering* 5 (2019): 624–636.

Index

Mechatronics for Complex Products and Systems: Project-Based Design Approaches for Robots, Cyber-Physical Systems, Digital Twins, and Other Emerging Technologies, First Edition. Zhuming Bi.

Companion website: www.wiley.com/go/bi/mechatronics

q

t

Printed and bound by CPI Group (UK) Ltd, Croydon, CR0 4YY
17/09/2026
14964315-0008